Hans-Peter Kohler
Population Studies Center
3718 Locust Walk
Philadelphia, PA 19104

GENETICS AND THE LOGIC of EVOLUTION

GENETICS AND THE LOGIC of EVOLUTION

KENNETH M. WEISS AND ANNE V. BUCHANAN

A John Wiley & Sons, Inc., Publication

Cover: Image provided by Ellen Weiss..

Published by John Wiley & Sons, Inc., Hoboken, New Jersey.
Published simultaneously in Canada.

Library of Congress Cataloging-in-Publication Data:
Weiss, Kenneth M.
Genetics and the logic of evolution / Kenneth M. Weiss, Anne Buchanan.
p. cm.
Includes bibliographical references and index.
ISBN 0-471-23805-8
1. Evolutionary genetics. I. Buchanan, Anne. II. Title.
QH390.W45 2004
572.8′38—dc22

2003014905

Printed in the United States of America

10 9 8 7 6 5 4 3 2

We wish to dedicate this book to our children, Ellen and Amie, for their forbearance, and for the inspiration they have continually given to us.

Apologia for advice not followed

And, freed from intricacies, taught to live
The easiest way; nor with perplexing thoughts
To interrupt the sweet of life, from which
God hath bid dwell far off all anxious cares,
And not molest us; unless we ourselves
Seek them with wandering thoughts, and notions vain.
But apt the mind or fancy is to rove
Unchecked, and of her roving is no end;
Till warned, or by experience taught, she learn,
That, not to know at large of things remote
From use, obscure and subtle; but, to know
That which before us lies in daily life,
Is the prime wisdom: What is more, is fume,
Or emptiness, or fond impertinence:
And renders us, in things that most concern,
Unpractised, unprepared, and still to seek.

J. Milton, Paradise Lost VIII: 182–197, 1667.

Contents

Preface

WHAT THIS BOOK IS ABOUT

A Philosophy of Biology

Our aim in this book is to develop some general principles to help describe the patterns to be found in the seemingly disparate facts about the diversity of life on Earth. It is an effort to assemble and digest observations made by naturalists and biologists from Aristotle to scientists publishing today—of organisms that live at temperatures above the boiling point of water and others that live in ice, organisms that fly and others that swim, those that inhale oxygen and those that expel it; how they are different and what they share. How can evolution, the single phenomenon that we invoke to explain how this endless diversity arose from one beginning, have produced it all?

Modern biological theory is thoroughly gene-centered, and this book is no exception. Genes are considered the essential storehouses of biological information and the mechanism through which evolution works. Thus, our specific interest in this book is in explaining the role of nucleic acids—DNA and RNA—in the evolution of complex organisms. At the same time, there is a danger in attributing too much to one cause, genetic or otherwise, or to one evolutionary process, or in considering the issues in such a detailed and itemized way that the broader picture gets lost. In this book we explore the ways in which an overly gene-based approach to biology can constrain our understanding of evolution.

Much of what we write about necessarily assumes evolution as its basic framework, that is, that organisms today have descended from ancestral organisms. But much of what we present considers alternative or supplemental general principles that we think are about as fundamental and ubiquitous in life as the core principles originally articulated by Darwin and Wallace. The theory of evolution was formalized as population genetics almost a century ago, but population genetics has little to say about the actual traits in organisms, how they are made, and how they evolve.

Natural selection is at the heart of the classical theory, but there is more going on than that, and we try to show what it is and where it might apply. Biology is forced to guess at the particulars of the evolution of traits and organisms because of millions of years of unobserved history that lies behind them. Natural selection is a rather generic explanation, which does not provide a very satisfying account of the particulars of the high degree of complexity found in organisms or even in cells themselves. To look at these, we consider aspects of life such as development, sensory systems, reproduction, and even perception. They illustrate some general principles that provide a remarkably consistent picture of processes involved in very disparate traits across the spectrum of life.

Organisms confront their world in a multitude of successful ways involving a comparable diversity of pathways to, or consequences of, complexity. Complexity requires that an organism have, for example, extensive mechanisms for communicating internally among its cells, and many such mechanisms have evolved in all branches of life, which we will discuss. Externally, organisms are surrounded by a wide diversity of information with possible relevance to their safety, reproduction, and food acquisition, and organisms have evolved many ways to use (or dismiss or get along without) that information, and we will discuss these. Indeed, the external environment contains "information" only when or if it is needed or used. Most of our attention in this book is on complex multicellular species, but we consider simpler organisms as well.

For reasons that probably go back to the way life first began, an elegant few strategies have been employed to confront the challenges of life (we do not imply conscious intent here). The word "logic" in our title refers to the way that the diversity of complex organisms has come about through a few general mechanisms that, along with shared history, enable a trait or developmental pathway or gene, once it has arisen, to be used, reused, and modified. Very similar characteristics and relationships are found among entirely different and/or unrelated genes across the living world. These facts make it much easier to understand complex nature than did earlier and simpler views of genes as each individually coding for a specific protein with a unitary function. Many of these attributes of life have been long known, though not always to all persons working in diverse areas of biology, or well integrated into their work.

We discuss specific genes throughout, but it is the relationship or process, not the detail, that counts. We also cover aspects of life not yet explained very well in genetic terms, but a major point is that one can predict the *nature* of those genes and processes, based on generalizations derived from what we already know. Such predictions are possible because the logic of life can be reduced to a small number of basic, ubiquitous principles. Nevertheless, a main point will be that this is not a prespecified system that follows necessary rules, or "laws of nature," the way formal mathematical logic does. Only in the broadest terms can there be a single theory of the contingent, largely chance-driven process that is the evolution of life. We can't prove that some mechanism other than the one we try to reconstruct might not have yielded the same diversity of life we see on Earth; that is the nature of retrospective analysis that we are stuck with in trying to understand the unobserved past.

We try to develop a broad and unifying sense of life and the ways that organisms live it. Our attempt is intended for any reader wishing to understand some of the most important generalizations to emerge from recent biological research. This is not only edifying—it is to us—but can also provide a guide for future work.

We hope especially to stimulate students learning about biology and evolution to see that there are broad principles at work in life that go beyond the one darwinian view so often taught. A theory helps us construct a consistent worldview but is always at risk of becoming a constraining ideology. Although we are involved in molecular biology ourselves, we seek to understand the unity of life in broader terms, compatible with the effort to reduce an understanding of biology to an understanding of genes and their action, but that always keeps its eye on the organism as a whole or on the phenomena that essentially rest on the interactions of molecules, cells, or even organisms.

Our Apologies

In this connection, it is certain that what follows will contain errors. Biology is a rapidly changing field, and facts are continually being amended or their importance reinterpreted, sometimes because of error and sometimes because of increasing knowledge. We are just two people confronting an enormous literature, and our own understanding will sometimes be flawed or we will have missed important papers; we will post errors and issues that we learn of on the http://www.wiley.com world-wide web page for this book. However, we think our general picture will be of some durability, and we hope that our attempt to go beyond the usually accepted principles of evolution, and to call some of those into question, will be useful and, if nothing else, thought-provoking.

We present some detail and technical material here but have tried to provide self-contained explanations; our intent is invariably conceptual rather than technical. Readers should be able to "read around" technical aspects that, because so much of the relevant genetics is of very recent vintage, are likely to be incomplete at best at this stage. We try to give a sense of what is known, with leads into the literature, without providing extensive lists of genes or pathways (we cite many excellent books, reviews, and scientific papers that do that). A reader interested in following up any particular points can easily find more about them through the literature and the internet—which would also help limit the damage that might be done by errors that we have made. If only because of the necessary lag time between writing and publication, no book can safely be regarded as definitive in detail, in a rapidly changing world.

A major risk in an era of exploding research and the sense of major discovery that now pervades genetics is that the firmness or importance of new results is probably overstated. However, we have tried to cite what seem to be reasonable interpretations of recent work that illustrate the generalities. We hope we have not been too restricted or parochial in doing so.

We have been unable even to approximate a thorough bibliography. As in the Technical Notes (below), internet web sources are so extensive and accessible that we think exhaustive citation is not as important for knowledge as it has been. We have cited primary literature to document our interpretation of various specific points, but as a rule we have preferentially cited recent reviews, texts, or convenient summary sources where we felt they would be useful, and/or that provide bibliographic entrée to the broader literature. Unfortunately, this nearly unavoidable way to handle information overload does undermine the proper assignment of credit for work and ideas because the authors of reviews are not always the sources of the material itself. We offer sincere apologies to many, many authors whose work we are aware of, staring us in the face from big piles on our floor, but that for practical reasons could not be cited.

Writing this book was a joint, interactive, and often grueling effort over several years as we tried to develop a credible understanding of fields entirely new to us, and to find the common threads among them. Although the illustrations new to the book were primarily done by one of us (AB), the writing itself was a joint, integrated effort in every respect regarding the ideas and the content.

If the ideas we present are interesting or stimulating to those who read this book, we will have succeeded, no matter how well our own particular views on life stand the test of time.

TECHNICAL NOTES

Gene Nomenclature

Genes are being discovered by the hundreds, often by automated means. The nomenclature system is somewhat undisciplined and not entirely consistent. In this book we have discussed results from work in most areas of biology, many of which have their own conventions for gene nomenclature, not always even internally consistent. Designation conventions change and seem likely to do so even more as an ever-larger set of species and their genes are identified, and genes are grouped ever more accurately and into more extensive phylogenies. Thus we have tried at least to be clear and consistent within the bounds of this book, to minimize distracting readers with confusing gene designations. We generally use italics for gene names (*Bmp4*), and corresponding standard font for their respective coded protein (Bmp4). This may be the single most consistent general aspect of nomenclature in the field. Our own consistency with these guidelines varies from strict to yielding to well-established conventions in various fields where a strict adherence would strike the informed reader as strange.

We have hopefully been clear about whether we are referring to a gene or to its product, although the distinction can usually be inferred in context. We try to identify relevant homologous genes among species, when they have very different names. Above all, while we undoubtedly have missed things and not been perfectly consistent, nomenclature should provide only a minimal distraction.

Bibliographic Support

Internet resources

We have not cited many internet URLs (worldwide web sites) in this book although the internet is a valuable resource for genetics. Readers who want to know more can usually use keywords to go right to major and minor resources for anything in the book and can follow up various issues by finding diagrams, DNA sequences, protein structures, technical descriptions, and even animations of many kinds and all levels. Unfortunately, the internet is a moving target, so that many URLs we would list here would be gone by the time a user wished to find them. The URLs we have cited seem to us to be likely to be relatively stable.

Reference citations

For the same reason, we have been especially sparing in our use of generic references. We usually give one or a few for broad topics. Similarly, it is utterly impossible to include all relevant technical references. The US National Center for Biotechnology Information (www.ncbi.nlm.hih.gov) provides many references and links, including PubMed (Medline) in which keyword searching can easily lead to the most recent literature. Readers should not rely on the accuracy of a conceptual survey such as ours, especially in an age in which so much is being learned so rapidly, and can be checked so easily.

Acknowledgments

We are pleased and grateful to acknowledge the help and guidance of a number of people as we wrote this book. Kris Aldridge, for an introduction into the literature on the brain, Mary Silcox for the tip on Linnaeus' classification of bacteria, Frances Hayashida for invaluable assistance with Adobe Illustrator, her provocative questions on the subject of every chapter and her delight in the answers (even though she's an archeologist), Ela and Janusz Sikora for cheering us on (even though they would prefer to talk about change by corrosion than by evolution), Nancy Buchanan for the photo of the flight feather from "the recent killing in the backyard," reminding us that nature can indeed be "red in tooth and claw," Bill Buchanan for his continual enthusiasm and support for this and all such endeavors, and Ellen Weiss for her photograph of Darwin's entangled bank (and her extra votes on the title).

We thank everyone, too numerous to mention by name, who gave us permission to use their figures or photographs, either as we redrew them, or the originals, and those we contacted for further clarification or expansion of their findings or ideas. People were invariably generous and helpful in guiding us to a better understanding of their ideas, whether or not we ultimately got it right or to their satisfaction.

And, we appreciate the tolerance of everyone who had to listen to us say for so long that we were "almost finished" with this book—all but our children were too polite to ask if it was really true this time. We thank Danielle Lacourciere and Rasa Hamilton at Wiley, for dealing with drafts rougher and more complicated than they probably expected. And last but certainly not least, we thank our editor at Wiley, Luna Han, for her patience and her belief in this project for more years than we would like admit.

PART I

Understanding Biological Complexity

Basic Concepts and Principles

In this section, we consider some of the general principles that characterize the nature and evolution of organized, functionally adaptive life on Earth. The mechanisms that determine the nature of organisms and the origin of the traits they possess can be approached at various levels of complexity. First, we will look at general principles. Inheritance is a vital component of diversified, specialized life, and we will consider just what it is that is inherited. We will then consider how that changes over time and relates to the processes we know as "evolution."

Genetics and the Logic of Evolution, by Kenneth M. Weiss and Anne V. Buchanan.
ISBN 0-471-23805-8

Chapter 1
Prospect: The Basic Postulates of Life

Natural history is the *descriptive* study of the natural world. The ultimate objective of science is to go beyond natural history to find generalizations, or explanatory *theories*, to account for our observations of nature. Theory enables us to explain a set of observations with fewer "bits" (a "bit" being equivalent to the answer to a single yes/no question) of information than are contained in the observations themselves.

The more dramatic the reduction in the amount of such information needed to account for observations and the more accurate the predictions we can make, the more *explanatory power* we credit to the theory. Predictive power is the gold standard for confidence in a scientific theory. The more sweeping and accurate the better, so long as the predictions are not vacuously vague. Newton's laws of motion, for example, apply broadly in the universe and are sufficiently accurate for many applications; Einstein's modifications are even more accurate and comprehensive.

Scientific theory involves many assumptions that may not always be stated. We assume that the facts of nature are objective and can be explained in *natural* terms, that is, without intervention of nonmaterial ("supernatural") factors. We also assume the universal validity of logical reasoning and mathematics. One of the most important assumptions that we make in building theory is that the fabric of causation in the cosmos is continuous and well-behaved, that facts are replicable—if we had the same conditions twice, we would have the same outcome. This may not be true in the ultimate sense (for example, if there is true randomness in the motion of atoms). More importantly for biology, our theory may assume replicability to a degree beyond what really applies, or, replicability may be the true state of Nature but our measurements too inaccurate. In fact, predictions and extrapolations can be almost completely inaccurate except in the short run, even for totally deterministic processes whose states or characteristics are not perfectly estimated (this phenomenon is sometimes characterized as "chaos" in the complexity literature).

The general belief among scientists is that we may not know the ultimate truth but that an ultimate truth does exist and that scientific methodology continually gets us closer to that truth. Philosophers of science debate whether this is actually so, noting that science is like other belief systems in resting on axioms—basic principles taken as givens and not to really be questioned. Indeed, science can be a kind of fundamentalism not unlike religion in its intolerance of challenges to its axioms. When, episodically, we become dissatisfied with the accuracy of this theoretical

---

Genetics and the Logic of Evolution, by Kenneth M. Weiss and Anne V. Buchanan.
ISBN 0-471-23805-8

edifice and an alternative explanatory framework is suggested, we experience what Thomas Kuhn called a scientific "revolution" (Kuhn 1962).

One rather curious basic assumption, the principle of parsimony (sometimes called "Occam's Razor"), states that nature is no more complex than it has to be. In scientific practice, this means that we assume that the simplest explanation for an observation is the *best* one. We implicitly accept that this also means the *truest*. But of course we don't know how complex nature really is or, in information terms, the degree to which any new theory could explain our current observations with fewer bits of information. This is a special challenge in biology because the biosphere is continually recreated through birth, death, and mutation in ever-changing environments. Unlike chemistry, we cannot replicate observations precisely at our will. Each new organism is unique, and life, unlike theory, does not always behave in the most parsimonious way. In the extreme, if life really were just as complex as our observations, then biology could not go much beyond descriptive "natural history."

Evolutionary biology both describes and predicts. The history of life is generally assumed to have been a one-time affair, whose specific events are unique, contingent (that is, depend on unique circumstances), and hence not replicable. Yet, each individual is a new test of the challenges of survival, and in that and other ways the living world continually replays the general principles of evolution. We find regularities, and these have led to a formal theory of evolution. Nonetheless, this has limited power because *specific* events in the future cannot be predicted the way one can predict the nature of a chemical reaction, for example. What can be "predicted" (or if we look back in time, "retrodicted") are patterns we might expect to see among descendants, based on postulated processes that affected their ancestors. A central problem is that in inferring how evolution produced what we see today we already know the outcome, so that much of what we do is to fit observations to theory rather than make truly deductive predictions.

One example of a very general prediction is that if different species share a recent common ancestor they will share more characteristics with each other than with species of more remote shared ancestry. If we could specify the extent of the similarity—say, in percent of difference between them on some scale—that specification could reduce the need to enumerate all the traits of each species. Linnaeus developed his systematic classification of life using morphological traits that he believed were important. The same idea can be extended to genes: related species will share genetic (DNA sequence) similarities to an extent that corresponds in some way to their phylogenetic history. This kind of divergence from a common ancestor was the basic idea underlying Charles Darwin's metaphoric tree of life (Darwin 1859) (Figure 1-1), an image that Alfred Russel Wallace also used to express the diverging nature of life, and one similarly employed by evolution's advocate in Germany, Ernst Haeckel, to show the nature of life diverging from "some one primordial form." (In this book for their symbolic utility we will frequently mention specific prominent individuals, but historians of biology have shown clearly that most advances have come from the work of many, famous and less famous).

Relationships previously characterized by Linnaeus have generally held up to studies of genetic data; morphology is not a bad guide to taxonomy. There are exceptions, but they usually involve subtleties, very ancient splits, or traits that can change easily or rapidly with relationships that can only be resolved with extensive amounts of DNA data. Although Linnaeus knew about bacteria (they were first seen microscopically in 1680 or so by Leeuwenhoek), he didn't understand them or their relationship to other living things and thus lumped them all into a category of miscellany that he called *Vermes*, in a class called *Chaos* (Magner 1994) (unrelated to the modern technical use of "chaos" referred to above). Sorting them out was left to future systematists. The complications are similar in nature to the complexity of nongenetic traits that have traditionally enabled debate among taxonomists.

In fact, genetic data are strikingly consistent with, and their characteristics were predicted by, darwinian principles, and it is significant that these findings were entirely independent of, and after, Darwin's formulation of his theory (in this book, we will use uncapitalized references, such as "darwinian," when discussing modified descendants of the original idea and capitalized references, such as "Darwinian," when discussing the specific notions of the person introducing them). Independent confirmation of theoretical ideas with new data is very important to the deductive aspects of science, and genetic taxonomy is an independent confirmation of Darwin. Of course, we know that morphological traits are affected by genes, so genetic data are not entirely independent; however, in a nonevolutionary world, for example, one made by a fixed creation event, there would not have to be any relationship between DNA sequence and morphological similarities.

If genes provide a kind of blueprint for life, genetic data should enable us to describe traits in different species or individuals with less information than is needed to describe each trait or individual separately. This is exactly the kind of reconstruction that Richard Owen and Georges Cuvier made famous in the early 1800s, when they used single bones to reconstruct whole animals, and why Thomas Huxley once exclaimed "A tooth! A tooth! My kingdom for a tooth!" (see Desmond 1994). Their theories were functional (not evolutionary): complex traits like a bone or

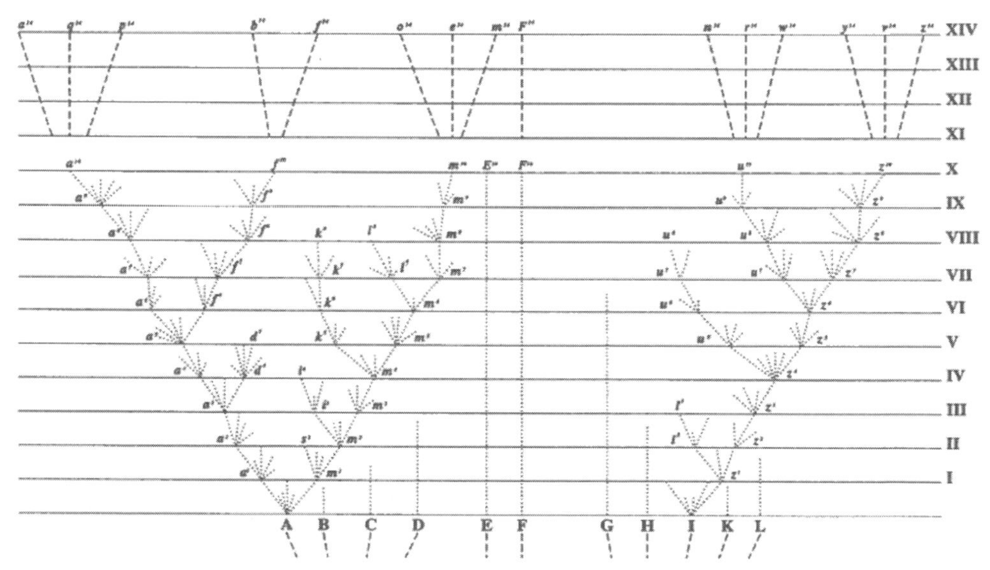

A

Figure 1-1. Trees of Life. (A) Darwin's from *Origin of Species*; (B) Haeckel's version from Haeckel (Haeckel 1906).

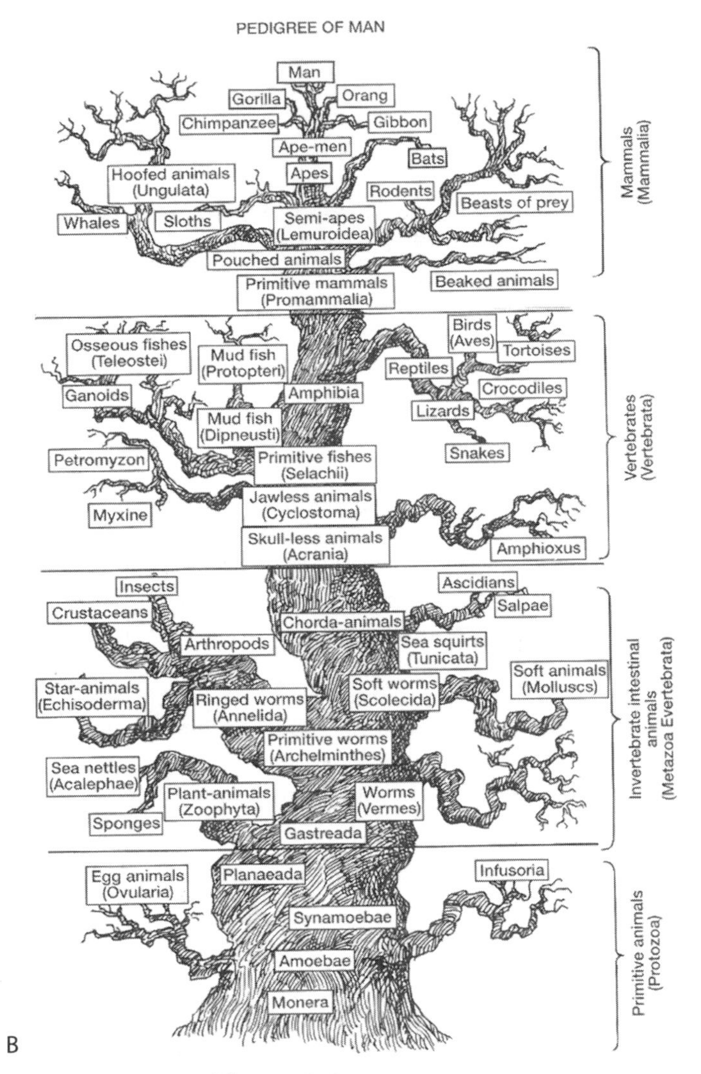

Figure 1-1. *Continued*

tooth reflect the function performed by the organism. A carnivore needs claws and teeth and speed, so to speak, and different carnivores share this general suite of characteristics. Genes provide similar kinds of relational information. One major purpose of this book is to ask how true are the simplifications that can be made from genes.

WONDROUS NATURE TO BE EXPLAINED

Naturalists, theologians, philosophers, and poets have written of their wonderment at the panoply of natural forms. Many have been struck by the adaptation of organisms to what they do in life; perhaps this insight alone is responsible for our modern view of biology. Different explanations for the origins of adaptation have been offered, but it is worth quoting one of the first advocates of an evolutionary view, the naturalist Henry Walter Bates, who described the following observations on the butterflies in Ega, on the Upper Amazon (Solimoens), hundreds of miles upriver from Manaus (Bates 1863):

> [They] vary in accordance with the slightest change in the conditions to which the species are exposed. It may be said, therefore, that on these expanded membranes Nature writes, as on a tablet, the story of the modifications of species, so truly do all changes of the organization register themselves thereon. Moreover, the same colour-patterns of the wings generally show, with great regularity, the degrees of blood-relationship of the species. As the laws of Nature must be the same for all beings, the conclusions furnished by this group of insects must be applicable to the whole organic world; therefore, the study of butterflies—creatures selected as the types of airiness and frivolity—instead of being despised, will some day be valued as one of the most important branches of Biological science.

This observation was made after Darwin and Wallace (a friend and co-Amazonian explorer with Bates) first publicized their views in 1858. In 1862, Bates laid out his view more formally in evolutionary terms, and it became known as *Batesian mimicry* (Bates 1862), the idea that tasty butterflies evolved to look like bitter ones that birds learn to leave alone. This is to this day one of the clearest cases of natural selection and is a textbook example cited to support the modern theory.

So what is this phenomenon called "evolution," and how is it that this one phenomenon can explain the diversity of life?

THE PHENOMENON OF EVOLUTION

The theory of evolution developed out of earlier ideas, some clearly anticipating what Charles Darwin and Alfred Russel Wallace would introduce to the world in 1858. The important concept was that the diversity of life, present and past, was not static and produced by externally derived creation events but was the product of historical *processes*, operating since some origin time on Earth—and *still* operating. One can view these as the biological version of the prevalent idea of a universally applicable natural law. Darwin himself left open how the whole process may have started, but biologists almost uniformly assume it was a terrestrial, strictly chemical phenomenon (this, too, is an *assumption* that, while not necessitated by specific knowledge, reflects the purely materialistic working world view of most scientists). The theory of evolution is elegantly simple and requires only a few basic elements. Darwin and Wallace introduced a few, to which several additional broad generalizations about the phenomenon of life can be added.

Darwinian Fundamentals

The basic postulates of evolution are simple and well known, but it is worth listing them: (1) organisms vary, (2) some of that variation is heritable from parent to offspring, and (3) there is population pressure on resources related to survival and reproduction; that is, organisms produce far more offspring than the environment can support. From any system with these general properties, the *fact* of evolution could be predicted, so that once they were clearly stated, darwinian phenomena are neither surprising nor really open to doubt. But this doesn't mean we could deduce any *particular* life form, even simple ones.

These basic principles can be summarized in Darwin's own phrase: "descent with modification." He deduced the consequence of persistent population pressure on resources: the variation that is able to reproduce more prolifically will be more commonly represented in future generations. He extrapolated this over long time periods, assuming it worked more or less consistently and gradually, to *hypothesize* that this *natural selection* explained the *adaptation* of organisms to their environment and accounted for the origin of new species from previous species over long time periods.

We should be aware of what these evolutionary premises do *not* say. They do not specify the resources under stress nor what it is that is heritable (or *how* it is inherited). Darwin did not specify correctly where new variation comes from, and there are widespread ideas in biology that are based on tacit additions to the basic principles (for example, some aspects of strong genetic determinism). We will see the implications of these assumptions.

We also don't need to argue about *whether* darwinian processes happen, as they are rather obvious and make very little in the way of specific assumptions. They not only apply to life but to any system built up of multiple, changeable units with competitive inheritance. But at the heart of Darwin's contribution to science was the theory that this is the process responsible for the transformation of species. Surprisingly, nothing in Darwin's premises necessitates species formation, even if his process adequately explained the form and structure of life. The separation into distinct species—which is what he was trying to explain—does not follow. At least one additional postulate is needed. This is (4) *sequestration*.

Sequestration and Diversification

Sequestration is implicit in Darwin's postulates. Life forms are isolated from each other, so that differences can accumulate between individuals. Darwinian variation does not immediately blend away (ironically, Darwin's mistaken idea of blending heredity was a problem for his theory, a fact that bothered him greatly). We use mating barriers between organisms to define "species." Even assuming they are genetically based, such barriers can be established by genetic changes having nothing to do with response to environments. In fact, whether adaptive or even random processes per se lead to new species has never been adequately proved as a generalization, and partly depends on our definition of "species."

Darwin was trying to explain the diversification of life into many species. In fact, he felt, and it is often argued, that *phylogeny*, or branching (divergent) speciation, is predicted by his evolutionary postulates. He developed his theory with the species question in mind, but adaptation does not by itself imply speciation. A global primeval soup could in principle evolve by changes in its chemical composition, energy, or some other cyclical processes diffusing through it over time. This does not constitute divergence among the states of life, except in the sense that there would be variation, as there is among readers of this book.

Evolutionary thinking predicts branching because descent with modification produces variation, and if that variation does not freely mix, then eventually reproductive exchange between the different branches becomes no longer possible. This is the essence of "speciation." Variation is sequestered within lineages, which accumulate increasing divergence over time. Because this process never ends, each lineage in turn diversifies. The result is a nested phylogeny.

It would seem from a superficial consideration of the similar nature of all cells that the basic machinery of life had developed before cells began to diverge. This assumes there was once only one cell population. Cells effectively isolate *very* localized packets of living matter from each other and from the surrounding "soup." Internally, the cell maintains the special conditions for using DNA to code for protein, a system almost certainly already present when organized cells evolved. Higher-level organization of life into multicellular organisms depended on this so that even within an organism there is local isolation of material.

Sequestration of material into cells, however, can never be complete. Even the first cells had to evaluate their environment and interact selectively with it (bring in nutrients, release waste, control ion concentrations and pH, and so on). Multicellular organisms require interaction and hence exchange of "information" among cells. Elaborate mechanisms have evolved for this, including partially permeable cell membranes, with mechanisms for transporting material across them, signaling mechanisms that work across cell membranes, and mechanisms for direct contact or transfer between adjacent cells.

DNA sequences, which will be described specifically in Chapter 4, are inherited across generations and thus, by nature, retain a trace of the past. Indeed, the sequestration of DNA from direct modification by the cell is one of the cornerstones of modern evolutionary theory, as we will see. However, DNA replication is not perfect or evolution could not have occurred, and if we have some external means of calibrating species history, such as known points in the fossil record, we can compare sequences of fundamental genes in representatives of the major branches of organisms to make educated guesses about what the ancestral cell type and its

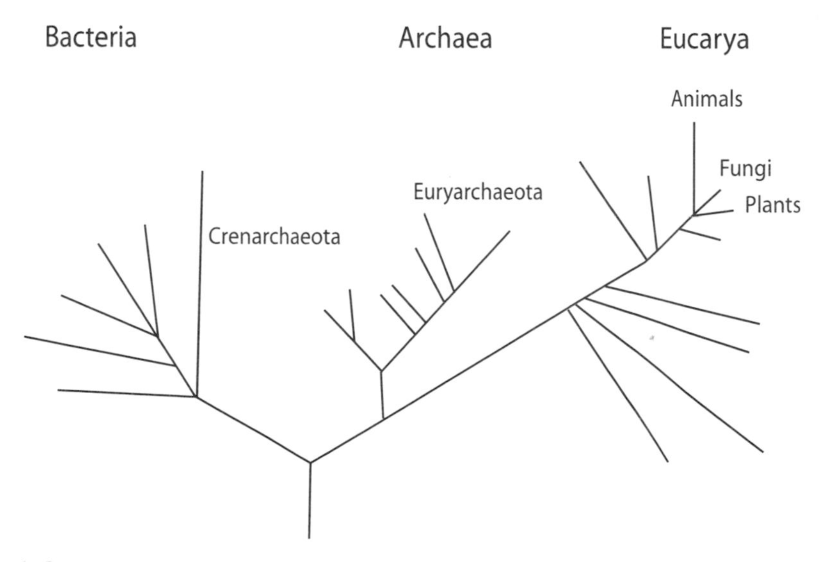

Figure 1-2. Tree of the major branches of life, based on ribosomal RNA. Redrawn from (Woese 2000) with permission. Original figure copyright 2000 National Academy of Sciences, U.S.A.

mechanisms may have been like (e.g., see Doolittle 1998; Doolittle 1998; Woese 2002). The process of accumulation of errors in DNA copying is highly stochastic (probabilistic); therefore, not all genes give precisely the same picture, so we have to aggregate data from many genes simultaneously. By grouping sequences that are most similar and roughly equating the amount of difference with time since common ancestry, we can reconstruct a hierarchical, treelike, representation of the history of life (e.g., Banfield and Marshall 2000).

The idea is based on the assumption that life had a single ancestry, here on Earth, represented by the trunk of our metaphoric tree. The tree of life reconstructed by genes presumably really is the tree of *cellular* life because basic biochemical mechanisms had to precede cells. Given a single origin of life, the principle of sequestration then leads naturally to diversification. Again, sequestration cannot be complete or we would never have aggregates of essentially similar cells that we call organisms or of essentially similar organisms that we call species. We will see, however, that this, like so many things in life, has important exceptions.

In addition to sequestration, three other aspects of life are so ubiquitous and fundamental that they should be added as generalizations about life as it happens to have happened on Earth. These are (5) *modularity*, (6) *duplication*, and (7) *chance*. Biological evolution could occur without them, but they have nearly comparable ubiquity and predictive power to the other postulates.

Modularity and Duplication

New structures from molecular to morphological are built by evolution from preexisting foundations. One of the most important and fundamental aspects of this is *modularity*. From molecules to morphology, we see variations on similar themes. These comprise separate modules or units from which more complex structures have been constructed. And one of the most important ways this has taken place is by the *duplication* of structures, with subsequent differentiation. The pervasiveness of duplication of structure has been known since systematic biology began and has only been reinforced by the history of discovery in physiology and molecular biology.

Modularity and Duplication Below the Level of the Cell

The modular nature of most of the basic biological molecules can be seen in Figure 1-3, which shows the chemical structure of nucleic acids, amino acids, and steroids. Variation on core structures as found in nucleic and amino acids was probably to a great extent a natural given, whereas variation in other molecules like steroids is at least to some extent manufactured by organisms. This is certainly true of protein families, as will be seen throughout this book.

The system of life today has been built on the modular nature of a corresponding concatenation of nucleic and amino acids into DNA/RNA and proteins. The nature of its ultimate origins is debated, but at some point biological information came to be stored in the form of the specific sequences, not the chemical nature, of these components. In particular, genetic coding is based on the *order* of concatenation of nucleotides in DNA and RNA, which has no chemical bearing on the *nature* of the protein being coded. The code for a given amino acid (see Chapter 4) is essentially universally used and has no bearing on the chemical nature of that amino acid nor on what that amino acid will do in a final protein. So it is in that sense a true *code*.

Nucleic Acids

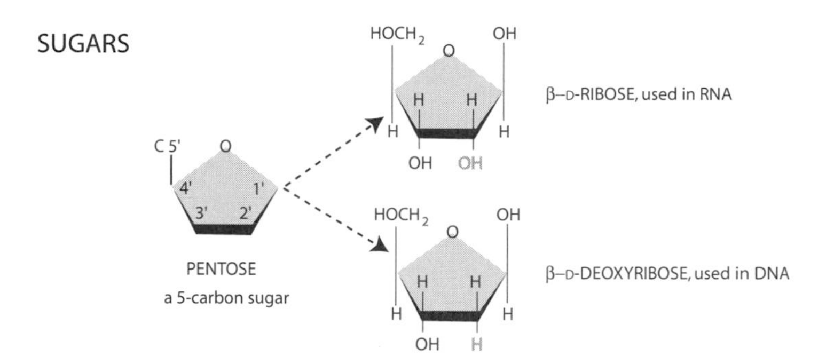

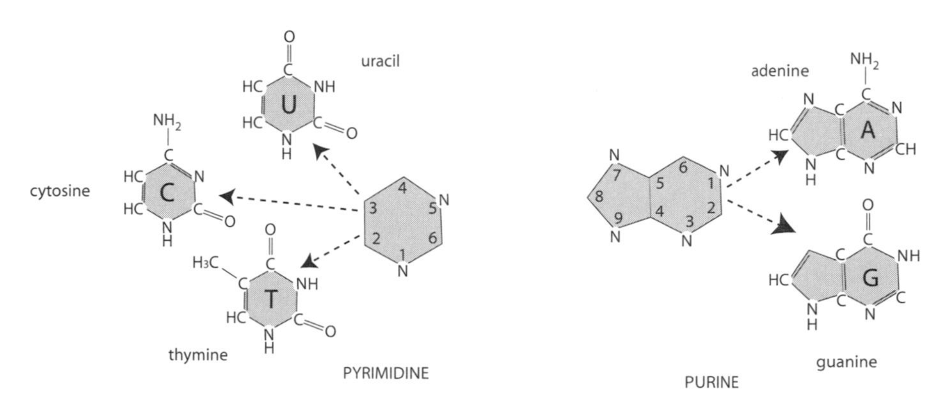

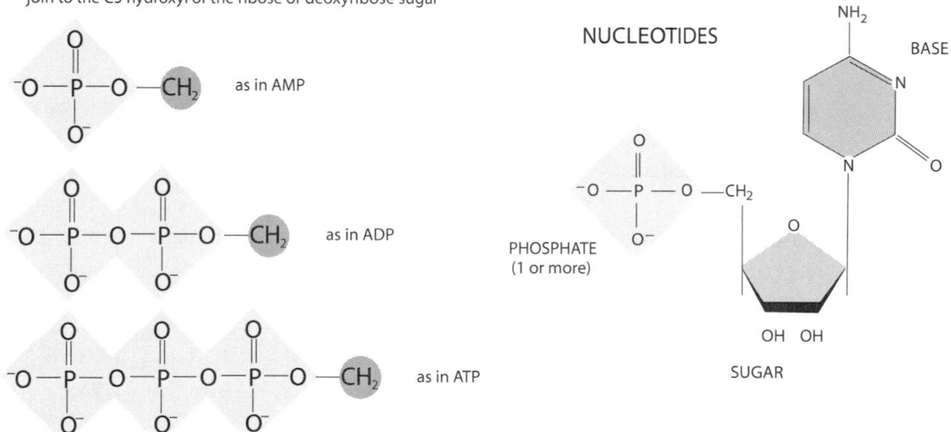

Figure 1-3. Modularity on basic chemical structure. (A) Nucleic acids; (B) amino acids, (C) Steroid molecules. (A) and (B) redrawn after (Alberts 1994).

Amino Acids

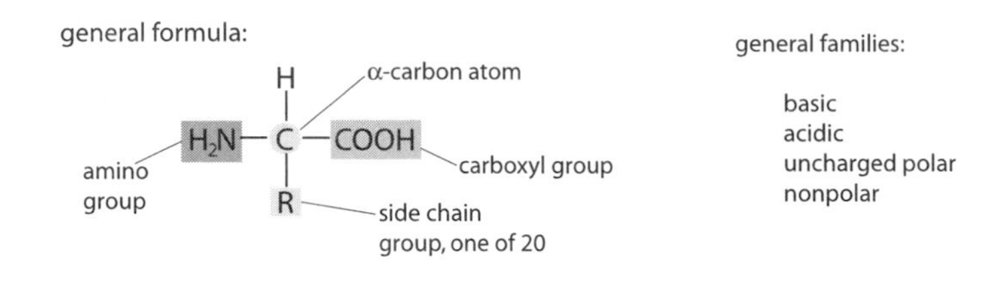

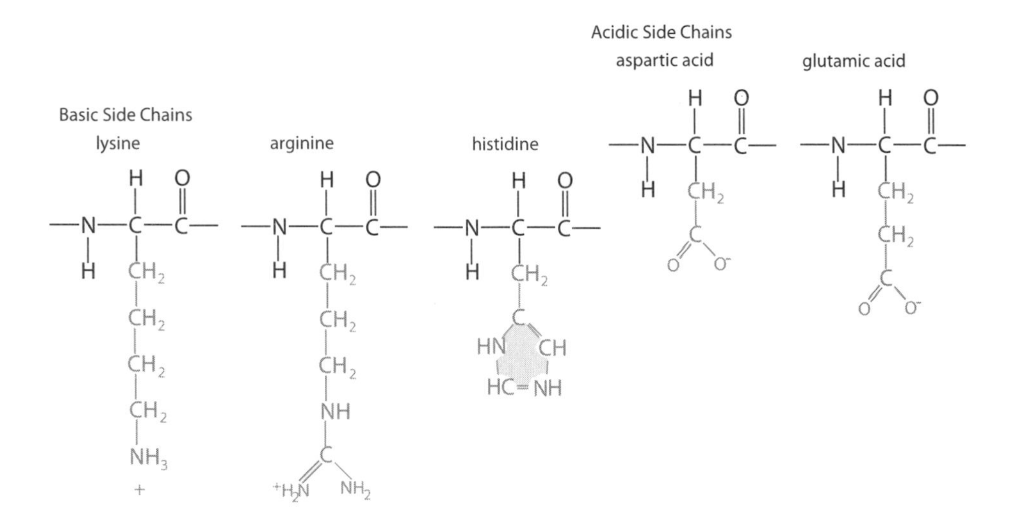

Figure 1-3. *Continued*

Much of what will be discussed in this book, and indeed in much of biology, is based on the elaboration of modular characteristics. Whether life had to evolve via modularity or whether some other form of aggressive energy-capturing self-replicating chemical system could have arisen from the conditions that existed when life began is difficult to say. But modular organization is certainly what happened and is so fundamental that one can surmise it would be inevitable.

Proteins and DNA/RNA are modular in that duplicate copies of the individual "beads" on these strings are used in the synthesis of new molecules. Indeed, new

Nonpolar Side Chains

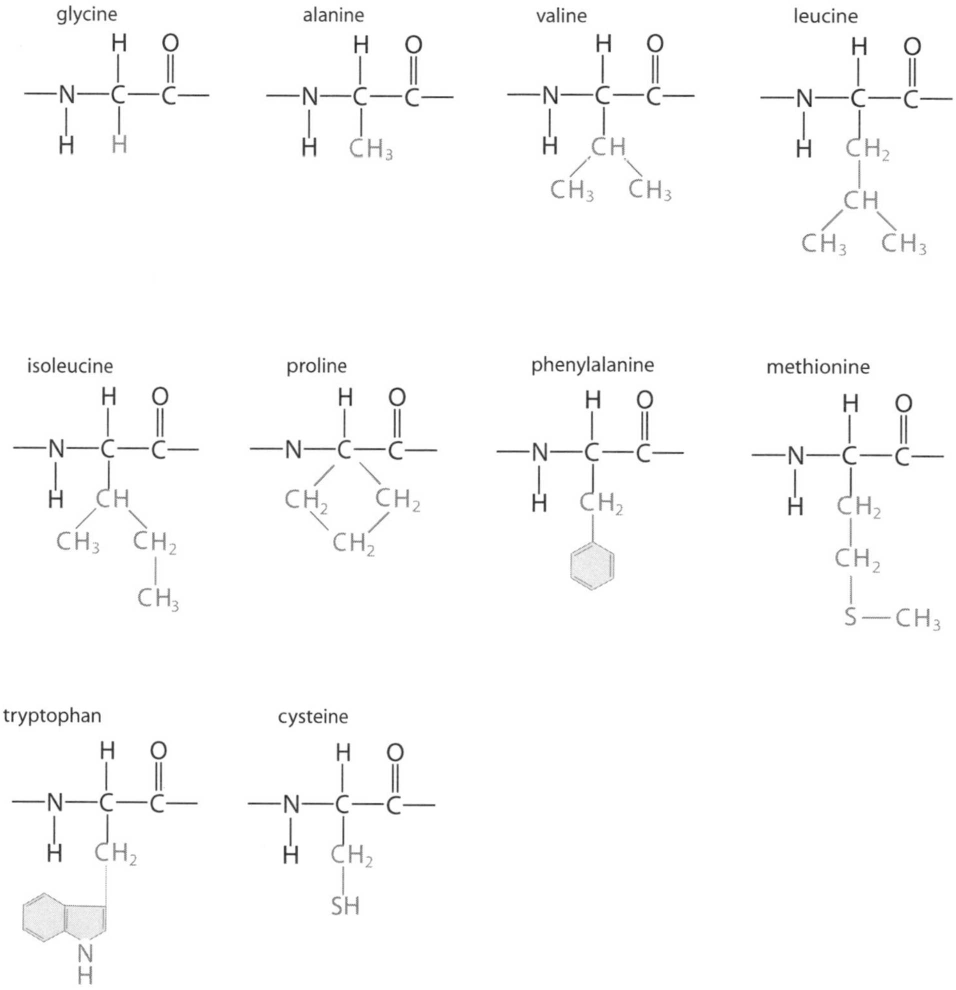

glycine

alanine

valine

leucine

isoleucine

proline

phenylalanine

methionine

tryptophan

cysteine

B

Figure 1-3. *Continued*

function also arises in a modular way and not simply by accretion. The molecules are occasionally modified when copied, and the copies can subsequently accumulate variation and modified or new function. This process can be termed *duplication with variation* and, as we will see, is a particular but important aspect of Darwin's major principle of descent with modification.

Duplication Above the Level of the Cell

Modular organization is related to sequestration as is seen by the important fundamental step of the evolution of cells, the modular units of which organisms are

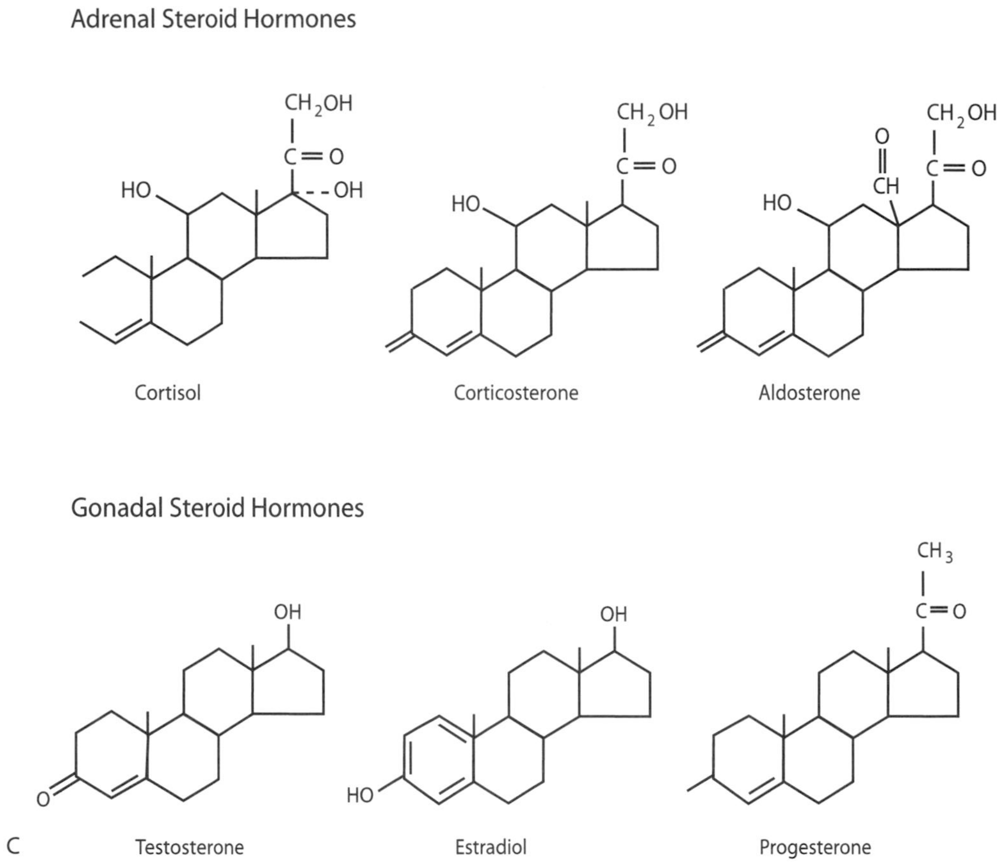
Adrenal Steroid Hormones
CH2OH
C=O
HO
OH
Cortisol
CH2OH
C=O
HO
Corticosterone
O
CH
CH2OH
C=O
HO
Aldosterone
Gonadal Steroid Hormones
OH
O
Testosterone
OH
HO
Estradiol
CH3
C=O
Progesterone
C

Figure 1-3. *Continued*

built. Cell division provided the mechanism for reproduction of the cell as an organism. Multicellular organisms are aggregates of differentiated cells that ultimately descend from a single cell (e.g., the fertilized egg). Thus, large complex organisms are built on a process of duplication with variation.

Organisms are modular in many ways beyond being aggregates of differentiated cells. Many if not most higher-level structures, like organ systems, are also modular (each themselves built up of cells, of course). A limited number of basic processes seem to be responsible for this hierarchical modularity, which we will review in later chapters. These processes are responsible for initiating very local cellular division and differentiation to produce individual organ subunits like leaves, flowers, intestinal villi, feathers, teeth, nephrons in kidneys, ommatidia in insect eyes, or vertebrae. Somewhat similar interactions may be responsible for the *branching*, a related but somewhat different process, that produces repetitive pattern in plants, lungs, blood vessels, and other structures.

A duplication strategy applies to physiological as well as morphological systems. The lipid (fat molecule) transport, endocrine (hormone), and immune systems, for example, are characterized by the interaction of slightly different products of related genes (the modules). How these various types of modularity arose and work will be discussed throughout this book.

Duplication and modularity were of course long known to be important in life but were not widely considered to be basic properties of evolution, because the phenomenon per se can occur without them. But early thinkers in the history of modern biology, for whom evolution meant the development of an organism, considered it; Goethe even likened the repetition of bones (as in vertebrae) to that in plants (leaves) (Richards 1992). The founders of biology were groping in the 19th century for such generalizations, or laws, of biological traits (at that time, essentially meaning morphology, because that was mainly what could be studied) and its variation, and for example how to explain the morphological changes through which embryos go, or the morphological or embryological similarities or differences (divergence) among species. Darwin used many of the same facts, but he accounted for them in terms of historical rather than developmental time. His idea provided a sweeping generalization about external processes that produce organismal diversity. After his and Wallace's exciting new idea, the central role of development, and the search for comparable internal generalizations, were somewhat shunted aside (e.g., Arthur 2002). They have returned.

Chance

Life as we know it depends on both precision and chance. The elegant precision of DNA replication gives life its predictability, generation after generation, but without chance mutations in genes, as well as the stochastic and error-prone nature of cellular processes in general, evolution would have no variation with which to work. We would all still be swimming in the primordial soup.

Darwin and Wallace had flashes of insight when they thought of the struggle for existence occasioned by overpopulation and the idea that the fittest organisms would prevail. This is the origin of the notion of natural selection, and both Darwin and Wallace saw selection as an immensely powerful tool for modifying organisms. To Wallace, and to some extent Darwin, the assumption was that the organized aspect of life is the product of selection, but in many ways this is as much a belief system for biologists as is that of fundamentalists who view the world of organisms as having been directly created by God. There are important reasons why we need to be careful about this view. One is our own human- and culture-dependent view of what it means to be "organized."

The second is a form of the *anthropic* principle, in essence, that things we see are not as improbable as they may seem, because had they not existed neither they, nor we, would be here to see them. Even in purely darwinian terms, whatever is here *had* to be "adaptive" in the sense of having been reproductively successful. Adaptations may seem highly refined, but among all the essentially infinite number of ways organisms might have evolved (even if by chance), *something* evolved. Thus, whatever is here is not as unlikely as we might otherwise think.

This of course cannot be used as an argument *against* natural selection; it should merely temper our after-the-fact reconstructions. However, the other side of causation is chance, and chance is pervasive and unavoidable—and often, as we will see, almost indistinguishable from causation. Darwin and Wallace were impressed by the universal ability of organisms to produce more offspring than could in the long run survive. Often, this means massive die-off. No matter how many acorns an oak tree produces or how many eggs a fly lays, most of the time populations are relatively stable in total size. Even a very small growth rate leads to major population size changes in short numbers of generations (a property of exponential growth). Thus, an oak tree on average produces one descendant tree.

This massive die-off means that differential survival based on the characteristics of the particular lucky acorn may over time favor that variant on the oak, but it doesn't imply that it will do so. In fact, it seems clear that life on Earth is much more awash in the disorder of chance than most biologists tend to acknowledge. At least, if the world is as remorselessly ruthless as the usual darwinian view holds, it is wanton rapine, directed to no particular end.

Actually, evolutionary geneticists recognize that much—perhaps most—of the genetic variation in the world not only arose by chance mutation but probably is barely if at all affected by natural selection: the amount and nature of variation changes over time by chance aspects of survival and reproduction alone. The same seems as likely of traits as it is of genes, as we will discuss in Chapter 3.

But What Is "Chance"?

Formal discussion of chance is a profound subject in cosmology about which there is no consensus; in this book we are not really concerned with whether true chance events occur in nature or whether they follow regular textbook probability distributions (e.g., binomial, normal, etc.). Instead, by "chance" we mean literal, or at least practical, unpredictability.

A probability distribution essentially provides a formula by which we can compute the relative likelihood of a given outcome in an experiment that can be repeated, like coin tossing. Sometimes observations can't be repeated in practice, but we still assume that underlying what we can actually observe is such a distribution. The Brownian motion of molecules is an example. But what about the "chance" that a given enzyme molecule meets a molecule of its substrate in a given cell? This may be a case in which we can at least assume that this probability could be specified in principle. In practice, this is not so clear: we can't really come close to specifying the probability because a cell is not a uniform fluid space, and there is no such thing as an archetypal cell in the practical sense, except perhaps in the somewhat artificial conditions of an experimental lab. Genes are expressed in cells based on regulatory mechanisms that, while specific, entail a substantial component of chance, but one that would be very difficult to predict. Each cell of a tissue of seemingly identical cells is somewhat different. What about the chance that a given wildebeest will be eaten by a lion or that a given human will have a given number of children during her lifetime? These questions seem somewhat simpler and more meaningful, and we can imagine replication or sampling distributions that could help us answer them, in principle. Many aspects of life and evolution, however, involve chance in a more profound sense. Biologists might try to estimate the probability that frogs could have evolved "by chance," for example, a question that is analogous to asking whether, if we could start life on Earth over again we would find the same outcome.

This is really a colloquial but scientifically misleading misuse of the term "chance" because there is no seriously meaningful sense in which the outcome of life represents a probability distribution in the sense of repeatable experiments. At best, the number of possible outcomes is so complex that it does not make real probabilistic sense to speak of the "chance" that frogs will evolve in the most stringent use of the criteria of science. At worst, our understanding is so rudimentary that the question itself makes too many assumptions that cannot be verified. In the end, we cannot operationalize this statement by specifying any practical way to make it testable. It is a metaphor.

Much of life is this kind of metaphor. It is important to realize that, although chance plays a role throughout life and its evolution, much of the time what that means essentially is unpredictability for all practical purposes. Some of the time—an unknown amount of the time—it means literal, ultimate true unpredictability as far as what we understand of the world today can tell us.

Implications of Sequestration and Modularity

Communication, differentiation, nesting and repetition are seen throughout life. They depend on controlled degrees of sequestration. In multicellular organisms, cells are organized into tissues of different types, then into organs or organ systems, such as the vertebrate epithelial, nerve, muscle, blood, lymphoid, and connective tissues, or, in plants the dermal, vascular, and ground tissues. Individual cells are bound together in layers in various ways and need modes of both communication and adhesion (and, for some cell types, such as sperm or egg or circulating blood cells, means of nonadhesion). Intercellular contact and communication allow the "blending" of isolated material to a limited or controlled extent.

The sequestration among cells in an organism is less than that between individuals in a species or between species. One of the most important of the mechanisms for reducing isolation of individuals within a species is sexual reproduction. Recombination among homologous chromosomes during meiosis breaks down some of the isolation of individual genes, and recombination of *genomes* (the entire set of genes inherited by an organism) in the formation of diploid zygotes allows diversity that would otherwise accumulate separately to be mixed, presenting greater variation and ability to respond to environmental circumstances among other things. Sexual reproduction allows individuals to be different and, for example, subject to screening by natural selection but also allows a limited and highly controlled amount of exchange that means that species can exist.

SOME BASIC ASPECTS OF THE CHEMICAL "LOGIC" OF LIFE

Here it is worth mentioning two facts that are fundamental to the way life works from a chemical point of view. Many other generalities apply, such as the properties of carbon-based life with the other major molecules (hydrogen, nitrogen, and oxygen). But here we are not referring to basic biochemistry but to the *logic* of evolution.

The details will be seen throughout this book. The first basic principle of the logic of life is that the four nucleotide bases, commonly denoted **A**, **C**, **G**, and **T**, chemically pair: **A** with **T** and **C** with **G**. As explained in Chapter 4, this is the essential fact in the "information" storage property of life.

The second major fact is that proteins, the other basic functional constituents of life and evolution, function by combining with other chemicals in the cell. From an evolutionary point of view, this in particular includes interactions of proteins with each other. This is fundamental to communication of all sorts, to the basic biochemical aspects of life, and to the way that evolutionary history and information are stored and used. An important example is the general phenomenon of one protein (known as a *ligand*) binding with another (known as a *receptor*) that has been evolved specifically to chemically recognize the ligand. We will see this phenomenon throughout the book.

Contingency

In the context of adaptation, sequestration, and the various levels of hierarchy seen across life from intracellular reactions to species evolution, it is important to keep in mind that these processes are *contingent*. That is, what happens now depends on the current state and not on previous or future states. The myriad biochemical reactions taking place within a cell are often localized within the cell, and each reaction depends on the current state, even if the raw materials of earlier states are still around or those prior-stage reactions are still occurring.

Two major points about this are worth raising in this context. First, life today depends on its history of evolution, and in that sense the nature of a cell or organism is dependent on its history as "written" in the genome (and in the nature of the environment around it, also the product of evolution). But what happens today is contingent on today's state, rather than some previous one. And any apparent longterm trajectory (e.g., teleology, or changes aimed at a certain distant adaptive end) is illusory except to the extent that the nature of things that are interacting today make it possible to predict what will happen tomorrow.

Beyond "tomorrow," it is generally agreed that life and its evolution are not specifically predictable, and the reason is contingency: what happens in the future depends on events that we view as purely chance, such as mutation, climate change, and the proverbial "acts of God" such as being struck by lightning. Similarly, it may be that many systems in nature are "chaotic," meaning that, even if they are totally deterministic, only perfect knowledge would allow *us* to predict future states with specifiable accuracy (a hypothetical example is the well-known one of a butterfly flapping its wings, unseen and unmeasured, that eventually leads to major weather changes). Whether such systems exist is itself essentially unprovable, but if it were to be shown that there really is nothing in nature that is purely chance—that is, that *all* the "laws" of nature are perfectly deterministic—would we have to back away from the view of nature as contingent? If so, everything really would be predictable from the beginning, a form of omniscience that would be truly God-like.

There are traits in some species that do not seem to vary, yet the species can be shown to harbor unexpressed variation (e.g., Dun and Fraser 1959; Gottlieb et al. 2002; Lauter and Doebley 2002) that can be revealed by changes in circumstances or other means—a kind of potential that is already there in the organism. Some aspects of life do not seem as chemically unlikely as previously thought. Functional proteins have been shown to arise rather easily even in random mixes of amino acid strings (Keefe and Szostak 2001), that is, they can be a kind of natural state.

Except for the stochastic element affecting whether two molecules come in contact, it seems generally true that the nature of molecular interactions is built into their atomic structure, that is, the principles of basic chemistry. Chemistry determines how DNA does its complementary base-pairing, the specific interactions of proteins or how enzymes interact with substrate molecules to affect chemical reactions. If there were no stochastic factor affecting whether molecules come into contact, the genome and other molecules within a cell would have essentialistic

TABLE 1-1.
Principles of Evolution.

Descent with modification:
1. Organisms vary
2. Some of that variation is heritable
3. There is population pressure on resources so that not all organisms survive to reproduce equally well (when differential reproduction is related to heritable variation we call that natural selection)

Duplication with variation:
4. Sequestration
5. Duplication
6. Modularity
7. Chance
8. Contingency

Basic chemical logic:
9. Complementary base-pairing in RNA/DNA
10. Protein-protein and receptor-ligand binding

properties such that life really would be the unfolding—the original, embryological development sense, of the term "evolution"—of the interactions latent in the molecular structure. In addition, life would be perfectly predictable from the day of the Big Bang onward.

If these kinds of molecular inevitabilities or chaos theory premises are literally true, it will have profound implications for cosmology, religion, and philosophical views of the nature of existence. However, based on what we know today, it seems just as true that the degree of any such predictability is so small, relative to the phenomena that we ourselves can observe, that evolution behaves for all practical purposes as if it were contingent and in that sense unpredictable.

CONCLUSION

In this book, we will be searching for generalities. We can think of these as the *logic* of evolution or of life, by which we mean the basic premises, properties, or processes, and their relationships to each other and to living forms. An important distinction is between how life *does* work, and how it might have *had* to work. Probably no one is really able to answer the latter question. That biology is not too far wrong in its assessment of the logic of life is shown by the fact that we routinely *use* the principles of evolution successfully to seek genes and understand processes across the entire span of biology. Using these principles allows us to account for a huge diversity of facts with only a modest number of basic units, processes, and principles. That is the goal of any science.

Chapter 2
Conceptual and Analytic Approaches to Evolution

The purpose of this book is to consider the genetic aspects of how various capabilities of complex life forms have come about, in the context of the basic postulates of evolution. There are many issues to keep in mind in the search for generalities. In this chapter, we consider various perspectives that may bear on this, given what we currently understand about both evolution and genetics.

SOME GENERAL CONCEPTS

UNIQUENESS AND GENERALIZATION IN A HISTORICAL SCIENCE

Evolution is an opportunistic process that builds solely on its present state. There is a large chance component in the environmental and biological variation that exists at any time and place. Evolution has been a one-time history. Each individual organism inherits, and must work with, the products of the events that happened in its unique *prior* history. Selection may mold organisms in a given direction *now* but does not—and cannot—aim toward anything in the future. Evolution does not require (and our theory does not even *tolerate*) any form of external force, as invoked by some religions, nor does it require internal animistic drive, invoked by Jean-Baptiste Lamarck and Henri Bergson (Bergson 1907), that directs organisms to evolve toward a particular *future* objective. Lamarck is today famously ridiculed for explaining evolution in terms of traits acquired by the striving of organisms toward some objective during their lives being inherited by their offspring.

But if evolution is contingent and without such directedness, can we expect to find generalizations or just local description? Might there be forms of biological necessity? Can we identify properties of life on Earth that should apply to life that might be discovered elsewhere, or are we simply describing history and calling it theory? In fact, general principles can be identified. Molecular and physical constraints and characteristics can be viewed as universal if we define life as varying,

Genetics and the Logic of Evolution, by Kenneth M. Weiss and Anne V. Buchanan.
ISBN 0-471-23805-8

self-replicating chemical interactions. Perhaps more specific to life on Earth are the intrinsic commitments of basing life on carbon and oxygen or on RNA and DNA and proteins.

To identify general principles, we make the core assumption of a single terrestrial origin, look comparatively at various forms of life to identify characteristics they share, and interpret them as mediated by shared historical processes. At a fundamental level, we compare gene coding mechanisms or biochemical usage among diverse species. To obtain ideas about higher-level phenomena, we can compare similar traits in multiple species—for example, forelimbs in mammals and birds. In such instances, we seek partial generalizations that apply to specifiable subsets of life. Many higher-order aspects of complex life are widely shared, even if the details vary, and a manageable number of broad generalizations are possible. Even so, there are usually exceptions because of the contingent, chance element involved in the historical phenomenon we call "life."

Is Evolution a Branching Process?

One of the most profound aspects of our notion of evolution is that it is a divergent or *branching* process. As shown in Chapter 1, the metaphor of a tree has been used as the fundamental image of evolution, at least since Darwin and Wallace. But is it correct?

It is true that mutation, natural selection, and other processes lead to *divergence*, or the accumulation of differences among different lineages descending from a common ancestor. If we look *forward* in time, the process produces a branching relationship as descendants of an individual, species, and so forth are isolated from each other, each acquiring unique new variation. Because of its ad hoc and complex nature and because of sequestration of the players, the process is, essentially, irreversible. If this is how life works, over time it would generate ever-proliferating diversity, with hierarchical, *cladistic*, or *nested* variation within branches sharing common ancestors.

In practice, we cannot really look forward in time, but must use the patterns seen among current individuals and species (calibrated by our understanding of mutational processes and the limited amount of geological and morphological information provided by the fossil record) to look *backward* in time. By grouping organisms with shared traits and assuming that to be the result of common ancestry, what we see involves the "coalescing" of today's forms into ever-fewer ancestral forms—ultimately going back to the common origin of life. This does not generally imply that there were fewer forms around at that time, except in a very general sense in relation to the very earliest times, because many past life forms will have left no descendants living today.

However, and especially if we attempt to infer the nature of the first life, the picture is less nested and treelike than the usual conception of evolutionary relationships. For example, reconstructing this history from DNA sequence data reveals inconsistent phylogenies among genes, suggesting that some systems have been transferred *horizontally* among long-separated branches. The recipient branch's original system is thus replaced in its descendants by the system transferred in from a different branch (Doolittle 1998; Doolittle and Logsdon 1998; Jain et al. 1999; Koonin et al. 2000; Ochman 2001; Ochman et al. 2000). As a result, today two groups *A* and *B* e.g. may appear to share a common ancestry based on studies of one particular gene, but groups *B* and *C* seem to have a common ancestor relative to another gene. The tree of ancestry is not the same for each gene. The extent of this phenomenon is unclear (Fitz-Gibbon and House 1999).

Essentially, horizontal transfer from one individual to a peer of a different species is a violation of the general darwinian postulate of variation transmitted *from parent to offspring*. Subsequently the immigrant gene, acquired horizontally, is passed by the recipient to its offspring in the normal way. Despite the occasional transfer of genes or sets of genes, one might nonetheless surmise that all of this reticulated ancestry at least goes back to a single ancestral *cell*. But even this may not be so.

An alternative to a single tree of all cells is that life evolved from a kind of communal pot of primordial soup, dished initially into rather imperfectly sequestered troughs, of incompletely different cell prototypes, and finally to more completely separated bowls. In the first stage of such a scenario, biochemical reactions took place communally, diffusing rather freely. As membranes or other barriers developed, some local, isolated specialization evolved tolerant to horizontal transfer (Woese 2000; 2002). Over time, local environments became more highly structured and organized and sequestration more important and impermeable—in the form of cells. Only by the latter stage were cell types among the major branches of life essentially isolated permanently from each other.

Woese (Woese 2002) questions when the effective isolation occurred, noting that basic gene replication mechanisms share little homology between archaea and other cell types. One possible example is DNA transcription and protein translation systems (the nature of these things will be described in Chapter 4), but mechanisms even more fundamental than that may also have transferred horizontally (Jain, Rivera et al. 1999). An incoming gene that cannot be used quickly becomes mutated into extinction, and it must be in the germ line of the recipient species to be transmitted (a major barrier to horizontal transfer in multicellular organisms). However, even among single-celled organisms, horizontal transfer more likely involves mechanisms that depend on the host having cell machinery compatible with that of the donor's.

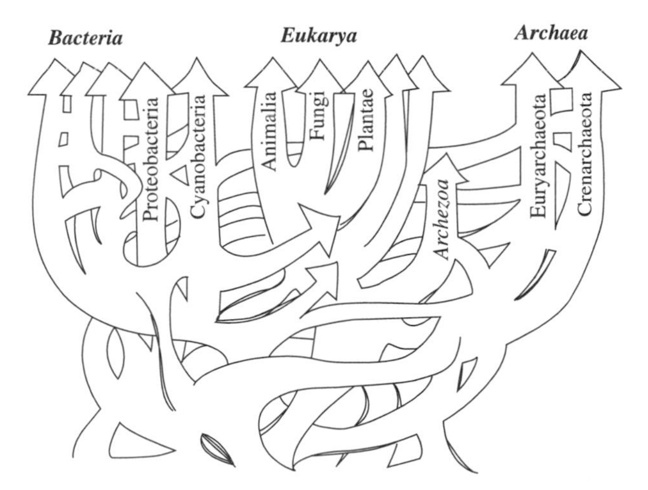

Figure 2-1. Tree of life showing horizontal transfer among major branches. Reprinted from (Doolittle 1999) with permission.

Eventually, according to this scenario, cells became too metabolically and genetically integrated across their diversity of functions to be tolerant to the importation of genetic mechanisms other than things basically self-contained. A "Darwinian threshold" is reached after which little horizontal transfer can occur, and cells and their descendants generate the standard kind of diverging, diversifying "tree" of relationships. There is, however, still debate about how much actual horizontal transfer has occurred, centering on, among other things, the analytic methods used to identify possible transfer products in present-day cells and species.

Eukaryotic cells (those with nuclei) are thought to have evolved by cell fusion (e.g., Hartman and Fedorov 2002) that introduced the subcellular organelles mitochondria and chloroplasts into a cell. This was initially a symbiotic relationship (a mutually beneficial arrangement between otherwise self-standing entities) but eventually the descendant cells and organelles were unable to survive without each other. Horizontal transfer may have been important in the evolution of algae, enabling them to engulf each other (or algal-bacterial transfer) and thereby acquire new function (Archibald et al. 2003). We also know that horizontal transfer at least occasionally still occurs, for example, by the insertion of viral genes into recipient genomes.

If there were complete sequestration, every cell lineage would be an entirely separate species. In fact, a number of normal mechanisms modulate sequestration and connect branches of life. Sexual reproduction within a species keeps the germ lines connected. This kind of transfer is easy and generally complete because donor (e.g., sperm or pollen) and recipient (egg or ovum) are so similar. Individual genes within species also are kept connected by recombination between an individual's incoming paternal and maternal genomes.

Life of course is *generally* cladistic, especially with regard to complex organisms. As the barrier to horizontal transfer became basically impermeable, species and their diversification in the usual darwinian sense became possible. This shows the importance of sequestration in biology.

All metaphors limp to some extent, but because various mechanisms of horizontal transfer exist, there is no single tree of life. Instead, what we see is reticulated, or interconnected, *networks* of historical relationships. This is true of deep aspects of species relationships, and also of gene relationships in shallower time.

Directedness in Evolution: Do Organisms Solve Problems?

It is difficult to write about evolution without using some kinds of convenient verbal shorthand, such as referring to existing adaptations as if they evolved *for* what they do today or to solve a problem confronted by the organism's ancestors. That is, we often tend to equate today's function with the selective forces of the past. Examples might be the ability to think, or fly, or digest cellulose. A cow's ancestors developed the ability to digest grass. However, no ancestral insectivorous mammal faced a field of grass and pondered how to digest it, the way we face a field of grass and ponder how to mow it. Organisms have no known way to develop heritable means to solve problems identified prospectively.

It is easy to think of environments as presenting problems for organisms to solve. But this can mislead us into Lamarckian thinking. The presence of the atmosphere makes flying possible, and flight has evolved many times. But birds' reptilian ancestors did not have to fly, as many contemporary land-bound reptiles demonstrate.

Nor has flight always taken the same form. An opportunity is not the same as a necessity. We can easily imagine opportunities that have not been taken.

Perhaps, genuine lamarckian mechanisms for directly producing heritable change in response to environmental circumstances will be discovered; some possible instances have been offered. There are examples to suggest that "evolvability" may exist, in that some organisms under stress respond by producing mutations, perhaps even in a context-specific set of genes (Caporale 1999; Fontana, 2002; West-Eberdard, 2003). The idea is that at some point organisms (and here we are speaking of single-celled organisms) had regions of DNA that were subject to mutations under, say, nutritional stress, and those mutations by chance led to tolerance of that stress and hence proliferation. However, the specific mutations themselves in this case are random, not directed to the specific need, and involve simple DNA changes rather than complex adaptations. In fact, organisms with a sequestered germ line are less likely to evolve such mechanisms because the mutations generated under stress would have to be inserted in the germ line and not just be in the body itself, yet it is the body that must survive the stress.

Our current understanding is that evolution is not *teleological*, that is, it does not work with future objectives in mind. We may some day discover lamarckian means of genetic evolution, but until then we have to hold to our view that mutational change is random relative to need. Neither directed, future-anticipating change nor the inheritance of acquired traits provides necessary explanations for the major functional characteristics of organisms. (Life may, however, someday evolve in a teleological way, if we develop genetic engineering methods to produce ends we envision in advance, such as sheep whose milk contains antibiotics useful for treating human disease.) The modern DNA theory of life, often called the Central Dogma, that a specific gene codes for a specific protein but that the gene's structure is not directly affected by how that protein fares in life (see Chapter 4), is the theoretical guarantee of this lack of lamarckian inheritance. However, this has to do in part with how we *define* heredity, and there are numerous examples of parent-offspring transmission of acquired traits—one being your ability to read this book.

What is a Trait? What Evolves?

It might seem strange to ask what a trait is, if the whole point of understanding evolution is to explain the diversity of traits in organisms. However, there have been and continue to be debates about what exactly it is that we refer to in this context. If traits are selected for or evolve, what are "traits"? Another way this has been put is this: What is the unit of selection?

This sounds simple but is not a trivial question. There is so much diversity in nature. Anything can be a trait. But if we want to relate our discussion to genes and adaptation, we need to know what we are considering relative to genes. Life cycle is a good example. This would seem to be directly related to the notions of darwinian fitness in the face of natural selection, since those who live longest or reproduce first might be declared the evolutionary winner. Does selection work directly on that or on the processes underlying the result? For example, much has been said of the notion that maximum lifespan is a characteristic of a species. This seems sensible, but does it imply there are genes for the timing of death? Does age at death evolve as a trait? Or is it just that causes of disease, that is, problems in cellular physiology, are screened and the net result is a statistical pattern of ages at death? This

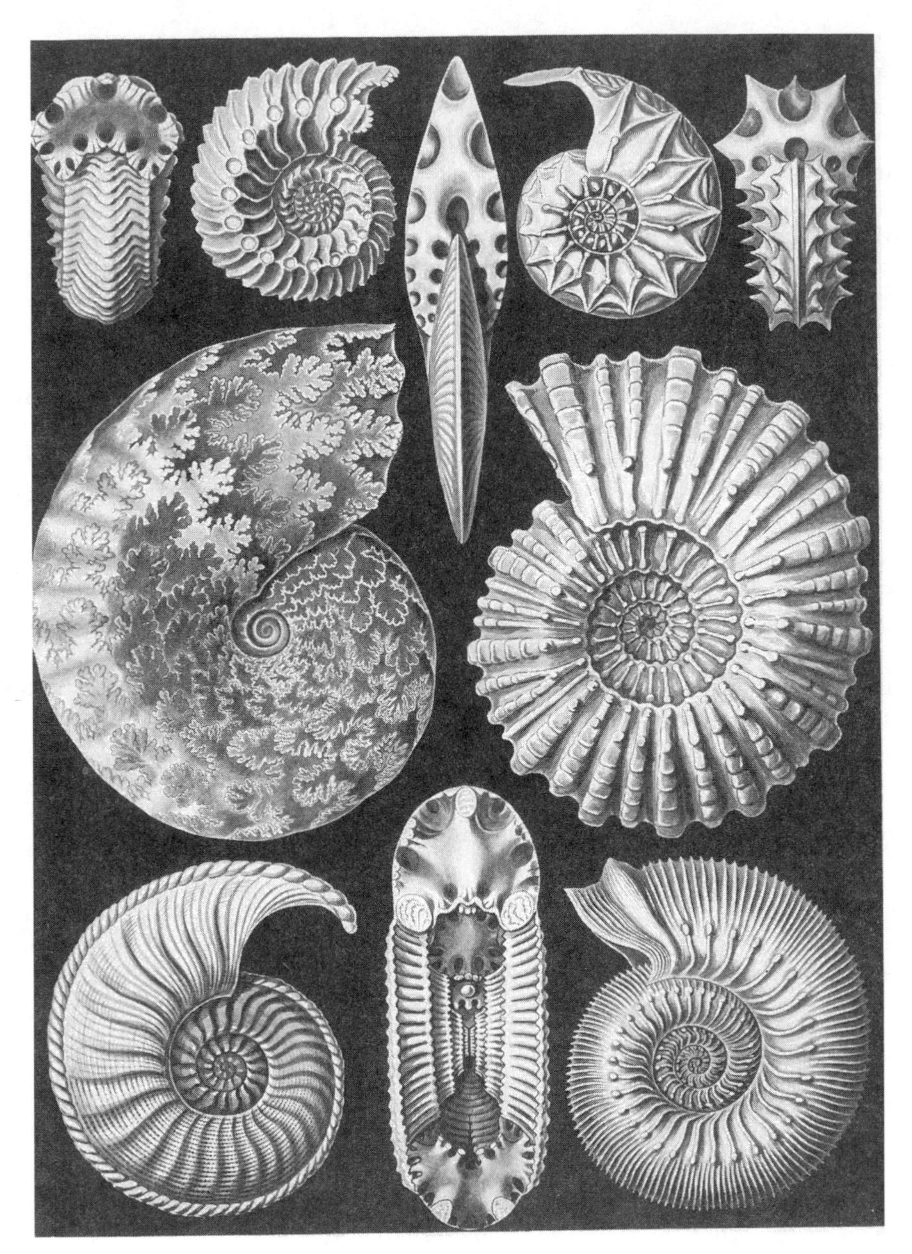

Figure 2-2. Ammonite shells drawn by Ernst Haeckel (a fine and avid artist) to represent diversity in nature. But what aspects are "traits" in the darwinian adaptive sense? From *Art Forms in Nature* (available in reprint as Haeckel 1899).

seems most likely (e.g., Finch and Kirkwood 2000); indeed, something we have already stressed is that there appears to be a huge component of chance even in life history events. This can easily seem to fly in the face of the universal appearance of adaptation in nature; but does it?

Is Adaptation a Profound or an Illusory Concept?

The lack of foresight in evolution leaves *us* as scientists (not organisms) with a problem because so many traits *appear* to have evolved to "solve a problem." Bats certainly fly! And it is not much of a misstatement that they evolved *to* fly or that they evolved because their ancestors *strove* to fly or that selection *favored* flight. Our theory holds that evolution is opportunistic. Selection screens variation that exists (by chance), favoring some functional variation that may or may not relate to flight but is useful at least in some way.

If an aspect of the environment remains relatively constant over long time periods, traits suitable for increasingly effective use of that aspect can be favored. Over time, the trait's evolution can continue in a generally consistent direction because variation that arises is screened by the same factors. This can effectively *canalize* (channel) evolution in a persistent way (Wagner et al. 1997). Biochemical constraints (see below) can limit what selection can achieve, and can contribute to what the prominent biologist William Bateson (Bateson 1913) referred to a century ago as "positions of organic stability." There is never any foresight involved, but a steady environment can lead to what *appears* to be directed evolution.

The resulting *teleological illusion* is what drove Lamarck, the Argument from Design, and numerous other responses to darwinian explanation. However, it can never be stressed enough that, if suitable variation had not arisen under particular circumstances, we would not have observed, for example, flying organisms today. Nothing we know about the mechanisms in biology suggests that flight arose through foresight or internal drive in any animal in the past or suggests that foresight or internal drive was necessary.

We are also somewhat trapped by the anthropic principle referred to in Chapter 1. *Every* organism we see today is the descendant of four or so billion years of *uninterrupted* success. Each has inherited genes that history blessed since the primeval soup. It cannot be otherwise. Critics note that, because of this fact, adaptive explanations verge on tautology because one can always invent an adaptive story that leads from past to present, and such explanations are sometimes applied so unconditionally by biologists as to be scientifically not much more meaningful than a collection of Just So stories because they cannot be verified. The issue might be ameliorated if we tempered these adaptive scenarios by keeping in mind that organisms clearly are not as finely tuned to their environment as is often casually assumed.

Similarly, our adaptive scenarios for complex traits might be tempered if we were obliged to specify *how* it happened. The human brain is considerably larger for our body size than the brains of other primates. We assume this is an adaptation for mental function. However, some humans have much smaller brains than others with no obvious defect (in behavior or, more importantly, reproduction), and genetic variation can seriously affect brain function without affecting brain size. It is easy to "explain" brain evolution by saying that a change of some very small amount in average size per generation (e.g., 1 mm^3) would be sufficient over many thousands of generations to increase brain size of the amount observed comparatively and from the fossil record. But how does 1 mm^3 of extra brain volume lead to increased reproductive fitness? If brain size did not increase in a gradual way, how did it increase? And why?

These important questions can be asked about most complex traits. For a scientific principle to have much meaning, or to be persuasive in a given circumstance, there should be some constraint on when and how it can be invoked. Adaptive explanations raise fundamental issues about concepts of causation in biology. This can be seen by the fact that, from Darwin to today, many biologists effectively assume that any complex trait is mainly the result of steady, gradual selection. (Rapid change by sudden mutation leading to a new level of complex organization has seemed impossible except for certain special cases, such as segmented, serially homologous systems like hands and feet or legs and wings in insects, in which the number of elements might under restricted circumstances change quickly and in simple ways we now understand at the gene level.)

There is no satisfactorily provable way out of the teleological illusion, but this has not shaken biologists into eschewing the making of adaptive scenarios, mainly because a good enough alternative material explanation for directed change does not exist. Religious creationists scurrilously misrepresent what biologists mean when they say that evolution is due to chance. But ironically, in insisting on adaptive scenarios, biologists share with religious creationists the belief that complex traits cannot arise just "by chance." However, we will suggest below that chance may be a more important factor in adaptive evolution than has been thought.

Natural Selection and the Species Question

One of the central concerns in the early stages of systematic professional biology in the late 18th and early 19th century was the "species question," that is, explaining the existence of the diversity of species, each suited to its way of life. Everyone knew that plants and animals were variable, and breeders could modify that variation up to a point. Domesticated species could be bred to change, but when the breeder's attention lapsed they seemed to "revert to type," and breeding never extended to the production of new species. Creationist explanations were weakening, as evidence from fossils, biogeography, and systematic, comparative, anatomic, and taxonomic studies accumulated. These studies showed that life was some type of historic phenomenon, and evidence showing that species *did* change and that new species arose by diversification from earlier species increased. But *how*?

Darwin and Wallace provided a general, codified, plausible, and in a sense *observable* mechanism—natural selection—by which species could in principle arise and change (Patrick Matthew also expressed the same argument clearly in 1831, but it went unnoticed in most subsequent priority credits because it appeared as only a brief comment in an appendix to a paper in a specialized book on naval arboriculture). But this lengthened the prevailing sense of time and made it a critical factor concerning relationships among species. Previously, when evolution referred to development, time was on the embryological scale. Charles Lyell, James Hutton, and other geologists had discovered slow processes by which the Earth's shape is changed, and that was an important factor lending plausibility to the ideas forming in Darwin's mind. His theory required a lot of time, and he was concerned that there might not have been enough for natural selection to mold the wonderful and complex diversity seen on Earth. He was convinced that the biblical estimates of the age of the Earth, roughly 6,000 years, were incorrect, but he thought "We have almost unlimited time . . . there must have been . . . millions on millions of generations" (C. Darwin, 1858, paper announcing evolution read before Linnean society; available on the public domain and web).

At the time, it was impossible to know just how many millions of generations the age of the Earth might truly have supported, and Darwin struggled with the problem. In fact, he thought that hundreds of millions (perhaps 300) of years would be required, and was highly discomfited by the British Royal Astronomer Lord Kelvin's estimate that the Earth was only about a tenth that old.

We now know that the Earth is much, much older, and that life has been here for several billion years (perhaps even 4 billion). But is that "long enough," not just for the evolution of cells or butterfly wings but for the evolution of *all* traits in *all* species, from leaves to language, without exception? Does 4 billion years make complex evolution, or adaptive explanations more or less plausible than some younger age? This is really a moot point, which is why Darwin could persist in his views despite unclear and sometimes quite contrary arguments about the age of the Earth. There is no real way to know how long is long enough for selection to have done its job. So long as we explain adaptation *conditional* on the *assumption* that life has evolved by natural selection on Earth, it *must* be old enough—it *was* old enough! Debates about this today are usually waged over contending adaptive scenarios. But so long as we accept the theory, there is no issue of the adequacy of time; instead, it is our job as biologists to use our theory to understand the details of *how* a given being evolved during its respective historic interval, which we document, for example, by molecular "clocks," calibrating time by the number of mutations that occur in DNA sequences compared among several species, and from fossil dates and biogeography. As we will see in Chapter 3, general mathematical and statistical theories have been developed for the rate and rapidity of selection in populations under various specified characteristics, and, while oversimplified, this yields an understanding of the way the process works in principle.

However, we will see in the next section that in some sense the ancient age of the Earth may actually make our reliance on natural selection *less* necessary than we have generally thought, which may change the kinds of reconstructions we should make or, at least, the need to invoke selection as much or as determinatively as we do.

SOME CONCEPTS OF EVOLUTION ARE USEFUL, BUT SHOULD NOT BE OVERUSED

Biological theory, like all theory in science, is an attempt to order the diversity of facts of the world. But the history of science consistently shows that the same "facts" can be interpreted in various ways and that what is chosen or accepted as fact is often culturally rather than, or as well as, objectively determined. An excellent and in some senses founding treatment of this now commonplace idea was presented by the Polish physician Ludwik Fleck (Fleck 1979), in the context of the way that the history of Western culture affected the drive to understand the causal nature of syphilis (Weiss 2003). That science is in part a product of its history and cultural context often means that alternative interpretations that we have not thought of or have minimalized or rejected might be of comparable utility. Better explanations might be rejected or not even considered because of such factors. It is interesting to consider some of the prevailing notions in evolutionary biology and how they may reflect the culture in which that theory developed.

Machine and Information Analogies About Organisms

In the 17th century, René Descartes promulgated the view that organisms were machines and could be understood in mechanical terms. Professional biology developed during the flowering of the industrial or machine age of the 19th century. Many important biological advances, particularly in genetics, occurred during the computer or information age of the 20th century; today, the prevailing view is that an organism is not just a machine but is the computable product of the information stored in its inherited DNA.

The machine analogy is in a curious way related to the 19th century's Lamarckian view of evolution. A machine *is* teleologically designed. It works because its parts are individually and independently manufactured in advance to serve a particular function. It is assembled from the outside and can be repaired part by part. Modifications can be introduced in various ways, some by chance perhaps, by purposive testing and experimenting *with some objective in mind*. Overall, the important characteristics of a machine are that it is *prospectively and purposively designed and manufactured*.

However, the same biology that views organisms in this way unambiguously denies that organisms have come about through teleological, lamarckian processes. Instead, an organism is the product of *contingent* processes, not de novo design. An organism develops from the inside, rather than being assembled by an external factory, and it evolved by modifications of that process, which works on the whole organism and not part by part and without any end objective in mind. We know that organisms are a patchwork of messy construction, yet we persist in analyzing them as if they can be decomposed part by part.

In a similar way, the computer age has led us to view genes as the sole blueprint for an organism, as if it were a simple storehouse of digital information for the assembly of an organism, a computer program for an organism. This metaphor has many problems (Kay 2000; Lewontin 2000). A program is modified not by overall natural selection but by debuggers that look for syntax errors and logical errors *relative to the preconceived function and built-in syntax rules*. Programs might be very different if all that we require of them is that they do *something*, as opposed to *this thing*. Selection works on organisms, not DNA, and we know that much of the DNA in this world is affected at most only weakly by selection (selectively *neutral* DNA; see Chapter 3). Nonetheless, much of modern biology is dedicated to the treatment of organisms as if they can be decomposed—and, for genetic engineers, repaired—gene by gene, just as we can execute a computer program step by step.

Of course, at some levels of approximation and for carefully chosen purposes, machine and information analogies work very well. Recently, however, we are learning that in important ways these analogies are frustrating at best and can be seriously misleading, and examples of these will be presented. Descartes did say that the body was a machine driven by the spirit.

Natural Selection: Cooperation as Well as Competition

Another element of the cultural context of the development of modern biology relates to ideas about the role of competition in evolution. We are not the first to observe that the formalized justification of competition is a core aspect of the industrial age in which evolutionary biology developed. It is often argued that were

Darwin not a wealthy industrialist in the world's Imperial power, he would not have produced his theory and it would not have been embraced by others. Nonetheless, for whatever reasons, although they differed in details, competition for scarce resources was central to the formulation of evolutionary ideas by both Darwin and Wallace. To both, the requisite force was the culling back of Malthusian overreproduction by an environment that favors those more "fit" for their circumstances. In the 20th century, it came to be widely believed this was to be explained not in terms of competition among individuals but by reduction all the way to a gene's eye view, in which the fundamental molecules of life—genes—are seen as "selfish", self-perpetuating units relentlessly trying to outdo each other and exploit each other for their own ends (e.g., see Hurst et al. 1996), most widely known through the uncompromising popularizations of this view by Richard Dawkins (Dawkins 1981).

We do not have to be stopped by the obvious fact that selfishness is a very human concept with many culture-specific nuances to agree that there is competition in nature and that those not fit to survive that competition are unlikely to do so. This was a natural way to view the world in the heady and dominant days of the British Empire and the competition and class hierarchies associated with industrialization and the growing organizational hierarchy of large, urbanizing societies. Selection does not screen on genes directly; however, it is true that over time some genetic variants become more common than others, thus winning a competition. But the historiographic explanation of evolutionary theory is weakened by the fact that Wallace was not a wealthy or socially conservative industrialist. Nonetheless it is fair to ask whether cultural circumstances affect how nature's observers interpret what they see; there are other ways to view the processes by which species might evolve. Competition is not all that occurs in nature.

Predation and Competition Between Species

As far as we know, life originated in some chemical mix in which the original organized molecules—whatever they were—increased by incorporating nearby molecules. Although initially life forms could have lived strictly by incorporating materials that were not previously involved in life or the decayed detritus of former life, the supply of such material would eventually have been exhausted. An important change occurred, in which living material depended on incorporating materials from other living organisms: the dawn of predation.

Obtaining nutrients from other living forms obviously became a successful way to do business and thus proliferated. It is easy to see how such a source of nutrition could be favorable, as the prey has been made to do the work of preparing the molecules needed by the predator (e.g., by making protein, carbohydrates, sugars, etc.). Consequently, a corresponding path to success emerged in the form of the ability to *avoid* being incorporated by another organism. Thomas Hobbes' war of all against all was on! Much of the structure of individual organisms and of ecosystems throughout the biosphere reflects billions of years of this arms race.

Recycling of dead matter by plants and many animals and, even better, capturing it in the form of living prey are major results of this history. Anyone reflecting on a meadow, forest, or pond (or on his own dinner plate) can see the importance and pervasiveness of this kind of interspecific competition. Systematic pressures of predatory or other limiting aspects of the environment can lead to increased organizational structure among living forms over time, and that is what Darwin and Wallace realized. It is easy to see how predation and other forms of competition that eliminate the sluggish could systematically produce organized traits in organisms *if* and *when* the traits of winner or loser are heritable. But this is not the whole story.

Cooperation and the Not-So-Selfish Gene

If we look at descendants over time, we can view those that increased in relative number as having acted in a way that served selfishly competitive ends. The notion of competition transformed scientific thinking by adding a dynamic process to a static worldview driven by religious dogma. However, much can also be learned by viewing things in terms of cooperation. Competition and cooperation are often viewed as antithetical, but biological processes almost by definition involve the interaction among molecules or other components of life. Success in life often depends on successful interactions. A purely selfish gene cannot even replicate itself. One can translate many of the phenomena now viewed in competitive terms into terms of cooperation (of course, as with competition, the entities need not be aware of what they are doing). A gene that gets along with others in its community—of molecules, cells, organisms, or even species—may have a better chance of succeeding. "Better chance" may be viewed as competition but "gets along" is just as necessary.

Within an organism, cells of different types are produced by a tree of descent from a common ancestral cell (the fertilized egg). These different cells perform vital complementary functions in very intricate ways. Without these functions, the organism perishes. Cooperation has been a basic strategy for building larger, complex organisms, as many chapters to follow will illustrate. Cells communicate with each other so that each differentiates in a way that is good for the organism. There is cooperative housekeeping, as nutrients and oxygen are delivered to all cells. Some cells protect other cells from damage by pathogens. Some communicate behavioral instructions from a central nervous system. At many stages in animal development, in the process known as *apoptosis*, cells self-destruct to make way, so to speak, for other cells. Animals require energy fixed by plants to survive. In plants, cells in roots deliver water to leaves.

Sexual reproduction has been viewed as fundamentally competitive at least since Charles Darwin wrote *Descent of Man or Selection and Selection in Relation to Sex* in 1871. Darwin quickly realized that competition to get or to avoid becoming food was not all there was to nature. Obtaining mates could also provide a strong selective force. Individuals of one sex choosing the "best" mate they can get or of the same sex competing with each other for access to the opposite sex, can in principle generate major phenotypic changes. This kind of competition can become exaggerated, a positive-feedback display race (Figure 2-3). Of course it is always difficult to infer after the fact whether this, or some more material type of competition, might actually have been responsible.

Beyond these kinds of organismal behavioral aspects of sexual selection, many modern darwinists peer into the intricacies of sexual reproduction to see all sorts of elements of competition—among individuals, among sperm, among fetus and mother, and so on (Hurst 1995; Hurst, Atlan et al. 1996; Parker and Partridge 1998; Partridge and Hurst 1998). Sexual reproduction has evolved many times, and the consensus reason is that this enabled species to be more diverse and responsive to changing environments, but it also has fundamental elements of cooperation. In addition, organisms from bacteria to humans manifest all sorts of cooperation,

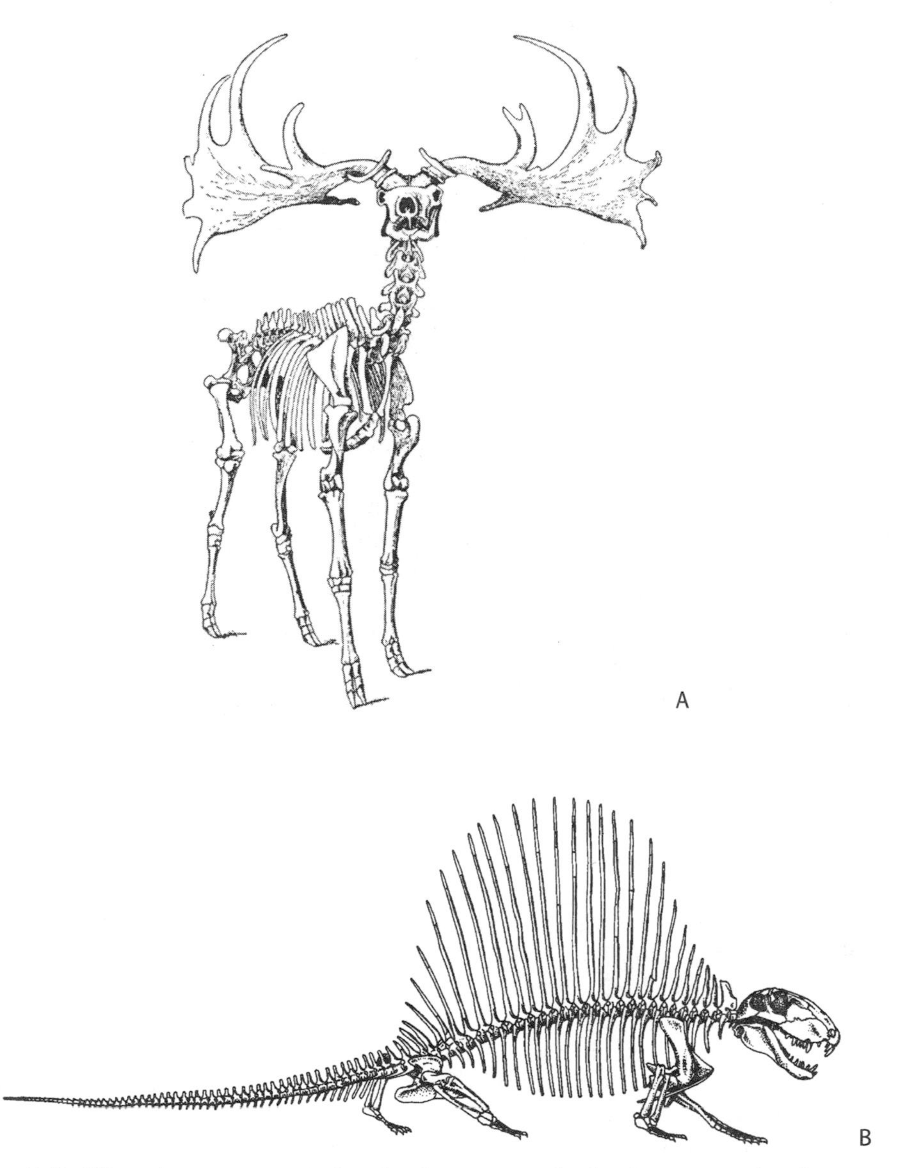

Figure 2-3. Were these exaggerated traits due to sexual selection, competing for mates, threatening predators, thermoregulation, or something not yet thought of? (A) Irish elk; (B) large-sailed *Dimetrodon.* (A) from (Romer and Price 1940); (B) from (Reynolds 1913).

including altruism and special dependency relationships such as that between plants and their pollinators, bees and their queen, or even bacterial cells and each other.

What is real here? A historiographic analysis can suggest that today, in the intensely competitive world of industrialized science, we view life through competitive lenses. If life must be viewed as competition for success in circumstances of limited resources, one would feel compelled to explain these phenomena as favoring genes leading to this behavior at the expense of less competitive genes. We see such interpretations at every turn.

However, from another point of view, this sounds quite forced. The best way to understand life in many cases is to view these phenomena in terms of their *driving* force—the cooperation that they are. Like Janus, the Roman god of change from past to future who has two faces, we should not ignore one view at the expense of the other, because that may blind us to understanding. How far we can take this is probably a matter of judgment and the viability of our interpretations. As we'll see in Chapter 8, prey can be viewed as cooperating with predators (by providing them food), as the essential nature of ecosystems. That cooperation is vital is easy to see: when the body dies, its perfectly normal cells die, but not the bacteria that were in the organism at the time.

Organismal Selection

Besides organized cooperation, there is at least one other way in which adaptive evolution can occur. Nature, like Hell in the play *No Exit* by Jean-Paul Sartre (Sartre and Gilbert 1947), can be portrayed as trapping organisms with each other, like people in a small room, for all eternity—or like throwing one piece of meat into an Roman arena and watching the lions kill each other to get it.

This is an accurate way to think of many circumstances in which species or local populations sometimes find themselves. But it is not the only way that life is lived, and there are ways to succeed other than by direct competition to the death. In many if not most situations, environments are complex. Organisms—even plants—actively seek environments in which they can survive, where their particular skills are suitable. A major characteristic of organisms is this kind of "plasticity" or adaptability and it is thoroughly integrated into their nature (West-Eberhard 2003). This is manifest by the great diversity of species of closely related animals (plants, fish, insects, mammals, etc.) found within the same lake or forest. The organisms have found many subenvironments in which to live.

As John McPhee (McPhee 2002) said offhandedly when observing the shape of the fins of American shad fish, "One look at that tail and you know that the fish is active in the middle of the water column and not sitting around on the bottom like a bullhead catfish, whose tail is so rounded it looks like a coin." In their preferred environment, they will meet and mate with like individuals. If migration is rare, or some mating or physical barrier arises, the individuals can become sequestered enough that they can diverge and eventually produce new species.

This is *organismal selection*, that is, active selection of the environment by organisms rather than passive selection of organisms by the environment, a reversal of the usual darwinian notion of natural selection. Over time, variation can accumulate, leading organisms to seek ever more specialized environments or to be more reluctant to leave the current one.

This self-selection may, but need not, initially involve genetically determined aspects of the organism. Divergence among populations that leads to subsequent speciation can then be aided by genetic changes that are unrelated to environment but provide mating barriers between individuals in the new and old environments. It may only take a very few such changes to create a reproductive barrier, and that is the definition of new species (Navarro and Barton 2003; Via 2001). Somewhat similar ideas have been expressed under rubrics like sympatric speciation or the Simpson-Baldwin effect, or "organic" selection, referring to the evolutionary effects of nongenetic transmission, that can eventually be canalized genetically (e.g., Hall 1999; Schlichting and Pigliucci 1998, and see Chapter 3). Though some of these terms have connotations from earlier disputes about evolution, sometimes offered in opposition to the notion of darwinian evolution, that need not be the case.

Of course, whether organismal selection, cooperation, classical competition, or just chance provides the best account for any given case is a separate question and sometimes difficult to answer.

Phenotypic Drift: Could Complex Traits Have Evolved by Chance?

As noted earlier, Darwin hungered for an old age to the Earth because that would provide enough time for natural selection to do its work. However, in an ironic way, an old age might relieve the necessity of invoking specific and highly discriminating natural selection to hurry along the process of adaptation. If the Earth is old, slower mechanisms can lead to organismal changes.

Defenders of creationist views of life accuse biologists of saying that creatures arose by chance. According to the alternative Argument by Design, finely tuned biological structures like the eye could not plausibly arise by chance. Biologists agree with this, and they never argued that adaptation occurs by chance in the sense that their opponents suggest. Biological organization is viewed as having arisen *systematically* by natural selection, but without any teleology because selection feeds on genetic change arising by chance *relative to the needs of organisms*. Selection is systematic, but opportunistic, working only on variation that happens—by chance—to be present.

Actually, classical darwinian explanations do not have to assume that selection is persistent and systematic enough to enable a complex trait to arise gradually from nothing. Instead, the evolution of such traits is assumed to be by a stepwise rather than via a continual teleological process. A complex state like bat flight or eyes with focusing lenses evolves through a series of earlier states. Each state itself became adapted by selection relevant to conditions *at each time and place*, unrelated to future states or needs. The variation that selection worked on at each stage had itself arisen by chance, without regard to states or needs. Because it only works with whatever is at hand and under the local circumstances at the time, evolution is known as a *contingent* process.

What the local selective reasons may have been at any stage is open to debate (or, perhaps, inherently speculative). Whatever their adaptive reasons for being, the earlier states, sometimes called *exaptations* (Gould and Vrba 1982) became available for selection that led to the next step, ending up in their final—that is, their present—use. For example, rudimentary wings useless for flight may have been selected for their value in thermal regulation, mating display, surface-gliding on streams (e.g., Thomas et al. 2000), and so on. Darwin called these traits "incipient stages," but that implies that they were functional rudiments of the present state, which is probably not the case for the evolution of novel function.

This view is essentially deterministic in that each stage is thought to be adaptive in the usual darwinian sense. But chance *not* aided by selection may have played a much larger role than usually assumed. Most organisms die for no reason related to their particular genetic makeup compared with that of their peers. Think of the millions of acorns whose unlucky fate it is to fall into shade or be eaten by squirrels.

Figure 2-4. Variation in sea shells coloration patterns, showing that patterns are highly plastic. Is this due to chance or adaptation? Courtesy H. Meinhardt.

Chance effects on reproductive success will affect heritable phenotypic variation, and the traits involved will thus change randomly over time. For example, their means, variances, or other characteristics will change in ways unrelated to their environment. Organisms might become larger, greener, rounder, and the like, by chance. Such *phenotypic drift* has nothing to do with natural selection except to the very general extent that the phenotypic changes must be compatible with successful reproduction.

Biologists are generally comfortable with the notion that a substantial fraction of *genetic* variation is due to chance aspects of reproduction from generation to generation (Chapter 3). In fact, random change of selectively neutral variation in DNA sequence has replaced selection as a theoretical baseline by which evolutionary dynamics are evaluated, and it can be shown theoretically that the effects of chance can even prevail over weak selection. But evolutionary biologists have been reluctant to apply the same view to phenotypes because of our long-standing adaptationist bias.

The working title of Darwin's *Origin of Species* was *Natural Selection* (1859), and he said his purpose was to show "that there is such an unerring power at work in natural selection." Wallace felt the same. Darwin later mused in *Descent of Man* (1872) that if he had overstressed the role of selection it was to show that creationist explanations were not needed to account for biological diversity. Selection provided a way out of creationist explanations.

Based on what could actually be observed in nature and domesticated species, adaptation could in principle occur—as Darwin repeatedly stressed—through the accumulation of *slight* variations, that is, slowly, gradually, and essentially deterministically. Even to Darwin, weak selection was the most prevalent kind, which is why he wanted the Earth to be old enough. These are just the circumstances in which phenotypic drift will occur and perhaps predominate. The general notion has been around for some time (e.g., Gulick 1872).

The difference between phenotypic drift and slow, local stepwise selection can be so little as to be philosophical rather than testable. In fact, darwinian selection is not even so parsimonious an explanation. It requires environmental factors that are steady, persistent, durable, and sensitive enough to mold organisms systematically in some way. Stepwise models help in some senses, but the meandering course of a series of exaptations is another way in which evolution is essentially "random." A stepwise scenario becomes nearly vacuous as an explanation if we take to heart the anthropic principle. It is at least as parsimonious to view chance as the ground state of phenotypic change and to invoke selection only when we have cause.

Energetics and Evolutionary Explanations: The Not-So-Thrifty Genotype

Characteristics of our age and culture include concepts of energetics, productivity, and efficiency. The evolution of biological traits is commonly explained the way we account for successful industry, as if natural selection can detect and favor subtle differences in form or physiology if they are more energy efficient. The idea is that efficiency would be favored because it costs less metabolic energy and hence less food and less struggle to acquire it. This essentially asserts that nature, like science, favors parsimony.

Energetic efficiency may be favored in many instances, especially in selection against very inefficient mutants, but this is not the only plausible view of life. Purging inefficiency in the face of mutations, organismal imperfections, and the sloppiness of the environment might levy a higher energetic cost on organisms than simply tolerating a degree of inefficiency. Such "noise" is pervasive in nature, and it is by no means clear whether natural selection is stringent enough to detect it. What reproduces, reproduces. Contrary to widespread notions of the genotype as a "thrifty" product of a prescriptive natural selection, *tolerance* of inefficiency may actually be a baseline—and more important—characteristic of evolution.

It is worth bearing in mind that it is we human interpreters who determine what is "efficient." One type of locomotion may be more efficient than another in terms of ground covered per calorie, but animals cover ground for particular purposes. Does a bird or bat expend fewer calories to catch a bug than a spider? Is it a mark of reptilian inefficiency that they cannot prey on big ungulates as a lion can? Clearly, energetics alone cannot be viewed as a very useful predictive or interpretive criterion. We should invoke energetic arguments only when there is a good, specific reason for them.

Three C's of Evolution

Evolutionary change involves the three C's of chance, competition, and cooperation in a way that is sometimes inextricable or even philosophically nondiscriminable. Different explanations can be offered for the same facts. For example, chance in small populations can have the same effect on variation as selection in larger ones. Afterward or even at the time, the empirical fact of change in heritable variation over generations can be translated as the result of chance, competition, or cooperation. A genetic variant that advances an organism's prospects has to be compatible with the cooperation among molecules and cells, and sometimes organisms, to succeed. If it succeeds relative to other variants at the same gene by leaving more copies in the next generation, this can be viewed as having out-competed the other variation and as "selfish." But the same result could be due to good luck.

In Chapter 3 we will see ways in which we may sometimes be able to distinguish between chance and selection, but this is usually a statistical criterion and not a direct observation of cause. Only in limited circumstances can we make convincing inferences of selection, and these usually require us to have a clear mechanism. But even classic cases of selection purportedly directly observed in action, such as industrial melanism in the peppered moth and the evolution of beak size and shape in the Galapagos finches, are not entirely clear-cut (Weiss 2002).

These points do not just relate to the degree to which our explanations are satisfying and consistent but can affect how we design experiments, draw generalizations, assess the role of genes in biological traits, and view Darwin's entangled bank of nature, the image with which he concluded his famous book (see Chapter 17).

GENES ARE INVOLVED IN EVERYTHING, BUT NOT EVERYTHING IS "GENETIC"

The Importance of Genes in Inheritance and Phenotype Determination

A century of unprecedented work has led to an understanding of the importance of genes as inherited material, as the molecule that stores the information from the history of life, and as determinants of the traits of organisms. Before genes were discovered and understood, it was difficult to explain inheritance and the evolution of organized traits. Genetics has become the central, theoretical organizing principle of biology.

However, recent work, and ideas that will be considered in this book, raise tempering questions about several aspects of the present view of genes. First, the connection between genes and traits is in many ways more indirect and subtle than most biologists have thought (or than many still seem to think). Second, although DNA is one of the most important and widespread constituents of new organisms, most of which begin life as single cells, some aspects of inheritance are not strictly based on DNA sequence. These include parental RNA and proteins, DNA packaging and modification, and other chemical characteristics of the cell. Many aspects of behavior and the construction of environments are also inherited (Chapter 3), and additional nongenetic aspects of inheritance may be identified in the future. Despite the undeniable and continued power of genetics to organize biological thinking and research and to account for the evolution of life, the degree to which variation, inheritance, or evolution should be described strictly in genetic terms is a more open question.

A Mendelian Illusion

Both Darwin and Gregor Mendel, the Moravian monk who first characterized the segregation of heritable traits with his experiments on pea plants, worked with very deterministic notions of "genes" (the term itself was not yet in use, of course), that is, with the heritable determinants of phenotypes. Darwin's theory of pangenesis turned out to be quite wrong, perhaps because he was thinking of a gradual change of continuous variation. Mendel, however, chose traits with simple transmission patterns that bred true in hybrids and that were essentially controlled by single "factors." That knowledge eventually led us to the genes themselves.

Mendel's discovery transformed biology, but in some ways we have become entrapped by the elegant simplicity of his choice of dichotomous traits manifesting "mendelian" inheritance. Many of the driving concepts of biology are built on this, and we still often pay little more than lip service to what goes on between the genotype an organism is born with and the phenotypes that develop in its lifetime. This is especially true of phenotypes not closely connected to the protein products of individual genes—the ones of most interest to Darwin and evolutionary biologists ever since.

In fact, the inner workings of life are far more complex than had been expected. In important ways, we attempt to force classical darwinian-mendelian theory in circumstances in which the fit is not so good. This does not mean that inherited traits do not involve genes, but genes are not always good predictors of traits, as will be seen in Chapters 3 through 5. Put another way, the mendelian inheritance of *genes* does not imply the mendelian inheritance of *phenotypes* (traits). Phenotypes are not inherited; organisms begin life as single cells with genes but not with arms, stomachs, or flowers.

Determinism, Reductionism, and Genetics

One of the fundamental aspects of most Western science is *reductionism*. The idea, attributed generally to influential thinkers such as Descartes and Francis Bacon, derives from our notions of empirical experimental design that the phenomena of nature can—indeed, perhaps should—be studied and understood part by part, ultimately all the way down to the most fundamental parts. This does not mean that each part acts independently, nor that we necessarily ever will understand all aspects of a trait. But it does assume that in principle we can come to a fundamental understanding of a phenomenon by isolating and analyzing its component effects. Just the way we disassemble a machine into its parts (discussed earlier).

The ultimate belief of reductionism is that the universe is (only) a space filled with matter and energy. If this view is true, then everything can in principle be "reduced" to, that is, ultimately explained in terms of (only), molecular and energetic phenomena. Biological phenomena, too, will ultimately be understood best in terms of the molecular biology of genes, the "atoms" of biological information (in some ways, biochemists would extend this even further down, of course). In this spirit, geneticists seek to study each trait in terms of genes, as separable causal elements. The objective is to explain a phenotype in terms of the effects of the individual genes that affect it—just the way we reassemble a machine from its parts.

Reductionism in biology works at various levels. Functional anatomists attempt to reduce traits like locomotion to the contribution of individual bones and muscles. To a psychologist, learning may be comprised of recognition, memory storage, memory recall, units of meaning, and ultimately to neurons and neurotransmitters. To an ecologist, an ecosystem consists of predators, prey, a food chain, etc.

Each biologist chooses his/her level of reduction. Some may choose to look only at the role of frogs in the biodiversity of a swamp without feeling compelled to try to explain the croaking of the frog in terms of its genes. Some wish to go farther and to "reduce" the croaking to hormone receptors, neuronal pathways, and the like. A biochemist may see this all as a problem in ligand-receptor binding, signal transduction, and gene regulation by action potentials of auditory hair cells.

Some biologists, although acknowledging that one can account for a frog in terms of chemicals, believe that reductionism cannot adequately *explain* the croaking. From this point of view, the phenomenon is an *emergent* one, that must be understood at its own level of organization. Field biologists may not even care to try to understand a bullfrog's croak and his mate's response in terms of DNA sequence data or hormone kinetics, a level of accounting in which they have no interest—any more than you might think the words you are reading could be understood by analyzing the chemistry of their ink and paper.

To some reductionists, higher-level studies barely count as important science, as they are too superficial. A common reason given is that we are not as good at making "operational" the study of complexity as we are at reductionist, experimental methods. The latter have a long history, and the triumphs of modern science and technology are the fruit. This view holds that higher-order phenomena are not *fundamental* and that eventually we *will* be able to predict "emergent" biological—or even cultural—traits by analyzing their components (e.g., Wilson 1998). Science does not allow nonmaterial causation, so how can an understanding of any phenomenon of nature not follow from an *adequate* understanding of its parts?

A reductionist perspective does not assert that causation is always one-to-one, but only that if we know all the actors, we will explain the play. An illustration of the issues involves gene action. Geneticists have long recognized that genes are often *pleiotropic*, that is, have many functions. Similarly, different genotypes can be found in individuals with the same phenotype. The genotype-to-phenotype relationship is often many-to-many in nature. Even if each component can be characterized in molecular terms, the overall effect of a gene on an organism, or of natural selection on a gene, may depend on the *set* of interacting constraints. We may not be able to predict the trait from any one of its components, but we should be able to do so from the set. An important question is when or how well we can ascertain or even define what that set is.

Arguments about reductionism are not new, and they are probably not resolvable, but the points are important in this book, whose aim is to understand the role of genes in how organisms manage their lives. But what it means to "understand" a phenomenon depends to a great extent on the question being asked.

Conservation, Variation, and Homology

With caveats raised earlier about horizontal transfer, evolution is generally diversifying over time. Based on this, a central organizing fact in evolutionary biology, indeed a key to what Darwin and Wallace contributed, is *homology*, the conservation of traits in descendants of a common ancestor. It is historical connectedness that differentiates evolution from creation. In the 20th century, as genes took the throne of the biosphere, the key element of homology tended to shift from shared traits to shared genes or DNA sequence elements. However, the more we learn about genes and how they are used, the more rather than the less elusive the concept of homology has become (e.g., Hall 1999; Wagner 1999; Wagner and Gauthier 1999).

For example, the limbs of tetrapods are considered to be descendants of fins in the common fishlike ancestor of these four-legged vertebrates, so limbs and fins are homologous in the classical sense. As expected, homologous genes are involved in limb development in different tetrapods. By contrast, eyes appear to have evolved independently in insects and vertebrates and have long been used as an example of *analogous* traits. However, developmental geneticists have recently discovered unexpected similarities in the mechanisms used to generate eyes in both groups of species—but only in some of the mechanisms and not all the same ones in all species. We will discuss this specific question in detail (e.g., Chapter 14). Here, the point is that recent discoveries of the genetic connectedness of diverse life forms show the need for a reformation of the very important homology-analogy question.

Ever since Mendel, and reinforced by the Central Dogma that one gene codes for one protein, the view that genes evolve "for" traits has been predominant. But this can be misleading in at least two important ways. On the one hand, unlike the eye situation, the genetic basis of similar, seemingly *homologous* traits sometimes turns out to be different (e.g., Hall 1999; Raff 1999; Wagner and Misof 1993; Weiss 2002; Weiss and Fullerton 2000). When this happens, the trait itself may be homologous in the traditional sense but not its underlying genes. On the other hand, it may be that a trait is produced by a developmental *process* that is completely conserved (homologous among species under comparison) but that the details of the process vary. For some traits with multiple elements, like body segments, the process and the nature of the trait may be conserved but the individual elements may not be homologous.

Another common view is that traits that have been shared since lineages diverged from a common ancestor have been conserved by natural selection. However, it is possible that the trait is conserved simply by the genealogical relationship. Even if selection is not acting, a trait will only change slowly between related lineages. Genes of humans and mice are around 80 percent identical on average, even in regions of the genome unlikely to have been seriously affected by natural selection. Perhaps of more potential importance, there are reasons to believe that in some instances, involving either particular chemical interactions or the interactions of many components, natural stable states or "attractors" may exist (such as "positions of organic stability" mentioned above). These may conserve a trait over long time periods or even provide an element of inevitability (e.g., Kirschner et al. 2000; Laughlin et al. 2000; Monod 1971; Morowitz et al. 2000; Schuster 2000)—in fact, as noted earlier, many aspects of life that seem complex may not be all that improbable in the first place (Keefe and Szostak 2001; Schuster 2000). Furthermore, a phenotype may be conserved but not its underlying genetic basis; the molecular structure of transfer RNA may be one example (Fontana and Schuster 1998). Similarly, as referred to earlier, natural chemical constraints would not need competitive natural selection in the usual sense, to be maintained. Indeed, once reached, it may be very difficult to get out of such canalizing constraints to try another way.

On Being a "Being"

For many centuries, the array of natural organisms was viewed in Western thought as a natural hierarchy, often referred to as the Great Chain of Being. The long-standing view was that a creation event had produced this natural order of life. In this book, we will see how the same sense of relationships was transformed by the idea of evolution. However, we will suggest ways in which, even from a genetic and evolutionary perspective, the notion of "being" has been too restrictive.

Biology is about beings, plural, not just the more philosophical phenomenon of "being"—the existence of life forms. The term is generally applied to organisms in the colloquial as well as scientific sense: a bird, ant, person, or birch tree. However, in many ways this is arbitrary: species, cells, and even genes can also be viewed as "beings." At these various levels of observation, the interactions within and among these entities are logically similar and, as we will see in various chapters, involve similar or identical genetic mechanisms. A broader, more flexible sense of the term "being" helps unify biology and make the origin of its complexity more readily understood.

WHY THESE CONCEPTUAL ISSUES ARE IMPORTANT

This chapter raised a number of issues, caveats, and perspectives that may impact our ability to understand the fundamental generalities of complex life. It is important to examine even the most basic of what are often tacit assumptions. Exceptions and alternative viewpoints can have considerable if sometimes unexpected merit, as we will try to show in various ways throughout the book.

The purpose of searching for generalizations about life is to be able to extrapolate from observations necessarily limited to only a subset of individuals, species, or model systems to as broad a scope as possible. Without observing everything, how far can we extrapolate? The answer is important for animal model work, agriculture, and biomedicine, as well as for understanding basic biology.

We would also like to reconstruct the unique history of life by taking specific account of the trace of past events left in DNA sequences or in their indirect manifestation in biological traits. For more than a century, this has been based on a strong branching model, but suppose this is not the right model? Suppose homology loses its clear-cut meaning before we get too far in the past, or life turns out not to have a single-trunked genealogy, or that genes turn out to be less determinative than has been thought.

In a historical field like evolutionary biology, the challenge of making retrospective evolutionary reconstructions is daunting. Unlike chemistry or physics, almost any general statement will have nontrivial exceptions, and for many biological observations there are multiple comparably plausible explanations (for example, chance, competition, and cooperation), and it is all the more important to put our notions to the test.

Classic darwinian theory has been exceedingly powerful at providing coherent explanations and has transformed thinking in biology essentially by equating similarity with genealogy. We caution against the overuse of the old Cartesian notion that an organism is a machine, but with linguistic irony our own evolutionary explanations are *fabricated*, in the literal sense. We make them up, hoping they are accurate re-*constructions* of a species' or trait's history. But we can rarely be too sure.

The historical connectedness of organisms and the consequent storage of historical information in genes provide unifying tools for understanding. But we can still ask important questions about how apparent order can emerge from the disorder of an undirected universe.

Chapter 3
Evolution By Phenotype: How Change Happens in Life

For the first part of the 20th century, the study of genetics was considered by many biologists to be separate from, or even irrelevant to, the processes by which darwinian evolution occurred. It did not seem possible that natural selection—*if* it was the mechanism of speciation—could work gradually, as Darwin had suggested, through the discrete particles of mendelian inheritance, whose known changes caused discretely different, or worse, grossly disruptive changes to the organism. However, as genetic understanding grew, it became possible to see how a unified theory of biology might work. By the 1930s, a group of leading biologists proposed what they referred to as the modern, or evolutionary, synthesis (Mayr 1982) that united the study of taxonomic relationships among species, the fossil record, and the theory of genetic inheritance into a single formal theory of evolutionary biology.

Before this time, it was difficult to have a rigorous, quantitative theory about the pace or nature (sometimes called the "tempo and mode") of evolution, and the theory was largely conceptual. But an assumption made by the modern synthesis, with widespread implications, was that genes are the fundamental elements of life—much as atoms are the units of chemistry and physics—and that evolution is to be explained in principle in terms of the processes of genetic change. The subsequent discovery of the nature and inheritance of DNA and its function as a protein and regulatory coding system greatly strengthened the gene-based view of life and provided a general research approach that predominates biology today.

Whatever the inherited material, if it is variable and particulate, so that each variant can be identified and not blend quantitatively with other variants, then the behavior of such variation over time and place can be quantified. If genes are the root units of biological causation, then the behavior of genetic variation over time will illustrate—and will *be*—evolution. The formal mathematical theory of evolu-

Genetics and the Logic of Evolution, by Kenneth M. Weiss and Anne V. Buchanan.
ISBN 0-471-23805-8

tion which describes this is called *population genetics*. Of course, organisms are more than genes, but to the extent that phenotypes (traits) can be ascribable to specific genotypes, it should be possible to subsume phenotypic evolution under the same theory. Here, we will present some of the basic principles of population genetics (thorough treatments are given by Gillespie 1998; Hartl and Clark 1997; Hedrick 2000; Lynch and Walsh 1998) and then discuss some of the subtleties that arise when considering that evolution works through phenotypes rather than directly through genotypes. We will refer to aspects of the nature of genes that will be explained specifically in Chapter 4 for readers not familiar with them.

EVOLUTION BY GENOTYPE: THE DARWINIAN POSTULATES FROM A GENE'S POINT OF VIEW

Population genetics is a kind of rigorous formalism about the evolution of genes. It can be done with no consideration of phenotypes and indeed many population geneticists specifically avoid dealing with the latter or with the connections between the evolution of genes and that of phenotypes. Population genetics is essentially a mathematical theory and hence is as rigorous as mathematics, which means that the theory and its usefulness depend on the values of various parameters, the degree to which they can be accurately known, and the degree to which the assumptions and formulas realistically reflect what goes on in life.

BASIC FREQUENCY CONCEPTS

The basic dynamics of variation over time can be described without being too specific about what is meant by "gene" beyond an identifiable, discrete, heritable unit of inheritance. We can assign a *relative* frequency to each of the alternative states, or *alleles*, that are found in a gene, *in some specified population of inference*; the latter can be a local deme or an entire species, as long as we realize that the analysis depends on this choice. Variation requires at least two alleles in our specified population.

Conceptually, an allele frequency can be viewed alternatively as the fraction of copies of the genes in the reference population that are of the specified type or as the probability that a randomly sampled gene from that population will be of that allele. Because the discussion is usually framed in terms of *relative* variation for a specific genetic unit in a specified population unit, by definition the frequencies of the alternative alleles in our population must sum to 1.0.

To present the logic of genetic theory, a simple situation is usually envisioned, in which the gene in the population has only two alleles that, ever since Mendel, have been conventionally labeled as A and a. We denote the frequency of A as p_A; in addition, because there are only two alleles, a must comprise the rest of the population, so its frequency must be $1.0 - p_A$. That is, $p_A + p_a = 1.0$. This can be seen schematically in Figure 3-1. It is important to keep in mind that this refers to the proportion rather than the number of copies of each allele *in the population of inference*. When working only with *samples* from that population, all of this refers to the characteristics of the sample, which if one is careful can be used as estimates of the true situation in the entire population.

Another important concept is the *genotype frequency*, again with regard to a reference population or a sample from it. Many species, including most animals, are

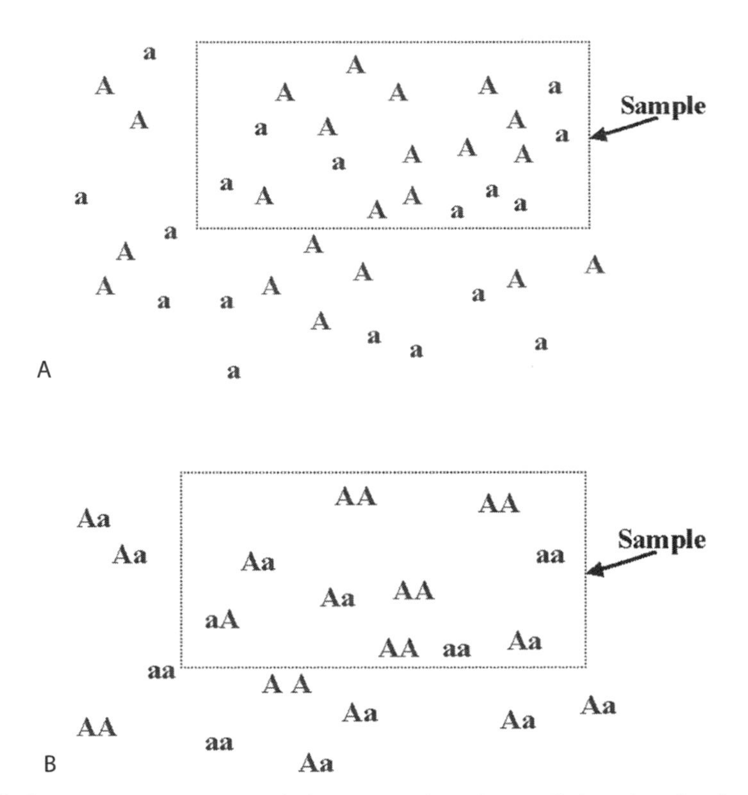

Figure 3-1. Frequency concepts. (A) Frequencies of two alleles, A and a, in a sample of 20 alleles from a population; $p_A = 12/20 = 0.6$, $p_a = 8/20 = 0.4$; (B) frequencies of genotypes $p_{AA} = 0.4$, $p_{Aa} = 0.4$, and $p_{aa} = 0.2$ in a sample of 10 individuals.

diploid, meaning that (like humans) they inherit a copy of each gene from each of two parents. This means that what happens to alleles in the population happens to them as they are carried around in pairs. The variants in a given individual are referred to as its *genotype*. (Some species like bacteria are *haploid* and have only a single copy of each gene; for them their genotype is the same as their allele. There are other ploidies in nature, but these need not be considered here to understand the basic principles.) In a diploid species in the simple case (Figure 3-1), the genotype must be a *homozygote*, *AA* or *aa*, with two copies of the same allele, or a *heterozygote*, *Aa*, with two different alleles. The relative frequencies of these genotypes can be denoted as p_{AA}, p_{Aa}, and p_{aa}. Again, because these are the only possible genotypes for this gene, $p_{AA} + p_{Aa} + p_{aa} = 1.0$.

Because a diploid genotype represents the contributions to the individual from each of two parents, the genotype distribution in a given generation is the product of the mate choice and reproductive pattern in the previous generation. Of specific interest is a baseline condition called *panmixia*, in which individuals choose their mates randomly relative to the genotype of mates choosing each other. When this occurs, the frequencies of the genotypes are just the product of the frequencies of each allele. Thus, the probability that a random individual has one A allele is p_A, and the probability that this individual has a second A is also p_A, so that we have $p_{AA} = p_A p_A = p_A^2$ and by similar reasoning $p_{Aa} = 2p_A p_a$ and $p_{aa} = p_a^2$. These are known as the *Hardy-Weinberg* genotype frequencies, and when they characterize a population it is said to be in Hardy-Weinberg Equilibrium (HWE) because, as has been shown mathematically, under idealized conditions the genotype frequencies will not change over time.

Mating is in fact typically strikingly close to random with respect to most genes because mate choice is unaffected by a mate's specific variants at most genes, and HWE serves as a baseline from which to judge observed genotype frequencies. If these differ in a statistically significant way from what is expected under random mating, there may be reason to investigate why this is so, and there are many possible reasons, including mating that *is* affected by the gene in question, or by natural selection.

These are the basic frequency measures needed in order to describe the essential concepts of evolutionary change from a gene's eye view.

Mutation: Change of State in the Genome

All genetic change ultimately comes about through mutation. Mutation can do many things that will be discussed in Chapter 4 and beyond. These changes can alter a gene's function, expression pattern, or structure. In some instances, new genes can be inserted from outside the individual, as for example when viral particles integrate into the genome. Such a change, if it occurs in the germ line and is transmitted to the next generation, constitutes horizontal transmission described earlier.

Frequency Change Over Time

Each new variant arises with allele frequency $1/N$, where N is the number of copies of the gene in the population of reference. (The frequency is $1/2N$ in a diploid species, since each of the N individuals in the population has two copies of the gene.)

Over time, alleles experience changes in their frequency. From one generation to the next, the proportion of a given allele may differ for several reasons, the primary one being chance. In a diploid species, for example, one of the two alleles that came together to form the parent randomly segregates into the germ cell to be transmitted to any given offspring. This is mendelian *segregation*. Which of these two *possible* outcomes occurs in any given case is inherently *probabilistic*.

Mendelian segregation introduces a fundamental element of chance in allele frequency change when individuals reproduce, and many other aspects of chance in survival, mate acquisition, or fertility affect whether that reproduction will occur in the first place. Under various assumptions, there are ways to quantify the relative probabilities of various outcomes of mendelian segregation in any family or in an entire population, and hence the distribution of possible allele frequencies in an offspring generation, if the frequencies in the parental generation are known.

The phenomenon of allele frequency change due strictly to chance is known as *random genetic drift*. We encountered the metaphor of "drift" earlier in discussing random changes in the distribution of phenotypes in a population. Frequency "drifts" randomly up or down over time, until a variant is eventually either fixed in or lost from the population, as illustrated in Figure 3-2.

Genetic drift is *inevitable*. Based on assumptions about the population, the expected (average, over many replicates could they occur) change in allele frequency from one generation to the next and the probabilities of any particular

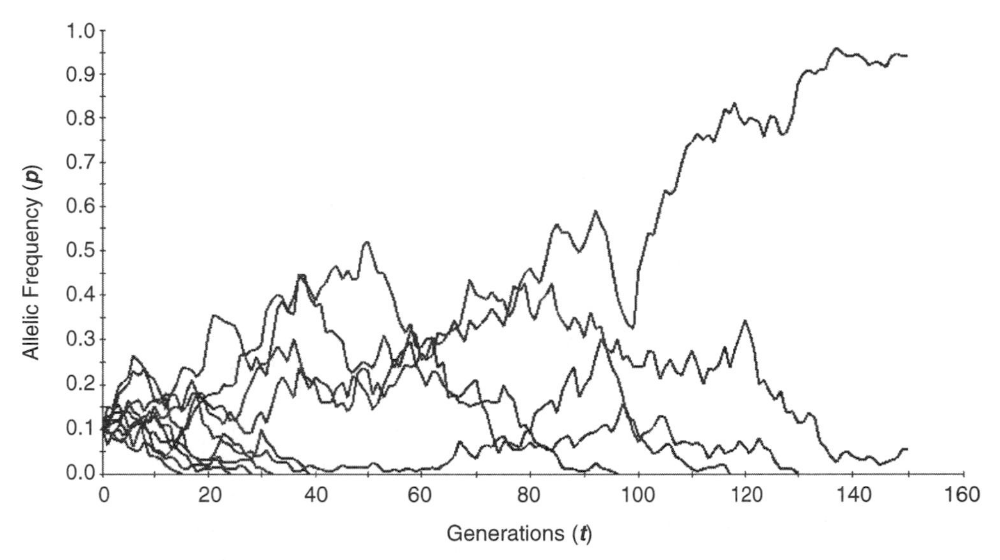

Figure 3-2. Schematic of genetic drift. Each line represents a separate "experiment" of an allele starting with a frequency of 0.10 followed over time in a population of size 100. In these computer-simulated data, frequency change is due to chance alone. Any frequency is achievable, but most new alleles quickly disappear while a few are lucky and rise to high frequency. In any population multiple alleles may be segregating, their relative frequencies drifting up or down over time, some being lost, new alleles generated from time to time by mutation. The fact that an allele has high frequency does not imply a history of adaptive natural selection having favored it. Usually, only statistical arguments can determine when that is most likely. Data simulated with Populus (Alstad 2003).

outcome can be specified. It is, however, not possible to specify what *will* happen in any given situation, Drift calculations are almost always approximations, but data often seem at least broadly consistent with the drift expectations—that is, as if drift were the *only* factor affecting allele frequencies.

Whether due just to chance or to other factors, allele frequencies can change substantially over time. If chancc alone is operating, descendant copies of any selected allele will *eventually* either drift to a frequency of zero (be lost) or to 1.0 (at which point it is *fixed*, that is, replaces all other alleles at the locus in the population). The reason is simple: there is always *some* probability of fixation or loss, and an allele can never return from the dead or change in frequency once fixed.

The *rate* of drift, or probability that fixation or loss will happen in any given time period, depend on population size and structure (mating patterns, subdivisions, and so on). This is because genetic drift is a kind of sampling from one generation to the next, and the variance of a sampling process depends on the sample size. In a large population the sample is large, and the sampling variance small, so that allele frequency changes little over a generation. However, actual species may be divided into small local populations in which drift can be important and rapid. Thus, *population structure as well as gene function are important determinants of evolutionary change*. This may seem to be a trivial consideration because species typically seem to have very large numbers of individuals. But they may not always have that, and in any case the Earth is old enough relative to population size that there has been time for drift to have been a major, if not *the* major, factor in genetic change during the history of life.

There is a continual flow of *substitutions* (one allele replacing another in the population) as new variation arises in each generation, and some fixation or loss of variants takes place. If the population size doesn't change, there is no in or out migration, and mutation rates stay the same, a steady state will arise between fixation, loss, and mutation. An ideal steady-state population will come to have a standing level of variation that depends on the relative strengths of these factors. However, and again and most importantly, the *specific* variants that are present, and their frequencies, will continually change—and will vary among local populations. Different genes and different populations accumulate differences over time in a statistically regular way.

Because this is a random process, it is impossible to say which of the alleles present today will be around at some point in the distant future. However, under idealized conditions, we can estimate various aspects of the process such as how long it would take for a given number of genetic differences to accumulate in copies descended from a single individual. One can thus look backward in time to estimate how long it has been since copies of a gene we see today shared a common ancestor. With some important technical caveats, genetic variation can serve as a *molecular clock* of evolution (e.g., Li 1997; Nei 1987). These are purely probabilistic phemenona, and population genetic models are at best crudely approximate; therefore, time estimates have a large statistical error. This can be reduced by looking at enough data. For example, the clock can be calibrated by external information such as experimental estimates of mutation rates and the age of relevant fossils representing the common ancestor of species being considered.

If the calibrating parameter values like population size change, then the timing and/or form of the numerical results will change, but the basic idea of a dynamic flux of variation holds. Effects can be estimated by making some simplifying assumptions such as that the population is in demographic equilibrium. Looking *forward*, a new mutation has very little chance of surviving very long and only about a $1/N$ chance that its descendant copies will become fixed in the population. But looking *backward*, from *any* gene, all copies extant today must be descendants of a single ancestral copy at some point in the past. This common ancestral copy is known as the *coalescent*. The existence of a coalescent is implied by the assumption that all life derives from a common ancestor and that mutation is a divergent, random process that generates new branches on the tree of descent from that common origin. Even if there is no single tree of life, as discussed earlier in regard to horizontal transfer of genetic material among species, copies of the different genes present today can in principle individually be traced back to a common ancestor, and molecular clocks are used to estimate when that was. This is true even if the individual genes in an individual each go back to a different ancestral copy, in a different individual, in a different place, and at a different time.

Common ancestry is perhaps the most important organizing factor in biology and was Darwin and Wallace's major contribution. Expressed in modern genetic terms, Darwin and Wallace's theory was based on branching divergence in genes produced by mutation aided by natural selection screening on allelic variation. Genetic drift will also bring about divergence, if not so systematically as Darwin's idea of continual natural selection. Even just by genetic drift, lineages (especially if sequestered by geographic isolation) will accumulate enough genetic differences over time that their members can no longer (or do no longer) mate with members of other lineages. This mating criterion is generally used to define "species." Darwin and Wallace provided a *process-based* explanation for the patterns of similarity among species that were already known. Population genetics tries to emulate that process, at least in regard to the genes that are involved.

Systematic Aspects of Evolution and Biodiversity

To say that mutation, mendelian transmission, and genetic drift all involve fundamentally random elements does not mean that they appear random because of inadequate data or insufficient knowledge on science's part about the underlying truth. Chance is an *inherent* aspect of life. Despite this, the prevailing view in biology is basically deterministic and is concerned with the molding of biological traits by natural selection. We discussed this phenomenon conceptually in Chapter 2, but here we can see how population genetic theory deals with it.

Differential Success: What Is "Adaptation"?

Everyone has an informal idea of natural selection. Usually it centers around the idea that the fittest organisms are those that are the most successful, that is, that reproduce the most. Over time, the population will comprise an ever higher frequency of these fittest organisms as they disproportionately leave descendants. From a genetic point of view, the *reason* they reproduce the most is not relevant so long as the outcome is systematic over time and, because genes carry information about organisms, that information persists and proliferates. We say that as a result the successful organisms have *adapted* to whatever was imposing the selective screen. Of course, the environment can change and new variation is always being introduced, so that the process may never reach an end. That is one aspect of the stepwise adaptation model described in Chapter 2.

Population genetics formalizes these notions by assigning *relative* selective or "fitness" coefficients to each genotype being considered. For example, if *Aa* and *aa* genotypes are lethal because of the effects of the *a* allele, these genotypes can be assigned a *fitness* coefficient (often denoted by w) of 0, and 1.0 is assigned to the *AA* genotype because in relative terms it is the most fit genotype. The corresponding *selective* coefficients, denoted s, are the complements of the fitnesses, ranging from 1.0 for traits with no chance to reproduce and 0.0 for genotypes experiencing no deficit in fitness (in some treatments, s is used for the relative beneficial or harmful effect on fitness of a specific allele rather than genotype, but this is a technical detail about the way population genetic models are used).

If in such a situation the *a* allele has frequency p_a at some time, we can compute how fast that frequency will change in the face of selection. As shown by a simple example in Figure 3-3, a favored allele increases steadily in frequency and will eventually become fixed in the population (that is, will be at a frequency of 1.0). There are many subtleties and various ways that selection may act, but this is the essence of the genetic theory of darwinian adaptive evolution.

Because selective coefficients are usually treated in relative terms, the fitness assigned to a genotype is measured not in terms of the inherent value of the organism but by comparison to the most fit genotype *at a given time and in the population of inference*. In the simple illustration of Figure 3-3, only one allele is ultimately

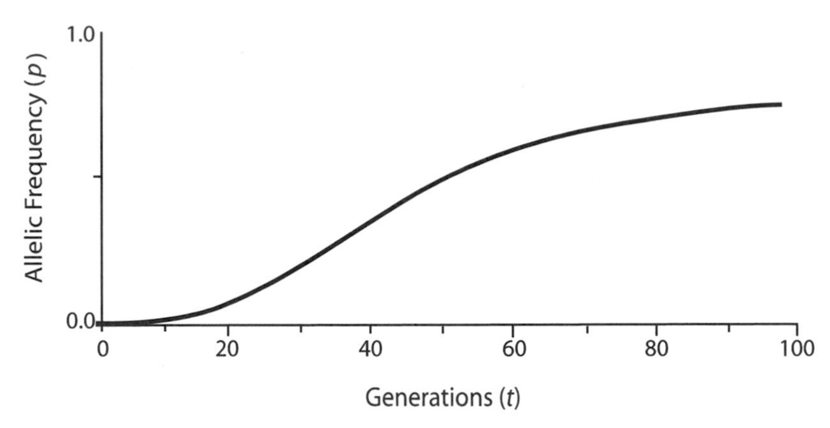

Figure 3-3. Effect of natural selection on allele frequencies. The increase over time of the frequency of an allele, *A*, in a population in which the *aa* genotype had a 10 percent disadvantage. These data are simulated by the method used in Figure 3-2, but here deterministic selection is simulated, with no factor of genetic drift. The allele's rise to high frequency in this case is due entirely to selection.

successful, but the theory doesn't imply "survival of the fittest" in the literal sense that in real populations only one genotype is successful while all others fail. There are typically many genotypes with comparable fitness, and even were there a single most-fit genotype, it can become disfavored the moment a new allele arrives by mutation or gene flow (incoming migrants). Something different may occur in other populations of the same species.

This is a formally competitive model: if an allele increases in frequency we can define it as having outcompeted other alleles in its given time and place. However, this says nothing about how that is brought about in real organisms, or what "competition" means in any particular case. And of course, having done better in this sense does not mean having been good or perfect.

Selection can act in many ways, including *balancing selection* (sometimes called *heterosis*, for heterozygote advantage), that maintain rather than exhaust variation by favoring or disfavoring multiple alleles in different genotypes. In the classic example, sickle cell anemia, the AA and SS genotypes are both harmful (one causes anemia, the other susceptibility to malaria), whereas the heterozygote AS genotype is better in both regards. A stable *balanced polymorphism* results over time, with unchanging allele frequencies that depend on the relative fitness values of these three genotypes. Other evolutionary strategies than allelic competition can generate heterogeneity in systems we will see later on, such as in immune resistance and olfaction, and we will see why that is important.

It is not unusual for individuals with extreme values for a trait to do less well in life than those near the average, presumably because the average is in some sense the result of an adaptive history. But this is not always the case, nor does it imply that variation in any individual gene associated with the trait is maintained in this way. Each trait and each gene are different.

Empirically, there are typically many alleles at a locus all the time, continually changing in frequency. Assigning fitness in relative terms, with 1.0 given to the best available genotype, means that the theory in this form does not specify whether the population grows or shrinks in absolute size as a result of selection. Population growth can be taken into account, and although it complicates the theory somewhat it typically does not change the way that relative fitness competition is at the core of the theoretical model.

What's In That "*s*"?

Genetic drift is omnipresent. Nonetheless, in classroom and some classical versions of population genetics (and essentially in Darwin's notions), selection coefficients are treated basically as inherent, deterministic properties of genotypes (this is what was simulated in Figure 3-3 as well). In truth, selection or fitness are probabilistic phenomena. An individual with the fittest genotype in our scheme might be struck down by lightning before reproducing. Does that mean that that individual's fitness was 0? The answer is a rather curious kind of "no" because we conceive of fitness or selection coefficients as applying to a genotype in a probabilistic sense, on average. That is, a genotype's average fitness does not mean that every individual with that genotype will have identical reproduction. In computer simulations, even if chance is built into models that include selection, we know the truth (a value of *s* is assigned to a genotype by the programmer). But it is much more ambiguous in reality.

What appears to be clearly selective may not be in the causal sense. Did that moth really get eaten because it did not have good camouflage and the bird saw it sitting on the tree trunk? Or because the bird noticed a leg moving, or smelled it—or chanced upon it? How can we know? These are not simple questions to answer, even in what seem like classically easy cases (Weiss 2002).

At best, the fitness of a genotype has to be estimated by observing many individuals who bear it and seeing how well they did. But if it were necessary to observe every individual event, and then determine which instances of survival, death, reproduction, or infertility "counted" specifically as being causally due to the effects of the specific genotype, rather than to other characteristics of the individuals, or to chance, we would be delving into epistemological quicksand. Usually, only when selection is consistent and rather strong can its effects be convincingly detected. However, it can be shown that the less that selection coefficients vary among existing genotypes in a population, or the more they appear to vary from one generation to the next, the more the frequency changes of the alleles will behave as if they were selectively neutral (affected by chance alone). With massive die-offs, it might seem that selection would be quite strong, and we use the evolution of antibiotic or pesticide resistances as examples. But massive die-offs in nature need not involve such selection, because they can affect individuals regardless of their genotypes. But massive die-offs, by their consequent reductions in population size, reduce the amount of variation available for selection to work with. The reason is that in a massive die-off much variation will be lost by chance.

Quantitatively, if we treat selection as a fixed property of genotypes, and apply it probabilistically, and if the absolute value of the product of population size and selective coefficient (Ns) is small, say much less than 1.0 or 2.0, then drift will predominate over selection in determining the future course of allele frequencies (e.g., Hedrick 2000). In fact, most genotypes most of the time in nature seem to be only weakly affected by selection, as discussed in Chapter 2, which essentially means that natural selection is not as discriminating or prescriptive as would be expected in the deterministic gradual adaptation described by a strictly darwinian view of life.

In general, it is difficult to appreciate the degree to which the fitness of various genotypes is empirical and contextual rather than inherent properties of the genotypes (Lewontin 2000; Schlichting and Pigliucci 1998). When genotypes are nearly neutral, their small selective coefficients are unlikely to be stable over time or across environments. Because they are formally treated as relative values of competition among peers, the coefficients depend not only on the constraints of the environment but also on the other genotypes in the population (and the genetic variation in *all* other genes in each individual).

After the fact, we might be able to estimate an *average* or net fitness for a given genotype over some time period, which could account mathematically for the observed net allele frequency change. However, this is not necessarily a good way to explain what happened to individual organisms in their individual lives over long periods of evolutionary time, much less *why* it happened. Yet this is important if we are to give biological meaning to a probabilistic s. For example, selection might work only through occasional episodes of intense screening or may only trim away genotypes associated with extreme phenotypes, otherwise leaving the field to drift. Except under unusually favorable circumstances, population genetic models of a given situation are basically schematic—sometimes even when the causal process seems quite clear (Weiss 2002).

With so much uncertainty in this system, or weak and perhaps changeable effects, it could require studies that were themselves on the evolutionary or whole-species scale to generate a satisfactory understanding (e.g., see Tautz 2000). This is consistent with the general contingent, step-by-step view of the evolution of complex traits. Although rarely put this way, it means that even when a trait is adaptively evolving, most of the time in most individuals most vital events are due mostly to chance (although, like every trait, every gene can be viewed as "adaptive" by definition, because it is here to be observed). And this also means tolerance of variation by the screen of selection, and that selection is correspondingly a less precise or prescriptive molder of traits.

The average or net fitness of a genotype does not say much about daily experience, but fitness does not just suddenly occur. The determinants of fitness act over the lifetime of an individual, and depend inherently on the age-specific aspects of survival and reproduction of organisms. These can be expressed in terms of the vital rate schedules, $l(a)$ and $f(a)$, the rates (or more properly, probabilities) of survival and reproduction, respectively, of organisms age a. Fitness is related to the product of these two, $l(a)f(a)$, over the lifespan, and these schedules have to be estimated specifically for each genotype. This must be done in the same population at the same time, and for a long enough time and large enough sample size to obtain useful estimates. In fact, these are highly probabilistic and not necessarily stable over time, even when the gene is having some effect. Does the genotype (or a trait with which it is associated) only affect reproduction or survival early in life? Or at certain life history stages such as larval or embryonic periods? Does it affect longevity?

Fitness also depends on successful mating in sexually reproducing species, which means that each individual's fitness depends on the genotype (and hence fitness) of its mate(s)—and this is especially complicated with overlapping generations, when mates are not chosen strictly from birth cohort peers.

The formal genetic theory of evolution by natural selection was famously articulated by one of the founders of population genetics, R. A. Fisher (Fisher 1930), and has been augmented by many others (e.g., Cavalli-Sforza and Bodmer 1971;

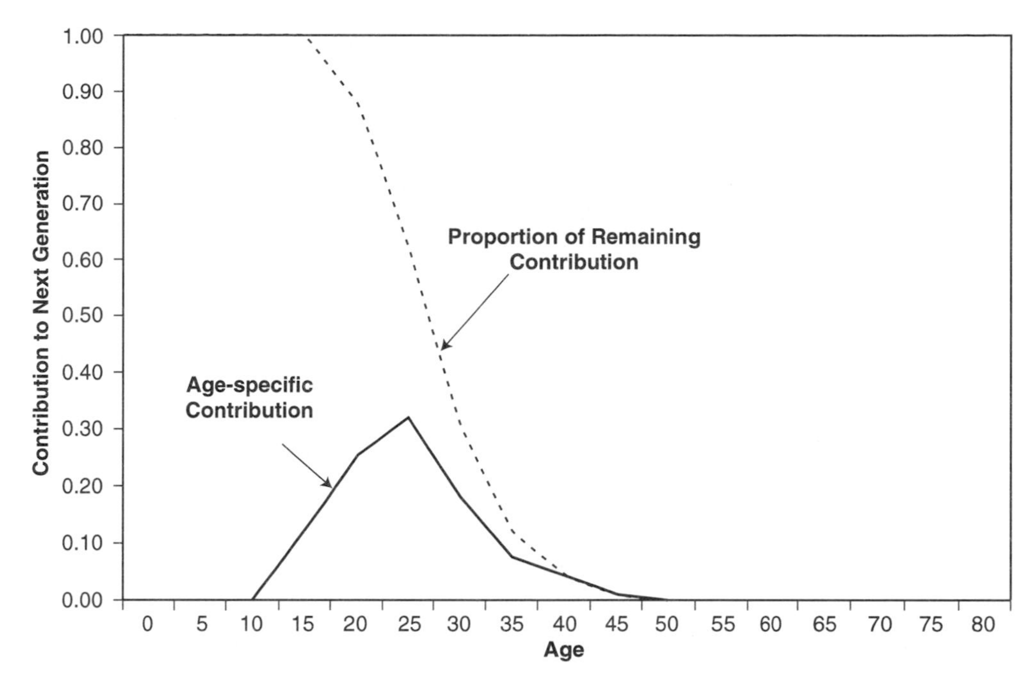

Figure 3-4. Schematic of human age-specific survivorship and fertility $l(a)f(a)$ schedules, showing age-related contribution to the next generation, and that the relative future contribution is greatest at puberty. The dashed line shows the proportion of remaining lifetime contribution as a function of age. Conceptually, each genotype would have its own such schedule, and that determines fitness or selection coefficients. In practice, these are usually weak inferences at best. The important point is that this is how fitness actually works on the ground and what must be estimated to truly understand its quantitative effects. It is worth keeping this in mind as we go through the diverse traits in nature and attempt to provide evolutionary explanations.

Charlesworth 1994; Crow and Kimura 1971). While we stress the problems with the concept to draw attention to the temptation to apply it uncritically or too universally, we can use it conceptually to see aspects of how selection might be expected to act. For example, using a term introduced by Fisher, an individual at a given age has a *reproductive value* that is determined by its genotype's vital rate schedules. This has been used to demonstrate the differential potential for selection to have an effect at various ages in the organism's lifespan (e.g., Crow 1958). This concept can help us understand how selection might work.

Figure 3-4 provides a schematic of human $l(a)f(a)$ schedules (Figure 3-4). Fertility is zero until puberty, then peaks for a while before gradually tapering off. Meanwhile, mortality rates increase from puberty onward. Overall, as shown by the dashed line, the amount of remaining reproductive potential is greatest at the beginning of adult life and declines thereafter. Events affecting young individuals can have a disproportionately great impact on reproductive success than events affecting them at later ages. So selection impacting young individuals will affect allele frequencies more rapidly than effects with later impact. Genotypes aiding in early reproduction more rapidly produce offspring than other genotypes, and their alleles outgrow their competitors over time. But late in life, differential mortality or fertility have little if any impact on the individual's total reproductive output. Even if a genotype leads somehow to an awful disease at such an age, it is not selected against in the evolutionary sense. To that extent, the alleles are selectively neutral. Thus, for example, one might in vain search for evolutionary explanations for the frequency of alleles that led to high late-onset cancer risk in humans. But an allele with high frequency that causes a serious childhood disease should have been eliminated by selection; if it isn't, one suspects some form of balancing selection has been at work.

The Illusion of Adaptation Revisited

It may now be easier to understand why in Chapter 2 we said that the notion of adaptation was in some senses a seductive illusion, if not a tautology. To be here today, an organism had to have been "adaptive"—sufficiently suited to its ancestral environments for an uninterrupted 3 billion year ancestry. What criterion can we use to infer anything beyond that? With genes, unlike phenotypes, we can at least define more directly specific units of evolutionary change, and this allows for somewhat more believable quantitation of effects.

Typically, we *define* selection as taking place when an observed pattern is unlikely to have occurred by chance alone. "Unlikely" is defined by choosing a statistical significance cutoff level, like 5 percent; we can use this convention to attribute to selection a pattern less likely than that cutoff value to have occurred by genetic drift alone. If the cutoff is 5 percent, that means we accept a 1 in 20 chance of an inference being wrong. Because drift is a sampling process, the probability that we estimate is only as accurate as our understanding of relevant evolutionary parameters such as sample size, population size, and the stochastic factors in birth, death, competition, and cooperation. These are rarely very precise, and the choice of a significance level is also, of course, entirely arbitrary. That this can be problematic can be seen in Figure 3-2. Consider the allele that became fixed. How can we tell whether it arrived at that frequency by selection or by drift?

In practice, indirect tactics are used. If we can make plausible estimates of the parameters, such as population size and mutation rates, then the amount of standing variation that would be produced by genetic drift alone can be estimated, and compared to that in a sample. If the time since divergence can be estimated (e.g., from the fossil record), population genetics theory can be used to estimate the amount of difference that would have accumulated by drift.

The amount of variation can be compared in different parts of the genome, searching for regions of reduced variability. There are biological reasons to think that some parts of the genome are more affected by selection than others. Figure 3-5 shows that this is reflected in the relative rates of accumulation of genetic divergence in different parts of genes observed between species. The relative amounts of standing genetic variation within species shows similar patterns. The reasons are the same: we expect and typically find less variation in regions with known function on the assumption that the effects of natural selection will generally prune variation that arises in such regions, relative to selectively neutral regions.

Particular attention is paid to the pattern of variation in regions that can most safely be assumed to have no function that could affect fitness. This variation then becomes the baseline. Regions showing less variation are assumed to have been constrained in some way. Likewise, parts of genes that are highly conserved (similar)

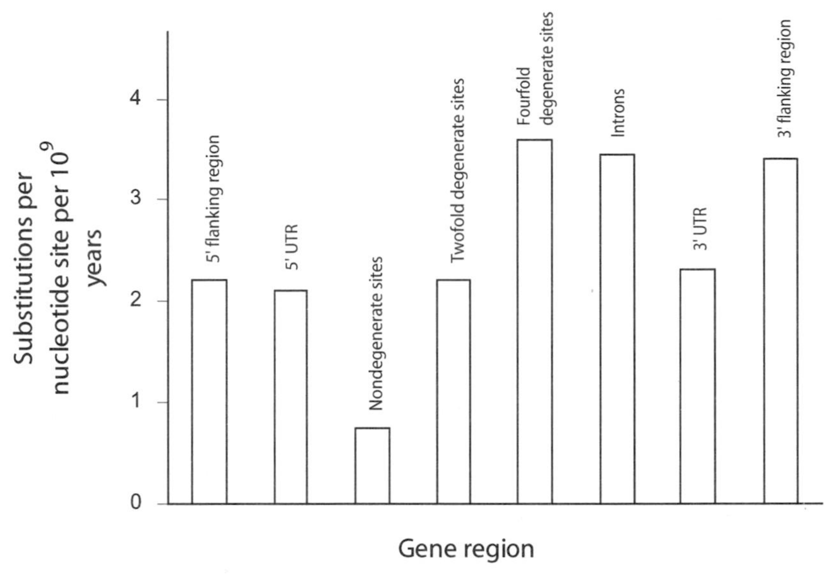

Figure 3-5. Relative rate of accumulation of variability in different parts of a stereotyped gene over evolutionary time. The units are the rate of nucleotide (basic DNA unit) substitution per 10^9 years. 5′ flanking, 5′UTR (untranslated region), and introns are not used to code for protein. 3′UTR is part of the protein coding mechanism. Degenerate sites are parts of DNA for which nucleotide variation has no difference on a coded protein. Fourfold degenerate sites are essentially freely mutable in that sense. Twofold degenerate sites, and nondegenerate sites will, if mutated, generally affect the coded protein. These regions are less variable. For details, see Chapter 4. Redrawn from data in (Graur and Li 2000).

between distantly related species are assumed to have been maintained by the effects of selection.

These indicators of the existence of past selection work because there are thousands of units that can be studied in a single analysis, so that the sample size at least approximates a replicable process (on the assumption that similar types of regions in different genes behave similarly). But the analysis is usually only statistical and approximate. It may indicate *that* selection has occurred, but nothing about the *particular* cause and may not support any particular adaptive story. There may be alternatives that do not involve selection, or involve it in some other way. A specific hypothesized scenario has to be supported with knowledge of specific function or by showing experimentally that a given DNA variant has an effect on fitness. But even this does not tell us directly whether the function that we observe today is the same as that favored by selection in the past.

The idea of "adaptation" is in some sense fundamentally philosophical. It is by a kind of idealism that fitness or selective coefficients are viewed as being permanent, inherent, or deterministic properties of the genes themselves rather than changeable and contextual phenomena. In experimental or agricultural selection, the theoretical principles can be observed and applied because we can control the situation to be, essentially, what is specified in the theory. Longterm adaptation, however, is too slow to observe directly. Most real selective coefficients are quite small, so that the evolutionary process is usually meandering, having a changeable opportunistic direction, and highly stochastic. The overall statistical analysis of DNA variation in identified functional elements generally suggests only rather weak average selection. Bursts of local intense selection do occur; but as a general rule, drift is the baseline characteristic of evolutionary genetic change.

At the Sequence Level, a Hierarchy of Structure

Population genetics is designed to explain and quantify change in genetic variation over time. It was initially developed from a consideration of genes simply as variable "things." Discoveries during the 20th century about what those things are like have revealed additional complexities and characteristics of variation.

Hierarchy of Change Due to the Modular Nature of DNA Sequence

As we will see in Chapter 4, DNA is organized as a string of modular units. This leads to the phenomenon of *molecular evolution* on which so much of life is based. To a great extent, new mutation is random with regard to its specific location in a gene, and, although some mutation occurs every time the DNA is copied during cell division, mutation in any *particular* location in any particular replication is rare. As a consequence, over time, mutations generate a nested or hierarchical pattern of variation. A central fact in life and its evolution is that this hierarchical accumulation of genetic divergence is what allows organisms to accumulate functional divergence, and it is in this sense also that traits as well as genes bear the history whose common origin it was Darwin and Wallace's insight to realize.

We can illustrate molecular evolution by a simple contrived example, in which a starting sequence of units, say **AACCC**, experiences a mutation that modifies the sequence to **AGCCC** in a descendant copy, and a subsequent change may change it further to **AGGCC**. Meanwhile, a different descendant copy of the original **AACCC** may mutate to **AACTC**, and so on. These sequences bear the trace of their history: the molecule evolves. That cannot be said of sugar or lipid molecules.

Figure 3-6 shows this kind of descent tree. Panel A shows the *demographic* history of the sequences we observe today. Regardless of its actual sequence details, each copy of a gene today has an ancestor, and all copies coalesce to a common ancestor in the past. Note that this is based on the *assumption* that evolution has proceeded as our understanding of life would have it. Panel B shows how mutational variation accumulates in descendant copies of the original sequence (the coalescent). If we assume this kind of process, the figure shows why we view the set of today's copies as having a nested, hierarchical pattern of relationships and population genetics theory allows us to make an educated guess as to the history of

Figure 3-6. The nature of gene sequence evolution. (A) Demographic (parent-offspring) ancestry of samples collected today, going back to the MRCA (most recent common ancestor) of the samples; (B) with no recombination, sequence relationships go back to a common ancestor (the coalescent) in a simple nested way; (C) with recombination (double arrow), segments of the sequence (in middle two samples) have separate paths back to the MRCA shared by all copies. Present-day samples in **boldface**, ancestral copies normal font. Lightning represents mutations.

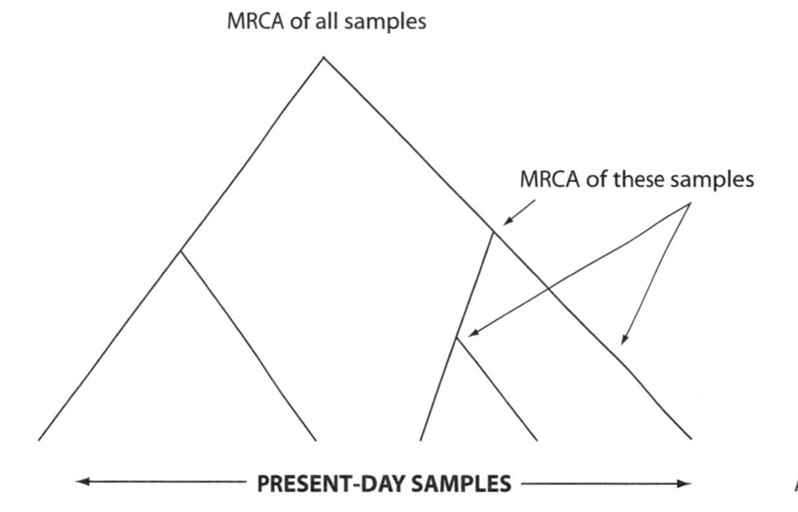
MRCA of all samples
MRCA of these samples
PRESENT-DAY SAMPLES
A

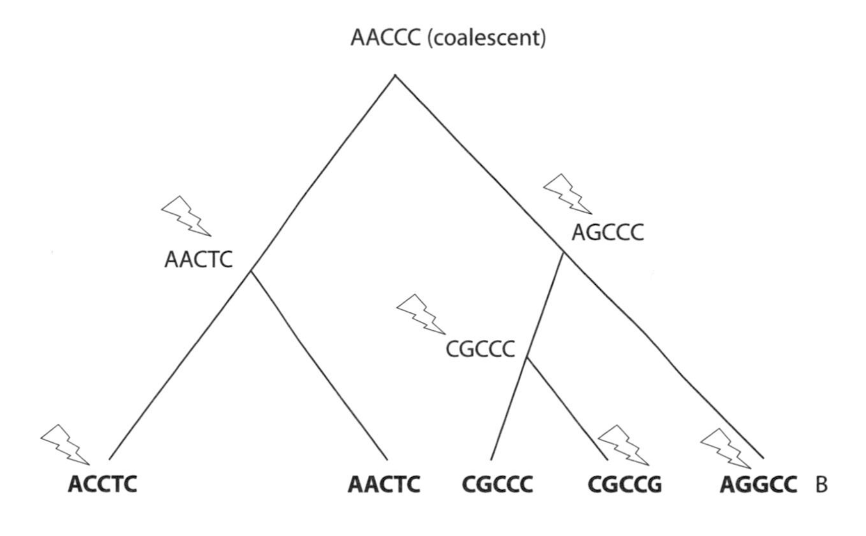
AACCC (coalescent)
AACTC
AGCCC
CGCCC
ACCTC
AACTC
CGCCC
CGCCG
AGGCC
B

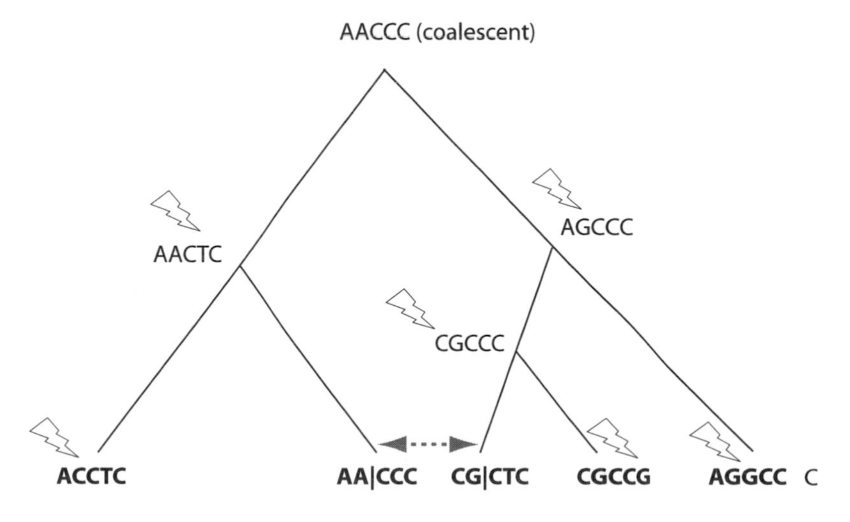
AACCC (coalescent)
AACTC
AGCCC
CGCCC
ACCTC
AA|CCC
CG|CTC
CGCCG
AGGCC
C

mutation (Li 1997; Nei 1987; Nei and Kumar 2000; Nordborg and Tavare 2002; Rosenberg and Nordborg 2002) and sometimes to reconstruct probable ancestral sequences. Viewed top down, Figure 3-6 shows an actual history, but looking bottom up it would have to be viewed as an inferred hypothetical history, because the ancestral sequences are not directly observed.

Panel C introduces the concept of *recombination*. Each generation, before reproduction in diploid organisms, the two copies of a chromosome align and occasionally exchange pieces. The figure shows that as a result different parts of the gene have different ancestral histories, though eventually going back to a common ancestor.

This type of analysis of DNA sequence data can be done to reconstruct the evolution of individual genes, or for multiple genes, sampled from various populations, to reconstruct species phylogenies (evolutionary histories) (Nei and Kumar 2000). Whole genes occasionally duplicate, and each copy then accumulates variation over time. The result is a set of divergent genes known as a *gene family*. Duplication events are a kind of mutation and the history of the set of genes can be reconstructed using coalescent concepts. We will see how this is used in many examples in subsequent chapters.

Within a species, we can examine the pattern of variation observed in different populations or different parts of its range. A specific mutation (e.g., the change from **A** to **G** on the right side of the sequence tree) is rarely recurrent, each change is likely to occur uniquely on a different sequence background, and to arise in a different part of a species' geographic range. As a result, DNA sequence variation is variegated over time and space. The elements of such a sequence are said to be *linked*, that is, colocated on the same chromosome, and the spatiotemporal nature of mutation produces statistical association between linked variants. For example, a given variant will only be found on related sequences (e.g., on the same branch of the sequence phylogeny in Figure 3-6B). The variation represented by the sites in a given branch is not independent because the sites have a shared history; we refer to this statistical association as *linkage disequilibrium* (LD).

When there is LD between sites in the genome, it means that knowing what is found in one site provides information about what is statistically more likely to be found in the other(s). Recombination breaks LD down over time because, as shown in Figure 3-6C, segments from different branches of the tree are exchanged, reducing the association between their respective variants and the consequent trace of history. Selection and population mixing or size instability can generate LD, and this and other subtleties mean that molecular evolution is often less straightforward to reconstruct than presented here.

Reticulated Variation

In nature, sexual reproduction is widespread and allows for the scrambling of the hierarchically nested variation that a history of mutation produces. In many animal and plant species, this means that individuals bear more than one copy of their species' genomes (humans have two—one copy from each parent). Scrambling of variation occurs because of recombination, or crossing-over, between homologous chromosomes during meiosis: prior to reproduction, the homologous (corresponding) chromosomes align and exchange pieces at one or more essentially random locations. This exchange reduces the LD among mutational variants along the chromosome, dissociating potentially important functional variation and allowing for new combinations of variation to occur together and to affect a trait or be presented to selection. The same phenomenon reduces the "hitchhiking" that can occur if an allele in one place on a chromosome is favored by selection or drift, raising the frequency of nearby alleles, so recombination enables each gene to be subject to its own history over time.

Sometimes the sequence on one chromosome is used as a template for correcting differences found on the other chromosome (e.g., Lewin 2000), a process called *gene conversion*. The effects on the resulting sequence relationships are similar to those of recombination.

This scrambling of variation produces reticulated rather than simple treelike branching relationships among gene sequences that constitute a kind of horizontal transfer between peer (contemporary) copies of a chromosome. This is different from horizontal transfer between species, although both create reticulated sequence relationships. Among other things, individuals in a species are more closely related (have more similar sequence, share a more recent common ancestor) in some parts of their genome than they are in others.

Recurrent mutation is another phenomenon that causes reticulation of sequence similarity relationships. In a sample, we are likely to identify multiple instances of the same allele. We say that these are *identical by state* (ibs). If mutation were strictly unique, all copies would also be *identical by descent* (ibd) from a single ancestral mutational event. Most mutations may be unique, but recurrent mutations generate ibs variants that are not ibd. Recurrent mutations are notably common in sequence repeat elements (Chapter 4). Descendant copies of each event will share ancestry with that event; however because we can only observe their present state, it is not always possible to sort out which are which. Nonetheless, we can accommodate most of these complications, and it is the generally unique nature of the mutational process that allows us to make plausible reconstruction of gene history and hence of gene evolution.

EVOLUTION BY PHENOTYPE: THE DARWINIAN POSTULATES FROM A TRAIT'S POINT OF VIEW

What we have described so far is a general quantitative theory of the inheritance and evolution of genes. The assumption is widely made that population genetics is the rigorous theory of evolution and that the study of phenotypes and their development is rather descriptive and ambiguous by comparison. If genes are the ultimate units of evolution, the theory of evolution really only need consider what happens to genes. This needs a bit of clarification.

Population genetics gives a plausible sense of constraints on values of various parameters like times of separation among species, mutation rates, selection coefficients, and typical population sizes. The theory is highly valuable when experiments can be done, as with bacteria or fruit flies, to study genetic change (and even relate that to the effects of experimental selection on traits). However, this may not provide much information regarding phenotypic evolution in nature, or about what genes do, especially over the long term, or how gene function comes about. Yet, evolution works through phenotypes (traits in whole organisms), not directly on genes, so there is much to learn by looking at evolution from a trait's rather than from a gene's eye view.

Darwin and Wallace gave us a general overarching theory of evolution by phenotype, but we still do not have a good formal theory of the relationships between genotypes and phenotypes beyond the simplest level. We can see some of the issues by reference to phenomena discussed in Chapter 2.

Some Slippage Between Gene and Trait

When organisms select local environments or in some other way associate with similar individuals, phenotypic variation can become partitioned, and this can in principle lead to speciation. We referred to various ways this might happen in Chapter 2 as *organismal* selection. One of the most important characteristics of organisms, especially complex organisms, is that they actively explore their environment and occupy favorable subenvironments that they find. Fish able to swim in strong currents may tend to stay in midstream, and weaker fish towards the margins, for example. What was a single initial population will become divided. Fish congregating (and mating) in a given part of the stream will be more alike (e.g., stronger fins), and there will be differences between groups inhabiting different parts of the stream.

Any kind of isolation of populations eventually leads to the accumulation by drift, mutation, and selection of genetic differences between them. Behavioral isolation may ultimately lead to the accumulation of variation that prevents mating should members of the two subdivisions encounter each other. Such changes may involve genes related to mating ability unrelated to the sorting trait itself (e.g., Rieseberg and Livingstone 2003). Variation in genes associated with the traits on which self-assortment was based will also be partitioned, in a way that is similar to the effects of competitive selection: reduce variation within populations because like individuals aggregate, increase it between them. This is a speciation scenario that is not about classical darwinian competition but can be difficult to discern from genetic data alone.

By a process known as *genetic assimilation* (Hall 1999; Waddington 1942; 1953; 1956; 1957; Wilkins 2002), something that is nongenetic can become genetic. If an originally acquired (environmentally or behaviorally produced) trait is favored by natural selection, mutations that increase the chance the bearer will have the trait can rise in frequency. It would be difficult to know from genetic data that such a process has occurred; yet it is not one initiated by genes. Canalization can lead a trait to become less variable in its environment, once the genes have been "focused" in this way. The extent of canalization depends on circumstances, that have recently been explored with a variety of theoretical approaches (Ancel 1999; Ancel and Fontana 2000; Fontana 2002; Siegel and Bergman 2002).

When there is phenotypic drift, a trait changes over time without the aid of selection. Genes associated with the trait will also change, but a rigorous or general kind of relationship between the two rates of change and the individual genes involved is not specifiable in practice. Although if we knew what the genes were we would find that their net pattern of variation today is consistent with selective neutrality, this tells us little if anything about the nature and timing of the phenotypic drift history.

What is needed is a good *phenogenetic* theory (if one is indeed possible), that is, for the relationship between genotypes and phenotypes: the number of genes involved, how those genes and their variation work, or the nature or extent of selection that may be operating over time.

When there is phenotypic drift, one consequence over time can be that a new "version" of the trait can replace a previous one. This is logically comparable to allele substitution by genetic drift. Although it is always chancy to cite examples retrospectively, an interesting possible instance of phenotypic drift might be the stridulation mechanisms observed in grasshoppers. A curiosity noted in observations of the Amazon by the prominent 19th century naturalist, W. H. Bates (Bates 1863) was that some species of grasshopper make their noises by rubbing asymmetric structures on their two wings, but others achieve the same kinds of objectives using legs or legs and wings, etc. Whether these chirping mechanisms could have evolved totally independently is debatable. More likely is that the behavior—chirping—has been shared by these related organisms since their common ancestor, and that one mechanism has gradually replaced another in different lineages. We might eventually identify the genes involved in each type of chirping behavior, but we would have to know a lot about the phenotypic history if we were to make proper sense of how the genetic differences actually evolved.

One thing we know about phenogenetic relationships is that for most traits a variety of genotypes can be found in individuals with essentially the same phenotype. Similarly, among individuals with the same genotype, we commonly observe a distribution of different phenotypes, presumably because of the effects of chance and the many-to-many relationship between traits and their associated genotypes. This is *phenogenetic equivalence* and can occur for various reasons, including redundant genes, alternative biological pathways, and additive (dose-like) contributions of many different genes to the trait.

Phenogenetic equivalence is very widespread and weakens our ability to predict underlying genotypes from knowledge of phenotypes and vice versa. Natural selection cannot see "underneath" the phenotype any better than we can and will be correspondingly weak at affecting genotypes by screening phenotypes. Even if selection were precisely prescriptive at screening phenotypes, genetic variants underlying equivalent phenotypes will evolve neutrally relative to each other (Clark 1998; Hartl and Campbell 1982). That is, there can be *phenogenetic drift* (Wagner and Misof 1993; Weiss and Fullerton 2000): some alleles may disappear, and new ones may appear. Some genes may lose their effect on the trait, while other genes may become involved. But the trait stays the same. Many examples of phenogenetic drift are being identified, in experimental species (Rutherford and Henikoff 2003), in developmental processes that will be of interest in this book (Gellon and McGinnis 1998; Kissinger and Raff 1998; Robert 2001; True and Haag 2001), and in many others.

Because phenogenetic drift is not restricted to selectively neutral traits, over time, classical darwinian selection can preserve a trait unchanged while its associated genes change, sometimes dramatically. Phenogenetic drift can have an important impact on our understanding of evolution and on routine work in biology. Conservation of genetic mechanisms is often tacitly assumed in research with animal models, for example, in extrapolating from one model species (e.g., mouse, *Arabadopsis*, zebrafish, *C. elegans*, fruit flies, yeast, *E. coli*) as well as to other unstudied species. Biomedical research depends heavily on this kind of extrapolation because we cannot experiment on humans. To the extent that phenogenetic drift occurs, these assumptions will be inaccurate. The common finding that genetic interventions have different effects in different strains of laboratory mice is but one disturbing example that is well known but conveniently ignored in the effort to lay out tidy phenogenetic scenarios.

The actual extent or impact of phenogenetic drift in nature is unclear, but given the prevalence of phenogenetic equivalence it must occur. Nonetheless, as we will see in subsequent chapters, there can also be extensive and strikingly deep phylogenetic conservation of phenogenetic mechanisms. Similar traits even among distant species often seem to share at least some aspects of their underlying mechanisms. Nobody now would suggest that complex traits like eyes or limbs or leaves, even in distantly related species, are controlled by entirely different mechanisms. But the conservation is typically not complete, and sometimes it does turn out to be fundamentally different. Explaining how such strong conservation has been maintained by selection in the face of phenogenetic drift and equivalence is not yet possible, though some ideas have been advanced and will be discussed below.

The unpredictability of evolution in terms of phenotypes can be seen in a different way. Many examples of alternative pathways are being discovered in experimental species and traits. Computer simulations of evolutionary processes (Yedid and Bell 2002) and replicate experiments in flies (Fry et al. 1995; Mackay 1995; 1996; 2001; 2001) and bacteria (Cooper et al. 2003; Lenski et al. 2003; Papadopoulos et al. 1999; Travisano et al. 1995) show that, from the same beginning, even with the same starting variation (as far as can be determined), different genetic bases may result even if selection generates the same resulting trait.

These various phenomena are logical extensions of the fact that variation in quantitative traits is contributed to by allelic variation at many genes simultaneously. To what extent we can formulate a general theory that goes beyond description of individual instances is not clear. But the absence of a good phenogenetic theory contributes to the difficulty in inferring from genetic data how evolution of organisms and their traits actually happened.

What Exactly is Inheritance?

Perhaps because of the predominance of the physical sciences through the 19th century, the prevailing worldview of science today is that the ultimate units of causation in biology must be chemical or molecular. Genes are molecules that provide the codes for protein sequences, and proteins form the basic chemistry of life. Genes also retain the sequence structure that has evolved over billions of years, and genes are universally important elements transmitted from organisms to their descendants. For these reasons, many in biology have come to view genes as *the* units of inheritance and, conversely, inheritance as the study of genes. The study of genes, however, is not the only way to view inheritance.

A new organism starts life as much more than a set of genes. A fertilized egg contains membranes, sugars, carbohydrates, messenger RNA (Chapter 4), salts, and so on. These components are not generally regarded as an important aspect of inheritance because their production is viewed as dependent ultimately on instructions from genes acting in the parent that produced the egg or sperm. This is only partly accurate. Genes themselves are highly modified by chemical packaging and the like, and their modification can vary and hence is an important component of inheritance. Genes are not really even self-replicating, because self-replication only occurs in the context of the complex machinery of a cell.

There are nonmaterial aspects of inheritance as well. There are many ways in which the environment of organisms is not just what they must adapt *to* but what they themselves *create* (Lewontin 2000; Turner 2000). These aspects of the parents' environment will also be inherited as part of the environment of their offspring. Humans are not the only organisms that directly and systematically transmit environments or behavior to their offspring. Language and culture are easy human examples, but similar characteristics apply to birds' nests, the local environment into which coral offspring are shed, the training to hunt that lion cubs receive, or the parental sounds that chicks respond to. Physiological effects can be directly transmitted transplacentally in mammals or indirectly via the nutritional or immunological environment an offspring is born into.

These traits are not directly genetic, but they are surely "biological" phenomena of animals and plants; they are meaningfully inherited, and they are widespread or even to some extent universal. They are as much the product of selection (and chance) as other traits; survival depends on nest–building abilities that birds have been passing to their offspring for countless generations as much as it does on their metabolism. The transmission patterns are different from those of genes; sometimes they are transmitted horizontally or depend on characteristics of the population as well as the parent. Nongenetic traits change, although not with the relatively specifiable characteristics of genetic mutation.

To some extent, it is rather arbitrarily that biologists have chosen to exclude nongenetic elements from the formal theory of inheritance and evolution. There are good reasons for making the distinction between genes and other aspects of life, but the boundaries blur and both should be considered (e.g., for perspectives on this, see Gottlieb 1997; Oyama 2000; Oyama et al. 2001). Population genetics generally treats nongenetic inheritance as an aggregate of "environmental" correlation in trait values observed among relatives, making very generic assumptions about these nonmendelian effects (e.g., that they have a normal distribution in the population) and treating them as temporary and of no interest to genetic evolution. Nonetheless, nongenetic inheritance affects the properties of organisms and populations, which can have direct, longterm effects on genetic evolution itself.

Elements of Evolutionary Phenogenetics

Darwin gave us a conceptual theory for the evolution of *phenotypes*, not genes, and we have developed a formal theory that applies to the evolution of genes. Population genetics and darwinian concepts would be subsets of a rigorous theory for the evolution of phenotypes and would be united by incorporating aspects of phenogenetics, such as the complex regulation of gene expression, the interaction among genes, the relationship to environmental factors, a more inclusive definition of inheritance, and an adequate accommodation for the role of chance.

Population genetics largely puts genes into a black box by being nonspecific about how they work, which in turn weakens our ability to understand the actual nature of evolutionary change even if we can examine its net results. Here we can mention a few things that might be considered, for example, constraints that might explain some of those aspects of phenogenetic relationships that seem to have been conserved over very long time periods.

Genes involved in fundamental biological processes, like basic cellular energy metabolism, are generally expected to be less variable and more slowly evolving than genes related to developmentally later and/or more specialized functions. The reason is that, once installed early in the history of life, too much has come to depend on them. However, there are exceptions, and even basic housekeeping genes do evolve.

Pleiotropic genes, genes with multiple functions, are expected to experience less change than genes with only a single function (like coding for a protein used only in one tissue). Genes related to embryological development can be highly pleiotropic, for example. Variation in such genes may be constrained because a mutation favorable to one of the gene's functions might be unfavorable to its other functions. Variation related to such genes is more likely to be found in aspects of its tissue-specific *expression* (usage) rather than in its coded protein product. But as we will see, the mutational and evolutionary dynamics of the elements that control gene expression are very different from those of protein-coding elements. We have theory for the behavior of pleiotropic genes under selection, but not the evolution of pleiotropy in the first place.

Along with some phenogenetic mechanisms, some other traits seem hardly to have changed from what is seen in fossils hundreds of millions of years old. Cockroaches and horseshoe crabs are famous examples. Some aspects of DNA sequences are also conserved among very distantly related species. What maintains the invariance? The usual explanation is that morphological conservation reflects the constraining force of adaptive natural selection. Yet this seems somewhat perplexing, because it is not as if there cannot be variation among crabs or cockroach relatives, or that their genomes have also remained frozen in time (they haven't). One contributing possibility is that, as mentioned in Chapter 2, longterm stability can occur when traits are canalized by developmental constraints, genetic commitments made in the past that are hard to modify. The organ systems of present-day species involve a complex network of genetic effects. These systems are the product of hundreds of millions of years of evolution, and their interactions seem to provide at least some canalization, or restraint on variation (Siegal and Bergman 2002). In a sense, they are so entrenched that they no longer need to be maintained by natural selection, or perhaps a better way to put it is that survival of the organism can tolerate little variation in the trait. But this is a weak general explanation, and there are exceptions.

Also mentioned in Chapter 2 was that some basic features of molecular interactions may simply, because of their chemical nature, remain in Bateson's "positions of organic stability" (Fontana and Schuster 1998; Kirschner et al. 2000; Laughlin et al. 2000; Monod 1971; Morowitz et al. 2000; Schuster 2000). It may be difficult or impossible for selection to alter these.

Genes that quantitatively affect patterning processes responsible, for example, for periodically repetitive traits like digits, leaves, hair, and the like may experience small mutational change with large phenotypic effect, very different from the Darwinian notion of gradual evolution. As will be seen, a substantial fraction of traits are patterned repetitively or in other similar ways. Darwin accepted the prevalent notion that nature makes no big jumps; to suggest otherwise is a kind of biological heresy. Population genetic models easily accommodate traits with a few states that can be altered by single mutations (like Mendel's peas) but generally treat complex trait evolution in a gradualistic way that does not provide a useful explanation of pattern traits.

The degree of change in phenogenetic mechanism over time is constrained by the availability of relevant genetic variation; this is hard to model without a better sense of the workings of the mechanism. Phenogenetic drift can only occur if alternative genetic mechanisms are available. Most traits probably have at least some phenogenetic equivalence, but it is not clear how to model the amount or its generation. Different alternative mechanisms may be subject to the effects of selection, drift, and mutation in different ways.

However, one important implication is that while selection may favor the same trait in different populations, it only works indirectly on genetic variation, so that in different populations selection can favor different genotypes that lead to similar phenotypes. That is, *phenotypic convergence* can lead to *genotypic divergence* between populations. An example is the diversity of genetic mechanisms by which humans in the Old World tropics have adapted genetically to malaria. Similar phenotypes do not imply similar genotypes.

CONCLUSION: DETERMINISM AND INDETERMINISM IN GENETIC INFERENCE

A trait can in principle be 100 percent genetically determined, in the sense that *if* we could replicate an entire genome *and* if we could ignore subsequent environmental and random effects, organisms would turn out to be identical. Each individual, however, is different in practice, making it impossible to view the inherited genome as "the blueprint" for life. We have experience with clones, twins, inbred animals, and other kinds of experimental evidence to show the degree to which a given genotype can determine traits (especially under similar environments). There is predictability, but it is not perfect and usually we only have observed the genes under a restricted range of environments, which can be deceiving (Lewontin 2000; Schlichting and Pigliucci 1998).

The issue of how deterministically genes drive evolution is more problematic, because evolutionary events cannot be replicated, so that the inference we can make from genetic data is usually indirect and statistical in nature. Common observations such as the strain-dependence of gene-inactivation experiments or the complex control of quantitative traits provide evidence that phenogenetic drift and phenogenetic equivalence are important aspects of evolution.

Phenogenetic drift illustrates this in a philosophically interesting way. Because new offspring must develop their final form from incomplete beginnings, often only a single cell, traits must be regenerated each generation. This justifies our search for the underlying "information" that makes this possible, and we attribute that information essentially to genes. From this perspective, the genes are real and a trait is their ephemeral product. But when a trait is preserved by selection while its underlying genetic basis changes, we can view the trait as real and its underlying genetic basis as an ephemeral servant.

We have raised and probably belabored some problems and incompleteness in current evolutionary genetic theory. The strengths of population genetics as a way to quantify the dynamics of genetic change and to reconstruct its history are well known. But it does not really address or predict the role of genes in what organisms actually do in life and how they came about. One purpose of this book is to try to identify some of the generalities that might be part of a phenogenetic theory of life.

PART II

Building Blocks of Life

A Genetic Repertoire for Evolving Complexity

Life is a complex, diversified phenomenon comprising species of many different kinds, some of them only very distantly related. Nonetheless, the basic system on which this diversity is built is common to all of life. A few principles and characteristics are shared, including the way biological "memory" evolves and is preserved in DNA, the cell, and a small set of basic genetic mechanisms that make it and all of life possible.

Genetics and the Logic of Evolution, by Kenneth M. Weiss and Anne V. Buchanan.
ISBN 0-471-23805-8

Chapter 4
The Storage and Flow of Biological Information

As discussed in Chapter 3, genes are generally given the place of honor at the core of biological and evolutionary theory. Genes are viewed as the carriers of biological "information." This metaphor may be somewhat overstated, perhaps fitting in too easily with the worldview of the computer age (Kay 2000), but given that caveat, in Chapter 4, we will describe the basic ways information is stored and used by DNA and related molecules.

GENETIC ASPECTS OF INFORMATION

BASIC GENETICS: A BRIEF PARTLY GUIDED TOUR OF DNA

The large macromolecules, the *nucleic acids* DNA and RNA, serve many functions. The vital structural property of both molecules is that they are *modular*: each is a string of concatenated nucleotide *bases*, of which there are four basic types: adenine (**A**), cytosine (**C**), guanine (**G**), and thymine (**T**) (or in a modified form as uracil (**U**) in RNA). These are linked by a phosphate-sugar backbone. Each of the two purines (**G** and **A**) easily forms a chemical bond with one of the two pyrimadines (**C** and **T/U**): **C** pairs with **G** and **A** with **T**. This *complementary base pairing* allows the molecule to take on higher-order three-dimensional structure, and these basic properties are used both by nature and by experimenters to manipulate DNA and RNA, including the important function of replication in which a DNA molecule is copied and passed down from one cell generation to the next. Indeed, much of life and of our research methods to *understand* life is based on complementary base pairing.

The most important characteristics of both molecules are that (1) the length of the concatenated string of nucleotides is chemically arbitrary and (2) the sequence of successive nucleotides can be arranged in any order. Clearly, both molecules are modular in nature and the subunits are variations on a basic chemical theme—and this is how and why it works.

Genetics and the Logic of Evolution, by Kenneth M. Weiss and Anne V. Buchanan.
ISBN 0-471-23805-8

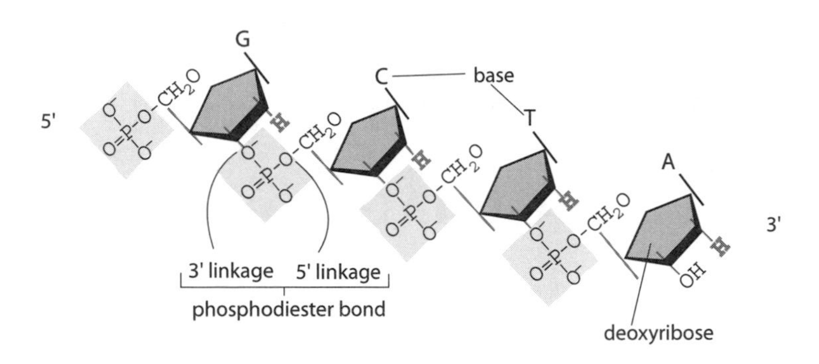

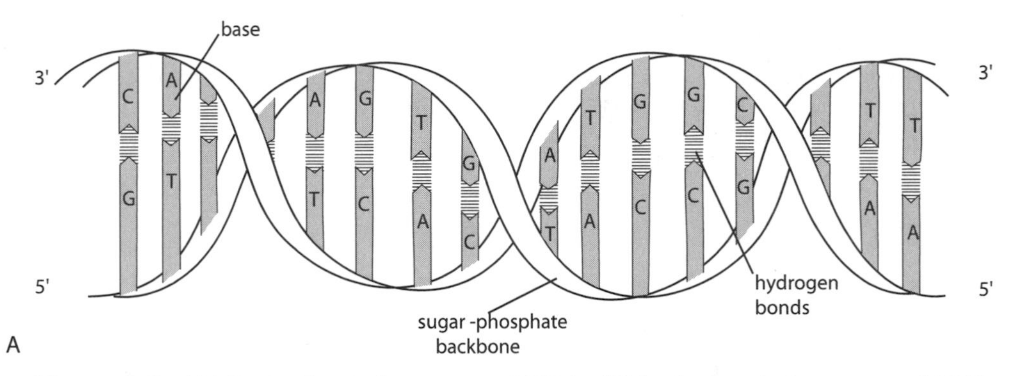

Figure 4-1. (A) Basic chemical structure of DNA; (B) basic chemical structure of RNA; (C) a tRNA molecule in natural conformation. Redrawn from (Alberts 1994).

DNA and RNA are illustrated schematically in Figure 4-1. DNA is double-stranded, with one strand having the complementary nucleotides to the other (e.g., **AAACGTA** would be paired with **TTTGCAT**). For physicochemical reasons, this arrangement leads to the well-known basic double-helix shape of DNA. However, the DNA molecule in a cell would, if stretched out linearly, greatly exceed the space available in the cell (e.g., in a human cell, the DNA would be more than one meter long); therefore, the molecule is wrapped and packaged into much tighter, more structured form.

The double-stranded nature of DNA has to do with its stability and its copying process; its information, however, is mainly contained in the sequence. RNA is usually viewed as a single-stranded molecule (Figure 4-1B), but autoannealing can occur by base pairing, as it folds upon itself, forming higher-order partly double-stranded structures, such as seen in a *transfer RNA* (tRNA) molecule (see below; Figure 4-1C). As we will see later in this chapter, although it has coding functions

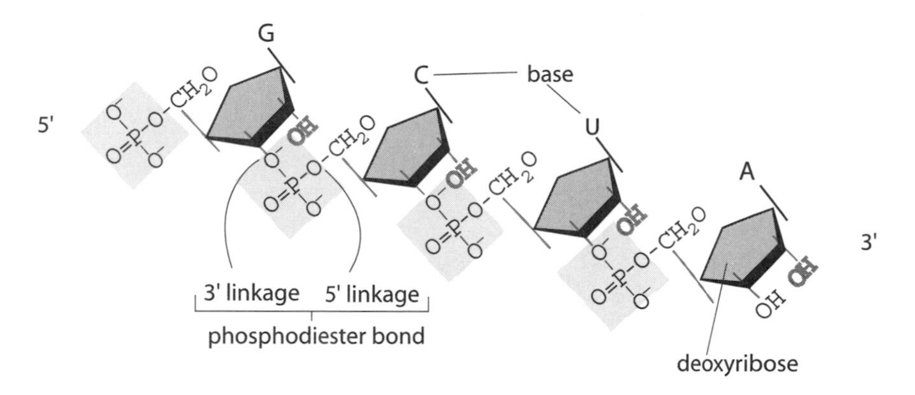

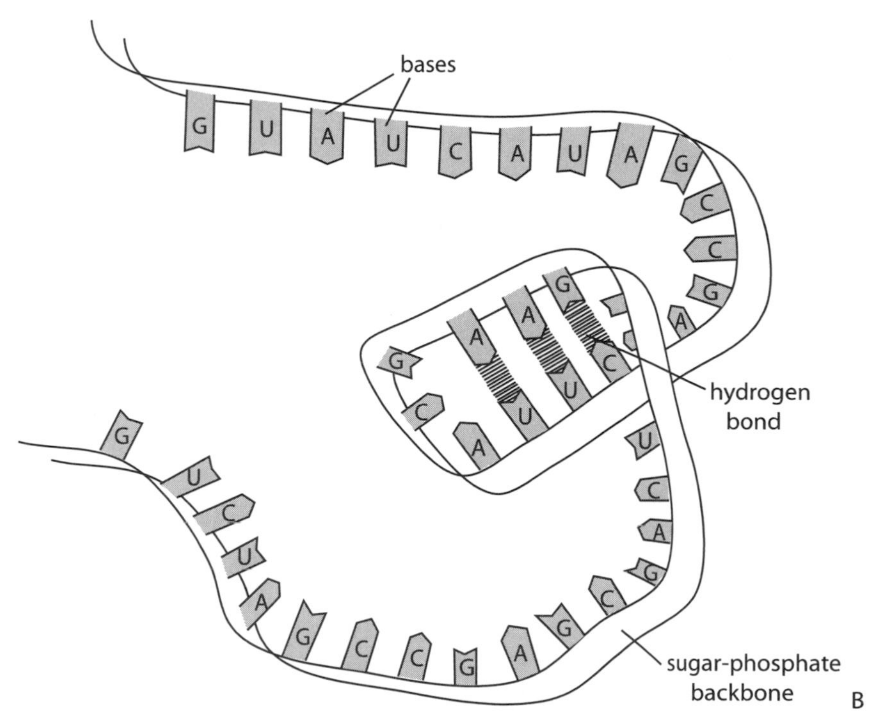

Figure 4-1. *Continued*

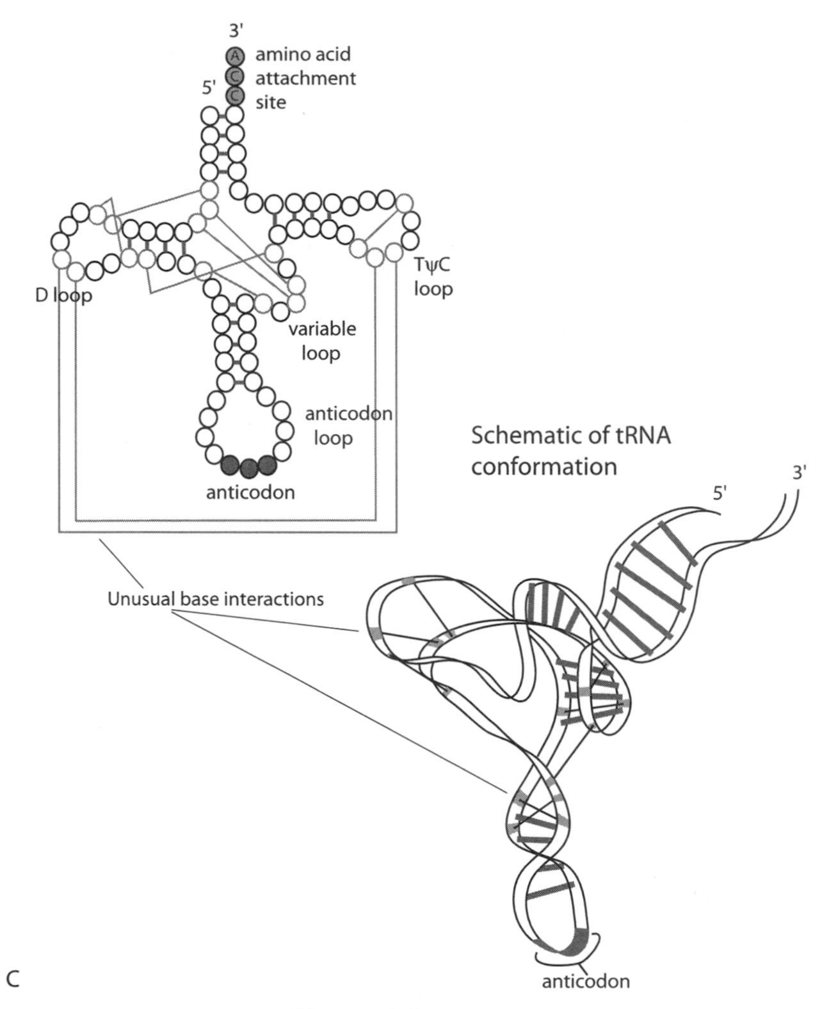

Figure 4-1. *Continued*

related to sequence as does DNA, RNA has many other direct functions having to do with the conformation it takes when autoannealing. RNA is considered a transitory source of information storage because, except for some groups of viruses, it is no longer the primary inherited coding material, for reasons to be explained below.

A segment of a DNA sequence (one strand only) is shown in Figure 4-2. DNA is more passive than RNA but is of much greater interest because of its role in the storage of inherited protein-coding information. However, like RNA, DNA is involved in other functions; these involve the contextual *use* of protein coding, as well as DNA's own maintenance, packaging, and replication during cell division.

Along with many other, simpler types of molecules such as carbohydrates, the diversity of life is characterized by a diversity of proteins. Proteins are modular concatenations of amino acids whose function depends on their specific sequence. There

GGAACTTGATGCTCAGAGAGGACAAGTCATTTGCCCAAGGTCACACAGCTGGCAACTGGCAGACGAGAT
TCACGCCCTGGCAATTTGACTCCAGAATCCTAACCTTAACCCAGAAGCACGGCTTCAAGCCCTGGAAACC
ACAATACCTGTGGCAGCCAGGGGGAGGTGCTGGAATCTCATTTCACATGTGGGGAGGGGGCTCCTGTGCT
CAAGGTCACAACCAAAGAGGAAGCTGTGATTAAAACCCAGGTCCCATTTGCAAAGCCTCGACTTTTAGCA
GGTGCATCATACTGTTCCCACCCCTCCCATCCCACTTCTGTCCAGCCGCCTAGCCCCACTTTCTTTTTTTTC
TTTTTTTGAGACAGTCTCCCTCTTGCTGAGGCTGGAGTGCAGTGGCGAGATCTCGGCTCACTGTAACCTCC
GCCTCCCGGGTTCAAGCGATTCTCCTGCCTCAGCCTCCCAAGTAGCTAGGATTACAGGCGCCCGCCACCA
CGCCTGGCTAACTTTTGTATTTTTAGTAGAGATGGGGTTTCACCATGTTGGCCAGGCTGGTCTCAAACTCC
TGACCTTAAGTGATTCGCCCACTGTGGCCTCCCAAAGTGCTGGGATTACAGGCGTGAGCTACCGCCCCCA
GCCCCTCCCATCCCACTTCTGTCCAGCCCCCTAGCCCTACTTTCTTTCTGGGATCCAGGAGTCCAGATCCC
CAGCCCCCTCTCCAGATTACATTCATCCAGGCACAGGAAAGGACAGGGTCAGGAAAGGAGGACTCTGGG
CGGCAGCCTCCACATTCCCCTTCCACGCTTGGCCCCCAGAATGGAGGAGGGTGTCTGTATTACTGGGCGA
GGTGTCCTCCCTTCCTGGGGACTGTGGGGGGTGGTCAAAAGACCTCTATGCCCCACCTCCTTCCTCCCTCT
GCCCTGCTGTGCCTGGGGCAGGGGGAGAACAGCCCACCTCGTGACTGGGCTGCCCAGCCCGCCCTATCCC
TGGGGGAGGGGGCGGGACAGGGGGAGCCCTATAATTGGACAAGTCTGGGATCCTTGAGTCCTACTCAGC
CCCAGCGGAGGTGAAGGACGTCCTTCCCCAGGAGCCGGTGAGAAGCGCAGTCGGGGGCACGGGGATGA
GCTCAGGGGCCTCTAGAAAGAGCTGGGACCCTGGGAAGCCCTGGCCTCCAGGTAGTCTCAGGAGAGCTA
CTCGGGGTCGGGCTTGGGGAGAGGAGGAGCGGGGGTGAGGCAAGCAGCAGGGGACTGGACCTGGGAAG
GGCTGGGCAGCAGAGACGACCCGACCCGCTAGAAGGTGGGGTGGGGAGAGCAGCTGGACTGGGATGTA
AGCCATAGCAGGACTCCACGAGTTGTCACTATCATTATCGAGCACCTACTGGGTGTCCCCAGTGTCCTCA
GATCTCCATAACTGGGGAGCCAGGGGCAGCGACACGGTAGCTAGCCGTCGATTGGAGAACTTTAAAATG
AGGACTGAATTAGCTCATAAATGGAACACGGCGCTTAACTGTGAGGTTGGAGCTTAGAATGTGAAGGGA
GAATGAGGAATGCGAGACTGGGACTGAGATGGAACCGGCGGTGGGGAGGGGGTGGGGGGATGGAATTT
GAACCCCGGGAGAGGAAGATGGAATTTTCTATGGAGGCCGACCTGGGGATGGGGAGATAAGAGAAGAC
CAGGAGGGAGTTAAATAGGGAATGGGTTGGGGGCGGCTTGGTAAATGTGCTGGGATTAGGCTGTTGCAG
ATAATGCAACAAGGCTTGGAAGGCTAACCTGGGGTGAGGCCGGGTTGGGGGCGCTGGGGGTGGGAGGA
GTCCTCACTGGCGGTTGATTGACAGTTTCTCCTTCCCCAGACTGGCCAATCACAGGCAGGAAGATGAAGG
TTCTGTGGGCTGCGTTGCTGGTCACATTCCTGGCAGGTATGGGGGCGGGGCTTGCTCGGTTCCCCCCGCTC
CTCCCCCTCTCATCCTCACCTCAACCTCCTGGCCCCATTCAGACAGACCCTGGGCCCCCTCTTCTGAGGCT
TCTGTGCTGCTTCCTGGCTCTGAACAGCGATTTGACGCTCTCTGGGCCTCGGTTTCCCCCATCCTTGAGAT
AGGAGTTAGAAGTTGTTTTGTTGTTGTTGTTTGTTGTTGTTGTTTTGTTTTTTTGAGATGAAGTCTCGCTCT
GTCGCCCAGGCTGGAGTGCAGTGGCGGGATCTCGGCTCACTGCAAGCTCCGCCTCCCAGGTCCACGCCAT
TCTCCTGCCTCAGCCTCCCAAGTAGCTGGGACTACAGGCACATGCCACCACACCCGACTAACTTTTTTGTA
TTTTCAGTAGAGACGGGGTTTCACCATGTTGGCCAGGCTGGTCTGGAACTCCTGACCTCAGGTGATCTGC
CCGTTTCGATCTCCCAAAGTGCTGGGATTACAGGCGTGAGCCACCGCACCTGGCTGGGAGTTAGAGGTTT
CTAATGCATTGCAGGCAGATAGTGAATACCAGACACGGGGCAGCTGTGATCTTTATTCTCCATCACCCCC
ACACAGCCCTGCCTGGGGCACACAAGGACACTCAATACATGCTTTTCCGCTGGGCCGGTGGCTCACCCCT
GTAATCCCAGCACTTTGGGAGGCCAAGGTGGGAGGATCACTTGAGCCCAGGAGTTCAACACCAGCCTGG
GCAACATAGTGAGACCCTGTCTCTACTAAAAATACAAAAATTAGCCAGGCATGGTGCCACACACCTGTGC
TCTCAGCTACTCAGGAGGCTGAGGCAGGAGGATCGCTTGAGCCCAGAAGGTCAAGGTTGCAGTGAACCA
TGTTCAGGCCGCTGCACTCCAGCCTGGGTGACAGAGCAAGACCCTGTTTATAAATACATAATGCTTTCCA
AGTGATTAAACCGACTCCCCCCTCACCCTGCCCACCATGGCTCCAAAGAAGCATTTGTGGAGCACCTTCT
GTGTGCCCCTAGGTAGCTAGATGCCTGGACGGGGTCAGAAGGACCCTGACCCGACCTTGAACTTGTTCCA
CACAGGATGCCAGGCCAAGGTGGAGCAAGCGGTGGAGACAGAGCCGGAGCCCGAGCTGCGCCAGCAGA
CCGAGTGGCAGAGCGGCCAGCGCTGGGAACTGGCACTGGGTCGCTTTTGGGATTACCTGCGCTGGGTGCA
GACACTGTCTGAGCAGGTGCAGGAGGAGCTGCTCAGCTCCCAGGTCACCCAGGAACTGAGGTGAGTGTC
CCCATCCTGGCCCTTGACCCTCCTGGTGGGCGGCTATACCTCCCCAGGTCCAGGTTTCATTCTGCCCCTGT
CGCTAAGTCTTGGGGGGCCTGGGTCTCTGCTGGTTCTAGCTTCCTCTTCCCATTTCTGACTCCTGGCTTTAG
CTCTCTGGAATTCTCTCTCTCAGCTTTGTCTCTCTCTCTTCCCTTCTGACTCAGTCTCTCACACTCGTCCTGG
CTCTGTCTCTGTCCTTCCCTAGCTCTTTTATATAGAGACAGAGAGATGGGGTCTCACTGTGTTGCCCAGGC
TGGTCTTGAACTTCTGGGCTCAAGCGATCCTCCCGCCTCGGCCTCCCAAAGTGCTGGGATTAGAGGCATG
AGCACCTTGCCCGGCCTCCTAGCTCCTTCTTCGTCTCTGCCTCTGCCCTCTGCATCTGCTCTCTGCATCTGT
CTCTGTCTCCTTCTCTCGGCCTCTGCCCCGTTCCTTCTCTCCCTCTTGGGTCTCTCTGGCTCATCCCCATCTC
GCCCGCCCCATCCCAGCCCTTCTCCCCCGCCTCCCCACTGTGCGACACCCTCCCGCCCTCTCGGCCGCAGG
GCGCTGATGGACGAGACCATGAAGGAGTTGAAGGCCTACAAATCGGAACTGGAGGAACAACTGACCCCG
GTGGCGGAGGAGACGCGGGCACGGCTGTCCAAGGAGCTGCAGGCGGCGCAGGCCGGCTGGGCGCGGAC
ATGGAGGACGTGCGCGGCCGCCTGGTGCAGTACCGCGGCGAGGTGCAGGCCATGCTCGGCCAGAGCACC
GAGGAGCTGCGGGTGCGCCTCGCCTCCCACCTGCGCAAGCTGCGTAAGCGGCTCCTCCGCGATGCCGATG
ACCTGCAGAAGCGCCTGGCAGTGTACCAGGCCGGGGCCCGCGAGGGCGCCGAGCGCGGCCTCAGCGCCA
TCCGCGAGCGCCTGGGGCCCCTGGTGGAACAGGGCCGCGTGCGGGCCGCCACTGTGGGCTCCCTGGCCG
GCCAGCCGCTACAGGAGCGGGCCCAGGCCTGGGGCGAGCGGCTGCGCGCGCGGATGGAGGAGATGGGC

Figure 4-2. DNA sequence of the coding strand of the human apolipoprotein E (*Apoe*) gene, in standard 5′ to 3′ orientation. GenBank accession number M10065.The core coding functions of DNA and RNA are determined by their *sequence*, that is, the identity of the successive nucleotides along the molecular chain. The specific sequence has specific implications, and the sequence information is used in many different ways in both DNA and RNA.

AGCCGGACCCGCGACCGCCTGGACGAGGTGAAGGAGCAGGTGGCGGAGGTGCGCGCCAAGCTGGAGGA
GCAGGCCCAGCAGATACGCCTGCAGGCCGAGGCCTTCCAGGCCCGCCTCAAGAGCTGGTTCGAGCCCCT
GGTGGAAGACATGCAGCGCCAGTGGGCCGGGCTGGTGGAGAAGGTGCAGGCTGCCGTGGGCACCAGCG
CCGCCCCTGTGCCCAGCGACAATCACTGAACGCCGAAGCCTGCAGCCATGCGACCCCACGCCACCCCGTG
CCTCCTGCCTCCGCGCAGCCTGCAGCGGGAGACCCTGTCCCCGCCCCAGCCGTCCTCCTGGGGTGGACCC
TAGTTTAATAAAGATTCACCAAGTTTCACGCATCTGCTGGCCTCCCCCTGTGATTTCCTCTAAGCCCCAGC
CTCAGTTTCTCTTTCTGCCCACATACTGCCACACAATTCTCAGCCCCCTCCTCTCCATCTGTGTCTGTGTGT
ATCTTTCTCTCTGCCCTTTTTTTTTTTTTTAGACGGAGTCTGGCTCTGTCACCCAGGCTAGAGTGCAGTGGCA
CGATCTTGGCTCACTGCAACCTCTGCCTCTTGGGTTCAAGCGATTCTGCTGCCTCAGTAGCTGGGATTACA
GGCTCACACCACCACACCCGGCTAATTTTTGTATTTTTAGTAGAGACGAGCTTTCACCATGTTGGCCAGGC
AGGTCTCAAACTCCTGACCAAGTGATCCACCCGCCGGCCTCCCAAAGTGCTGAGATTACAGGCCTGAGCC
ACCATGCCCGGCCTCTGCCCCTCTTTCTTTTTTAGGGGGCAGGGAAAGGTCTCACCCTGTCACCCGCCATC
ACAGCTCACTGCAGCCTCCACCTCCTGGACTCAAGTGATAAGTGATCCTCCCGCCTCAGCCTTTCCAGTA
GCTGAGACTACAGGCGCATACCACTAGGATTAATTTGGGGGGGGGTGGTGTGTGTGGAGATGGGGTCTG
GCTTTGTTGGCCAGGCTGATGTGGAATTCCTGGGCTCAAGCGATACTCCCACCTTGGCCTCCTGAGTAGCT
GAGACTACTGGCTAGCACCACCACACCCAGCTTTTTATTATTATTTGTAGAGACAAGGTCTCAATATGTTG
CCCAGGCTAGTCTCAAACCCCTGGCTCAAGAGATCCTCCGCCATCGGCCTCCCAAAGTGCTGGGATTCCA
GGCATGGGCTCCGAGCGGCCTGCCCAACTTAATAATATTGTTCCTAGAGTTGCACTC

Figure 4-2. *Continued*

are 20 amino acids, each with its own molecular characteristics, and this diversity allows proteins to be highly diverse in their own functional characteristics. Proteins can be hundreds of amino acids long. Because each amino acid has its own particular chemical characteristics, the characteristics of the concatenation that makes a particular protein can be very diverse indeed. It is for this reason that proteins are considered to be the basic biochemical difference between animate and inanimate matter.

Proteins carry the biological information, so to speak, for the many chemical reactions that take place in life, but the term "information" is usually not applied to them because they are generally ephemeral molecules that are destroyed or recycled but not inherited. Instead, for every protein, there is a DNA sequence that directly corresponds to its amino acid sequence—and the former sequence *is* inherited. Because of its modular structure, we can see how mutation and the processes that subsequently determine their frequency, reviewed in Chapter 3, generate sequences that are a cumulative reflection of their history; that transmitted legacy accounts for our viewing DNA as bearing the essential information in biology.

This same system is used throughout the entire biosphere: in present-day trees, bacteria, mushrooms, trout, and humans, as well as 3.5-billion-year-old stromatolites and 100-million-year-old dinosaurs. Indeed, we would not define as "life" anything earthly that did not in some way involve this system. This single modular system of building blocks (nucleotides) evolved from a very rudimentary system to be able to incorporate open-ended functions and to preserve the information required to carry the diversity of functions of all life, across all generations.

Over billions of years, by growth in size and change in sequence, and 3 to 4 billion years of *uninterrupted* copying and descent, "endless forms most beautiful and wonderful have been, and are being, evolved," as Darwin (Darwin 1859) stated, describing the history of life on Earth (see Chapter 17). He did not know about DNA, but he sensed the unity and common ancestry of all life which we now know is due in large measure to the incredibly flexible sequence-variation storage properties of these modular molecules.

Digital Storage of Biological Information

The Information Structure of DNA

DNA can be a very long molecule, and we often refer to it in terms of kilobases (kb, or thousands of sequential bases) or megabases (Mb, millions of bases). If we were to look along a molecule of DNA from end to end and observe the nucleotide sequence, it can superficially appear to be random; for example, knowing one nucleotide might provide essentially no power to predict the next nucleotide in line. But in a strange and interesting way, there is rich and complexly structured order in the sequence, only some of which can be discerned from the sequence alone. To identify, much less understand the patterning, chemical or biological function studies are required. For example, proportions of **CG** vs. **AT** pairs can vary considerably from one region to the next; one would have to scan a lot of sequence, like that shown in Figure 4-2, to identify pattern reliably and to infer that regions rich in **CG**s have particular meaning—in vertebrates protein-coding regions tend to be characterized by high **CG** content.

Figure 4-3 shows schematically the kinds of patterns that can be found in DNA sequence; each identified element varies both within and between individuals and species. In a virus, the DNA (or RNA) of a few thousand base pairs is a single molecule packaged within a protein coat. In bacteria, and in organelles (mitochondria and chloroplasts) within cells, the DNA is in the form of a closed circle or ring.

In eukaryote cells between cell divisions, DNA is found in the nucleus, wrapped periodically around a complex of proteins called *histones* and coiled into even more complex, compact form. At some points, some sequence motifs related to the incorporation sites of the histones are present. Several separate DNA molecules—the *chromosomes*—are in each cell. This set of chromosomes is referred to as the *genome*. Each species has its own number of chromosomes; a stereotyped display of the lengths and structures of the chromosomes in a species is called a *karyotype*.

The total length of each chromosome is tens or hundreds of millions of base pairs. Each end of each chromosome comprises a variable number of roughly similar (but also somewhat variable) sequence repeats, for example, $(\mathbf{TTAGGG})_n$ in vertebrates, where n refers to the number of adjacent copies. These are *telomeres* and function to protect the chromosome ends from enzymatic degradation. Somewhere along the sequence another motif (also variable in sequence and location among and within

Figure 4-3. Stereotypical chromosome structure. Bead representation of various elements labeled as in key. "Junk DNA" is a colloquialism for DNA currently of unknown function.

species and characterized by complex patterns of sequence repeats) is found; this is the *centromere*, which is used to separate chromosomes into daughter cells during cell division (Sun et al. 2003; Tyler-Smith and Floridia 2000). Some of this DNA has been copied and inserted into this region from elsewhere in the genome, perhaps facilitated by repeat sequences such as **CAAAAAGCGGG**, and in flies, at least, there are also many short **AATAT** and **TTCTC** elements in centromeres. This structure may allow the DNA to loop out from its chromosome packaging and fold up upon itself, to provide an attachment of the spindle fibers that separate the chromosomes in cell division.

Along the sequence, numerous runs of the same nucleotide (e.g., **TTTTTTTTTT** . . .) of variable lengths or sets of tandem (adjacent) short nucleotide motifs, like $(\mathbf{ATTT})_n$ or $(\mathbf{CA})_n$, are present. Some of the repeat copies are exactly or nearly exactly the same in different (but more or less randomly spaced) locations along a chromosome. Many of these repeated motifs have been catalogued (Vossilenko 2003). (See Table 4-1 for a sampling of these elements.) Depending on the motif or species, the repeat motif may occur tens or hundreds of thousands of times in the genome. When scanning different individuals of the same species in their corresponding chromosomal regions, one finds variation in the number of times a motif is repeated. In fact, the two sets of chromosomes in a diploid individual vary considerably in this respect. Thus, an individual may have **CACACA** on one chromosome and **CACACACACA** on the other *homolog* (copy of the corresponding chromosome), in a given location. These highly variable length repeats are called *microsatellites* (depending on the details, they are also known as short tandem repeats, or minisatellites). More than 50 percent of the human genome, in fact, consists of repeats (Lander et al. 2001). The proportion of the genome that consists of repeat regions varies considerably between species, however, from around 1 percent in some species to over 50 percent in others. The high variability in repeat numbers, within species as well as individuals, raises an important but perhaps little-appreciated point that there is no single length of the genome for a species or even in the two copies within an individual.

Some of these repeated elements have complex structures, such as a particular sequence at each end of a short (few hundred or fewer base pair) stretch of DNA, sometimes the sequence is in inverted order at the two ends, and sometimes such elements occur multiple times in tandem. Variation in short elements appears to be due to error in DNA replication during cell division. Some elements, however, are *transposable*, that is, have mechanisms known to make pieces of DNA move around among chromosomes from time to time, or move from parasites like viruses into a host's chromosomes. Repeat element sequences often resemble some other functional element that was captured in some way and subsequently distributed in copies around the genome. The *Alu* elements in primates are an example (Mighell et al. 1997); they are about 300 base pairs long and are distributed in hundreds of thousands of places in human (and other primate) genomes. The core sequence suggests that the *Alu* is a transposable element, including what once was a small RNA gene (Mighell, Markham et al. 1997). Sequence comparison shows a hierarchy of variation as if these elements episodically insert copies of themselves around the genome, accumulating mutations in the interim.

Clearly, selection has tolerated the presence of this repetitive DNA. It may be the harmless detritus of imperfect DNA repair or replication or of random insertions due to viruses or other "unintentional" processes manipulating DNA in the

TABLE 4-1.
A Sampler of Some Sequence Repeat Elements.

Tandemly repeated DNA
Minisatellites such as di-(($\mathbf{TA}$)$_n$) tri- (($\mathbf{CAA}$)$_n$) or tetranucleotide (($\mathbf{GAGA}$)$_n$) repeats
Microsatellites (10–40 bp) and
Macrosatellites (3–20 kb)

Telomeres
Long arrays of **TTAGGG** repeats

Centromeres
A complex of highly variable duplicate repeats of variable length, nature, and origin, often **A–T** rich, elements often having inverted repeat ends

Interspersed repeats
Short *I*nterspersed *N*uclear *E*lements (*SINES*), generally evolved from small RNA species, usually tRNA, but also 7SL cytoplasmic RNA
Alu repeat family (primate specific); ~1,500,000 copies in the human genome each about 280 bp long, usually flanked by 6–18 bp direct repeats
B1, B2 (mouse); ~150,000 copies in the mouse genome 140–190 bp long
Mariner (Mariner-like) elements, about 80 bp long, two inverted 37 bp regions, flanked by TA dinucleotide
*L*ong *I*nterspersed *N*uclear *E*lements (*LINES*) (mammal specific); called "retroelements" because they are related to retroviruses and retrotransposons
L1 element (Kpn repeat), generally 6–8 kb long, but as small as 500 bp

Transposable elements (*TE*s)
Transposable elements with Long Terminal Repeats 1.5–10 kb
Retroposons derived from RNA and transposed to DNA via cDNA
DNA transposons, transposed directly from DNA to DNA

Transposable repeat elements
*M*iniature *I*nverted-Repeat *T*ransposable *E*lements (*MITE*) (eukaryotes) 80–500 bp, terminal inverted repeats (TIRs)
Maize transposable elements; Ds/Ac; *Ac* is 4563 bp long, and contains 11-bp terminal inverted repeats; *Ds* are truncated versions of *Ac*. *Ds* requires *Ac* to move, and is called "nonautonomous;" *Ac* can move without *Ds*, and is called "autonomous"
P elements (*Drosophila*), 2907 bp, terminal 31 bp inverted repeat

Whole Gene or Cluster Duplication
Tandem gene duplications (e.g., *Hox*, globin, olfactory receptors, immunoglobulin, and R-genes)
Gene cluster duplications (e.g., 4 clusters of *Hox* and associated genes)
Whole genome duplications, for which there is some evidence in vertebrates

cell. These repeat sequences can have functions; for example, they may be used to help the packaging or replication of the genome (e.g., see papers in Caporale 1999). Regardless of their origin, some appear subsequently to have experienced mutation that, in their new context, affects protein coding and expression. This is interesting because the Japanese puffer fish (*Fugu rubripes*) has basically the same genes as other fish but a substantial scarcity of repeat (and other noncoding) elements. Likewise, despite the highly structured nature of centromeric DNA, cell division can apparently take place without it (Amor and Choo 2002).

Genes have arisen historically through duplication events. Functional pieces of genes, or whole genes, are occasionally duplicated or transposed during meiosis. Gene duplication produces a family of related genes. They may remain in tandem on the same chromosome or may be inserted in other chromosomes. Gene families constitute another type of repeat element in the genome; the number of members of a gene family can range from only a few to hundreds. As with other duplicates like *Alu*, the individual copies can accumulate subsequent mutational variation. Gene duplication is a fundamental characteristic of evolution (Ohno 1970).

Figure 4-4 exemplifies the origin and evolution of gene families with a famous example, the *Hox* genes, that are involved in many aspects of embryological pat-

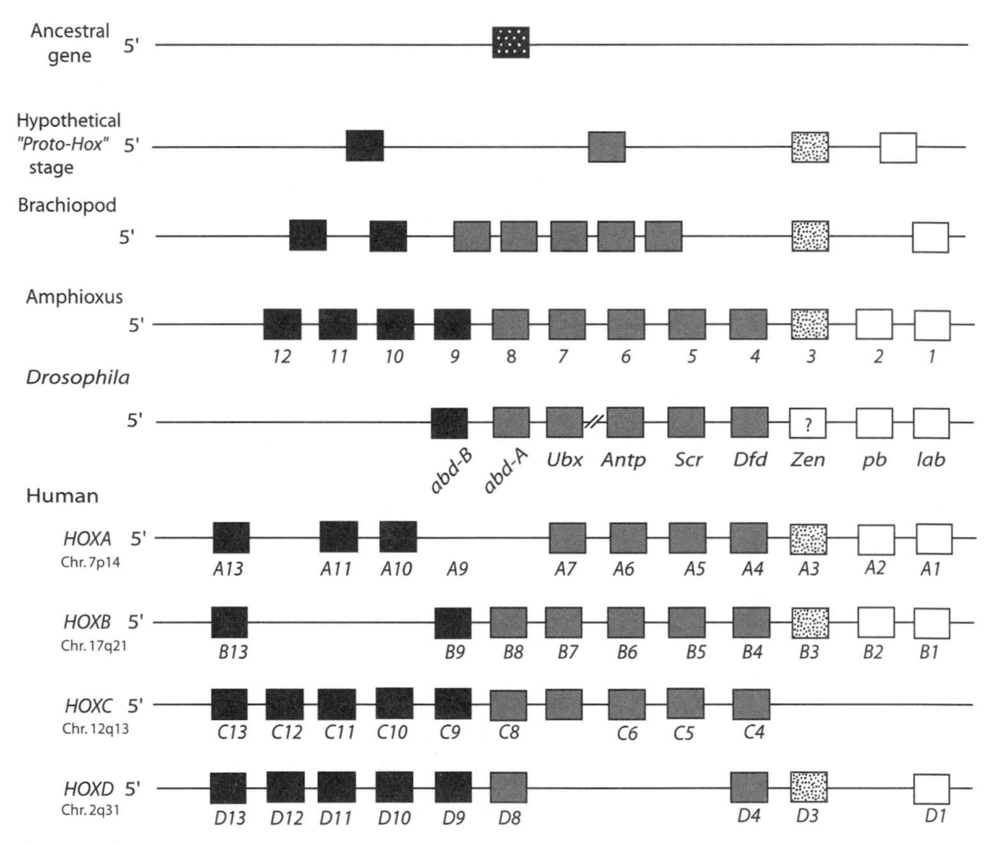

Figure 4-4. Evolution of the *Hox* subfamily of homeobox-containing genes involved in developmental patterning (see Table 7-5 for a description of this family). From an original gene, a small set of "proto-*Hox*" genes evolved by duplication. From these, subsequent duplication has created chromosomally linked clusters in invertebrates and early vertebrates. These clusters continued to gain and lose genes by tandem duplication and the clusters themselves were duplicated on at least two occasions (perhaps more in fish). Shading indicates likely homologies, that is, genes thought, based on sequence comparisons, to be descended from a specific common ancestor. Gene names are shown for the vertebrate human and *Drosophila* clusters and for the stem chordate *Amphioxus*.

terning. Figures of the evolution of the *Hox* genes are so often published that perhaps everyone is aware of them and they have become trite. But the example is important because the discovery of these genes had a transforming effect on biology. As will be seen in later chapters, not only are these genes used in corresponding structures in very diverse animals, but they are used in a way that was striking when first discovered. The *Hox* genes persuasively, and dramatically, showed the continuity of animal life and the much greater than expected homologies of structure and process across the animal world.

Hox genes and their action were first identified in patterning of the major fruit fly body axis. That was remarkable enough, as it was one of the first examples of complex patterning to be understood genetically. But then homologous genes were found in vertebrates. In addition, the gene arrangement structure was similar to that in flies, and indeed vertebrates have four separate clusters that resulted from major cluster-duplication events. Comparable sharing has now been extended to most groups of animals, and indeed following these discoveries many more instances of deep conservation of genes have been found, to the extent that it is now perhaps expected rather than surprising. We will see examples in several subsequent chapters.

With some understanding of what is generally found in a molecule of DNA, we can now look briefly at some of its major sequence-based elements that serve to *code* for various functions. The biology of DNA itself (its replication, packaging, and the like in cells) encompasses core biological traits but is not of primary interest in this book; therefore, we next consider the role of DNA in coding for the production of other substances.

Genes: Coding for RNA

Located at various places in the chromosomes of every species are the codes for the various types of RNA molecules. These include the tRNA molecules that "transfer" (carry) a specific amino acid for use in protein assembly, the *ribosomal RNA* (rRNA) molecules that are major constituents of *ribosome*s, where the tRNA-borne amino acids are concatenated, and the "small nuclear" snRNA molecules that participate in a variety of functions such as splicing mRNA and attending to telomeres (chromosome ends). These genes code for RNA that is itself directly functional, depending on its nucleotide sequence and autoannealing conformation.

RNA *transcription* is achieved by enzymes that move along a specific one of the two DNA strands at a given location (the two strands locally separate for this to happen); the enzyme "reads" the sequence incorporating (ribo)nucleotides one at a time into a chain by complementary base pairing to each nucleotide on the DNA template strand. The resulting RNA molecule is identical in sequence to that of the other "coding" DNA strand, except that ribonucleotides that are chained together to make an RNA molecule have a slightly different sugar than DNA and **U** replaces the **T** of DNA. Different polymerase (chain making) enzymes are used to transcribe different types of RNA; the control of which polymerase is used is encoded in the flanking sequence on the chromosome that represents a physical binding site (known as a *promoter*; see below) for the enzyme.

DNA is the "permanent" codebook. Once the RNA is transcribed, it leaves the DNA and proceeds to go about its business, while the DNA template—the "master" code—remains intact and can reanneal into a stable double-stranded state. When a cell divides, for example, the whole intact DNA molecule is copied and passed to

A. Lysine tRNAs,Various Species

```
Consensus             --GCCCGG-TA-GCTCAGT-CGGT--AGAGC-ATCAGACTTTT-AATCTGAGG--GTCGTGGGTTCGA-GTCCCACGTTGGGCGXXX

Archealglobus fulg    GGGCTCG--..-.....GCCA..C--.....-GACGGG..TTT-..CCCGTCG--GTCGCG...T.AA-AT..CGTCGAGCC.G
Bombyx mori           --GCCCGGC..-.....GT-C..T--.....-ATGAGA..CTT-..TCTCAGG--GTCGTG...T.GA-GC..CACGTTGGG.G
C elegans             --GCCCGGC..-.....GT-C..T--.....-ATGAGA..CTT-..TCTCAGG--GTCGTG...T.GA-GC..CACGTTGGG.G
Chicken               --GCCCGGC..-.....GT-T..T--.....-ATGAGA..CTT-..TCTCAGG--GTCGTG...T.GA-GC..CACGTTGGG.G
Chlorella virus k2    --GCCCGTC..-.....GT-C..T--.....-GCCAGA..CTT-..TCTGGTG--GTCGTG...T.GA-GC..CACGATGGG.A
Drosophila            --GCCCGGC..-.....GT-C..T--.....-ATGAGA..CTT-..TCTCAGG--GTCGTG...T.GA-GC..CACGTTGGG.G
E coli                -GGGTCGT-..-.....GT-T..T--.....-AGTTGA..TTT-..TCAATTG--GTCGCA...T.GA-AT..TGCACGACC.ACCA
Haemophilus influ     -GGGTCGT-..-.....GA-C..T--.....-AGCGGA..TTT-..TCCGTTG--GTCGAA...T.GA-AT..TTCACGACC.ACCA
Helicobacter pyloris  -GACCCGT-..-.....GC-T..T--.....-AATTCC..TTT-..GGAATGG--GCCGTT...T.AA-AT..AACACGGGT.A
Human                 --GCCCGGA..-.....GT-C..T--.....-ATCAGA..TTT-..TCTGAGG--GTCCAG...T.AA-GT..CTGTTCGGG.G
Loligo bleekeri       --GCCCGGC..-.....GT-C..T--.....-ACGAGA..CTT-..TCTC-GG--GTCGTG...T.AA-GC..CACGTTGGG.GCCA
Methanococcus jan     GGGCCCG--..-.....GTCT..C--.....-GCCTGG..TTT-..CCAGGTG--GTCGAG...T.AA-AT..CTTCGGGCC.G
Methanotherm fer      GGGCCCG--..-.....GTCT..T--.....-GCTTGG..TTT-..CCAAGTA--GTCGCG...T.AA-AT..CGTCGGGCC.G
Mouse                 --GCCTGGA..-.....AT-T..T--.....-ATCAGA..TTT-..TCTGAGG--GTTCAG...T.AA-GT..CTGTTCAGG.G
Mycoplasma capric     -GACTCGT-..-.....GC-C..T--.....-AACTGG..TTT-..CCAGTGG--GTCCGG...T.GA-AT..CCGACGAGT.ACCA
Rat                   --GCCCGGC..-.....GT-C..T--.....-ATGAGA..CTT-..TCTCAGG--GTCGTG...T.GA-GC..CACGTTGGG.G
Treponema pallidium   -GGGCCAT-..-.....GT-T..T--.....-AACAGA..CTT-..TCTGTGG--GTCGCA...T.GA-AG..TGCATGGCT.A
Wheat germ            --GCCCGTC..M.....GD-D..T--.....MGCAAGG..CTTT..CCTTGTGMGDMCGTG...Y.GMAGC..CACGGTYGG.GCCA
Xenopus laevis        --GCCCGCA..-.....GT-C..T--.....-ATCAGAC.TTT-..TCTGAGG--GTCCAG...T.AA-GT..CTGTTCGGG.G
Phage T5              -GGGTTGC-..-.....AC-T..TTT.....-ACTGGT..TTT-..ACCATAG--GTTACA...T.GA-GT..TGTGCAACC.ACCA
```

B. Human Serine tRNAs

```
Consensus         GTAGTCGTGGC--CGAGTG-GTTAAGGCGATGGACTTGAAATCCATT-GGGGTT-TCCCCGC---GCAGGTTCGAA-TCCTGCCGACTACGXXX

Hum Serine1       .T..TCG...C--CG.GTG-GTT.AG..G.T.....TGA..T..A.T-G.GGTT-.CC.CGC---..A...TCGA.-...T.CCGA..ACG
Hum Serine2       .T..TCG...C--CG.GTG-GTT.AG..G.T.....AGA..T..A.T-G.GGTT-.CC.CAC---..A...TCGA.-...T.CCGA..ACG
Hum Serine2.2     .T..TCG...CACCG.GDGMGDD.AG..G.Y.....TGA..Y..A.TMG.GGTMC.CC.CGMC--..A...TCGA.-...T.CCGA..ACG
Hum Serine3       .T..TCG...C--CG.GTG-GTT.AG..G.T.....AGA..T..A.T-G.GGTC-.CC.CGC---..A...YCGM.A...T.CCGA..ACGCCA
Hum Serine4       .T..TCG...C--CG.GTG-GTT.AG..G.T.....TGA..T..A.T-G.GGTT-.CC.CGC---..A...TCGA.-...T.TCGG..ACG
Hum Serine5       .G..ATA...C--AC.ATCTGGT.CT..A.C.....CGA..T..G.CGA.CCGCG.TT.GGGTGT..G...TCAA.-...C.CTAT..CCT
```

Figure 4-5. Conservation and variation in tRNA. (A) Alignments showing conserved (dots) and variable elements in the DNA sequences coding for lysine tRNA in various species; (B) alignment showing variation among the multiple copies of genes coding for serine tRNA within the human genome. Consensus (generally, the most common nucleotide at each position, even though no gene has this actual sequence) shown above each part. Bold face triplets are the mRNA-binding anticodon (this shows the redundancy of the coding system—see below).

descendant cells. In contrast, RNA is much less stable and is often enzymatically destroyed.

Multiple Functions via Modularity in RNA

A tRNA molecule folds upon itself via complementary base pairing into a general "cloverleaf" shape that is characteristic of tRNAs. Over evolutionary time, a considerable amount of variation on this structure has accumulated, although the basic shape is retained. The sequence at one end of a given type of tRNA determines which amino acid will attach there, and a sequence of three nucleotides at the other end of the (folded) tRNA molecule, called the *anticodon*, determines its specificity to a particular *codon* in the assembly of polypeptide (protein) chains.

The genomes of individual organisms contain multiple copies of codon-specific tRNA genes, some of them clustered on their respective chromosomes. Whatever other variation may have occurred, each codon-specific tRNA has a recognition triplet in the proper location (relative to the folded tRNA molecule) and the sequence structure that enables it to carry its specific amino acid. The "same" (amino-acid-specific) tRNA within or among organisms can vary in many details of its sequence, as can be seen by aligning them. Because of the redundancy of the coding system (see below), different recognition triplets that specify the same amino acid can be found in the appropriate positions on their respective tRNA molecules. Thus, within the flexibility of the cloverleaf conformation, there is considerable variation that serves this variation-on-a-theme function (amino acid specificity).

From DNA to RNA to Protein: A Digital Coding System for the Diversity of Life
The most familiar role of DNA is as a code that specifies the structure of a protein, the role to which the word "gene" was first applied in regard to DNA. The diverse functions of life are brought about by the chemical properties of the 20 different amino acids (listed in Table 4-2); their number and order determine the interactions of a protein with other molecules. These properties depend on how it is folded, chemically modified, and combined with other molecules (including other proteins). Many proteins function as complexes of several polypeptides, each coded by region is of DNA that may or may not be located close to each other on the same chromosome. This simple logic accounts for the diversity of function observed in the biosphere.

Amino acid specification is based on a three-nucleotide genetic code, as listed in Table 4-3. Because there are 20 amino acids but only four nucleotides, at least three nucleotides are required to specify 20 different amino acids. However, because there are $4^3 = 64$ possible different three-nucleotide codes and the code is *redundant*, most amino acids are specified by more than one triplet, in ways that probably reveal the evolutionary history of the coding system (see below). Within a protein-coding segment of DNA, each successive three-nucleotide stretch comprises a codon, whose sequence specifies a single amino acid in the coded protein. Most amino acids are specified by more than one codon (code redundancy), and three codons specify

TABLE 4-2.
Amino Acid Single Letter Designation Code.

A—Alanine (Ala)
C—Cysteine (Cys)
D—Aspartic Acid (Asp)
E—Glutamic Acid (Glu)
F—Phenylalanine (Phe)
G—Glycine (Gly)
H—Histidine (His)
I—Isoleucine (Ile)
K—Lysine (Lys)
L—Leucine (Leu)
M—Methionine (Met)
N—Asparagine (Asn)
P—Proline (Pro)
Q—Glutamine (Gln)
R—Arginine (Arg)
S—Serine (Ser)
T—Threonine (Thr)
V—Valine (Val)
W—Tryptophan (Try)
Y—Tyrosine (Tyr)

TABLE 4-3.
Genetic Code, Showing the Amino Acids Coded for By Three-letter Codons.

First Letter of Codon	Second Letter of Codon: T			C			A			G		
T	TTT	Phe	(F)	TCT	Ser	(S)	TAT	Tyr	(Y)	TGT	Cys	(C)
	TTC	Phe	(F)	TCC	Ser	(S)	TAC	Tyr	(Y)	TGC	Cys	(C)
	TTA	Leu	(L)	TCA	Ser	(S)	TAA	STOP		TGA	STOP	
	TTG	Leu	(L)	TCG	Ser	(S)	TAG	STOP		TGG	Trp	(W)
C	CTT	Leu	(L)	CCT	Pro	(P)	CAT	His	(H)	CGT	Arg	(R)
	CTC	Leu	(L)	CCC	Pro	(P)	CAC	His	(H)	CGC	Arg	(R)
	CTA	Leu	(L)	CCA	Pro	(P)	CAA	Gln	(Q)	CGA	Arg	(R)
	CTG	Leu	(L)	CCG	Pro	(P)	CAG	Gln	(Q)	CGG	Arg	(R)
A	ATT	Ile	(I)	ACT	Thr	(T)	AAT	Asn	(N)	AGT	Ser	(S)
	ATC	Ile	(I)	ACC	Thr	(T)	AAC	Asn	(N)	AGC	Ser	(S)
	ATA	Ile	(I)	ACA	Thr	(T)	AAA	Lys	(K)	AGA	Arg	(R)
	ATG	Met	(M)	ACG	Thr	(T)	AAG	Lys	(K)	AGG	Arg	(R)
G	GTT	Val	(V)	GCT	Ala	(A)	GAT	Asp	(D)	GGT	Gly	(G)
	GTC	Val	(V)	GCC	Ala	(A)	GAC	Asp	(D)	GGC	Gly	(G)
	GTA	Val	(V)	GCA	Ala	(A)	GAA	Glu	(E)	GGA	Gly	(G)
	GTG	Val	(V)	GCG	Ala	(A)	GAG	Glu	(E)	GGG	Gly	(G)

Note the redundancy of the code, that is, an amino acid can be coded for by more than one codon.

the instruction *stop*, that is, the end of the amino acid coding sequence. As can be seen from the table, the redundancy is mainly in the third position. The level of code degeneracy (e.g., whether two or four nucleotides in a given position code for the same amino acid) was what was shown in Figure 3-5 concerning the relative amounts of diversity in genes. Thus, the bar for variation found in two-fold degenerate sites refers to sites like the third position in the codons that specify the amino acid lysine, which can be an **A** or a **G**.

A functional gene is more than a protein-coding region. The structure of a typical gene is shown schematically in Figure 4-6A. By convention, relative position in DNA is denoted 5′ and 3′ (in a left-to-right diagrammatic layout), referring to the location number of the carbon atoms in the sugar ring that links the nucleotide "before" and "after" the position being considered, respectively. In the 5′ DNA flanking a coding region, sequences are present that carry a different kind of information—that for the regulation of the gene in question (discussed below), not for protein structure. The major features in the immediate gene-flanking region include various generic sequences used in gene transcription (for example, the "**TATA**" box is a sequence element containing **TATA**), a site where the transcription starts, and an **ATG** codon (for methionine) that is the *start* signal for the translation of the code into protein (seen in the second exon of the *Apoe* gene in Figure 4-6B). This is followed by a series of consecutive codons, called an *open reading frame* (*ORF*), which specifies a corresponding sequence of amino acids.

Similar to the production of rRNA and tRNA, transcription of a gene is brought about by an appropriate RNA polymerase that binds the promoter DNA itself (along with a complex of other regulatory proteins, themselves coded for by other genes) and then copies one strand of the locally unwound double helix of DNA into a single-stranded complementary molecule of *messenger RNA* (mRNA). Transcription ends when the polymerase encounters a *stop* signal (one of several particular codons, shown in Table 4-3) in the DNA.

Most eukaryotic genes are interrupted by sequences of variable lengths that do not code for amino acids and typically may not code for anything or even have a function. The coding regions are called *exons*, and the intervening sequences are called *introns*. An exon ends when an appropriately recognized **GT** . . . is reached, and this is followed by intronic sequence until an appropriately recognized **AG** . . . is reached. After that, another exon may be encountered. The details, especially of the DNA flanking these splice donor and acceptor sites, are somewhat variable. Different proteins have different numbers of exons, a number that is the result of history but generally seems not to be constrained by anything biochemical; some genes have no introns, whereas others can have tens of them.

The freshly transcribed RNA is a temporary structure whose introns must be excised and consecutive exons *spliced* together by chemical machinery that recognizes the **AG-GT** signals. The RNA is further altered by the addition of a methyl cap on one end and a poly(**A**) tail (long string of **A**s) at the other to enable translation into protein. All of these modifications are functional, that is, affect chemical reactions at later stages, and/or the stability and processing of the RNA itself. The molecule is now mature mRNA. It is a messenger because it embodies the codon sequence that will be used to specify the amino acid sequence of a protein.

The mature mRNA is transported from the cell nucleus to the cytoplasm for *translation* into polypeptide (the corresponding system is somewhat different in prokaryotes, which have no nucleus, but the basic coding mechanism is similar). Protein synthesis is catalyzed by ribosomes, which are composed of multiple rRNA subunits (each transcribed from a locus in the chromosomal DNA), and other constituents. The mRNA serves as template; moving through the ribosome three nucleotides at a time, tRNAs whose anticodon is complementary to the mRNA

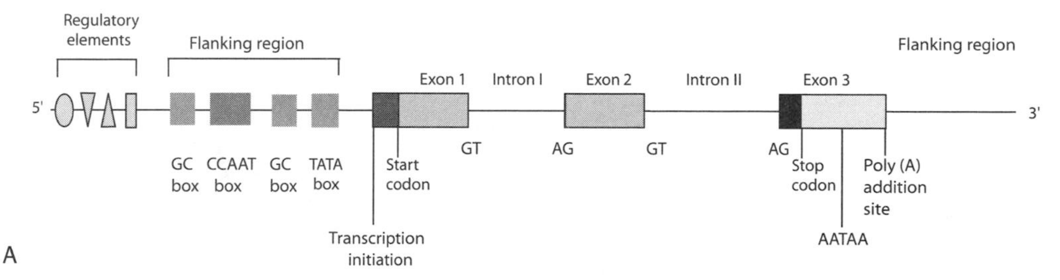

Figure 4-6. (A) Stereotype of a typical gene, including flanking regulatory regions, introns and exons, and so on. The number of exons varies among genes; (B) these components identified in the *Apoe* gene shown earlier in Figure 4-2, also indicating the location of repeat elements.

GGAACTTGATGCTCAGAGAGGACAAGTCATTTGCCCAAGGTCACACAGCTGGCAACTGGCAGACGAGAT
TCACGCCCTGGCAATTTGACTCCAGAATCCTAACCTTAACCCAGAAGCACGGCTTCAAGCCCTGGAAACC
ACAATACCTGTGGCAGCCAGGGGGAGGTGCTGGAATCTCATTTCACATGTGGGGAGGGGGCTCCTGTGCT
CAAGGTCACAACCAAAGAGGAAGCTGTGATTAAAACCCAGGTCCCATTTGCAAAGCCTCGACTTTTAGCA
GGTGCATCATACTGTTCCCACCCCTCCCATCCCACTTCTGTCCAGCCGCCTAGCCCCACTTTCTTTTTTTC
TTTTTTTGAGACAGTCTCCCTCTTGCTGAGGCTGGAGTGCAGTGGCGAGATCTCGGCTCACTGTAACCTCC
GCCTCCCGGGTTCAAGCGATTCTCCTGCCTCAGCCTCCCAAGTAGCTAGGATTACAGGCGCCCGCCACCA
CGCCTGGCTAACTTTTGTATTTTTAGTAGAGATGGGGTTTCACCATGTTGGCCAGGCTGGTCTCAAACTCC
TGACCTTAAGTGATTCGCCCACTGTGGCCTCCCAAAGTGCTGGGATTACAGGCGTGAGCTACCGCCCCCA
GCCCCTCCCATCCCACTTCTGTCCAGCCCCCTAGCCCTACTTTCTTTCTGGGATCCAGGAGTCCAGATCCC
CAGCCCCCTCTCCAGATTACATTCATCCAGGCACAGGAAAGGACAGGGTCAGGAAAGGAGGACTCTGGG
CGGCAGCCTCCACATTCCCCTTCCACGCTTGGCCCCCAGAATGGAGGAGGGTGTCTGTATTACTGGGCGA
GGTGTCCTCCCTTCCTGGGGACTGTGGGGGGTGGTCAAAAGACCTCTATGCCCCACCTCCTTCCTCCCTCT
GCCCTGCTGTGCCTGGGGCAGGGGGAGAACAGCCCACCTCGTGACTGGGCTGCCCAGCCCGCCCTATCCC
TGGGGGAGGGGGCGGGACAGGGGGAGCCCTATAATTGGACAAGTCTGGGATCCTTGAGTCCTACTCAGC
CCCAGCGGAGGTGAAGGACGTCCTTCCCCAGGAGCCGGTGAGAAGCGCAGTCGGGGGCACGGGGATGA
GCTCAGGGGCCTCTAGAAAGAGCTGGGACCCTGGGAAGCCCTGGCCTCCAGGTAGTCTCAGGAGAGCTA
CTCGGGGTCGGGCTTGGGGAGAGGAGGAGCGGGGGTGAGGCAAGCAGCAGGGGACTGGACCTGGGAAG
GGCTGGGCAGCAGAGACGACCCGACCCGCTAGAAGGTGGGGTGGGGAGAGCAGCTGGACTGGGATGTA
AGCCATAGCAGGACTCCACGAGTTGTCACTATCATTATCGAGCACCTACTGGGTGTCCCCAGTGTCCTCA
GATCTCCATAACTGGGGAGCCAGGGGCAGCGACACGGTAGCTAGCCGTCGATTGGAGAACTTTAAAATG
AGGACTGAATTAGCTCATAAATGGAACACGGCGCTTAACTGTGAGGTTGGAGCTTAGAATGTGAAGGGA
GAATGAGGAATGCGAGACTGGGACTGAGATGGAACCGGCGGTGGGGAGGGGGTGGGGGGATGGAATTT
GAACCCCGGGAGAGGAAGATGGAATTTTCTATGGAGGCCGACCTGGGGATGGGGAGATAAGAGAAGAC
CAGGAGGGAGTTAAATAGGGAATGGGTTGGGGGCGGCTTGGTAAATGTGCTGGGATTAGGCTGTTGCAG
ATAATGCAACAAGGCTTGGAAGGCTAACCTGGGGTGAGGCCGGGTTGGGGGCGCTGGGGGTGGGAGGA
GTCCTCACTGGCGGTTGATTGACAGTTTCTCCTTCCCCAGACTGGCCAATCACAGGCAGGAAGATGAAGG
TTCTGTGGGCTGCGTTGCTGGTCACATTCCTGGCAGGTATGGGGGCGGGGCTTGCTCGGTTCCCCCCGCTC
CTCCCCCTCTCATCCTCACCTCAACCTCCTGGCCCCATTCAGACAGACCCTGGGCCCCCTCTTCTGAGGCT
TCTGTGCTGCTTCCTGGCTCTGAACAGCGATTTGACGCTCTCTGGGCCTCGGTTTCCCCCATCCTTGAGAT
AGGAGTTAGAAGTTGTTTTGTTGTTGTTGTTTGTTGTTGTTGTTTTGTTTTTTTGAGATGAAGTCTCGCTCT
GTCGCCCAGGCTGGAGTGCAGTGGCGGGATCTCGGCTCACTGCAAGCTCCGCCTCCCAGGTCCACGCCAT
TCTCCTGCCTCAGCCTCCCAAGTAGCTGGGACTACAGGCACATGCCACCACACCCGACTAACTTTTTTGTA
TTTTCAGTAGAGACGGGGTTTCACCATGTTGGCCAGGCTGGTCTGGAACTCCTGACCTCAGGTGATCTGC
CCGTTTCGATCTCCCAAAGTGCTGGGATTACAGGCGTGAGCCACCGCACCTGGCTGGGAGTTAGAGGTTT
CTAATGCATTGCAGGCAGATAGTGAATACCAGACACGGGGCAGCTGTGATCTTTATTCTCCATCACCCCC
ACACAGCCCTGCCTGGGGCACACAAGGACACTCAATACATGCTTTTCCGCTGGGCCGGTGGCTCACCCCT
GTAATCCCAGCACTTTGGGAGGCCAAGGTGGGAGGATCACTTGAGCCCAGGAGTTCAACACCAGCCTGG
GCAACATAGTGAGACCCTGTCTCTACTAAAAATACAAAAATTAGCCAGGCATGGTGCCACACACCTGTGC
TCTCAGCTACTCAGGAGGCTGAGGCAGGAGGATCGCTTGAGCCCAGAAGGTCAAGGTTGCAGTGAACCA
TGTTCAGGCCGCTGCACTCCAGCCTGGGTGACAGAGCAAGACCCTGTTTATAAATACATAATGCTTTCCA
AGTGATTAAACCGACTCCCCCCTCACCCTGCCCACCATGGCTCCAAAGAAGCATTTGTGGAGCACCTTCT
GTGTGCCCCTAGGTAGCTAGATGCCTGGACGGGGTCAGAAGGACCCTGACCCGACCTTGAACTTGTTCCA
CACAGGATGCCAGGCCAAGGTGGAGCAAGCGGTGGAGACAGAGCCGGAGCCCGAGCTGCGCCAGCAGA
CCGAGTGGCAGAGCGGCCAGCGCTGGGAACTGGCACTGGGTCGCTTTTGGGATTACCTGCGCTGGGTGCA
GACACTGTCTGAGCAGGTGCAGGAGGAGCTGCTCAGCTCCCAGGTCACCCAGGAACTGAGGTGAGTGTC
CCCATCCTGGCCCTTGACCCTCCTGGTGGGCGGCTATACCTCCCCAGGTCCAGGTTTCATTCTGCCCCTGT
CGCTAAGTCTTGGGGGGCCTGGGTCTCTGCTGGTTCTAGCTTCCTCTTCCCATTTCTGACTCCTGGCTTTAG
CTCTCTGGAATTCTCTCTCTCAGCTTTGTCTCTCTCTCTTCCCTTCTGACTCAGTCTCTCACACTCGTCCTGG
CTCTGTCTCTGTCCTTCCCTAGCTCTTTTATATAGAGACAGAGAGATGGGGTCTCACTGTGTTGCCCAGGC
TGGTCTTGAACTTCTGGGCTCAAGCGATCCTCCCGCCTCGGCCTCCCAAAGTGCTGGGATTAGAGGCATG
AGCACCTTGCCCGGCCTCCTAGCTCCTTCTTCGTCTCTGCCTCTGCCCTCTGCATCTGCTCTCTGCATCTGT
CTCTGTCTCCTTCTCTCGGCCTCTGCCCCGTTCCTTCTCTCCCTCTTGGGTCTCTCTGGCTCATCCCCATCTC
GCCCGCCCCATCCCAGCCCTTCTCCCCCGCCTCCCCACTGTGCGACACCCTCCCGCCCTCTCGGCCGCAGG
GCGCTGATGGACGAGACCATGAAGGAGTTGAAGGCCTACAAATCGGAACTGGAGGAACAACTGACCCCG
GTGGCGGAGGAGACGCGGGCACGGCTGTCCAAGGAGCTGCAGGCGGCGCAGGCCGGCTGGGCGCGGAC
ATGGAGGACGTGCGCGGCCGCCTGGTGCAGTACCGCGGCGAGGTGCAGGCCATGCTCGGCCAGAGCACC
GAGGAGCTGCGGGTGCGCCTCGCCTCCCACCTGCGCAAGCTGCGTAAGCGGCTCCTCCGCGATGCCGATG
ACCTGCAGAAGCGCCTGGCAGTGTACCAGGCCGGGGCCCGCGAGGGCGCCGAGCGCGGCCTCAGCGCCA
TCCGCGAGCGCCTGGGGCCCCTGGTGGAACAGGGCCGCGTGCGGGCCGCCACTGTGGGCTCCCTGGCCG
GCCAGCCGCTACAGGAGCGGGCCCAGGCCTGGGGCGAGCGGCTGCGCGCGCGGATGGAGGAGATGGGC
B AGCCGGACCCGCGACCGCCTGGACGAGGTGAAGGAGCAGGTGGCGGAGGTGCGCGCCAAGCTGGAGGA

Figure 4-6. *Continued*

GCAGGCCCAGCAGATACGCCTGCAGGCCGAGGCCTTCCAGGCCCGCCTCAAGAGCTGGTTCGAGCCCCT
GGTGGAAGACATGCAGCGCCAGTGGGCCGGGCTGGTGGAGAAGGTGCAGGCTGCCGTGGGCACCAGCG
CCGCCCCTGTGCCCAGCGACAATCACTGAACGCCGAAGCCTGCAGCCATGCGACCCCACGCCACCCCGTG
CCTCCTGCCTCCGCGCAGCCTGCAGCGGGAGACCCTGTCCCCGCCCCAGCCGTCCTCCTGGGGTGGACCC
TAGTTTAATAAAGATTCACCAAGTTTCACGCATCTGCTGGCCTCCCCCTGTGATTTCCTCTAAGCCCCAGC
CTCAGTTTCTCTTTCTGCCCACATACTGCCACACAATTCTCAGCCCCCTCCTCTCCATCTGTGTCTGTGTGT
ATCTTTCTCTCTGCCCTTTTTTTTTTTTTAGACGGAGTCTGGCTCTGTCACCCAGGCTAGAGTGCAGTGGCA
CGATCTTGGCTCACTGCAACCTCTGCCTCTTGGGTTCAAGCGATTCTGCTGCCTCAGTAGCTGGGATTACA
GGCTCACACCACCACACCCGGCTAATTTTTGTATTTTTAGTAGAGACGAGCTTTCACCATGTTGGCCAGGC
AGGTCTCAAACTCCTGACCAAGTGATCCACCCGCCGGCCTCCCAAAGTGCTGAGATTACAGGCCTGAGCC
ACCATGCCCGGCCTCTGCCCCTCTTTCTTTTTTAGGGGGCAGGGAAAGGTCTCACCCTGTCACCCGCCATC
ACAGCTCACTGCAGCCTCCACCTCCTGGACTCAAGTGATAAGTGATCCTCCCGCCTCAGCCTTTCCAGTA
GCTGAGACTACAGGCGCATACCACTAGGATTAATTTGGGGGGGGGTGGTGTGTGTGGAGATGGGGTCTG
GCTTTGTTGGCCAGGCTGATGTGGAATTCCTGGGCTCAAGCGATACTCCCACCTTGGCCTCCTGAGTAGCT
GAGACTACTGGCTAGCACCACCACACCCAGCTTTTTATTATTATTTGTAGAGACAAGGTCTCAATATGTTG
CCCAGGCTAGTCTCAAACCCCTGGCTCAAGAGATCCTCCGCCATCGGCCTCCCAAAGTGCTGGGATTCCA
GGCATGGGCTCCGAGCGGCCTGCCCAACTTAATAATATTGTTCCTAGAGTTGCACTC

Legend:

SEQUENCE: Short repeat element

SEQUENCE: Alu repeat

SEQUENCE: Exon, untranslated regions

B SEQUENCE: Exon, coding region including signal peptide, mature peptide, stop.

Figure 4-6. *Continued*

triplet carry their amino acid to the site, where it is concatenated to the previously joined amino acid, thus producing a polypeptide, the precursor of a functional protein. This codon-anticodon matching is how the sequence of DNA specifies, through mRNA as a temporary information carrier, the sequence of amino acids, and this is why the DNA is referred to as a modular, digital code.

Correspondence Between Coding and Coded Units Enables the World of Unconstrained Biological Complexity

This system is built up of modular units at all levels and depends on that fact to make it open ended and flexible. A mutation that replaces one nucleotide by another in a codon may change the amino acid that is specified: for example, **GCA** specifies the amino acid alanine; an **A→C** mutation, resulting in **GCC**, will have no effect on the protein because the latter is also a codon for alanine; this is known as a *synonymous* mutation. However, if the **G** mutates to **A**, the resulting **ACA** codon specifies threonine, and hence is a *nonsynonymous* mutation because it changes the amino acid that is coded by this piece of DNA.

Modularity does not imply independence among the constituent parts, however. A change in one can have dramatic effect on the others. An insertion or deletion of a nucleotide will alter the *reading frame*, throwing all subsequent amino acids out of order. Thus, removing the second letter (h) of the following sentence, while preserving the number of letters in each word, destroys its meaning completely: "The fish swam well" becomes "Tef ishs wamw ell." The following single-letter *substitu-*

tions make relatively little change: "The fish swim well" and "Thy fish swam well." The following change also makes sense—but with quite a difference: "The dish swam well."

With a modular amino acid coding mechanism, matched by the modularity in tRNA, a system evolved by which a sequence of amino acids could be assembled and "remembered" across replication, cell division, or organism reproduction. The number of ways a polypeptide can be put together via different amino acid sequences is essentially limitless, and life evolved to build structures and catalyze reactions around the kinds of interactions that depend on the chemical properties of folded, modified polypeptides. This constrains, but also enables, the huge diversity of life and form on Earth.

RNA Regulating Gene Expression

If it is true in physics that for every action there is an equal and opposite reaction, there is something like this in genomic information as well. If mRNA represents activation of a coding signal, the same signal can be dampened by *antisense* RNA, which can bind to mRNA by complementary base pairing. In Chapter 7, we will see that important means to do this do in fact exist. Of course, were this "reaction" truly equal and opposite, nothing in life would get accomplished. But then, biology is not the same as physics!

RNA is also used in an interesting and direct physical way in the control of chromosomal inactivation during sex determination in mammals, by affecting expression of genes related to functional differences between males and females. This will be described in Chapter 8.

DNA: A Protected Code

The beauty of the DNA/RNA system is that it solves many functions by the one phenomenon of base pairing. The double-stranded nature of DNA and its packaging make it a stable molecule resistant to damage. When a cell divides, the entire complement of DNA must be replicated so that each daughter cell receives a complete copy. To achieve this, the two strands, call them A and B, separate and are matched nucleotide-by-nucleotide by a new concatenation of nucleotides. This generates two new double helices, AB and BA (and these are chemically identical—the order we write them in is strictly arbitrary). Uncorrected copying errors (mutations) can occur in the replication process, which is vital to evolution, but basically the DNA is preserved for posterity.

Meanwhile, RNA is the local, contextual, temporary tool for making polypeptides. A given gene can be transcribed many times to make for abundant protein, but the mRNA itself is not replicated and is eventually degraded. RNA transcription is based on base complementarity and so is the tRNA anticodon system. Thus all these functions are based on the same base-complementarity system. While this is general and applies between strands, the linear sequence along strands is used to carry the *specific* coding information.

The "Central Dogma"

The discovery of the logic DNA→ mRNA→ protein as a fundamental characteristic of life led to the formulation of the "Central Dogma," referred to in Chapter 2. This states that this information flow goes only one way: there is no flow back of information (that is, to DNA sequence) that reflects life experience directly in a heritable, lamarckian way. The causal arrows go only one way. Evolution unfolds but does not involute back upon itself.

The Central Dogma asserts an aspect of biological sequestration—of information—at the heart of life and evolution that, unlike other kinds of sequestration, is widely and almost uniquely considered to be complete and inviolate. Whether exceptions exist or will be found is uncertain; we know that DNA does many other things, some more codelike than others, and, as we will see, the nature and level of sequestration between those aspects of the genome and the organism are more variable and incomplete. They have to do, for example, with the regulation of gene transcription. The "Dogma" is fundamental to the genetically centered view of life that provided a powerful prescription for research that transformed biology. It is an elegant system and is the core of molecular biology now being done, and was one of the most important discoveries in biology in the 20th century.

Information Change: Genetic Mutations

We introduced the general notion of mutation in the last chapter; functional aspects of genes can be altered in several important ways. As illustrated by the alterations in the fish sentence above, mutations can alter the amino acids for which a gene codes, can interrupt or even destroy the coding system and make the protein wholly functionless, or can alter the amino acids in subtle ways that make small changes in the action of the coded protein. But because of the redundancy of the codon system, a high fraction of single nucleotide substitution mutations do not even change the amino acids coded for; all they do is change the particular tRNA species that is used to supply that amino acid. Sequence-altering mutations can involve simple nucleotide substitutions, deletions, or insertions, rearrangements of chromosomal fragments, or the transfer of repeat elements into or out of a region.

As well as affecting coding sequences, mutations can also affect regulatory sequences, creating or destroying them, altering their location relative to the gene itself, or altering the binding efficiency of regulatory factors. The sequence itself can be affected, or change can occur in the way that the nucleotides are modified (e.g., methylation of stretches of adjacent **C**s and **G**s flanking mammalian genes). These various changes can have subtle or dramatic effects, depending on the particular gene or context.

Gene function can also be altered by the way DNA is packaged, which can make a gene more, or less, available for transcription in a given cell or context. The determinants of the location of histones or other packaging elements are currently not well understood and may depend on factors other than just sequence signals that could constitute a kind of nonsequence-based inheritance.

Another type of mutation occurs by horizontal transfer of genes among individuals or species. It appears to be rare, especially today, with a few major exceptions. One is the transfer of plasmids among bacteria, for example, to confer resistance to natural (or human made) antibiotics. Viral DNA survives mainly by horizontal transfer because it functions only in the context of a host cell and its genome; viral transfer can cause cancer or other diseases but is of longterm evolutionary interest relative to hosts only if it integrates into the germ line rather than just somatic cells.

As we have noted in several places, we know of no truly lamarckian mechanisms by which a mutation occurs in order to bring about a specific adaptive change in a protein. As we noted in Chapter 3, however, aspects of nongenetic inheritance should perhaps be more seriously included in evolutionary biological theory. Some gene-*related* changes may be inherited, such as gene-expression levels that may affect traits like blood pressure, the stimulus for which can be transmitted across the placenta. But this is reversible in different environments, and does not permanently alter DNA, and hence is not lamarckian in the genetic sense.

CODING CELLULAR DIFFERENTIATION: INFORMATION REGULATING INFORMATION

Except for simple organisms, the number of genes in a given species is somewhat a matter of speculation, even when the complete DNA sequence is known. This is partly because of the multiple splicing of mRNA and overlapping transcription that are sometimes found. The definition of a "gene" also changes as we continue to learn of ways new functions are encoded in the genome. Even with the simplest classical definition of genes as protein codes, a complex organism typically has tens of thousands of genes.

This raises an important biological question. Why, if essentially all cells in an organism contain the entire genome, whatever the number of genes that includes, do individual cells differ in their structure and behavior? The answer is that this is mainly because, in any given cell or cellular context, only a subset of these genes are used. Genes are turned "on" or "off" (are expressed or repressed) in a highly controlled active way that is affected by circumstances. This control is mediated through DNA sequences, another form of information in the genome—information that controls the use of coding information. How is this done?

Transcription Regulation

The key is *cis*-regulation, whose information signals are found in sequence on the same chromosome (hence, *cis*) as the protein-coding gene (Figure 4-7) (Davidson 2001; Davidson et al. 2002). Usually the regulatory signals are in the nearby 5′ flanking region. Gene expression is also regulated by the local state of the DNA in terms of its packaging or chemical modification. Typically, a gene is expressed when a complex of specific regulatory proteins including an appropriate polymerase is bound to promoter sequences just upstream (5′) of the transcription start site.

This is a generic signal found in corresponding places in most genes, but specific gene expression is affected by other sequence elements flanking (and sometimes within) a gene; these are known as *enhancers*, *repressors*, or other *response, or regulatory, elements* (REs), which are bound in the appropriate cellular context by a variety of regulatory or DNA-processing proteins to make the DNA accessible to the polymerase. "Binding" means that the chemical structure of DNA with a particular nucleotide sequence—the RE—fits the charge and shape of a binding domain of the regulatory protein.

The Logic of Gene Transcription

Gene transcription begins with RNA polymerase binding to the promoter site, a short sequence about 25 nucleotides upstream of a gene's transcription start site. Sites like those labeled **TATA** and **CAAT** (Figure 4-6A) are found in about the same location flanking most eukaryotic genes. The rate of transcription is modified by the binding to enhancer sequence elements by DNA-binding proteins known as

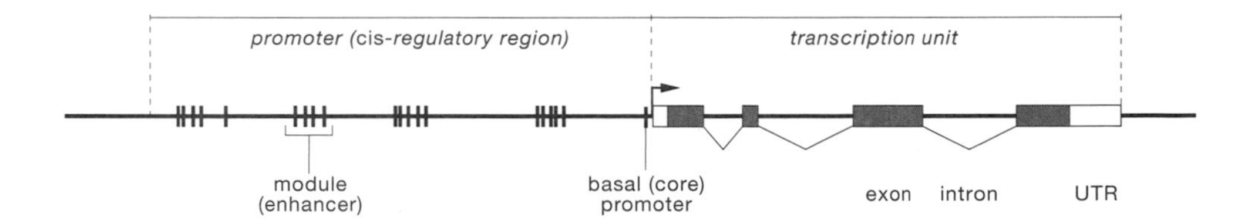

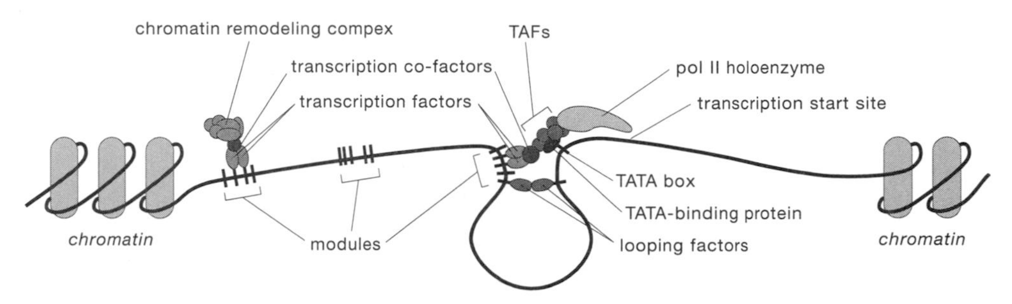

Figure 4-7. Basic *cis*-regulation of gene expression. Top line shows schematic organization of regulatory regions in relation to a gene. These can vary in copy number, number of elements, number of repeats and their arrangement relative to the gene. Bottom of figure shows the nature of protein complexes in a chromosomal region of a gene being expressed. Various components of the regulation process, including Transcription Activating Factors (TAF), are shown generically. Reprinted from (Wray et al. 2003) with permission.

transcription factors (TFs). These DNA-binding regulatory factors are encoded by genes elsewhere in the genome.

REs are typically very short (e.g., 5 to 10 base pairs) sequences, often assembled into regulatory modules or cassettes. This is shown schematically in Figure 4-7. A module typically contains binding sites for four to eight TFs, but there is no fixed number or arrangement or even location relative to the gene. A cluster of such elements is sometimes jointly called a *locus control region* (LCR), and can be quite distant from the gene itself. The TFs binding in such a region are often from different gene families. When in place, the configuration of TFs attract and enable the polymerases and other basal RNA transcription machinery.

Many regulatory proteins typically assemble in the regulatory region of a gene when it is to be expressed and interact in a variety of ways, including direct protein-protein as well as protein-DNA binding. Products of the *Sox* family of genes produce bends or loops in DNA near a gene, to enable other regulatory proteins to bind there. The number, location, or specific sequence details (e.g., binding affinities) of the REs can quantitatively affect expression levels. There are other sequence elements, for example, insulators that prevent multiple neighboring genes from being expressed at the same time.

Table 4-4 provides some examples of TF recognition sequences. This sequence information is not unambiguous, in that related TFs may recognize similar binding motifs, so there can be crossreaction; for example, many *Hox* genes (that is, their coded proteins) recognize motifs like **ATTAAATTA**. POU domains recognize sequences like **ATTTGCAT**. Some leucine zipper TFs bind **CCAAT**.

TABLE 4-4.
Known motifs in yeast.

Known motif	
Factor	Motif
ABF1	RTCRYnnnnnACG
UME6	TCGGCGGCTA
CBF1	RTCACRTG
NDT80	TCGGCGGCTDW
REB1	TTACCCGG
MCM1a	TTWCCCnWWWRGGAAA
SWI6	ACGCGT
PHO4	CACGTG
MBP1	ACGCGTnA
SWI4	TTTTCGCG
DAL81	GATAAG
RPN4	TTTTGCCACC
MSN2	CCCCT
MSN4	CCCCT
PDR1	CCGCGG
ESR2	AAAAWTTTT
MIG1	CCCCRSWWWW
MIG1b	CCCCGC
BAS1	TGACTC
GCN4	ATGACTCAT
GAL4	CGGnnnnnnnnnnnCCG
HSF1b	TTCTAGAA
ESR1	GATGAG
MET31	AAACTGTGGC
AFT1	YRCACCCR
TEA1	CGGnCGG
PUT3	CGGnnnnnnnnnnCCG
HAP2	TGATTGGC
RAP1	ACACCCATACATTT
LEU3	CCGGnnCCGG
MCM1b	YTTCCTAATTWGnnCn
INO4	CATGTGAAAT
INO2	CATGTGAAAT
GLN3	GATAAK
ADR1	GGAGA
FKH2	TTGTTTACST
FKH1	TTGTTTACST
RLM1	CTAWWWWTAG
SWI5	KGCTGR
HAP1	CGGnnnTAnCGG
XBP1	MCTCGARRRnR
MAC1	TTTGCTCA
TBF1	TTAGGG
MSE	TTTTGTG
STE12	RTGAAACA
DIG1	RTGAAACA
MET4	TGGCAAATG

TABLE 4-4.
Continued.

Known motif	
Factor	Motif
HAP4	TnRTTGGT
SMP1	ACTACTAWWWWTAG
ACE2	GCTGGT
YAP1	TTACTAA
CIN5	TTACTAA
RME1	GAACCTCAA
HAC1	CAGCGTG
GCR1	GGAAG

Source: (Lander, Linton et al. 2001).

Like the codon system for amino acid specification, the recognition sequences are *degenerate* or redundant; that is, the RE sequences for a given TF can vary. An example is given in Table 4-5 for the "paired" domain of *Pax6*, a gene involved in the development of eyes and other structures. These enhancer sequences may not have equal binding efficiency, but all are recognized by a given TF protein. The nominal or "consensus" binding sequence for the (vertebrate) *Pax6* double-bHTH paired domain is shown on the rightmost column ("N" means any nucleotide). *Pax6* also has a homeodomain (a separate bHTH domain) and binds **TAAT(T/C)(A/C/G/T)(A/G)ATTA** (simplified somewhat from Callaerts et al. 1997). See Table 7-5 for a general description of the different TF classes that bind these domains.

Unlike the codon system, however, in which redundancy is amino acid specific, variation in recognition sequences can lead to binding by different TFs. Variation in the copy number or location of REs can also contribute to variation in the strength of gene expression and hence have qualitative or quantitative effects. This kind of variation is shown in Figure 4-8 for REs for *Pax6* among various lens protein genes. Figure 4-9 shows variation in RE location among related genes used in rhombomere (hindbrain) segmentation.

The location of REs also varies among genes and among species for the corresponding gene and can be on either side of the gene, within the gene, near the gene, or tens of kilobases away.

This variation is clearly constrained by selection because there are so many similarities, but the pattern also shows the tolerance that evolves. In life, the functional meaning of REs is determined not just by their sequence but by their context-specific combinations. For example, a 30-base pair region called DC5 is needed to express a lens protein in the eye in chick embryos; this works if particular *Sox* and *Pax* class transcription factors jointly bind the 5′ and 3′ region of this enhancer (Kamachi et al. 2001).

Because RE sequences are short, they can be erased or generated relatively easily by chance in any DNA sequence. That is, any mutation has a substantial probability of generating a new RE, turning an existing RE into a different one, or altering the binding efficiency of an existing one. Mutations can also erase a binding site. Over reasonably short periods of evolutionary time, it appears that the blinking on

TABLE 4-5.
Variation in binding site sequences for the TF Pax6 ("paired" domain).

Pos	A	C	G	T	Consensus
01	15	7	6	10	**N**
02	21	9	3	10	**N**
03	10	9	10	18	**N**
04	8	14	9	16	**N**
05	3	2	4	38	**T**
06	2	0	1	44	**T**
07	3	29	1	14	**C**
08	40	5	1	1	**A**
09	3	39	0	5	**C**
10	1	0	44	2	**G**
11	1	36	7	2	**C**
12	23	2	1	21	**W**
13	1	4	0	42	**T**
14	2	13	26	3	**S**
15	40	1	6	0	**A**
16	14	11	15	7	**N**
17	2	4	3	37	**T**
18	1	0	20	25	**K**
19	13	17	9	4	**N**
20	14	8	4	6	**N**
21	4	12	3	9	**N**

The first column is the position along a *Pax6* binding site as experimentally determined; the next 4 positions give the frequency of each nucleotide in this position observed among 47 tested sequences bound by this TF. The last column gives the predominant nucleotide (if any). N indicates no predominant nucleotide, S is **C** or **G** predominates, W is **A** or **T** predominates.
Source: Transfac data base: http://bioinformatics.weizmann.ac.il/transfac/)

and off of REs in this way can alter the number and nature of REs, or their arrangement, near a given gene. Therefore, even if the coding part of a gene is not altered, its usage may be.

Sequence bears other sorts of expression-control information. **CpG** regions (**CpG** rather than **CG** to indicate successive nucleotides along a strand rather than **C-G** base pairs between strands) located just 5′ to a gene can be enzymatically methylated to make the region unavailable for binding by regulatory factors. Histone modification by acetylation or methylation appears, respectively, to allow or suppress access by transcriptional machinery and hence to affect gene expression (e.g., Bird 2001; Struhl 2001).

Linkage relationships can present a kind of arrangement-based information that allows clusters of genes to be coordinately regulated via the information in shared REs. An LCR can be the initiation point for a process that moves down the chromosome, for example, by starting a competitive or sequential binding process that activates downstream genes. Or each gene can have its own copy (or copies) of the same RE(s). There are many examples, including coregulated, corelevant cellular housekeeping genes (Lercher et al. 2002). However, linked genes may be expressed together without being regulated by the same mechanism, and not all the genes

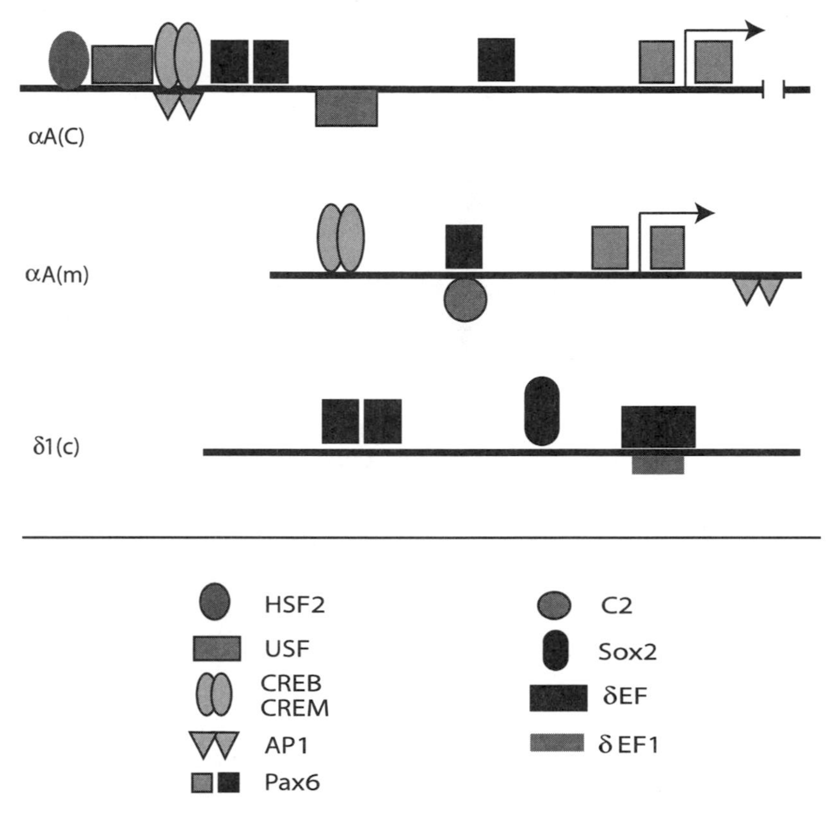

Figure 4-8. Variation in enhancer binding patterns in α A(C) chick, α A(m) mouse, and δ 1(C) chick lens crystalline genes (that make the lens protein in these species). Symbols show the location of REs for various TFs. In particular, note the variable location and number of the *Pax6* sites, in these genes that are each expressed in developing lens tissue. The broken arrows indicate RNA transcription start positions. Modifed from (Cvekl and Piatigorsky 1996).

coexpressed in a given cell are phylogenetically or even functionally related (Caron et al. 2001; Spellman and Rubin 2002).

Similar REs may be, but are not necessarily, involved in coregulating genes in the same cells. For example, expression of *Hox*, *immunoglobulin*, and some color vision and olfactory genes is closely tied to their linkage relationships, but the alpha and beta hemoglobin genes are coregulated in erythrocytes by essentially unrelated mechanisms (Hardison 2000; 2001). Typically, only one alpha and one beta gene from each cluster of related genes is active in a given cell, and similar statements apply to opsin, immunoglobulin, olfactory, and other linked, related genes. Sometimes several, but not all, of the linked genes are expressed in a given context but others of the genes in other contexts; this is the case with the *Hox* developmental genes.

In addition to the mechanisms of *cis*-regulation by *trans*-acting TFs (that is, TFs whose own gene is on a chromosome other than that of genes it is regulating), there is also evidence for some true *trans*-regulation, that is, by sequence elements on one strand interacting with those on another. For example, there can be direct interac-

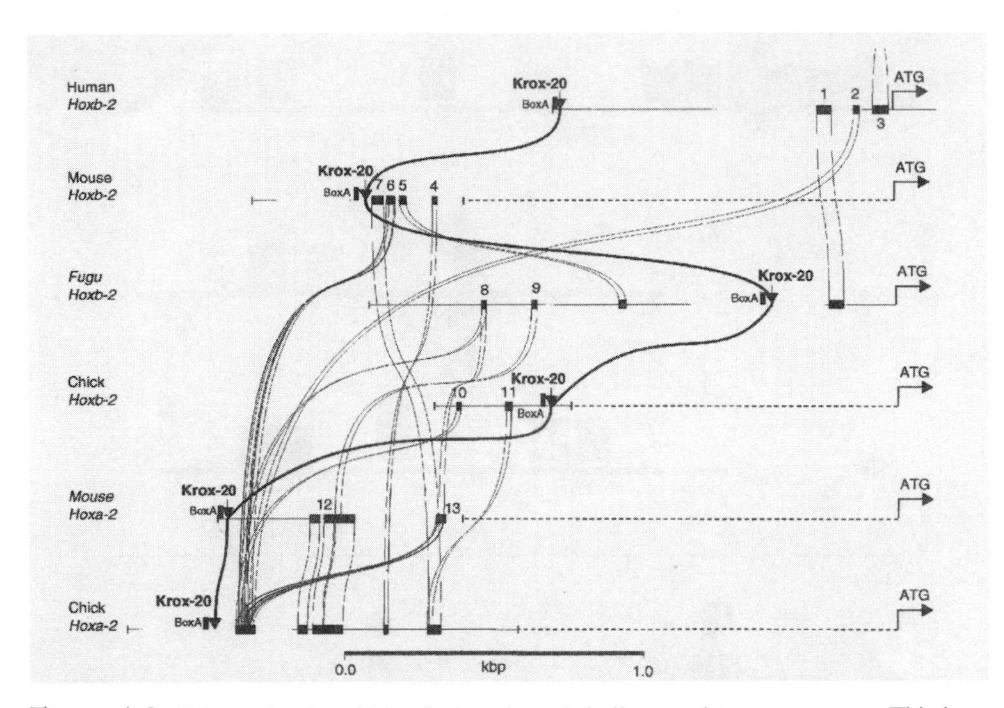

Figure 4-9. Schematic of variation in location of similar regulatory sequences. This is an alignment of the hindbrain r3/r5 segments enhancer elements of group 2 Hox genes from mouse, chicken, pufferfish, and human. Curves connect homologous blocks (solid squares) of sequence in the regulatory domains. The binding sites for the transcription factor *Krox20* (indicated by a solid triangle) and the BoxA REs (indicated by a solid rectangle) are situated at similar locations with respect to the start codon **ATG**s in all six fragments and joined with a thicker curve. The solid boxes (blocks 1–13) represent short stretches of at least 70 percent sequence identity dispersed along the enhancer elements of all four species. Regions of the mouse and chicken genomic DNAs between the **ATG**s and the enhancers that had not been sequenced are depicted with a broken line. Broken arrows are RNA transcription start positions. Reprinted from Nonchev et al., 1996, with permission. Copyright 2000 National Academy of Sciences, U.S.A. For details consult this source. See Chapter 16 on hindbrain segmentation genes.

tion of mRNAs from different homologs or genes. Direct contact-based influence of a sequence on one chromosome on a gene on another chromosome occurs regularly in at least some plants (called *paramutation*), and similar phenomena have been seen in at least some animals (e.g., Hollick et al. 1997). Paramutation often is associated with inhibiting gene expression. Although these phenomena are rare, we will see peculiar instances of gene regulation (e.g., in olfaction) where they may have an important application.

Evolving, But Not Treelike

One of the keystones of evolution, in the usual model, is that it generates divergence among descendants relative to their common ancestor. We have seen how mutation generates this kind of relationship among gene sequences and how gene duplication will generate a tree of divergent sequence relationships among the members of a gene family. Interestingly, this is not quite how RE sequences evolve. A given TF recognizes a variety of RE sequences that are variations on a common theme (e.g., Table 4-4). They have relationships constrained perhaps by selection related to the binding efficiency, location near a regulated gene, and so on. But because they are at least largely generated by local mutation, not duplication and translocation, there need be no orderly tree of evolutionary relationships even among the enhancer sequences used by the same gene.

A Protein is More Than a Polypeptide

Genes are usually portrayed as coding for proteins, and a simple schematic of this relationship is given by Figure 4-10. This shows the modular nature of the coding logic, but it is not entirely accurate. Genes code for polypeptides—strings of amino acids. Polypeptides are not mature proteins. Functional proteins often comprise several polypeptides—from the same gene (homomultimers) or different genes (heteromultimers). Its shape gives a protein its ability to interact with some other specific compound, but a polypeptide does not fold into this shape without help, and other enzymes (that each must itself be coded by a gene) fold, modify, repair, protect, transport the polypeptide, and/or complex it with other polypeptides, to form the active protein. Many active proteins have *signal peptides* or other domains that must be enzymatically cleaved for the protein to be activated (or secreted from the cell, and so on, as needed in each case).

This processing machinery largely consists of other proteins, which means that, like the TFs that regulate the expression of a protein, these processing enzymes are also the product of genes. In general, although much of the information for a protein's function is in a sense contained in its amino acid code, a gene does not really code for a protein.

EVOLUTION OF THE DNA SYSTEM OF LIFE

DNA Evolved From an RNA World

Because of the diverse enzymatic roles now known for RNA, it is thought by many who speculate on the origin of life that some kind of "RNA world" preceded the DNA-based coding system (Gilbert 1986; Joyce 2002; Maynard Smith and Szathmary 1995). This was suggested by the discovery in the 1980s that RNA has enzymatic properties that include catalyzing some aspects of RNA replication (e.g., Cech 1986; 1990), and many additional active functions of RNA have since been discovered. Indeed, RNA can fulfill all the elemental functions of life as we know it (self-replication, catalysis of biochemical reactions) (Altman 1990; Cech 1990; Doudna and Cech 2002; 2003; Joyce 2002). Today, we see the remnants of this earlier history in the various fundamental uses that small, nontranslated RNA molecules still play. tRNA, rRNA, and snRNAs described earlier are examples. However, there is no clear consensus about the relative roles of protein, DNA, RNA, or other components, nor the stages through which this early evolution went. For our purposes, it is enough that the basic functions are essentially universal, so that how *organisms* manage has to do with how they use this system rather than how the system originated.

Similar questions are asked about the evolution of the DNA-protein coding system. This is a modular system, and it has been suggested that it evolved from a

```
START
M   K   V   L   W   A   A   L   L   V   T   F   L   A   G   C   Q   A
atg aag gtt ctg tgg gct gcg ttg ctg gtc aca ttc ctg gca gga tgc cag gcc

K   V   E   Q   A   V   E   T   E   P   E   P   E   L   R   Q   Q   T
aag gtg gag caa gcg gtg gag aca gag ccg gag ccc gag ctg cgc cag cag acc

E   W   Q   S   G   Q   R   W   E   L   A   L   G   R   F   W   D   Y
gag tgg cag agc ggc cag cgc tgg gaa ctg gca ctg ggt cgc ttt tgg gat tac

L   R   W   V   Q   T   L   S   E   Q   V   Q   E   E   L   L   S   S
ctg cgc tgg gtg cag aca ctg tct gag cag gtg cag gag gag ctg ctc agc tcc

Q   V   T   Q   E   L   R   A   L   M   D   E   T   M   K   E   L   K
cag gtc acc cag gaa ctg agg gcg ctg atg gac gag acc atg aag gag ttg aag

A   Y   K   S   E   L   E   E   Q   L   T   P   V   A   E   E   T   R
gcc tac aaa tcg gaa ctg gag gaa caa ctg acc ccg gtg gcg gag gag acg cgg

A   R   L   S   K   E   L   Q   A   A   Q   A   R   L   G   A   D   M
gca cgg ctg tcc aag gag ctg cag gcg gcg cag gcc cgg ctg ggc gcg gac atg

E   D   V   C   G   R   L   V   Q   Y   R   G   E   V   Q   A   M   L
gag gac gtg tgc ggc cgc ctg gtg cag tac cgc ggc gag gtg cag gcc atg ctc

G   Q   S   T   E   E   L   R   V   R   L   A   S   H   L   R   K   L
ggc cag agc acc gag gag ctg cgg gtg cgc ctc gcc tcc cac ctg cgc aag ctg

R   K   R   L   L   R   D   A   D   D   L   Q   K   R   L   A   V   Y
cgt aag cgg ctc ctc cgc gat gcc gat gac ctg cag aag cgc ctg gca gtg tac

Q   A   G   A   R   E   G   A   E   R   G   L   S   A   I   R   E   R
cag gcc ggg gcc cgc gag ggc gcc gag cgc ggc ctc agc gcc atc cgc gag cgc

L   G   P   L   V   E   Q   G   R   V   R   A   A   T   V   G   S   L
ctg ggg ccc ctg gtg gaa cag ggc cgc gtg cgg gcc gcc act gtg ggc tcc ctg

A   G   Q   P   L   Q   E   R   A   Q   A   W   G   E   R   L   R   A
gcc ggc cag ccg cta cag gag cgg gcc cag gcc tgg ggc gag cgg ctg cgc gcg

R   M   E   E   M   G   S   R   T   R   D   R   L   D   E   V   K   E
cgg atg gag gag atg ggc agc cgg acc cgc gac cgc ctg gac gag gtg aag gag

Q   V   A   E   V   R   A   K   L   E   E   Q   A   Q   Q   I   R   L
cag gtg gcg gag gtg cgc gcc aag ctg gag gag cag gcc cag cag ata cgc ctg

Q   A   E   A   F   Q   A   R   L   K   S   W   F   E   P   L   V   E
cag gcc gag gcc ttc cag gcc cgc ctc aag agc tgg ttc gag ccc ctg gtg gaa

D   M   Q   R   Q   W   A   G   L   V   E   K   V   Q   A   A   V   G
gac atg cag cgc cag tgg gcc ggg ctg gtg gag aag gtg cag gct gcc gtg ggc

T   S   A   A   P   V   P   S   D   N   H
acc agc gcc gcc cct gtg ccc agc gac aat cac tga
                                            STOP
```

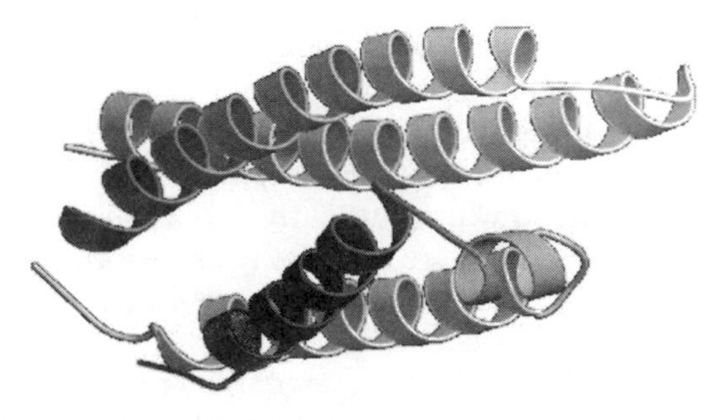

Figure 4-10. Modular nature of the information transfer system illustrated with human *Apoe*. Single letter amino acid sequence codes (translation of mRNA) sit above the three letter codons which specify them. Below is the protein structure, showing the *Apoe* lipid binding domain only (for the "e4"allele). Based on GenBank accession number M10065.

simple beginning (de Duve 1991; Maynard Smith and Szathmary 1995; Trifonov 1999). It seems very likely that a more direct relationship between RNA and protein existed before the conversion of mRNA to an intermediary message bearer. The most likely scenarios that have been suggested have this evolving from primitive RNA-amino acid interactions, perhaps related to amino acid synthesis, and there are relationships between the similarity of codon sequences, the genealogical relationships among the tRNA sequences, the chemical properties, and biosynthetic relationships of their corresponding amino acids (Maynard Smith and Szathmary 1995; Ronneberg et al. 2000). In this sense, the coding system can be viewed in adaptational terms as being selected to minimize the impact of mutational error on protein characteristics (Freeland and Hurst 1998; Freeland et al. 2000; Haig and Hurst 1991).

Today, the coding system is nearly universal. There are a few exceptions (for example, there are slight changes in the codons used by mitochondria, the organelles in cells that are used for energy metabolism), but its core elements seem to go back to a single beginning.

A Digression on Selection: Genetic Load

We have raised several points about the potential illusory aspects of adaptation and that selection is generally a probabilistic process (even if the common notion is more deterministic) whose characterization in nature can be elusive. We've suggested various reasons why darwinian selection by fitness competition may not need to be invoked as universally or strongly as is often done to account for evolution. But, as so often happens, there is a reverse problem as well. When we look at genetic variation across life, we may have a problem providing an explanation for *too much* evidence for selection. It is clear from the patterns of relative variation (shown in Figure 3-5) that functional regions of genomes vary much less within and between species than regions for which either there is no known function or strong evidence only for weak function.

Pseudogenes are the clearest case, and provide an approximate baseline for the rate of accumulation and amount of standing sequence variation to be expected when selection is not acting at all. A pseudogene is a gene, or duplicate copy of a gene, that no longer has the regulatory sequences needed for transcription; it is a sequence like a gene or a former gene that is clearly "dead" today and hence without function and whose variation is invisible to selection—a truly neutral sequence. By contrast, splice junctions, first and second positions in codons, regulatory sequences, and regulatory sequences are systematically less variable, and this constraint on variation is explained today as the result of selection. That is entirely plausible, indeed, the only good explanation until or unless some chemical error correction mechanism is found that applies specifically to such regions of DNA.

However, the order of magnitude of the amount of selection required to maintain this reduced variation is worth considering: for every 10^4 genes there are about 10^2 amino acids (and hence codons), in each and the variation in the first two nucleotides is constrained relative to the third, most redundant position—amounting to roughly 20,000,000 constrained sites per 10,000 genes, not counting enhancers, splice junctions, start signals, polyadenylation sites, **TATA** boxes, and so on. From today's estimates, the human genome has somewhere between 30,000 and 50,000 genes. This means something on the order of magnitude of 100,000,000 constrained variation sites. Yet, in the long term, each human has on average only one surviving child (two per couple) each generation.

Not only are codons less variable than noncoding sequences, presumably because of selection related to function, but there is sometimes a further *codon bias* related to the availability of the alternative tRNA genes, based on codon redundancy, that use the possible codons specifying each amino acid. The genes for alternative tRNA molecules for a given amino acid vary in their number of copies in the genome. Codon bias is a slight statistical excess of codons for the most abundant of the alternative tRNAs. Codon bias has been found in screens of many species, although at least one search in mammals found no statistically meaningful evidence for codon bias (Urrutia and Hurst 2001) and the bias is not found uniformly among genes.

The subtle and basically statistical conservation of so many parts of genomes raises the problem known as *genetic load*. Genetic load became a critical issue in biology when protein electrophoresis methods made it possible to characterize variation in detail. How is such variation maintained? There would seem to be simply too much variation to be sustained by the amount of overreproduction needed to screen out mutational variation in the millions of conserved sites by mechanisms of darwinian selection. It was this problem that led to the development of the neutral theory of evolution (sometimes for this reason called "nondarwinian evolution"), because selectively neutral variation places no reproductive burden on the organism. The neutral theory holds that most variation most of the time is not affected, or barely affected, by selection and proposes neutrality and genetic drift in finite populations as the baseline for evolutionary population genetics.

The genetic load problem has been largely ignored recently but was never very satisfactorily solved. Even if we accept that introns and so on are neutral, the remaining subset of the genome that we take as constrained in variation involves a huge potential amount of selective screening. If selection is illusory in some ways and very difficult to document in nature, there is also strong evidence for pervasive selection. Several explanations have been suggested to reconcile this seeming paradox.

Species with high rates of reproduction, in which each individual produces thousands or even millions of eggs or pollen grains, can sustain a lot of selective loss, since on average an individual in a species needs to only just replace itself so that only one of these eggs or pollen grains will be successful (populations are not typically growing in size). Think again of the number of acorns shed by a single oak tree over its lifetime, only one of which on average will become a new tree. Bacteria, too, can reproduce rapidly and have smaller genomes and may be able to sustain extensive but subtle selection. But the same kinds of constraints on variation across the genome are also seen in vertebrates, who are comparatively slow reproducers. This suggests but does not prove that there may be common mechanisms not wholly related to high or low reproductive potential.

One possible way to account for ubiquitous selection is that gametes even in slow reproducers are generated in profusion and may be screened even before fertilization (prezygotic selection), at least for genes related to cell division and normal metabolism, which gametes must do successfully.

Another possibility is that selection is not even approximately gradual the way Darwin suggested; the idea of selection against a variant because of a single trait is probably also too simple. Because of pleiotropy, selection on several different functions of a gene could constrain its variation so that there may be many ways to select against a given variant. Furthermore, because of *epistasis*, or the functional interaction among genes, any given instance of selection may simultaneously affect many genes. Individual variants might be harmless in themselves but become harmful in rare circumstances or in combinations with many other variants in the genome, so that a selective event removes many of them at a time. Theory for epistatic effects of this kind of process has been advanced (e.g., Eyre-Walker et al. 2002; Keightley and Hill 1983) but does not seem able to account adequately for the amount of constrained variation seen across genomes.

We assume that mutations occur more or less randomly with respect to function and that selection can only purge them after they occur and in comparison with contemporary alleles in the population. Unless understanding of evolution is missing something basic, there is no known way for mutations in specific regions of a gene to be prevented from happening. For that to be possible, parts of codons would have to be specifically protected from mutation; perhaps this could somehow be achieved by DNA packaging in areas to be transcribed, but it is not something generally known as yet.

No matter what, it seems clear that for most variation of this kind the average selection coefficient has to be very small, probably so small as to be immeasurable even in large samples at any given time. In practice, most of these variants are probably essentially neutral most of the time. How to test some of these notions about genetic load is unclear, and the issue this raises is whether or to what extent there is something else important and widespread going on in life besides selection that we may not yet have stumbled upon.

Part of the problem applies generally to evolution. It is difficult to understand things that happen so slowly that we can hardly perceive them. In a famous Amazonian travelogue, H. M. Tomlinson noted that a dense tropical rain forest seemed eerily silent and almost lifeless as one moved through the dark floor under the remotely high canopy. But if time were accelerated, we would have seen the movements in that "war of phantoms . . . we should have seen . . . the greater trees running upwards to starve the weak of light and food, and heard the continuous collapse of the failures, and have seen the lianas writhing and constricting, manifestly like serpents, throttling and eating their hosts" (Tomlinson 1912).

IF DNA IS A BOOK, HOW DOES IT GET READ?

The metaphors are overstated, sometimes misleadingly so, but DNA is widely characterized as the Book of Life—a repository of the information of biology and the program from which organisms are computed. But this is very different from a book a human would write for the same reason that an organism is not truly a machine. A book or machine is assembled with a purpose in mind, from parts derived from all sorts of other knowledge. An organism is built from within, and its DNA must function that way.

We have described many ways that DNA fills a number of physiological roles in the cell. Only some of these are related to protein coding itself. Looking at the genome from without, as we have done, shows its proliferating, diversifying, modular nature. Remarkably, one might say that there are too many modular units to make any sense (biological or otherwise): any stretch of DNA contains nothing but sequences identical to codons. Every three base pairs of every enhancer has a codon sequence. Similarly, because of the tolerance of TFs for variation in their RE sequences, and the fact that they are short, the genome is also nearly saturated with possible REs. Even a given 10-mer (that is, stretch of 10 consecutive nucleotides) occurs every 4^{10} nucleotides on average, about 1 Mb or about 3,000 times in the human genome. In fact, the critical regions of REs are often shorter than 10 base pairs. This also means that any stretch of sequence is simultaneously saturated with wholly overlapping codon and enhancer sequences.

Even if we only consider exons, the same codons are repeated millions of times in the coding parts of the genome, in no detectable order from the point of view of the sequence itself. Likewise for RE sequences. This is not an illusion because, for example, it is the functionally open nature of polypeptides, and hence of codons, that makes the diversity of life possible. As a consequence, one could view DNA sequence naively and say that it would be laughable to think it actually contained any information at all! Yet, these short otherwise essentially random motifs carry the *most* important biological information in the genome.

Thus, if DNA is so densely packed with information, how can an organism open and read the book? We don't have a very good answer to this, other than to say that it happened incrementally over eons of time. Also, because of cellular continuity from one generation to the next, the apparatus for reading the genome properly may always be changing as cells do what they have to do, but is never entirely missing. How scientists learned to read genomes (to the extent that we can) is more easily explained.

Scientists Learning to Read

By what authorization do we turn Figure 4-2 into Figure 4-6B? We do not learn about DNA just by examining DNA. For example, it is not possible even to identify all the genes in a given genomic DNA sequence. There is no known signature by which all genes can be identified, and indeed it is not even clear what should constitute a "gene." If our understanding of the sloppy, contingent nature of evolution is accurate, there undoubtedly is no single signature or definition.

Instead, we have let organisms (sometimes enslaved by us experimentally) teach us their workings step by step. But it is important to understand the degree to which what we ask of organisms affects what we see and how we interpret what we see. Thus, our knowledge is built on observation filtered through and directed by theory. The theoretical understanding that DNA codes for mRNA that codes for proteins was discovered through experiments, but the information now can be applied to entirely new sequences.

For example, it is possible to separate mRNA from DNA in cells. We can then use mRNA sequences to determine the sequence of the corresponding exonic DNA sequences in the genome. The exons are separated by introns, which are excised when a gene is transcribed and thus are not part of the messenger RNA, but computer-based sequence alignments can be done to find the sections of genomic DNA that code for the successive exons that assembled to form the RNA. This shows us where the gene is in the genome, and, since only exonic sequence is included in mRNA, we can identify exon-intron boundaries by finding how an mRNA sequence is interrupted in the genome. This in turn makes it possible to identify the sequence characteristics of exon-intron boundaries and transcription start/stop sequences in the genome. (We can also work backward to DNA by first identifying the amino acid sequence of a protein, which can be done chemically).

Of course, the Central Dogma of DNA → RNA → protein is only partly correct. DNA does many other things. Once we knew that genes were selectively expressed in ways affected by DNA itself, we could identify the sequence elements (REs) that are responsible. mRNAs are informative of gene regulation because genes used in the same tissue may share some regulatory mechanisms. For example, once we find their genomic location, we can search the chromosomal regions of coexpressed genes to see whether they share any sequence elements in their flanking regulatory regions. Genes and their flanking regions can also be inserted in expression systems (letting bacteria, flies, or even mice express our test region); systematic deletions can identify sequence elements necessary for particular expression patterns. Then, because our theory is that response elements are bound by transcription factor proteins, we can use these elements experimentally to "trap" proteins that are bound to them in cultured cells and use this in turn to identify the transcription factor genes themselves.

There are developing databases of known RE sequences (pdap1.trc.rwcp.or.jp/research/db/TFSEARCH.html). It is clear that simple catalogs of REs cannot suffice to identify or characterize regulation from bioinformatic (sequence data) approaches alone. Some success with such computerized searches has been reported, however, and candidate sequences have been experimentally confirmed (e.g., Bonifer 2000; Bussemaker et al. 2000; Chiu et al. 2002; Dermitzakis and Clark 2002; Dermitzakis et al. in press; DeSilva et al. 2002; Hardison 2000; 2001; Michelson 2002; Pennacchio and Rubin 2001; Spellman and Rubin 2002).

The theory of evolution instructs that related organisms share similar traits because they share common ancestry. The same is true of genes. Through our understanding of evolution and the nature of DNA and mutation, the amount of sequence variation in regions that are functionally important to selection will be constrained, relative to the amount of drift that occurs in less stringently functional elements. This allows us to identify homologous regions of genomes of one species, to find regions that resemble the sequence of known genes in other related species. Similarly, because of our theory that genes arise by duplication, we can find new genes by looking for different sequences within the same genome that are similar to those of genes we have identified, and we can guess—often rightly—about what types of cells the new genes will be expressed in from the expression pattern of the known gene and the notion that related genes have related function.

We can align sequences from different species, first anchoring the alignments by known regions such as homologous genes. Then, we can search the aligned regions for conserved sequence shared between them. Since theory suggests that highly conserved sequence is likely to have some function, an analysis of the candidate sequence can show us what that function is. This approach can be used to identify at least some of the control elements that are conserved between the tested species (Schwartz et al. 2000). Alignments can identify previously known elements and other good candidates. Phylogenetic methods may be useful (e.g., Blanchette et al. 2002; Blanchette and Tompa 2002; Chiu, Amemiya et al. 2002; Dermitzakis and Clark 2002; Shashikant et al. 1998) in the search for conserved REs in subsets of related species that share a particular trait (e.g., type of teeth or limb or leaf structure).

We read the book of life by applying external information—viewing life from outside, whereas organisms have to do it all from within. They were also "designed" from within, that is, through evolution by phenotype, bit by bit over billions of years.

It easy to see why we have trouble stereotyping the function of sequence too tightly, or thinking too deterministically about how it all works, or that it must be as neatly organized and regular as something we might design from the outside. Nonetheless, the various methods described have allowed scientists to peer rather successfully into the private business of every species we choose to look at.

Chickens and Eggs—Literally

Because this all occurs in *biological* context, elements that on the surface do not appear patterned or modular or meaningfully repeated are in fact densely filled with absolutely vital information. The trick is not the sequence alone but in its context along the chromosome in living cells because that is what determines what the sequence means.

So then, what is primary? DNA or its context—the chicken or the egg? Life may not exist without DNA, but DNA by itself would just sit there, inert. Nor does DNA control the expression of proteins or their structure. That only happens in the natural cellular context of proteins and other elements already in a cell (which is why viruses are not "alive" by themselves).

TFs themselves are proteins, which means that, if they are actively regulating genes in a cell, their own regulating mechanisms must be active in the same cells. This shows the rolling circle of developmental regulation because these TFs must thus be activated by other TFs, and so on. The mechanism that activates each TF in a cell must either have been active in that cell's mitotic ancestors or must be stimulated in the new cell by developmental signals coming into the cell from outside.

Biologists give primacy to DNA and say that the egg only makes a chicken because it starts out with DNA. But this reasoning, that the proteins that make all this possible are encoded in the DNA, is inextricably circular. This chain of events goes back to the organism's first cell(s), and hence to its parent, and . . . as fleas on fleas on fleas on dogs . . . ad infinitum. To break the circle, and settle the ultimate chicken-or-egg argument, we would have to go back 4 billion years—to the hypothesized RNA world. If the original "soup" comprised chemical reactions between RNA and amino acids, then there was neither chicken nor egg: the code developed by the addition of function to elements already interacting. Neither came first in time or importance. Nor has one been able to function without the continued presence of the other for the consecutive billions of years of life on Earth.

It is basically for our own *practical* reasons, to provide a research and interpretive strategy, that today we assign evolutionary primacy to DNA: the circle of life continually rolls through a DNA component, and that codes for the proteins that do everything else (including regulate DNA). This means that DNA is the entry point for new heritable variation, and to that extent stream of evolution is then mediated through this process. But that perspective on life is a *decision* of the culture that is science, not the only way we might view life, if we chose to feature other of its aspects. As discussed in Chapter 3, the flow of genetic information it is not even the exclusive element of inheritance.

CONCLUSION

We have seen some of the many ways in which what is conventionally and conveniently called information is contained within the coding sequence of DNA and how this is used to produce protein and serve other functions. There are probably many other ways in which information is stored in nucleic acid sequence, some perhaps quite vital but simply not yet known. Although the information metaphor is inexact or even misleading, understood in their contexts, the coding functions of DNA are certainly fundamental to life, and we will see the importance of this as well as the modular nature of genetic information throughout this book. In particular, we can now look at some of the basic elements of the way genotypes are related to phenotypes.

Chapter 5
Genotypes and Phenotypes

We have seen that the genome provides a highly elaborate code, largely based on its nucleotide sequence. We have also seen how the code works, how it evolves, and the indirect nature of the relationship between the code and phenotypes in an organism. In Chapter 3, we explained why having a genetic theory of evolution does not automatically provide a satisfactory phenogenetic theory, that is, a theory that explains the evolution of phenotypes. Here, we will try to give a sense of what is known about the internal connections between the genome and phenotypes. How are genes used, and how precise is the "code"? Does it actually code for phenotypes? In what senses can we read the organism from its code? How do we determine these relationships?

Phenogenetic relationships are the product of nature's single de facto criterion, that of the screen of evolutionary survival. Science looks at these relationships through different filters, however. In addition to wanting to understand evolution, we may wish to understand the genetic basis of disease or develop better agricultural products. These are not necessarily "natural" measures of a trait relative to its evolutionary origins. We may or may not be able to use our chosen filters to infer how phenogenetic relationships evolved, but we have been able to further our understanding of the relationship of the genetic code to phenotype.

THE RELATIONSHIP BETWEEN GENOTYPES AND PHENOTYPES

A number of important aspects of the relationship between genotypes and phenotypes relate to both the biology of the relationship and the problems we face in trying to identify and understand the underlying genes (Weiss 2003c). Some of these relationships are shown in Figure 5-1. The terms and components shown in this figure are those commonly used in the study of human disease, where we try to infer genetic causation. However, the same issues apply to traits studied in nature or experimentally (after all, humans are studied in nature). The difference is that in experiments we can control some of the variables better or more explicitly. However, the overall idea is to identify genes associated in some way (mechanistically or in terms of variation) with a trait of interest to us.

Genetics and the Logic of Evolution, by Kenneth M. Weiss and Anne V. Buchanan.
ISBN 0-471-23805-8

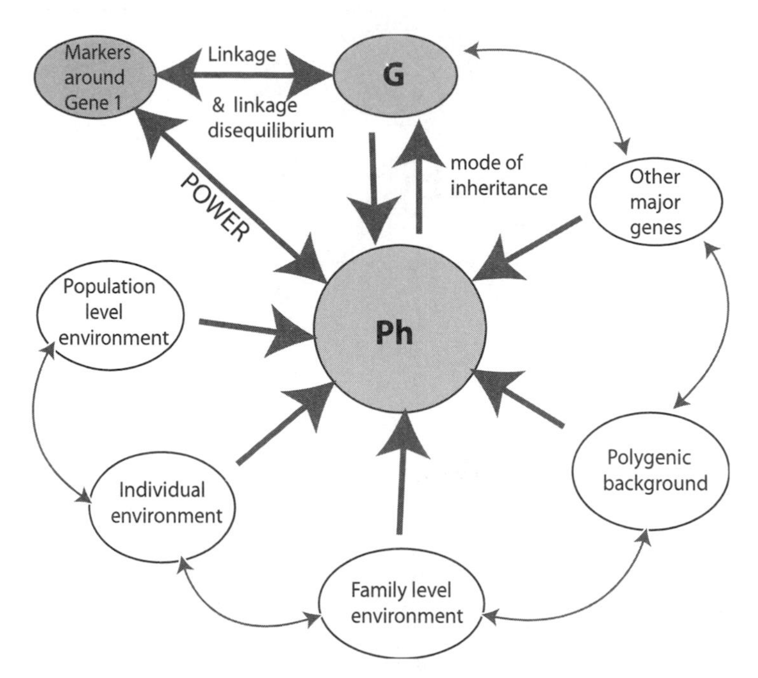

Figure 5-1. Gene mapping and Genotype-Phenotype relationships. **G** designates a gene that is assumed to have major effect on the phenotype, **Ph**. Other factors are also involved as indicated. To find the chromosomal location, or to "map," **G**, we rely on a sufficient predictive power from **Ph** to **G**, as well as a statistical association between nearby genetic markers and **Ph** (see text). After (Weiss and Terwilliger 2000).

The primary relationship of interest is shown in the vertical arrow between the phenotype **Ph** that we actually identify or measure and some important genotype **G** that we wish to find. Associated with this are many other genes (the two ovals on the right) that contribute individually, in small interacting sets, or as a "polygenic" aggregate of unspecified numbers of individually small contributions.

Environmental factors—known or unknown—contribute in various ways. We often model them as an aggregate with some tractable behavior, which can be separated into various subaggregates: the effects of factors shared by siblings or other specified sets of relatives, factors unique to the individual, and cultural factors shared by the whole population.

We can refer to the number of genes, their interaction, and perhaps frequency relationships of their alleles in a population or species as the *genetic architecture* of a trait. An important point to make right away is that the genetic architecture is not inherent in the genome. Genes do a lot, but not everything, and the same allele or gene may do different things in different individuals, even within the same species; this is another way of expressing the many-to-many **G-Ph** relationship discussed in Chapter 3.

Penetrance Concepts for Alleles and Genotypes

The effects of genes on an organism can be assessed in many ways, although typically we are interested in some particular trait, such as the level of expression of a specific protein like insulin, or in more indirect genetic traits, such as wool quality, blood pressure, or flower morphology. We can ask the two genetic questions: (1) what is the mechanism of action of the genes? and (2) how is variation in the genes related to variation in the trait? Much of this book will concern the former question, but our interest at present is in questions of the second type. We can define the probability that a particular phenotype will be found in an individual with a particular genotype as the *penetrance* of that genotype.

For qualitative traits (ones with a discrete number of possible states, like presence/absence or the number of petals on a flower), the penetrance is the probability that a specified state will be found in an individual with the specified genotype. For quantitative traits, like blood pressure, which can take on an infinite number of possible values across some range, the penetrance of a genotype (sometimes called the expressivity) is the probability of a given value of the trait in bearers of the genotype. Actually, this is more properly referred to as a probability density, but since the probability of any *exact* trait value—like being 2.000001 meters tall—is zero, in practice we divide the possible values into intervals of a certain width, such as being 1.95 to 1.99 meters, 2.00 to 2.04 meters, etc.

Penetrance is inherently a *statistical* concept that is tied to a particular sample. Only in the limit (extremes) is a trait never or always observed (penetrance equal to 0 or 1, respectively). The extent to which this statistical nature is due to observational error and finite samples or to inherent chance factors in the actual **G-Ph** relationship is a major question with no single (or simple) answer, and clearly varies from situation to situation. The penetrance distributions provide the *spectrum* of genetic effects on the phenotype, at least relative to the alleles and genotypes being considered.

Gregor Mendel dealt with selected traits in crosses of hybrid pea plant strains. He deliberately chose traits and strains in which each parent strain was qualitatively different, meaning there were only two causal factors, one in each parent (one with strong and the other with weak effects). Only decades later was the nature of the factors, alleles at an individual gene, understood. Because of the simple situation he studied, Mendel got the general principles of genetic inheritance right. In fact, much of the 20th century, the century of genetics, was spent working out the details of the theory and designing experiments to understand genes in which there were basically only two alleles, or at least relatively simple relationships like Mendel's. In natural populations, or experimental organisms, the standard way of viewing things is to think of one state as more common and new states that arise as usually harmful; thus the alleles came to be called normal, or "wild type," and "mutant." These terms are still with us, although they somewhat misportray the situation outside the laboratory, where things are more variable and more complicated. Mendel might not have been able to reach the conclusions he did, if he had had to work with wild peas (Weiss 2002b).

For complex traits, like weight, stature, or blood pressure, the possible role of genes could until rather recently only be estimated crudely and indirectly. Relationships in trait values could be observed among individuals in specifically structured samples, such as sets of parents and offspring, controlled crosses among inbred strains, twins, or parents and offspring raised under conditions of artificial selection. This observed relationship was compared with the expected relationship that would arise under theoretical assumptions about how genes would affect the trait if they contributed quantitatively to it *in aggregate*. The genetic effects of many individually unidentified genes could be considered in aggregate, for example, as if the alleles contributed additively to the trait. In experimental situations, selection could be done by the investigator to follow the changes in the nature and amount of variation over many generations and then could be related to the model of underlying aggregate effects.

As long as one was satisfied to study effects and not specific genes—which was all that was possible for most of the 20th century anyway—this largely empirical approach to genetics was an effective way to study quantitative traits. Essentially, the underlying assumption, was that the many genes were individually inherited like Mendel's pea genes, and that the alleles at each gene, although not identifiable, were related to the trait in the same *kind* of way. Artificial selection, either on the farm or in the laboratory, is more deliberate and systematic than selection in nature, but it does mirror the natural way that selection works on phenotypes in whole organisms.

Advances in genetic technology in the past quarter of the 20th century made it possible to identify genes directly and thus to identify their alleles and assess the association of their observable variation with traits of interest. It is now fairly straightforward to do this for simple traits, in which genotypes have high penetrance—that is, are directly associated with the measured phenotype. However, we have also learned how favorably artificial Mendel's experiments were. In general, there are multiple alleles at a typical locus in natural populations (including Mendel's traits in wild peas).

For a diploid species, multiple alleles means many possible genotypes. A locus with m alleles has $m(m + 1)/2$ possible genotypes. For example, 2 alleles can form 3 genotypes (**AA**, **Aa**, and **aa**), 3 alleles 6 (**AA**, **Aa**, **aa**, **A**$\boldsymbol{\alpha}$, **a**$\boldsymbol{\alpha}$, and $\boldsymbol{\alpha\alpha}$), and 100 alleles can produce 5,050 different genotypes. Unlike Mendel's and many experimental situations, natural alleles are not equally frequent. Based on Hardy-Weinberg proportions, the genotypes will have comparable variation in frequency. If the three allele frequencies at a locus are 0.7, 0.2, and 0.1, the six genotypes will have frequencies 0.49, 0.28, 0.04, 0.14, 0.04, and 0.01.

Many possible genotypes are too rare to be found in a given population, and many may never be found at all (for example, if an allele has a frequency of 0.001—not unusual for a rare allele—then its homozygote will have a frequency of 0.000001, or one per 1,000,000 people). In addition, not all alleles that exist at a locus in a species will necessarily exist in any given population, and rare alleles usually will be found to be geographically localized near the area where the mutation that generated them originally occurred. Thus, not all possible genotypes, and hence phenotypes, will occur to be screened by evolution, certainly not in every population of the species. In addition, unlike Mendel's situation, even at a single locus, there is often a quantitative relationship between genotype and phenotype rather than a simple mendelian penetrance probability. An allele that is potentially favorable only in some genotypes may never get the chance to shine.

As we discussed in Chapter 3, new alleles arise as mutations mainly unique at the DNA level, at least in terms of the variants in their nearby chromosomal background. Because of the effects of genetic drift, migration, and natural selection, the distribution and even the identity of many alleles will vary among populations, and the penetrance may be affected by variation at *other* loci (often referred to as the genetic "background" relative to the gene(s) in question).

This complex of variation is consistent with the ability of experimental breeding systems, as found in agriculture and model systems like *Drosophila*, to respond to selection in the laboratory and also provides the fuel for natural selection to act on almost any trait. However, the amount of variation also generally means a relatively loose or nondeterministic relationship between genotypes and phenotypes. Over time, the variation can accumulate and change, especially between populations as they diverge into separate species or for other reasons cease to exchange mates.

One implication of quantitative genetic models, which is routinely confirmed by studies that look at the genes involved in complex traits, is that there is extensive phenogenetic equivalence, resulting in different genes in different individuals or populations being associated with the same phenotype (e.g., blood pressure). This provides fuel for phenogenetic drift. When we address the question as to how various biological traits are produced genetically, it may be that the answer differs among species even when they have shared the same trait since they shared a common ancestor.

How We Know: Mapping and Inference Issues

The elusive and statistical nature of penetrance relationships and the potential allelic and genotypic complexity even of individual loci in nature should be kept in mind when we think about genetic causation. This is especially so when a trait of interest is many steps removed from the direct action of a specific gene product. The safest, and most general way to conceive of the role of genes in life is to assume a priori that any gene or system of genes that we may wish to consider may be associated in some way with any phenotype. If natural selection can't effectively screen on genotypes via a given phenotype, we can't expect to predict the genotype from the phenotype (Weiss and Buchanan 2003).

When we attempt to or identify genes associated with a particular trait, for each genotype in the system, **G**, we can express penetrance as the probability, **Pr(Ph|G)** that some phenotype, **Ph**, will be found. Of course, a system of genes may empirically be found *not* to be related to the phenotype; the evidence would be that the penetrance functions were identical (that is, to statistical accuracy) for all genotypes under consideration. The red and green visual pigment (*opsin*) genes, for example, have to do with color vision, but all genotypes in those genes probably are associated with identical distributions of body weight or insulin levels. However, we have many times been surprised at the degree of pleiotropy in natural organisms, a major problem in contemporary biology. Apolipoprotein E was originally studied for its lipid-transporting effects, but it turns out to have important neurological function; its alleles may even be related to the ability to recover from childhood head trauma, it is associated with Alzheimer's disease, and it seems to be involved in the embryological development of the brain.

These issues are important in both a "forward" causal and a "reverse" inferential sense. In the forward sense, they relate to the way in which individual genes affect traits. This reflects the biology of an individual and the products of evolution. In particular, we generally use a model that has genes "causing" traits, following the downward **G** → **Ph** path in Figure 5-1, based on the role of genes as protein codes. To evaluate the effects or associations of genotypes with phenotypes, we should be careful to consider the range of environments in which a genotype might find itself, including the background genotypes. This is impossible, actually, because we have no way to know that range except approximately. But it is an important point of caution (Schlichting and Pigliucci 1998).

The "reverse" approach is also important; this has to do with how we *infer* the role of genes, or identify genes, associated with a trait we wish to understand, using natural variation to map the chromosomal location of genes that affect the trait. What we would like to do is to go from **Ph** directly to **G**, but that requires that we first *map*, or find the gene in its chromosomal location.

To map a gene, we try to take an indirect route through the correlation arrow to *genetic markers* in Figure 5-1. These are variable sites all along the genome, each with known chromosomal location. We genotype these markers in all sampled individuals. If the model (the main causation we hypothesize) is reasonably close to being true (*in the sample we decided to collect*), then at markers chromosomally near to **G** there may be alleles in LD with causal alleles at **G**. If so, then when we find such a marker allele in an individual we are also likely to find a causal allele at **G** in the same individual. That is, we find association between the marker and the phenotype **Ph**; because we know the location of the marker, we now know chromosomally where to look for **G**.

There can be many a slip twixt **Ph** and **G** in this situation, ranging from statistical sampling variation, to our choice of sample, to our actual causal guess. Biologists often assume something about the forward sense, how genotypes predict phenotypes, but then use it in the reverse sense, assuming how phenotypes predict genotypes; too often, that "something" is highly oversimplified. The challenge of genetic inference is a subject beyond the scope of this book and dealt with in many places (Millikan 2002; Sham 1997; Weiss and Buchanan 2003; Weiss and Terwilliger 2000). However, it is appropriate to make a few comments here because this concept affects how we will interpret the evidence that exists for traits that we will consider.

Mapping Qualitative and Quantitative Traits

Mapping is an empirical approach to finding genes, that is, it is a technique that makes use of naturally existing or experimentally arranged phenotypic variation to find associated genetic variation, rather than the direct experimental manipulation of known genes. Traits can be treated as qualitative or quantitative, but they are approached in logically similar ways.

In mapping a qualitative trait, our statistical analysis searches for associations between the presence, or probability of presence, of a trait quality with specific underlying alleles or genotypes. The idea here is classically mendelian—directly related to Mendel's experiments with peas. We use genome-spanning markers to find the gene involved. If more than one gene can generate a qualitative outcome, things become more complicated but are conceptually related.

In mapping a quantitative trait, however, the idea is to find mendelian segregation, but of alleles or genotypes that contribute to a quantitative trait measure. Usually, this means finding genotype-specific shifts in the mean or variance of the trait measure. This is a dose-like concept, in which one gene does not make or break the trait. A locus whose allelic variation in a given sample contributes to statistically meaningful fractions of the observed phenotypic variance is called a Quantitative Trait Locus (QTL). It does not "cause" the trait in the general way that we conceive of a particular allele causing wrinkled vs. smooth peas.

What Mapping Genes Tells Us and Doesn't Tell Us

When we do mapping studies, the idea is to find genes that contribute quantitatively or qualitatively (if probabilistically) to a trait. This is an important objective because, without identifying the genes, we can't understand what they do. However, at this stage of knowledge, much of what we do remains in the black-box category.

In many if not most instances, especially for complex or quantitative traits, variation in genes identifiable by mapping methods accounts for only a fraction, often a small fraction, of the apparently genetic variation in the population. Referring to Figure 5-1, we search for **G**, but we know that there may be other genes (again, the ovals on the right) with identifiable effects. That is, most of the phenotypic correlations among relatives remain unexplained after we statistically remove the effects of the already identified genes. As shown in Figure 5-1, this may be due to the effects of shared environmental factors of various kinds; however, even when genes are involved, the residual genetic correlation is treated perforce as a polygenic aggregate in the figure. When many genes vary but each contributes only a small amount to variation in a trait, there is almost inevitably a large amount of phenogenetic equivalence in the trait, and the concept of genetic causation becomes almost inherently elusive if not actually ephemeral (not to be exactly repeated in the next generation of organisms).

What mapping studies do is to pick out the genes whose alleles happen to be playing a relatively important role *in our particular sample*. We sometimes say that, although many factors are at work, we are identifying the "rate limiting" ones. In a metabolic pathway, one substrate is converted by a particular enzyme into the substrate for another enzyme and so on, until the original materials are converted into the phenotype being measured. If in our sample, alleles at one of these genes reduce the substrate concentration for subsequent steps, that allele may have a greater effect on the phenotype than alleles at other genes in the pathway. When this is replicated among samples, populations, or species, it may be that the gene in question really is more important in the normal range of variation than the others (or, perhaps, the least buffered by alternative pathways).

Such a rate-limiting step may appear as a QTL in a given sample; however, we may have a misplaced sense of inherency of genetic effect. If the same alleles are missing or alleles at other genes are newly present in subsequent samples, or environments have changed, the original gene may not have the same kind of effect. For these reasons, the generic term "genomic background" is invoked; that is, one strain of a model system responds differently to a particular genetic manipulation than do others (or families differ in the genes responsible for a trait, as detected by mapping studies).

Gene mapping can be done in various ways not described here, but the idea is generally the same—to find genes not already known to affect a trait. But mapping is an imperfect tool for identifying all such genes or characterizing the genetic architecture.

ORGANIZATIONAL ASPECTS OF THE GENOME: MODULES, MODULES

A major theme of this book is that the genomes of complex organisms essentially comprise a huge collection of scattered, sequence-nested, gene families and other repeat elements (e.g., Edgell et al. 1997). This modularity is closely connected to the relationship between phenotypes and genotypes.

Proteins Function in Modular Ways

Proteins are molecules whose shape, charge, miscibility, and so on determine their interactions with other molecules with which they come in contact. A protein typically interacts with a limited number of the thousands of other compounds in a cell, and these have been screened by evolution so that they are not just chance specificities, although a certain amount of chance determines which interactions happen in a given cell.

Different regions of proteins have specific functions. These regions are often due to distinct physical domains of the protein. Some physical parts of proteins are stable in the neutrally charged lipid membrane of a cell, where the active protein resides. Some regions bind specifically with other compounds to serve enzymatic function (to bring about or speed up a reaction involving those compounds). Still other regions of some proteins interact specifically with another copy of the protein itself to form homodimers or with other (often related) proteins to form heterodimers.

Proteins take a few basic shapes. Some form what are called beta sheets, a flat conformation, whereas others form helical shapes. These depend on many factors, which are shared generically among proteins. For example, sulfide bridges can connect cysteine amino acids in the polypeptide to hold it in a folded conformation. This is a complex subject, the point being that proteins are functionally modular beyond the simple fact that all proteins are basically chains of different combinations of the same set of amino acids.

The Origin of Parts of Genes: Modular Coding for Protein Domains

We saw in Chapter 4 that much of the repeated structure of the genome is due to the duplication and rearrangement of whole genes or even groups of genes.

The evolutionary origin of genes is as interesting as the prior evolution of coding and codons. New genes provide opportunity for new function. Genes as whole units appear largely to evolve by duplication, but what is the source of their internal structure? This is important to gene function and its evolution. A few important facts help us here. First, genes in eukaryotes typically contain coding exons interrupted by noncoding introns; second, exons often correspond to the functional domains of the coded proteins. In addition, exons with very similar sequence can be found as parts of very different genes.

This led to the idea that the exons were functional or evolutionary units in their own right and that genes have arisen by the shuffling around of individual exons. There is considerable debate between this exon-shuffling "introns early" view and the contrary introns-inserted "introns late" view of the origin of genes. The latter is plausible because bacteria do not possess introns; if introns arose early, how did bacteria lose theirs? One way to examine the question is to see whether exons tend to be broken at natural codon boundaries (that is, are multiples of 3 base pairs in length) rather than having their terminal codons interrupted by introns. A statistical bias in favor of unbroken coding units has been found (Gilbert et al. 1999), favoring the exon-shuffling view. However, it may be that not all genes have arisen by exon shuffling and that only some, such as those associated with some early aspects of multicellularity, may owe their origins to this process (Patthy 1999).

The idea that exons can be shuffled around, and the ubiquitous evidence for duplication events in the genome, is consistent with important aspects of not just modular but repetitive structures within genes. Many examples are known in which the phenotype directly reflects repetitive internal gene structure.

The important animal structural material, collagen, is built of long chains of interconnected collagen subunit proteins. Collagens have repeat amino acid motifs that allow the molecule to form long fibers. Apolipoproteins have repeated cyclic domains that allow the protein to wrap around and hence provide an internal core that sequesters lipid molecules so that they can be transported in the aqueous environment of the bloodstream. Calcium-binding proteins provide another example of internal repeats (Kawasaki and Weiss 2003).

The source of the repeated coding sequences clearly seems to be duplication. Important evidence for this is that sometimes the copies have nested sequence structure, reflecting episodic duplication events and subsequent mutations. Thus a general initial core sequence **TCGGACGAGGC** can be duplicated as **TCGGACGAGGC**TCGGACGAGGC (using font to indicate different copies), then experience mutation to **TCGGACGAGGC**TCGGACGAG*t*C, and later duplicate to **TCGGACGAGGC**TCGGACGAG*t*C***TCGGACGAGGC****TCGGACGAGtC*. These various ancestral and descendant elements may be found in the same gene and/or in related genes, so that the history of the events can be reconstructed.

The duplication example here is hypothetical, but the core sequence used to illustrate it has been suggested as an ancestral core apolipoprotein gene sequence (Luo et al. 1986; Rajavashisth et al. 1985). The example is hypothetical because these genes are old, and the actual variation that has accumulated has been complex and considerable. The reason is that, for genes with internal repeat motifs, it is the protein properties of the repeat domains and not their codons or specific amino acids that count. Because there is codon redundancy, and some amino acids are chemically similar to others, there can be substitutions in the DNA sequence that are compatible with keeping a similar overall protein structure (that is, phenogenetic drift).

The Origin of Genes: Gene Families and What They Do

Duplication events can produce copies of entire genes and their regulatory regions. Such events, although unlikely in any given meiosis, are a characteristic of life and have occurred regularly over evolutionary time (Ohno 1970). Through gene duplication, a single ancestral gene can spawn many descendant genes, creating a gene family. The related genes may remain in tandem array on the same chromosome or may be scattered around the genome or both. Examples are shown in Figure 5-2. Initially, if the required regulatory sequences and coding domains are copied intact, the organism has redundant genes with the same function. Subsequently, mutation can inactivate the gene (producing a pseudogene, detectable by its sequence relationship to the ancestral gene until further mutation eradicates the similarity). Or, mutation can modify the function of one or both genes, through the usual evolutionary processes of chance and selection.

Like the duplication of domains within a gene (or of dispersed repeat elements), nested sequence structure can arise as mutations occur in different members of a gene family, which themselves may then become the parents of new genes through subsequent duplication. The nested sequence relationships can be used to reconstruct the history of these events.

Human Globin Genes

Consensus	MXLSXXDRTXVXAXWXKXGXXAGXYGTEALERXFLSFPXTKTYFPHFDLSPGSAQVRAHGXKVADALXXA
	10 20 30 40 50 60 70
Theta Globin	.A.SAEDRALVRAL.K.LGSNVGVYTT.A...T..AFPA.....S.L..SP..S.VRA..Q..AD.LSL.
Zeta Globin	.S.TKTERTIIVSM.A.ISTQADTIGT.T...L..SHPQ.....P.F..HP..A.LRA..S..VA.VGD.
Alpha Globin	.V.SPADKTNVKAA.G.VGAHAGEYGA.A...M..SFPT.....P.F..SH..A.VKG..K..AD.LTN.

Consensus	VXXXDDXPXALSALSXLHAXXLRVDPVNFKLLSHCLLVTLAAHXPADFTPAXHASLDKFLSXVSSVLTSKYR
	80 90 100 110 120 130 140
Theta Globin	.ERL..LPH...A..H...CQ.....AS.Q..G........RHY.GD.SPALQ.SL....SH.ISA.VSE..
Zeta Globin	.KSI..IGG...K..E...YI.....VN.K..S........ARF.AD.TAEAH.AW....SV.SSV.TEK..
Alpha Globin	H.AV..MPN...A..D...HK.....VN.K..S........AHL.AE.TPAVH.SL....AS.STV.TSK..

A

Consensus	MTG-FDSLVXSLXS--ISSS-VFH---PP-S---ASP-G----------L--PGXXSAPD-SYS--SSYG----HPXG--------Y-H-GS---HAS------YSPKSQY--G-
Dlx1	MT--MTTMPESLNS-PVSGKAVFMEFGPPNQQMSPSPMSHGHYSMH--CLH-SAGHSQPDGAYSSASSFS----RPLG--------YPYVNSVSSHASSP----YISSVQSYPGS
Dlx2	MTGVFDSLVADMHSTQITASSTYHQHQQPPSGAGAGPGGNSNSSSSNSSLHKPQESPTLPVSTATDSSYYTNQQHPAGGGGGGASPYAHMGSYQYHASGLNNVSYSAKSSYDLGY
Dlx3	MTGVFDRRVPSIRS--------------------------------------GDFQAP---------------FPTS---
Dlx4	--
Dlx5	MSGSFDRKLSSILT-DISSSLSCHAGSKDSPTLPESTVTD------------LGYYSAPQHDYYSGQPYG----QTVN-------PYTYHHQFNLNGLAGTG-AYSPKSEYTYGG
Dlx6	----MTSLPCPLPD-RGASNVVFPDLAPALSVVAAYPLG----------LS-PGTAASPDLSYS--QSYG----HPRS--------YSHPGPATPGDS------YLPRQQQLVAP

Consensus	AAAAQSRA-----ESP-----DEPESEVLEXGEVRVNGK--GKKVR**KPRTIYSSLQLAALNRRFQQTQYLALPERAELAASLGLTQTQVKIWFQN**KRSKFKKLLKQGSXPLEQSP
Dlx1	ASLAQSRL-----EDPG---ADSEKSTVVEGGEVRFNGK--GKKIR**KPRTIYSSLQLQALNRRFQQTQYLALPERAELAASLGLTQTQVKIWFQN**KRSKFKKLMKQGGAALEGSA
Dlx2	TAAYTSYAPYGTSSSPV---NNEPDKEDLEPEIRIVNGK--PKKVR**KPRTIYSSFQLAALQRRFQKTQYLALPERAELAASLGLTQTQVKIWFQN**RRSKFKKMWKSGEIPTEQHP
Dlx3	AAMHHPSQ-----ESP-----TLPESSATDSDYYSPAG-------**AAPHG-YCSPTSASYGKALNPYQYQYHGVNGSAAG----------------**YPAKAYADYGYHPYHQYG
Dlx4	-AAAQTRG-----DDT-----DQQKTTVIENGEIRFNGK--GKKIR**KPRTIYSSLQLQALNHRFQQTQYLALPERAELAASLGLTQTQVKIWFQN**KRSKFKKLLKQGSNPHESDP
Dlx5	SYRQYGAYR----EQPLPAQDPVSVKEEPEAEVRMVNGK--PKKVR**KPRTIYSSYQLAALQRRFQKAQYLALPERAELAAQLGLTQTQVKIWFQN**RRSKFKKLYKNGEVPLEHSP
Dlx6	SQPFHRPA-----EHPQ---ELEAESEKLALSLVPSQQQSLTRKLR**KPRTIYSSLQLQHLNQRFQHTQYLALPERAQLAAQLGLTQTQVKIWFQN**KRSKYKKLLKQSSGEPEEDF

Consensus	GASXSALSAS SPA-LPASWD--------------SASGKGSSXPSXSYXAS--FL-----SWYHSAS------------QDAMQQPQLM--------------------
Dlx1	LANGRALSAGSPP-VPPGWNP------------NSSSGKGSGSSAGSYVPS--YT-----SWYPSAH------------QEAMQQPQLM
Dlx2	GASASPPCASPPVSAPASWDFGAPQRMAGGGPGSGGGGAGSSGSSPSSAASA-FLGN--YPWYHQASGSASHLQATAPLLHPSQTPQAHHHHHHHHHAGGGAPVSAGTIF
Dlx3	GAYNRVPSATSQP--AFSWP--------------------------LYREG--FR-----RLSTSPC------------QNARSWPPL
Dlx4	LPGSAALSPRSPA-LPPVWD-------------VSASAKGVSMPPNSYMPG--YS-----HWYSSPH------------QDTMQRPQMM
Dlx5	NNSDSMACNSPPS--PALWDTSSHS--------TPAPARNPLPPPLPYSASPNYLDDPTNSWYHTQNLSG---------PHLQQQPPQP---ATLHHASPGPPPNPGAVY
Dlx6	SGRPPSLSPHSPA-LPFIWG----------------LPKADTLPSSGYDNSH-FG-----AWYQHRS------------PDVLALPQM

B 
(Homeobox in bold)

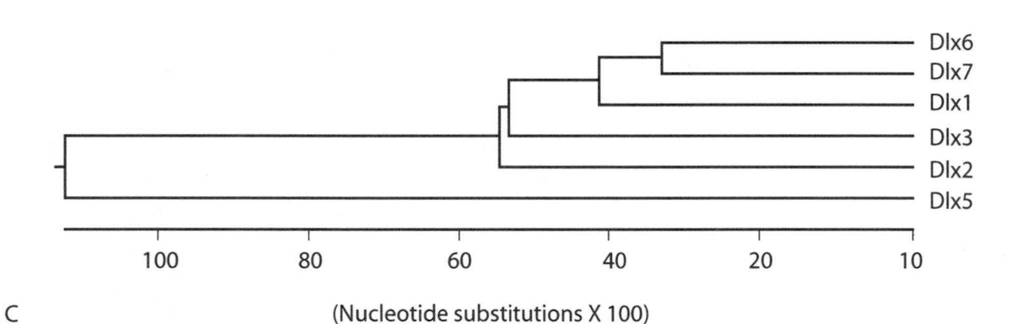

C (Nucleotide substitutions X 100)

Figure 5-2. Amino acid sequence comparison of gene family members within a single species' genome. (A) Alignment of human globin proteins (components of hemoglobin); (B) alignment of the mouse *Dlx* gene family, involved in embryological development (head, limbs, etc.); (C) phylogenetic tree of the *Dlx* genes based on degrees of sequence similarity, relationships that reflect the genes' history of descent by gene duplication from a common ancestral gene. Here, the topology of the tree also reflects duplication of two-gene clusters (see text).

In some instances, a cluster of genes duplicates (perhaps as part of a large chromosomal or even genomic duplication event), as in the case of the *Hox* genes. The genes within the original cluster will have accumulated some sequence divergence because they arose from an initial gene duplication event. Subsequently, the copies of these genes in the duplicated clusters will accumulate their own mutations. However, the *orthologous* genes between clusters (those corresponding to the gene in a given position in the original cluster) will be closer in sequence than the *paralogous* genes (those within each cluster, relative to each other). This is because the cluster duplication was subsequent to the individual gene duplication. We infer that this has happened when the orthologous genes are found together in a tree of comparative gene sequences (Figure 5-2C shows this for the six *Dlx* transcription factor genes that arose as a single pair followed by cluster duplication). The pairs of genes *Dlx1/2*, *3/7*, and *5/6* are each located on a separate chromosome, these having arisen by cluster duplication. This is shown by the fact that *Dlx1/7/6* and *Dlx2/3/5* form separate branches on the tree of sequence relationships. These relationships occur because the origin of the first pair was earlier than the cluster duplications. At any stage, additional individual gene duplications can occur in each cluster.

Transcription and other regulatory factor genes evolve just as other classes of genes do. There are numerous TF gene families, whose members are sometimes found in linkage groups, but these may also be distributed as isolated single genes around the genome. The genes diverge in their sequence details and hence in their specific and protein recognition properties. This allows members of a TF class to attain different function after they are produced by gene duplication, again just as other genes do. The result today is a regulatory tool kit with sets of similar tools; for example, related TFs may recognize related but not identical enhancer sequences (REs). We can make analogy with the different socket wrenches, screwdrivers, and so on found in a mechanical tool kit. As discussed in Chapter 4, there is a corresponding littering of the genome with potential response elements (REs) that these regulatory proteins recognize.

Another common way in which modularity works to achieve complex results is by the interaction of gene products. An important class of such interactions allows cells in one part of an organism to affect others. This allows an organism to be more "whole" or to respond to environmental situations in an organized way. Thermal homeostasis, locomotion, and diurnal or seasonal reproductive or growth cycles all involve such interactions. Genes code for molecules that travel to other cells where they complex with molecules in those cells to affect gene expression and hence cell behavior.

Overall, a combinatorial model of gene regulation explains complex cellular behavior, differentiation, and development. That is, in a given situation a specific set of regulatory factors activate or repress an appropriate set of genes. There are tens to hundreds of factors available in complex organisms. The number, location, and/or binding efficiency of the specific sequences that occur in and around a gene provide a kind of quantitative regulatory power, a "tuning knob" for gene expression (Trifonov 1999). Similar quantitative effects might occur by other kinds of copy number variation, such as in redundancy due to multiple crossreacting members of gene families.

This regulatory system also allows for variability and flexibility in evolution. We have mentioned that the genome is saturated with potential RE sequences. Regulatory sequences are short—often only four or five base pairs—and they can arise in multiple copies in one location and multiple locations in the genome by mutational chance more easily than by the highly coordinated duplication and translocation events that might otherwise be needed to change gene expression.

The members of a gene family usually retain some common aspects of function, or at least the general nature of their function (Marshall et al. 1994). The combination of constraint and variation allows for the evolution of complex structures. Members of a TF family may retain the property of being a TF—activating other genes—but do so in unrelated contexts (e.g., forming vertebrae, limbs, or gut; cell adhesion or immune reaction; regulating ion transport in and out of cells in muscle; or neurotransmission). This allows related genes to contribute to related but not identical structures, such as fore- and hind wings or mouth parts, antenna and legs on insects, anterior and posterior vertebrae, and so on. Whole new mechanisms need not be invented for each such modification.

OTHER ASPECTS OF INFORMATION AND CONTROL

Much more is in an egg than DNA. There are thousands of chemical constituents without which the egg could not live or function, much less duplicate to begin the development process. We tend to view the chemicals in the egg as the products of the parents' biochemistry, driven by *their* genes, and hence genetic at its root. The reason is that the biochemicals (in particular, the proteins that synthesize or use them) are exhausted and replaced (coded for by the DNA, which remains intact). In cell division, DNA but not protein or RNA is *directly* replicated. Still, DNA itself cannot be used or copied without those elements.

However, a number of elements are inherited that are not directly coded in DNA sequence. These include the way in which DNA is packaged and chemically modified. Sequence-specific modification, such as methylation of some **CpG** sequences, occurs in the parents and differently between males and females. This is known as *imprinting* and affects the expression of genes early in embryonic life (and, in some instances, throughout life: there are human diseases that depend on whether a mutant allele was inherited from the father or the mother). Other aspects of chromatin modification may affect gene expression and, in turn, the range of phenotypes presented to selection (Sollars et al. 2003). These traits do not depend on the DNA sequence of the genes themselves, but on modifications that affect their expression; the sequence itself is transmitted faithfully.

Every time a cell divides it is subject to mutation. This applies to somatic (body) cells as well as to the germ line. An organism is a mosaic of slightly variable genotypes in its different cells.

CONCLUDING REMARKS

Genes are not the only elements of inheritance but broadly defined seem to be the single most important information-carrying system. The genome, however, is not just a digitally arranged set of beads on a string coding for amino acids determining the traits that make up an organism.

During the previous century, our conceptual approach to genetics was heavily influenced by Mendel, who provided a fundamental understanding of how *genes* were inherited. However, he used traits very closely tied to individual genes to show that inheritance was particulate in nature. Whatever our definition of a gene finally ends up being, this discovery led to a century of unprecedented progress in understanding the behavior of *genes*. However, we may have followed Mendel's conceptual world too closely, by overinterpreting the inheritance of genes to be equivalent to inheritance of *traits*. There is conceptual trouble as a result because this has led to thinking of a gene for or per trait and hence of traits as simple products of unique genes.

Mendelian thinking does work very well for carefully chosen traits and for those aspects of natural variation that are at the extremes of the penetrance functions. Much if not most phenotypic inheritance does not seem to be of that type. The greater complexity provides a greater challenge to understand but also probably provides greater natural stability for traits.

The modular nature of the genome is of tremendous help in explaining the origin, nature, and arrangement of genes and for reconstructing the history of duplication and rearrangement events between species. Gene families with related and partially overlapping function, and similar characteristics or regulatory elements, provide understandable ways to account for the "evolvability" and developmental nature of complex organs and systems.

But accounting for something like a wing or a peach does not really explain what it "is" because there is so much complexity in the myriad pathways between a fertilized egg cell and a wing or a peach. We can use the machine analogy to identify parts and connections; at least up to a significant point, we are already learning tremendous amounts about the inner workings. Overall, the most important generalization may be the combinatorial use of related pathways and genetic "parts" that are also used in other structures as will be discussed in subsequent chapters.

The power of complementary base pairing in DNA and RNA leads to an essentially open-ended diversity of possible function. Floating all around a cell are thousands of DNA or RNA sequences, providing many possibilities of complementary hybridization or protein-DNA/RNA binding; this means that opportunistic evolution can in a sense "create" new function if some sequence turns out to be recognized in one of these ways (which can affect transcription or translation). The complex mix of molecules looking for partners raises an additional issue—how to *avoid* uncontrolled binding and/or inhibition of these many elements. This is a largely unsolved problem for science (but not for organisms, who have figured it out).

Modularity is so fundamental a principle of life that new manifestations are likely to be found, and these may be entirely different from sequence-specific modularity. Such modularity may involve the conformation of DNA near expressed genes, the shape of chromosomes as a key or code for particular differentiation, or any number of other characteristics and conformations of genes. A whole different set of simple processes can explain the nature and evolution of segmented traits with variable numbers of repeated similar structures, like petals, vertebrae, or toes. We will see many examples.

Modularity, gene and gene family evolution, and the like provide a beautiful satisfaction—these few general characteristics of genomes and their "strategies" for producing complex traits apply widely across both the animal and plant worlds, providing elements of a unified view of the way life uses the information stored in genes.

Chapter 6
A Cell is Born

"*Omnis cellula e cellula*"—all cells come from other cells. This is an insight we now take for granted, but when it was made by one of the founders of modern pathology, Rudolf Virchow, there was still uncertainty over the origin of new life. The uncertainty was a continuation of the age-old debate over spontaneous generation. Virchow's observation was one of the most important contributions to modern biology in the 19th century and is the cornerstone of the Cell Theory, the unifying idea in biology that all living things are composed of cells, that cells are the basic unit of life, and that cells come only from preexisting cells. Indeed, it is only in this way that the deeper point, that all life comes from other life, settled the long-standing question of spontaneous generation. Before the invention of microscopes and key experiments that followed, the appearance of organisms arising from dead matter (maggots in dead meat or widespread fermentation phenomena) suggested to many scholars that life could arise totally on its own. Today, we assume that that happened only at the beginning, *before* cells had evolved.

In Chapter 6, we will describe some of the basic characteristics and functions of cells with the objective of showing how, within this fundamental unit of life, the same kinds of primary phenomena of controlled sequestration and modularity, as well as many functional elements for "information" handling presented earlier, are used in the systems to be discussed in the remainder of the book. These components of the cell are largely familiar; the genetic concomitants are rapidly being identified. The organization of biochemistry into the compartments we call cells is an important, perhaps *the most* important, aspect of life on earth, since everything subsequently is essentially an elaboration of cellular life. If primary biological molecules are frequently modular as we have seen, and perhaps even come from different points on a modular cycle of basic energy metabolism, then the cell can be viewed as a very much higher level of intricate, modular physically and chemically hierarchical organization, internally, as well as the fundamental sequestration phenomenon that enabled more complex life to evolve its diversity of function and adaptation.

Cells show that complex organization has required highly orchestrated cooperation at every level—within as well as among cells, tissues, and organs. Even single-celled bacteria, completely self-sufficient as isolates, can enhance their survivability by massing together into sheets of cells that, among other things, communicate, pass genetic material, increase their antibacterial resistance, and better survive periods

Genetics and the Logic of Evolution, by Kenneth M. Weiss and Anne V. Buchanan.
ISBN 0-471-23805-8

of starvation (Hoyle and Costerton 1991; Miller and Bassler 2001; Pratt and Kolter 1999).

THE BASIC CHARACTERISTICS OF A CELL

A cell is a pool of controlled chemical composition bounded by an outer membrane. It is the main structure of undifferentiated single-celled organisms and the essential building block of highly complex multicellular animals composed of many different cell types. A cell (1) sequesters biological resources relative to the outside world, and also internally; (2) maintains the necessary concentrations of chemical components, pH, and so forth; (3) localizes, transports, exports, and imports select molecules; (4) uses selected, controlled, context-dependent subsets of its genes and controls whether gene products are kept local, for use by this cell, or are sent outside the cell; (5) allows differentiation from the surrounding medium, and (6) provides a building-block mechanism by which life can evolve more complex traits.

Although there is extensive variation in cells found in the biosphere, generalizations can be made about them and hence about life, a fact basic to our understanding of how life works and evolved. These generalizations have placed at least some constraints on what has evolved and on what can or will evolve in the future. All known organisms (except very primitive "life" forms like viruses and prions, which, although not cells themselves, depend on cells to replicate and to continue to exist) are composed of one of two basic cell types, *prokaryote* and *eukaryote*, shown schematically in Figure 6-1. Figure 6-2 provides a detail of the cell membrane. See Table 6-1 for details of the structure of prokaryotes and eukaryotes.

All cells use the DNA-RNA coding system for replication and for coding proteins. All cells are bounded by lipid membranes, although the composition of the membrane varies across cell types. All cells use ATP (adenine 5′-triphosphate) as their source of energy for carrying out metabolic processes. They do this by glycolysis, the anaerobic (in the absence of oxygen) breakdown of glucose into lactic acid with a net gain of ATP molecules, or by photosynthesis, the conversion of sunlight into energy either nonoxidatively or oxidatively with a net gain of many more ATP molecules than glycolysis yields. Actually, as we currently understand early life, the evolution of these processes both depended on and drastically altered the Earth's atmosphere, and knowledge about how particular cells make ATP and carry out metabolic processes such as movement, synthesis of various cellular constituents, and the like tells us something about the evolution of life itself. Because much of cellular physiology ultimately rests on basic energy metabolism and that is highly constrained at the molecular level, some of the basic nature or constraints may be predictable solely on chemical grounds (Morowitz et al. 2000).

Beyond these generalities, prokaryotes and eukaryotes are quite different. Prokaryotes, the simpler of the two, are generally much smaller. Two distinct groups of prokaryotes are known: bacteria and archaea. Bacteria are single spherical or rod-shaped cells, found living in almost every known niche—in the water and on, above, and deep below the surface of the earth. Often found protruding from the cell wall are *flagella*, which allow the bacterium to move, and *pili*, which are flagella-like, but are involved in the transfer of genetic information between cells or in anchoring the cell to other cells. Archaea have some features in common with Eukarya and some with bacteria. They are often found in extreme environments—intense heat, cold, salinity, alkalinity, acidity, and the like.

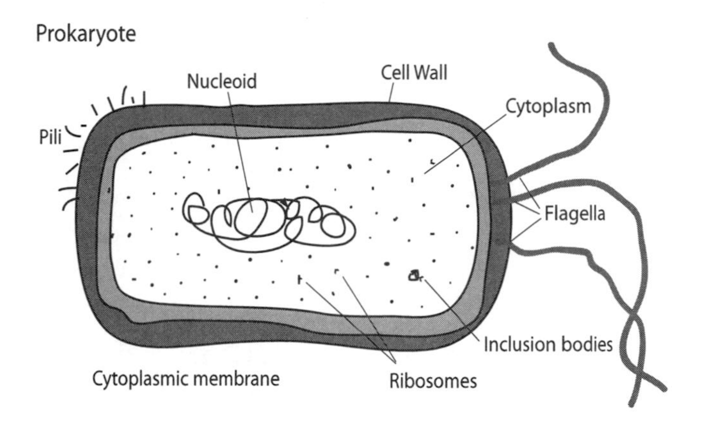

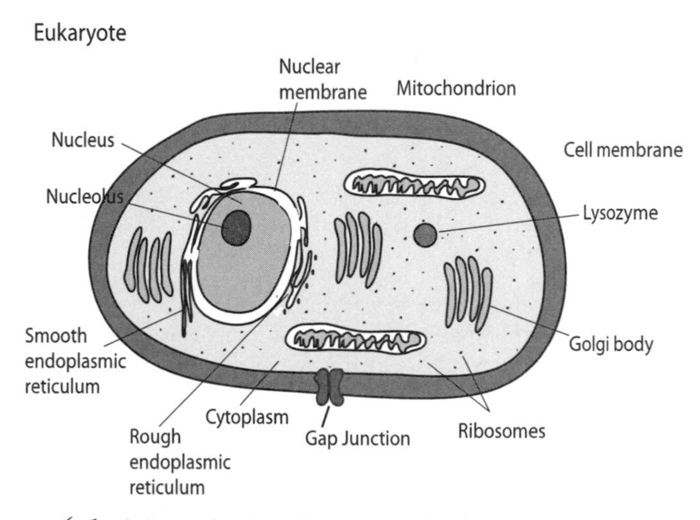

Figure 6-1. Schematic of prokaryote and eukaryote general structure.

Prokaryotes typically are bounded by a tough *cell wall.* However, among eukaryotes, plants, fungi, and algae do have cell walls, whereas animal and protozoan cells lack them. Bacterial cell walls are complex, but generally are composed of a type of petidoglycan called *murein*, a complex polysaccharide, whereas archaeal cell walls are made of protein, glycoprotein, and carbohydrate or *pseudomurein*, but never petidoglycan. Plant cell walls are made of cellulose and other polymers, and fungal cell walls are different still. The common characteristic of these cell walls, even if they differ in composition, is that they provide rigid structural support for the organism.

Within the cell wall is a plasma membrane, composed of lipids, mainly phospholipids and cholesterol, and proteins (see Figure 6-2). In bacteria, the lipids are saturated or monounsaturated. The lipids in archaeal membranes are quite different from those of either bacteria or eukaryotes. Among other things, they are polyunsaturated. The specific components of these membranes are probably what allow archaea to live in the extreme conditions in which they are found. In all cells,

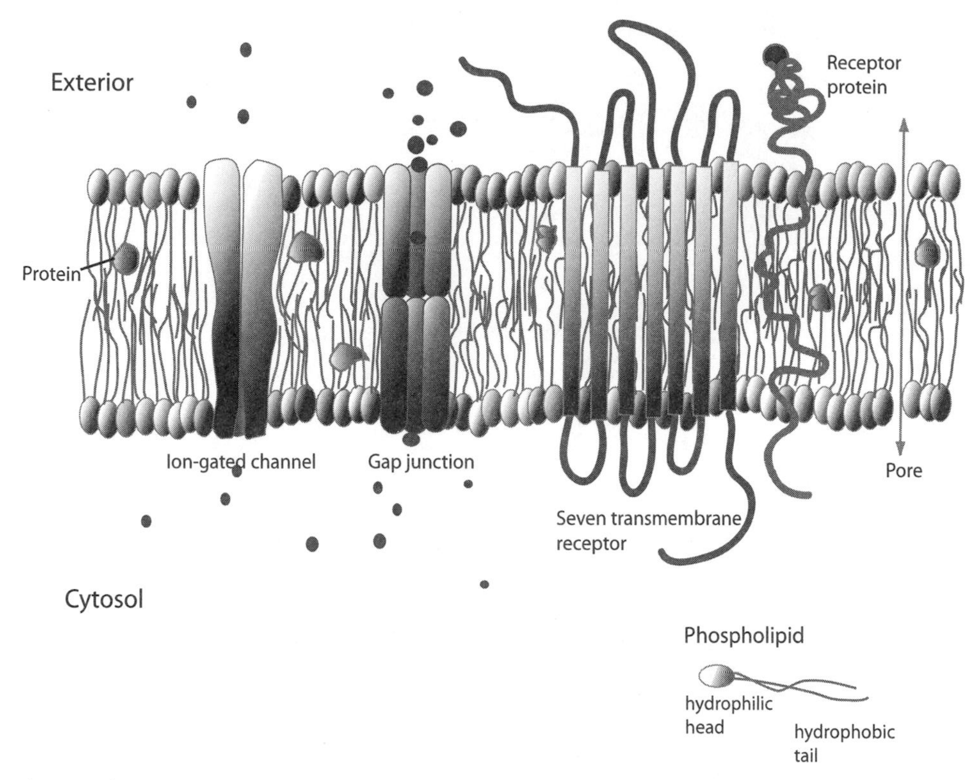

Figure 6-2. Schematic of constituents of prokaryote and eukaryote cell membrane/wall. The elaboration of the original bilipid membrane now enables cells to undertake a complex variety of functions, and provides sophisticated levels of sequestration within the cell itself.

no matter what their composition, the plasma membrane encloses the genomic material, proteins, and small molecules, all of which float free in the cytoplasm of prokaryotic cells, but are encased in a variety of organelles in eukaryotes.

In prokaryotes, the genome consists of one circular molecule wound tightly to form the *nucleoid*. Also in the cytoplasm are *ribosomes*, which are composed of ribosomal RNA and associated proteins and which translate RNA into proteins; *inclusion bodies*, which are globules of starch, glycogen or lipid, and stored nutrients; and *metachromatic granules*, which are storage sites for phosphate. Some bacteria also have *magnetosomes* in their cytoplasm, which help them orient in their environment.

Many of the metabolic processes of the prokaryotic cell take place on the cell membrane, including energy metabolism and waste removal. Other processes, such as protein synthesis and glycolysis, occur in the cytoplasm.

An important class of substructures in prokaryotic cells, and occasionally in lower eukaryotes, is *plasmids*, small circular extra-chromosomal DNA molecules found in many bacterial and archaeal cells. They contain genes for proteins that a bacterium can generally function and replicate without, although plasmids are also replicated when the cell divides. They provide functions like virulence factors, which are important for pathogenesis in some bacteria, and they can code for antibiotic resistance,

TABLE 6-1.
Components of Prokaryotic and Eukaryotic Cells.

Cell wall is a rigid structure that surrounds a plant, bacteria, fungus or alga cell. It maintains the structural integrity of the cell, and resists osmotic pressure. Although its structure depends on the species of bacteria or plant, it is generally multilayered. It does not regulate transport into and out of the cell, as the cell membrane of an animal cell does, but instead allows the passage of anything that fits through its pores.

Plasma membrane is the outer boundary of a cell. It is a continuous sheet of phospholipid molecules, with proteins studded throughout that serve to regulate transport of ions and chemicals into and out of the cell and receive and coordinate hormonal and other signals from adjacent cells or the animal's bloodstream.

Flagella and pili are often found protruding from the cell wall of prokaryotes. Flagella are involved in cell mobility, and pili are involved in the transfer of genetic information from one cell to another, or in anchoring the cell to other cells. Protists have flagella as well and their own form of anchoring device.

Cytoplasm is the aqueous contents of the cell, excluding the contents of the nucleus of eukaryotic cells, and is the site of most of the chemical activities of the cell. In prokaryotes, the genomic material floats freely in the cytoplasm. In eukaryotes, other membrane-bounded organelles are found here.

Nucleus is the largest organelle within the eukaryotic cell.

Nuclear envelope bounds the nucleus, is composed of two lipid bilayers, and is punctuated by nuclear pores through which transcribed DNA (mRNA) travels to the cytoplasm for translation into proteins by ribosomes.

Inner membrane gives the nucleus its shape.

Outer membrane is joined to the inner membrane at the nuclear pores, and is in contact with the rough endoplasmic reticulum, also joined at the nuclear pores.

Nucleolus is the site of assembly of ribosomes. Nucleoli form on specific sites on specific chromosomes, called "nucleolar organizer regions," which contain multiple genes for ribosomal RNA. The two ribosomal subunits are produced here and transported into the cytoplasm through the nuclear pores. The assembled ribosome is too large to reenter the nucleus, so stays in the cytoplasm, often attaching to the rough endoplasmic reticulum, to carry out synthesis of proteins.

Chromatin, complex of DNA, histones, and nonhistone proteins, of which chromosomes are made.

Chromosome, a long molecule composed of DNA and associated proteins. In eukaryotes, the chromosomes are packaged in the nucleus; in prokaryotes, they float freely in the cytoplasm. The number of chromosomes in a genome is species-specific. Bacteria have one circular chromosome, whereas humans have 23 pairs of linear chromosomes.

Endomembrane system of eukaryotic cells is composed of organelles with a single membrane or membranes connected via tiny vesicles, with perhaps a common origin. This system greatly increases the membrane area of the cell. The organelles of this system are as follows.

TABLE 6-1.
Continued.

Endoplasmic reticulum (ER) which maintains structural continuity with the cell membrane, is continuous with the nuclear envelope, and synthesizes and transports lipids and membrane proteins. There are two kinds of **ER** in the cell; the **rough ER** is a flattened sheet, studded with ribosomes carrying out protein synthesis, and the **smooth ER** is more tubular with no attached ribosomes, but with major lipid synthesis function and for the storage and release of calcium ions into the cytosol, where calcium catalyzes many processes. However, the function of **smooth ER** varies from tissue to tissue;

Golgi apparati (GA) play an active role in the synthesis, modification, storage, sorting, and secretion of the cell's chemical products. Lipids and proteins produced by the ER pass through the **GA** to the other organelles in the endomembrane system.

Lysosomes are made by the Golgi apparati, and are responsible for intracellular digestion of particles like other organelles that are no longer functioning properly, food molecules, foreign particles like bacteria, or antigens.

Endosomes are one of the bodies in the endocytic pathway. They carry newly endocytosed materials to the lysosomes for degradation.

Microbodies

Peroxisomes destroy hydrogen peroxide, a potentially dangerous byproduct of cell metabolism.

Glyoxysomes, in fat-storing tissues of germinating plant seeds, contain enzymes that start the conversion of fats into sugars.

Vacuoles (plants) store enzymes and waste products, as well as provide the turgor pressure that allows plants to stay upright. These organelles can comprise up to 90 percent of the volume of a plant cell.

Mitochondria, about the size of bacteria, are bilipid membrane bounded, and are the energy producing organelles found in all eukaryotic cells. The inner membrane usually contains many folds, or *cristae*, which give the organelle its characteristic appearance in cross-section. Communication between the mitochondrial matrix and the cytoplasm takes place through *porins*, junctions through the outer membrane. Mitochondria convert energy from the oxidation of foodstuffs into ATP, or adenosine 5′-triphosphate, a nucleotide present in every cell and the principal carrier of chemical energy. Mitochondria have their own genomes, which are circular chromosomes floating in the mitochondrial matrix, and they replicate by binary fission.

Ribosomes float free in the cytoplasm, or, in eukaryotes, some are attached to the endoplasmic reticulum and are responsible for the translation of mRNA into proteins. Mitochondria and chloroplasts have ribosomes for cDNA translation, but they resemble bacterial ribosomes more than they do those of eukaryotes.

Cytoskeleton, in eukaryotes is a highly flexible set of protein filaments that extends throughout the cytoplasm and is responsible for the shape of the cell, cell movement, cytokinesis (the division of the cytoplasm after the nucleus has split during cell division), and the organization of the organelles within the cell.

Chloroplasts (plants), like mitochondria, are energy producing organelles in plant cells. Unlike mitochondria, they convert sunlight, rather than food, into ATP. Again, as mitochondria, they are thought to have been introduced into cells early in the evolution of eukaryotes by endocytosis.

---

which is essential to bacteria that live in environments awash with natural antibiotics (that is, this predated by aeons the human development of antibiotics, in the battle by other organisms to protect themselves against bacterial attack).

The more complex types of cell, eukaryotes, are also bounded by a lipid bilayer membrane. The bulk of the genomic structures in eukaryotic cells are further sequestered within a *nucleus* encased in its own membrane that controls molecular transport in and out. Eukaryotic metabolic processes are carried out within and between membrane-bounded organelles floating in the *cytosol* (the contents of the cell minus the nucleus and organelles). Eukaryotic chromosomes are as a rule linear and vary greatly in size and number; as in prokaryotes, they are tightly wound but, like archaea but not bacteria, in association with periodically spaced protein complexes called *histones*, which package the DNA so that it will fit into the cell (see Figure 6-3).

Additional organelles in eukaryotic cells include the ribosomes, where protein synthesis takes place, as in prokaryotes, the *Golgi apparati*, involved in transport of macromolecules between organelles, the *endoplasmic reticulum*, which carries out the synthesis and transport of lipids and membrane proteins to the cell membrane for transport out of the cell, *endosomes*, *lysosomes*, and *peroxisomes*, which are involved in intracellular transport, digestion, and maintenance of the proper concentration of specific chemicals, the *cytoskeleton*, which consists of protein filaments and most importantly *actin filaments*, involved in movement of the cell, *microtubules*, which are most likely the primary organizers of the cytoskeleton, *intermediate filaments*, ropelike structures that are thought to be responsible for the

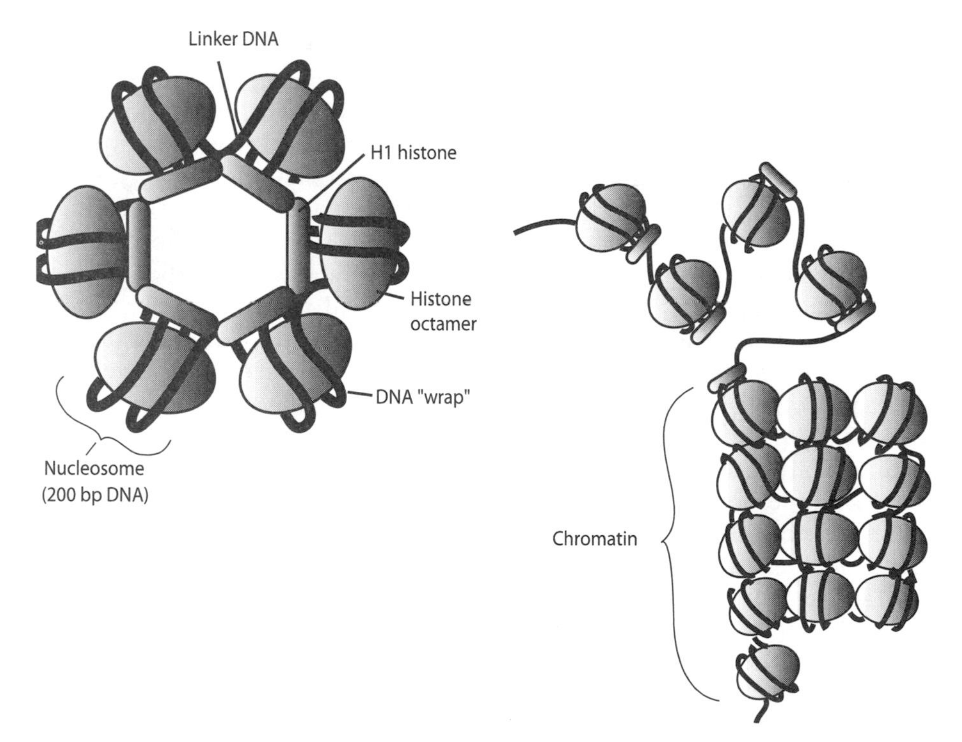

Figure 6-3. General characteristics of DNA packaging.

TABLE 6-2.
Genome Size and Number of Chromosomes in Selected Eukaryotes.

Organism	Genome size (megabases)	No. Chromosomes (2n)
Fruit fly	100	4
Mosquito	30	7
Housefly	900	6
Rice	400	12
Frog	3100	13
House mouse	3454	20
Wheat	1650	21
Human	3400	23
Tobacco	3800	24
Carp	1700	50–52

Genome size data from: http://www.cbs.dtu.dk/databases/DOGS/, http://mcb.berkeley.edu/labs/cande/pages.dir/genomesizes.html; chromosome numbers from: http://www.kean.edu/~breid/chrom2.htm, http://www.wikipedia.com/wiki/Chromosome.

mechanical strength of the cell, and *mitochondria*, where cellular respiration takes place—the production of energy by the combining of oxygen with food molecules. Mitochondria actually contain their own DNA, a circular molecule of about 16,000 bp. Listed in Table 6-2 are the genome sizes and chromosome numbers for a selection of eukaryotic organisms.

The volume of eukaryotic cells is so large, relative to the cell membrane, that an additional role of the organelles is to expand the membrane area of the cell to allow controlled and rapid transport of waste products, newly synthesized proteins, and other products of cell metabolism into and out of the cell. Molecules move around rapidly and passively encounter each other with reasonable probability, but cells take a surprisingly active role in locating components where they are needed on a microscale, or in actively moving them around. They also *chaperone* molecules that protect them from degradation or interaction by other molecules in the cell on their way to the right location. Such mechanisms help fold proteins properly, prevent them from agglomerating with other proteins, and so on. Together, these organelles comprise the *endomembrane system.* The system includes the nuclear envelope, the endoplasmic reticulum, the Golgi apparati, lysozymes, endosomes and peroxisomes, the plasma membrane, and various microbodies and vacuoles. The membranes can be in direct physical contact, such as that of the endoplasmic reticulum with the nuclear envelope, or connected by the transfer of cellular constituents via tiny vesicles. There is no equivalent system in prokaryotes, as the cells have no membrane-bounded organelles; all of the communication with the environment takes place directly across the plasma membrane.

The endomembrane system of eukaryotes allows these cells to take up macromolecules from the environment by *endocytosis* and deliver them to lysosomes for digestion. The resulting metabolites are delivered directly into the cytosol. The membrane system also regulates the delivery of newly synthesized proteins and the like to the exterior via *exocytosis.* Each molecule follows an elaborate pathway on its way to the outside of the cell, and it can be modified each step of the way and stored for delivery to the cell surface at the appropriate time. The exocytic pathway extends from the endoplasmic reticulum to the Golgi apparati out to the cell surface, whereas the endocytic route is from the plasma membrane inward to the endosomes and lysosomes. If the membranes of the various organelles are not contiguous, and thus able to pass molecules from one to the next directly, *transport vesicles* engulf the molecule and bud off from the membrane of the first compartment and fuse with the membrane of the next to deliver the molecule directly into the acceptor organelle, meanwhile managing to maintain the separate identity of each of the organelles involved.

An additional organelle in plant cells is the *vacuole*, which stores enzymes and waste products; it also provides the turgor (fluid) pressure that allows plants to stay upright. Vacuoles can take up much of the volume of a cell, sometimes as much as 90 percent. Higher plant cells contain additional partially autonomous organelles, called *plastids*, which have various functions, such as synthesis and storage of food, as well as the production of, among other cellular constituents, most amino acids and all fatty acids. In animal cells, fatty acids are made in the cytosol. *Chloroplasts* are the most commonly found plastid; they contain chlorophyll, giving green plants their color, and produce energy from sunlight by photosynthesis. Another is the *amyloplast*, which transforms glucose into starch and stores it within its membrane.

Mitochondria and plastids are membrane-encapsulated structures that contain their own DNA (mammalian mitochondrial DNA uses a coding system somewhat different from that of nuclear DNA) and are thought to be the descendants of once free-living bacteria that were endocytosed into cells as parasites several billion years ago (Margulis 1970). In this regard, incorporation through horizontal transfer, they are analogous to plasmids in bacteria, other horizontally transferred elements referred to in earlier chapters, and probably more to be discovered.

After millions of years of cohabitation, a combination of their own and the cell's gene products has evolved to control their activity, and the cell and these organelles now completely depend on each other. Although these organelles are similar in that mitochondria perform respiration-driven ATP (energy) synthesis and the synthesis of ATP is driven by light in chloroplasts, phylogenetic analysis of mitochondrial DNA and bacterial genomes suggests that, of currently available bacterial genomes, mitochondria are closest to the present-day *Rickettsia prowazekii* bacterium (which causes typhus in humans) and perhaps descended from its ancestor (Andersson et al. 1998), whereas chloroplasts appear to be descendants of oxygen-producing photosynthetic cyanobacteria (Alberts 1994).

FUNCTIONAL LOCALIZATION IN EUKARYOTIC CELLS

The nucleus is a highly organized organelle, with a nuclear matrix that controls the location of the chromosomes themselves, of transcription factors, transcriptionally competent or silent genes, and so forth, allowing the cell to precisely regulate the timing of gene expression (Lemon and Tjian 2000; Stenoien et al. 2000). At the same time, proteins move through the nucleus by passive diffusion, so that the encounter of any given molecule with its binding site also has a random component (Misteli 2001; 2001).

At the subcellular level, gene transcription, DNA replication, recombination, or DNA repair all depend on proteins having access to the target DNA. As described earlier, DNA wraps around a core of histones to form nucleosomes, and the entire genome packaged into a series of nucleosomes constitutes chromatin. To carry out their role in the cellular machinery, transcription factors, repair proteins, and so forth must gain access to the nucleosome-bound DNA—to do so, they must outcompete nucleosomes for binding sites. That is, the conformation of the packaged DNA must be changed, and it is *chromatin-remodeling complexes* that do this, harnessing energy from ATP hydrolysis to loosen DNA-nucleosome bonds or to reposition nucleosomes, thus allowing proteins access to the DNA (King and Kingston 2001; Lemon et al. 2001).

In addition, transcription seems to occur in specific locations, called *transcription factories*, within the nucleus (Hume 2000; Navarro and Gull 2001). These are sites of active RNA polymerase II-mediated transcription; about 2,500 such sites are in every cell. Whether a specific protein-encoding gene winds up in the right place in the nucleus at the right time seems to be a partly random process (Hume 2000; McAdams and Arkin 1999). Gene expression seems to be digital—it happens or it doesn't in a given cell. That is, in some experimental situations, stimulating the expression of a gene causes it to be expressed in more cells, not necessarily at a higher level in a given or in every cell (Hume 2000). The average level of transcription varies among genes, cells, and circumstances (there can be many more copies of mRNA for some genes than for others in a given cell). Whether transcription of a specific gene happens in a specific cell seems to be probabilistic; however, when it is "time" for a gene to be expressed, the essential thing is that it happens in enough cells in a tissue or organ or organism to matter. Similarly, the level of expression appears to be based on various aspects of the *cis*-acting regulatory elements (number of copies of a given RE, its specific recognition sequence, location relative to the gene, etc.), although the use of these elements, too, has a stochastic component. Gene expression is a population-level event—the product of transcription in many cells in a tissue, not a single-cell event. Therefore, although it is stochastic at the cell level, at the population level, gene expression is highly predictable. As long as the tissue overall does what is needed, the organism can suffer the failure of some of its cells to do so (though this doubtlessly is dependent on how many spare cells there are—not many in creatures like *Caenorhabditis elegans* with only a restricted number of cells in total).

Structures such as ion channels and receptors for neurotransmitters are on the cell surface, sometimes localized into subregions to control the level of possible cross-reactions that could confuse signals or other information transfer or undo local conditions from their intended state. Supramolecular islands of receptor complexes or ion channels are produced by aggregation via mechanisms such as shared protein-protein binding domains or protein carriers (Sheng 2001; Sheng and Lee 2001; Sheng and Sala 2001). Similarly, cells often have chemical as well as shape polarity, with location-specific and/or cell-surface concentrations of proteins, asymmetric secretion, and the like modulated by a variety of trafficking processes known as *constitutive cycling* (Royle and Murrell-Lagnado 2003).

Ion Channels and Osmoregulatory Units

One of the most important aspects of sequestration brought about by the packaging of biochemical materials within protective membranes was that it allowed the cell to control internal chemical conditions. This meant keeping some basic chemical properties under control relative to the surrounding (presumably ocean-like) environment. Much of this is achieved by shuttling elements into and out of the cell. Some adjustment can be made by *carrier proteins* that bind some kinds of molecules, carry them through the membrane, and release them on the other side. In this case, the surrounding environment is the womb of cells, with shared characteristics so that a more tightly controlled barrier is important. In particular, ions affect the pH of the fluid in which cellular reactions take place, and pH is an important aspect of how those reactions work. Cells can titrate (adjust) the relative concentrations of many aspects of their internal environment relative to their external concentrations—for example, adjusting the salt concentrations inside vs. outside the cell. However, cells need means to manage the passage of ions and small molecules through their uncharged cell membranes. Several ways to do this have evolved, including *facilitated diffusion* and *active transport.*

Facilitated diffusion of ions takes place through water-filled pores or *ion channels* in the cell membrane. These pores are formed by proteins or groups of proteins that connect the cytosol of eukaryotic cells to the exterior. They are narrow, highly selective, and *gated*, that is, they open or close in response to a signal from the environment. Some ion channels are *ligand gated*; others are *voltage gated*, *mechanically gated*, *light gated*, or *stretch activated*, but in all cases they allow only some ions to pass only under specific conditions.

More than 100 types of ion channels are known, used in different cellular contexts and formed from channel-specific proteins coded for by different sets of genes. Neural cells are a good illustration of how ion channels function. The charge in a cell is related to the relative concentration of positive and negative ions on each side of its cell membrane. When the relative charge difference across the cell membrane reaches a critical point, ion channels are activated and a pulse of electrical charge moves along the cell, or a burst of neurotransmitter molecules is released, to be picked up by receptors on adjoining cells and sent further along the neural line.

Voltage and ligand gating are the most frequently found gating mechanisms. Voltage-gated channels open and close to ion transport depending on the membrane potential of the cell, that is, the relative internal charge. When the membrane potential is not in equilibrium or the concentration of ions inside the cell is too high or too low (the cell is *polarized*), Na^+, K^+, and/or Ca^+ ion channels open, admitting or releasing ions and depolarizing the cell. Each type of channel preferentially allows transmission of one particular ion type (for example, Na^+, K^+, Ca^{2+}, or Cl^-) and responds to the relative concentration of that ion inside the cell. Transport is usually passive, with ions flowing along the charge difference.

Ligand-gated channels respond to small signaling molecules received by receptors in the cell surface that protrude either into the extracellular matrix or into the cytosol. There are a few main classes, each activated by a particular neurotransmitter molecule or class of molecules (e.g., glutamate, GABA, acetylcholine). Each class has its own structural characteristics and behavior, but always receipt of the ligand toggles the channel to open or close and thus regulate the transmission of ions.

Ion channels serve many important functions within the cell, including maintenance of osmotic balance, neurotransmission, and muscle cell function. Voltage-gated channels are more passive respondents to the external environment (and are present on single-celled organisms) compared with ligand-gated channels, which require a ligand typically produced elsewhere in the organism as an intercellular signaling phenomenon. In this sense, a multicellular organism actively controls the environment to which its cells respond, by evolving its own environmental constituents—the ligands—to provide organism-specific cues. This is how a chain of nerve cells can carry a message or how a nerve cell can activate some other type of cell.

Ion channels of different types can be present in multiple copies on the surface of a given cell. However, they can work by activating similar intracellular molecules. This raises the question as to how specificity can be achieved and why the cell is not awash in chaotic cross-reactions among different channels with different function. One way seems to be that given channel types are clustered together in floating "islands" in the cell membrane, which are actively located by the cell into specific regions of its surface. This sequestration mechanism, which can even localize and compartmentalize functional aspects of a given cell, is produced by evolution.

Small hydrophilic molecules, such as some sugars, pass readily through cell membranes by facilitated diffusion through transmembrane proteins that create pores filled with water through which the molecules diffuse following their concentration gradient.

Active transport is the pumping of ions or molecules through the cell membrane, *against* their concentration gradient, using a protein transporter and an energy source, usually ATP. For example, because the H^+ concentration of gastric acid is about 3 million times that of the blood parietal cells in the stomach, very low pH gastric acid is secreted into the stomach via active transport. Very different gradients of K^+ and Na^+ concentrations in and outside of animal cells are maintained by a specific kind of active transport called the Na^+/K^+ pump, and in mitochondria (the energy factories of the cell), the H^+ pump actively pumps hydrogen atoms into the intermembrane space of the organelle for the process of making ATP.

Osmosis

Water passes through lipid membranes by *osmosis*, from the level of higher concentration to the level of lower, and thus maintains the fluid balance of a cell with its environment, preventing the cell from shrinking or bursting. Osmosis is a specialized form of passive transport, although water concentration can be affected by the active transport of solutes across a membrane.

The nature of osmosis can be illustrated by its use in the production of urine in the waste-disposal process in vertebrate kidneys, where much of the animal's waste is removed as solutes, highly concentrated by the kidneys. One quarter of the blood volume pumped by every heartbeat passes through the kidneys where it is filtered through a very complex network of blood vessels; both waste and useful molecules are passed into the urine. This "early" urine is then refiltered through the nephrons and sent into the tubules where much of its initial water volume is released, by osmosis, into the interstitial spaces around the tubules, again following a concentration gradient in the interstitial spaces that goes from low to high. The urine gets more concentrated as it passes through the tubules; osmotic flow increases along the concentration gradient to pull more water from the urine on its way to the center of the kidneys. The concentrated urine is then passed to the ureters for transport to the bladder; the final volume of urine going to the bladder is approximately one percent of its initial volume with the remainder of the fluid returning to the blood.

Most of the solutes are recaptured as well, for reabsorption and reuse. This concentrating of the urine saves an animal the size of an adult human 2–3 liters of water a day.

CELL REPLICATION

Replication by Fission

Perhaps the single most important characteristic of life is its ability to replicate. This will be the topic of Chapter 8, but a few comments can be made here. Today, biomolecules replicate mainly via the elaborate DNA-RNA-protein coding system, which is an intracellular mechanism. This can be direct, as in the repeated production of the same protein, or indirect, as in the production of a biomolecule through the action of enzymes (proteins coded for by genes).

An important stage in evolution was, of course, the acquisition of the ability of the whole cell as an organized structure to replicate. In bacteria, this is done by *binary fission*, that is, by the division of a cell into two "daughter" cells. Prokaryotes replicate quickly, usually asexually, dividing into two identical cells that each contain a copy of the entire original genome and hence are genetically identical (except for any mutations that might occur during replication).

In binary fission, the bacterial cells first replicate their ribosomes, enzymes, and other cell components and duplicate their chromosome. When the replication process is complete, the two chromosomes anchor to different parts of the cell wall, and new cell wall material lines up down the cell midline, one chromosome on each side, and the cell splits in two. Under most normal conditions for prokaryotes, the new daughter cells contain approximately the same nongenomic material that was in the parent cell; in some special instances, however, cell division is not completely symmetric, and the two daughter cells may differ.

Replication is much more complex in eukaryotic cells because they are larger and because they must replicate multiple chromosomes, including the chromosome within the mitochondria (and within plastids in plants), as well as all the membrane-bounded organelles. Eukaryotic replication is a multistage process, from *interphase* to *mitosis* to *cytokinesis*. Interphase is the longest stage of the cycle, during which the cell prepares for replication and the DNA in the nucleus is replicated. Mitosis includes the process of nuclear division up to the point of cell division, cytokinesis, when the cell splits into two genetically identical *daughter* cells—cells that each contain the same chromosome complement. There are variations on these themes, as is always to be expected. One, for example, is the syncytium of some muscle, insect eggs, or other single large cells with many nuclei. Syncytia can form by incomplete cell division, or by the fusion of cells, and so forth.

Replication is not Always Symmetric

Contrary to what is typically thought, simple binary fission may be singular in nature. In particular, the two daughter products of eukaryotic cell division are often, perhaps usually, different from each other. Indeed, this asymmetry is at the foundation of the hierarchical nature of development and of functional specialization among tissues in complex organisms. The daughter cells may look similar physically, and they do typically both inherit the complete genetic complement. Often, however, the daughter cells do not look alike, and genes can be used differently in each.

A ready example of physical asymmetry is when a mammalian egg cell is produced (three smaller cells, known as "polar bodies" are also produced in two rounds of cell division). Less widely appreciated, however, is that a differentiated organism depends fundamentally on genetic differentiation among cells that often takes place at or as a part of cell division. For example, many differentiated animal tissues arise from a layer of *stem cells*, partially differentiated cells that divide and produce one daughter stem cell, and a second cell that begins a sojourn of differentiation to become, for example, a sperm cell, an intestinal lining cell, or a skin cell. In mammals, hematopoietic cells divide into, among other things, red blood cells that eventually get rid of their nucleus and thus no longer have nuclear genes (and concomitantly limited lifespan and physiological capabilities).

Asymmetry of gene expression in daughter cells occurs quite regularly during embryogenesis and in some senses is the basis of organogenesis and morphogenesis. Cells sharing a common ancestral cell turn on, or off, genes that were repressed (or expressed) in the former. Some of this, at least in vertebrates, is regulated by gene-specific chemical modification (e.g., methylation of **C**'s in **CpG** pairs), which sometimes affects only one of the two homologous chromosomes.

An important concept in embryogenesis is that of the morphogenetic "field," in which some cells produce, and others receive, molecular signals. Although signals can pass in various ways among cells, so that a cell can change its genetic program in the resting phase of the cell cycle (and hence, develop gene expression differences unrelated to cell division), clearly a cascade of events differentiates cells from stage to stage that, at least in some instances, requires asymmetric behavior of cell division products.

Prokaryotes, too, have mechanisms for the equivalent of sexual reproduction and the transfer of genetic material from one cell to another, thereby altering the symmetry of simple binary fission. *Conjugation* is the one-way transfer of genetic information, usually plasmid but occasionally chromosomal, from a donor bacterium to a recipient via the pili. *Transformation* is the process by which bacteria take up unpackaged DNA from the environment. Only cells that are "competent" can receive DNA in this fashion. Cells can be made competent, however, via a routine procedure in molecular labs where introduction of foreign DNA into bacteria is fundamental to recombinant DNA technology. In *transduction*, viruses move DNA from one cell to another.

A somewhat different type of asymmetric cell division, used across the eukaryote realms, results in the production of germ cells (sperm and egg). It is the two-stage process known as *meiosis*. Beginning with a diploid precursor, the final functional product is haploid, containing only one copy of each chromosome (spermatogenesis yields four haploid sperm cells, and oogenesis yields one haploid functional egg and three polar bodies destined to die). There are variations on this theme. Some plants, for example, are tetraploid rather than diploid, but the essentials of the logic and net result of their meiosis are the same. Although traditionally thought of as unique compared with the common notion of evenly dividing cells in mitosis, from the foregoing discussion it is clear that meiosis, although relatively rare, is but one of many asymmetric aspects of cell division.

RELATIONSHIP OF CELL STRUCTURE TO THE CONCEPT OF A TREE OF LIFE

Exactly what counts as the oldest fossil evidence of life is somewhat controversial. However, many accept that the very earliest evidence includes structures known as *stromatolites*, which are made of layers of bacteria-like organisms of various types (Schopf 2000). Details are controversial with such old and often ambiguous material, but the evidence suggests that at least some stromatolites radiometrically date to about 3.85 billion years ago. Because the Earth is only around 5 billion years old, these fossils suggest that cells were already quite elaborate relatively early in the history of life. Furthermore, they attest to the robust nature of the early cellular "designs" because some stromatolites are indistinguishable from microbes alive today. This fact is sometimes used to treat time as if it were compressed, and the cell as we know it as an inevitable or even automatic consequence of biochemistry. Nonetheless, we need to keep things in perspective and not assume that reaching this stage was very early in the history of life on earth: it has been guesstimated that perhaps up to 0.5 billion years were required from the time of the first self-replicating molecules that we might call "life" to the appearance of the cells fossilized in these stromatolites, suggesting that life began on Earth around 4–4.3 billion years ago.

The most striking thing (from the perspective of biologists) that happened in the subsequent 3.85 billion years was the evolution of the much larger and more complex eukaryotic cell and the proliferation of multicellular plant and animal life. Of course, it is difficult to judge by just how much unicellular (or more simple viral) life has changed since its external shapes were first fossilized. As will be seen throughout this book, the elaboration of multicellular life was extensive, but it again and again employed basic mechanisms, and very similar if not identical biochemistry, that presumably were present in early eukaryotic cells, in whatever form, or kind of organism, that first arose.

The classical idea of a universal Tree of Life is based on a single origin of cells and generally divides all the forms of life found today into three main domains: Eucarya, Archaea, and Bacteria. The latter two branches comprise unicellular prokaryotes, and the former, the Eucarya, includes organisms from tiny protists (unicellular nucleated eukaryotes) to the most complex organisms on Earth. Because prokaryotes are less complex than eukaryotes, it had long been thought that the universal ancestor of all life must have been a simple prokaryotic cell, from which eukaryotes descended by adding structures over time.

As discussed in Chapters 1 and 2, current phylogenetic evidence built on DNA sequences from a diversity of genomes suggests that this unitary view of life is at least partially wrong. Primordial life may have consisted of a kind of primitive bacterial *biofilm*, not a single cell, with genomic material encased in permeable membranes that were not very efficient at sequestering the cell's contents from the external world. Lateral exchange of genetic information between cells and the acquisition of characteristics from anywhere in the biotic mass probably occurred much more readily than is possible between cells with the enclosed genetic systems that subsequently evolved. The primordial cells probably lacked sophisticated translation mechanisms, and it is likely that proteins were simple and short and that replication was inaccurate. This type of simple beginning could account for the evolution of the codon system, of genes by exon shuffling, and of the transition from RNA to

DNA discussed in earlier chapters. Woese calls these primitive cells *progenotes* (e.g., Woese 1998). According to some views, as we have noted, there was so much horizontal transfer of information among progenotes that the idea of a single ancestral life form cannot be correct. See Figure 2-1 for a tree supporting this view (Doolittle 1999).

Stromatolites are prokaryotic organisms that most closely resemble modern cyanobacteria (Schopf 2000), single-celled, photosynthetic organisms that contain a blue pigment in addition to chlorophyll and that live singly or in colonies. Their descendants or closely related organisms still thrive; cyanobacteria are commonly found as the green scum on a stagnant pond. Interestingly, if indeed these earliest life forms were cyanobacteria, this would mean that photosynthesis had already evolved more than 3.5 billion years ago. Green plants use photosynthesis to synthesize carbohydrates from carbon dioxide and water, using light as the source of energy, thus suggesting that water and carbon dioxide were present very early in Earth's history (Schopf 1992, 1994, 2000).

Oxygen is the byproduct of most forms of photosynthesis; various geological formations from around 3.75 billion years ago show evidence that oxygen was already in the atmosphere, probably being produced by cells capable of photosynthesis (Schopf 1992, 2000). This implies that life already existed long before 3.75 billion years ago, perhaps at least another 0.5 billion years before (Schopf 1992). If the Earth was formed roughly 5 billion years ago, this would mean that life began relatively soon after the environment cooled from its extremely hostile early state. Fossil evidence shows that eukaryotic cells had arisen around 1 billion years ago. This is a minimum time estimate, based on the size of the cells in microfossils; the organelles that comprise eukaryotes, which would be definitively diagnostic, generally do not preserve as fossils. This represents a huge gap in time between the appearance of prokaryotes and eukaryotes, but the evidence shows that this evolution happened before, rather than as part of, the evolution of complex differentiated organisms: the first eukaryotic fossils are of apparently single-celled organisms. Given the successful persistence of prokaryotes to the present day, it is curious both that single-celled eukaryotes would evolve and that they would take so long in doing so.

Modern genomic techniques show that eukaryotes are not simply prokaryotes with more complex cellular machinery added on. Indeed, as we have seen in Chapter 2, comparisons of the DNA sequences of different components of eukaryotic organisms, compared with bacteria and archaea, suggest that eukaryotes are chimeric, the result of a fusion between (or at least sharing the intrinsic characteristics of) archaebacteria, eubacteria, and a cytoskeleton-bearing prokaryote (Doolittle and Logsdon 1998; Katz 1999; Vellai and Vida 1999). Similarities between genes involved in transcription, DNA replication, and translation in eukaryotes and archaea suggest that the genetic mechanisms for these functions in eukaryotes came from archaea; other genes, however, are closer to those of bacteria, and still others are too divergent from either lineage for their ancestry to be determined. Phylogenetic analysis of mitochondria and chloroplasts suggests that these organelles were once free bacteria that entered the cell as the result of an endosymbiosis event, the phagocytosis of a bacterium by perhaps an archaebacterium. An endosymbiotic origin for other eukaryotic organelles has also been suggested (Margulis 1970). There is also recent evidence of other levels of gene transfer, including between eukaryotes and prokaryotes and even between eukaryotes (Andersson et al. 2003; Archibald et al. 2003).

Evolution of Complexity

Multicellular organisms appear to have arisen first in the Vendian period of the late Precambrian, about 600 million years ago (Gerhart and Kirschner 1997), and 3–3.5 billion years after the development of the first single-celled organism. The reason for the immense time lag between the development of single and multicellular organisms is not specifically known, but we can make educated guesses, keeping in mind that (like Darwin) we are somewhat blinded by, among other things, the rather minimal fossil record. The standard default assumption is that multicellularity evolved because increased complexity was favored by natural selection for some reason(s). However, any such selective advantage did not mean that single-celled life became obsolete. The extent to which characteristics are shared between contemporary single and multicellular organisms suggests that single celled life was probably pretty much as sophisticated then as it is today. Single-celled organisms predominate on Earth today and are at least as diverse as any kind of multicellular organism.

However, if no *inherent* advantage is associated with multicellular life, at some time(s) and at some place(s), opportunity(ies) must have arisen at which cells that aggregated gained some sort of local advantage—and we should keep open the possibility that the first instance(s) were due to chance and became assimilated or committed over time, subsequently ramified by selection and drift. Because this did not put single-celled life out of business, a default presumption is that this represented a new kind of niche rather than obsolescence of the old one.

Initially, multicellular life probably occurred by little if anything more than the failure of dividing cells to fully separate or by the adhesion of similar cells. Some primitive multicellular life today is little more than this; bacterial biofilms are an example: they consist of sheets of bacteria that can survive as single cells but that under some natural circumstances can mass together to form a film with different characteristics and different genes expressed compared with their single-celled planktonic counterparts (Prigent-Combaret et al. 1999).

COMMON CELLULAR FEATURES OF MULTICELLULAR LIFE

Once cellular differentiation became possible (whether it happened many times or only once), an advantage or some resource was eventually gained by the use of *both* intercellular signaling and signaling between *different kinds of cells*. As a general guide to cellular diversity as an index of complexity, prokaryotes manifest a very small number of different cell types. Plants have about 30-40 cell types, and higher animals like ourselves have about 150–200. To achieve whatever advantage there is to becoming complex in this way, cells had to become able to differentiate and, at the same time, to communicate with different cell types. This may be why multicellular life took so long after the origins of single-celled organisms to arise. Today, several classes of activity are important across the spectrum of multicellular life. Most of these clearly have their origin in unicellular times.

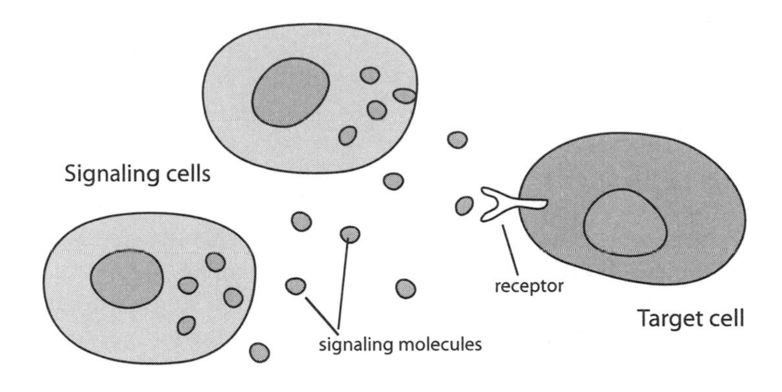

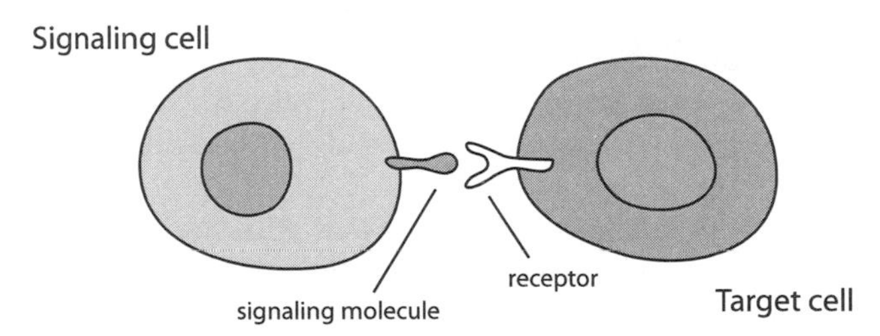

Figure 6-4. Schematic of cell signaling and information transfer mechanisms.

CELL-CELL SIGNALING

Cells in multicellular organisms communicate via hundreds of kinds of signaling molecules, using many pathways (not all of which would necessarily be found in the same cell type). Figure 6-5 schematically shows some of these pathways. Most of these molecules are ligands secreted by a signaling cell and received by a specific receptor protein on a target cell; the receptor-ligand binding then activates a response inside the cell. Signal molecules can be secreted by active transport mechanisms through the cell membrane into the extracellular space to be picked up at either short or long range by target cells, or the signal can be tightly bound to the surface of the producing cell and transmitted only to target cells that come into contact with the signal.

Some small signaling molecules, like retinoic acid, thyroid hormone, vitamin D_3, and steroid hormones, are hydrophobic and diffuse through cell membranes but then are bound by specific *binding proteins*, at which time their information is used to activate or repress specific genes. However, most *signaling factors* (SFs) cannot directly enter a cell because the lipid membrane, being uncharged, is largely impermeable to water-soluble molecules or because the molecules are too large; therefore, the information they carry is transmitted by the binding of the SF to a receptor on the cell surface, which communicates the occurrence of the event to the cell.

Signaling can occur at any distance from a cell, from intracellularly, to a few to thousands of cell diameters distant, or indeed even among distant organisms. A variety of terms are used to indicate the distance relationships, although as usual

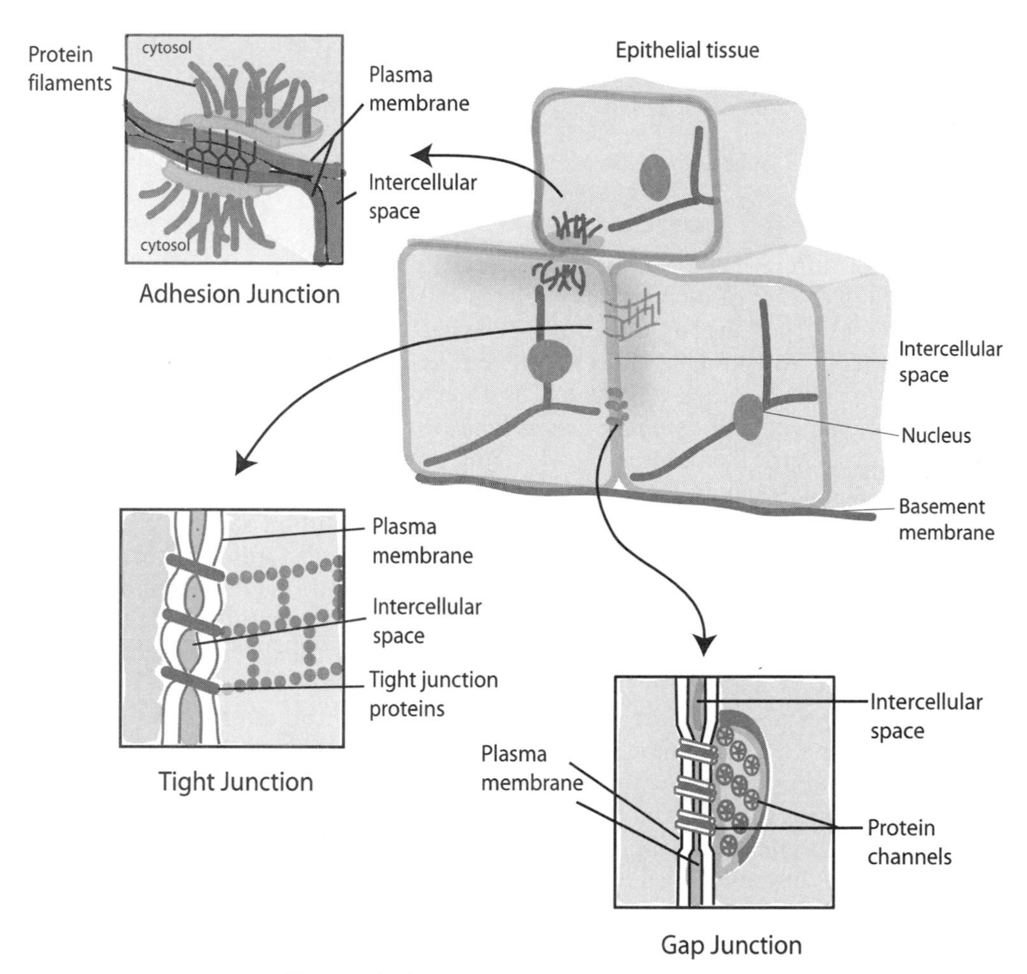

Figure 6-5. Some types of cell junctions.

they are not unambiguous and can overlap. Short-range signaling between adjacent cells is called *paracrine signaling.* The transmission of signals between specialized sites, called synapses, in the cell for long-range signaling, such as along nerve cells, is *synaptic signaling* (the role of ion channels in this process was referred to briefly above). *Endocrine signaling* in the form of a *hormone* molecule secreted from endocrine cells into an animal's bloodstream or the sap of a plant or from one cell to its neighboring cells or even for intracellular use targets cells anywhere in the organism. Hormonal signaling from one organism to others works through *pheromones*, specific molecules released into the air or water for which conspecifics have receptors.

These kinds of signaling mechanisms pass messages between different types of cells, but cells can also send signals that are bound by their own receptors or those of the same type of cell. This is called *autocrine signaling* and is important in development, as it allows all cells of a single type to respond identically to the same differentiation signals. Other signaling is transmitted through shared "pores" among adjacent cells (see below). In this case, whatever the signaling molecule is, and whether it simply diffuses through the pore to the adjacent cell or is actively transported, when inside, it works as it would within the cell in which it was produced. Signaling to neighbors directly via their surface constituents is known as *juxtacrine* signaling.

CELL JUNCTIONS

A number of different cell-cell and cell-extracellular matrix pores form the foundation of the network of interactions between cells or between a cell and the extracellular matrix. They are of three main classes: *occluding junctions*, which seal together cells in an epithelial sheet to prevent leakage through the sheet; *anchoring junctions*; which attach adjacent cells to one another or attach cells to the extracellular matrix; and *communication junctions*, which regulate the passage of chemicals or electrical impulses between the cytoplasm of two cells. These are illustrated schematically in Figure 6-5.

Perhaps the most important tissue to develop in the evolution of complex animals was the *epithelial* sheet of cells. Tight sheets of epithelial cells form the skin and line the digestive system, body cavities, organs and glands, and one of their important functions is to serve as a selective sequestering barrier, preventing the leakage of fluids from one side to the other. Occluding or *tight junctions* help create that barrier by sealing together the cells in the sheet.

Anchoring junctions connect filaments of the cytoskeleton of one cell to that of another, allowing a large number of cells to function as a structural group. They are especially abundant in tissues that undergo stress, such as skin and muscle. Three forms of anchoring junctions have been described: *adherens* junctions, *desmosomes*, and *hemidesmosomes*. A number of genes that code for these structures are known, in part because they are associated with skin diseases, such as skin fragility and carcinomas. Adherens junctions and desmosomes both play roles in cell-to-cell connections, whereas hemidesmosomes help to connect the basal surface of a cell to the adjacent connective tissue.

Gap junctions, *chemical synapses*, and *plasmodesmata* are *communicating junctions*. Gap junctions are found in most cells in most tissues and in almost all animals, both vertebrates and invertebrates. Gap junctions are channels in cell membranes formed by two neighboring cells that allow the two cells to communicate, to share cytoplasmic ions, small regulatory molecules, and macromolecular substrates, selected principally by size. Cells can readily share small molecules, passing them from cytoplasm to cytoplasm via gap junctions, but they cannot share larger molecules, nucleic acids, or proteins in this way.

Gap junctions are *gated*, that is, sometimes open and sometimes closed (like ion channels), and transient; they do not necessarily exist for the life of the cell. They are important for the normal functioning of organs such as the heart or the intestine, which requires constant calibration of ion concentrations. Gap junctions seem to be important in development as well; the existence of these communication channels allows a group of cells to function as a whole (Kumar and Gilula 1996; Wilson et al. 2000). In very early embryos, cells are electrically coupled to each other, and this is maintained by gap junctions. Figure 6-6 illustrates several signaling mechanisms.

Higher plant cells have a communication mechanism similar to gap junctions called *plasmodesmata*. These, in fact, are the only form of intercellular communica-

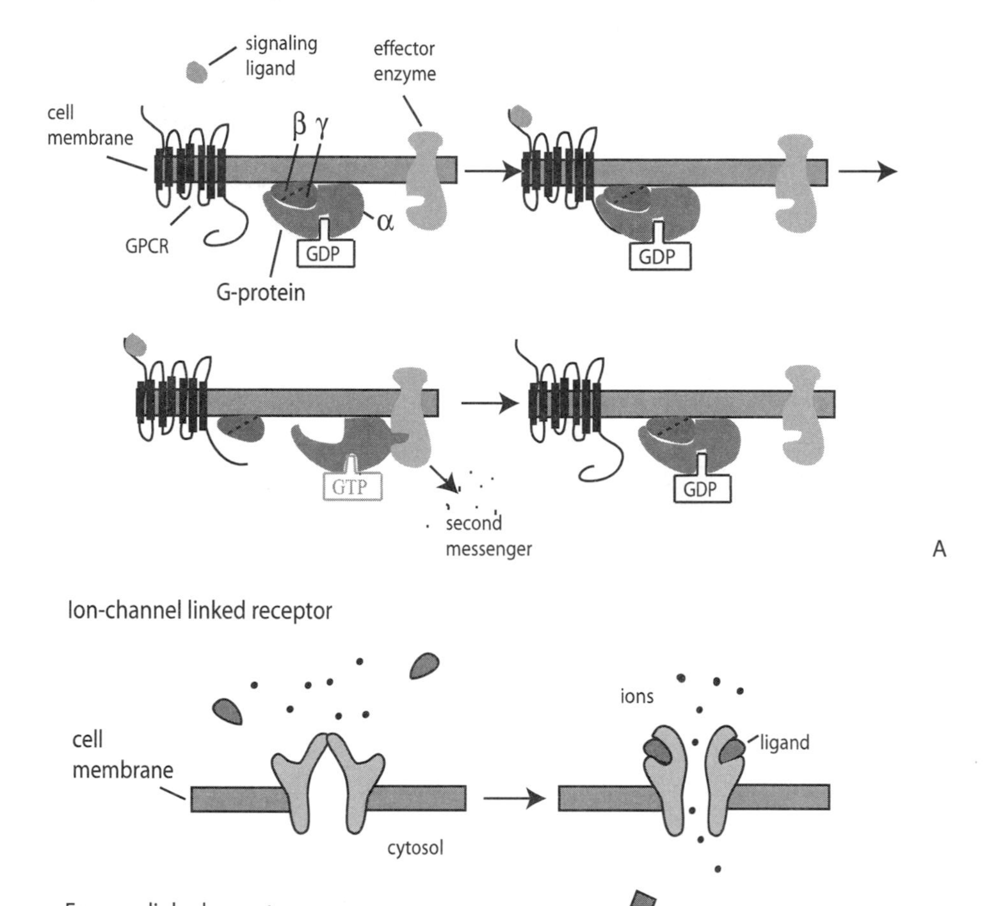

Figure 6-6. Schematic of some communicating pores in cells. (A) G protein-linked receptors; ion-channel linked receptor; enzyme-linked receptor; (B) plasmodesmata in plants.

tion plants have because, with their rigid cell wall, they have no need of anchoring junctions. Plasmodesmata connect every cell in a plant with its neighboring cells, allowing the passage of cytoplasm from cell to cell. In a sense, then, plant cells connected in this way form one mega-cell containing many nuclei.

The outer membranes of bacteria, mitochondria, and chloroplasts are permeated by pore proteins called *porins* that function similarly to gap junctions, although they are structurally very different. They allow the passage of small molecules through the membrane by passive transfer.

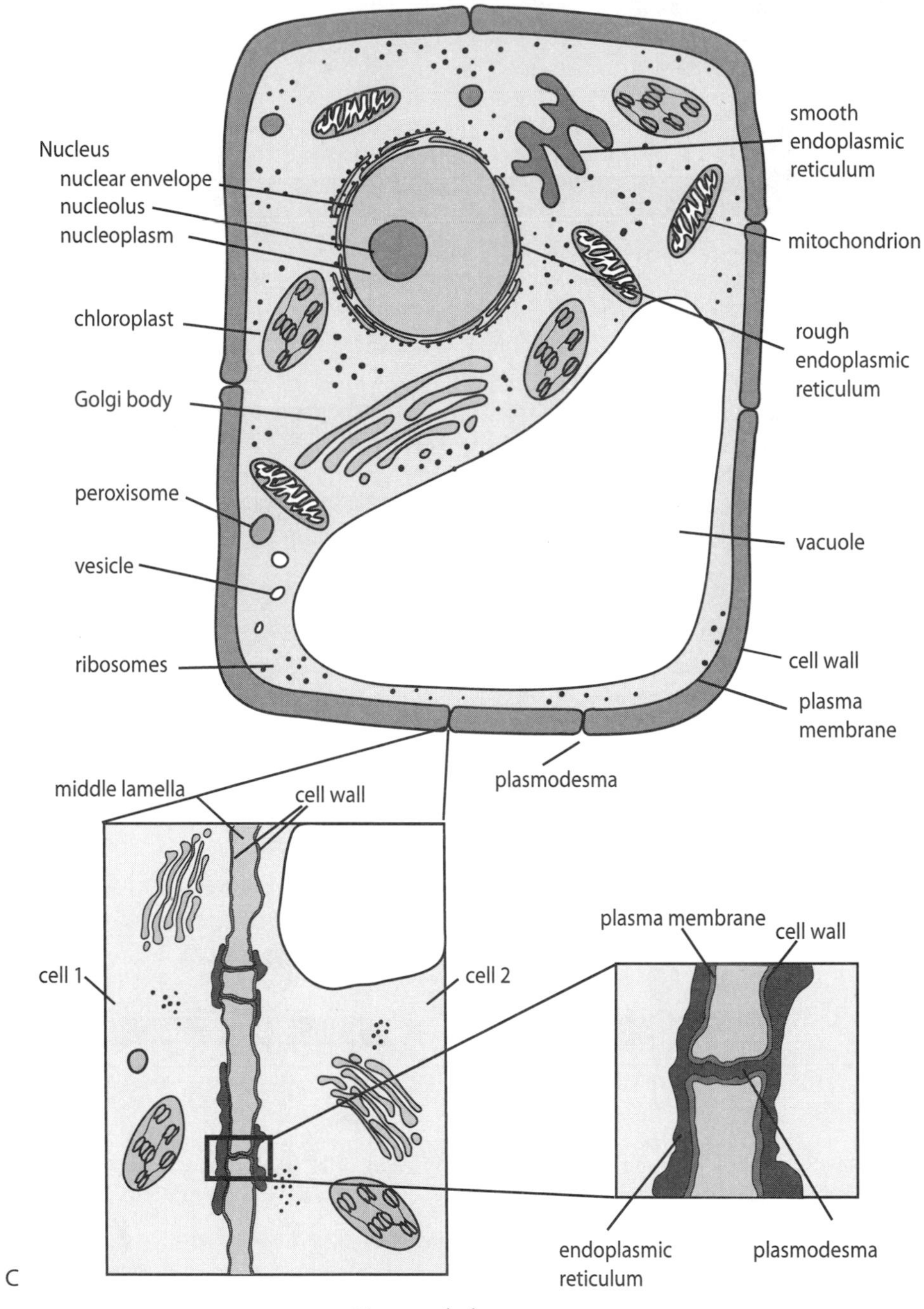

Figure 6-6. *Continued*

But a hole in a cell is a dangerous thing! In fact, one ancient part of our immune system, the *complement* system, functions mainly to poke holes in invading bacterial cells. The cell leaks its guts out, so to speak. In general, cells don't want drafty open doors, yet they must communicate. This explains the diverse and highly elaborated means animal and plant shave for getting things into and out of themselves and of communicating with other cells without becoming too vulnerable.

SOME CONCLUDING GENERALIZATIONS ABOUT MULTICELLULAR LIFE

The organizational advances necessary for the cell as we know it to exist and predominate in the biosphere were many and included a cell membrane and internal structures of some kind in which controlled reactions could take place. Such evolutionary steps alone resulted in single-celled organisms such as bacteria and protozoa, which still comprise more than half of the total biomass on Earth. Of course, we do not know how much more "advanced" current unicellular organisms are versus those that look so similar in the 3.85-billion-year-old mud; however, in terms of external appearance they are very similar, which is how we can even recognize the earliest fossils. The extent and durability of their success show that evolution does not automatically favor complexity or overwhelming change. Nonetheless, as with conserved fossils like horseshoe crabs, there *is* evolution at the DNA level and this is why deep phylogenies—despite their horizontal connections—can be constructed. Appearances can in some sense be deceiving.

Despite this conservation of form and basic physiology, some circumstance somewhere was favorable to or tolerated the evolution of multicellular life in some way, and it proliferated into diverse forms, doing things far more complex than what unicellular forms do (although the latter are found in essentially all the same niches, and more).

Perhaps for a long time, adhesion and interaction among individual cells were rather fleeting or incidental and not necessary for survival, or for survival only in a simple physical ganging-together sense. Eventually, size and coordinated organization provided advantages over simple free-floating biomolecules or single cells with ephemeral interactions. It is interesting to speculate about what the advantage of cooperation between cells was, which extended from occasional interaction to complete mutual dependency.

Indeed, many cells currently undergo programmed self-destruction (*apoptosis*) during development and modeling of tissues. Generally, in this process, a cell activates a series of internal biochemical reaction steps that lead to its death: its nucleus shrinks, the cell condenses, and it fragments without spilling any of its cytosolic contents; the fragments then are quickly phagocytosed by macrophages or other nearby cells. Apoptosis is usually actively "ordered" (signaled) by other cells, and the ill-fated cell cooperates by suicide, the sincerest form of cooperation.

Apoptosis is in some ways logically like many other kinds of apparently altruistic behavior in life. Sterile ants, self-sacrifice of individuals to protect their group, and the abscission layers that allow leaves and fruit to separate from plants are other well-known examples. These are signal-driven or in other ways "programmed" into the behavioral or genetic repertoire of the organisms concerned; the bases for abscission or apoptosis and some aspects of sterility in social insects are at least partly known. Coordinated communication is often involved. As noted earlier, there are debates about whether every such action must have evolved through selfish competition, but we understand the choreographed interactions in life at least as well by examining what is achieved by cooperation between the cells, organisms, or within the community in the process.

Unicellular organisms may be more pervasive and more ominous for us, but much of what we as humans find most interesting in the world today has to do with the function and evolution of multicellular organisms. The evolution of these inter- and intra-organismal dependencies has led to the complex traits that are the main subject of this book. From this point of view, cells are reasonably stable and uniform entities, whose common general properties are used in a diversity of contexts that make complex organisms what they are and made it possible for complex organisms to exist.

However, we do not want to leave the impression that the workings of cellular machinery are elegant and precise. It must be noted that not all cells of the same kind even in the same individual are identical. Many of the mechanisms of the cell cycle are error-prone and can introduce errors of many kinds at any stage. DNA or RNA polymerases are inaccurate to varying degrees, gene transcription and translation are imprecise, and cell division can go awry. A number of enzymes that repair replication errors, mitotic error, or errors caused by environmental degradation (for example, ultraviolet UV exposure) are ubiquitous in cells. Endonucleases, exonucleases, proteases, tumor suppressors, which we will discuss further in Chapter 7, perhaps even the immune system itself, because it recognizes and destroys foreign proteins, all keep the very imprecise cellular functions in line. Still, error (that is, deviation from the norm) happens.

In bacteria, perhaps 1 in every 10,000 amino acids or more is misincorporated into proteins as they are being synthesized (Bouadloun et al. 1983; Kurland and Gallant 1996; Kurland 1992). Perhaps, about 20 percent of newly synthesized proteins are degraded because they are misfolded or prematurely terminated or otherwise mis-synthesized. Another 10–20 percent of new proteins are found associated with molecular chaperones, which suggests a role for these molecules in the folding of proteins during synthesis (Wickner et al. 1999). These quality-control mechanisms recognize exposed hydrophobic regions in the polypeptides; proteases degrade them, whereas chaperones fold them properly.

In a very real sense, because cellular mechanisms can be so imprecise, the evolution of elaborate repair mechanisms to clean up the nontrivial errors of cellular functions is an essential facet of life as we know it. In fact, perhaps two percent of the energy of a cell goes into error repair (Scriver 2002). However, errors are not always repaired, or are not reparable by the tools of the trade, and thus fall outside the acceptable range of noise in the system. They may fall far enough outside the range of acceptability that they cause disease, including cancers and neurodegenerative diseases such as Alzheimer's, Parkinson's, Huntington's, and Creutzfeld-Jakob, that develop when cells do not rid themselves of misfolded proteins and they aggregate into amyloid fibrils (Wickner, Maurizi et al. 1999). As with many other biological systems, there is some tolerance of errors in cellular mechanics, but too much error left unchecked can be momentous, if not disastrous.

In Milton's *Paradise Lost*, Satan is accompanied by hell hounds of Rumor, Chance, Tumult, Confusion, and Discord, dedicated to frustrating the good intentions of God with respect to humans at every possible turn. This is supposed to lead to the human struggle to be good, to be vigilant and to overcome. Similarly, error and variation are central to evolution, were at the heart of Darwin's central notions, and have been a fundamental force that has kept the gods of selection ever on the alert during the long evolution of all of the pathways we have discussed in this book.

Chapter 7
A Repertoire of Basic Genetic Mechanisms

Observing the very similar chemical constituents of diverse life forms, biologists long ago realized that life (on Earth, at least) is a single subject. This edifying realization did not require knowledge of anything about genes and was basically understood even by Aristotle and others in ancient times. However, we now have genetic knowledge, and it confirms the unity of life in exquisite detail and adds whole new sets of phylogenetic relationships that go beyond and in some ways are independent of those of species.

Genes play a variety of roles. The differentiation of complex organisms into a variety of tissues depends on the fact that cells typically only express a fraction of their genes in any given context. This requires mechanisms to regulate context-specific gene expression, which as we have seen largely resides in the genome. Despite the diversity of genes and mechanisms, general statements can be made that apply broadly, from single-celled organisms to plants and animals. Chapter 4 described generalities concerning the ways DNA carries various kinds of "information," Chapter 5 described ways in which the modular nature of the genome is used to produce complex function, and Chapter 6 described many of the diverse characteristics of cells and their behavior. Table 7-1 provides a few reminders, summary pointers, and general principles. In this chapter, we will discuss some of the types of genes that have arisen in terms of mechanisms of gene action.

Our objective is to identify different kinds of gene action and to describe a few salient features that can help identify generalizations about phenogenetic relationships. In each case, there is a diversity of genes, always involving at least one gene family (usually several). It would not be possible (for us) to enumerate all of these, but there would also be no point to that. They share the general characteristics of gene families: divergent sequence, divergent but related function, and so forth. Online references are available and easily located by keyword searching; they include sequences, functional and evolutionary aspects, protein structure and

Genetics and the Logic of Evolution, by Kenneth M. Weiss and Anne V. Buchanan.
ISBN 0-471-23805-8

TABLE 7-1.

Some basic pointers, reminders, and generalizations about gene functions.

1. Proteins are modular in structure, characterized by discrete, identifiable *functional domains*, which often correspond to their exon-intron structure.
2. Genes arise by duplication, or function domains may be introduced to a gene by translocation of exons from other genes. These events are individually rare but over evolutionary time have been very important.
3. Many gene products are characterized by internal repeated protein motifs that probably arose by tandem duplication.
4. Gene family members diverge in function over time, but these divergent functions may retain some similarities or redundancies, buffering the organism against mutation and enabling us to reconstruct their descent from a common ancestral gene.
5. Similar functions may be achieved by genes with no detectable evolutionary relationship.
6. Phylogenetic relationships among genes in a gene family reflect their history of gene duplication. There can be substantial differences in the numbers of members of gene families, especially among distant taxa, but even among more closely related ones.
7. It is not easy to define exactly what a "gene" is because genetic activity is so diverse, and genes can be variably spliced.
8. Only a small fraction of genes code for classic structural proteins or metabolic enzymes.
9. A large number of genes code for products to manage the processing of other genes, including transcription, translation, and chaperones that help fold, protect, or transport proteins in the cell to their appropriate destination transport.
10. Many genes code for RNA molecules that have direct biological rather than coding function, including regulation of the production of other genes.
11. A large number of genes are involved in communication among cells or between cells and the outside world, enabling cell specialization, homeostasis, and development to occur.
12. Gene family relationships are sometimes related to the arrangement of genes on chromosomes because duplicates sometimes stay in tandem linkage relationships with each other; but gene family members can also be translocated to other chromosomes.
13. Syntenic linkage arrangements of genes (that is, location together on the same chromosome) may, but need not, be related to their coordinated expression.
14. Many if not most genes are expressed in many cell types and have multiple, often unrelated functions. In many instances, the functions can be shown to descend from some common ancestral function (like cell adhesion).
15. To the extent that differentiation is the vital key element of complex organisms—or even of the existence of single cells—gene regulation is as important as gene coding.
16. Genomes are filled with regulatory or response elements (REs), sequences whose chemical qualities "attract" transcription factors that bind to those sequences and help express or repress the gene. These are functional units, though not usually called "genes" and that can be more variable in sequence, number, and location than the coding genes that the REs regulate.
17. Regulatory sequences, like most other functional (and indeed many nonfunctional) elements of DNA structure, can be grouped into related sequences that have some internal variability. Except when found in tandem arrays, these short (approximately 5–15 bp) sequences appear to arise repeatedly by de novo accretion and sequence evolution.
18. There are many genes and mechanisms for policing or preventing errors in a cell. These protect proteins when the cell is under stress, degrade misformed proteins, degrade or repair DNA or RNA, or perform other similar error-moderating functions.

TABLE 7-1.
Continued.

19. Almost everything that goes on in a cell is affected by chance or environmental events. As a result, no two organisms are complete twins or completely symmetric, cells even of the same type don't express exactly the same genes at the same time (see Hume 2000), and a cell never divides into identical daughter cells. Stochasticity, including the often unpredictable effects of gene-environment interaction, has driven the evolution of various systems but can also drive the pathways themselves.
20. Many processes of importance to organisms are not genetic, and nongenetic inheritance can be important in the short as well as the long term.

gene-evolution diagrams, and tools for analyzing the details of any example. Interested readers can use these as they go along through this and subsequent chapters.

SOME GENERAL TYPES OF GENES

The Central Dogma of biology, that genes code for protein, led to the general notion that genes specify individual structural proteins and enzymes that control basic physiology. The idea (not always explicitly stated) was that an organism is built up of separate identifiable functions and that one gene coded for one function. We know this is an overstatement, and one that can be quite misleading, but it is a view that has persistent effects on biology and biological research. The traditional kinds of genes, perhaps those most easily understood, code for extracellular structural proteins like collagen, intracellular structural proteins like actins and spectrins, carrier molecules like albumin or hemoglobin, or enzymes that trigger reactions like insulin or hormones, as well as for catalysts of basic energy storage and release or catalysts of reactions in the production and manipulation of basic biomolecules such as nucleic acids, amino acids, carbohydrates such as sugars and starches, cellulose, lipids, steroids and vitamin D derivatives, and so on. Other genes have fundamental roles in controlling DNA packaging and replication and have many general cellular housekeeping functions. Large families of genes exist for these purposes. Such "end-product" genes have each traditionally been associated with a biological function itself and, hence, from an evolutionary point of view, the targets of natural selection. Of course, not one of them acts alone, nor without a chain of cofactors related to their use, protein maturation, expression, and the like.

Such functions are vital to life, and many must have been present in the first cells. But essential as they are, these generic functions are not the focus of this book. We are concerned with the intricate specialized functions of differentiated organisms and the many ways the sequence in DNA works to serve these biological functions. Their discovery has broadened our understanding of how integrated organismal complexity is achieved and evolved and deepens the unifying view that genes are the core functional elements of life. But while the phenogenetic relationships being discovered display logical regularities, they are generally anything but simple.

Genes are discovered in many ways, sometimes by identifying coding sequence and sometimes via effects of mutation or studies of function. One reason that genes can be challenging to classify functionally is that as soon as we think we know what a gene does, other uses and expression may be discovered. Genes originating with one function can evolve others that initially may be related but eventually become at best distantly so. One way this is a problem in scientific practice is that genes are often named for the function for which they were first discovered, which can be very misleading when, as so often happens, diverse additional functions are identified; often, the "original" function is minor or misperceived (e.g., the gene called "eyeless" is not a gene evolved to make—much less remove—eyes but is part of a class of general developmental genes).

Keeping in mind that genes are not necessarily "for" any particular thing (even within a given species), we can at least generally classify them into various functional types, as shown in Figure 7-1. As the figure makes clear, a rather small fraction of genes appear to have "function" in the classical sense. Although not shown in the figure, for each functional group there is a common theme: one or more families of genes descended by gene duplication from a common ancestor, usually with widespread sharing of members of the family among plants, or among animals, or both, and typically also with single-celled organisms. What are some of these types of genes?

Processing Genes

A large class of genes codes for proteins involved in the processing of mRNA or of polypeptides after they have been translated. The processing functions include transport, packaging, error repair, scavenging, activating, inactivating, and modifying proteins or RNA. One interesting example are the heat-shock proteins, often called *chaperonins*. These genes were discovered because they are expressed under

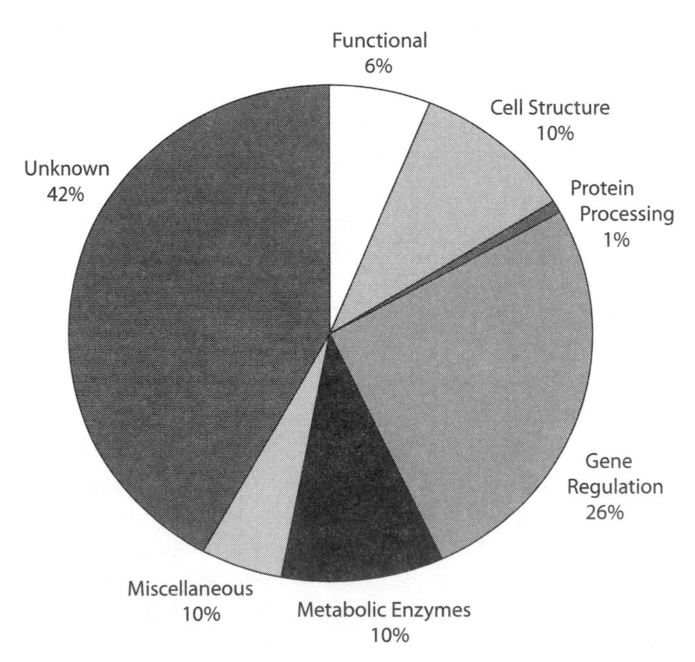

Figure 7-1. Distribution of human gene types based on annotation of human genome sequence. Definitions are rather imprecise and some genes overlap in function. For up to date details see the Gene Ontology website (Gene Ontology Consortium 2003).

conditions of stress. Their intracellular functions include the folding and protection of proteins (normally and/or to resist stress of a high-temperature environment, hence "heat shock") and detecting unfolded protein or infection. These genes are found in all major life forms. Perhaps, chaperonins are (or were) also evolutionarily primitive intercellular signaling molecules, as they can be found outside of cells and adhere to various cell-surface receptors.

Cell Adhesion and Related or Descendant Function

Multicellular life involves cell-to-cell contact in many ways, as described in Chapter 6. Some of this is achieved with members of a large and ancient family of *cell adhesion molecule* (CAM) genes. These typically involve a single transmembrane domain and an extracellular domain that binds to a molecule on an adjacent cell, holding the cells together. Some function as homo- or heterodimers (that is, two copies of the same gene product or one or more copies of each of more than one gene product forming a single functional unit). They may also require a cofactor element such as calcium (e.g., cadherins). Desmosomes and adherens junctions that stitch cells together are formed with CAMs.

Most cells in multicellular organisms have CAMs on their surface, so that they can be properly arranged in relation to other cells. This gives the organism its physical architecture. However, not all CAM binding is between cells; *integrins* can bind to extracellular matrix, for example. *Selectins* and integrins in some situations facilitate cell shape or movement along a substrate (which can be another cell or a blood vessel, for example).

This ancient class of genes has been involved in the evolution of many functions that would not now perhaps be thought of as simply involving architectural adhesion, but their properties are recognizable and the DNA/protein structure shows the common ancestry. Developing and migrating neurons form bundles that depend in part on a combinatorial cadherin "code," in which cells adhere if they express the same specific set of cadherin gene products. This helps organize neural function (Redies and Puelles 2001). There are also other neural-specific CAMs (called N-CAMs).

These latter genes are members of a large family of immunoglobulin-like CAMs. As will be seen in Chapter 11, the immune system deals with internal damage and foreign molecules by recognition and binding processes that seem to be descendants of simpler cell adhesion phenomena. The immune system recognizes and handles infected or inflamed cells using the CAMs selectins as well as integrins. As things have evolved today, antigen-antibody binding goes beyond cell-cell binding to involve molecular fragments derived from invading cells or molecules. In vertebrates, immunoglobulin-like CAMs include antibody and histocompatibility genes involved in self-non-self recognition.

CAM genes are scattered over the genome, sometimes in large coordinately regulated clusters as in the immunoglobulin and T-cell receptor clusters in mammals.

Cell Environment Control

As described in Chapter 6, cells have pores that enable them to adjust their internal conditions relative to that of adjacent cells and the extracellular space around them. This allows them to keep their internal conditions under control for their particular needs by regulating what goes into and out of the cell. These pores are modular in that they are controlled by different families of genes with different evolutionary histories, although they share the type of functional role they play.

Ion Channels

Ligand-gated and voltage-gated ion channels are separate types of structure that were described in Chapter 6. Both are formed by complexes of proteins that make controlled passages into the cell. The ion channel may be composed of a single protein with multiple domains that pass through the cell membrane or of homo- or heteromultimers of various channel proteins. Voltage-gated channels typically involve four subunits, with the main subunit having multiple (e.g., five or six), similar transmembrane complexes, arranged in a circular fashion around the pore. Changes in the molecular configuration of the channel complex open or close the channel to ion flow. Whole families of genes code for the ion channels that preferentially transmit specific ions. That is, there are related genes that specify channels for Na^+ or K^+ or Ca^{2+}.

Ligand-gated channels also have multiple copies of components (often 5), including multiple membrane-spanning elements around the pore, along with receptor and response domains. There are three known superfamilies of ligand-gated channel genes (cys-loop receptors responsive to GABA and other signaling molecules, ATP responsive, and glutamate activated). Each class is coded by its own gene family, whose members serve different physiological functions but via shared mechanisms and channel-structure characteristics.

Gap Junctions

Like ion channels that face intercellular spaces, gap junctions are pores that connect cells directly to each other. The gap junction is formed by a circular complex, a *connexon*, composed of four to six protein subunits called *connexins*, which can be repeats of the same subcomplex. What can pass through a gap junction is determined by which gene family members code for the proteins of which it is comprised. In invertebrates, the proteins are coded for by members of the *innexin* gene family; interestingly, these are not homologous to connexins at the DNA sequence level, and they also appear to be unrelated to the CAM cell-contact genes.

Genes to Keep Things in Balance

If genes are generally thought of as making something, an obvious issue that arises is what happens when enough is enough. Getting rid of something can be as important as making more of it. Excess levels can be toxic, and genes related to development or homeostasis may be very damaging if expressed at the wrong times. At some point, a structure has to *stop* developing. Not surprisingly, a large number of mechanisms have evolved whose function is to adjust the balance of constituents in a cell, by removing or modifying things that are already there. Some of these involve repressor REs near genes, but others are genes specific to the purpose. As a generalization, almost anything produced in a cell is regulated in one or more of these ways.

Complex genetic mechanisms exist to destroy incoming bacterial parasites, and internal wounds can be repaired in part by first destroying or removing damaged cells. This can be achieved by genes that code for products that degrade DNA or RNA or protein or that break down cell membranes or cell surface components.

One large class of such genes codes for *proteases* (sometimes called peptidases). Proteases typically cut proteins at the ends or internally or at particular amino acids. They are expressed in particular contexts and may serve special functions, such as the breakdown of extracellular matrix proteins. Plants, animals, and even bacteria have proteases, and there are several families of protease genes, each sharing a specific cutting mechanism (for example, that cut proteins using serine residues in their enzyme structure), which presumably characterized their common ancestor (Proteases 2003). There are four basic, ubiquitous classes (serine, cysteine, aspartic, and the ancient metalloproteases found even in bacteria, which use a metal ion, usually zinc, as a cofactor).

Proteases can degrade proteins no longer needed within a cell, or they may be secreted to work outside the cell, for example, to destroy extracellular matrix as tissue remodeling occurs during development or cell movement. Proteases can be used in defense against incoming proteins, or in the attack as in the breakdown of a host's proteins by toxins introduced by a parasite or predator (e.g., mosquitoes or snakes).

Corresponding to these and with analogous kinds of function are various RNases and DNases that break down RNA and DNA, including nucleases to break down nucleotides, *endonucleases*, and *exonucleases*, which cut DNA internally or at its ends, and a variety of gene products that help control cell growth, detect and repair mutations, and other functions related to DNA/RNA integrity and use. A well-known class of genes is one that codes for *restriction endonucleases*, enzymes that are used by bacteria to destroy incoming DNA (e.g., from viruses). This is a large class; each endonuclease specializes in cutting DNA at a particular short sequence. There is a large repertoire of restriction sites recognized by the products of this group of genes, which humans have appropriated for our own daily use, in molecular genetics laboratories around the world, to cut DNA at specific places of our choosing.

We have mentioned apoptosis, or programmed cell death, which occurs in numerous situations such as tissue remodeling during the complex spatiotemporal processes of development. A variety of genetic means are used for this purpose, including the degradation of mitochondria, cellular protein, or DNA or signals released by a cell that are recognized by cytotoxic mechanism in the organism's immune system. The mechanisms can destroy or disable the cell or prevent it from going through the mitotic cell cycle.

Another diverse set of genes is the *cytochrome P450* (*CYP*) genes (the name has to do with their means of discovery and isolation in the laboratory). The products of these genes use iron to oxidize and inactivate classes of toxic compounds; the human liver uses these gene products to breakdown environmental toxins, for example. There is speculation that this huge class of hundreds of genes evolved in the competition between animals and plants—plants producing toxins to avoid being eaten and animals evolving CYP genes to enable them to keep nibbling.

Not surprisingly, in the checks and balances of life, every destructive action can also be overdone, and the amount, timing, and location of degradation activity is controlled by cells; defenses such as protease inhibitors protect against exuberant degradation. This is another manifestation of the widespread homeostatic mechanisms in complex organisms.

GENES FOR REGULATING OTHER GENES

RNA "GENES" AND THE REGULATION OF RNA

mRNA is an indirect code bearer for the translation of protein, but there are more direct RNA-based coding mechanisms. One mechanism for posttranscriptional gene silencing (known in these acronym laden days as PTGS) is *RNA interference* (RNAi). Stretches of DNA that include *antisense* sequences that are complementary to parts of specific mRNA,are transcribed into RNA that folds into a double-stranded (dsRNA) form. As currently understood, the latter is recognized and cleaved by an RNase (RNA-degrading nuclease) enzyme in the *Dicer* gene family, releasing small interfering RNA (siRNA) fragments of 20–25 bp. The siRNA has antisense sequence to part of the target mRNA and together with a protein complex called RISC (RNA-induced silencing complex) binds the target, making it thus partly double-stranded so that it is degraded by RNases (Banerjee and Slack 2002; Cerutti 2003; Grosshans and Slack 2002; Hannon 2002; Sharp 1999; Tijsterman et al. 2002; Zamore 2002) (see Figure 7-2). There may be an intermediate step in which the dsRNA is amplified to many copies so the inhibition can be more effective in the cell. These chains of events are diverse within and between species and involve carrier and facilitator proteins coded by a gene family called *Argonaute*.

Another and related mechanism involves similar processing machinery to activate small temporal RNA (stRNA). The roughly 70-bp stRNA transcript self-hybridizes to form a stem-loop structure that is inactive until a short segment

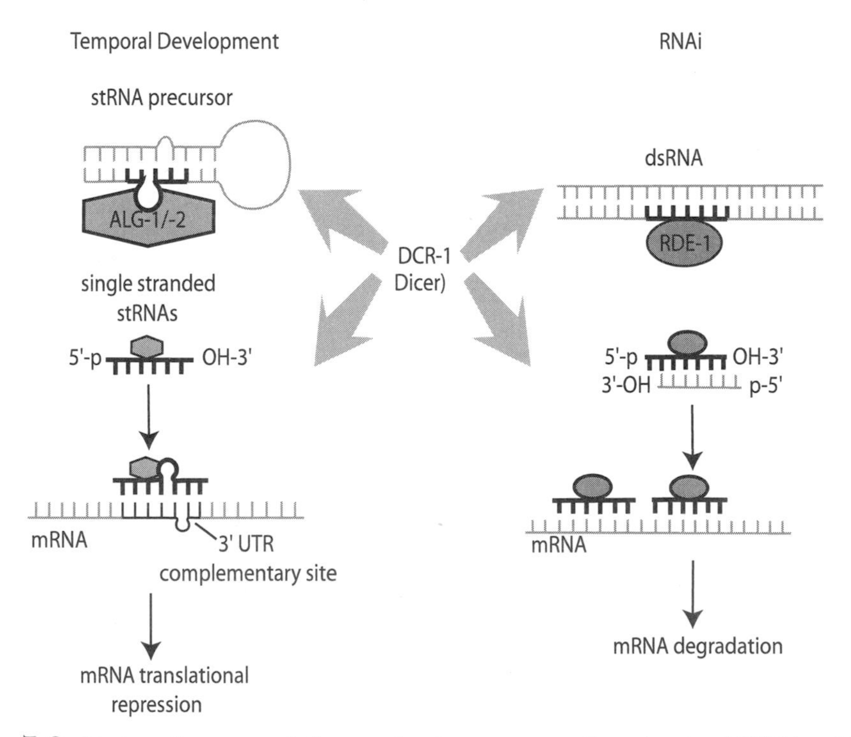

Figure 7-2. Basic antisense regulating mechanisms as currently understood. RDE and ALG are proteins that participate in the process. Redrawn after Banerjee and Slack, 2002.

(about 20 bp) is excised. This piece is diced to single-stranded antisense fragments that bind to 3′-UTR sequences in target mRNA. The binding does not appear to cause the RNA to be destroyed, but instead prevents translation, which effectively down-regulates the target gene. Some stRNAs discovered early were the *Let7* and *Lin4* genes in the nematode *Caenorhabditis elegans*, but there are probably many more. In the nematode, these genes are used to regulate heterochronic aspects of development, that is, highly timed expression of each of these genes is required for the transitions through the four stages of larval development, ensuring that patterns of developmental-gene expression and tissue differentiation occur at the right times and places. A *Let7* gene may block translation of specific mRNAs, that have *Let7*-complementary sites in their 3′–UTRs. In one proposed scenario, a temporal chain of interactions appears to occur: a target of *Lin4* inhibits *Let7*, which had been blocking translation of *its* target, *Lin41* mRNA; the latter is an inhibitor of the final adult developmental moult. These genes are widely conserved among animals and seem to have similar developmental function in *Drosophila melanogaster* (Banerjee and Slack 2002; Grosshans and Slack 2002; Sempere et al. 2002), in which *Let7* is temporally regulated by the steroid molting-related hormone *ecdysone*. However, the two stRNA genes and their critical sequences appear to be evolutionarily unrelated. There are numerous other genes of similar function already known, and likely many more to be found.

The inactivation of mRNA may be involved in quick changes of gene expression that can be important in some stages of development or cell differentiation. RNAi also may clean up dysfunctional RNA in the cell and may even affect chromatin or DNA modification (Cerutti 2003). RNAi involving homologous mechanisms is found in animals, plants, fungi, and protozoa, with the usual diversity of related gene family members, reflecting the ancient age of this system, which means that what we see today in terms of function (or the function in which a system was first discovered by us) may not be its original function(s). RNAi may, for example, have originally served an immune function because plants and other species activate RNAi when invaded by infectious agents like viruses; at least some plant viruses appear to have counteracting genes (Tijsterman, Ketting et al. 2002). This would be a mechanism functionally analogous to the use of restriction enzymes by bacteria to degrade incoming DNA, although RNAi works by complementary base-pairing, whereas restriction enzymes bind and cut double-stranded DNA at specific short sequences. Like restriction enzymes that molecular biologists use to cut and manipulate DNA in the laboratory, RNAi is proving to be useful for experimentally interfering with gene expression in the laboratory, by introducing selected antisense fragments into cells at particular times. It is likely that many more genes will be found related to the large RNAi system.

Searches of genome sequences and cDNA libraries have also revealed numerous natural antisense RNA transcripts (NATs) that are complementary to mRNA sequences (see www.hgmp.mrc.ac.uk, and Cerutti 2003; Lehner et al. 2002). Many imprinted genes—their DNA chemically modified to inhibit transcription factor (TF) binding and hence expression—appear to have associated antisense transcripts as another way to inactivate them. A considerable fraction of *cis* antisense RNA sequences are transcribed from the antisense strand of regions of genomic DNA coding for the gene (that is, so the result is complementary to the mRNA being transcribed from the template strand); the antisense sequence is thus automatically available for use so long as the appropriate transcription start signals are available on this complementary strand. There are also a substantial number of possible *trans* antisense transcripts located elsewhere in the genome (Lehner, Williams et al. 2002). The importance of these, if any, is not known but may indicate elaborate *trans* activation/regulation mechanisms.

Many types of cells have biochemical asymmetry, or polarity, which can be produced by differential distribution of mRNA. For example, asymmetric polarity in the egg cell is fundamental to fly development. In *Drosophila*, the protein coded by the *Bicoid* gene binds to a short 3′-untranslated recognition sequence in the mRNA of the *Caudal* gene. This inactivates the latter. Initially, caudal mRNA is uniformly distributed in the egg, and *Bicoid* is more concentrated at the anterior end (determined by the mother when the egg is produced). The result is an asymmetric concentration gradient of functional Caudal protein, which establishes locally different environments for the expression during the first stages of differentiation of the egg (Rivera-Pomar et al. 1996; Sauer et al. 1996).

ADAR

Other *RNA editing* phenomena affect the translation of a gene's amino acid code. Besides mRNA splice variation, there is at least one other known pre-mRNA editing phenomenon, adenosine deaminase acting on RNA (ADAR), which has been observed in both invertebrates and vertebrates. What ADAR does is change the sequence of particular codons so that the mRNA is translated to contain a different amino acid from that nominally coded in the gene itself (e.g., Reenan 2001). In the target regions, the DNA **A** is changed to a nucleotide inosine (**I**), which for chemical reasons the ribosomes read as **G**, because **I** chemically pairs like **G**. This change is an active enzymatic process; although the system is only recently discovered, an understanding of the targets in immature mRNA that will be converted is developing. No simple reason, however, is yet known regarding why such a mechanism would have evolved.

Most known examples of interference with translation were stumbled upon, rather than deliberately sought, because this is a phenomenon strange to the usual theory of the evolution and use of DNA. Among the known instances are interesting effects on neural function, suggesting that ADAR evolved in regard to some aspects of behavior. ADAR can alter the properties of ion channels in neural tissue and have major behavioral effects on flies (Reenan 2001). Because the degree of ADAR activity may vary, or may even be context-dependent, an element of behavioral flexibility may be introduced via ADAR. However, different uses may yet be discovered, and ADAR activity has been found in other tissues.

SIGNALING

It has been said that only a small percentage of the soldiers in an army actually do the fighting, whereas the rest are there to make the fighting possible by taking care of supplies, maintenance, food, administration, medical care, and the like. Biology is similar, as reflected in Figure 7-1. A majority of genes are used primarily in support of the functional end-products that, as we noted at the beginning of the chapter, comprise the usual image of what genes are. Among the most important support services, as in an army, is to communicate "instructions" by signaling among the many components of an organism.

Examples are legion and include all aspects of life. As will be detailed in Chapter 11, the detection and destruction of an infectious pathogen involves a chain of interactions in which one type of cell is signaled by another to go into action: cell-surface immunological receptors on mammalian white blood cells bind to fragments of pathogens presented on the surface of other immune-system cells. Antibodies secreted by B cells bind these, and once bound, the cell is marked for destruction by the complement system or neutralization by phagocytes. Macrophages and neutrophils have surface receptors that bind to circulating bacteria to target them for phagocytosis.

Metabolism involves signaling, as for example the use of insulin to stimulate cells to bring glucose in from the blood system, or apolipoproteins to stimulate the liver to take in cholesterol (and not make so much new cholesterol). Hunger and satiety responses are triggered by circulating signals. Chromophore molecules bind to photoreceptors, changing them when subsequently hit by photons of a particular energy to which they can respond. Steroid hormones and retinoic acid diffuse into cells where they become bound to specific binding proteins. Airborne odorant molecules are captured by binding proteins and transported to cell surface olfactory receptors. Small molecules diffuse between cells through gap junctions and through cell membrane ion channels.

In plants, diffusing auxins cause cell proliferation to promote upward or lightward growth or establish a dominant meristem to control plant shape. Abscissic acid causes the closure of leaf stomata and promotes seed dormancy by inhibiting cell growth, both reactions to dry conditions, for example. The gas ethylene causes fruit ripening (and is produced by ripe fruit) and is a ligand for transmembrane kinase receptors for the plant or neighboring plants (see below).

Similarly, development and differentiation depend heavily on the production, distribution, and detection of signals. Such signals are responsible for the establishment of polarity and segmentation and induce major cascades of differentiation.

A variety of gene-coded signaling factors (SFs) diffuse from cells to trigger these processes. Table 7-2 provides a description of some of the mechanisms. Signals can travel in various ways or for various distances. Cytokines or growth factors enter the circulation and travel around the body to stimulate growth, but some such signals are carried by circulating cells and only released locally, for example, to stimulate wound healing or immune responses. At the other extreme are pheromones that carry information between organisms, with similar kinds of effects and using very similar mechanisms. In a sense, this makes the members of a population of organisms a kind of large single super-organism, the connecting medium being air rather than blood or sap. We think of organisms as being different entities rather than serial homologs of each other the way vertebrae are but the communication processes are essentially the same. This is but one way in which the nature of what constitutes a "being" is more open than usually considered.

Passive Signaling

We mentioned in Chapter 6 that some small signaling molecules can diffuse through cell membranes directly into the cell. Steroid hormones are examples. These are nonprotein molecules. Once inside the cell, the signal molecule usually complexes with some binding factor, and this then affects gene expression or other aspects of cell biology. The action of various kinds of nuclear receptors is described in Table 7-3.

TABLE 7-2.
A sampler of receptor-mediated signal transduction mechanisms.

The logic of receptor-mediated signal transduction is similar among many classes of signaling factors and receptors. Here are some of the variations that exist on this theme. Note that gene names are historical; they have diverse, sometimes opposite function from those for which they were originally named. Also, we list only representative stereotypes; other examples are known and probably others have not yet been discovered. For example, there are several classes of growth factor that work in different ways; we only enumerate a few of them, because the basic ideas are similar.

1. *FGFs* (fibroblast growth factors) are a class of around 20 diffusible mitogens (stimulants of cell division) that have numerous splicing isoforms (mRNA equivalents, that for example, make proteins that contain different complements of exon units). FGFs are ligands for a corresponding set of *FGF receptors* (*FGFR*s). Ligand binding is aided by cofactors (heparin sulfateproteoglycans, or *HSPG*s) that complex with the FGF and the FGFR. The intracellular FGFR domain is a tyrosine kinase. The extracellular part of an FGFR has several immunoglobulin-like domain loops, because of their origin by gene duplication from the cell-adhesion gene family that includes antibody and many other types of genes. However, the homology does not imply functional similarity. These receptors often are activated when two adjacent copies jointly bind ligand molecules. Other cell adhesion molecules may serve as ligands. FGFs stimulate growth and differentiation in many contexts and tissue types, mainly in embryos.
2. *Wnt* signaling factors (called *Wingless*, in flies) are ligands for serpentine receptors in the *Frizzled* class (see Table 7-4). The receptor is ligated to an intracellular *Disheveled* gene product, which transduces the received signal through a pathway that releases a β-catenin molecule to bind with and activate a TF (e.g., *Lef/Tcf*). Different pathways respond to different members of the *Wnt* family, through different second messenger systems, and there are specific *Wnt* inhibitors (Niehrs 2001).
3. *TGFβ* (transforming growth factor) signals are diffusible signals that include the BMP (bone morphogenetic protein, called Dpp in flies) signaling factors. TGFβ dimers bind between copies of two variant types of TGFβ receptors. The type II receptor phosphorylates serine or threonine residues on the adjacent type I receptor molecule. The latter then phosphorylates and activates intracellular Smad proteins, intracellular signal mediators that dimerize to form active TFs in a complex set of pathways (von Bubnoff and Cho 2001).
4. *Hedgehog* gene products are diffusible signaling molecules that travel only a few cell diameters. Hh proteins are ligands for cell-surface receptors of the *Patched* gene class bound to another protein called Smoothened. Receipt of the Hh ligand by this Patched-Smoothened complex leads to changes in Smoothened whose intracellular domain activates an intracellular transcription activator Ci. When Smoothened is quiescent, a modified Ci protein acts as an inhibitor of the same genes. *Hedgehog* signaling may work in relay fashion, with cells receiving signal and secreting Hedgehog protein of their own to travel the next few cell diameters.
5. *Notch* class proteins reside in the cell surface and can receive several ligands, including proteins from the *Delta* (*Jagged*, *Serrate*, or other) class, which are bound to the membranes of adjacent cells. The binding event enables protease cleavage of an intracellular domain of the Notch protein. The cleaved fragment migrates to the nucleus where it complexes with and activates a TF (e.g., of the CSL class). Another consequence is that the Notch-presenting cell may be inhibited from producing its own Delta. *Notch-Delta* signaling generates zones of expression (cells presenting Notch) and inhibition zones in surrounding cells. This mechanism is used in neural structures, broadly defined, in vertebrates and invertebrates, often involving repetitive patterning.

TABLE 7-3.
Nuclear receptors and how they work.

Nuclear receptors (NRs) are ligand-dependent TFs that direct targeted gene expression when bound with steroid and thyroid hormones, retinoids, vitamin D, and other ligands. Ligand-bound NRs by and large interact with transcriptional coactivators, whereas ligand-free NRs interact with transcriptional corepressors. Ligand-binding significantly affects the distribution of receptors within the cell, both within the cytoplasm and the translocation of receptors from cytoplasm into the nucleus because ligand-binding releases chaperones and exposes nuclear localization signals (NLSs) (Black et al. 2001).

NRs activate gene expression by binding to a short DNA sequence, often **(A/G)GGTCA**, called a *hormone response element* (HRE). HREs are typically located in the promoter region flanking the coding region of the responding gene. This HRE is often present in two copies, either head to head as palindromes, as direct repeats, or tail to tail as inverted palindromes. The NR binds either as a homodimer or as a heterodimer, and the orientation of the HRE pairs determines to which HRE the receptor dimer binds.

The typical NR includes A/B, C, D, E, and sometimes F domains, each with a separate role in ligand binding and gene activation. C and E/F are conserved in basically all family members, whereas the A/B and D domains vary (Aranda and Pascual 2001). The structure of a typical nuclear receptor is diagrammed in Scheme 7-1. The A/B domain contains the ligand-independent transactivation domain. Region C recognizes specific DNA sequences. Linker region D connects the DNA-binding domain to the ligand-binding domain, E/F.

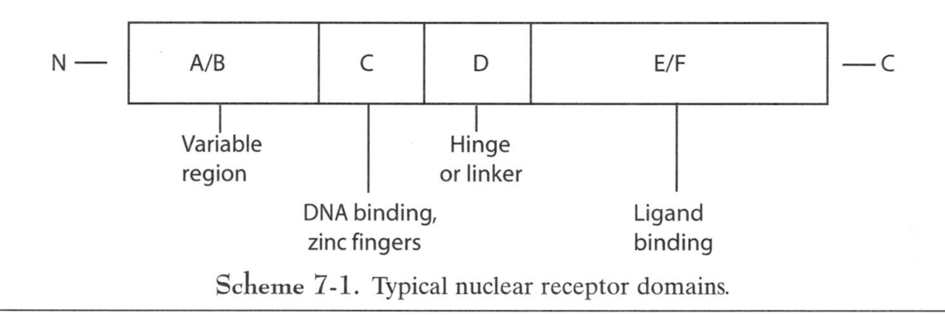

Scheme 7-1. Typical nuclear receptor domains.

Receptor-mediated Signaling

Larger signaling molecules cannot diffuse through the membranes of receiving cells and instead serve as ligands for specific receptors protruding from the surface of cells sensitized thereby to receive the message. This is true of a variety of growth and signaling factors as well as of protein hormones like insulin and adrenaline. In fact, message exchange among cells is so important that cell membranes are often littered with receptors of many kinds.

Cell surface receptors have one or more ionically neutral (uncharged) domains of the right length to reside stably within the thickness of the cell membrane. Transmembrane domains typically evolve by duplication events and often coincide with exon-intron boundaries in the receptor gene. The extracellular domain of the receptor is usually the ligand-binding site, and an intracellular domain of the receptor protein is altered by the physical event of a ligand binding to its extracellular domain. The changes in the intracellular part trigger second-messenger events, described in Chapter 6.

There are three general classes of cell surface receptor proteins, defined by the manner in which they transduce signals into the cell. We have already discussed some of them and their attributes. *Ion-channel-linked receptors* transmit inorganic ions between electrically excitable cells. *G protein-coupled receptors* (GPCRs) are a large family of transmembrane proteins (more than 2,000 have been identified) found in all eukaryotes, although fewer have been found in plants than in animals; aspects of this family are given in Table 7-4. The table shows the diversity of known function fulfilled by the various classes of these receptors, with class subdivision based on sequence phylogeny, indicating some, but not total, relationship between function and evolutionary origin. These are seven-transmembrane spanning helical proteins (e.g., see below). Binding of a ligand (signal molecule) to the extracellular domain causes a conformational change in the intracellular domain, which activates the G protein bound there. The G protein dissociates from the receptor and carries the signal to an intracellular target, which can be an enzyme or ion channel. Many transmembrane signaling mechanisms have evolved and will be discussed more fully below. *Enzyme-linked receptors* act either as enzymes directly or in association with enzymes.

A particularly important class of receptors, including GPCRs, that we will see more of in later chapters is described in Table 7-5. These are called seven-transmembrane membrane receptors (7TMRs) or "serpentine" receptors because their seven domains "snake" through the cell membrane. Many of these are receptor tyrosine kinase (RTK) genes, meaning that the signal receipt triggers the intracellular phosphorylation of tyrosine residues on responding proteins. Other kinase-based receptors phosphorylate serine or threonine amino acid residues on their target proteins. The phosphates are transferred from ATP molecules by kinases and removed by *phosphatases* to deactivate the protein. Phosphorylation produces structural change in a protein, causing it to bind or release another molecule. Protein phosphorylation is a predominant means by which activity is controlled inside the cell. It has been estimated that perhaps one-third of the proteins in a mammalian cell are phosphorylated at any given time, and that 2–3 percent of a eukaryotic genome encodes kinases (Hunter 1995). The initial response to ligand binding in turn activates *second messenger* molecules inside the cell. Through cascades of subsequent interactions, activated messengers are transported to the nucleus where they serve to activate (or repress) the expression of specific genes.

As would have to be the case, signaling genes have corresponding receptor genes. The two families coevolve in the face of mutation, so the signal and its receptor mechanism still work, and there can be cross-reactivity among them; thus, a given SF can bind to the product of more than one receptor of the same class. The binding may not be as efficient, but this cross-reactivity provides some functional redundancy, perhaps at the loss of some specificity. This is an example of the partial nature of functional modularity or sequestration that is so prevalent in life. The presence of cross-reactivity also suggests that during evolution the SF and receptor duplication events do not need to be tightly coordinated: mutational divergence of duplicate SF genes can lead each eventually to bind preferentially to one of the evolving, duplicating family of receptors.

TABLE 7-4.
G Protein-Coupled Receptor Superfamily.

	Group					
Family	I	II	III	IV	V	VI
Receptors related to rhodopsin/ β-adrenergic receptor	Olfactory, adenosine, melanocortin, cannabinoid, several orphan receptors	Serotonin, α- and β-adrenergic receptors, dopamine, histamine, muscarinic, octapamine, and orphan receptors	Bombesin/ neuromedin, cholecystokinin, endothelin, growth hormone secretagogue, neuropeptide Y, neurotensin, opsins, tachykinin, thyrotropin-releasing hormone, and orphan receptors	Bradykinin invertebrate opsins, and orphan receptors	Angiotensin, C3a and C5a, chemokine, conopressin, eicosanoid, fMLP, FSH LH and TSH, galanin, GnRH, leukotriene, P2 (nucleotide), opiod, oxytocin, platelet activating factor, thrombin and protease activated, somatostatin, vasopressin, vasotocin, and orphan receptors	Melatonin and orphan receptors
Receptors related to the calcitonin and parathyroid hormone receptors	Calcitonin, calcitonin gene-related peptide, corticotropin-releasing factor, insect diuretic hormone, and orphan receptors	Parathyroid hormone/ parathyroid-related peptide receptors, orphan receptors	Glucagon-like peptide, glucagons, gastric inhibitory, peptide growth hormone releasing hormone, pituitary adenylyl cyclase activating peptide, vasoactive intestinal peptide, secretin, and orphan receptors	Latrotoxin and orphan receptors		

TABLE 7-4.
Continued.

Family	Group					
	I	II	III	IV	V	VI
Receptors related to the metabotropic glutamate receptors	Metabotropic glutamate receptors	Extracellular calcium ion sensor receptor	GABA-B receptor	Putative pheromone receptors		
Receptors related to the STE2 pheromone receptor	STE2-α-factor pheromone receptor					
Receptors related to the STE3 pheromone receptor	STE3-α-pheromone receptor					
Receptors related to the cAMP receptor	*Dictyostelium discoideum* CAR2 to CAR4 receptors					

From: (Conn and Means 2000). These groupings are provisional because the phylogeny of the *FTMR* genes is still tentative.

TABLE 7-5.
The special case of the Serpentine receptors.

A particularly interesting and important class of cell-surface signal receptors are those known as the seven-transmembrane G protein-coupled receptors (7TM-GPCRs, or 7TMRs). There are hundreds of genes in this class, involved in diverse aspects of information transfer to cells. As shown in Scheme 7-2, these receptor molecules contain seven helical transmembrane domains, with additional extra- and intracellular domains.

Transmembrane information transfer that relies on secondary messengers often involves G protein cascades. G protein–coupled cell surface receptors (GPCRs) are a large and ancient class of seven transmembrane receptors, cellular signaling mechanisms found in many different kinds of receptor-ligand interactions in a wide range of cell types and organisms, including single-celled yeast (the receptors for yeast mating factors are an example) (Alberts 1994). In vertebrates, close to 2,000 GPCRs have been identified, including more than 1,000 for odorant and pheromone detection alone. About 1,100 GPCRs are known in *C. elegans* (Bockaert and Pin 1999).

When GPCRs are compared at the amino acid level, by phylogenetic relationship and the type of ligand to which they can bind, they cluster in at least six main families; GPCRs from the different families show little sequence homology at the DNA level (GPCR 2003; Bockaert and Pin 1999). These tentative groups are itemized in Table 7-4. Class I GPCRs constitute the largest and includes many small peptide or weakly hydrophobic organic molecules. This class is basically for chemoreception and chemotaxis and includes olfactory and taste receptors, for which the ligands are weakly hydrophobic, organic molecules, or small peptides. Class II GPCRs antedate the protostome-deuterostome divergence, and this class comprises receptors for biogenic amines (acetylcholine, catecholamine, and indoleamine) largely related to neurological function. Class III genes include small neuropeptide receptors and receptors for chromophores (vertebrate visual pigments); these are the opsins used in light sensing. Class IV genes are insect and molluscan opsins, which bind to small odorant molecules of various types. Class V genes are receptors for small hormone molecules of various kinds that are involved in immune chemotaxis and blood-clotting; these genes are also involved in the secretion of steroid hormones. In vertebrates, some similar neurological and endocrine peptides are mediated by single-transmembrane receptors. Class VI GPCRs are members of the *smoothened/frizzled* family of vomeronasal receptors.

Additional GPCRs continue to be found and classified. No amino acids are conserved among all the classes of these genes, and few are conserved even among many. Within a given class, a relatively small number of sites determine the receptors' specificity to ligands.

GPCRs set into motion a chain of events that alter the concentration of intracellular signaling molecules, which in turn affect the behavior of other cellular components. Most G protein cascades involve alterations in the cellular concentrations of cAMP, cGMP, or Ca^{2+}, which induces activation of an effector molecule that produces a response specific to the receptor and cell type. This system is common in but not restricted to sensory perception; regulation of heart rate, contraction of smooth muscle, and secretion of hormones or growth factors are examples of other systems that use this mechanism. The effects can be achieved in various ways, including the regulation of membrane ion transport channels, transmission of signal pulses to adjacent cells as in the transmission of nerve impulses from retinal neurons, or the regulation of transcription factors. Similar mechanisms are fundamental or even ubiquitously used for the transfer of cell-differentiating signals in both plant and animal development (Chapter 8).

TABLE 7-5.
Continued.

Extracellular workings: A G protein cascade is activated when a ligand binds to a receptor (see Scheme 7-2). As 7-spanning transmembrane proteins, the free NH_2 terminus of the protein remains outside of the cell, and the COOH terminus floats in the cytoplasm of the cell. The receptor-ligand binding causes a conformational change (phosphorylation) in the three-dimensional structure of the receptor, and this results in the transfer of signal from the receptor to a G protein, which is in the plasma membrane of the cell.

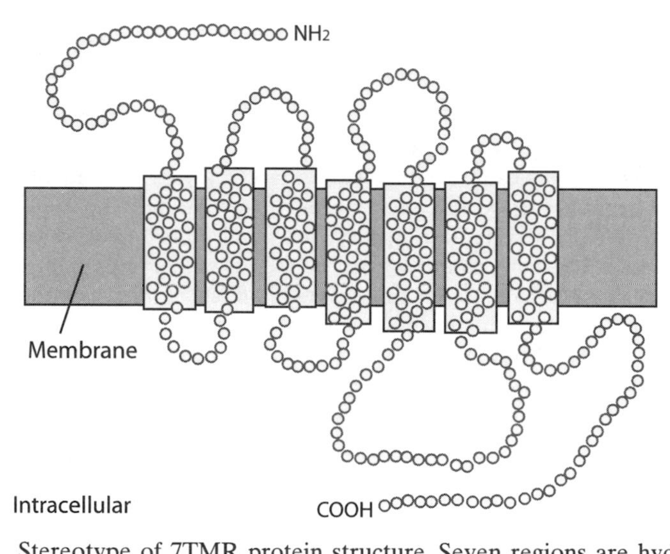

Scheme 7-2. Stereotype of 7TMR protein structure. Seven regions are hydrophobic and insert stably in the cell membrane. The intracellular domain is involved in signal transduction. Signal may be received by the extracellular domain(s) or in some instances occurs in the intracellular or membane domains. Specific amino acid variants in the appropriate domains affect the nature of the ligand (or light energy frequency) that triggers a response, and hence is how divergence and specialization in function is brought about.

Intracellular workings: G proteins are *heterotrimers*, proteins with three subunits: α, β, and γ. The structure of the α-subunit usually defines the type of G protein. The α-subunit binds a guanine nucleotide (thus giving this class of proteins their name). When the guanine is in the form of GDP (guanine nucleotide *di*-phosphate, rather than *tri*-), the α-subunit binds with the βγ-subunits. When the nucleotide becomes a GTP (guanine nucleotide *tri*-phosphate), which is induced by the receptor-G protein interaction, the G protein is activated, and the α-subunit is freed from the complex (Lewin 2000).

What the α-subunit does next depends on the class of G protein-coupled receptor; despite the large number of GPCRs, there are only a few basic ways in which they affect signal transduction. Some GPCRs regulate the activity of enzymes, and some affect the concentration of cyclic nucleotides or calcium inside the cell, with the effect of activating or inactivating ion channels. Others directly activate or inactivate ion channels, but the effect of G protein activity involved in sensory transduction is to alter the ion permeability, and thus the excitability, of the membrane of the target cell.

TABLE 7-5.
Continued.

Of the GPCRs that modulate cyclic nucleotides, some regulate *adenylate cyclase* activity. Activation of adenylyl cyclase leads to the production of cAMP as a second messenger, and cAMP in turn leads to the activation of ion channels. Odorant receptors are an example of this type of GPCR. Photoreceptors are a class of GPCR coupled to the G protein *transducin*, which activates phosphodiesterase, an enzyme that causes a drop in the level of cGMP, which leads to the closing of sodium and calcium channels and *hyperpolarization* (an increase in the negative charge) of the cell.

These are general, stereotypical descriptions. Not all aspects of all of these many genes have been carefully checked. Some things are reasonably well-studied, however. The ligand binding characteristics differ among genes and among classes and will be discussed in several places in this book. The point is that receipt of an odorant by an olfactory receptor GPCR is "wired" to the olfactory part of the brain, whereas the receipt of a photon by a chromophore GPCR ligand is wired to the visual system. We have given this brief summary of this class of genes because its members are used in so many diverse ways relevant to this book and because it is central to communication between a cell and its outside world. But it is the logic and modular nature of the various uses of these common signal-related molecules that is of importance here. For biochemical details, seek appropriate sources.

Regulation of Gene Expression

Signaling between cells may affect a variety of functions. Usually this means changes in gene expression in the receiving cell, which implies that the signal causes the activation or inactivation of one or more TFs. There are several TF gene families (Table 7-6), characterized by their DNA binding domains of their coded protein. The most famous of these is the homeobox after which a major TF class is named. The homeobox is a region coding for a 60-amino acid homeodomain (so called because of the effect discovered in the 1980s when mutations in these genes produced famous replicated or altered segmental structures in flies). Each of the many subclasses of homeobox genes shares one or more variants in the homeodomain, and they bind to somewhat different response or regulatory element (RE) sequences; however, as a rule, this region of the class of genes is much less variable than other active domains, and there is a lot of sharing of the binding sequence (e.g., **TAAT**) among homeobox genes; the other domains are more varied and have different functions, not all of which are known (Alberts 1994; Transfac 2003).

CHAPTERS 4–5–6–7: COMPLEX TRAITS REVISITED

Signaling and gene regulation are so fundamental to complex organisms that we can expect there to be an elaborate set of mechanisms, and there are. We can now look at them in light of concepts covered in this and previous chapters. The importance of the modular organization of the genome and the various information encoded in DNA sequence is clear.

The major classes of genes for signaling appear to have evolved from multiple essentially independent beginnings. However, the *logic* of signaling mechanisms

TABLE 7-6.
Characteristics of the major transcription factor classes.

Transcription factors (TFs) share the generic property of binding to specific DNA sequences (which are often palindromic so the binding of both strands can occur in anti-parallel ways). But each family of TFs varies in detail in how its DNA binding is accomplished. Scheme 7-3 shows some of the major mechanisms (this section is not exhaustive but again is intended to suggest the modular nature and logic of the process and the diversity of agents that bring it about).

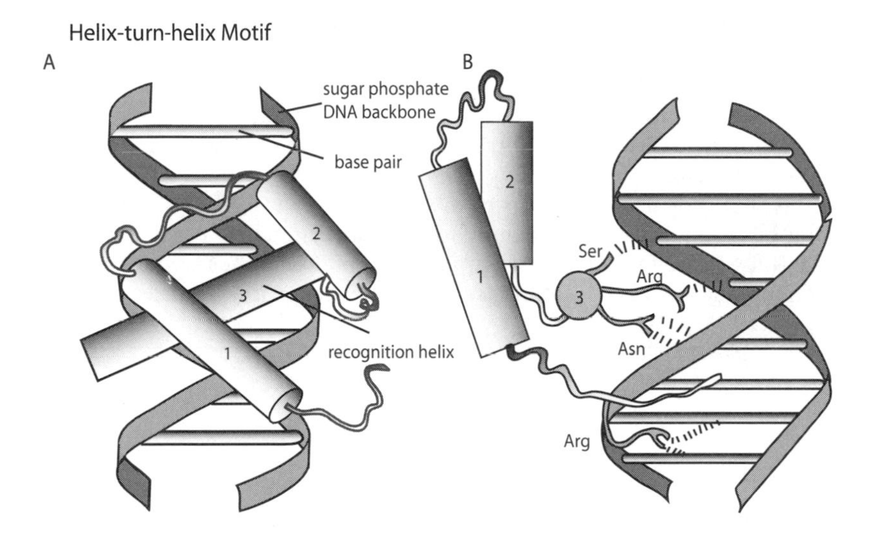

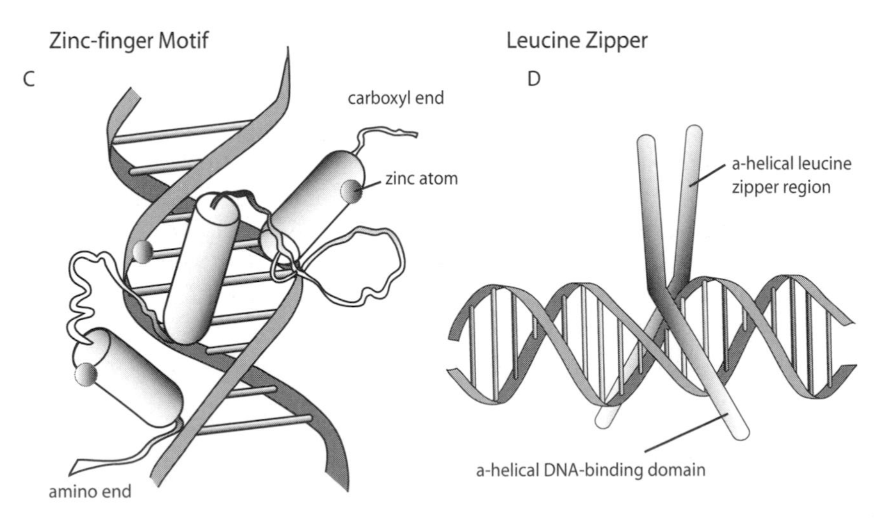

Scheme 7-3. DNA binding motifs. Helix-turn-helix motif; (A) homeodomain bound to its specific DNA sequence; (B) specific amino acids of helix 3 (recognition helix) make contact with the major groove of the target DNA at specific sites; (C) zinc-finger motif: this motif uses zinc fingers of specific amino acids to recognize DNA binding sites; (D) leucine zipper motif; two α-helical DNA-binding domains join to form a leucine zipper dimer that binds to a specific DNA sequence.

TABLE 7-6.
Continued.

Homeodomain TFs (coded for by homeobox genes) have a basic 60 or 61-amino acid helix-turn-helix (bHTH) protein domain that, depending on its amino acid sequence, binds specific DNA sequences; the third helix extends into the major groove of the DNA that it "recognizes." Amino acids in the NH_2-terminal portion of the homeodomain also contact the bases in the minor groove of the DNA double helix. A class called *POU* TFs also has a second DNA-binding domain. *Lim* and *Pax* genes comprise two other homeodomain gene classes. The *Hox* class is the most well-known; they are involved in neural differentiation, anatomic axis specification (e.g., main anterior-posterior, or proximal-distal appendage axes), and head and eye development.

Zinc finger TFs form domains that include loops made by the binding of spaced pairs of cysteine and histidine around a zinc molecule, leaving a projection ("finger"). Between such fingers is a sequence-specific DNA-binding region. *Zf* genes include the *Sox* family that bend DNA to assist in the assembly of regulatory protein complexes, sex differentiation, cartilage development, and hormone and retinoic acid signal reception. They are part of a large number of chromatin-associated genes with a High Mobility Group (HMG) box, a DNA-binding domain that recognizes a consensus sequence **(A/T)AACAAT**. There are many *Sox* genes, and several classes of *HMG* genes that are not closely related to this family, which may or may not include a zf domain.

Leucine zipper TFs are formed of two polypeptides joined by regularly spaced leucine residues (the "zipper") adjoining a DNA-binding region of basic (positively charged) amino acids.

Basic helix-loop-helix (*bHLH*) TFs form dimers in each of two helical domains, that are separated by a loop; DNA binding is by a region of 10–13 basic amino acids at the end of the first helix in each of the polypeptides. The two molecules may be of the same or from a related bHLH gene. bHLH genes are important in a variety of neural and sensory developmental functions, and include the *MyoD* genes used in muscle development.

MADS box genes regulate many functions but are particularly prominent in flower morphogenesis (the acronym is the initial letter of the first four family members identified). Along with a variety of other functional domains, these genes have a characteristic 57 amino acid DNA binding motif, and a domain for forming homo- or heterodimers.

Other classes of TFs, such as the "paired" group, have one or two DNA binding domains that are variants of these classes. The TF families are old and have accumulated family members through duplication events in long phylogenetic lineages; there are therefore numerous members overall, that vary among species. For example, there are tens of genes in the *Hox* homeobox superfamily (39 in humans). Typically, representative homologs in the animal TF families are found across the animal world.

(e.g., receptor-ligand binding leading to TF activation and *cis*-regulation) is generic and development and its evolution are now frequently viewed in terms of regulatory pathways (Carroll et al. 2001; Davidson 2001; Wilkins 2002). A regulatory *pathway* or *circuit* goes from inducing signal to TF activation and hence an activated (or repressed) developmentally *downstream* target gene. Development is hierarchical in that one circuit can activate subsequent circuits—for example, when the first

TF activates a second TF that then activates subsequent genes. A set of genes regulated by the same circuit has been called a *battery* of genes. Interactions among contemporary circuits or batteries comprise a regulatory *network*. It might be worth noting that *meshwork* could be even more evocative, because two-dimensional networks of regulation are connected to each other through time, with some circuits active and others missing at different times, or with a pathway having different uses at different times, and effects feeding back upon themselves, depending (presumably) on what else is active at each time.

An indicator of the basic importance of this kind of genetic logic is that SFs, their receptors, and TF classes are widely represented in vertebrates and invertebrates, sometimes even plants and bacteria. It is possible in many instances to identify specific orthologs among gene family members in very distantly related groups, such as flies and mammals. Homologous interaction pathways are often deeply conserved, such as between a given ligand, its receptor, and/or the TFs it activates. In addition, such conservation may effect similar function (such as activation in neural or gut tissue), revealing surprisingly deep homologies in basic animal or plant functions.

Members of TF families may be linked in chromosomal clusters that have been conserved for long evolutionary time periods, and the conserved linkage arrangement is related to the control of the expression of the genes in the cluster. However, this is not an automatic need for TF regulation, because multiple TFs, including members of the same families as those in clusters, are chromosomally off by themselves.

Regulatory Control Occurs in Different Ways

Some TFs are autoregulatory; that is, once expressed they enhance their own continued transcription—copies of their own RE sequences are located 5′ to the TF genes themselves. Just as apoptosis and proteases remove cells or substances in a variety of remodeling or repair contexts, gene expression can also be down-regulated or inhibited.

Inhibition can be achieved in various ways. One way is pretranscriptional, when other regulatory repressor proteins bind areas around a gene to prevent activation by TFs. Chemical modification of the regulatory region has the same effect, preventing access or binding by TFs. The DNA cannot be opened (e.g., histones removed) to permit access by an activating TF. Earlier, we described posttranscriptional regulation by RNAi. There is also posttranslational inhibition. For example, bHTH TFs (Table 7-6) bind REs as heterodimers. The TF protein has separate dimerization and DNA binding domains. There are genes that code for truncated proteins that lack the DNA-binding domain; they form normal dimers, but the result cannot bind DNA properly, which prevents or inhibits the expression of the gene a normal dimer would activate. Similarly, there are receptors that lack one of their domains, and hence reside in the membrane where they receive signal, but the receipt is not transmitted internally to the cell (e.g., Kroiher et al. 2001).

Expression of a particular gene or cell-specific set of genes depends on the required set of TFs and/or inhibiting factors being present at the same time in the cell, meaning that the appropriate set of receptors, second messengers, and the like must also be expressed in the cell, along with the TFs themselves. Cells may require more than one signal for a given action to take place and thus all the requisite receptors may need to be triggered; this means that the extracellular as well as the intracellular environments must be appropriately prepared. Thus, itemizing a pathway by itself is somewhat of an illusion of simplicity.

An Embarrassment of Riches, or Neo-victorian Beetle Collections?

A description of the basic logic of signaling and gene-regulation mechanisms, along with a catalog of their major classes, does not do justice to what actually is going on within a cell. The cascades of regulation that bring about even relatively simple aspects of complex traits are themselves complex (e.g., Carroll, Grenier et al. 2001; Davidson 2001) and often involve parallel shunts or redundancy. Regulatory and signaling pathways can be investigated by a host of experimental in vivo and in vitro methods, including cell culture, animal models, and the direct manipulation of individual genes or regulatory sequences. Experiments dissect pathways component by component in a way to reveal their individual effects (*under the particular laboratory conditions being used*). As shown in Figure 7-3, regulatory pathways can be intricate, and also, the apparent complexity of a pathway can depend on the method used to study it. For example, in part A, the pathway seen from normal development of relatively simple model organisms can seem quite straightforward. In part B of the figure, the results of different experiments, in different cell types under different conditions, in more complex organisms, can make the "same" pathway appear more complex, because the latter approach reveals more of the phenogenetic equivalence (redundancy, interconnectedness) of "the" pathway.

Experiments with individual known pathways can be supplemented with genomic approaches that identify which of the entire set of an organism's genes are used in a given cell type. With the use of various technologies, it is becoming possible to identify and quantify all the genes that are being expressed in a given cell type. This *expression profiling* can identify, for example, genes whose expression changes quantitatively or qualitatively before and after cells are subjected to nutritional stress or those that are expressed differently between tumor and normal cells from the same tissue (e.g., Eisen et al. 1998; Tamayo et al. 1999). Expression profiles are useful when the situation is relatively simple or well-controlled experimentally, but they can also reveal the quantitative and complex nature of cell behavior. Redundancies and other interactions may be difficult to discern from an expression list (even with quantitative data) alone.

Expression profiles can reveal whether genes with coordinated expression under the tested circumstances are (1) in coregulated chromosomal clusters, (2) coregulated members of gene families, or (3) scattered genes coregulated because they have related function (e.g., networks of TFs and downstream expression cascades). Co-regulation within clusters can be achieved in various ways, and genes on different chromosomes that are coordinately expressed can use the same or different regulatory mechanisms. That is, this can be due to similar REs, or different REs whose coordinated usage evolved; in some instances, phenogenetic drift is probably responsible, if different genes have had a similar common expression but have evolved different regulatory mechanisms. The alpha- and beta-globin genes, and different means of coordinate regulation of *Hox* genes in vertebrates probably represent instances of the former. But the function could have been achieved by recruiting different, otherwise unrelated genes to some new function.

In some ways, we have in recent years been presented with a sudden embarrassment of riches. There is a danger in taking the details too seriously, mistaking

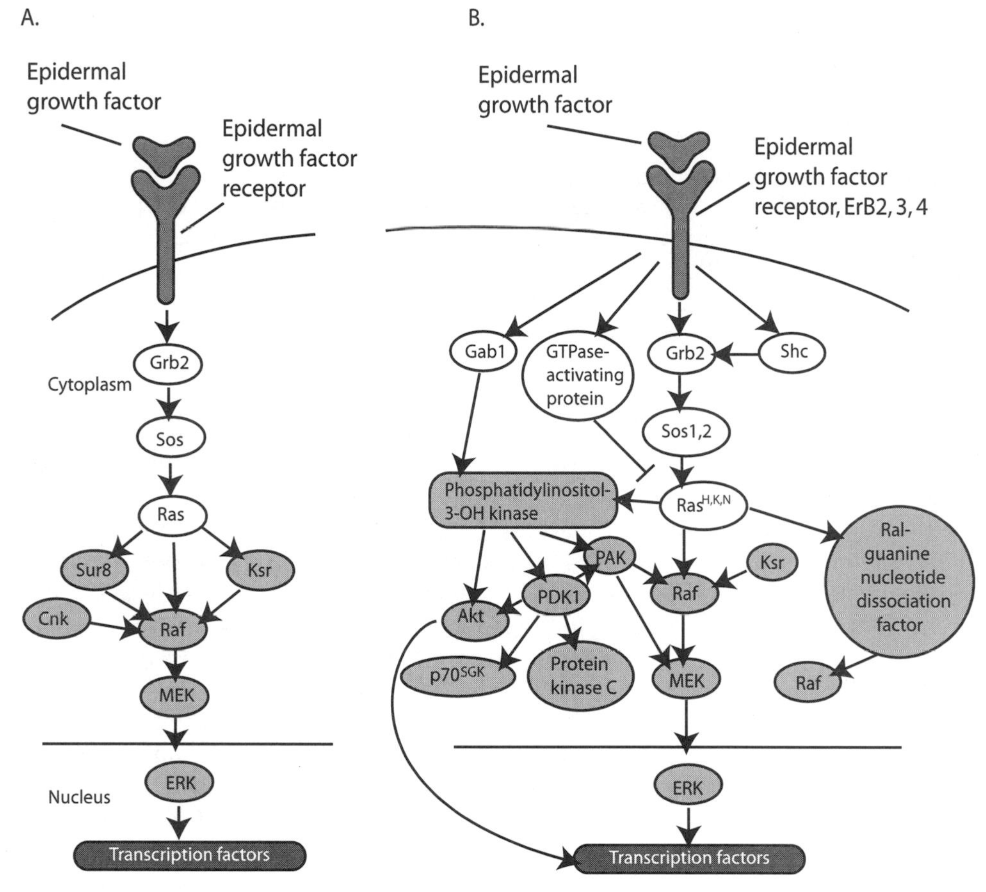

Figure 7-3. Two views of the *Ras* regulatory pathway that responds to growth-factor signal. (A) Genetic view; (B) view from molecular cell biology. Ovals indicate individual factors whose specific names are unimportant here. Grayed ovals indicate factors classified as "effectors;" white ovals as "intracellular mediators" of the signal. Redrawn from (Downward 2001) with permission; see original for details and terms, text for explanation. Original figure copyright Nature Publishing Group 2001.

enumeration for understanding, underestimating phenogenetic equivalence and its importance or prevalence, and viewing genotype-phenotype relationships as being more prescriptive and unitary than they are in nature. In this sense, this new work resembles the stage of biology in the 19th century, when naturalists traveled around the world collecting the beetle specimens that were then dispersed throughout the museums of Europe. This led to an accumulation of species lists that we tend today to view as merely descriptive natural history. Until we have a better ability to synthesize and digest the information from gene expression studies, the difference between then and now may be less than we think.

One problem is that manipulation of gene expression in the laboratory is almost always artificial and can even make systems seem misleadingly complex (or blind investigators to alternative pathways) (Downward 2001; Niehrs and Meinhardt 2002). This is illustrated by comparing Figures 7-3A and 7-3B, and that does not even consider variation within species. Sequence variation in all the elements, from TF protein sequences and regulation to the REs of regulated genes and their surrounding sequence and chromatin structure, can lead to quantitative variation in traits, while experimental studies in development generally focus on the invariant aspects of mechanisms (at least, as seen in model animal strains). Experiments cannot in a practical sense test all combinations of variants, and strain-by-strain variation has raised caution flags that should not be ignored.

Most pathway diagrams seen in the literature are partial, even when they appear as 7-3B, in part because they do not include all the other pathways required to activate the components that are shown. More thorough diagrams can be intimidating in complexity (Davidson et al. 2002; Davidson et al. 2002) even if the pathways shown are an exhaustive list and even if they are all correct and always operative. Individual studies report experiments on pieces of networks, often those known by experience with natural or randomly engineered mutation, but then specifically followed up experimentally. This can sometimes give the impression that the tested genes are the only genes of importance, but this can be highly misleading (expression profiling shows this as well).

Reports of only subsets of genes are understandable for practical reason, but often they consist of a mixed bag of genes, including receptors, TFs, some second messengers and the like. One can view these as akin to throwing darts randomly at a complex diagram, and then studying the result as if those were the only important genes. These facts should be kept in mind in reading this book. Many examples will follow in which we present these kinds of partial sets of genes, and in discussion of other phenomena we simply omit pathway details that do not apply to the logic of the system we are discussing.

Considering the complex nature of such cascades one can see the many points where mutation can introduce variation. Nonetheless, despite all these caveats, there is considerable and tractable order. Some broad generalizations have already emerged, and we can consider some of them and what they mean.

A Basic Functional Toolkit

The logic of *cis*-regulation, messenger-transduction, modular genome organization, and combinatorial expression provides a powerful new way to understand how new function arises and complex organisms are assembled: a small number of components can generate a huge number of combinations, and a new combination can in principle add novelty without destroying existing traits in an organism, for example, by only affecting specific enhancer locations, not a TF itself.

Pleiotropy Aplenty

The same TFs, SFs, receptors, and REs are typically used in numerous developmental or histological contexts in an organism or even at different times and different ways during the ontogeny of a single tissue or organ. This in a sense is what is implied by the term "toolkit", but there is an important implication: if expression control is combinatorial and each gene has many uses, we can't expect to understand the development of a given structure until we know most or all of the regulatory factors involved.

Pleiotropy can appear to involve changing sets of functions, and phenogenetic drift can bring that about by substituting a new regulatory mechanism for the original one. A well-studied example is the Stripe 2 enhancer region responsible for expressing the developmental gene *Eve* in the second of the many stripes in the very early fruit fly embryo. *Eve* activity is similar between closely related species of fly, but the REs and TFs involved bear only limited homology (Carroll, Grenier et al. 2001; Ludwig et al. 2000; Ludwig and Kreitman 1995; Ludwig et al. 1998). However, the multiple uses of a gene or pathway in organisms today can be an indicator of phylogenetic relationships among those uses: they may have diverged from some common ancestral function.

Development is Functionally "Arbitrary"

Just as the codon system for specifying amino acids is completely arbitrary relative to the chemical properties of the proteins being coded for, regulatory DNA sequences are arbitrary relative to the function they enable (Weiss 2002d; 2002e). There is nothing about regulatory genes or RE sequences that has anything directly to do with the function of a regulated protein. There is nothing about the chemical nature of SFs or their receptors that is related to the physical or functional properties of the limbs, guts, feathers, teeth, collagen, or chitin, whose production they induce. Regulatory sequence information *is* highly specific because ligands only fit particular receptors or TFs only recognize particular REs, but the same signaling machinery can operate with any downstream function. As shown in Figure 7-1, a large fraction of the genome is indeed of this logically necessary but functionally arbitrary type. This again is implied by, or a consequence of, the modular, repetitive, combinatorial use of the same elements for sequestered differentiation that is deeply built into the nature of life.

Hierarchies of Regulatory Circuits, Batteries, and Networks

Some TFs are known as "selectors", meaning that they activate a hierarchical cascade of change in the expression of other, developmentally "downstream" genes. This is comparable to the notion of an embryological "organizer" that we will see in later chapters. Another term for the idea is "master control" genes, but it can be misleading to think of genes so metaphorically in terms of our own culture's social structure; rarely if ever is one gene the master of a whole organ in a very meaningful sense. Selectors cause a major differentiation commitment or branch point. They activate a series of other TFs that in turn activate still more TFs that ultimately alter the expression of structural genes, enzymes, and the like. *Hox*, *Pax*, and *MyoD* genes are examples. Different selectors work at different levels or stages of embryological development (Carroll, Grenier et al. 2001). But they don't act alone, and, typically, other factors (perhaps controlling their own downstream effects) are also involved.

We will elaborate further on this in Chapter 9, but here we can indicate the uses of linkage relationships among selector genes. The members of the *Hox* class of homeobox TF genes have become an archetype of high-level selectors in which linkage arrangements are important. The *Hox* genes are linked in chromosomal clusters (shown in Figure 4-4), whose individual genes are expressed in a colinear, sequential, cumulative fashion in cells from anterior to posterior regions of the early embryo, respectively. That is, the gene at the 3′ end of a *Hox* cluster is expressed earlier in development, in cells in the most anterior parts of the embryo. A bit later, as the next most posterior region of the embryo begins to develop, the next most 5′ *Hox* gene in the cluster also becomes expressed (that is, along with the *Hox* gene that is already "on"). This continues down the line of genes in the cluster to the most 5′ gene, when the most posterior anatomic segments develop, which is when most or all of the genes are being expressed. Because the latest gene to be activated as more posterior segments differentiate has its effect on those segments, even though earlier (more "anterior") genes may still be expressed, the effect is known as "posterior prevalence."

Figure 7-4A repeats part of Figure 4-4 to show the gene clusters in *Drosophila* and in mammals, and is another oft-used figure because of its great importance in the recent history of developmental biology and genetics. The colinear summed expression pattern sets up combinatorial sets of *Hox* transcripts that effect gene expression cascades that differ locally along an anatomic axis. The figure shows the similarity in this aspect of the expression pattern that is shared between vertebrates and invertebrates, first documented in a famous paper by Lewis in flies (Lewis 1978) and later shown in vertebrates.

Essentially, the same summed colinear expression system is used to specify the primary AP axis in species across much of the animal world, although the details of what the various combinations achieve vary. However, the *Hox* cluster genes in vertebrates are also used in the development of many other structures, including the segmentation of the gut and limbs. Thus, a *Hox* gene is not a "vertebra" gene, but can be viewed as a selector gene, or in some contexts a patterning selector gene in several structures. The use is functionally arbitrary as noted above and not inherent in the genes themselves, and can only work because of the contingent, hierarchical, sequestered nature of development (e.g., limb buds are not the same kinds of cells as are early vertebral precursors).

Not all pathways are as deeply conserved as the *Hox* system, nor rely on conserved linkage relationships. Even this classic system is subject to all the sources of variation that one might expect; in fact, the more intensely it is studied, the less completely conserved it seems to be (e.g., Levine et al. 2002), including the variety of ways the genes in the cluster may be regulated to coordinate their spatial and temporal domains of expression (Kmita and Duboule 2003). Nonetheless, much is conserved, and similar situations apply to a number of other differentiation hierarchies, such as the role of *Pax6* in eye development (Chapter 14), or *MyoD* in muscle development (Gerhart and Kirschner 1997). Figure 7-4B shows the general logic of a selector gene, and the way that REs are used to start and proliferate a cascade of downstream effects on gene expression.

A regulatory gene that acts early in helping trigger differentiation cascades may be widely expressed in an organism; some TFs seem to help regulate a high fraction of all the genes in the genome, as seems to be the case with the *Ftz* and *Eve* genes in *Drosophila* (Liang and Biggin 1998). However, the spatiotemporal distribution of most regulatory genes is restricted at least somewhat, and like a Venn diagram of the embryo, the set of regulatory signals activated in a given cell is assumed to be correlated with the functional genes activated in that cell. It is assumed that there is a unique combination for each function, or at least for each context in an organism; although this is far from proven, it is central to the present view of differentiation by sequestered combinations of *cis*-regulating factors.

A critical aspect of this that enables so much combinatorial pleiotropy to work is that gene activation states can be mitotically heritable. This is a most important sequestration factor. Once a differentiation cascade has been initiated, a cell and its descendants in the organism can become strikingly autonomous, either self-

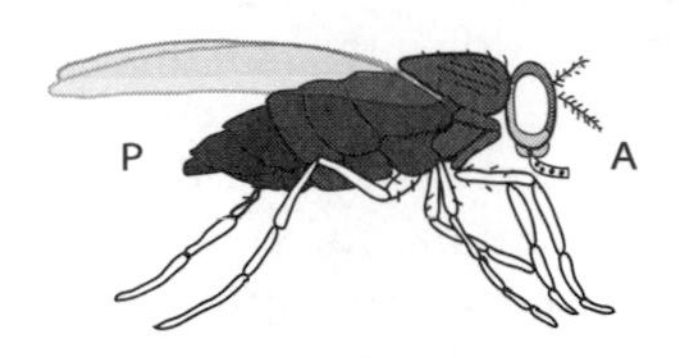
P
A

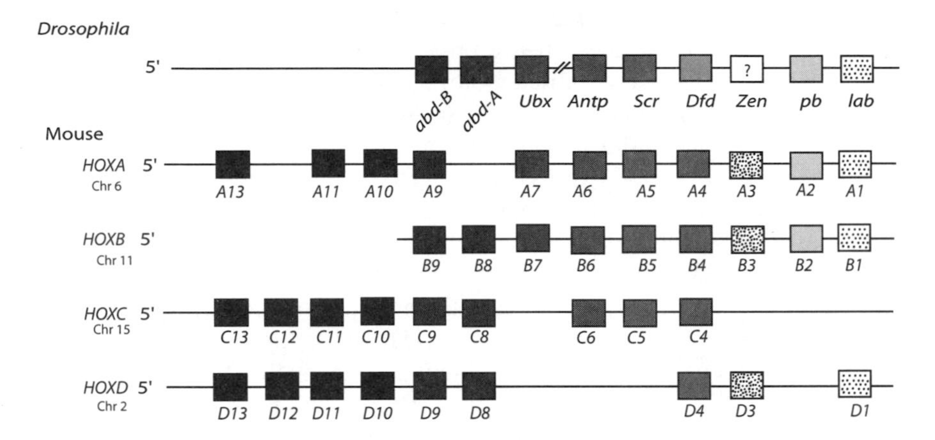
Drosophila
5'
?
abd-B
abd-A
Ubx
Antp
Scr
Dfd
Zen
pb
lab
Mouse
HOXA 5'
Chr 6
A13
A11
A10
A9
A7
A6
A5
A4
A3
A2
A1
HOXB 5'
Chr 11
B9
B8
B7
B6
B5
B4
B3
B2
B1
HOXC 5'
Chr 15
C13
C12
C11
C10
C9
C8
C6
C5
C4
HOXD 5'
Chr 2
D13
D12
D11
D10
D9
D8
D4
D3
D1

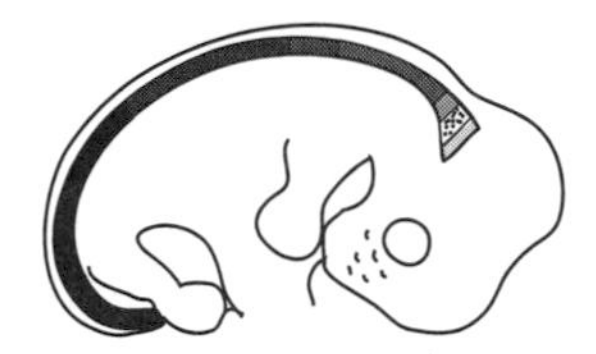

A

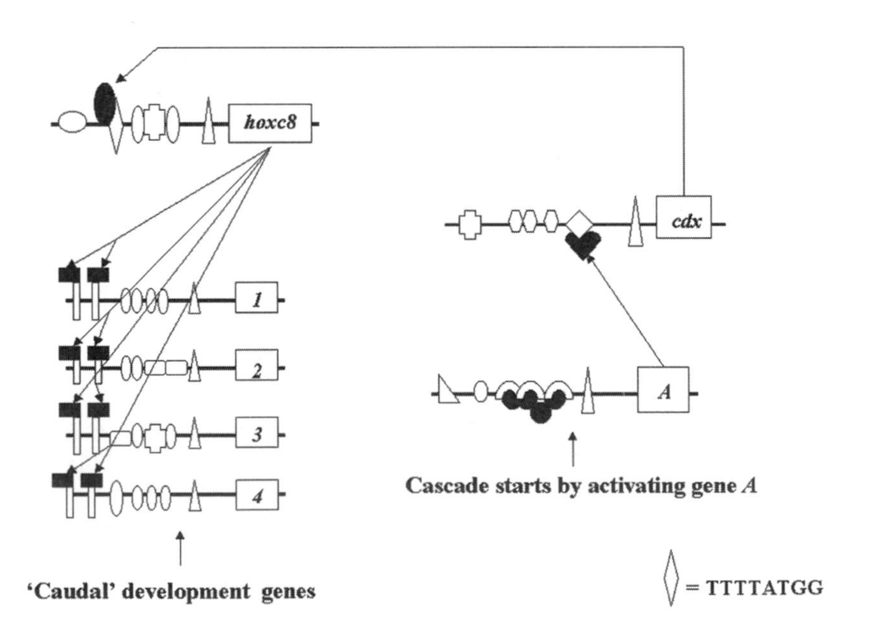
hoxc8
cdx
1
2
3
4
A
Cascade starts by activating gene A
'Caudal' development genes
= TTTTATGG

B

sustaining or changing in a contingent way. The state of differentiation is not washed out by being blended with that of neighboring cells. One cell or few cells can be the progenitor of a whole complex structure even if isolated from the rest of the organism (e.g., grown in vitro). As cancer shows, and may be important to many normal traits as well, a mutant somatic cell can generate a clone of cells with altered function. Cellular commitments are achieved in various ways; for example, the differentiation state can affect chromatin structure that leaves some genes open for TF binding, whereas others are closed to regulatory machinery (Davidson 2001; Davidson, Rast et al. 2002; Davidson, Rast et al. 2002; Lieb et al. 2001).

In the chicken-and-egg discussion raised in Chapter 4 about RNA and DNA, the regulation of an organism is in this view necessarily hierarchical, all kick-started by parental mRNA and proteins produced in the zygote. The increasingly committed nature of cells that occurs during development is the result. The new zygote is but another form of single-cell precursor to an autonomously developing tissue cascade. Of course, this is hierarchical with respect to the zygote but not the process itself. The process is circular, since it was the same genome that made the maternal mRNA and will eventually make that of the germ cells of the next generation.

Because sequestration is necessarily incomplete, cells are able to continue differentiating in contingent, context-dependent ways, but in some cases can re- or dedifferentiate. The degree of reversibility that occurs in nature varies. In a mature animal embryo, most cells are unable to differentiate into more than a limited set of tissue or cell types; artificial cloning from somatic (body, as opposed to germ) cells in mammals, for example, requires special treatment of the cells to dedifferentiate them. However, the nature of commitment depends on species and even context, and not all cells are fully committed to some end state. Reptiles can regenerate entire tails or limbs, and humans can regenerate blood vessels and skin, from cells that are partially differentiated even in adults. Cells in a developing brain may in some instances be entrained to behave like cells amongst which they intercalate (see Chapter 15). Plants are rather different in this respect, with much more flexibility built into the basic way they do business (for example, in some, the tip of any branch able to form flowers).

Implications of Pleiotropy in Complex Regulatory Circuits

If the pleiotropic use of a limited set of regulatory genes makes complex evolution easier to explain, it entails the potentially serious problem that mutation in a TF could disrupt *all* of its different functions simultaneously, which could be a disaster

Figure 7-4. Aspects of regulation by and of genes. (A) *Hox* combinatorial expression figure; similar anterior-posterior colinear expression affects both invertebrates and vertebrates; gray shade is used to indicate the gene class and the most anterior location of its expression along the axis where the gene has its predominant effect, as shown by corresponding shading in the two types of animal; for example, darker-shaded genes are expressed posteriorly and affect caudal regional development; (B) schematic of regulatory circuits, hierarchies, and batteries, illustrated by a schematic of the use of the TF *Cdx* to regulate *Hoxc8* that in turn is a selector for many other genes, in patterning early caudal vertebrate development; the diamond symbol represents an enhancer sequence that flanks *Hoxc8* (e.g., see Shashikant et al. 1998).

for the organism (e.g., Losick and Sonenshein 2001; Struhl 2001). However, because transcription factors work via regulatory elements, there is a ready escape: mutations that add, delete, or modify a particular RE will only affect the expression of its nearby gene and hence only that particular use of the TF that recognizes the RE.

This provides room for tremendous evolutionary flexibility, but there is also a potential flip side. If the same TF is used in many contexts, an organism is vulnerable to a high degree of multiple jeopardy. A given TF may use hundreds of REs scattered across the genome in its different functional contexts. Relying on one sequence in too many ways raises the likelihood that one or more of these REs will be mutated, and that could be harmful. There are at least two ways to protect a gene against this kind of mutational regulatory "noise" (Sengupta et al. 2002). One is to flank it with multiple copies of the relevant RE; Figure 4-7 provided an example of enhancers upstream of lens protein genes (this shows between-species variation, but the principle is the same). As was shown in Table 4-5, another mutational buffer is for a TF to evolve tolerance for variation in its recognition sequence. Not all the sequences are equally efficient at binding the TF, but they are at least recognized.

However, this same robustness that buffers a gene's enhancers from being erased by mutation makes it more difficult for evolution to remove that gene from the TF's set of regulated genes—it could take many mutations to delete all the relevant REs for a given gene to completely destroy the binding site (as opposed just to making it less efficient and, for example, lowering the expression level of the gene). This could tie functions together over evolutionary time once the use of REs for a common TF is established, yielding suites of traits that evolve together, at least for a period until mutational variation accumulates. During that time, traits might appear to share adaptive constraint, and could possibly be wrongly interpreted as of apparent canalization of traits, of selectively constrained variation, evolutionary stability, or range of response to environmental changes (e.g., Siegal and Bergman 2002). This is speculative, but a balance between robustness to mutation and stabil-

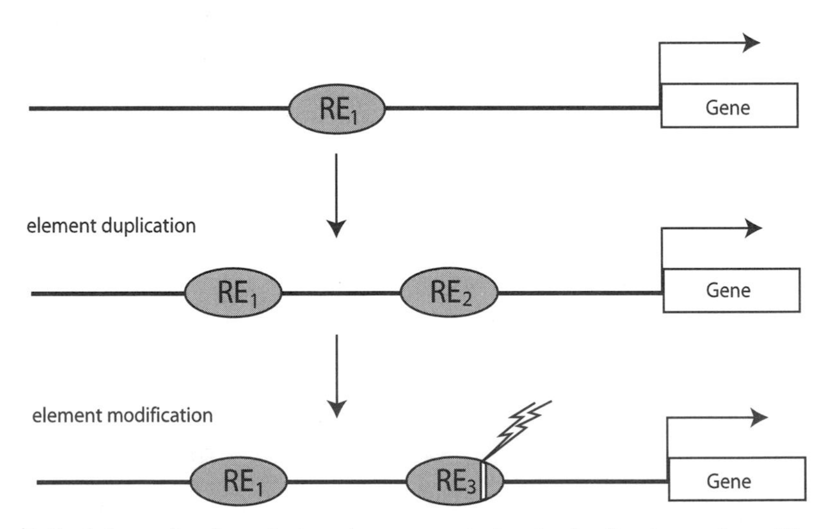

Figure 7-5. Schematic of regulatory element evolution by duplication and modification by mutation (lightning). Short elements can also be created or destroyed by mutation in otherwise random sequence. Bent arrow indicates start and direction of gene transcription.

ity on the one hand and evolutionary flexibility on the other may be the consequence of the network or mesh-like nature of developmental and homeostatic controls.

CONCLUSION

In the preceding overview of the ways DNA encodes various kinds of biological information, the relationships between genotypes and phenotypes, the multiple functions in cells, and the various classes of genes that have evolved to serve these functions, we have tried to outline a general picture of the working tools with which biological traits are made. There are important broad generalizations that are seen repeatedly.

Biological processes use a limited number of genetic regulatory factors and their interactions. The diversity of mechanisms has been elaborated by gene duplication from a few ancestral starting points and by multiple uses of the same factors and pathways. This generic logic rests on overall principles of evolution and/or biochemistry, which can perhaps be attributed in a general way to the modular nature of coding, the fact that so many biological reactions involve the interaction among proteins (including ligands and their receptors), and the nature and ubiquity of cell membranes. All of this is constrained by the core of basic biochemical reactions from which so many of the constituents are derived and upon which it all depends.

A few of the "classic" kinds of genes, like single enzymes or structural proteins, are used only in one context for a direct function. Hemoglobin and tooth-enamel proteins are examples. But we need a different kind of explanation of specificity for the broader set of complex functions. Are signaling factors Fgf4 or Shh, which we know are expressed in limbs and teeth (and many other structures), genes "for" teeth or limbs? Is the distal-less homeobox TF an insect wing spot or leg or tooth gene? Is the answer simply "yes," or are there better ways to think of it? In a meaningful sense, many aspects of a biological trait are perhaps better defined not by the specific genes that bring it about but by their interactions. Many if not most of the signaling pathways have both positive and negative feedback components, which can be viewed as providing a stabilized interaction tool (Niehrs and Meinhardt 2002) that can function under a variety of circumstances.

There is degeneracy in RE sequences for a given TF. There is also similar regulatory degeneracy (sometimes referred to as promiscuity) in developmental signaling. The regulatory toolkit comprises sets of gene family members that are able to substitute for each other experimentally or evolutionarily—and this probably means that there is considerable cross-reaction among SFs and their receptors under normal conditions. For many traits, this leads to the many-to-many phenogenetic relationship described earlier.

The genes in the regulatory tool kit are also arbitrary relative to traits they control. This is reminiscent of the arbitrary coding nature of DNA. The complex nature of gene interactions is such that we can typically no longer associate a gene with a single function (Greenspan 2001). The action is in the specifics of interactions, and this also provides malleability and robustness. Just as a given type of tRNA transports the same amino acids to any protein whose message has the appropriate codon-anticodon match, a Hox or Ffg protein can regulate any gene that has the appropriate flanking RE sequences. In each case, there is binding specificity but functional arbitrariness.

Thus, the biological traits that have been the focus of evolutionary biology (limbs, flight, and the like) are the very specific end-stages of complex developmental processes, but much of how the traits get here is genetically arbitrary. This is very different from the view of evolution that has predominated since Darwin and of molecular biology since the modern synthesis and Central Dogma. Throughout the remainder of this book, we will see these principles in action.

PART III

An Internal Awareness of Self

Communication within Organisms

The purpose of this section is to review how some fundamental challenges in information transfer within a multicellular organism have been met, and how the mechanisms involved today evolved from simpler precursor states. First, we will look at the ways in which multicellular life has come about. We will look at various ways a modest number of basic "strategies" that are used have evolved, and how genes are used to achieve them. These strategies have been employed throughout the biosphere.

A major—if not the major—stage in the evolution of life as we know it was the evolution of the cell. Cells are more than gene-translating factories and are the essence of more complex organisms, constituting a fundamental organizing aspect of life. For an organism to function as a coherent whole, there must be an organized division of labor and communication among cells. We will look at how that occurs among nearby cells and then at the processes that allow long-distance communication among cells within an organism.

Finally, an organism develops generally from a rudimentary beginning, such as a single cell, through a highly orchestrated process of development, and we are beginning to learn how that works. A major aspect of that is the vital problem of reproduction, which itself has many meanings. Similarly, many if not most organisms are able to maintain viable states by responding to changes in their environment. We will describe general aspects of development as well as genes and the regulation of their expression that makes this responsiveness possible.

An important point is that within an organism communication largely involves specific *signals that are received by cells specifically looking for them. Another problem of dealing with the internal world actually derives from the external. Organisms are always subject to being invaded by microorganisms that could do them harm. They have developed a variety of ways to cope.*

Genetics and the Logic of Evolution, by Kenneth M. Weiss and Anne V. Buchanan.
ISBN 0-471-23805-8

Chapter 8
Making More of Life: The Many Aspects of Reproduction

Reproduction is so fundamental to life that it is one of the characteristics by which we *define* life. As well as the generation of new life in the short term, in the long term, reproduction provides an opportunity for heritable change of various kinds, and hence for evolution. A key view of 20th century biology was that life is basically a nucleic acid information phenomenon, in the dual senses that biological information flows from DNA to RNA to protein (the Central Dogma), but not the other way, both in an individual and across generations. Thus an important primary form of reproduction, *replication*, is molecular copying, of RNA for individual genes and of DNA for chromosomes. Replication of other molecules of life, such as amino acids, steroids, lipids, carbohydrates, and nucleic acid components themselves and reproduction of organisms follow as a result. We've seen in earlier chapters that this view can be tempered by other ways in which interited change can come about, but they all involve reproduction in some fashion, and an important element is that in each case the heritable aspect of the new iteration is sequestered from direct modification by its parent. A new generation has a life of its own.

We usually associate the term "reproduction" with the production of new organisms from a parental generation. But this same process is essentially the way a single cell differentiates and divides into a multicelled organism. Among those cells, genetic information flow is also a unidirectional evolutionary microcosm. This includes somatic mutation that induces gene structure change and the various mechanisms that affect gene expression within each, partially sequestered, lineage of cells. There are many interesting subtleties in the processes of reproduction, which we discuss in this chapter.

THE PHENOMENA OF REPLICATION

Multicellular organisms have various ways of producing a next generation. Some, like bacterial biofilms, are essentially aggregates of otherwise free-standing cells that

Genetics and the Logic of Evolution, by Kenneth M. Weiss and Anne V. Buchanan.
ISBN 0-471-23805-8

reproduce by individual cell division. The biofilm does not reproduce as a single entity. Slime molds are another class of cells that normally live as individual single-celled organisms, although under some circumstances (to be described in Chapter 12), the cells are induced to come together to form multicellular aggregates, taking on many properties of multicellular organisms. At this *slug* stage, they have internal organization, signaling, various sensory behavior, cells that undergo apoptosis, and the like (e.g., Bonner 2000), and these reproduce basically as integrated organisms in which not all cells contribute directly to the reproductive process. Some simple multicellular animals, like sponges, can produce small aggregates of cells that shed as primitive larvae to become new organisms. A specific cell type is generally the precursor, but these are located throughout the parent's body. In plants, many or even most cells have the potential to produce an entirely new plant on their own, as even small pieces can regenerate complete new plants.

Most complex animals and plants have a specialized form of reproduction that we tend to treat as a standard for reproduction, although it is by no means the only or "best" form. In these, a type of cell, the *gamete*, is used only for reproduction. The organisms begin life as a single cell and grow through mitotic cell division. In many organisms, a lineage of *germ line* cells is set aside, usually early in development, and reserved for gamete production, while the rest of the organism's functions are carried out by other lineages of *somatic* (body) cells. Somatic cells may be replaced, lost, or even actively killed by apoptosis during life, and they can undergo natural selection that can even take the life of the organism (that is what cancer is). But although they are the product of a form of reproduction, and "beings" within the higher-order being known as an organism, somatic cells make no contribution to the next generation. Genetically, gametes are produced by meiosis and have half the chromosome complement of the somatic cells (e.g., are haploid vs. diploid). Typically, reproduction occurs by the union of gametes from different individuals, but self-fertilization can occur and some species reproduce by *parthenogenesis*, in which the organism's genome is fully represented without a separate fertilization stage; there is wide variation in the ways this kind of reproduction occurs in nature.

Somatic cells experience mutation just as germ cells do; however, although mutations accumulating in somatic cells may affect the organism (and cancer and other changes associated with aging are examples), only those in the germ line affect the next generation. Hence, in organisms reproducing via single gamete cells, this is where the evolutionary memory lies. Of course, somatic mutations may have an evolutionary impact if they determine whether a particular organism's gametes are transmitted.

One theory explaining the early and active sequestration of primordial germ cells is that they can escape the chromosomal modification by methylation that is used to control differential gene expression in somatic cells. Before transmission, gamete genomes in some species (including humans) are systematically imprinted, sometimes differently in males and females, by methylation at specified sites on the chromosomes or by other means, affecting how and/or when genes transmitted from that parent are used in the offspring. Another common characteristic of germ line lineages is to undergo fewer cell divisions during the life of the individual (Buss 1987); we may think of this as protecting the patrimony from mutational damage, but it is not always true. Sperm-producing cells in mammals undergo many cell divisions, with evidence showing that this results in correspondingly more mutations occurring through the male lineage (e.g., Crow 2000).

However, there is no one rule about this. Plants and many other organisms do not even sequester a special germ line; flowers, for example, reproduce via sexual reproduction but can be produced on hundreds of different stems. Some species, including many plants, are hermaphroditic, that is, the individual contains, and functions as, both sexes. And not all reproduction occurs through single gametes.

Most cells in multicellular organisms toil their lives away at the evolutionary service of the few cells that will contribute to the next generation. However, in sexually reproducing organisms, only half the genome is transmitted to a given offspring anyway. The prevailing view is that, for evolution to produce individuals who make this genetic sacrifice, the cell and organism must gain some selfish advantage in exchange. A general explanation is that somatic cells are helping to advance the reproductive cause of their genetically identical germline kin. This is only partially true because accumulating mutations make each somatic cell different. We might thus expect furious natural selection among the varying cells in our bodies all the time, but, although there is much machinery to interfere with aberrantly behaving cells, there is no master genome controlling this. In regard to reproductive self-sacrifice, germ cells from the same organism are, at least, a somatic cell's closest genetic relative but no way has evolved for a really superior somatically mutated cell to make a germline copy of itself (though in principle something like this could occur in plants).

We can take a similar view of the colonies of certain organisms like ants and wasps, in which most individuals do not even contribute gametes. This has been seen as a theoretically instructive example of how sterility (e.g., in drones or workers) could evolve: their reproductive self-sacrifice leads to the high reproductivity of the queen they protect, to whom they are closely related by a special chromosomal system called *haplodiploidy* (see below). But if this is compatible with evolution it is not inevitable: termites and naked mole rates are similarly eusocial, forming colonies with one breeding female, a few breeding males, and the rest nonbreeding, but these species are diploid, not haplodiploid.

Because of their locked-in reproductive fate, some biologists have suggested that it is meaningful to consider such colonies as organisms themselves. The idea has been around for a long time (Bateson 1894; 1913). Indeed, in many ways, our own somatic cells can be likened to the drones in a beehive, and our germ line to the queen. What an ant, or the set of drones, are to a hive may not be so different from what a liver is to a human. A major difference is that the individual ants are not stuck together, whereas, for example, individual nephrons in a kidney are. We could even consider a *species*, like humans, as a single large superorganism, with various sets of cells, aggregated in hierarchical ways and exchanging genetic variation over time. Genetically, this is not far-fetched at all because the responsible mechanisms are similar in nature.

Sexual reproduction is a nearly ubiquitous part of life and has evolved countless forms, many presumably having arisen independently. The most plausible evolutionary advantage leading to this is provided by the added variation produced by recombination of homologous chromosomes from different individuals. The partial sequestration made possible by sexual reproduction makes a species more able to respond to changes in the environment, while keeping the species from devolving into a set of totally sequestered lineages.

That so many multicellular organisms reduce their reproductive activity to the shedding of only single cells raises the question of whether sexual reproduction is still a remnant of single-celled life. After all, reproduction could in principle occur by fusing many cells rather than one, and among more than two organisms, taking advantage of somatic mutations and natural selection within each organism's life history. This would allow a kind of semi-lamarckian evolution to take place: somatic cells used in reproductive fusions would be screened by selection for those cells that behaved well, including those carrying favorable somatic mutations. That can't happen now in lineages long committed to meiosis-based sexual reproduction, and problems could arise in terms of making a unified cellular tree of differentiation and development, or of self/nonself immune recognition. But different mechanisms might have evolved, and plants (for example) don't have a simple unified tree of development (see below).

Of course, some organisms do reproduce by shedding sets of somatic cells, and the rather fluid, context-specific nature of aggregation in otherwise single-celled organisms forming biofilms and slime molds shows that something of this kind can occur (Bonner 1998). Still, for whatever reason, the single-cell route has been repeatedly successful; whether it is "best" or not, it works and may be easiest to evolve. These are facts of importance to the definition of an individual from evolutionary perspectives (Buss 1987) and affect the tempo of evolutionary genetic change and perhaps also some aspects of its mode of action.

Viruses contain nucleic acids that have normal protein-coding function, but they are not alive and replicating when free-standing and must infect cellular organisms to reproduce. There, they shed their protein coat, allowing their RNA or DNA (depending on the type of virus) to interact with relevant enzymes in the host cell and to be copied. New viral coat proteins are coded for by the viral genome and translated into protein by the host cell. Multiple copies of viral DNA/RNA, and viral coat proteins, can circulate within the cell; thus a given virus can be assembled from components that did not all derive from a single parental ancestor. Viruses can exchange genes by recombination within the cell before new particles are formed, further adding to their repertoire of variation. Another way viruses reproduce is by integrating their DNA (or DNA copies of their RNA) into the host genome. If this occurs in somatic cells, the viral genome is inherited somatically; in a germ cell, the viral genome is transmitted to the next host generation via the horizontal transfer referred to in Chapter 2.

Bacterial plasmids, chloroplasts, mitochondria, and many chromosomally integrated genes are examples of the past or present infectious transfer of parasitic DNA that can be replicated as units within a cell and inherited through mitosis and meiosis.

THE SPECIAL PHENOMENON OF SEX DETERMINATION

We have described sexual reproduction from its general genetic point of view, which is a source of shuffling of chromosomes (and recombination during meiosis). But sexual reproduction requires much more than mechanisms for the production of male and female reproductive cells (sperm, egg, pollen, ovule, and correspondingly different cell structures in yeast and bacteria). Making gametes is just the beginning: they (that is, the organisms that carry them) have to find each other. This burden clearly reflects the importance of whatever advantage sexual reproduction must have. It has led to so many differences in behavior and morphology that sexual dimorphism can be the greatest single source of variation among individuals in a species.

There is extensive variation in the genetic basis of sex determination. Even bacteria and yeast occasionally reproduce sexually, often induced to do so by nutritional or other environmental changes that we tend to characterize as stress. Morphological changes occur to denote "male" and "female" cells, which fuse to form a temporarily diploid organism, in which recombination can occur. In the case of bacteria, a small chromosomal element known as the fertility factor (F) can transfer between bacteria. F+ cells form short projections called *pili* on their surface, in a process called conjugation that joins F+ and F– cells; genetic information can then transfer through chromosomal recombination (see panels A-D in Figure 8-1A) (e.g., Griffiths et al. 1996; Lewin 2000; Suzuki et al. 1998). On rare occasions, the F element is incorporated directly into the bacterial genome and can thereafter be transferred to other bacteria in subsequent conjugation.

There are many variations among species, but the basic mating systems of yeast involves a "cassette" system. Figure 8-1B shows this for the brewer's yeast *Saccharomyces cerevisiae* (e.g., Haber 1998). A gene in a chromosome III location called MAT codes for mating-related proteins. Depending on the mating type, one of two flanking genes, *HMLα* or *HMRa*, replaces the gene currently at MAT. Chromatin structures symbolized by gray ovals repress these genes, but at the appropriate time in the life cycle, a cut is made at MAT. Then, controlled in part by flanking

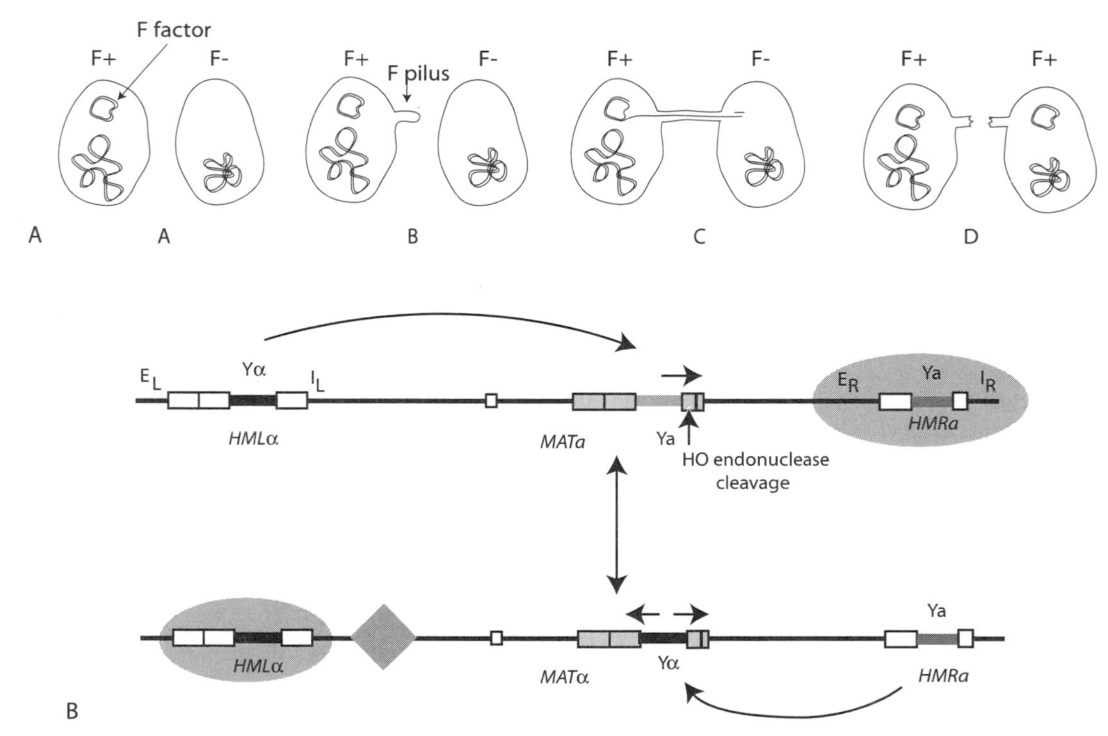

Figure 8-1. Some different kinds of mating systems in single-celled organisms. (A) Mating in bacteria (e.g, *E. coli*), stages A–D; (B) mating in yeast (*S. cerevesiae*). Separate lines show switching from each of two states. For explanations see text. (B) redrawn from (Haber 1998).

regulatory sequences (E_L, I_L, E_R, and I_R) recognized by DNA-binding regulatory proteins, and a repressor switch (symbolized in the figure by a gray diamond), a copy is made of the derepressed donor gene, and the copy is inserted at MAT replacing what was there. A subsequence within *HMLα* or *HMRa* (Y_α or Y_a) codes for genes determining the different mating type. Haploid yeast of complementary mating types unite to form a diploid reproductive cell. The genetic mechanism for this switching, and the conditions under which diploid and haploid states occur or persist under different environmental circumstances, are highly variable among yeast species.

Many animals have sex determination in which males and females have different chromosomal complements. In mammals, females are homogametic and have two structurally identical X chromosomes, whereas males are heterogametic, with one X and one Y. The remaining chromosome sets, called *autosomes*, are structurally identical in each sex, although they may be differently modified in their respective somatic or gametic cells (see below), and of course may bear different alleles at any locus. Birds and nematodes have reversed male/female heterogamy.

Some species (e.g., some plants) produce diploid hermaphrodites, but closely related species may have obligatory sex-specific morphology and go through a haploid stage. In the haplodiploidy of social hymenopterans (ants and bees), males are haploid and come from unfertilized eggs, whereas workers and queens come from fertilized (and hence diploid) eggs. Queen development is environmentally induced by feeding larva on a special substance known as "royal jelly."

Factors in the environment can determine sex in some vertebrates. In many reptile species, such as turtles, sex determination is temperature-dependent. However, these environmental effects appear to work via genetic pathways similar to those used in descendant vertebrates, so that experimental sex-hormone manipulation can override the effect of temperature. From their presence in diverse living reptiles, we might infer that sex hormones already existed ancestrally for some purpose and may have been recruited to take over the role previously played by temperature, becoming regulated by genes linked to sex chromosomes.

Many plants can be male, female, or hermaphroditic. Flowering plants do not sequester a separate germ line, and cells at the shoot meristems (tips) can differentiate into male and female flowers, or flowers can be produced that bear both stamens and carpels, depending on patterns of regulatory gene expression. There can be location-specific sex determination, such as male tassels atop and female flowers along the axis of corn. Once committed, the designated cells undergo meiosis that is similar, although not identical, to meiosis in animals. Although some plants can self-fertilize, others have mechanisms that force them to crosspollinate using, for example, insects, birds, or wind. Darwin wrote an entire monograph on the devious ways this is achieved by orchids (Darwin 1862).

Some species form two sexes without a specific set of sex chromosomes. This can be achieved by quantitative polygenic contribution of genes on several chromosomes or (as in some reptiles) through environmental effects.

Some General Genetic Issues

It is very interesting that, as important and pervasive as sexual reproduction is, its mechanisms are so highly variable and rapidly evolving. The mechanisms for sex specification differ greatly at the molecular level. Some species have sex chromosomes, but they work in different ways in different species. The genes on the sex chromosomes are responsible for some aspects of sex determination, but they typically induce developmental cascades using genes all over the genome to effect the wealth of differences in size, morphology, and behavior between males and females that go well beyond their respective gonadal differences.

Some of these effects are qualitative, but others are quantitative (e.g., Griffiths, Miller et al. 1996; Raff 1996). In mammals, the mechanism is dichotomous and sex-linked. The *SRY* regulatory gene, a member of the *HMG/Sox* chromatin-associated DNA binding protein family (see Table 7-6), binds and bends DNA in a critical region of the Y chromosome, presumably providing access to transcription factor (TF) activity that initiates a cascade to turn a gonad into a testis. Downstream activation includes a related gene, *Sox9*, which is involved in male determination in mammals and other tetrapods (but also having unrelated developmental functions). The sex-determining cascade also involves X-linked genes. However, germ cell differentiation and gonadal and morphological sex differences are not inherently connected (and there are developmental genetic differences, even between humans and mice).

By contrast, sex determination is more quantitative than dichotomous in flies (Schutt and Nothiger 2000). In a zygote with two X chromosomes and two sets of autosomes (A), the ratio of X- and A-linked gene products is roughly 1.0, a state that activates a TF, *Sex lethal* (*Sxl*), to initiate a female-producing downstream cascade of differentiation. The latter activates a gene known as *Transformer* (*Tra*) that activates the gene *Doublesex* (*Dsx*), which leads to female development. These appear to involve sex-specific splicing variants (e.g., Graham et al. 2003). In XY males, the X-to-A dosage ratio is less than 1.0, inactivating *Sxl* and producing a male. However, not all flies use the same control system for activating *Dsx* (Dubendorfer et al. 2002; Graham, Penn et al. 2003), suggesting that the lower-down mechanism was established initially and its control assumed by various genes during arthropod evolution. In many fly species tested, *Sxl* is not even involved in sex determination (it is not true that once you've seen one fly you've seen them all, nor that all flies are like *Drosophila*). In the species *Ceratitis capitata* the X-to-A ratio appears not to be involved either, but a separate Y-linked factor called *M* is.

The widely studied animal model nematode *Caenorhabditis elegans* also uses a dosage-sensing mechanism, with activity of the autosomal gene controlled by the sex-chromosome dosage. However, the controlling sex-determining gene, *Xo1*, is not homologous to *Sxl*, nor does the downstream cascade of genes that are on or off differently in male and female development bear much if any homology with the fly mechanism. Indeed, recent work suggests that even the clear-cut "mammalian" mechanism is not so clear-cut because mice and humans use *SRY*, and some other genes thought to be fundamental for sex determination, in different ways (Ostrer 2001; 2001). Furthermore, although they have a similar *SRY* mechanism, marsupials can be hormonally induced to form either sex, and gonadal development is determined by mechanisms separate from the sex-determining ones (Renfree and Shaw 2001).

The diverse sex-determining mechanisms appear to have evolved independently many times. However, some organisms, vertebrates and yeast for example, seem certainly to have had sexual reproduction ever since they shared a common ancestor;

therefore, the differences in mechanism seen today have probably arisen by phenogenetic drift rather than via independent evolution from nonsexual forms.

Sex-Specific Chromosome Modification

Functions in the early zygote, from the single-cell stage onward, may be dominated by conditions inherited from the mother. This is especially true in species whose fertilized egg receives little besides DNA from the sperm. The nutrients and mRNA required for early development are produced by the mother. In mammals, maternally and paternally derived chromosomes arrive differently marked by *imprinting*, by methylation of the DNA (e.g., the **C**s in **CpG** pairs) flanking the 5′ start site of many vertebrate genes, which prevents promoter binding by TFs and thus represses transcription. In the formation of gametes, the parent resets its respective imprinting pattern.

How has this curious arrangement evolved? One explanation offered from a strongly darwinian behavioral-evolution perspective is that the fetus is a kind of parasite on the mother. It is related to the mother genetically but only by half, and she and the fetus compete for resources. The fetus could sap her of so much energy that she would not be able to reproduce again, which is against her interests, and conflicting with this is her clear interest in the health of the fetus.

Genes expressed in the trophoblast cells that form the extraembryonic tissues including the placenta are predominantly maternally derived, leading to a different possibility, that she is less likely to detect the implanted fetus as a foreign invader and destroy it. Sperm-derived chromosomes are generally more heavily imprinted and not expressed until somewhat later, and then in the cells that will form the embryo itself. Some imprinting signal does persist in the zygote, but differential imprinting cannot affect most autosomal genes, or at least not for long, because individuals typically show the expected equal inheritance proportion and effect of their autosomal alleles from both parents; for example, offspring from ***AA*** × ***aa*** matings typically manifest both ***A*** and ***a*** allelic effects. The known exceptions are notable for their rarity.

Gene Dosage and Related Issues

Diploid organisms must be able to function genetically in a haploid state, during early development as just mentioned, and the haploid germ cells must also be viable. However, heterogamic sexual reproduction raises an additional *gene dosage* problem. To the extent that gene products coded by autosomal and X-linked genes interact in a particular cell type, a female mammal has a double dose of autosomal and X-specific genes but a male has only one-half dose of the latter. In fact, this kind of dosage difference determines sex in flies, as described earlier.

In some species, there is upregulation (increased expression) of genes on the single chromosome in heterogametes, whereas in others, including mammals, there is downregulation. This is achieved by random X-inactivation in females. During embryogenesis, in each cell in females, one of the X chromosomes is inactivated so that the cell only expresses genes from a single of its two X chromosomes. This is achieved by an interesting mechanism. A gene called *Xist* is transcribed from an X-inactivation center, and the resulting *Xist* RNA "paints" (covers) that X chromosome (but not the other X chromosome in the cell) to block its genes from transcription at positions that may be signaled by the location of dispersed repeat elements (Bailey et al. 2000; Brockdorff 2002; Plath et al. 2002). The result is that, in males and females, protein interactions work at the same dose level. The genetic mechanisms used to inactivate the entire X in this process were already available as part of the general mechanisms used in selective gene silencing in differentiating cells.

Because of the random and differently timed nature of inactivation, a female is somatically mosaic, with approximately half of each tissue using the maternal and half the paternal X-linked genes. Furthermore, because the inactivation pattern is mitotically inherited, the time at which a given inactivation event occurred in development determines the nature and size of the downstream clone—the developmental lineage—of cells expressing the particular active chromosome. This long-known principle has been shown to be incomplete, however, in that some genes do appear to escape inactivation. Interestingly, humans and mice differ considerably in the details of their *Xist* systems (e.g., Daniels et al. 1997). In addition, both the X- and Y-linked copies of some genes are active in males; the gene *Amelogenin* that is involved in dental enamel production is an example, with an active copy of the gene on both X and Y chromosomes.

Sex-linked genes in some organisms, one example being *Bombyx mori* (silkworm), appear not to be dosage compensated, showing that, clearly, interactions among gene products can be adjusted. Both male and female gametes are similar and in some species essentially identical. Some parent-specific marking persists in life so that alleles identical in sequence function differently if they were inherited from the maternal or paternal line. A few instances of this have been discovered in inherited disease, when the genes were examined carefully. Angelman syndrome is a set of conditions including developmental, speech, and other behavioral disturbances that arises when maternally imprinted information in a region of chromosome 15 is lost through various mechanisms such as aberrant meiosis. Alternatively, if paternal imprinting pattern in that chromosome region is absent a different disorder, Prader-Willi syndrome results; Prader-Willi has very different symptoms, predominated by morbid obesity. Several other known diseases are preferentially inherited through one parental line, again for presumed imprinting reasons.

Variation Everywhere

Variation is everywhere in the sexual reproduction arena. And an arena it is, because of the behavioral complexities required for the two sexes to find each other. Add to this the variation involved in making different kinds of gametes, not to mention their delivery by various mechanisms as well as the diversity of genital organs, gestational systems, and the huge repertoire of associated appearance, behavior, and morphological differences between males and females. Indeed, it is too simple to think of sex as dichotomous (males and females). There are, for example, plants that make both sexes and may be self-fertilizing, reptiles that can be either sex depending on how hot they are (or were as eggs), and fish that can respond to environmental conditions by changing their sex. Many if not most aspects of sexual morphology are variable among closely related species (e.g., Renfree and Shaw 2001). We also have culturally based ideas about *gender*, that is male and female behavior, but we know that there are exceptions, both in our species and many others.

More importantly perhaps is that as cultural beings who monitor our own societies closely, we know that there are all sorts of morphological as well as behavioral gradations related to plumbing as well as sex-related morphology and gender. These are not all highly correlated and hence are likely to be genetically unlinked. Not all this variation is dysfunctional in reproductive terms, and while in many or most species mating behavior involves stereotypical components, wide ranges of subtleties in morphology and behavior are compatible with successful reproduction. This probably indicates the kind of variation with which evolution works as sexual mechanisms change over time.

Finally, all evolutionary explanations have a Kiplingesque Just So element in them. Bdelloid rotifers illustrate this in regard to sex. These are microscopic aquatic animals that have been around for a long time and are phylogenetically diverse. They reproduce entirely by parthenogenesis (either by laying eggs or "hatching" already formed young) (Judson and Normark 2000; Welch and Meselson 2000). How this "violation" of so important an evolutionary rule as having sexual reproduction (and yet surviving quite well) occurred is not known, but of course even to propose this as a conundrum is to build in assumptions about the importance of sexual reproduction.

THE PACE OF REPLACEMENT

Organisms approach reproduction in various ways. Evolutionary biologists have sometimes characterized these in two general categories as *r* and *K* "strategies." This nomenclature derives from an equation for the dynamics of population size, relating the rate of change in size of a population now of size *N*, to its intrinsic growth capability of the population (*r*), and the carrying capacity (maximum sustainable population) (*K*) of the local environment:

$$\Delta N = rN(K - N)/K$$

This is the logistic growth equation. At low relative population size, $K - N$ is large, $(K - N)/K$ is nearly equal to 1 so that growth occurs nearly at the maximum capacity (a factor of *r* per generation); but as population size nears the carrying capacity, *K* nearly equals *N*, so that $(K - N)/K$ is nearly zero, and growth tapers off.

In this conceptualization, a species that reproduces rapidly is said to be using an *r* strategy. Its offspring mature quickly and shed large numbers of gametes. Insects and fish that produce thousands of eggs are examples. They can tolerate massive die-off of their young and still have net successful reproduction during their lifetime. The parent typically does not invest much care or effort into the maturation of the offspring, but there are exceptions (for example, among mammals, mice are rapid reproducers, but they also nurse their young until self-sufficiency). The race for success is run by producing many offspring, increasing the odds that at least some will survive. Animals lower in the food chain, whose lives tend to be risky and short, often reproduce rapidly and can be more adaptable. Any individual offspring is, more or less, expendable in terms of the parents' evolutionary interests. On average, in a stable environment, there is no net population growth after all this struggle, but an *r* strategy is able to recover more quickly from losses in numbers.

In contrast are the lumbering *K* strategists. Elephants, albatrosses, and some trees would be examples. So are humans and other primates. These organisms are, so to speak, near their carrying capacity or at least their ability to respond in terms of growth. They reproduce only irregularly, after a long juvenile time period, or only produce a small number of slow-growing offspring. In some cases, there is a need for longterm parental care so offspring can survive to reach reproductive age. Carnivores high in the food chain are fewer in number than their prey and may need more training for life. Because their reproduction is slow, maximizing the probability of survival of each individual offspring is important. A lost offspring is devastating to its parents' net reproductive output.

Mathematical theory has been developed to show how these reproductive behaviors might evolve genetically through the action of natural selection. Clearly, however, there is no one way to succeed in reproducing, nor is any way restricted to particular taxa, nor is a clade of taxa necessarily restricted to one way of doing business. But we should remember that while populations may fluctuate in size, no matter what their "strategy," in the long term the net reproduction on average for members of a population is just at the replacement level: one offspring produced per individual. In Chapter 3 we described various aspects of the way selection might act in regard to life history, to differentiate success among genotypes. The only invariant and merciless principle is that if it works, it works.

A QUICK AND INCOMPLETE TOUR OF ANIMAL AND PLANT DEVELOPMENT

A major hallmark of multicellular life is differentiation. Cells reproduce, and a single cell becomes a complete organism. Much of the work done on this subject before and even since our recently increased ability to document genetic mechanisms experimentally has concerned the determination of overall body plans. Multicellular organisms are quite diverse, but in fact there are only about 35 body plans associated with the major animal phyla (e.g., Raff 1996) and an additional set of basic plant forms. There is a similarly limited repertoire of basic patterns of structural organization of systems within an organism. We will see in the next chapter that a limited repertoire of developmental processes brings this about, helping to explain how the great diversity of life has been achieved.

Once a basic system has become highly integrated, early in development, it can be *canalized* or its general features retained, a term introduced by CH Waddington (Waddington 1942; 1957; Wilkins 2002) that we have several times mentioned. A view fully compatible with classical darwinian theory is that, once entrenched, so much of later development depends on the early establishment of a body plan that the latter is difficult to change. This leads to evolutionary and developmental stability, although at the expense of flexibility (e.g., Siegal and Bergman 2002). Other ways of buffering include heat shock (or chaperone) proteins that can serve as a "capacitor" against harmful environmental or mutational effects, epistatic feedback (Rutherford and Henikoff 2003; Wagner 2000), and redundancy. In Chapter 7, we discussed the way that tolerant enhancer elements can contribute to these kinds of stability over individual as well as evolutionary time.

Balancing selection, that is, selection against all extreme phenotypes (e.g., too big, too small) can facilitate the evolution of such stability. Of course, no foresight is involved in any case. Transmitted chromatin modification that affects gene expression is a way in which genetic variation can exist in a population but not be expressed, perhaps easily "reawakened" under stress (if the chromatin modification is itself variable or released by fortunate mutation). Selection need merely maintain the suitable forms in their respective environments. Later, crosses between what appear to be invariant strains, or other kinds of selection, can release the constraints on this variation so that it becomes expressed. Various effects of this kind have been observed in animals and plants (e.g., Dun and Fraser 1959; Lauter and Doebley 2002; Sollars et al. 2003).

Here, we will attempt a cursory survey of the basic aspects of multicellular organization that these processes have produced. Examples, illustrations, and even animations are plentifully available on the internet, and there are several excellent texts concerning morphological development and its genetics (Gilbert 2003; Raff 1996; Wilkins 2002; Wolpert et al. 1998). Tudge (Tudge 2000) has provided an excellent compendium of animal form, and there is an excellent resource on known evolutionary relationships among the creatures on Earth (Maddison 2003).

Animal body plans include various types of symmetry from essentially none (as in hydra and sponges) to radial (coelenterates), to pentameral (starfish and other echinoderms), to bilateral (ourselves). Plants establish radial symmetries in their roots, shoots and flowers, and branching symmetries in the arrangement of leaves on stems or veins within leaves. These symmetries define or reflect anatomic axes, most notably the anterior-posterior (AP) axis from head to tail and the dorsal-ventral (or dorsoventral, DV) axis from back to front and lateral or left-right or proximal-distal (or proximodistal, PD) when viewed relative to the midline; in plants, correspond to the main vertical and radial axes and the longitudinal axes of shoots and roots.

Evolutionary explanations of body plans usually relate to their current function. For example, the segmentation of the vertebral column enables animals to become long and flexible, and yet rigid from head to tail, and to have separate structures in different places along the way. The bilateral symmetry of our eyes and ears, the forked tongue in snakes, and the antennae in insects allow perception of the world in stereo. Of course, keeping in mind the step-by-step nature of complex trait evolution, these may or may not be the original selective advantages.

Not all body plans are as organized or simply explained. Bacterial aggregates form from free-floating single-celled organisms that, under environmental conditions that favor their congregation, can emit chemical signals that induce others nearby to change behavior and join together with no fixed kind of plan (Bonner 2000; Shapiro JA 1998). By aggregating on a surface, to which they adhere through the secretion of various polysaccharide or other substances in which they also become covered, the bacteria act in a sense as a single entity. The secreted adherent film can serve to concentrate trace nutrients, attract commensal organisms to process waste, or provide or recycle nutrients. The adherence may cause the walls to thicken, protecting the individual cells.

Different microstructures in the aggregates affect microenvironments, the flow of and access to water, response to antibacterial agents, and the like. In a sense, by organizing fluid flow in their environment, bacterial colonies become vascularized. More remarkably, it has been discovered that bacteria use different sets of genes when living in biofilm aggregates than when planktonic (free-floating), with major changes in the protein composition of their cell walls. More complex and diverse are ways that bacteria aggregate and accomplish things they could not do alone

(Armitage 1999; Hoyle and Costerton 1991; Miller and Bassler 2001; Parsek and Greenberg 2000; Pratt and Kolter 1999; Prigent-Combaret et al. 1999); sometimes, different species cooperate in this regard (we are anthropomorphizing in discussing these things as if they were done on purpose).

ANIMALS

Here we consider general characteristics of multicellular organization, and in Chapter 9 we will consider the genetic mechanisms that bring this structure about. First, we should repeat what has been noted by countless other biologists: the range of multicellular organization cannot be arrayed in a hierarchical Great Chain of Being graded from worst to best. The simplest organisms are still with us and, if anything, seem far less likely to become extinct than most complex organisms (who may, in fact, be at an overall disadvantage and more vulnerable to extinction in the long run). If there is one multicellular structure that is "better" in this sense, it might be the human brain because it is the only structure that seems to make it possible to exterminate so many others, around the globe, at least in the short run.

Starting Simple

Very simple organisms that can live as isolated single cells but can also form associations in various ways appear to have multiple evolutionary origins (Bonner 1988; 1998; 2000; Buss 1987). The simplest organisms that we think of as truly multicellular, such as sponges, have little organized structure or cellular differentiation. Most are referred to as *diploblast*, because they have an organized outer and inner cell layer but little cellular material or organization in between. Choanoflagellates are single-celled organisms consisting of a cell body and a collar of microvilli through which water is filtered, surrounding a single flagellum used for propulsion. Some choanoflagellates aggregate to form what may resemble an early form of metazoan life.

Sponges have only a few distinct cell types: a protective external layer; a ciliated internal layer that moves water through the organism; and a middle layer composed of freely migrating archeocytes that scavenge food particles and can differentiate into all the other cell types, as well as cells that lay down an extracellular matrix of spicules (Sponges 2002). A sponge is a hollow structure pierced by numerous incurrent and excurrent pores through which water flows. There is species-specific variation in shape, which implies that there is also underlying developmental programming, and specialized functions such as contractile ability and color variation. However, sponges do not develop from a single cell in the highly orchestrated hierarchical way we will see is typical of more complex animals. Their lesser level of tissue commitment and organization can be seen by the ability of disaggregated cells to reaggregate into a normal whole, something "higher". Cells in vertebrate embryos have similar reaggregation properties but only early in embryogenesis (in a way that may be informative about evolution and mechanisms of early development) (Steinberg 1998). Environmental conditions can affect sponge morphology, reflecting their rather loose level of genetic programming.

Sponges can reproduce asexually by shedding buds or "gemmules," small groups of cells that float away to anchor at a new location (Darwin used the term to hypothesize ubiquitous rudimentary and rather lamarckian units of inheritance). Even the simple sponge can also differentiate into haploid sperm and eggs and undergo fertilization and the shedding of ciliated larvae. Reproduction can even be seasonal. Thus, without much real histology or organized form, some developmental complexity occurs.

Coelenterates (or Cnideria) include corals, jellyfish, and hydra. These species are diploblasts but make only a few specialized cell types, including the stinging nematocysts. They are typically bag-shaped with a single oral opening but can have internal septa, and they have more species-specific body plans than sponges. Coelenterates reproduce asexually as sedentary polyps or via sexual, free-swimming forms known as medusae; their life cycle often includes both forms. Corals are aggregates of polyps, whose association is so highly constrained that it is somewhat moot whether they should be considered separate individuals. They exemplify organisms so integrated with each other and with the environment that they themselves construct that it is difficult to consider the organism as only what is within each unit (e.g., Turner 2000). Some forms branch in a simple treelike way as they grow, with branches occurring in a more or less random pattern but affected by external conditions. Coral polyps grow on acellular (corallite) stalks that occasionally divide into two similar units, each forming a new stalked polyp, overall resulting in a treelike structure that approximates a fractal pattern.

More Complex Animals: Basic Characteristics

Much of our knowledge about the remaining forms of animal life is generalized from the intense study of only a few model organisms (e.g., mice, fruit flies, nematodes, sea urchins) and a few special traits (e.g., eyes, early embryonic stages, limbs), usually studied for historical reasons (because so much is already known) or convenience (transparent embryos, short lifespan, easy to manipulate or raise in the laboratory). But the broad picture is likely to be representative of the general array of "strategies" used.

These "higher" forms of animal life essentially are committed to multicellular life derived from a single starting cell. Subsequently, a highly orchestrated four-dimensional dance of development in space and time takes place. Differentiated organ development includes a few simple processes: make a ball; change its shape by differential, asymmetric growth; fill space by local division of similar cells; pinch it; make a local bulge outward or indentation inward; close or open a hole in the ball forming tubes; and split growing areas into two or more branches. Cells then differentiate to produce specific substances such as extracellular materials that provide structural rigidity or other properties, perhaps comprising dead cells or protein debris (e.g., hair, lenses), or may become mineralized and quite hard, as in shell, bone, or cartilage. Some cells secrete particular functionally important products such as hormones or digestive enzymes.

Despite sharing these basic characteristics, sometimes in a highly programmed way within a given species, there is great variation in most aspects of development among species, sometimes even among closely related species. A few rounds of rather unspecialized cell divisions generically referred to as *cleavage* turn the single fertilized egg into a simple primary shape, generally a hollow ball (*blastula*). But there are various basic patterns, including very regular and symmetric division in echinoderms, many asymmetric patterns, and the initial formation of a syncytium, a large single cell that contains many nuclei not surrounded by nuclear membranes or cell walls (in many insects) (Gilbert 2003).

Commitment of cells quickly occurs in two basic ways that have been defined by manipulation of embryos, or more recently, of gene expression. *Polarity* is also established early and involves basic axes of symmetry mentioned above. In insects like *Drosophila*, polarity is established through gradients of maternal gene products in the syncytium. Vertebrates first separate a yolk or nutrient section from a section that will generate the embryo itself. Then, the AP and DV axes of the embryo form, instructed by signals produced by centers known as *organizers* that affect nearby cells that proliferate as growth extends away from the organizer region. For example, Spemann's organizer is a region of cells capable of establishing the early AP and DV axes in a vertebrate (the homologous structure has different names in different species), as is shown by its removal or surgical relocation in an embryo and by mutations that alter or prevent it from developing.

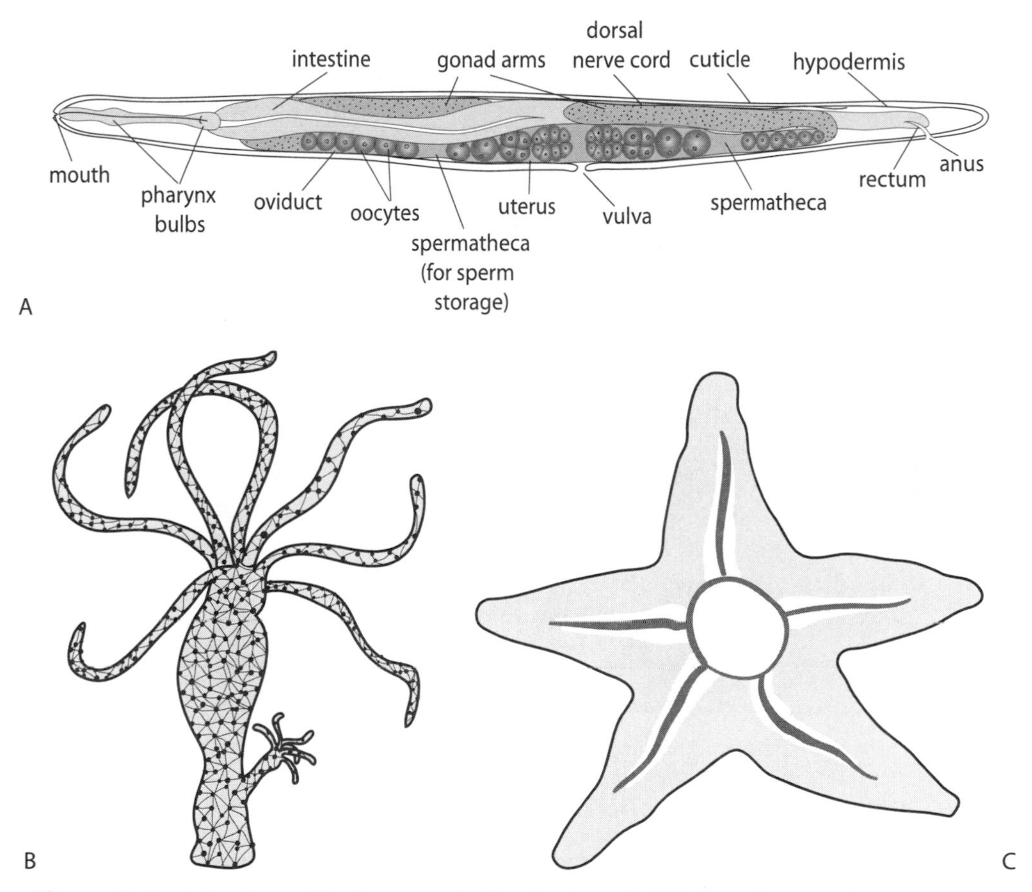

Figure 8-2. A bestiary of basic anatomical body plans: (A) Nematode, experimental model system, *C. elegans* with tissues identified (for use below); (B) hydra; (C) starfish illustrating radial symmetry; (D) fossil fern from Carboniferous age around 350 million years ago (Mya); (E) fossil trilobite from the Cambrian (Burgess shale) about 500 Mya; (F) Richard Owen's famous archetype of a standard, shared vertebrate body plan. Also see the frog, fly, and plant body plan figures below. Sources: Fern, (White 1899); trilobite, (Walcott 1918); (figures D–F reproduced from Moore 1987; 1989, with permission).

Figure 8-2. *Continued*

The notion of an organizer is somewhat imprecise, but in a general sense there are organizers from head (see Chapter 15) to tail (Agathon et al. 2003) and many places in between, and in subsequent chapters we will see how some of them work.

The notion of an organizer, like others in biology that use a term borrowed from everyday life, should be understood with circumspection. The term is used to identify signals in a practical sense and does not mean that these cells somehow come from the outside, as the word "organizer" often implies in other uses. Developmental organizers result from prior cells in a continuum of a developmental process.

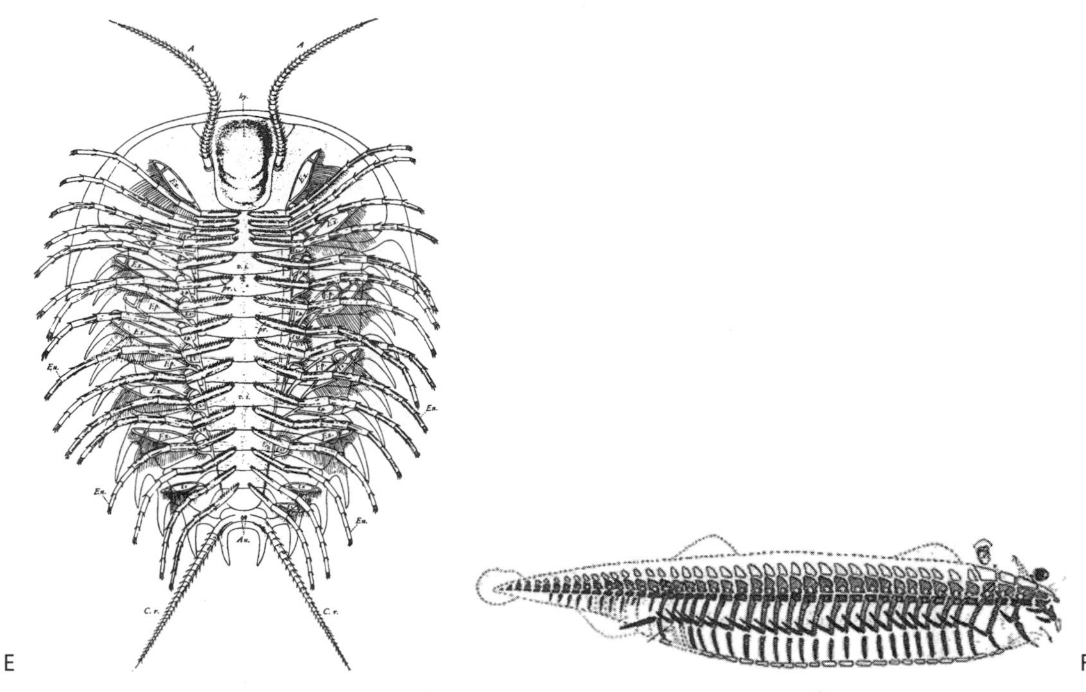

Figure 8-2. *Continued*

Pentameral animals such as echinoderms (e.g., starfish) have an additional kind of axis, comprising the five "arms" arrayed around the central body location; they have a central radial axis and a single, repeated PD axis going out along each arm. The genetic and developmental homologies with linearly patterned animals are not yet well understood (Popodi and Raff 2001).

Cells in different regions of the blastula become committed to three basic lineages that are the precursors of different sets of tissues in the final organism. Broadly, *ectodermal* cells will generate the outer covering and nervous system, cells that make up the *mesoderm* will become muscle and connective tissue, and *endodermal* cells will become the gut and its associated organs. Organisms formed from these three basic tissues are referred to as *triploblastic.* These first steps in the hierarchy of histological differentiation are highly varied in detail, even if the result is somewhat similar. Figure 8-3 shows a selection of developmental tissue "fates," in different classes of animals, that form in a hierarchy of commitment (e.g., see Carlson 1999; Gilbert 2003; Wake 1979). By "fate" is meant that descendants of the tissue in question differentiate into a unique hierarchy of cell or tissue types. This is a manifestation of the hierarchical, histological sequestration that is the major outcome of development, which is why the pattern can be drawn cladistically (in a treelike diagram of descent and differentiation). Commitments are made at different stages of development; these are not yet well understood, but some late-developing organ systems may already be *preprogrammed* in cells at very early developmental stages (the cells await a subsequent proper environment to begin elaborating their program).

The formation of these layers involves a process known as *gastrulation* by which the committed cells take their relative positions. Gastrulation in most animals involves the movement of cells from the outer cell layer to the inner, a process dependent on cell division and migration. Basically, through inpouching, the formation of holes in the outer layers or simply the expansion of dividing cells into the space inside the embryo, a layer forms that will develop into an inner pouch or tube that will become the gut—essentially a tube running through the animal and open to the outside world. Mesodermal cells are caught between the outer ectodermal covering and this inner tube. The higher animals are classified as protostomes (basically, mollusks, worms, insects), and deuterostomes (chordates), depending on how the gut is formed.

Among the most important invertebrate model species is the nematode *Caenorhabditis elegans*, which develops from a single cell into an adult with a stable 959 cells; lineages of all 959 have been entirely traced through a stereotypical developmental journey. Each organ or structure develops from a specific, very small number of cells, and gastrulation and other basic processes are correspondingly simple.

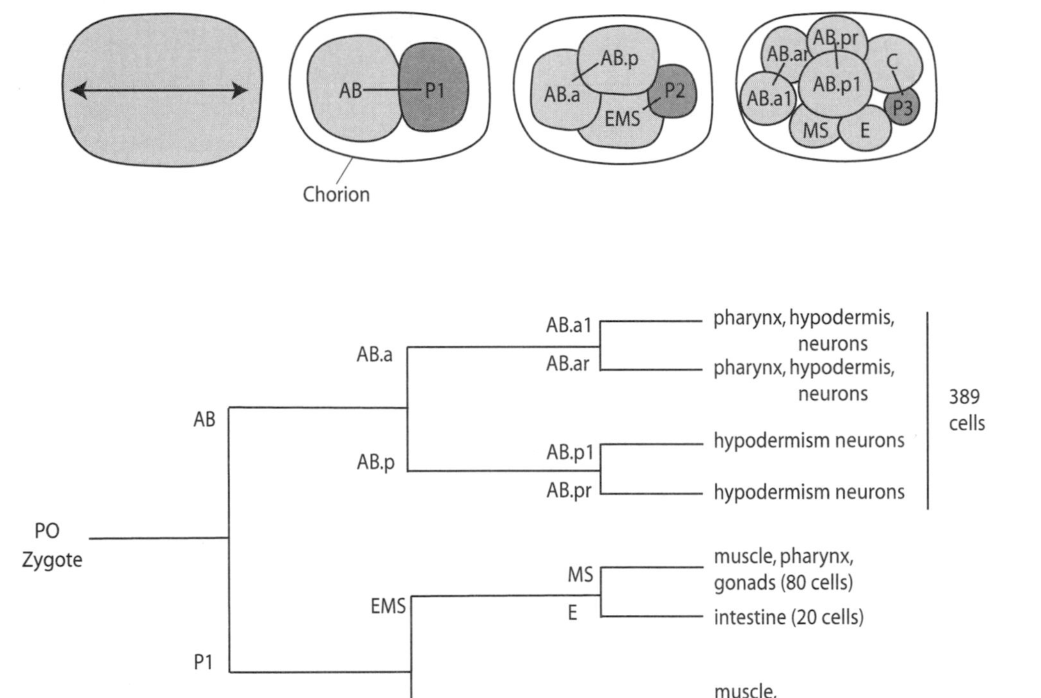

Figure 8-3. Embryological fate maps. (A) The development of the nematode *Caenorhabditis elegans* from a single cell to 558 differentiated cells (these will further proliferate to a final total of 959 cells, but without further change in cell type); (B) the tissues descended by hierarchical differentiation of the three germ layers in vertebrates (example is from humans): endoderm, mesoderm, ectoderm; neural crest interactions not specifically identified.

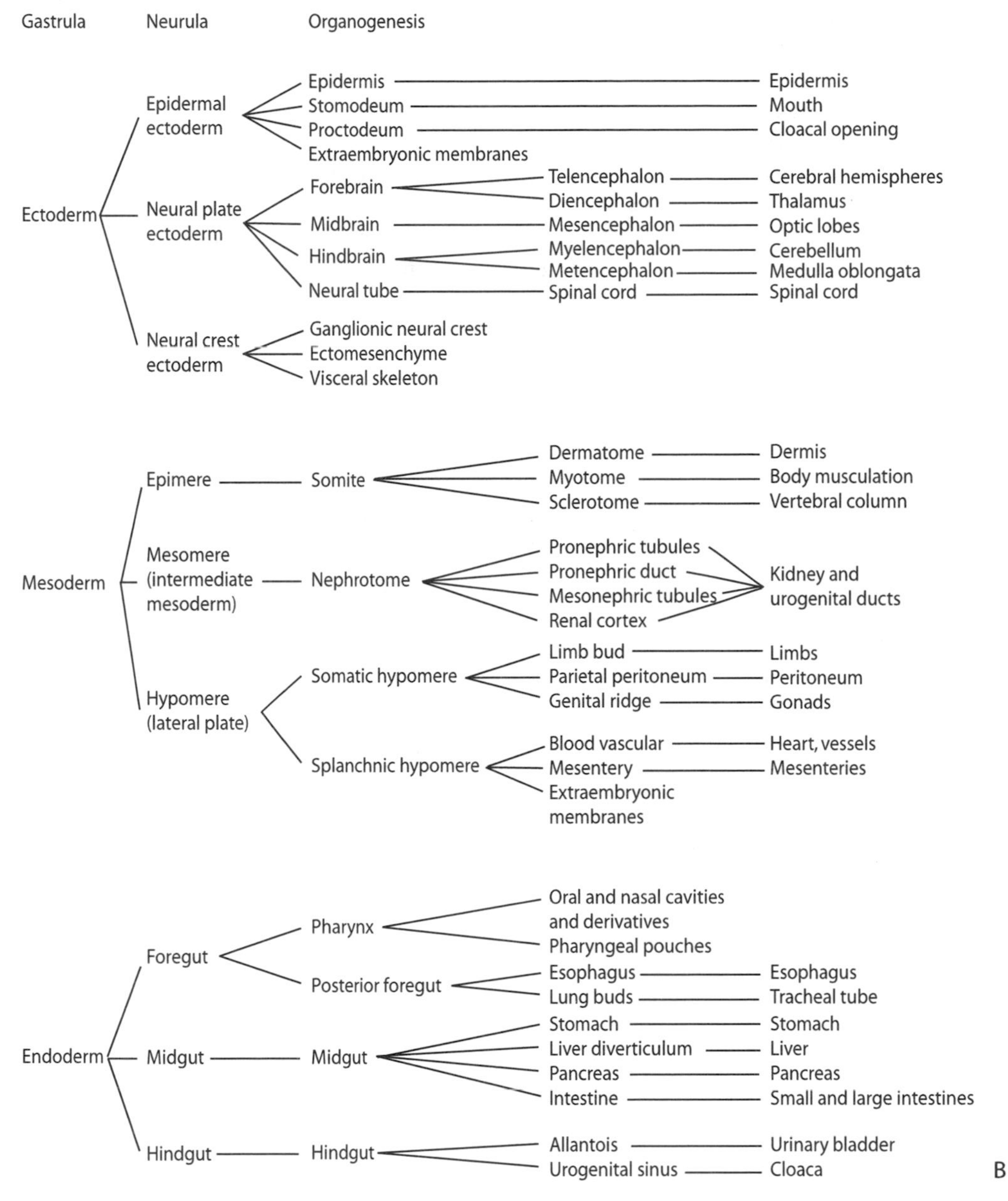

Figure 8-3. *Continued*

Regarding arthropods, more is probably known about *Drosophila* than any other (Lawrence 1992). Cell walls form around the nuclei in the syncytium to form a hollow structure along which there quickly appears a series of segmental structures that differentiate along the AP axis as the embryo grows. Local organ primorida called *imaginal disks* develop at specific locations along the axis. The imaginal disks themselves serve as organizers for cellular subsets that are precursors of the various parts of the antenna, mouth and tailparts, legs, wings, sensory bristles, respiratory spiracles, and the like. At appropriate times, ingrowth or outgrowth starts the process of organogenesis from the imaginal disks, and these structures themselves develop AP and DV axes, and then develop PD axes as they grow in relation to the body of the embryo.

In vertebrates, developmental processes are more complicated, although the basic principles are similar (Gilbert 2003). Figure 8-5 shows one pattern, as seen in *Xenopus*, the laboratory frog. The basic cell types (e.g., gut, muscle, photoreceptor, neural) are similar to those in invertebrates like *Drosophila*, but vertebrates have to deal with some other complications. They are large, which is made possible by having an internal rather than external supporting skeleton (though in their early evolution they were covered, protected, and supported by an external scale-like skeleton), they have more complex organ systems, and they tend to live longer and have more renewable tissues (especially organs lined by epithelia, tissues with ever-dividing *stem cells* that renew locally differentiated special cells like those of the gut and skin), and they have more elaborate immune systems. Large internalized bodies entail thermal, mechanical, and respiratory demands and the circulatory and neural apparatus that goes with them.

Vertebrates, like arthropods, are organized with an external wrapping around a hollow digestive tube, having organ outpockets such as the pancreas. Also like arthropods, which are only distantly related, they have a longitudinal hollow nervous

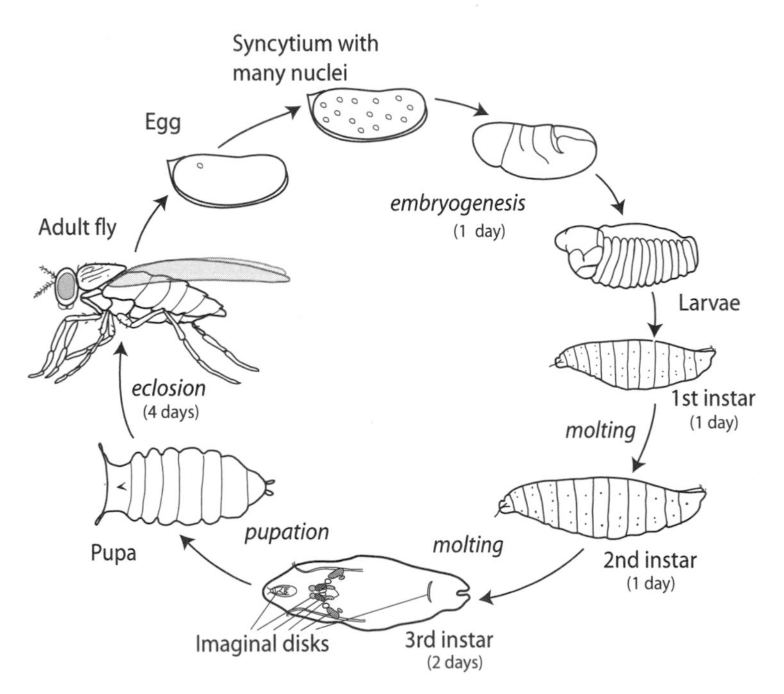

Figure 8-4. Basic stages in insect development; *Drosophila*, from fertilized egg to embryogenesis, larval stages, pupation and eclosion, or metamorphosis, to adult fly. In the important transition from the syncytium to the cellular embryo, the nuclei in the former locate around the periphery of the egg, and membranes form to separate them into separate cells. The embryo forms stripes of gene expression (not shown) that are precursors of the axial segmentation that is important from the early embryo stage through to the highly segmented adult.

system, with the brain at one end and differentiated branching along the AP axis, referred to as an "inverted" body plan. The vertebrate circulatory, gut, and nervous systems, however, develop with more complex turning and folding, by differential growth at various places along the axis, versus the corresponding system in arthropods, which is more or less straight along its major axis. In both groups, local signaling centers or organizers guide the development of structures like the distal tips of developing limbs and the midbrain and hind brain regions of the central nervous system, teeth and feathers, and the like. The neural axis is dorsal in chordates but ventral in arthropods, and the digestive systems have reverse relative locations. There is healthy debate about what homologies if any might be involved, as will be discussed in Chapter 9.

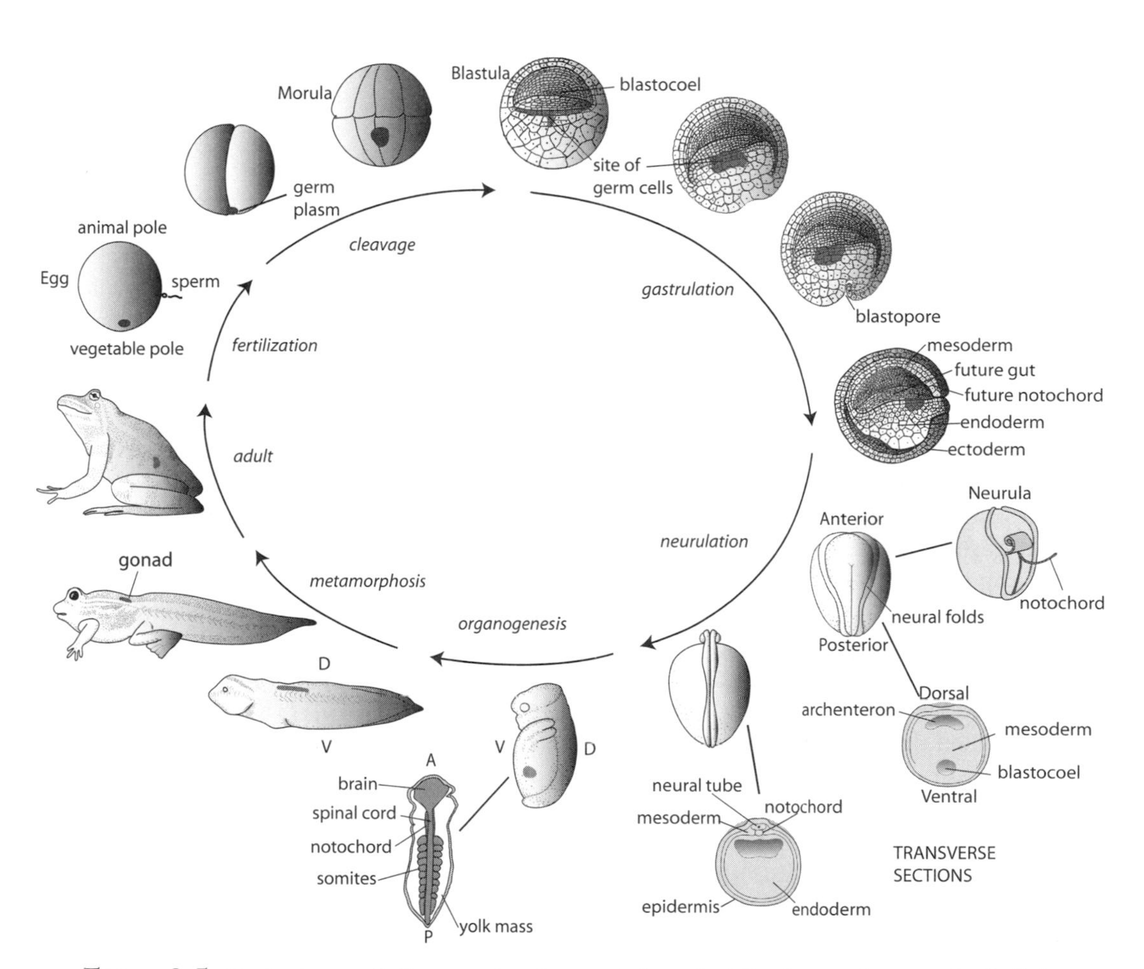

Figure 8-5. Basic stages of *Xenopus* development, from fertilization to cleavage (cell divisions that produce the blastula, a spherical layer of cells that surround a fluid-filled cavity), gastrulation (invagination of cells from the blastula to form the gastrula, a two-layered sphere consisting of ectoderm and endoderm surrounding an archenteron that communicates with the exterior via the blastopore that will form the gut), neurulation (embryonic stage at which the neural tube develops), organogenesis and metamorphosis to the adult frog. Shaded spots indicate site of development of germ cells and gonads at each stage.

Among the defining characteristics of vertebrates is a cell lineage known as the *neural crest* (NC), which derives from the early dorsal neural ectoderm. The NC cells migrate from positions along the dorsal midline, intercalating through mesodermal tissue to reach various locations where, in contact with overlying ectoderm, inductive tissue interactions trigger the initiation of various organs, like hair, feathers, mammary gland, teeth, and others (NC cells migrating from the head region are sometimes specially referred to as Cranial Neural Crest, or CNC). These inductive *epithelial-mesenchymal interactions* (EMI) are location and context specific and are responsible for so many important characteristics of vertebrates that some researchers give the NC a special place as a fourth primary germ layer (Hall 1999; Hall 2000), making them "quadroblastic." Among other structures, the segmental patterning of parts of the head have been correlated with EMI and NC migration (but the role of EMI is not entirely clear; see Chapter 9).

At least one reason EMI is important in vertebrate development is that many organs can develop only after the embryo has reached a certain stage of complexity. At that stage, however, it may be difficult to initiate a dispersed structure because the tissue bed is too large for effective signaling. For example, teeth cannot develop until there is a pharynx or jaws to develop in, taste buds cannot develop until there is a tongue, and so on. The use of NC cells and EMI allows tissues to be preprogrammed for certain developmental fates, but then not be activated until an appropriate organizational level. Neural crest cells can then come into contact with the inductive tissue located where the structure will develop. Other organ systems form in vertebrates by inductive interactions that involve mesoderm. Other forms of differentiation occur in cells programmed in some home location but that then circulate freely, as in the blood and lymphatic systems. Consistent with the notion that early program is important in later development, invertebrates establish basic regional differentiation very early in their development.

In addition to the widespread presence of epithelia, vertebrate organs typically contain both primary "functional" tissue, called the *parenchyma*, often in the form of repeated units, like pancreatic islets, nephrons in the kidney, osteons in bone, hairs, taste buds, or ovarian follicles. The functional tissue in each unit secretes hormones or enzymes, filters blood, transmits neural signals, and so on. There is usually also supporting tissue, called the *stroma*, such as Schwann or glial cells to protect neurons, or a cellular matrix to hold and structure the organ. Invertebrates are also organized in a similar patterned way, for example, with discrete eye subunits including sensory elements of various kinds that will be described later (not to mention the joints for which arthropods were named).

Many if not most organ systems have periodic, modular, or segmented, organization, with multiple regularly spaced and perhaps regionally differentiated subunits. Some systems do not segment linearly but ramify—literally—by physical branching, as in the nervous, vascular, and respiratory systems. Some organ systems, like limbs in tetrapods, are both segmented and branching.

It is easy to think of development as having to do with morphology, but much if not most of what goes on in life is "virtual," that is, is physiological and has to do with chemical interactions rather than physical structure. Some of these take place within the cell, but some take place in the internal circulating fluid, like lipid transport and respiratory gas exchange—in some ways comparable to those of the external environment, to the extent that some believe that the environment should not be considered as separate from the organism itself (Turner 2000). Metabolic differentiation at the gene level involves many of the same characteristics found in genes responsible for physical systems: the use of multiple subunits often involving related gene products, which interact with each other. The processes of differentiation of physiological systems are often hierarchical in time or cell lineages and demonstrate homeostatic interaction with the environment, in response to signaling such as of hunger, hormones, pheromones, exogenous infections, and the like.

PLANTS

Plant life histories differ from those of most animals in several basic ways; in part, their system is different because the rigid cell walls prevent the kind of migration so important in animal development. Many plants have life cycles that include diploid and haploid stages, perhaps in alternating generations. Although not reproduction in the sense we have already described, because everything stays connected, some plants nonetheless effectively propagate via *rhizomes*, part of their root systems, that push up through the soil to start a new above-ground plant. Plants that reproduce via single fertilized cells do not sequester a separate germ line, and gametes are independently produced in many places (different flowers) on the same plant. Plants also continue to undergo differentiating development and morphogenesis throughout their lives, as stem and root meristems retain the ability to develop new structures at their ends because they maintain a core of pluripotent stem cells, resulting in plants being able to respond to their environments in ways that are less stereotypical or fixed than most animals.

Plant embryos also generally form three basic tissues, known as *dermal*, *ground*, and *vascular* tissues, but this is not established by gastrulation. Plants have an active vascular system for the upward and downward transport of nutrients, but it forms differently from these tissues (Figure 8-6). Like animals, plants use a limited repertoire of basic processes, including growing, elongating along the stem and root axes to form branches, and differentiating some stem tips into flowers, leaves, or other specialized structures. Thus plants have a vertical axis (separately behaved in stems and roots), as well as corresponding axes along stem and root branches.

The First Few Cell Divisions

Not surprisingly, early plant development varies among the many plant species that have been studied (Weigel and Jurgens 2002). Under normal conditions, the fertilized zygote undergoes a few rounds of development to form initial root-tip and leaf structures, thus providing the means of survival through its early days. The new plant quickly develops a few primary cell types or layers, which lead to the apical root and apical shoot meristems. Development occurs radially around the meristems. Several primary tissue layers form distinct parts of the plant: dermal tissue forms the outer protective layer, vascular tissue forms the hollow fluid transport tubes, and ground tissue grows in between these layers.

The Basic Plant Body Plan

The two meristems grow to become roots and stems, respectively. Cells behind the growing meristem divide to allow elongation of the shoot or root. At intervals determined by growing conditions and the genotype of the plant, buds of meristem tissue

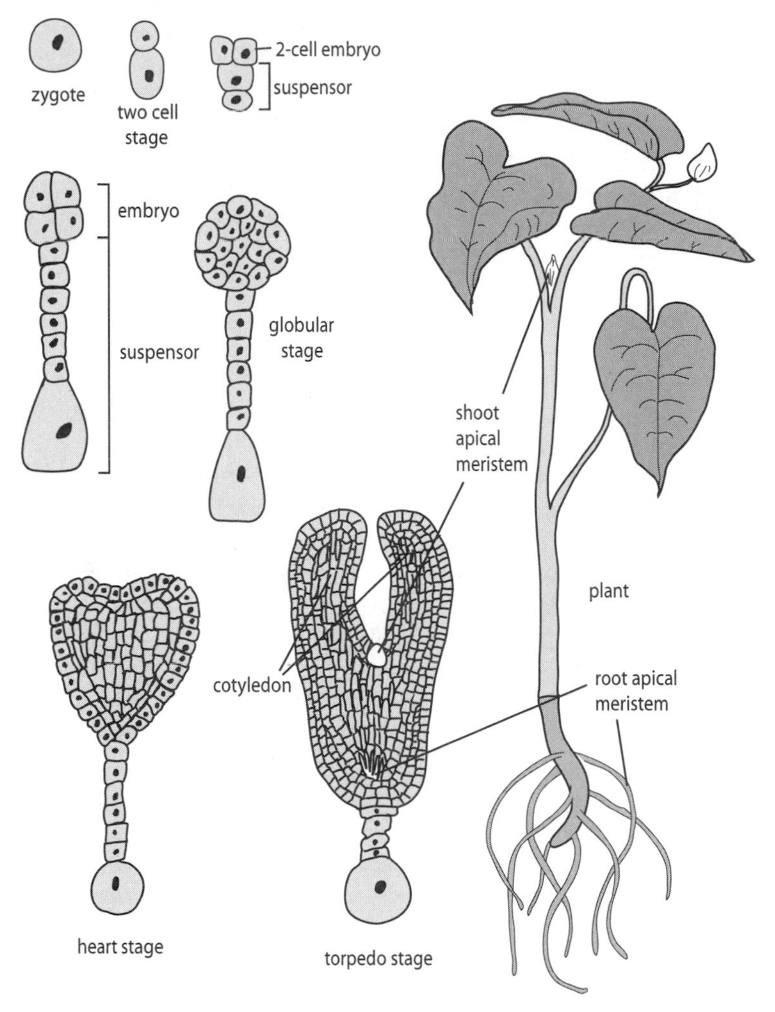

Figure 8-6. Schematic of plant development. Figure depicts the stages of a dicot; monocot development is basically the same.

separate laterally, to initiate branches. In the shoot, each such branching normally develops into a leaf and a small meristem, which itself can generate a new branch or shoot (depending on species, conditions, and the like). Signaling from the primary shoot meristem may control the timing of development of other shoot meristems, maintaining "apical dominance" of the primary shoot. An interesting feature of plant development is that it achieves a kind of spiral symmetry as new shoot meristems are formed at relatively species-specific angles of rotation around the developing apical meristem. Furthermore, plants have hierarchical branching symmetry, with each new shoot or root having radial and PD axes of their own. At points determined by internal and external conditions, a shoot meristem will differentiate into a specialized structure, a flower. This can be male, female, or both. Plants vary in whether they have only one, both, or "bisexual" flower types. Root meristems also send off lateral branches at intervals.

The notion of "stem" cells in animals and plants is similar, and there are at least some genetic similarities shared between stem cells in the two kingdoms (Benfey 1999; Laux 2003; Weigel and Jurgens 2002). This suggests that in the common ancestral period, multicellularity may have been a widespread sometime trait among otherwise single-celled organisms, perhaps due to the effects of gene(s) (today with homologs *Zwille* in plants, *Piwi* in animals, of those currently identified) related to basic cell division or adhesion. In any case, the properties of stem cells are similar in the two kingdoms, but plant stem cells are essentially totipotent in adults, able under appropriate signaling to generate all the plants stem, or root tissues, unlike animal cells which are "stem" for a tissue but not all tissues (Laux 2003; Weigel and Jurgens 2002). It is this that makes each stem and root rather independent, unlike the case in animals where each tissue can further differentiate, but generally only down its respective fate-map pathways. A comparable notion in animals would be the cells that give rise to each hair or limb. But in a more substantial way, each branch of stem or root is like a separate organism (see below). Plant cells have less rigid developmental fates in that differentiated cells can more easily dedifferentiate to become stem cells.

Plants are simpler in their developmental patterns than animals because plants do not develop as many organs or as much interstructure communication as animals may. Plants do not have to deal with mobility and hence have simpler communication systems. Plants can respond to predation, but not in ways that require the degree of movement flexibility and hence muscular or nervous system complexity of animals. This does not mean that plants are all alike or are simple or that they have little sense of their environments. Plants do differentiate many different cell types and tissue structures, they have to maintain rigidity often at great body sizes, and they have to maintain mutational integrity over very long lifespans. They must be able to find nutrients, light, and water (and mates) and must be able to protect against infection or predation while remaining immobile (or, for floating plants, without real control over their mobility). Plants have distribution systems (although they are not closed as are the circulatory systems of vertebrates) and internal as well as interindividual communication. Although the systems differ in detail, there are many fundamental similarities in the molecular mechanisms by which they are achieved. Plants sense many aspects of their environments and integrate that information to "make decisions" relevant to their kind of life. To achieve this, plants also have evolved a variety of diverse ways of branching (Sussex and Kerk 2001), and developmental fate maps, even if they are not implemented in as unitary way as animals' are (Jurgens 1994).

SOME ADDITIONAL GENERAL PRINCIPLES

The foregoing attempted to give a general, if superficial, picture of the basic patterns of animal and plant development and form, stressing the processes or phenomena that are repeatedly observed among species or within individuals during their development. So brief a catalog gives a poor reflection of the diversity of mechanisms, and we have to be careful about how we interpret the data, perhaps especially in regard to what appears to be widely conserved developmental characteristics because they often are quite different when examined in detail. This is clear when multiple species (sometimes even closely related species) are compared.

Not Everything We See Is "Necessary"

Simpler organisms have no clear-cut body plan, some can develop a whole new organism from any cell or at least from cells located throughout their body, and some have no clear-cut "embryo" stage. Plants have an organizational "plan" and have essentially totipotent cells throughout their bodies. In the case of animals with classic body plans, the hierarchy of development and phylogenetic similarities led biologists to form a general idea of how development works, and this then became intimately related to the study of evolution (Richards 1992). Going back to the leading embryologist Karl Ernst von Baer in the early 1800s, it had been shown that there is typically less phenotypic divergence early and more divergence later in development, consistent with a hierarchical dependency of later structures upon earlier foundations. Before the discovery of evolution, the idea was that similar animals developed their adult forms by modification of a largely identical early form. Around Darwin's time, the changes among species were widely considered to have come about by sequential building of new structures upon *existing* adult stages over the generations, a process called "terminal addition." This explained how complexity arose or evolved. Essentially, natural selection was treated as if it worked only on adults.

The famous and classic example has to do with the stage of vertebrate embryos known as the *pharyngula*, at which gill arches, somites (prevertebral segments), dorsal nerve cord, and early limb buds were present. Ernst Haeckel (see Figure 8-7), Darwin's energetic and prolific defendant and science-popularizer in Germany, developed his well-known recapitulation theory of morphological evolution and called it the Biogenetic Law, that Ontogeny (development) Recapitulates Phylogeny. This theory held that in development an organism passes through the adult stages of its ancestral species, adding on later stages leading to its current more advanced form.

Figure 8-7. Ernst Haeckel. 1899 drawing by F. von Leubach, in Haeckel 1906.

In Haeckel's famous figure, diverse vertebrates are shown passing through pharyngula stages, which are essentially the same, presumably representing their ancestral state as a kind of primitive gilled fish (Figure 8-8B). An "hourglass" model developed over the years in which the different forms of vertebrate cleavage had been canalized to produce a stereotypical pharyngula common to vertebrates, which then diversified later in development to the varying adult morphologies. We know now (and in truth it was known to Haeckel) that the pharyngula stages are not as identical as depicted; even more importantly, however, the earliest stages from cleavage onward are in fact quite diverse among vertebrates, and indeed the supposed stereotypical traits vary in their timing of appearance (Bininda-Emonds et al. 2003; Raff 1996; Richardson et al. 1997; Richardson and Keuck 2001; 2002). Nonetheless, there are similarities, even if they were exaggerated, and they provide important evidence for shared ancestry.

As a digression, historians and philosophers of science have many times remarked on the ubiquitous, perhaps even necessary, element of public advocacy (or even propaganda) that is necessary for a controversial view to gain acceptance. One can go further and note that one of the problems in science is that there is, and always has been, a whole lot of "fudging" going on: theory is routinely defended by selective use, presentation, collection, and manipulation of the evidence, ignoring some facts and stressing others, redefining terms, and retrospectively and conveniently reinterpreting data. This cannot be justified formally, but it in many classic cases was necessary, in the face of imperfect data or measurement (or understanding), in order to reach a general understanding of things. Gregor Mendel's supposed cheating in his analysis of pea plants is a famous example (Weiss 2002b), and so it was with the classical Copernican revolution in astronomy that occurred despite many totally wrong inferences and facts incompatible with the theory as proposed at the time. Many theories are wrong, but even correct theories typically require commitment to a framework even in the face of such evidence. Darwin's wrong ideas about inheritance and the problem of the age of the Earth did not prevent recognition of the overall truth of common ancestry.

Even in related animals that end up with similar adult morphology, developmental stages *subsequent* to the pharyngula can vary greatly (Raff 1996). There are several well-studied examples of this: sea urchins, anurans (frogs), and ascidians (primitive chordates including tunicates, or sea squirts). From genetic phylogenies of contemporary species, among closely related species with similar adult forms, some are found to pass through an intermediate juvenile or larval stage, whereas others do not. Furthermore, the similarity of larval stages across the phylogeny suggests that it probably was the ancestral state (at least, it is an ancient state). However, a larval stage has been dispensed with in several subsets of the descendant lineages. Figure 8-9 shows the "pluteus" larval stage in sea urchins. Several frog lineages have subsets of species that pass through an immature, tadpole stage. The converse situation also appears to be true: among sea urchins, there are genetic differences in how *similar* developmental stages are achieved (Kissinger and Raff 1998; Nielsen et al. 2003; Raff 1996; Raff 1999). Comparable statements appear to apply to the development of flowers among angiosperm species (Kramer and Irish 1999; Ma and dePamphilis 2000).

It is clear that there has not been an absolute, or clear-cut, advantage to going through one or the other form of development—a ladder of developmental sophis-

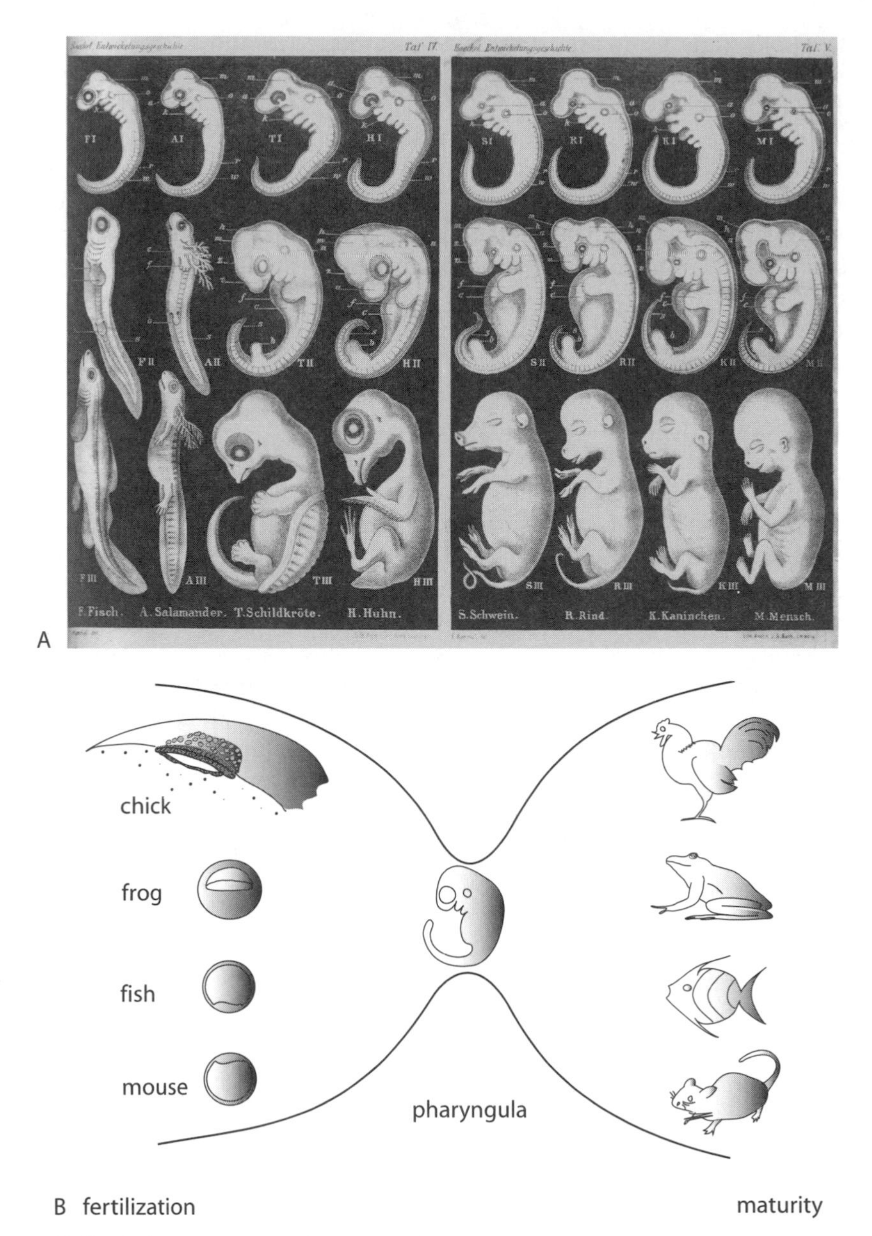

Figure 8-8. (A) Ernst Haeckel's famous figure showing divergent adult stages from similar pharyngula stages; (B) developmental constraint notion, that divergent blastula stages among vertebrates all pass through a common pharyngula stage and then diverge again to the adult form. (A) photograph courtesy Michael Richardson; (B) modified after Richardson, Hanken et al. 1997 with permission.

tication or improvement is not evident. Direct and indirect developers with very similar morphology are doing well today. A larval stage is not a necessary aspect of a frog's solving the "problem" of becoming an adult; some do, some don't, and axolotls stop at the larval stage. If the idea of evolution by natural selection is correct, then it must be that under some local environmental conditions, when the

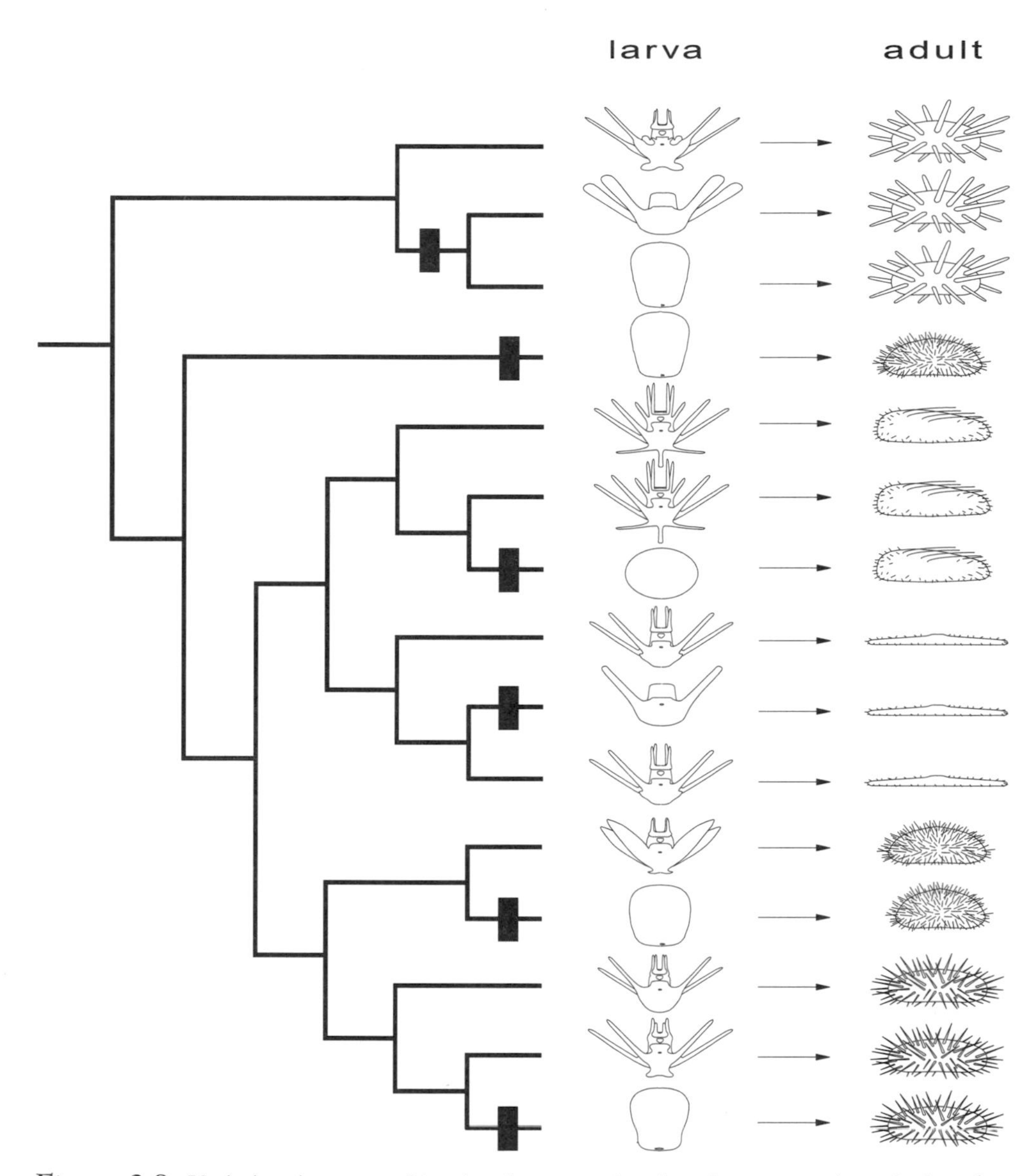

Figure 8-9. Variation in sea urchin development showing the repeated evolution (or repeated loss) of the pluteus larval form in different branches. Modified from (Wray et al. 2003), reprinted with permission.

appropriate mutations were present, one route of development provided an advantage of some sort.

Repeated evolution of a structure is plausible if the mechanism is not too intricate. A potential alternative compatible with modern theory is that the structure was present in the common ancestor, and was repeatedly lost. A variation on this theme would be that the changes occurred by phenotypic drift: that both developmental strategies existed in a population because of some tractably simple genetic mechanism, and one variant replaced the other by chance (or, was not needed and its loss was tolerated by the screen of natural selection). When we see something spotted throughout a phylogeny, such as wings in some insect clades, tadpoles and optic cups in some amphibians, pluteal stages in some sea urchins, and some aspects of olfaction or color vision we face these explanatory possibilities.

Growth, Heterochrony, and Allometry

One of the persistent questions in the evolution of morphology or development is how what appear to be complex changes can be brought about in a practicable way in the time available and by the simple process of natural selection. It has long seemed—correctly, it appears—that it can't be that each variation on a theme is due to the arrival of a new gene specific to the purpose. For example, it would seem to be impractical for every stripe on a tiger, or vertebra, or cusp on a tooth to come about by the action of a new gene.

Discussion of this topic goes far back in the history of biological thought (Schlichting and Pigliucci 1998), but the best-known systematic attempt to explain the phenomenon of changes in shape in modern scientific terms belongs to D'Arcy Thompson (Thompson 1917), who stressed concepts of *allometry* (size differences in the same structure that may or may not be linear in all dimensions), *heterochrony* (changes in timing of the same structure), and *heterotopy* (changes in the placement of a structure in an organism) during development. His idea was to explain the structures in organisms in terms of the universal laws or properties of physics, rather than invoking biology-specific explanations. Whatever the proximate mechanism (which we would attribute to genes, cell characteristics, structure proteins, and the like), biological structures achieve their variation due to physical constraints and temporal aspects of their growth. A number of authors have recently dealt with these general topics (e.g., Calder 1984; Gould 1977; Hall 1999; Raff 1996; Schlichting and Pigliucci 1998; Wilkins 2002; Wolpert, Beddington et al. 1998). The close study of developmental timing has shown this to be a source of variation in vertebrate body plan, in contrast to the notion of a conserved pharyngula stage in vertebrates mentioned earlier. Timing seems to be simple to change, but can have what appear to be complex effects: a powerful tool for evolution.

In modern genetic terms, we seek to identify molecular mechanisms that will bring about these ends and to understand their relative importance. The appeal of an allometric or heterochronic approach is that in principle it can provide a means by which rather simple molecular processes can account for a wide and otherwise perplexingly complex diversity of shape or other higher-level complexity. Shape can be substantially changed simply by changes in the speed of or the differential timing of developmental processes like tissue growth, the onset of segment-formation, and so forth. This can be achieved by mutations in the dynamics of gene expression or the interaction of products of the same genes, without a need for new genes or mechanisms to arise. This may be a major means of morphological evolution (e.g., Carroll et al. 2001).

In fact, there are many examples of substantial differences in the timing and pattern of growth and development among related species, showing how variable this can be or how weakly correlated it is with phylogeny (e.g., Chipman et al. 2000). However, neither heterochrony nor allometry explains all morphological variation. For many changes, even among closely related species, more profound genetic mechanisms seem to be responsible (e.g., Raff 1996).

A famous example of the work of such mechanisms in evolution is the idea that humans are an example of *neoteny*; that is, we retain juvenile morphological proportions, particularly in brain size, as adults; this enables us to have relatively large heads compared with our closely related primate ancestors or contemporaries. Of course, we aren't really just grown-up babies (even if we act that way), but the differential shape and growth history resembles that of an arrested maturity.

Many aspects of development within a generation are responsive to environmental conditions, and some may be somewhat heritable genetically (e.g., mating type in yeast) or epigenetically (a well-nourished mother may produce a larger egg), a point that cannot be stressed enough (Lewontin 2000; Schlichting and Pigliucci 1998). If the environment persists, the effect is effectively heritable, even though not in genetic terms (Chapter 3). Many such factors, including effects on chromosome packaging (akin to genetic imprinting, e.g.), may remain to be discovered. Removal of the relevant environment may lead to a reversion of the organisms to their former state, but it may be that successful survival or mating will come to depend on the "environmental" trait. If so then if mutations arise and are favored by selection, genetic assimilation can occur (Hall 1999; Waddington 1953; 1956; 1957; Wagner 2000; Wagner and Misof 1993). Then the trait becomes inherited in the genetic sense. For example, by being bigger, an animal might be able to utilize a different food source, which could lead to selective favoring of other traits related to that change: animal gets bigger because the egg was larger because climate was favorable for its parents; because of larger size, animal can eat larger seeds than rivals from less favorable parental environment; genes for eating larger seeds hitchhike to high frequency as a result.

From Cells, to Organisms, to Communities, to . . . "Gaia"?

There are numerous ways in which cells aggregate to form organisms, or organisms aggregate for common purposes, or ways that cells are connected physically, logically, or temporally or by social aggregation. This suggests that an appropriate notion of an "organism" should be broader than is usually considered. As human organisms that sequester a germ line, we tend to think that the "individual" in our usual sense is the essential unit of evolution, the very definition of a living being. But this is rather arbitrary (e.g., see Buss 1987).

The nature of association among cells carrying genotypes within a "species," defined as those that can potentially reproduce together, is variable. Cells within our bodies aggregate necessarily, and only the germ line connects us cellularly to other humans. But trees and sponges do not sequester a germ line, and the reproductive cells of species like some annelids change during the individual's life. Is there a major distinction in regard to the logic of evolution between the cells in our various organs and the ants in a hive? Bees can be viewed as having a restricted, sequestered, specialized germ line, in the form of the reproductive cells of the queen. This differentiation is induced, for example, by royal jelly, but is that conceptually different from the induction of our germ line by diffusible hormones? Our limbs develop more or less autonomously relative to other parts of our body, but is this so different from the "limbs" of a hive—its individual bees—relative to each other?

Sexually reproducing organisms are not completely separate units of evolution (except hermaphrodites). But how completely separate are the different humans in a population? We are cellularly connected internally by the tree of our developmental life histories and externally by our reproductive history. We communicate among disconnected cells in our own bodies by use of circulating free cells or signaling molecules and to others by the air using pheromones and their receptors and through transmitted vibrations (sound) received by a different kind of receptor, and so on. As we will see, the mechanisms are very much the same even at the fundamental level of genes. We have cellular dependence in our bodies, but the bodies (and hence cell aggregates) of our offspring depend on others in their population, certainly until they reach maturity. It is possible to take a more seamless view of biological connectivity, all through the common phenomenon of cell division. This is much the notion Bateson was trying to express long ago (Bateson 1894; 1913) in the sense of different individuals being extensions of a single organism.

Some of the subtleties, and perhaps the reasonableness of our suggestion that the notion of the individual organism is not so clear as is usually thought, can be seen in plants (Halle 1999). Because of the relative independence of the different shoot meristems, plants can shed gametes from different locations rather than having a single sequestered genome. Somatic mutation can lead these different reproductive units to become genetically variable over the many stems and long life of a tree. Further, the individual units compete with each other during the life of the plant. The individual stems draw sustenance from a common root system (which, however, has its own independent repetitive units), but they act in other ways as individuals.

We can extend the same ideas even more broadly to include the self-reproducing, homeostatic communities we call ecosystems. They have "organic" behavior, and even the very diverse species that make up an ecosystem have common ancestry, if very ancient. There is also complex communication among the organisms (e.g., via sounds, smells, and in many other ways). They eat each other, but how different is that from apoptosis or other aspects of degrading and recycling of its constituents that routinely occurs in complex organisms?

Notions of these sorts have been taken to their global extreme by a few investigators (Lovelock 1979; 1988; Lovelock and Margulis 1974). They suggest that it is productive to consider the entire global biosphere as a single homeostatic, self-evolving superorganism. This idea has been called *Gaia*, after the Greek goddess who drew the living world forth from Chaos. The Gaia idea has sometimes been used to argue that the Earth is an almost self-aware system that tries to maintain harmonious balance, resisting the pressure toward disorder from the forces of entropy (e.g., discussed by Turner 2000). Actually, within the confines of modern empirical science, roots of the Gaia hypothesis can be found in the work of a founder of modern geology, James Hutton (Hutton 1788), who viewed the world as a closed physical system subject to the physical laws of nature.

The basic idea is simply that, through various feedback mechanisms, the diverse forms of life maintain on Earth a stable physicochemical environment that would not be possible were the planet inert and life-free. Unlike passive physical structures, living forms are homeostatic, and their interactions depend on their evolutionary connectedness (for example, species consume each other in part because the prey is made of constituents similar to the predator's and hence reflect what the latter needs). In evolutionary terms, an excess of some resource can stimulate the evolution of organisms to fill it, and extinction can be viewed in some senses as a way to remove imbalance. At least, any multicomponent system with inputs, outputs, and interactions can achieve regularities of structure or various states of homeostasis or equilibrium (or can spin out of control, but that has not yet happened to the biosphere). No mystic or conscious component need be invoked in considering the biosphere as a single interacting system.

Indeed, because evidence suggests that all life is cellularly connected through evolution, and currently interconnected in numerous ways, the question is whether we gain scientific insight by thinking of the biosphere as a unit. Certainly, we do this in reconstructing phylogenetic trees from DNA sequences or in using the biochemical pathways found in common to try to infer the origins of life. And many ecologists have thought about the overall energenetic aspects of life, such as its relationship to the conservation of energy and thermodynamic efficiency, as a criterion for selection and for the evolution or balance of ecosystems, food chains, and the like. It is valuable, at least, to remove conceptual constraints in trying to understand how organisms manage to get through life.

CONCLUSION: MANY WAYS OF "LIVING THE DREAM"

The great biologist Francois Jacob supposedly said that the "dream" of every cell is to become two cells (so said his friend Jacques Monod) (Monod 1971)) (if DNA can dream, would it be of the ecstasy of base pairing?). In his classic book written shortly after the DNA coding system had been worked out, in which he had a major role, Monod considered that all of life is about the single "project" of reproduction. He refers to this as *teleonomy* to try, perhaps only partly convincingly, to avoid invoking teleology. The driving dream of reproduction is fulfilled in many ways; but rather than the project of life being *to* reproduce, it may be more accurate to say that a core *characteristic* of life is *that* it reproduces.

What is done to be big or complex are but different aspects of the same processes of basic cell biology and replication and of differentiation that started when they were small and simpler. The three to four billion-year-old unbroken membrane and its contents continually ooze off buds that, when they stick more or less together we call the "development" of an organism, when they separate we call the "reproduction" of a new organism, and when they no longer join cells during their life history we call "speciation." But these are all forms of variation on one, long, connected largely branching process. This is a remarkable fact, and it makes the evolution of life as we see it much more believable.

The single phenomenon "make more of life" has been a conserved trait of all the lineages of life forms that have survived to the present. But *how* making more is done varies tremendously. There are many ways to become long, large, mobile, complex, or to make progeny. We presume that the growth, development, and reproductive phenomena we have discussed at so many levels came about through the agency of random genetic change and natural selection. In modern biology, we are not satisfied that we understand a phenomenon until we understand it at the genetic level. This is but one way, and perhaps not even the best way to understand life and its development (Keller 2002). But our major purpose in this book is to understand the role of genes in the nature and evolution of life, and we now turn to what is known about the genetic basis of these phenomena, as they relate to development. We will see that genetic mechanisms, like the phenotypes to which they are related, may be conserved for long time periods, but are also quite fluid.

Chapter 9
Scaling Up: How Cells Build an Organism

A single cell can develop, more or less on its own, into a full adult organism. This basic phenomenon was noted nearly as long ago in the history of Western thought as we can go back, by Aristotle. In *On the Soul*, he wrote, "It is a fact of observation that plants and certain insects go on living when divided into segments; this means that each of the segments has a soul in it identical in species, though not numerically identical in the different segments, for both of the segments for a time possess the power of sensation and local movement." William Bateson (Bateson 1913) likened organismal growth to the basic process of cell division, and Bonner (Bonner 2000) suggests that the beginning of complex organisms was the failure of cell division products to separate, presumably in the early ocean.

A single cell somehow contains within it all the requisite "information" to go beyond its first division. Here, we will describe how a repertoire of a relatively few basic and logically simple processes is widely used to produce organismal diversity. Much of this chapter may seem like gene name-dropping. Listing genes is not very informative, and we are most likely to have found only what we know how to find, not every gene involved in the trait or process of interest, but the results do make sense. Not only are the same processes used and reused, but very similar genes or genetic mechanisms are involved over and over again to bring those processes about. These repeatedly used processes include what we would expect: signaling factors and receptors, transcription factors, activating and inhibiting effects. By mixing the order, frequency, and timing of their use—that is, the repeated contextually varying expression of subsets of the same set of genes—an organism can bring about its own transformation from a simple beginning to an adult.

HOW CELLS MAKE ORGANISMS: BUILDING MORE WITH LESS

In a way, cell division was a fundamental trick in making more of life. The mechanisms by which the constituents of a cell replicate and a cell divides are related to

Genetics and the Logic of Evolution, by Kenneth M. Weiss and Anne V. Buchanan.
ISBN 0-471-23805-8

the way cells adhere, move, or cooperate (e.g., Bonner 1998; 2000). Thus materials in spindle fibers are similar to those used in flagella or cilia, and various speculations can be made about how both cell division, with its rules for the distribution of materials into daughter cells, and other multicellular functions evolved. Much of what happened in the rest of evolution is a use and reuse of the many things that were needed for basic cell biology and division.

One ultimate question that we touched on earlier is *why* cell division has led to the aggregation of cells, and their dependence on each other, to the point that only some of them are even able to copy themselves into new colonies (individuals). We can imagine biological challenges or potential opportunities that single cells could not answer very well. Autonomy can be liberating in many ways, but we can imagine (and observe) circumstances in which large aggregates of cells can better withstand environmental changes or acquire resources. Bacterial aggregates appear to have this effect: bacteria in biofilms are much more difficult to destroy with antibiotics than planktonic bacteria, for example, due to a number of antibiotic resistance mechanisms that have recently begun to be elucidated (e.g., Stewart 2002; Stewart and Costerton 2001). Directed mobility enables organisms to pursue potential food if there is a local shortage (or avoid becoming somebody else's food).

On Being Big

It has been said that there is usually an open ecological niche at the top of the size range (e.g., Bonner 1998). Several times in evolutionary history a new branch of organisms has evolved a subset of lineages of ever-larger species, a premise sometimes known as Cope's law (see, e.g., Bonner 1988). Thus the first fish, land creatures, dinosaurs, birds, etc. may have been small, but larger species ensued. Yet, to reiterate a point we have made before, small is beautiful, too.

Large species have experienced several major die-offs, so it may be precarious to sit atop the size pyramid even if it gains you an empty niche when you first move in, but there are still a lot of large creatures around. Perhaps a kind of arms race develops among aggregates of cells (that eventually are reproductively isolated enough to become different species). Multicellularity has arisen many times in many different ways. Even when "bigger" was evolving, small members of the same lineages also typically did very well. Tiny reptiles scurried through the giant legs of dinosaurs. Single-celled and other simple organisms are thriving still and may be what is left standing at biological Armageddon.

Being large and complex implies a division of labor, which in turn requires specialized cells or tissue structures that can communicate with each other, but must be kept chemically sequestered so they can *be* specialized. This means carrying messages and nutrients to them and porting waste and messages from them, which usually means formalizing communication. It means protection against micro- and macro-attack and processing information from the outer world as well as among parts of the inner world. As a result, we see specialization for respiration, digestion, neural control, immune resistance, sensation, and the like (Bonner 1988). A large organism must also be sufficiently supported physically. Water can do this (internally or externally), and large organisms like seaweed or octopuses have very little skeletal structure. But most large creatures, land and sea, plant and animal, have skeletal structures of some kind (and so do most small creatures). Plants have rigid cell walls, and animals have external (arthropods) or internal (vertebrates, some gastropods) skeletons.

We cannot answer the general question of why cells originally began to aggregate, but we can address the question of how that aggregation led to something more than the sum of its parts: a differentiated organism. In this chapter we will first look at some of the general differentiation "strategies" and then at the repertoire of their use in the development of complex organisms. Much of the discussion will relate to morphology, but physiological systems can be similarly characterized in terms of their biochemical pathways.

Differentiation Is Hierarchical

Differentiation is a contingent process, in that what a cell does next depends on its current state. Because of the sequestration of gene expression patterns within cells, and the mitotic inheritance of those patterns, an organism develops a hierarchy of different parts. Cells and their descendants in this sense are partially autonomous, in a nested way, over time. However, they are not completely sequestered, and they may be directed to change their state by circulating signals from headquarters, or from the environment. Hierarchical differentiation means that if cells of a given type are needed in a particular context, the local precursor cells must provide them by division and changing gene expression, and if signaling is to be a part of this process, the cells must be prepared to be able to detect the appropriate signal.

It is a kind of developmental dogma that differentiated cells rarely revert to undifferentiated cells. In vertebrate organ systems, there are beds of stem cells that are committed to a particular organ but have not yet undergone terminal differentiation. When the time is right they differentiate into a stem cell and a descendant cell that undergoes this final state change process. These decisions are made during development but also throughout life as part of homeostasis or response to circumstances. In some cell lineages, like those of most vertebrate neurons, the final differentiation is thought generally to occur only once (mature neurons cannot undergo further mitosis). Some stem cells give rise to the panoply of different types of descendants, as in the generation of the diverse blood cell types. The epithelia that line most vertebrate organs, a layer of stem cells, undergo terminal differentiation repeatedly as needed. For example, cells in crypts between the villi that line the gut continually divide to replace the villus cells as they slough off into the lumen. Figure 9-1 shows an example of hierarchical differentiation of cells that form the central nervous system, a cellular rather than tissue fate-map.

The idea in animals generally is that differentiated cells within an organism have made too many gene expression commitments—often meaning modification of their DNA or its packaging—to go backward. This is analogous to Dollo's "law" for species: evolution does not reverse itself (Marshall et al. 1994). But unlike species, the cells in an animal retain the genome and under some circumstances can reverse their differentiation, whereas once mutations have occurred in a species it can't really regenerate its prior state. The recent engineering of cloned individuals derived from chromosomes obtained from differentiated somatic cells demonstrates that this can be done under artificial conditions (such as somatic chromosomes being placed in enucleated host undifferentiated cells to be stripped of the changes that differentiation had already made in it). Of course, like species, somatically derived chromosomes may bear some somatic mutations, a minor caveat to their

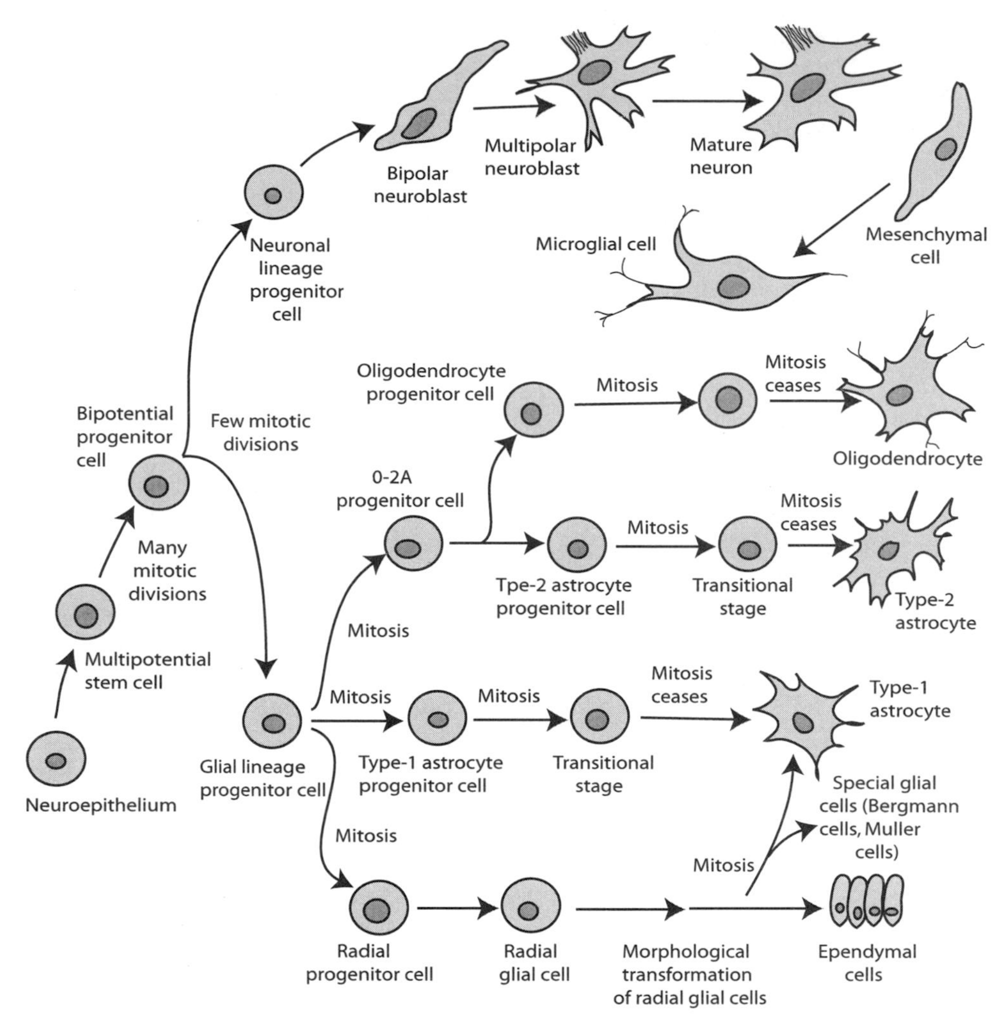

Figure 9-1. Cell lineages in the developing central nervous system, showing hierarchical cytological differentiation from stem cells. Modified after (Carlson 1999).

differentiation capabilities for most genes (though not in all, as will be seen in future chapters).

Cloning is one example of a growing realization that animal cells may not be as wholly committed as had been thought, but the generalization about normal context-dependent differentiation, to whatever state, seems generally to hold. And as we also noted in Chapter 8, plants are much less rigidly committed to their cell fates. Nonetheless, the use of hierarchical differentiation to produce sequestered, differentiated cells is a fundamental aspect of the development of complex plants and animals.

Organisms Are Built by Duplication and Repetition

William Bateson (Bateson 1894; 1913) wrote extensively and prominently on the problem of reproduction and replication, in particular the *meristic*, or repetitive, nature so common to biological traits (Carroll et al. 2001; Carroll 2001). As he said (Bateson 1886), "greater or less repetition of various structures is one of the chief factors in the composition of animal forms." We have cited Bateson many times in this book because of his interest in repetitive traits. He coined the term "genetics" and was an early advocate of Mendel's work on heredity, but he did not think repetitive traits could be accounted for in mendelian terms, one of the reasons that he did not think natural selection could explain evolution. We now know that genes can indeed be responsible for such traits, and we are finding the genes, as this chapter will illustrate.

In fact, the living world surrounds us with duplicate structures. Vertebrae and fingers are examples of this, but the same applies to leaves and petals, starfish arms, insect antennae and legs, worm segments, feathers, hair, and scales; branches in nerves and arteries; pancreatic islets, intestinal villi, and taste buds. Examples of meristic structures (that have repeated segmental elements) can be seen in Figure 9-2. Duplication of structures is clearly a fundamental aspect of complex organisms, and we have already stressed the repetitive nature of the underlying genetic and biochemical systems themselves.

Morphogenesis is Stimulated by Inductive Signaling

Because an organism is not just a simple linear array of different structures, but a complex three-dimensional structure assembled from a simple beginning, each component can only begin developing when the proper context exists. Organs are typically built when cells capable of the appropriate differentiation cascade are *induced* to begin differentiating by some form of context-specific signaling. We introduced examples of this in Chapter 8.

The Spemann organizer is an area on an early vertebrate embryo that patterns axial polarity, and an area between the future mid- and hindbrain is an organizer for brain regionalization. Epithelial-mesenchymal interaction (EMI) involving overlying epithelium and migrating NC cells was mentioned in Chapter 8 as an important characteristic of vertebrates. EMI is responsible for inducing sites where teeth will form along vertebrate jaws; subsequently, local spots of signaling factor (SF) release called *enamel knots* serve as organizers for cusp development within teeth. These are examples in which the Fgf signal spreads from its source to induce cell growth in surrounding cells.

All the cells in a tissue field may be capable of differentiating in some particular ways, such as to initiate tooth development, but only those that receive the SF signal will be induced to respond. Surrounding areas may then be inhibited from forming the structure. In arthropods, corresponding to vertebrate organizers, the imaginal disk precursors of wings and other structures are organizers for those structures. In plants, cell differentiation of pluripotent meristem cells is induced by cell-cell interactions and hormones in response to external conditions such as day length, temperature, water and so forth. Shoot and root meristems in plants can be viewed as containing organizers, and some of the signaling and indicator genes that maintain the source of stem cells will be described below (e.g., Benfey 1999; Laux 2003; Weigel and Jurgens 2002).

Inductive signals may come from adjacent to distant cells, as communication from one organism to another via pheromones or behavioral cues such as displays, or simply the presence of conspecifics, or from environmental information (olfaction, vision, immunity). The use of terminology for interorganismal communication that

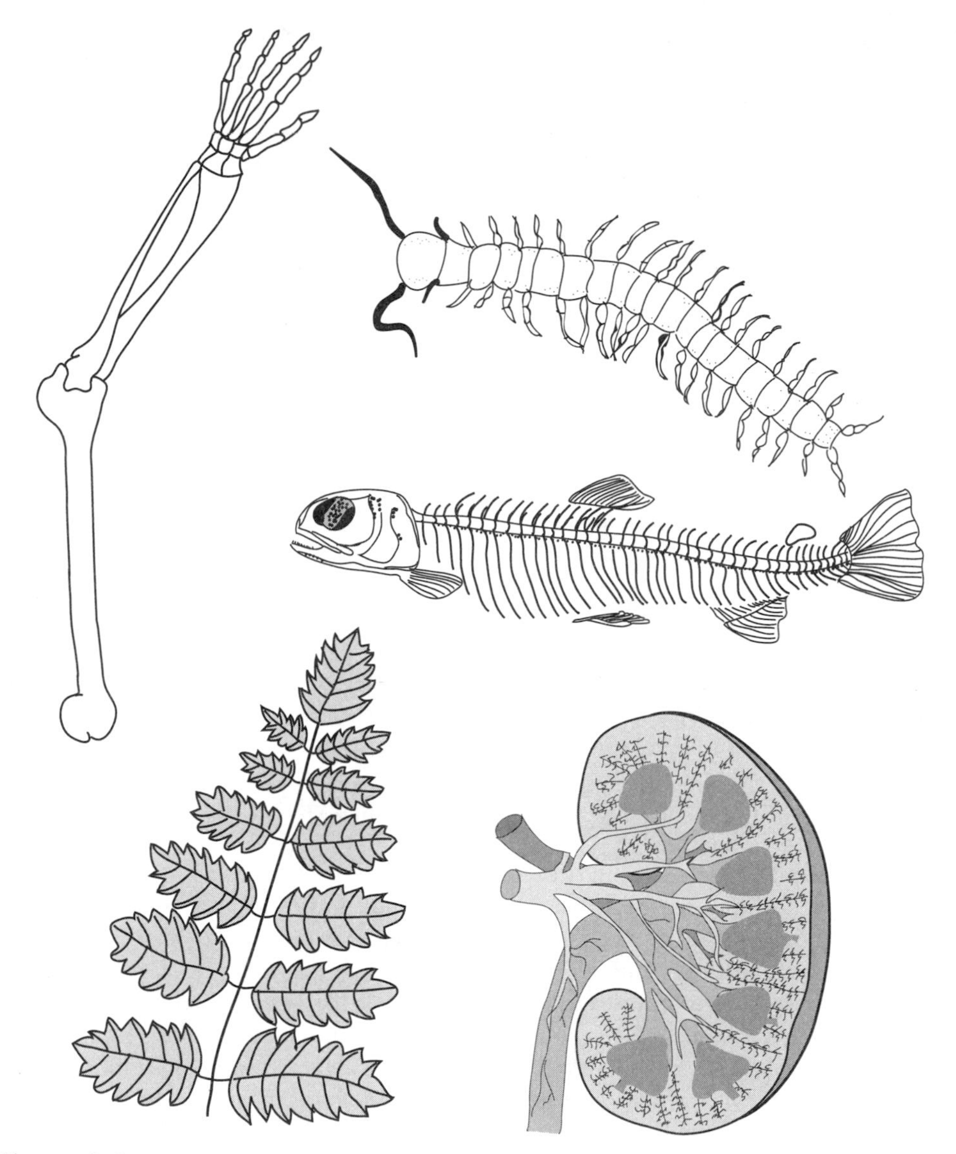

Figure 9-2. Duplication of structures: whole-body serial homology in centipede and fish; nested repetition of structures in plants; segmental nature of tetrapod limb; structural repetition within an individual organ (primate kidney).

is similar to that for cell-to-cell communication is justified because similar genes or at least information transfer mechanisms are involved.

Position in a Region Can Be Specified by Signaling Gradients

Local differentiation can be specified by concentration gradients of extracellular SFs (in this context known as *morphogens*). Morphogens can be protein products, such as those from the *Hedgehog* gene family, or other small, diffusible molecules like retinoic acid. The cells across an otherwise undifferentiated tissue field interpret their position by sampling the local concentration of the morphogen (see, e.g., Tabata 2001; Teleman et al. 2001; Wolpert 1969; 1981; Wolpert et al. 1998), using appropriate receptor mechanisms. Gradients of concentration can occur if a substance is produced in a localized source and physically diffuses away from it along the tissue.

A concentration gradient can form either simply as a passive ink-in-water process or by the binding of diffusing signal molecules to extracellular receptors removing them from the flow, by being diminished by active degradation by a factor produced at another source, and probably in other ways. Signals may move in various ways, including by passive diffusion or even by being relayed cell to cell (see, e.g., Figure 9-3). Gradients can induce gene expression changes if the signal concentration exceeds a threshold (Figure 9-3A), for example, by binding a large enough number of cell-surface receptors or by leading to the binding of enough REs flanking a target gene to induce its expression.

A classic example of gradient signaling involves establishing anterior-posterior (AP) polarity by the diffusion of transcription factors (TFs) within the early *Drosophila* egg (Figure 9-3B). Maternally deposited *Bicoid* mRNA diffuses from the anterior, and *Nanos* mRNA from the posterior ends (the early stage of the embryo is essentially a single cell with many nuclei on its inner side). Initially, mRNA from the genes *Caudal* and *Hunchback* are distributed rather uniformly throughout the egg and are translated in the cytoplasm. At the anterior end *Bicoid* message is translated, and Bicoid protein diffuses toward the posterior end. As described in Chapter 7, Bicoid protein binds to and inhibits the translation of *Caudal* mRNA, generating a gradient of increasing Caudal protein posteriorly establishing the first gradient. A corresponding Hunchback protein gradient is established because Bicoid protein binds to REs upstream of Hunchback, inducing expression of that gene. At the same time posteriorly, *Hunchback* mRNA is bound by Nanos protein, which is concentrated there.

As cell walls develop in the syncytium forming the cellular blastoderm, the intracellular environments become more sequestered, autonomous, and precisely controllable. Hunchback, Bicoid, and Caudal proteins are involved in expressing the "gap" genes that are the first zygotic genes to be expressed in an axially segmenting pattern, as the larva develops (of which more below).

A second example of a gradient patterning mechanism controls dorsoventral (DV) patterning. There, *Bmp*-class SFs (Dpp and Screw) are produced along with the proteins Tolloid in the fly blastoderm (Srinivasan, Rashka et al. 2002) (Figure 9-3C) (terminology and nomenclature can be complicated; *Dpp* is a fly relative of the *Bmp* class of *TGFβ* SF genes, Bmp being the name coined for vertebrate genes). Sog protein diffuses dorsally from a ventrolateral source in the embryo which at this stage consists of a layer of cells surrounding an inner extracellular space. Tolloid (Tol) and Tolkin (Tok) proteins bind and degrade the Sog protein as it diffuses dorsally, and another protein, Dynamin (Dyn) retrieves undegraded Sog molecules.

Together, these extracellular interactions establish a Sog concentration gradient, decreasing from the lateral source to the dorsal sink for this protein. Since Sog protein binds Bmps so they cannot serve as receptor ligands, and there is more active Sog laterally than dorsally, there is an inverse DV Bmp gradient. Its source is dorsal, and Sog serves as a sink. The resulting Bmp gradient affects gene expression differences that compartmentalize the dorsal embryo, as we will see below in a broader context.

A third example of morphogenic gradients patterns the vertebrate limb. At specified sites along the side of the embryo, diffusible Fgf8 protein induces (or at least is associated with) the initial outgrowth of limb buds. Anterior-posterior (AP) limb polarity is initially established by the induction of the SF Shh (a vertebrate *Hedgehog* family gene product) in a posterior (caudal, or tailward) part of the future limb known as the *zone of proliferating activity* (ZPA). ZPA-derived Fgf4 diffuses to induce the asymmetries associated with the posterior side of the limb. The organization of the distal part of the limb begins in the Fgf4-negative part of the limb bud. As the limb develops, regional variation in signaling molecules helps establish combinations of *Hox* gene expression, in roughly summed-sequential ways, along different limb axes, as will be discussed below.

These are tidy stories, but similar phenomena need not be due to the same mechanism, even within the same animal, even if a similar logical means is used. For example, gradients of *Dpp* activity are also used in AP patterning of the *Drosophila* wing imaginal disk. But in this case, unlike DV patterning in the early embryo, the gradients are established not by interaction with degrading proteins, but because a separate protein, called Thick-Veins (Tkv) inhibits the rate of Bmp diffusion away from its source (Srinivasan, Rashka et al. 2002) (Figure 9-3D). It does this because Bmp molecules are tightly binding ligands to Tkv and this stops their further

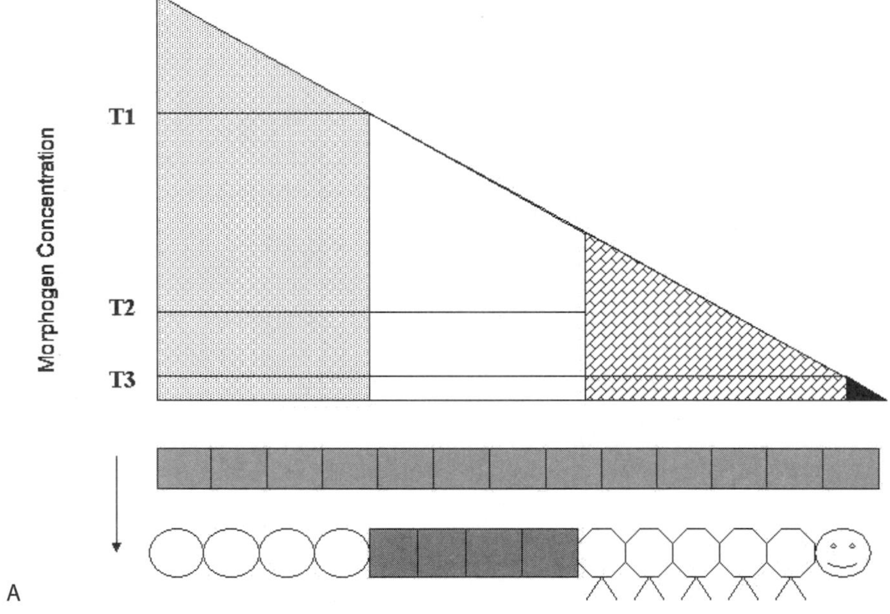

Figure 9-3. Gradient patterning mechanisms. (A) Multiple concentration thresholds (T1, T2, T3) established across a tissue field affect local cell gene expression, inducing regional differentiation (bottom) of previously identical cells (middle, induction indicated by arrow); (B) AP patterning in the *Drosophila* syncytium; (C) DV patterning in the fly cellular blastoderm: product of the *Sog* gene (+) diffuses dorsally intercepting Dpp (gray) diffusing laterally from its dorsal source. *Tol*, *Tok*, and *Dyn* gene products inactivate or intercept Sog molecules dorsally; (D) AP patterning in the Drosophila wing imaginal disk: Dpp (gray) gradient spreads asymmetrically from a source because the concentration of its ligand Tkv (+) is greater posteriorly; cell-to-cell movement of Dpp is aided by Dyn, not shown. For C & D see (Srinivasan et al. 2002).

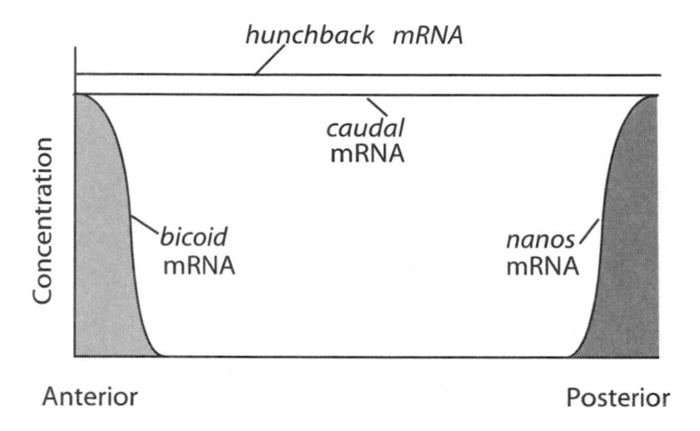

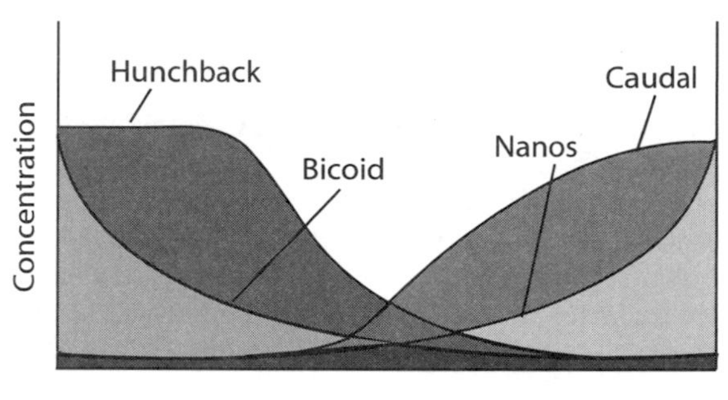

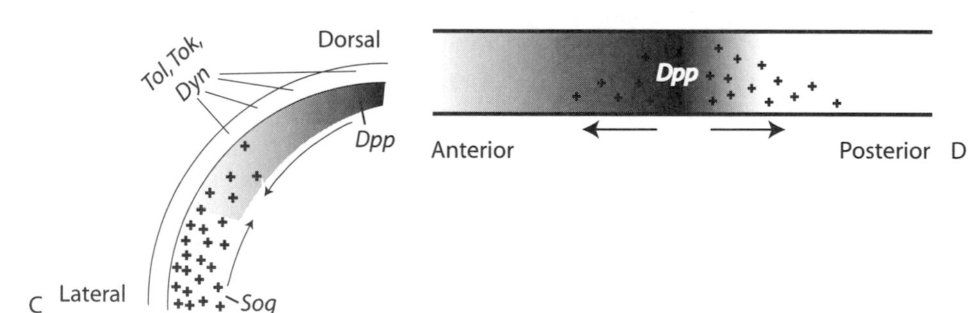

Figure 9-3. *Continued*

diffusion, and in the mainly cellular tissue of the wing disk, Dynamin here acts to move Dpp from cell to cell to help it diffuse from its source.

Figure 9-3C described the workings of a simple concentration gradient mechanism, for DV patterning in the fly. This establishes the dorsal dominance of Dpp signal that, among other things, inhibits neural development, which takes place more ventrally. This is all part of a broader DV gradient patterning mechanism, that results in regionalized DV areas that develop into the major tissue fates in the fly body plan. This process is worth describing briefly to show the nested or layered

Dorsal inhibits *Tolloid, Dpp, and Zerknüllt*

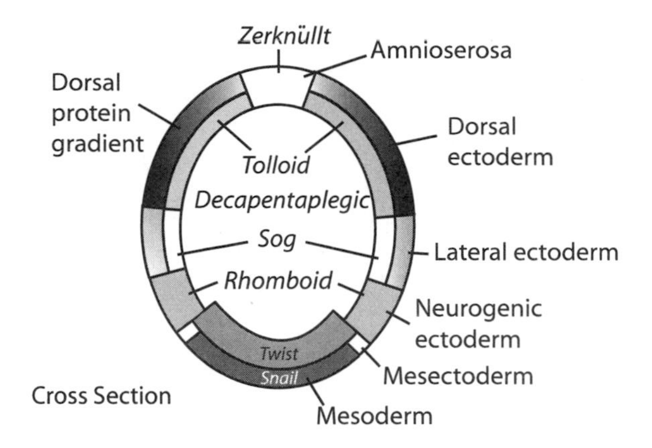

Dorsal activates *Twist, Snail, Rhomboid*

Figure 9-4. Gradient mechanism for dorsoventral body plan patterning in *Drosophila*. The center figure shows a cross section of the early embryo, identifying locations of gene expression as well as (right side) the tissue fates of the various regions. Above the figure is shown schematically that Dorsal protein inhibits genes that must be expressed for dorsal tissue fates; below the figure is a schema of gene induction by Dorsal, of *Twist* and *Snail* expression to induce mesodermal ventral fates, and inhibiting *Rhomboid* whose expression (ventrolaterally) is associated with neural fates. For details, see text and (Gilbert 2003).

nature of patterning and the numerous ways it can be brought about (Figure 9-4, and for details see Gerhart and Kirschner 1997; Gilbert 2003).

Early DV polarity involves a cascade of effects beginning with a protein, called Gurken (that is, coded by the *Gurken* gene) whose gradient is provided by the oocyte nucleus, which by the process of oocyte formation is located dorsally within an ovarian follicle that also contains within it 15 other cells, called *nurse cells* that supply mRNA and other materials to the oocyte (for details see Gerhart and Kirschner 1997; Gilbert 2003; Lawrence 1992). Gurken protein inhibits ventralizing signal from forming in the dorsal site of its (Gurken's) production. Another protein, Dorsal (i.e., a TF encoded by the *Dorsal* gene), in part provided by the follicle to the oocyte, is initially distributed throughout the egg. However, Dorsal's effect is to inhibit dorsal structures, so that it must form higher concentrations in the ventral side (the gene is so named because in its absence the entire embryo takes on dorsal fates). A series of events involving 11 known genes causes the translocation of Dorsal protein to the ventral nuclei in the syncytium.

As indicated in Figure 9-4, dorsal tissues form at the lowest concentration of Dorsal. Dorsal is complexed with a gene product, Cactus (not shown in the figure), and is activated only when released, which occurs via the Nudel protein whose concentration is highest in the ventral part of the egg, because Gurken protein, which is dorsally concentrated, represses *Nudel* transcription. Thus, a DV gradient of active Dorsal TF is established; via a series of threshold-level effects such as those shown in Figure 9-3A (where the cartoon at the bottom suggests AP patterning), Dorsal acts as a selector for differential gene expression initiating the major embryonic tissues. Ventrally, where the Dorsal concentration is greatest, it activates genes including *Twist* and *Snail* that form mesoderm and undergo gastrulation when external tissue moves inside the embryo to form the gut and other internal structures. Just lateral to that, the lower Dorsal concentration enables the gene *Rhomboid* to be expressed, and so on. On the dorsal side of the embryo, where Dorsal concentration is least, dorsalizing genes including *Tolloid*, *Zerknüllt*, and the familiar *Dpp* are activated. We saw in Figure 9-3C how Sog and other proteins, whose expression is induced as a result of the Dorsal gradient, then interact to establish the Dpp gradient, that in turn inhibits neural development, that occurs in cells in the region shown in Figure 9-4.

The Dorsal gradient mechanism is interesting in that it is a TF not a signaling factor, showing that it is the gradient, and not the type of gene that establishes it, that is important in the logic of patterning. In this case also, depending on the circumstances, the Dorsal protein appears to activate some genes and inhibit others. The same TF can have both kinds of effects, depending on the REs to which it binds near a target gene—for example, what else must bind there, or must be prevented from binding there.

Development Is Plastic . . . and Other Caveats

The patterning results just discussed are among the many phenomena that have been carefully studied experimentally. Many if not most of the genes involved were first found in mutant flies, and even named for those mutant effects, and their expression was then documented and manipulated to identify their effects. But traits are often studied first in terms of expression patterns of known genes, and it is important to be careful about interpreting expression patterns because they can be misleading: expression of a gene in a tissue or developmental stage does not automatically mean critical function at that stage. For example, the importance of EMI in pharyngeal development in vertebrates seemed somewhat perplexing because pharyngeal region patterning seemed to have arisen in evolution before neural crest. Experimental ablation of NC cells in the chick to prevent their migration did not disrupt early pharyngeal arch patterning (Veitch et al. 1999). Thus, although NC may be involved, normal development does not entirely depend on it. In a similar way, it is a common experience that experimental inactivation of developmental genes in mice does not affect a structure in which the gene is expressed and/or in the way the expression pattern might suggest. Unfortunately, it is not practicable to do experimental manipulations of all gene expression patterns, especially in large or long-lived species.

The *Bicoid* and *Dorsal* stories are classics in the discovery of morphogenic gradients. However, organisms are robust and the actual effects are subtle. Bicoid gradients can be manipulated experimentally in *Drosophila*, and the result shows that to a considerable extent the embryo somehow compensates for experimentally induced aberrant dose levels and develops more or less normally (Driever and Nusslein-Volhard 1988; Namba et al. 1997). This is a point to stress something we have noted earlier, the role of chance even in the biochemical concentrations of cellular constituents. Organisms must be resistant to such changes. Bicoid concentrations have been found to vary considerably among embryos (Houchmandzadeh et al. 2002), probably buffered by scale-information built into the Hunchback concentration.

This variation may be a reflection not just of chance variation in transcription rates but of variability even within species in the structure and sequence details of enhancers more generally (Wray et al. 2003). We referred in earlier chapters to the fluidity and variability in enhancer binding location, number, and sequences.

But there is more. This early patterning might seem to be quite a fundamental aspect of development, one hard to change, but a fruit fly does not all insects make. Fruit flies are "long-germband" insects, in which the blastoderm (the part from which the embryo develops) takes up most of the egg and the body segments are specified simultaneously by processes such as those we have described. By contrast, "short-germband" insects develop more structures outside of the egg cell itself, sequentially during development. Recent work on the short-germband beetle *Tribolium* (another popular model arthropod species) shows that two other genes, *Orthodenticle* (*Otd*) and *Hunchback*, play the *Bicoid* role in establishing AP patterning (Schroder 2003). Whatever the common ancestor, there has been a substitution of mechanism for a conserved basic polarity-establishing phenomenon, a kind of phenogenetic equivalence and perhaps a manifestation of phenogenetic drift.

Genes from all four *Hox* clusters (denoted *HoxA*, *HoxB*, *HoxC*, and *HoxD*) are expressed in mouse embryonic development and are thought to be required for pattern formation. Many examples of effect, including homeotic shift (of identity of segments) have been observed in natural or experimental mutation of these genes. But the picture is not always so straightforward. As one of several known examples, mice experimentally lacking their *HoxC* cluster genes establish normal patterning (they have subsequent other problems, so the genes, which are used in many tissues, do have important function) (Suemori and Noguchi 2000).

SOME BASIC DEVELOPMENTAL PROCESSES AND SOME GENES THEY USE

We have identified a few generic principles by which spatial and temporal differentiation takes place. We can now describe how these same principles are employed in a repertoire of basic developmental processes that characterize much of development in animals and even plants. The same sets of genes appear again and again in different developmental contexts, even in the same organism. Many or even most regulatory pathways have homologs in vertebrates as well as invertebrates, or even in animals and plants—the kind of deep conservation referred to in Chapter 3 and elsewhere. These are often used in logically similar developmental patterning across diverse species, and in fact this pathway homology has helped establish homologies between traits that had been thought for more than 150 years (since Darwin) only to be analogous, that is, similar in function but independently evolved. Indeed, deep conservation and multiple use of genetic mechanisms have seriously challenged the concepts of analogy and homology themselves.

Primary Axis Specification

An embryo quickly establishes structured asymmetry (polarity) along at least one and usually two or three anatomic axes. Subsequently, many new axes or asymmetries are established, often in a nested way. Position along a temporal or physical axis is correlated with the expression of different genes or gradients of the same genes. Primary axis specification is fundamental and, once laid down by evolution, has stubbornly retained many of its essential features.

We have described how concentration gradients establish aspects of axes of the fly embryo. Homologous genes are used in vertebrates in an interesting way that relates to long-standing questions about the evolutionary relationships between the two major groups of animals—and showing what was, when discovered, surprising evidence of our common ancestry. As shown in Figure 9-5, the relative DV positions of the central nerve axis and gut are oriented similarly relative to early expressed genes, even though these signals are "inverted" between vertebrates and invertebrates relative to the DV axis. Proteins coded by the homologs *Chordin* in vertebrates and *Sog* in invertebrates bind the homologous proteins from *Bmp4* and *Dpp*, respectively, preventing the Bmp4/Dpp signal from being received by local cell surface receptors. Product encoded by the homologs *Xolloid/Tolloid* inactivates Chordin/Sog, allowing Bmp4/Dpp signal to be received by the receptors inducing neurogenic ectoderm. In a sense, the default cell fate of ectoderm is neural, so this antagonistic relationship prevents ectoderm in nonneural regions from becoming neural. Thus the *Chordin/Sog* genes "protect" a region from becoming neurogenic ectoderm. These are quantitative inductive interactions and have been shown experimentally to work across the vertebrate-invertebrate divide (e.g., vertebrate genes having their same effect in invertebrate embryos). Sometimes a conserved series of pathway interactions is found in both groups though performing different functions, but sometimes there is functional homology that reawakens notions of the unity of animal form that had gone out of acceptability a century or more ago.

It was hypothesized in the early 19th century by Geoffroy St Hilaire that invertebrates were "inverted" vertebrates (or vice versa). Figure 9-5A shows a forced example to position an invertebrate—a cuttlefish, and a vertebrate—a bird, suggested by two little-known authors, trying to show that the two types of animal were of essentially the same design; this figure was drawn by investigators who themselves had been drawn into a famous heated debate in 1830 between Geoffroy and his former colleague Georges Cuvier, about the extent to which the similarities were true (Appel 1987). Although there are clear similarities between vertebrates and at least some invertebrates (mouth and head at one end, anus and reproductive organs at the other, limbs branching out along the side, and so forth), at that time the homology of this overall organization, much less the skeletal, neural, digestive, and other systems was not clearly established. The uncertainty was probably made worse by the fact that this was before Darwin and hence before common ancestry was a serious explanation of body-plan similarity (Richards 1992).

Even after Darwin, and despite these overall differences, the known ancestral forms and fossils were such that it still seemed fanciful to suggest that a vertebrate was an inverted insect. However, the kind of genetic patterning data recounted above breathed surprising new life at least into the general idea. It seems clear that today's deuterostomes (chordates) and protostomes (worms, arthropods) have dorsoventrally inverted body plans relative to each other, both in terms of morphology and the expression of genes associated with corresponding structures (Figure 9-5B). In both groups, *TGFβ* genes (*Bmp/Dpp*) repress, but antagonists (*Chordin/Sog*) locally counteract that signal to permit, neural commitment. How this works in flies was described above; in vertebrates the combined action of the developmental signal β-*catenin* and *Nodal*-related proteins activates *Chordin* in the midline, repressing *Bmp* expression and allowing neural development (De Robertis et al. 2000; Gerhart 2000; Nielsen 1999).

Does this indicate that St Hilaire was right after all (Holley et al. 1995)? Clearly this seems to be so in the sense that the conserved but inverted anatomic relationships are matched by conserved but inverted gene induction patterns. However, an inversion would require one of the two forms to be ancestral, and rather than an inversion per se the common ancestor might have had a less dorsoventrally specific body plan, or have been intermediate in other ways between what we consider today to be two distinct body plans; the means of expression of these homologous genes in the two groups is different and some primitive animals have less distinct DV orientation than either vertebrates or arthropods (Gerhart 2000). Modifications over evolutionary time of the pattern of migration of cells that will take neural fate, relative to the locations of the mouth at the blastopore stage, could provide an explanation without requiring a real and more complicated inversion event (Fitch and Sudhaus 2002; van den Biggelaar et al. 2002).

Since the discovery of this conserved aspect of polarity, which made a wonderful story of the resuscitation of a prescient kind of guess that had been long ridiculed, many more kinds of homology have been observed between vertebrates and invertebrates. It is probably fair to say that major elements of most corresponding systems, and many if not most basic cell types, share at least some homologous gene expression. Table 9-1 provides a number of these, and Wilkins (Wilkins 2002) presents a thorough discussion of the prominent examples known to date. Of course, the concept of homology relative to specific structures is challenged by the fact that many of the shared genes, such as the *Hox* and many other TF and SF genes, have numerous sites of expression, making it likely that some similarities might occur even by chance, an implication of the powerful "strategy" of evolution to use and reuse a limited toolkit.

Darwin would be pleased, because this confirms his ideas about common ancestry, but with entirely new types of data. However, it is tempting to overstate the homologies. Geoffroy's idea of "inversion" included the exoskelton of invertebrates corresponding to the internal skeleton of vertebrates (e.g., a vertebra corresponding to the exoskeletal elements of an arthropod limb). We know that *Hox* and other genes like *Bmp*s and *Distalless* TFs are used in both, but they really are not very similar. Even if homologous pathways are used in similar ways, they are embedded in different overall organizational contexts. Thus the fly wing and vertebrate forelimb are both limbs, and their overall polarity is established by some strikingly

Figure 9-5. Comparison of stereotypical body plans for vertebrates and invertebrates to show the idea that they are inversions of the same overall plan. (A) Alleged similarity between vertebrate (bird) and invertebrate (cuttlefish) sent to St Hilaire in the 1830s by otherwise unknown authors Meyranx and Laurencet, and published by his opponent Georges Cuvier (Couvier 1830); (B) vertebrates and invertebrates have a common body plan in many ways, but with a 180 degree dorsoventral rotation. Top row is invertebrate, bottom vertebrate. Left are correspondingly oriented copies of an image of Geoffroy's inversion hypothesis by Wilder (Wilder 1909), right are transverse sections of the body plans of each animal type showing the major basic structures; to the right of that are genes involved in early dorsoventral patterning showing their conserved usage. For more homologies see Table 9-1, and see text.

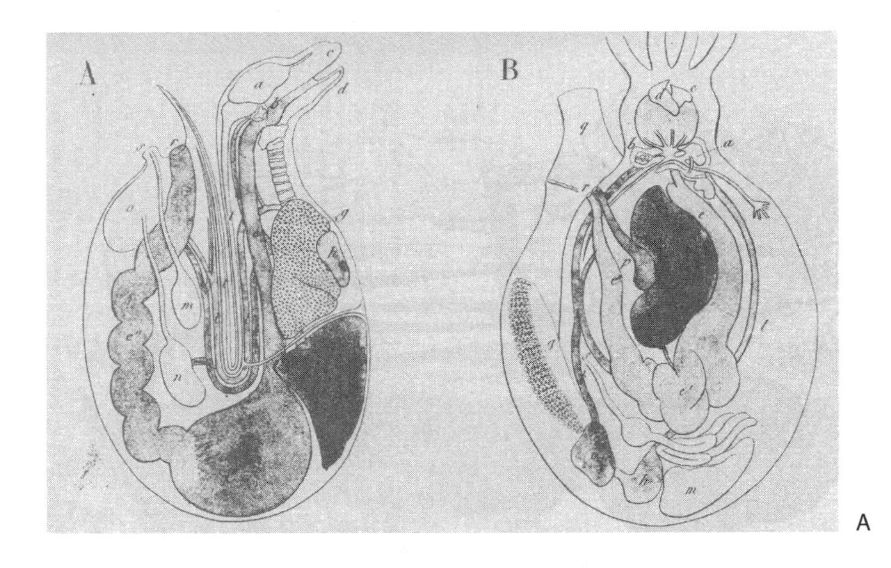

A

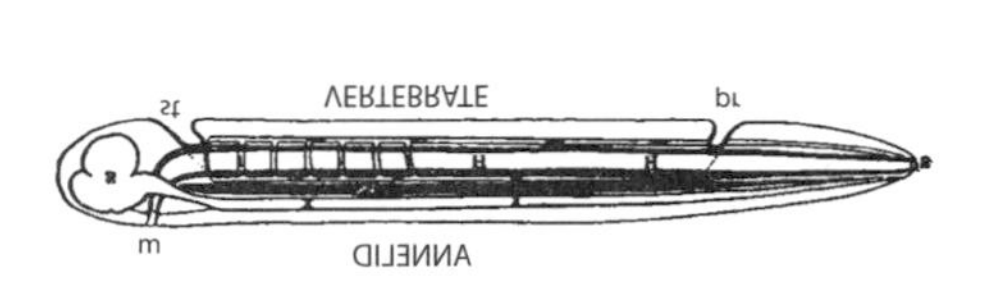

ANNELID
m
st
VERTEBRATE
pr

Fig. 140. Reversible diagram illustrating the Annelid theory. Reversible designations, applying to both forms; *S*, brain; *X*, nerve cord; *H*, alimentary canal. Designations applying to Annelid only; *m*, mouth; a, anus. Designation applying to Vertebrate only; *st*, stomatodæum; *pr*, proctodæum; *nt*, notochord.

ANNELID/ARTHROPOD

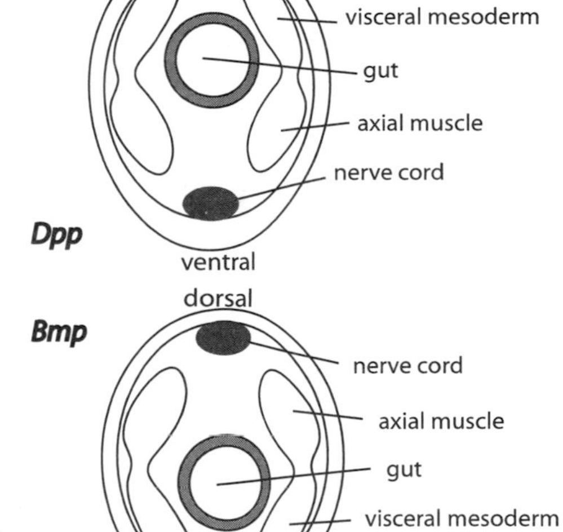

CHORDATE (VERTEBRATE)

B

TABLE 9-1.
Some shared gene expression patterns suggests genes that may have been responsible for patterning and development in the common ancestor of bilateran animals (vertebrates and invertebrates).

AXIS OR TRAIT	SHARED GENE EXPRESSION
Anterior-Posterior	Summed combinations of *Hox* gene expression specifies region along the axis; Segmentation behind CNS: *Hairy, Engrailed*
Dorsoventral	*Sog/chordin* dorsal, *dpp/TGF*β ventral
Gut	Para-hox genes: Anterior, *Lab/Gsx*; central, *Bcd/Xlox*, and perhaps an Antp/Ubx-like gene; posterior, *Cdx/Abd-B.*
CNS	*Otd, ems, Hox*
Photoreception	*Pax6, opsin*
Chemosensation	*Odorant receptors*
Body axis outgrowth	*Dll/Dlx, Fgf*
Heart	*Tinman/NK2.5*
Mesodermal structures	*Snail, Twist, Slug*

See (Arendt and Nubler-Jung 1994; Campbell 2002; Carroll, Grenier et al. 2001; De Robertis and Sasai 1996; Gerhart 2000; Scott 1994; Wilkins 2002).

similar gene hierarchies and inductions; but they are very different in their details: your legs and a fly's are not the same.

Repeated Structures Produced by Periodic Patterning Processes

Many traits—plant and animal—have periodic (repeated) and/or hierarchically nested patterns. What might be responsible at the gene level? Modular structures can develop Roman candle fashion by being generated from a source region and growing away from that region; the continued replacement of reptile teeth is an example. Combinatorial expression of TFs like the *Hox* genes can be responsible for initiating different cascades in the repetitive elements along an axis. However, neither combinatorial enhancer binding nor signal concentration gradients can explain all periodic patterning. For traits like hair, scales, feathers, intestinal villi, leaves, or fly sensory bristles or ommatidia (individual units of the compound eye), it is not plausible that each individual unit would be programmed by its own unique gene combination. There are too many units, and they are also often too similar to each other. Some other patterning process must be responsible.

In one such process the individual elements develop from initiation sites periodically spaced along an axis or in a tissue field. Once an initiation site becomes committed for the appropriate differentiation cascade, the unit can proceed autonomously without communication among the individual units; for example, the structure may develop when isolated experimentally.

Bateson developed an "undulatory" theory of life (Bateson 1894; 1913; Hutchinson and Rachootin 1979; Webster et al. 1992; Webster 1992; Weiss 2003a). He likened repetitive systems to interference patterns seen in wavelike phenomena being studied intensely according to field theories of late nineteenth century physics (Bateson had some notions derived from his time, that are not far from a kind of vitalistic inherency of built-in patterning, but we need not accept all of someone's views to see where perceptiveness may lie). Diffusion usually establishes simple gradients, as described earlier, but in 1952 the famous computer scientist Alan Turing (Turing 1952) suggested a *reaction-diffusion* process in which the interaction between two chemicals with different diffusion characteristics could establish spatially heterogeneous and/or repetitive patterns. Essentially, interference waves are established. In the archetypal simple model, the pattern can be described by two differential equations, one for the dynamics of change of concentration in space and time of each factor.

In a basic activator-inhibitor system of that type (Figure 9-6A) the interacting elements are SFs. One, the activator, catalyzes its own production in cells that detect it and also induces the production of a second signal that inhibits the activator. Both diffuse from sources of production across a cellular field. In this basic model, pattern formation requires that the inhibitor diffuse faster than the activator. This single *process* can in principle generate a series of spaced initiation sites in which an organ element will develop (that is, sites where the activator exceeds some expression threshold), surrounded by inhibition zones where no structure develops. The nature and stability of the resulting pattern depends on the production and turnover rates of these key substances (or on how many of them there are, and so on) (e.g., Bar-Yam 1997; Meinhardt 1996; 2003; Murray 1993).

That such processes were involved in biological patterning was suggested by the striking similarity between mathematical models and computer simulations and a host of observed traits including natural pattern traits such as the locations of hair, feather, teeth, butterfly spots, fish coloration, seashell and mammalian fur patterning, and others (see, e.g., Figure 9-6B and Kondo 1995; Asai 1999; Meinhardt 1996; Meinhardt 2003; Meinhardt 2000; Murray 1993; Nijhout 1991; Salazar-Ciudad 2002; Jung 1998). Very different patterns like butterfly wing patterns, fish stripes, and seashell coloration can be generated by similar processes worked in different ways. At the same time, small changes in the characteristics of a single process can generate very different patterns—just what one would want of a system that can evolve, can generate complex outcomes, but is itself not so complicated.

As one example, the mantle edge of a seashell is also a kind of linear *progress zone* in relation to shell color patterns. The shell grows out from the mantle forming a hollow spiral (hollow except for the organism itself). The mantle constitutes a line of cells *along* which periodic waves of different color generation are produced by a reaction-diffusion-like process. Wavelike color differences form along this line but then grow out and away from the mantle to become the new part of the shell. As the linear process "unrolls" from the mantle like a window shade, a two-dimensional color pattern results (Meinhardt 1996).

Simulations by various authors attempting to model different living patterns have used variations on the basic equations, and the process may never be exactly as specified by simple differential equations. But the nature, or *logic* of the patterning processes is probably similar to what is modeled. In fact, processes of this general

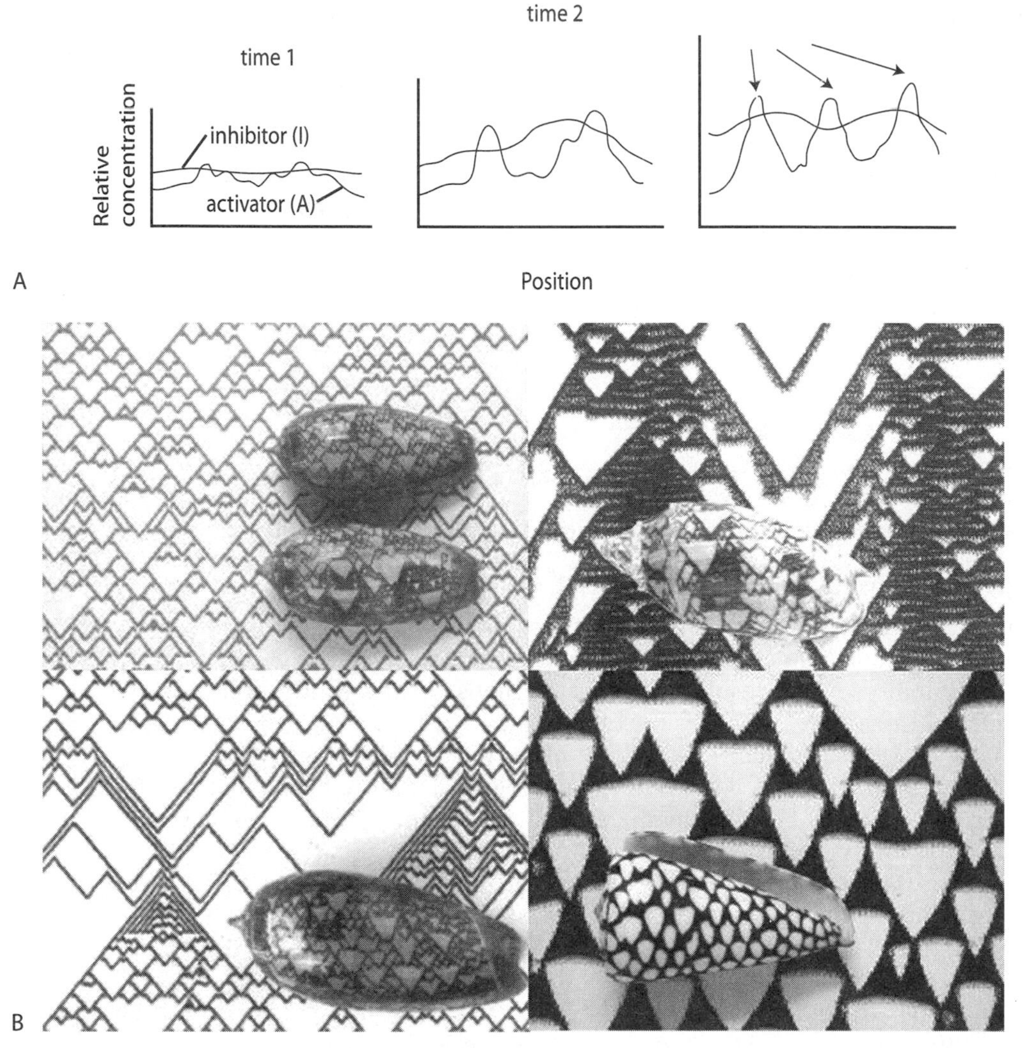

Figure 9-6. Reaction-diffusion systems can generate repetitive patterns. (A) The principles of a basic model shown at sequential developmental times; the horizontal axis represents position along a layer of otherwise similar cells. Cells produce an activator, **A,** that catalyzes its own continued production. **A** diffuses across the tissue, where cells detecting it are stimulated to make **A** as well as a rapidly diffusing inhibitor, **I**. Cells whose receptors detect **I** inhibit their production of **A**. Where the concentration of **A** relative to **I** exceeds a threshold level, gene switching occurs leading to activity peaks (arrows) where structural units, like feathers, teeth, or tooth cusps form. (B) Seashells real and simulated. Courtesy H. Meinhardt. For model see (Meinhardt 1996; 2000; 2001; 2003).

type seem likely to be responsible for a large fraction of the organization of complex organisms. There has been direct molecular demonstration of reaction-diffusion-like patterning in the use of *Fgf* and *Bmp* signaling as activators and inhibitors, respectively (Hogan 1999; Jernvall and Jung 2000; Jernvall and Thesleff 2000; Jung et al. 1998; Jung et al. 1999). Here we add the qualifying terminology, "reaction-diffusion-like" to indicate that these repetitive patterning systems vary in most of their details, the nature and number of interacting agents, their diffusion mechanisms, and the like.

In this kind of *dynamic patterning*, the frequency and location of the waves, peaks, or zones of expression are determined not by the specific ingredients of the system but by its dynamic *parameters*: timing and intensity of expression, number of cells, diffusion rates, receptor/enhancer binding efficiencies, and the like. Combinatorial expression is also involved, because it is the combination of activator(s) and inhibitor(s) and the cascade of gene expression they trigger that specifies the zones of growth and inhibition, and while there may only be quantitative differences between the two, combinatorial differences consequent to the signaling process account for the structural differences.

While nested patterns may involve several patterning signals, there is not a separate gene or even gene combination for each hair, feather, or tooth cusp. Instead, each unit is produced by a reinvocation of the *same* developmental cascade. The dynamic interaction of SFs determines when, where, and how many such instances will occur. There are genes for the process, but not "for" each individual component. The units *and* their intervening regions are part of a single trait.

For repetitive organ systems, an initial process called *prepatterning* occurs, which involves the priming of genetic switches that are "remembered" in the cells (or their descendants); the "enabled" genes are activated only at some subsequent time and tissue context. That is how they "know" to express *receptors* for the SFs they will have to be able to detect in order to be induced. Traits like dentition or taste buds, mammalian coloration patterns, and perhaps feathers in birds or respiratory spiracles along the sides of insects are examples. EMI may be the triggering phenomenon, when migrating NC cells in the jaws encounter inductive signals from specific locations in overlying prepatterned or prepared epithelium.

Feathers and hair are arrayed along what first appear as linear axes and then become more spatially distributed. Extensive studies have been done on the patterning of chick feathers (Figure 9-7) (Jung, Francis-West et al. 1998). Lines of comparably pluripotent cells expand in the epithelium, ultimately forming stripes, detectable experimentally by the presence of SFs including Shh, Fgf4, and Ptc (from the *Patched* gene). The stripe then organizes into local "condensations" of cells, or slightly thickened *placodes* that are the initial indication that a structure will develop. Primordial placodes serve as autonomous local signaling centers or organizers, which interact inductively with underlying NC-derived mesenchyme to differentiate into feathers. Zones of activation that then generate surrounding zones of inhibition are typical of periodic patterning processes. SFs Bmp2 and Bmp4 are produced and diffuse away from the feather placode, where Bmp receptors on surrounding cells receive these ligands, and are inhibited from forming placodes, thus producing inhibition zones around each placode. Bmp antagonists like Noggin or Follistatin in the placode prevent self-inhibition. Much the same pattern is responsible for hair patterning (Jung, Francis-West et al. 1998; Oro and Scott 1998) and as noted earlier teeth as well (Jernvall and Jung 2000; Weiss et al. 1998).

This general and rather simple process is nested in various ways. As the inhibiting signal diminishes with distance from an initiation site, new placodes appear. The original stripe differentiates into periodic feather buds along a line, but the diffusion process then produces new feathers laterally, to produce the two-dimensional array. Different types of feathers arise even among adjacent locations, and in dif-

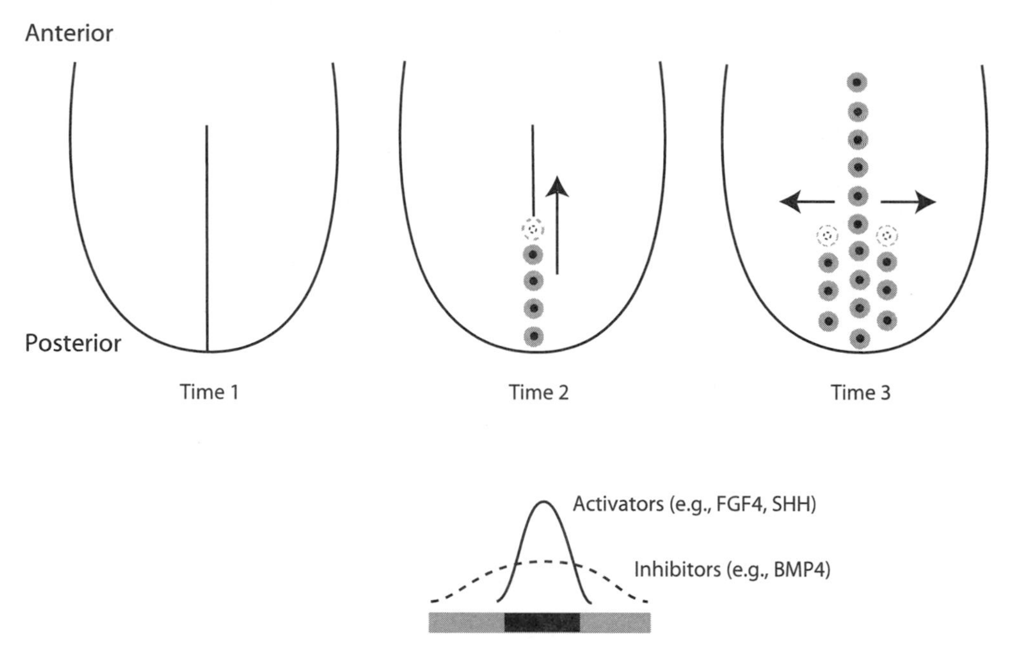

Figure 9-7. Feather patterning reaction-diffusion schematic at standard times during chick development. Original signal develops along a line, The signal then develops activation spots where initial feather buds arise. As the process continues, diffusing activator and inhibitor signal leads to the formation of additional rows lateral to the original one. Bottom figure shows the activator-inhibitor gene system with gray-shade that corresponds to the top figure. Redrawn after (Jung, Francis-West et al. 1998).

ferent parts of the bird (see Figure 9-12), much as different color patterns are generated in different parts of a mammal's fur or different types of vertebrae or segments appear along the AP axis of vertebrates and invertebrates, respectively. Hair, fish scales, mouse facial sensory whiskers, and mammary glands are patterned structures that develop similarly.

Although elegant experiments have shown the reaction-diffusion-like nature of this process, knockout and other experimental manipulation of these signals also show that they are not all required for successful patterning. This suggests that other factors are involved (e.g., Chuong et al. 2000). Of course they must be: as we have noted in many places, at the very least the cells must be primed to express the SFs, receptors, and signal processing machinery and to make the appropriate parts of chromosomes available to respond to the signal.

In vertebrates, the oral epithelium expresses the SF *Fgf8*, but this signal becomes inhibited by zones of *Bmp4* expression, and local Fgf8 foci remain that will become tooth-forming placodes along the upper and lower jaws (and/or elsewhere in the oral cavity, depending on species). Other genes, including the SF *Shh* and the TF *Lef1*, are also expressed in the initiation sites. The first known indicators are organizer-like zones of SF production that are similar to those just described for feathers in birds (indeed, hair and teeth can be made to grow in similar areas in some experimental manipulations in mice) (Jernvall and Jung 2000; Jernvall and Thesleff 2000; Thesleff et al. 2001).

Specific genes in the underlying mesenchyme (e.g., TFs *Msx1*, *Msx2*, *Pax9*) whose presence allows the NC-derived cells to respond to these epithelial inductive signals and others (e.g., Bmp2) appear to inhibit laterally adjacent mesenchyme from participating, thus focusing the mesenchymal response into a local responding center under the epithelial initiation site, in which a tooth will form. Subsequently, the local signaling centers within each tooth, the enamel knots (EKs) referred to earlier, appear and secrete similar (though not entirely identical) combinations of SFs. Interestingly, the EKs themselves do not express receptors for the factors they secrete, but the surrounding cells do (Kettunen et al. 2000); this stimulates cell division and downgrowth around the EK, which eventually undergoes apoptosis, but not before secondary EKs repeat this process to generate other cusps. The dentition has several axes: along the jaw, the breadth and width of individual teeth, and the crown-to-root vertical axis. Each is produced by signaling systems, sometimes involving the same genes. In the end, the pattern can be modeled closely by reaction-diffusion-like models for the array of teeth along jaws (Kulesa et al. 1996), and for periodic cuspal variation within teeth (Salazar-Ciudad and Jernvall 2002).

Feathers and teeth are among the model systems that illustrate what are much more widespread phenomena (there are many similar examples in insects and other animals and the equivalent in plants). In the case of teeth, experimental evidence suggests that the epithelial tissue layer has been prepatterned long before the initiation sites appear (Weiss, Zhao et al. 1998). Taste buds have comparable spatial patterning on the tongue, and direct experimental evidence suggests that these locations are prepatterned as early as gastrulation (Barlow and Northcutt 1998; Northcutt and Barlow 1998).

There are good reasons why this early preparation might be expected, which also illustrate why differentiation is hierarchical: a tissue like hair or feathers cannot form until there is skin to form on. Upper and lower teeth develop in anatomically separate regions of the jaws that suggest that prepatterning may have occurred very early there as well (Weiss, Zhao et al. 1998) and we referred above to evidence that the limb may also be prepatterned before it becomes a physically distinct structure. Another indicator that prepatterning occurs is that closely related animals can achieve similar results even though they are of very different size (such as the stripes on a Bengal tiger and a small tabby house cat, or teeth in an elephant and a mouse). One might say that this solves the allometric problem: the same basic mechanism is likely to be involved across the species size range, and the initial patterning event is likely to occur at a developmental stage when the embryos of all the species are more or less the same size (e.g., Murray 1993). The early stage can be much earlier than the first manifestation of the actual trait in the embryo. Then later, growth differences can generate structures of size appropriate to the species and these can be allometrically stretched or distorted by the growth factor (a subset of SF class gene products) patterns.

In a regionally patterned organ system, it is typical that the simplest and most global aspects of structure arise first and then more complex structures are produced. Axial region-specific expression of *Hox* genes is induced by gene- and region-specific enhancers (e.g., Shashikant et al. 1998). Once a region is specified, differentiation of hindbrain segments, mouthparts, vertebrae, and limbs occurs (Figure 9-8). Thus the *Hox* system is involved in the regionalization of major AP zones like thorax and abdomen, but these have multiple nested structures within

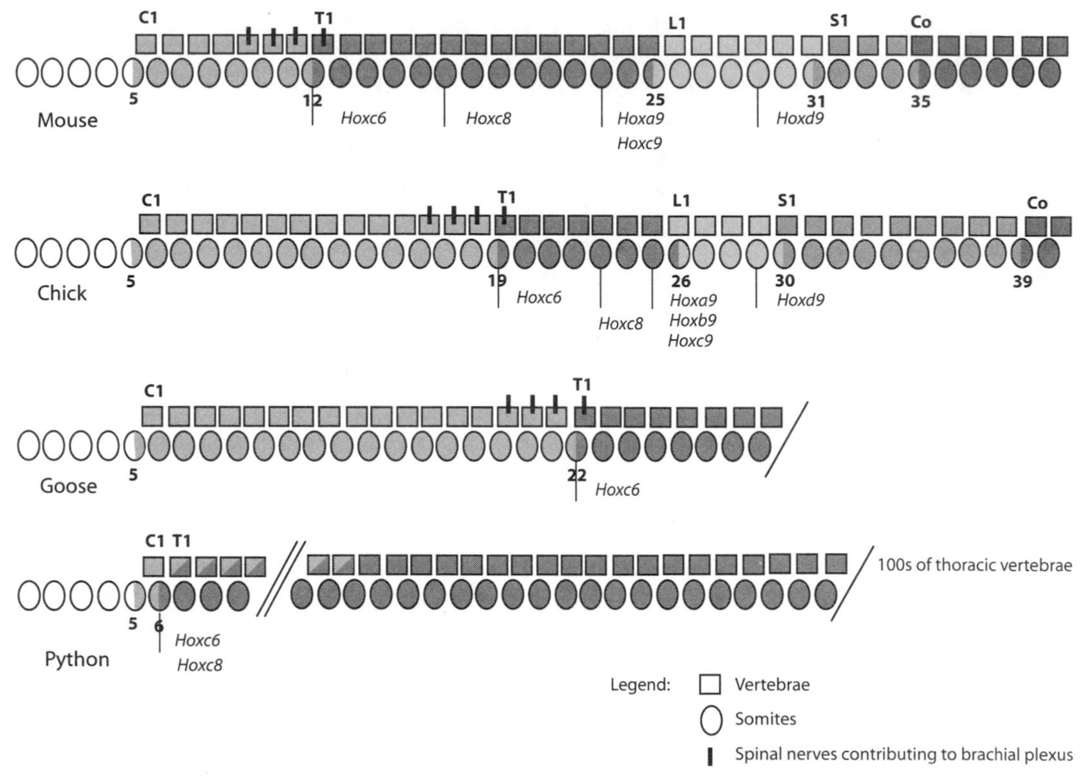

Figure 9-8. Functional conservation of axial *Hox* expression boundaries in vertebrates. Despite differences in detail, such as the number of vertebrae, gene expression patterns along AP axis are conserved relative to major homologous boundaries in the organism, such as the origin of nerves to the brachial plexus (that services the forelimbs). For corresponding body regions in other animal taxa, see (Carroll, Grenier et al. 2001). Redrawn from (Burke et al. 1995).

them (vertebrae and ribs, for example, that themselves vary and/or have patterned subunits). The homologous boundary of *Hox* gene expression switches shown in the figure indicates that an earlier broader patterning process is in place as the *Hox* system engages. Such layering of what may be rather similar kinds of patterning process may be involved in the structure of most meristic traits (and that means a high fraction of all traits in complex organisms).

Each of the roughly 800 ommatidia in the compound eyes of fruit flies has a basically fixed number of cells that are programmed, or patterned, by a reaction-diffusion-like activation-inhibition system. Each has about 20 cells of seven different types, including eight photoreceptor cells, seven of which are arranged hexagon-like around the other (known as R8); six surrounding accessory cells; and four lens-generating cone cells. The eye develops from an imaginal disk of about 20,000 cells. The cells in the eye imaginal disk are apparently equivalent and equipotent (e.g., all expressing the *Eyeless/Pax6* TF). These cells are induced to form ommatidia by a sweep of activation-inhibition activity induced by an indentation called the *morphogenetic furrow* that moves wavelike from posterior to anterior across the disk.

As the furrow passes it induces cells to express *Hh* (a *Hedgehog* gene family homolog of the vertebrate *Shh*), which then induces *Dpp* (the *Bmp* homolog) in the furrow to drive it anteriorly. In the R8 cell, the SF Boss (*Bride of sevenless*) is released and induces the *Sevenless* gene in the neighboring R7 cell in the little prephotoreceptor cluster. This signal to R7 initiates a cascade of differentiation of the other photoreceptors and surrounding supporting cells. The Wingless SF is also active in the disk as an inhibitor expressed ahead of the morphogenetic furrow, which helps establish anatomic polarity in the eye. Again the result is a somewhat elaborate pattern of structures generated by a single and relatively simple quantitative, nested signaling process.

Growth and Inhibition in Patterning

By themselves, pattern-generating processes might produce continually changing patterns, but the system can be stopped in various ways including apoptosis and mineralization (e.g., calcification of teeth and vertebrae, the moving away of shell material from the active mantle cells). One mechanism that establishes inhibition zones surrounding initiation sites is the *Delta-Notch* system (Table 7-2). This is used in the establishment of sensory bristles and ommatidia in insects and in feathers, teeth, and other structures in vertebrates. In the case of bristles (whose structure and nature will be described in Chapter 12), activation zones arise in a set of comparably prepared cells; cells expressing high Delta levels inhibit surrounding cells, and a few of those cells become the precursor of bristles. The successful cell (if being a bristle is more "successful" than being the surrounding tissue!) arises by chance.

The similarity of biological patterns to patterns simulated by computer (noted above) justified the inference that activation-inhibition processes were responsible for such traits in the real world. This was confirmed with the examples we have discussed. However, as so often happens, evolution frustrates what experience excites. The early *Drosophila* egg manifests stripes of expression of various genes, including *Even-skipped* (*Eve*) (the gene was named for the effect of its mutants on the stripes of early fly embryos). This seemed a clear example of a reaction-diffusion system, yet experimental work has shown that there are separate stripe-specific *Eve* expression mechanisms (Carroll, Grenier et al. 2001; Ludwig et al. 2000; Ludwig and Kreitman 1995; Ludwig et al. 1998). There is no rule without an exception—or is it that there is no exception without a rule?

Important Aspects of Quantitative Patterning

Turing's and other similar processes can be modeled mathematically in terms of the relative properties of interacting factors. For many years this was thought to be fanciful but rather meaningless (or even mystical and nonscientific) biological theorizing, because it could not be operationalized (proven in experiments, for example). As we have seen, that is certainly now done successfully. In a sense these processes can be viewed as "mathematical" in that it is not necessary to know what the factors are. If they have the specified properties, the effects are seen automatically. This can be said whether the process is by diffusion signaling, mechanical interactions, or temporal or physiological interactions among molecules rather than among physical cells. The result will be pattern; it can be ever-changing, divergent, or steady state depending on the parameters of the interactions.

This is an example of what we mean by the *logic* of gene use and mechanism in biology. We referred earlier to the functional arbitrariness but logical necessity of mechanism, meaning that it did not matter for the specific structural attributes of a trait what patterning process brought it about. If we think of processes in terms of their logic, it is the interaction of entities that is the process and, in a sense, not the entities themselves. Unfortunately, mathematical verisimilitude does not imply identity. There may be multiple mathematical ways to generate a similar outcome, which can be expressed one in terms of the other. In terms of the logic of patterning, it may not matter which equations are used; but for discovering the actual mechanism in any given case it matters considerably how many and which genetic factors are involved—and this becomes especially important in trying to understand homologies in species or processes we can't study directly from ones we have. Phenogenetic drift shows this as well.

But not all repetitive patterns can be simulated by a single set of equations, and in fact, meristic structures raise all sorts of subtle questions about homology (e.g., Hall 1999; Wagner 1996). The usual description of meristic traits is that they are represent *serial homology*. The notion there is that each unit is a separate use of the same process, not among species but among regions of the same embryo.

Reaction-diffusion-like processes can generate repeated structures as described, and it is tempting to explain any such pattern by these elegant processes. When true, each iterate of the structure (e.g., each hair or scale) is part of a single *continuous* process. However, by a different route, each unit might more literally be a repeated but *independent* use of the same process, something more akin to the idea of homology between structures. As with reaction-diffusion processes, the same genes would be used in each unit. Mutations in those genes would therefore affect all units, demonstrating the unity of the overall system.

Another way to generate repeat units is for the same genes to be used, but to be invoked by separate means in each part of the embryo. In such case, unlike a standard reaction-diffusion patterning system, the separate domains could be sequestered from each other. For example, the regulatory region of a gene that is involved in the cascade that develops each unit can have different enhancer cassettes, each responding to different TFs and in a specific compartment. This is the case, for example, with the *Eve* gene in *Drosophila*, where enhancers for different TFs control expression in different stripes in the egg and this helps establish subsequent AP segmentation. Here, mutation within the regulatory region will affect only those units that use the specific enhancer that was mutated.

There are probably instances of all three of these phenomena, and the latter two require that in some way a prepattern is laid down so that the repeated invocation of the same expression cascade takes place. Local reaction-diffusion like processes may be involved.

The patterning of butterfly wing spots seems to be a good example of these phenomena (e.g., Carroll, Grenier et al. 2001; McMillan et al. 2002; Nijhout 1991; Nijhout 1994). Wings are divided into cells (compartments, separated by veins), within which quantitative activation-inhibition patterning involving the development of expression sites ("organizers") of the genes *Dll* and *En* and others, leads to a ray of *Dll* that grows from a stripe of expression along the distal wing margin to the middle of each wing segment. There, it becomes a focal location of expression for a color spot. Other genes, involving the veins separating the segments, participate, and the genes include some of our familiar early polarity-specifying genes (*Dpp*, *Wg*, *Rhomboid*, and others) (Keys et al. 1999; McMillan, Monteiro et al. 2002).

Artificial selection experiments in which butterflies were bred to have larger or more, or fewer or smaller spots, have found that the effects are correlated among spots in the same individual (Beldade and Brakefield 2003; Monteiro et al. 1994). The effect of selection was to favor whatever combinations of genes might affect spotting activity. The result shows that in a sense the wing spots together constitute a single trait. However, other experiments have shown that mutations can affect (e.g., delete) individual spots within the row of spots (each in a different wing compartment), without affecting the rest of the spots (Monteiro et al. 2003). This reveals a strong modular element of control than normally achievable by a single reaction-diffusion-like process, and suggests that a complex enhancer was mutated. Some prepatterning process must, however, still be present, even if moved earlier in development.

Here we have an example of multiple components generating a serially homologous trait. The interpretation of the various results is not yet unambiguous, but begins to show how a complex meristic patterning process can evolve and vary (and scales are developmentally related to sensory bristles, and the focal spot of *Dll* is produced by the coopting of an evolutionary prior use of *Hh* signaling to separate anterior and posterior wing compartments) (Carroll, Grenier et al. 2001; Keys, Lewis et al. 1999). It may not be a simple, single patterning process. But it is a series of familiar processes, occurring apparently separately, within the cells of the wing, whose nature and origin are as we would expect.

OUTPOUCHING

Most animals are hollow tubes relative to their overall AP, DV, and left-right symmetries, but they have structures like limbs and antennae that grow outward from placodes or imaginal disks along the main axis, as we described above. A variety of new patterning signals are then expressed in spatiotemporal order, to establish the new main proximodistal (PD; sometimes called mesiodistal) axis as well as DV and AP axes within the outgrowing structure.

The same genes can be involved in multiple axes even within the same structure. A good and well-studied example is the formation of tetrapod limbs (Carroll, Grenier et al. 2001; Davidson 2001; Gilbert 2003). Genes from all four *Hox* clusters are used first to establish the overall AP axis. Then, locally, *HoxD9* and *10* are expressed as the most proximal region, or *stylopod* (e.g., humerus), develops in the outgrowing limb bud in a summed-sequential way, roughly from anterior to posterior. Along the PD axis, *HoxA9–13* are expressed. In the formation of the next segment, or *zeugopod* (e.g., radius, ulna), *HoxD10–13* are expressed in an AP summed-sequential combinatorial manner. Finally, as the *autopod* (hand) forms, *HoxA13* as well as *HoxD10–13* are expressed again in an AP manner as the digits form (in an axis separate from the PD axis, or in a modified, curved single axis that wraps around the end of the limb to induce the digits).

In each case, *Hox* expression is associated with the formation of cartilage models that later ossify as the bony elements. *HoxC* genes (along with other known genes) are expressed differently between forelimb and hindlimb. In this way, the same genes expressed at different times control different aspects of the hierarchical, regionalized, multiple-axis morphogenesis (see Figure 9-9).

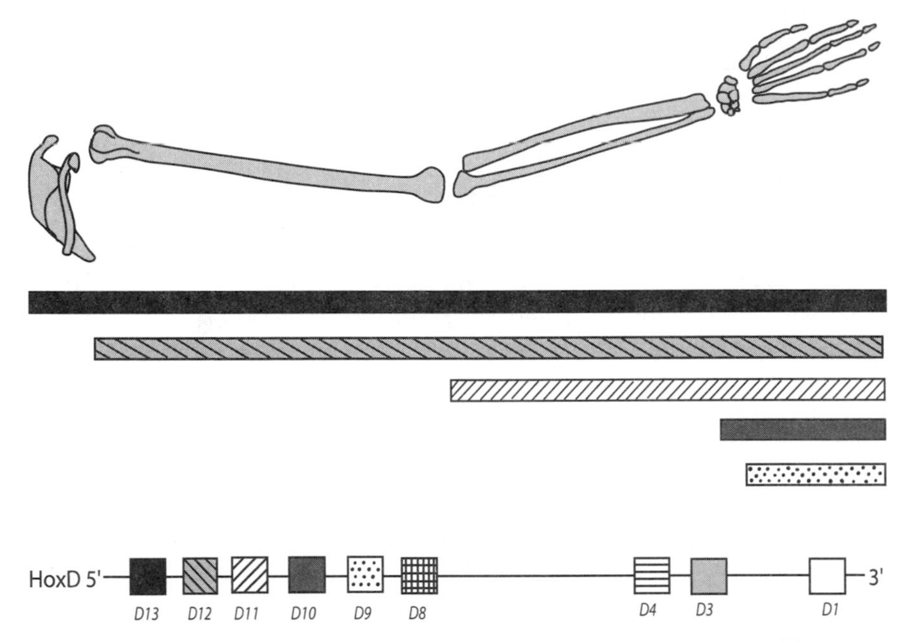

Figure 9-9. Combinatorial *HoxD* gene expression during development and its correspondence to sections of the adult tetrapod limb. Redrawn with permission from (O'Day 2003).

An interesting aspect of the vertebrate limb has to do with how its PD axis is formed and the *Hox* expression patterns are initiated. A structure called the *apical ectodermal ridge* (AER) forms as a kind of line that caps the distal edge of the limb bud. The AER has been thought to be a dynamic generative structure, or *progress zone*, so that as cells divide under the AER, the longer their precursors have been in that region the more they are prepared to make the segments that they do (stylopod, zeugopod, autopod). This seemed to be similar to the model of seashell patterning along a linear progress zone described earlier. Unfortunately, like so many stories of this kind, continued research shows that the story is not so tidy. The AER had been thought to be perhaps a stripe of reaction-diffusion-like process too slow to repeat before the limb becomes mineralized or stops growing for other reasons, but recent evidence suggests that the AER is already prepatterned in miniature, perhaps like the dental arch or tastebud patterning of the tongue, the three major limb segments ready to be "revealed" in the process of growth (Chiang et al. 2001; Duboule 2002; Dudley et al. 2002; Sun et al. 2002).

Wings grow out from the main body axis in flies and butterflies, representing a different kind of limb. AP coordinates are specified by sequential activation of genes called *Engrailed* (a homeobox TF), and the SFs *Hedgehog* and *Dpp*. Only the posterior part of this region expresses *Engrailed*, which induces *Hh*. Anterior cells express the *Ptc* receptor for Hh signal; receipt of this signal induces a line of *Dpp* expression. DV compartments are separated by dorsal expression of the *Apterous* homeobox TF, which induces expression of *Serrate* and *Fringe* components of the *Notch* signaling system; *Notch* receptors are expressed across the whole region but only receive Serrate/Fringe in the posterior part, and that induces in turn a stripe of the *Wg* (*Wingless*) SF expression.

Wg, like Dpp, diffuses in both directions from this line, inducing different subsequent expression leading to compartment-specific differentiation. Some of the SFs work across only a few cell diameters, such as Hh, but induce others that can travel more (e.g., up to about 20 for Dpp), in a kind of relay system. Two TFs, *Vg* and *Sd*, are expressed in all wing cells and enable other wing-specific differentiation to occur. When these two genes, along with compartment-specific TFs, are activated in the same cell they induce further compartment-specific gene expression, all resulting in compartment-specific combinations of expressed genes.

Branching

Budding occurs when a cell or cluster of cells in a tissue begins to proliferate locally and grow outward or inward from the initial tissue layer. Branching occurs if secondary buds form from a primary, and if this occurs repeatedly over time a treelike structure results. The processes are not unique; for example the hand can be viewed as a branched structure coming of the original budlike structure of the limb. We think first of branches and roots in plants, naturally, but similar patterns occur in animals. Branching can occur from an initiation site outward, as in plants, or inward as in the branching lobes and bronchi of lungs. Other examples include the branching of vascular systems, and the development of nephrons in the kidney, the ductal structure in mammary glands, and nerve networks. The branching of the major nerves of the human nervous system is shown in Figure 9-10A, and the circulatory system in Figure 9-10B.

Some branchlike organs are produced by a process known as *clefting*. Ingrowth of cells from the periphery of a bud can form a wall that divides the bud in two. Along with the formation of buds and branches, these processes can generate many of the structures we see in complex organisms (Hogan 1999). One can view the hand as a clefted as well as a branched structure. A pad forms first, in which the digital ray cartilages form, but then apoptotic processes cleft the spaces between the digits (but not in webbed species).

Branching systems arise in various ways. Some begin developmentally as a single initial indentation or sac (rather than a placode), which divides into a nested hierarchy of internally descendant branches. The first or major divisions of branching systems are often highly stable among individuals, as in the left and right lobes of the mammalian lung or the major arteries, veins, and nerves (which is why anatomy students have to memorize them). Eventually, these systems divide into more numerous and/or highly variable branches like capillaries (which is why anatomy students *don't* have to memorize them).

Some systems, like the circulation, are closed (branches must connect so that blood can circulate), whereas others, like the lung, are open-ended (branches can diverge ever farther from the original trunk). The mammalian circulatory system forms as an initial plexus of connected vessels and then proliferates by sprouting new vessels and collapsing or destroying others to form the large, deeply branched, but closed system in the adult. Vessels can form "on demand," induced by the release of signals from tissues starved of oxygen—including tumors—so that a branching pattern can often best be characterized by the *process* that generates it rather than by a map of the final structures.

Some deeply branched systems produce ever more numerous and smaller branches that resemble a *fractal* phenomenon. This means that the branching *pattern* is essentially the same no matter at what point you observe it. The developmental process continues to generate new copies of that pattern, though in many cases of ever-smaller absolute size. Mathematical models can mimic such structures, and give a compelling explanation of the type of continual process that generates the pattern, but some patterns that look as if they "must" be fractal really are not literally so (Metzger and Krasnow 1999). Instead, what is most useful is to say that, like the periodic patterning discussed above, a process is at work rather than a specific program for each element in the system. Indeed, because new branches can be induced at least in some systems (blood vessels, plants), this must be so for them.

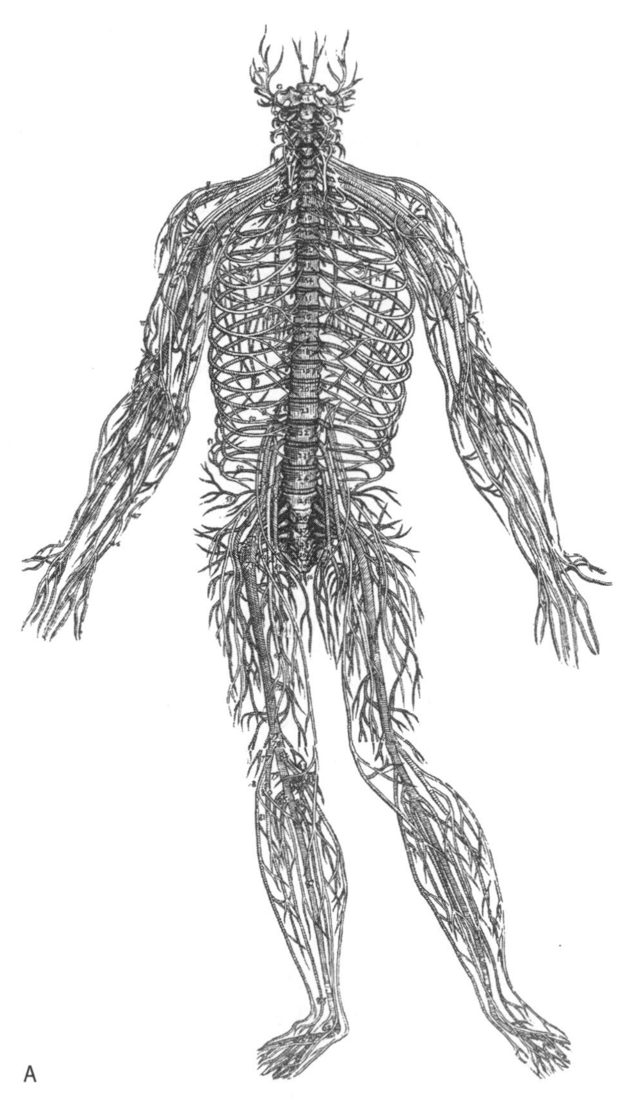

Figure 9-10. Branching patterns are important and widespread in life. (A) Branching of the major nerves in humans; (B) branching of the venous drainage system in humans. There are corresponding branching structures in the arteries and lymph ducts. From Andreas Vesalius' classical drawings of 1543, that helped introduced the modern era in anatomic studies (Vesalius 1543).

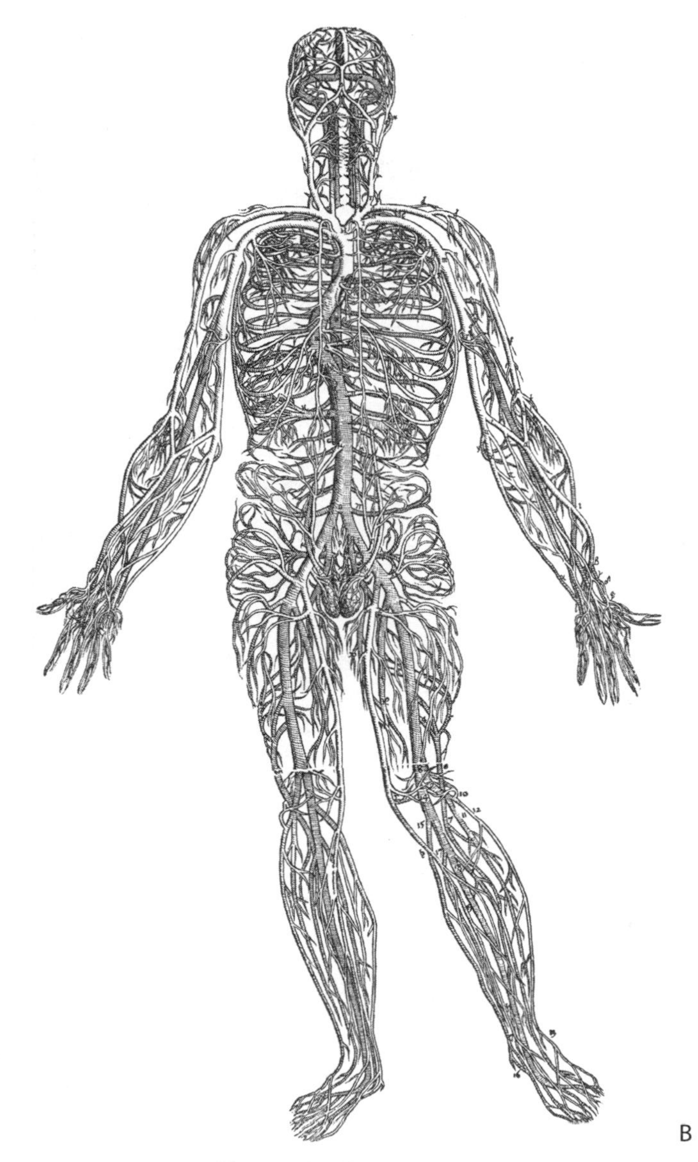

Figure 9-10. *Continued*

That the result has a fractal dimension (a numerical way to characterize the richness of the branching pattern) would be an incidental finding rather than something built into the process: there seems to be nothing about fractal geometry that is physiologically necessary per se.

Along an insect body, tracheal sacs form and invaginate to begin a branched tracheal system that carries oxygen to the internal tissues. The openings are known as *spiracles*. Once located along the embryo, each sac generates a structure of six primary, about 25 secondary, and hundreds of tertiary branches. There is considerable variation among species, but in the "archetypal" condition (i.e., in species like *Drosophila* where this has been specifically studied) the entire tracheal structure is formed by movement and morphological changes in an initial population of cells rather than by the cell growth by which mammalian bronchial trees develop. It is in this sense an autonomous developmental unit. It is adaptable, however, in that eventually the smaller branches grow toward signals produced by oxygen-deprived cells and connect with branches from other trees.

In *Drosophila*, spiracles initiate in 20 sites of expression of the *Trachealess* TF gene along the midgestation fly embryo, involving about 80 cells each (what locates the spots is not known) (Davidson 2001; Metzger and Krasnow 1999). Trachealess and Tango protein (bHLH class TFs) expression presumably regulates downstream genes necessary for tracheal sac development. A cascade of expression of the same genes occurs in each sac, which develops autonomously, by growing inward, forming secondary and tertiary branches, each of which appears to be a module involving similar genes.

The secreted *Branchless* (an *Fgf* homolog) is activated in five additional locations surrounding the initial *Trachealess* expression zone; expression appears to be delimited by determinants of the overall AP and DV axes relative to the sac (Metzger and Krasnow 1999; Samakovlis et al. 1996; Sutherland et al. 1996). *Breathless Fgf Receptor* homologs on nearby cells receive the Branchless signal, triggering a cytodynamic cascade that guides the migration of cells to their budding locations in each of the five primary branch sites. Meanwhile, a second secreted factor represses *Fgf* genes to create an inhibition zone surrounding the cells forming the primary branches. *Branchless* turns off in the cells as they move along the branch but then switches on again in the distal cells as the next round of branching takes place. Cells moving toward the Branchless signal express branch-related genes not found earlier, as well as a second inhibitor that confines the location of the secondary branching. Terminal branching again involves Branchless signaling but also oxygen-sensitive signaling; together, they generate the more variable subsequent branching that is needed to respond to local tissue oxygen requirements. Tens of downstream SF and TF genes that affect the different steps in this process have been identified by mutation analysis (that is, mutations lead to aberrant development).

The process of tracheal ramification is, however, not simply a repeated invocation of the same branching signal (Metzger and Krasnow 1999). At different stages along the tree different genes are expressed, and mutational analysis suggests that different branches within a tree are controlled by specific genes (although a common set of *Fgf* pathway genes are also expressed). In the latter sense, as with so many similar systems, subsequent branching may be more like a single process analogous to reaction-diffusion processes, a type of mechanism that can generate roughly fractal branching. At least, although there may be some differences along the developing tracheal tree, hundreds of branch-specific gene signals are not required. In having formed from a single primordial set of cells rather than by mitotic growth, the tracheolar system can be viewed as the unfolding of a program somehow latent or prepatterned in those cells.

Homologous genes are expressed in mammalian tracheal and bronchial branching, which goes through 6 to 20 or so generations depending on the size of the animal. The major initial stages (trachea to two primary bronchi, etc.) are essentially the same among individuals and species. *Fgf10* and its receptor are involved in triggering branching in the end of a growing tracheal bud, with inhibitory signal (perhaps of *Shh*) expressed in the center, so that the outer sides grow but the center does not (Figure 9-11). Experiments have indicated that mesenchyme is important in providing patterning cues, but the epithelial layer is also needed. As with flies,

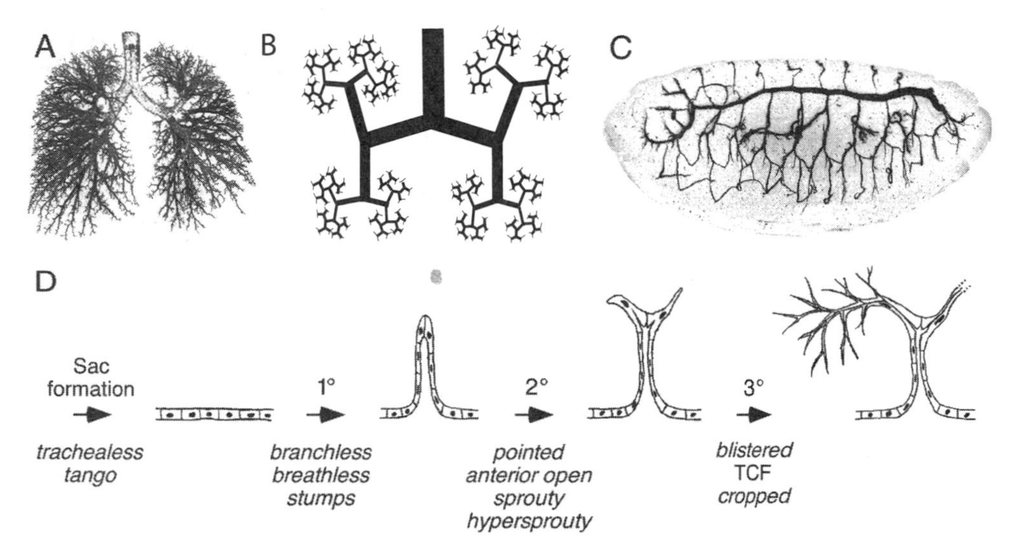

Figure 9-11. Animal budding and branching patterns. (A) Mammalian lung; (B) a fractal-like pattern generated by computer resembles natural budding and branching of this sort; (C) similar branching in the tracheole system of insects; (D) the steps in forming the tracheolar branching as in C are shown along with gene expression specific to each. Modified from (Metzger and Krasnow 1999), reprinted with permission.

the lung budding process is basically autonomous (can be achieved in isolated cultures of bud primordium), and the same signaling is reused in subsequent branches. Receptors are expressed more diversely than the cells secreting Fgf, which is expressed in a temporally dynamic pattern in the mesenchyme. Buds grow toward the nearby source region at any given time. As elsewhere in the embryo, Bmps are antagonists of Fgf, and in the lung they appear to inhibit growth and limit branch formation. Mesenchymal differentiation produces supporting structures that are more complex than in insects and involves further gene cascades.

A simple computer-generated fractal pattern is shown in the center of Figure 9-11. This has been said by many authors (but not those of the paper from which the figure was derived) to represent the kind of branching divergence found in traits like lungs where, for example, such a process generates the maximum surface area packed into a given volume. The figure shows only the first few branching iterations of what would be a space completely packed with ever more nested versions of exactly the same branching (what the term "fractal" refers to). The idea is a good metaphor, perhaps representing the "objective" of a branched lung to pack as much surface into a given volume as possible, but the lung is not exactly fractal nor symmetric. The left and right human lungs have two and three lobes respectively and are more different when seen in cross section (though not easily seen in the format of Figure 9-11), and this is a standard pattern not a chance difference.

Nested Complexity by Budding and Branching

Branching and budding can go together, and feathers provide an interesting and well-studied example (Yu et al. 2002). Feathers have three levels of branching: rachis or main axis to barb, barb to barbule, and barbule to cilia or hooklike structures that hold feathers together. As noted earlier, feathers form by a periodic patterning process that sets up initial buds separated into rows and columns by inhibition zones. Each of the buds then branches internally, then branches again (Figure 9-12), in a nested hierarchy that, unlike tree or lung branching, leads to different structures at each level. The hierarchy of branching provides multiple opportunities for diversity in final structure, which is manifest in the world of (and within single) birds. Not only is each feather differentiated by structure, and by color, but the array of feather types is different on different regions of the bird, and there are different types of feathers patterned within the regions as well, such as down, contour, and flight feathers. Each has its own modified morphology (and the whole complex can be repeated over space).

Key facts in branching morphogenesis are signals that specify (1) the location of the original bud, (2) the location and number of subsequent branches, and (3) the size, shape, and final histological differentiation of the structures at the ends of the branches (Hogan 1999; Metzger and Krasnow 1999). Feathers involve EMI and are also closely related in this respect to nonbranching structures, including hair and teeth and scales before that). It is not a surprise that despite such differences we find developmental friends including *Bmp2*, *Bmp4*, *Shh*, and *Noggin* (a *Bmp* inhibitor) involved in feather production (Chuong, Patel et al. 2000; Jiang et al. 1999;

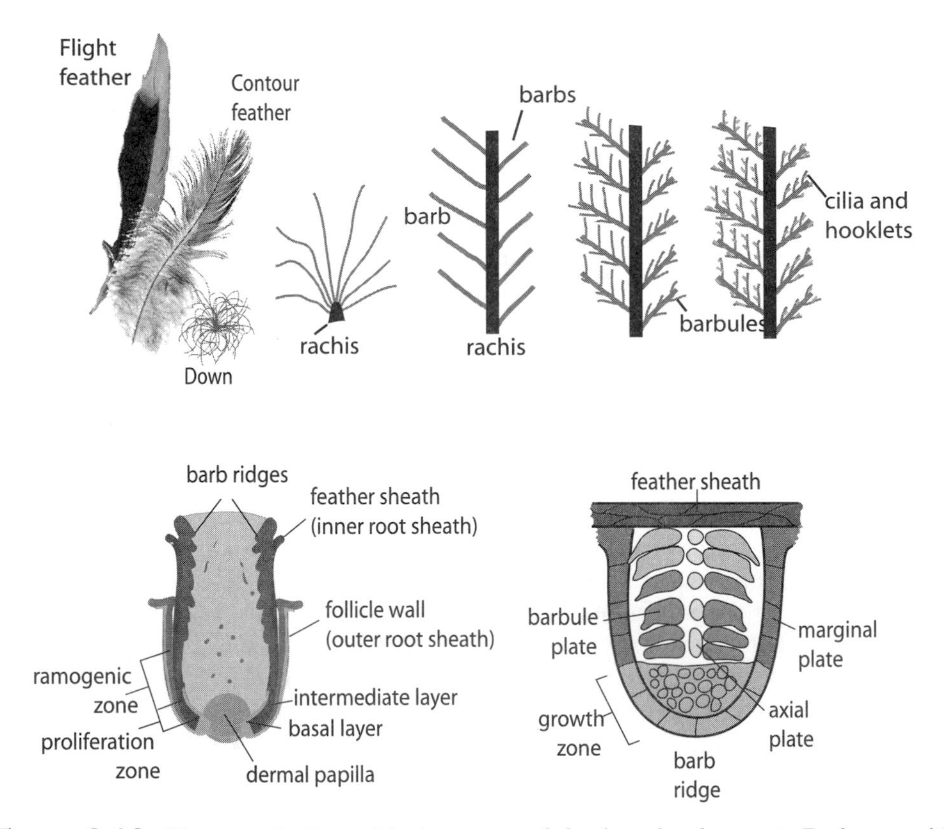

Figure 9-12. The nested, hierarchical process of feather development. Redrawn with permission from Nature (Yu, Wu et al. 2002) copyright 2002 Macmillan Publishers Ltd.

Jung, Francis-West et al. 1998; Metzger and Krasnow 1999; Patel et al. 1999; Yu, Wu et al. 2002).

Shared gene use is no surprise, but some aspects of the presumed gene action are interesting. Although Bmp4 inhibits tooth initiation and may be involved in inter-feather inhibition zones, it appears at the initiation sites within feathers. Bmp2 is a Bmp4 antagonist in teeth but has a generating effect in barbule generation. Shh is here not a growth stimulator but instead appears to be involved in apoptosis that removes tissue to make spaces between barbs. We noted above that Shh may have inhibitory effects in bronchial branching. There may be signaling gradient effects in feathers as well as switchlike effects. Overall in these and other systems, a nested, repeated use of the same genes occurs in different and sometimes at least partly opposite roles.

"Exploratory" Branching

Branching development may be "random" or fractal or stereotypical, but it can also be "exploratory" (e.g., Gerhart and Kirschner 1997). NC cells migrate through mesoderm to peripheral locations where, if they get there, they are stimulated to proliferate, aggregate, or differentiate by SFs emitted by overlying ectoderm of the appropriate type. Angiogenic factors secreted by tissues (including tumors) induce differentiation of angioblasts to form new blood vessels. Vascular branching at the local level seems largely random. Endothelial (vessel lining) cells present Fgf and other surface receptors, and chemotactic growth occurs as the cells proliferate to follow concentrations of growth factor ligands produced by oxygen-deprived cells. Tracheogenesis in insects may "seek" local oxygen-deprived cells, but in vertebrates oxygen is provided by the circulation rather than directly by the lungs; this means that vertebrates must supply lung branches with ample vessels.

In plant branching, which has some similar exploratory characteristics, the meristem acts as an organizer, as mentioned above. Central meristem cells express the homeobox TF *Wus* that maintains meristem status, but induces expression of a secreted protein, Clv3 in adjacent cells. These cells are maintained in undifferentiated state by the TF *Stm* (*Shoot meristemless*), that represses a cascade of expression of differentiation genes (including TFs called *As1* and *Knat1*) (e.g., Laux 2003; Weigel and Jurgens 2002). As the meristem grows upward these cells are displaced peripherally, away from this repressive signal, lose their *Clv3* expression, and become capable of differentiating into primordial cells for a new branch, flower or leaf (as part of the repetitive induction of competent states, the central cells in a flower regain *Clv3* expression). There are new genes here, but also some homologs. *Zwille/Pinhead* in plants are homologous to *Piwi* and its relatives in animals, and both help maintain stem cell state (Benfey 1999). This brief description is mainly from the *Drosophila* of the plant world, the mustard relative *Arabidopsis*, but similar and/or homologous mechanisms are found in other plants, differing, of course, as the plants themselves differ. Even where the mechanisms are not homologous, however, there is nothing new here in that the logic of the processes is similar to those in animals.

Leaves are also internally branched. They provide rich venation for the cells through diverse patterns among species from reticular patterns in broadleaf plants to parallel veins in many grasses. As with lung or other branching, there can be a hierarchy of branching order and branch (vein) size. Reaction-diffusion-like mechanisms seem plausible, perhaps involving transport of signaling factors like auxin, or other interaction gene products (Dengler and Kang 2001). That a quantitative kind of process is involved in leaf shape is suggested by experiments showing that changes just in the timing of expression of a single gene, *Knox1*, can modify a leaf from a simple to a complex shape (Bharathan et al. 2002).

These various patterns are exploratory in the sense that growth takes place without a predetermined autonomous plan, and without a prepatterning process, and can occur in response to local factors. Indeed, there can be great variation in a trait among individuals with the same genotype (e.g., inbred identical plants or animals), even if the environmental conditions are essentially the same. The genetic program is the mechanism for the branching, not the branch.

Flower Segmentation

Plant apical meristem tissue corresponds in some senses to the apical tissue in animal organ buds, but the analogous *process* does not involve homologous genes. Plant differentiation decisions are controlled more by external than internal conditions, including temperature, light, and humidity. Time since last branching also has determining power. Plants grow and bud in response to signaling molecules (generally small nonprotein molecules). Diffusible substances, including auxins and cytokinins, act as hormones to trigger these differentiation processes (see Chapter 10). Basic branching patterns in plants are shown in Figure 9-13. As with other examples in life, the apparent complexity of variation is probably brought about by relatively simple modifications of a basic process (Sussex and Kerk 2001). One of these appears simply to be in the differential timing of the development of axillary (side) branches off the main ("dominant") apical meristem; this appears to be controlled by hormones as will be seen in Chapter 10. Gene mapping studies (Chapter 5) have been done and some QTLs (candidate chromosome regions) have been found, that is, genes that quantitatively affect branch pattern and proliferation. One gene, a TF called *Teosinte-branched 1* (*Tf1*), related to tissue proliferation, appears to affect branching architecture in maize by suppressing lateral branching (Sussex and Kerk 2001).

In some cases, particularly of flower differentiation, a cascade of transcription factor expression follows (Ng and Yanofsky 2000; Weigel and Meyerowitz 1994). The stereotype of this cascade is the "ABC" system (Figure 9-14), in which three different sets of genes in classes denoted A, B, and C are expressed in different positions around the meristem (looking at it end-on). The genes in the model species in which the system has been studied are mainly plant members of the *Mads* TF family, and the system is very similar in spirit to the combinatorial use of members of the *Hox* family in axial patterning in animals.

Essentially, a meristem is induced to produce flowers by meristem identity genes including *Leafy*, *Unusual Floral Organs* (UFO), and *Apetala1, 2* (Ng and Yanofsky 2000). The meristem is arranged in roughly concentric whorls in which different A, B, and C genes are expressed. Cells expressing the A genes *Apetala1* and *Apetala2* alone lead to differentiation of sepals (green leaflike parts) around the outside of the inflorescence. Expression of the C gene *Agamous* leads to carpel (female part) formation in the inside. A + B gene expression generates flowers, whereas B + C generates stamens (male parts). Mutually antagonistic interactions among these and

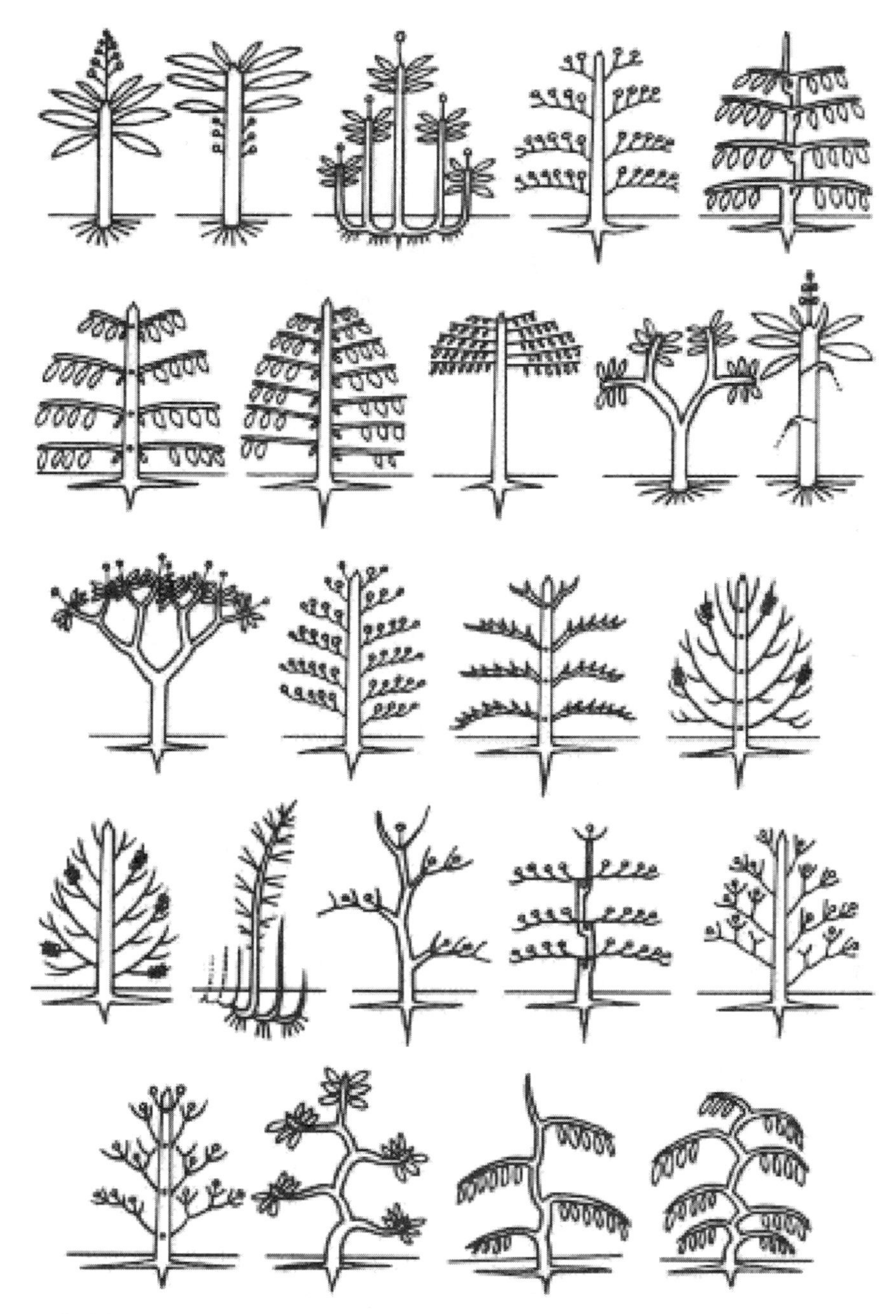

Figure 9-13. Twenty-three basic branching patterns in plants. Reprinted from (Halle 1999), with permission from Elsevier.

other genes generate the different expression zones early in the developmental cascade, which allows differentiation to occur. Experimental manipulation of these genes has produced homeotic mutations, which shows the combinatorial nature of this system of identity determination. Plants like maize with separate male and

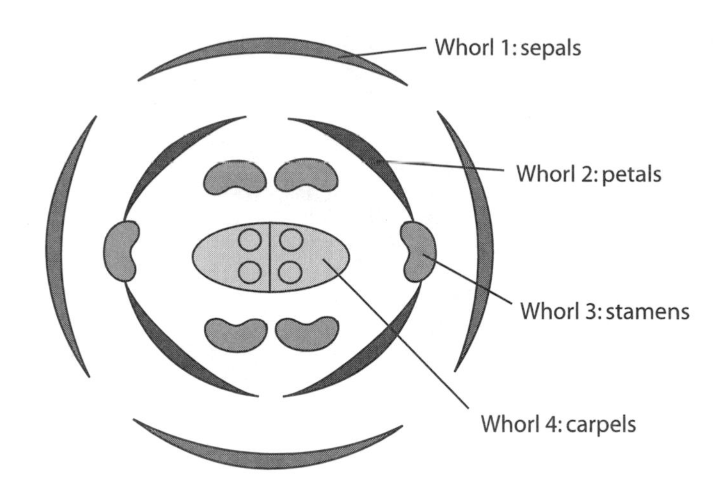

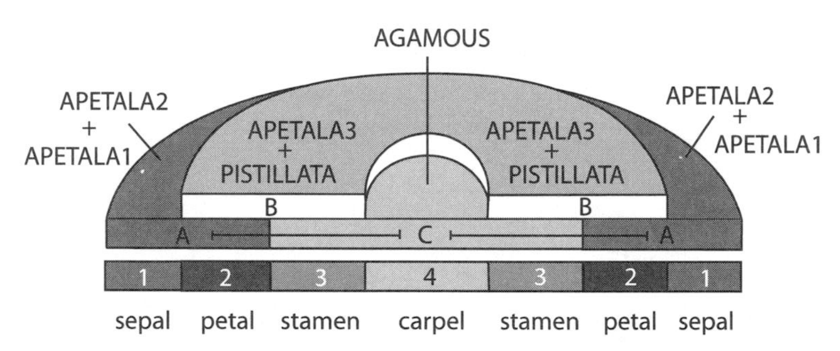

Figure 9-14. The basic ABC model of combinatorial gene expression and the development of flower organs. Three classes of homeotic genes specify the identity of the floral whorl components: sepals, petals, stamens, and carpel organs. Class A genes (dark grey) specify sepals, classes A and B (white) specify petals, classes B and C (light grey) specify stamens, and class C alone specifies carpels. For details see (Ng and Yanofsky 2000).

female flowers use homeotic gene differences to produce single-sex flowers (e.g., Ma and dePamphilis 2000), and similar changes can be brought about experimentally with the *Silky1* B-class gene homolog.

A general combinatorial system involving many of the same ABC system *Mads* and other TF genes is conserved among angiosperms (flowering plants), but the details, specific genes, and functions of specific genes vary (Kramer and Irish 1999; Ma and dePamphilis 2000). As with other examples noted earlier, such as tadpoles or pluteal larvae, phenogenetic drift seems to have been at work during the 130 million or so years of angiosperm evolution: either some flower structures have evolved repeatedly, each time recruiting new genes, or an original angiosperm flower has lost structures in different lineages while phenogenetic drift has modified genes or gene use in what is otherwise a generally conserved ABC system. As in the other systems we have discussed, gene duplication with divergence and overlap in function has also played a role.

Sculpting by Cell Death

Organs are sometimes said to be "sculpted" during development—but the metaphor is somewhat inapt, because sculptors work from the outside. Michelangelo may have "seen" *David* within a solid block of newly cut Carrara marble, but David did not make himself from within that stone. Perhaps we should view even the simplest embryo as much more a master sculptor than Michelangelo. One trick is that the embryo first makes a rather crude form of a structure and then forms the elegant final version by selectively removing unwanted material. This is done by apoptosis, using various pathways mentioned in earlier chapters.

Even some life cycles of bacteria involve autolytic cell death. Cells can die if they are starved of some nutrient or growth factor. Neural connections in the brain are formed through "exploratory" growth of some cells, leading others to follow them; the guide cells die once connections among the latter are made. Neurons can die if they fail to reach cells in an appropriate target tissue, a process triggered by the failure of receptors on the neuron to be bound by ligands secreted by cells in their destination. As mentioned earlier, vertebrate limbs form as paddlelike structures, but once the cartilage segments that will be the bones in the digits form, the tissue between the future fingers or toes dies away (when this doesn't happen in humans, anomalous webbing can result). In a somewhat different way, deciduous plants shed their own leaves by sealing them off from nutrient supplies when faced with stress such as drought or the cold of winter.

Apoptosis occurs in two basic ways. The death sentence can come from the outside. Cells presenting apoptosis-related receptors on their surface undergo a cascade of self-destruction when a ligand (Tissue Necrosis Factor, Fas, or others) is received. This type of externally triggered mechanism is involved in the targeted destruction of infected cells by the immune system. *Bmp* signaling can also lead to the degradation of cells during development, as occurs in some reaction-diffusion patterning. Apoptotic genes include the oncogenes *p21* and *p53* (named for their protein size), which antagonize genes that promote cell growth, preventing cells from overgrowing their normal tissue structure. In one common pathway, signal reception activates a member of the *Caspase* (cysteine-aspartic acid protease) family whose proteolytic activity digests the materials in the cell by anomalous phosphorylation or dephosphorylation of cellular proteins affecting their function. For example, the phosphorylation regulator *Pten* modifies second-messenger molecules. Apoptosis can be achieved by disrupting the normal cell cycle, so that the cell can no longer divide and eventually withers.

Apoptosis can also be managed by internally derived pathways. Some of these also involve *Caspase* genes. In one example, an *apoptosome* containing various proteins forms when internal cellular damage releases Apaf1 from an Apaf1/Bcl2 complex bound to the mitochondrial surface membrane. Damaged mitochondria also release cytochrome *c*, which complexes with Apaf1 and Caspase9, to activate a cascade of Caspase-based proteolysis that cleaves proteins in the cytoplasm. Apoptosis can also destroy chromosomal DNA, another effective way to stop a cell in its tracks. Essentially, the logic of apoptosis is to use external signaling or internal cues to produce compounds that effectively mutate (interrupt) existing pathways in the cell, and as expected there are many ways to achieve the net result of cell death.

Timing and Asymmetric Growth

We referred to spatial and temporal asymmetry in Chapter 8 as another set of ways of making a complex organism by simple mechanisms that affect scale and shape (Calder 1984; Carroll, Grenier et al. 2001; Hall 1999; Raff 1996; Thompson 1917; Wilkins 2002). The timing, level, and persistence of growth factors are responsible for these phenomena. Simple beginnings can be amplified hugely in terms of general growth, leading to considerable differences even among closely related species. We referred earlier to the likelihood that some pattern is laid down very early in the embryo and that this can explain how very similar traits like teeth or color stripe patterns can be generated across a large size range of species so closely related that they clearly must use essentially the same patterning mechanism (e.g., similar stripes in very small and very large cats; teeth in voles and elephants). No new duplicate genes or new mechanisms are required, only differences in their expression.

Timing is a relevant phenomenon—when an activity starts, how long it goes on, and its intensity during that period. Hormone- or induction-receptive periods are established in which a signal can be interpreted properly. This is true even at a higher level of trait complexity, such as language learning ability, songbird song learning, rat olfactory pathways, puberty, developmental patterning, and blossoming/leaf abscission. What determines the sensitive times? This is unknown at present for most examples, but some genes are known that relate to periodic events in development.

Chronobiology is a rapidly growing field in genetics. Various genes, including *Clock* and *Chairy*, are known to be involved in expression timing, the latter having 90-minute activity cycles in some systems. Melanopsin is involved in calibrating light-related cycles.

CONCLUSION

Diverse biological processes in diverse species use a limited number of genetic regulatory factors and their interactions, raising questions about how local specificity is established or how the requisite complex combinatorial patterns are controlled in a stable way. These processes include ligand-receptor binding for signal transduction that alters gene expression, gene regulation by the binding of *cis* enhancers by TFs, message transduction mechanisms such as the protein kinase networks, and combinatorial use of a limited set of factors to establish numerous unique contexts.

The fingerprints of gene duplication are all over these generalizations, so that in a real sense the diversity of mechanisms is derived from only a few ancestral instances. The widespread presence of the same mechanisms across the biosphere justifies referring to them—and in particular, their "logic"—as basic principles of life and shows that not much is around that is really new, in the regulatory sense, compared with the diversity of final function. Relatively simple mechanisms are used to make relatively complex and variable structures. However, the same network is not always preserved to have the same function across taxa. So it is not the specific regulatory network that counts so much as the product of that network—the final phenotype.

None of these phenomena is specifically predictable as a fundamental of life from the formal theory of evolution or its supporting theory of population genetics, which in this sense do not provide a very useful theory for development. *Fgf4* and *Shh* are expressed in limbs and teeth and many other structures, *Distal-less* in insect wing spots and vertebrate teeth. There is nothing about these genes that is physically tooth- or limblike, or even mouselike (because the same genes do similar things in many other species). As we noted in earlier chapters, regulatory genes constitute a toolkit whose own identity is purely arbitrary relative to what the tools are used to make. A *Hox* or *Fgf* gene can be involved in the regulation of any gene that has the appropriate flanking enhancer sequences. In a meaningful sense, a trait is not really due to the genes that bring it about, but to their interactions.

Phenomena like periodic patterning affect a diversity of traits, and even simple coloration has widespread uses including camouflage, mate attraction, age grade or sex marking, and the like. Similar developmental phenomena have varied functional effects, such as the functions of different types of feathers or hair (display, flight, thermal protection). Jointed locomotor structures, suckers in octapi, tentacles (pentameric symmetry), protective scales, worm segments, cactus spines, and leaves and their internal structures all reflect similar processes, some but not all due to the action of homologous genes.

Development was one of the key factors in Darwin's thinking, for example, in explaining embryonic similarities as due to common ancestry and using development as a kind of recapitulation of prior evolution. Indeed, the word "evolution" (which literally means "unfolding") had previously applied to what we call development. However, for many decades in the twentieth century geneticists considered developmental biology to be but a crude and empirical business of little generalizability or rigor (Gilbert et al. 1996) that could not be accounted for by the one-by-one evolution of genes via selection of allelic variation that was taken to be the real theory of evolution. Fortunately, this is no longer true, but we still need to make progress in reconciling the evolution of the more organismal phenomena of development and the more particulate perspective of evolutionary genetics. It is not enough simply that they are not inconsistent with each other.

Chapter 10
Communicating Between Cells

The specialization in cells and tissues that multicellularity allows depends on mechanisms for regulating and integrating the function of the different cells and organs. One cell must be able to influence the physiology of another, on the basis of its context or conditions. In a sense, this is what makes what we term an "organism" as a coordinated biological entity possible. The organism coordinates many functions that involve communication between neighboring cells or among more distant cells or even communication from a single location to everywhere else in the organism. The functions involved are often programmed, specific, or stereotypical rather than responses to open-ended situations. In this chapter, we will describe ways by which complex organisms communicate internally to achieve these ends.

Intercellular communication or at least its basic mechanisms probably arose before multicellularity, because, as we have seen in earlier chapters, single-celled organisms are capable of communicating with each other. These mechanisms include cell surface receptors, G proteins, protein kinases, and the like, many homologous mechanisms used by the cells of more complex organisms.

A number of cell interactions play a role in the timing of gene expression and changes in the biochemistry of the cell to regulate growth, development and homeostasis, and the like. These interactions can be triggered by environmental cues that can induce cellular functions such as photosynthesis or the running of sap in plants, hibernation in animals, or diapause (biological dormancy) in insects or by intercellular signaling via molecules like nutrients such as sugars or ions or via hormones, organic substances that in minute concentrations have defined effects on cellular activity.

Two basic regulatory systems control cell-modulator interactions in animals. One is a nervous system, and in more complex organisms a central nervous system (CNS), which sends electrochemical signals to peripheral systems and receives electrochemical messages from the periphery. This occurs in a relatively specific way to connect the two endpoints of the communication. The other, in which communication is more diffuse, is the endocrine system and its analog in plants. The endocrine system consists of a number of glands throughout the body that synthesize organic

Genetics and the Logic of Evolution, by Kenneth M. Weiss and Anne V. Buchanan.
ISBN 0-471-23805-8

substances, polypeptides or steroids, which are released into the bloodstream or extracellular space and circulate to target organs whose cells present specific receptors, stimulating a variety of functions, one of the most important of which is regulation of gene expression. Given the genetic message mechanism, the sending cells must be prepared to use the sending mechanism under appropriate circumstances, and cells that need to respond must be prepared to receive the signal.

NEURO-IMMUNO-ENDOCRINOLOGY

The vertebrate CNS and endocrine systems interact with each other, and current knowledge suggests that the two systems are not as functionally, mechanistically, or evolutionarily distinct as had long been believed. Until recently the nervous system was viewed as a regulator of brief, rapid, localized responses, whereas the endocrine system slowly initiated more prolonged and generalized responses. However, understanding of the overlapping functions of components of these distant regulating systems has led to a more synthetic view, completely consistent with a more unified view of life and the common origin of systems and genes, as well as of species. No new phenomena are involved, but familiar ones are.

Nerve cells are secretory, in that they release chemical *neurotransmitters* at synapses to trigger fleeting neuronal responses; inhibitory cells ensure that the triggering signal can be received only briefly. However, some nerve cells are specialized for translating neural signals into chemical rather than electrical stimuli. These *neurosecretory* cells release *neurohormones* into the bloodstream at *neurohemal organs*, the synapse between the axon of the neurosecretory cell and a blood capillary, or so close to their target that they need never enter the bloodstream at all. Additionally, the endocrine system regulates the function of the nervous system, and the nervous system controls the blood flow and secretory activity of endocrine glands, because those glands are innervated and their cells receive neural signals. Neurohormones can travel to distant target cells.

The signaling molecules of the nervous, endocrine, and immune systems are often the same. Immune function is affected by hormones, and the endocrine system is affected by cytokines, mediators of the immune response (Chapter 11). Immunologically reactive cells have recently been found to secrete hormones that had previously been thought to be synthesized only in the brain or pituitary gland (Ojeda and Griffin 2000), and excretions from glands long thought to have exclusively endocrine function, such as norepinephrine from the adrenal medulla, are now known to be neurotransmitters as well. So, although we will concentrate in this chapter on the action of hormones, the isolation of the endocrine system in this way is an artificial classification that ignores the essential interactions between the different systems, interactions being discovered and characterized very actively today.

An important point about cell signaling is *that* it happens, and that this, rather than *how* it happens, is what natural selection has acted upon. Understanding the distinction can help explain many things in biology, such as why the G protein-coupled receptor signaling cascade is involved in so many signaling pathways in tetrapods and arthropods but so few in plants and the fact that kinases are important to signaling in all organisms but the particulars of the amino acid that gets phosphorylated or the shape of the transcription factor that results are irrelevant to the general process and what natural selection "sees." Cellular communication mechanisms can vary even though the regulated trait has remained similar.

Hormones can be divided into two general classes. *Regulatory* hormones are involved in the control of metabolic and homeostatic processes, whereas hormones that control the onset of irreversible physiological switches are referred to as *developmental* hormones. The latter basically trigger events such as molting cycles or reproduction. However, the difference is in function, not mechanism of action. The general modes or logic of action of plant and animal hormones are similar and involve processes we visit many times in this book.

But what is a hormone? The classic definition is an organic substance produced in an endocrine gland and released into the bloodstream or the sap of a plant in very small amounts for transport to distant cells or organs prepared to receive it and where it affects physiological activity via both genomic and nongenomic pathways. However, in fact, hormones need not travel long distances, nor need they be released into the bloodstream or hemolymph or extracellular space for transport to distant cells. As described in Chapter 6, they may act at various distances, for example, paracrine function being local and autocrine or intracrine function affecting the hormone-producing cell itself.

Hormones might not even be synthesized in a gland traditionally seen as part of the endocrine system—in mammals adrenocorticotropin (ACTH), β-endorphin, prolactin, and luteinizing hormone-releasing hormone (LHRH) are all produced by immunologically reactive cells, and hormones are released by cells lining the stomach and small intestine, among many other examples. Some growth factors work in a nondiffusible way, when the membrane-anchored growth factor on one cell binds to its receptor on the membrane of an adjacent cell (as, for example, the *Delta/Notch* signaling system).

The more we know the more life defies categorization, but if we need a more inclusive definition of a "hormone," it might be simply an endogenously synthesized organic substance that, when bound by a receptor or binding protein, triggers a specific cellular response. Yet even this is not quite right, because there are other endogenously produced substances—for example, SFs, that induce changes in gene expression to alter cell physiology—that are not classed as hormones. The connotation of the term "hormone" is bound up, like so much else, in history and the traditional notions that predate modern molecular biology. Ultimately, given the blind nature of evolution by phenotype and the striking connectedness of life reflected in the way genes and genetic mechanisms evolved by divergence from a common ancestor, it is likely that any kind of grouping that we might conceive is an overlay of our own construction for our own convenience, rather than a discretely programmed part of nature.

PLANT SIGNALING

In plants, as animals, growth and development are controlled by responses to environmental or inter- or extracellular triggers. Environmental triggers of plant responses include features of light signals such as quality (red light activates different genes than blue light), duration, direction, and quantity, changes in temperature that can alter the fluidity of cell membranes, chemical signals (air pollutants, phytotoxins, or elicitors from other organisms), changes in water availability and thus cell turgor, insect or pathogenic attack, which can trigger the immune response, gravity, available nutrients, proximity of other plants, wind, and other variables.

Intracellular signals, generally activated by receptors that bind a variety of extracellular signals, include second messengers such as inositol triphosphate (IP_3), which binds to and activates a calcium channel in the endoplasmic reticulum, and Ca^{2+}, an important mediator of the activity of many proteins inside the cell, and metabolites such as sugars and glutamate (an amino acid). Cellular activity is also driven by intercellular communication. Electrical signals, small molecules (<800 Da), and some mRNAs can pass between cells for a short distance via the plasmodesmata. Finally, plant hormones, which are released from a cell into the *apoplast* (the extracellular space) and transported to other cells, are important in altering gene expression and physiology.

Like animal hormones, plant hormones are compounds synthesized in the plant that at low concentrations elicit physiological responses either in the cell in which they are synthesized or in other cells when translocated. Plant hormones are transported from cell to cell through intercellular spaces or through vascular bundles, the xylem and phloem. Plant hormones have synergistic and antagonistic effects; they inhibit and stimulate other hormones. These signals thus "circulate" in the organism, but this is not the rapid, closed circulation that we are familiar with in vertebrates, because the upward and downward flows are essentially independent and there is nothing corresponding to a capillary bed to connect them.

A number of classes of phytohormones have been isolated (see Table 10-1). *Auxin*, *cytokinin*, *gibberellin*, and *brassinosteroids* are, by and large, growth stimulators, and *abscisic acid* and *ethylene* can generally be called growth inhibitors, whereas *salicylic acid*, *jasmonic acid*, *systemin*, and *oligosaccharins* are synthesized in response to stress and initiate plant defenses. Another class of hormone, the *phytoecdysteroids*, is noxious to nonadapted insects and herbivores and probably evolved as a plant defense. Generally, the growth stimulators and inhibitors affect cell division, elongation, and differentiation, but their effects are modulated by the type of target cell, the life cycle stage, interaction with other hormones, and hormone concentration, so that hormones that are usually categorized as growth stimulators can, depending on the cellular context, sometimes inhibit growth, and vice versa.

At the level of the cell, hormones primarily induce, and sometimes inhibit, gene expression and enzyme action, although some hormones act via nongenetic pathways, such as by being cofactors in reactions among existing constituents of the receiving cell. The mode of action of phytohormones is most easily characterized through the study of mutant plants, whose phenotype seems to be the result of over- or underproduction of a hormone. The biosynthetic pathways of a number of plant hormones are not yet understood, primarily those of hormones for which mutants have not been available for study.

These core signaling pathways are widely shared among plant taxa and used over and over in many different plant developmental pathways and in the maintenance of homeostasis, and most have a long evolutionary history, of gene duplication and diversification, as they are found in plants as old as the primitive mosses (McCarty and Chory 2000; Schumaker and Dietrich 1998).

Auxins are involved in a large diversity of developmental processes, in response either to environmental cues or to intracellular stimuli. Synthesized in shoot apical meristems (differentiating tips of branches), young leaves, and embryos, they induce cell division, promote elongation, and inhibit growth of lateral buds (a phenomenon known as apical dominance), and early in development they transcriptionally activate a set of early genes that mediate processes such as determination of the

TABLE 10-1.
Plant Hormones.

Hormone	Place of Synthesis	Action
Growth regulators		
Auxins	Shoot apical meristems, young leaves, embryos,	Cell division, promotion of elongation, inhibition of flowers, fruits, and pollen, growth of lateral buds, axial growth, induction of tropism, vascular patterning, and differentiation, lateral organ outgrowth in root and shoot
Gibberellins	In immature seeds, the root and shoot apical meristems and in young leaves	Stem elongation, germination, growth of some fruit, development of male sex organs in some flowers, control of juvenility in some plants
Cytokinins	Most in roots, in angiosperms of flowering plants	Cell division, shoot and root formation, apical dominance, xylem formation, leaf senescence, solute mobilization, cotyledon expansion
Brassinosteroids		Regulate gene expression, stimulate stem growth, inhibit root growth, promote xylem differentiation, retard leaf abscission
Abscisic acid	Fruits, root caps, mature leaves	Inhibits cell growth, promotes seed dormancy, involved in opening/closing of stomata when leaves wilt
Ethylene	Anywhere, but primarily apical buds, stem nodes, senescing flowers and ripening fruits, wounded tissues	Promotes ripening
Defense Mechanisms		
Salicylic acid		Disease resistance, control of heat production, essential to systemic resistance response
Jasmonic acid/ Methyl jasmonate		Help regulate plant growth, defend against fungi, involved in leaf senescence
Systemin		Disease resistance
Oligosaccharins		Disease resistance, induction of SRR

bilateral axis of the developing embryo (Christensen et al. 2000; Ouellet et al. 2001). Later in development, auxin is involved in vascular patterning and differentiation and lateral organ outgrowth in the root and shoot (Christensen, Dagenais et al. 2000), cell division, phototropism, gravitropism, and gene induction.

High levels of auxins stimulate the production of ethylene, which essentially counteracts auxin activity. Leaves remain attached to the stem as long as there is auxin moving from the leaf blade down the petiole (leafstalk). When auxin is interrupted, as the leaf begins to senesce, the cell walls of the cells at the base of the petiole (the *abscission zone*) undergo dissolution and the leaf falls off (Cleland 1999). This is called abscission, and it also happens in fruit when the seeds cease transporting auxin through the fruit pedicle. However, the route by which auxin reaches the abscission zone is important in determining the speed with which senescence occurs. Auxin reaching the abscission zone from the tip promotes abscission, whereas auxin reaching the structure from the opposite end inhibits it. (Abscission is a major example of a plant process that depends on interaction between different hormones to catalyze the reactions that break down the cell walls. Once bound by their intracellular receptors, ethylene stimulates the synthesis of the enzymes involved in destruction of cell walls and abscisic acid speeds the required senescence. Gibberellin inhibits abscission by promoting growth rather than senescence.)

Three *Auxin* gene families have been characterized; the *Aux/Iaa* gene family, the *Gh3* gene family, and the *Saur* gene family (Ouellet, Overvoorde et al. 2001). The *IAA*, or indole-3-acetic acid, gene family encodes the predominant group of auxins in higher plants. *Arabidopsis*, for example, has at least 25 *Aux/Iaa* genes (Rouse et al. 1998). IAA is synthesized mainly in the apical meristem shoot tips, young leaves, and developing fruit. A number of pathways seem to be involved, and IAA can be catalyzed by endogenous plant enzymes as well as by enzymes produced by plasmids of plant pathogens. The endogenous production is not yet well understood (Zhao et al. 2001), although it seems to be dependent on the amino acid tryptophan or the breakdown of glycosides, carbohydrates.

Cytokinins (CKs) promote cell division and are involved in shoot formation, root formation, apical dominance, xylem formation, leaf senescence, solute mobilization, root growth, and cotyledon ("seed leaf") expansion. They are produced in growing areas such as meristems and are derivatives of adenine. CKs interact with auxins in mitosis and the initiation of shoot and root primordia. The antagonistic action of auxin versus cytokinin is involved in apical dominance (predominance of one meristem relative to others in the plant) and xylem development, that is, the process depends on the relative amounts of these hormones. A similar antagonistic interaction determines the relative growth of roots, shoots, and callus (the undifferentiated tissue that grows at the edge of a wound) (Zhao, Christensen et al. 2001). A high IAA/CK ratio promotes root growth, and a lower ratio promotes growth of the callus and shoots.

The CK signaling pathway is not yet well understood, although it has been proposed that it is a two-component signaling system. Two-component systems control signal transduction pathways in many microorganisms but have only recently been recognized to be important in plant signaling (Hwang and Sheen 2001; Inoue et al. 2001; Moller and Chua 1999). The two components include a histidine protein kinase that detects the signal and transmits it to the second component, the response regulator, which mediates the plant response. The signal is transmitted by phosphorelay—that is, the protein kinase autophosphorylates at a histidine residue, taking the phosphate from a donor ATP molecule, and then transfers the phosphate group to an aspartate group on the response regulator. This activates the regulator, generally a transcription factor, inducing it to repress or stimulate expression of a target gene. A cytokinin receptor, a histidine kinase, has been identified in *Arabidopsis*, *Cre1* (*Cytokinin response 1*), but the regulator is not yet known. Presumably it will be a transcription factor (Hwang and Sheen 2001; Yamada et al. 2001).

Gibberellins are a group of more than 100 related compounds, all of which contain a chemical structure known as a "gibbane ring" (Cleland 1999). No plant synthesizes all of the gibberellins, but all plants seem to have more than a few. These hormones are produced in immature seeds, in the root and shoot apical meristems, and in young leaves. Gibberellins promote stem elongation, induce enzymes involved in germination of grass seed, and promote the growth of some fruit, the germination of some seeds, the development of male sex organs in some flowers, and the control of juvenility in some plants. Their effects are enhanced in the presence of auxins.

Less is known about the gibberellic acid (GA) signaling pathways than those of other phytohormones. GA synthesis is catalyzed by the enzyme copalyl-diphosphate synthase (CPS), which is encoded by *Gai1* (*Ga-insensitive*). The *Gai* genes were first identified in a plant with reduced GA sensitivity. *Gai1* mutant plants do not synthesize GA and are extreme dwarf, have reduced male fertility, and germinate poorly (Moller and Chua 1999). Other GA mutants lack GA repressors; the *Spy (Spindly)* mutant has a phenotype similar to that of wild-type plants that are overexposed to Ga. *Rga* (*Repressor of Ga*) represses the GA signaling pathway in *Arabidopsis*. It encodes a protein of the *Gras* family, which includes *Gai*. The repressor activity of *Gai* and *Rga* can be inactivated by GA and is triggered by the upstream gene, *Spy*.

Brassinosteroids are key regulators of plant responses to light. They are steroid-like compounds, related to cholesterol as are animal and insect steroids, that elicit growth responses by regulating the expression of genes associated with development. They stimulate stem growth, inhibit root growth, promote xylem differentiation, and retard leaf abscission. Their effect on stem cell elongation and xylem differentiation are auxin-mediated processes. This class of hormones works by binding to a brassinosteroid receptor on the cell surface, which induces a serine/threonine receptor kinase signaling pathway. Brassinosteroids have recently been found to be involved in inducing resistance to disease in higher plants as well (Nakashita et al. 2003).

One brassinosteroid receptor gene has been identified in *Arabidopsis*: *Bri1* (Wang et al. 2001). Homologs to *Bri1* have also been found in *Arabidopsis*, but their role in brassinosteroid signal transduction, if any, is not yet known. *Bri1* encodes a leucine-rich repeat (Lrr) transmembrane receptor kinase. Structurally, the repeat region contains 25 Lrr's, interrupted by a 70-amino acid island, a transmembrane domain, and a cytoplasmic serine/threonine kinase domain. Because plants do not have nuclear steroid receptors, any genomic response a plant steroid hormone initiates must begin at the cell surface. The *Bri1* receptor ligand-binding domain, therefore, is extracellular and ligand binding initiates signal transduction via the cytoplasmic kinase domain of the receptor, which autophosphorylates serine and threonine residues. Other components of this signaling cascade have not yet been identified (Li et al. 2001; Mussig and Altmann 2001). Even the nature of the ligand is still unknown (Wang, Seto et al. 2001).

Ethylene, derived from the amino acid methionine, is a gas that promotes ripening. It can be produced anywhere in a plant but primarily is produced in apical buds, stem nodes, senescing flowers, and ripening fruit as well as wounded tissues. As a gas, ethylene diffuses readily to other cells in the same plant and even to nearby plants, so this hormone need not be produced endogenously to trigger ripening, although the extent to which plants receive or respond to signal from other plants is debated.

Ripening is a series of events leading to senescence and, ultimately, rotting. Ethylene initiates the cascade by inducing the breakdown of cell walls, changes in pigments, and formation of flavor compounds. It is a factor in leaf and fruit abscission and facilitates the withering and death of petals of flowers, among other effects. Many of the processes now known to be due to ethylene were once attributed to auxin, because ethylene is often produced as a response to high concentrations of the former.

Ethylene synthesis is controlled by the ACC (1-aminocyclopropane-1-carboxylate) synthase family of enzymes. ACC is the precursor molecule to ethylene. The genes that convert ACC to ethylene comprise multigene families in different plant species. Expression of ACC synthase is induced by stimuli that lead to increased ethylene production—wounding, senescence, fruit ripening, etc. This suggests that the rate of ethylene production depends on ACC synthase activity. Oxidases also seem to be involved in regulation of ethylene production (Johnson and Ecker 1998).

Like the cytokinin signaling pathway, the ethylene pathway also shares homology with bacterial two-component systems. Here, the receptor is a histidine kinase that autophosphorylates an internal histidine residue of target proteins, and the response regulator includes a conserved aspartate residue that is the recipient of the phosphate group from the histidine kinase (Johnson and Ecker 1998; Stepanova and Ecker 2000). Unbound ethylene receptors activate a negative regulator, Ctr, a protein kinase in the *Raf* family, which induces a cascade that ultimately leads to the repression of the *Ein2* (*Ethylene insensitive*) gene. But when ethylene is bound to its receptors, *Ctr* is inactivated, releasing *Ein2* repression and leading to activation of the ethylene response gene (Moller and Chua 1999; Ouaked et al. 2003).

Abscisic acid (ABA) inhibits cell growth but, like so many biomolecules, has multiple functions. ABA levels, for example, rise during times of environmental stress; thus ABA is sometimes called a stress hormone. ABA promotes seed dormancy and is involved in the opening and closing of stomata when leaves wilt. A 15-carbon acid, ABA is made in fruits, root caps, and mature leaves in response to environmental signals. Water stress, or freezing temperatures, for example, result in massive and rapid ABA synthesis. Movement of ABA is through phloem and xylem and by diffusion between cells. ABA seems to be antagonistic to auxins, cytokinins, and gibberellins.

Five *Aba-insensitive* genes, *Abi1* through *5*, have been cloned in *Arabidopsis* and shown to have homologs in at least two other plant species. Some components of the *Aba* pathway are even found in animals (Gampala et al. 2001) *Abi1* and *Abi2* encode serine/threonine kinases with various downregulatory functions in genetic responses to cold or drought and regulation of ion channels, and *Abi3*, *4*, and *5* encode genes for various transcription factors (Gampala, Finkelstein et al. 2001).

Apical dominance, the control of development by the main shoot meristem, characterizes the branching and differentiation of many plants, but not all. In general, branching seems to depend on relative amounts of CKs, ABA, and IAA, with relatively high CK:IAA ratios affecting and/or reflecting the location and relative growth rates of nodes that develop in axial locations on shoots, and the pattern is then modified by ABA (Sussex and Kerk 2001). We noted earlier that CK:IAA ratios seem to be important in root development as well. The distribution of relative levels of these hormones may function as a kind of quantitative patterning process, possibly reaction-diffusion-like, in which simple alteration of the parameters (the quantitative levels of the hormones), are responsible for the diversity of branching patterns shown in Figure 9-13. As discussed in Chapter 9, the genetic mechanisms responsible for these quantities are not yet identified.

Another group of plant hormones appears primarily in response to wounding and pathogen attack. *Salicylic acid* is involved in disease resistance and control of heat production in some species and is essential to the systemic resistance response. *Jasmonic acid* (JA) and *methyl jasmonate* (MeJA) help regulate plant growth, defend against fungi, and are involved in leaf senescence. MeJA has been found to be synergistic with ethylene and activates enzymes involved in defense against wounding. JA is probably confined to the cell in which it is produced in most cases. MeJA can act between cells, as well as between plants by signaling between infected plants and unaffected plants. Although controversial, it has been suggested that volatile MeJA, released into the air by a plant under pathogen attack, can induce increased resistance in nearby unaffected plants (Dolch and Tscharntke 2000; Karban et al. 2000; Preston CA et al. 2001). If true, this is another instance of altruistic cooperation that many biologists would argue requires specific evolutionary explanation. MeJA also promotes tuber formation and storage protein formation.

Systemin is involved in disease resistance, although how ubiquitous it is as a hormone is not known. It acts by inducing proteinase inhibitors throughout the plant. Action may be via the synthesis of JA, which acts intracellularly as a second messenger in the induction of proteinase inhibitors. Plant cell walls are composed of a mixture of complex carbohydrates. When these break down, small pieces are released, some of which have biological activity. These are *oligosaccharins*, and they are produced during pathogen attack and signal other cells to prepare to defend against attack.

As will be discussed in Chapter 12, one way that plant defenses are triggered is via gene-for-gene resistance pathways, the rapid defense response a plant has when its *R* genes are induced by a pathogen's *Avr* genes. This hypersensitive response (HR) involves apoptosis of plant cells in contact with the pathogen. The precise mechanism of induction of this response is not well understood, but it is hypothesized that R proteins "guard" plant proteins that are the targets of attack by Avr proteins ("guardees") and that the HR is triggered by Avr-guardee interactions (Glazebrook 2001).

The HR triggers the systemic resistance response (SAR), which protects the entire plant and is long-standing. The SAR involves the signal molecule salicylic acid and the expression of a characteristic set of defense genes. Some, but not all, response pathways use the signal molecules JA or ethylene, and cross-talk between the pathways is required by some responses. JA and ethylene mutants, either in the production or the perception of these hormones, are more susceptible to disease than their wild-type counterparts. The mechanism by which they are involved in defense responses, however, is not clear; it may be via a synergistic interaction with genes in various defense pathways (Ellis and Turner 2001). Many of these genes can be induced by the application of JA in the absence of wounding or attack (Reymond et al. 2000), showing the importance of this hormone in the transduction of signal and the initiation of defense responses.

ARTHROPOD HORMONES

INSECTS

Insect hormones control a broad range of processes in the insect life cycle, both developmental and physiological. These include the metabolism of carbohydrates and lipids, the maintenance of water balance, including excretion of water after a blood meal, stimulation and inhibition of the circulatory system and the firing of muscles, growth, including molting and metamorphosis, diapause (a period of suspended growth or development), apoptosis, reproduction, caste determination (which occurs in social insects, such as bees and ants, whose fate as queen or worker is determined not by genes but by differential feeding and pheromone exposure at the larval stage), and aspects of behavior during molting and migration, reproductive and other social behaviors, and response to pheromones.

Insects produce two general classes of hormones, as do vertebrates, lipid or steroid (lipophilic) hormones and polar (hydrophilic) peptides, and their mode of action depends on how the hormone interacts with the cell membrane (peptide hormones are coded directly by genes; the other forms are the product of enzymatic reactions for which the enzymes, but not the final hormone product, are genetically coded). Being hydrophobic, the lipid hormones pass readily through the cell wall to bind with receptors in the cytoplasm or nucleus of the cell. The hormone-receptor complex in turn binds with specific DNA sequences to initiate, enhance, or inhibit gene expression. The peptide hormones, however, do not easily pass through the cell membrane but instead bind with receptors on the cell surface, which transduce the signal to second messengers inside the cell, often via a G protein cascade. Target cells are not continuously receptive to hormone stimulation. There are "hormone-sensitive periods," and although hormones circulate throughout the hemolymph, exposing all tissues to the same levels of hormone at the same time, the tissues are not all equally receptive. What controls the timing of the sensitivity of a tissue is not known, but clearly both the signaling and receiving cells need to be prepared, implying a prior element of differentiation in these two, often distant, locations in the body.

Insects synthesize hormones in two distinct organ systems: endocrine tissues and neurosecretory cells. The glandular endocrine tissues are specialized for the synthesis and release of hormones. The most important endocrine glands in insects are the *prothoracic glands*, which secrete ecdysteroids during development, and the *corpora allata*, where the juvenile hormones are produced and secreted. The ovaries and testes of many adult insects also produce ecdysteroids.

Endocrine Tissues

The gross morphology of the prothoracic glands varies widely throughout insect phylogenies, both in size and in location in the body, but the cellular structure is quite uniform (Nijhout 1994). As in vertebrates, probably because of shared ancestry, the distinction between the nervous and endocrine systems is not clear-cut; homologous nerves of the CNS in all species generally innervate the prothoracic glands, and some of these are neurosecretory and are involved in regulating secre-

TABLE 10-2.
Insect Hormones.

Hormone	Site of Synthesis	Function
Prothoracicotropic (PTTH)	Neurosecretory cells in the brain	Stimulate secretion hormone of ecdysone
Ecdysteroids		Promote growth, control molting, embryonic development
Ecdysone (an ecdysteroid)	Thoracic glands	Induce apolysis, cell division, degradation of old cuticle, production of new
Juvenile hormones	Corpora allata	Development, metabolism, behavior
Bursicon	Abdominal ganglia	Control of hardening of new cuticle

tions from the prothoracic glands (Nijhout 1994). These glands secrete, but do not store, *ecdysone*, a steroid that promotes growth and controls molting. In most insects, the prothoracic glands undergo apoptosis during the metamorphosis of the larva to adult stage and ecdysone is then no longer synthesized and released.

Molting is essential if an insect is to change size or shape because the insect's hard outer shell, or cuticle, cannot accommodate growth by expanding (in contrast to the endoskeleton of vertebrates, which grows along with the individual). The process is complex, involving the formation of a new cuticle *within* the old one while the old one is digested and the proteins reused. Therefore, the new cuticle must be protected against degradation by the digestive enzymes and at the same time remain pliable enough to expand when the old cuticle is shed (Nijhout 1994). Chemical assault to interfere with normal molting is one insecticide strategy.

Molting cycles are triggered by the secretion of the *prothoracicotropic hormone* (PTTH), which is produced by neurosecretory cells in the brain. The only known function of PTTH is to stimulate the secretion of the molting hormone ecdysone by the two prothoracic glands in the thorax. These two hormones trigger every molt, whether larva to larva or pupa to adult. Levels of a different set of hormones, the *juvenile hormones* (JHs), control metamorphosis.

The corpora allata are a pair of small glands found along the main vessel in the neck of most insects and are attached to the brain by a nerve that passes through the *corpora cardiaca* (additional small glands just anterior to the corpora allata). The corpora allata produce JHs. Innervation of the corpora allata is by nerves that conduct impulses, as well as by neurosecretory neurons. In some insects the corpora allata are also neurohemal organs for some neural secretions from the brain. The brain and these associated neurohemal glands form the *brain-retrocerebral neuro-endocrine complex*, a control, synthesis, and excretion system that is the most important neuroendocrine organ in the insect endocrine system.

JHs serve both as regulatory and developmental hormones and have a role in every aspect of insect life, from development to metabolism to some behaviors. Three major forms of JH are known; some insects secrete only one, others two or all three. After secretion from the corpora allata, JHs bind to the *juvenile hormone binding proteins*, which increase the solubility of the hormones in the hemolymph and protect them from degradation. JHs are lipid hormones, but their mode of entry into the cell, and subsequently into the nucleus, is not yet definitively known (Davey 2000). It may be that juvenile hormone binding proteins chaperone JHs into the cell and regulate their binding to JH-specific receptors once there.

As in plants, insect hormones have an antagonistic or complementary relationship with each other and highly tissue-localized, specifically timed gene expression. The presence of JHs prevents a juvenile insect from becoming an adult (thus the nomenclature) by suppressing secretion by the brain of hormones involved in molting and metamorphosis. The process is complicated and depends on JH-sensitive periods during the molting cycle. Basically, when the insect is JH sensitive and JH is present, the insect does not molt. If JH is absent, the developmental stage changes. This generally depends on the secretion of ecdysone having already initiated the next molt. Different sections of the epidermis have different JH-sensitive periods. The onset of JH-sensitive periods is independent of presence or absence of JH, and it is during this period that a cell can be committed to a specific developmental fate. Usually, it requires the action of *ecdysteroids*, however. JHs are pleiotropic and play different roles at different stages in the life cycle (see below).

Ecdysteroids are a family of steroids that includes ecdysone and its analogs and metabolites. These compounds are found in insects and crustaceans and in some plants as phytoecdysone. Nearly 100 of these compounds have been described in insects and other arthropods, and more than 200 have been isolated from plants (Sadikov et al. 2000). These hormones promote growth and control molting and play a role in embryonic development. Ecdysteroids are lipid hormones and act by enhancing or inhibiting gene transcription. The hormone-receptor complex usually binds a G protein inside the cell to initiate the signaling cascade.

Ecdysone is produced by the thoracic glands and is a relatively inactive *prohormone* that becomes active when converted by the fat body and epidermal cells into *20-hydroxyecdysone*, the most important molting hormone in insects (Nijhout 1994). Ecdysteroid action is the same at all stages of insect life; these hormones act on epidermal cells to induce apolysis (the first step in the synthesis of a new exoskeleton), cell division, degradation of the old cuticle, and production of the new. As with other hormones, ecdysteroid concentrations are affected by the concentration of other hormones. In this case, ecdysteroid secretion depends on the pattern of PTTH secretion in the brain.

When the new-stage insect emerges from the old cuticle of the previous stage, initiating a new *instar* (the phase between molts) in a process called *ecdysis*, the new cuticle must harden or *tan*. This is controlled by a neurosecretory hormone called *bursicon*, except in the higher Diptera in which it is controlled by a set of neurosecretory hormones called *pupariation factors.* The principal source of bursicon is the abdominal ganglia, and the hormone is released into the hemolymph from the abdominal previsceral organs, although it is synthesized throughout the nervous system, including the brain.

The ecdysone signal is transduced by the *ecdysone receptor*, a heterodimer formed by the joining of two nuclear receptor proteins, EcR (Ecdysone receptor) and Usp (*Drosophila* Ultraspiracle receptor), a homolog of the vertebrate nuclear receptor RXR (Retinoid X receptor) family of ligand-dependent transcription factors (TFs) (Arbeitman and Hogness 2000; Ghbeish et al. 2001; Mouillet et al. 2001). The RXR receptors have two signature domains, the DNA binding and ligand binding domains. The ecdysone receptor requires not only heterodimerization for DNA binding, as do the RXR family of receptors, but also for ligand binding, a characteristic it does not share with the RXR receptors. Through alternative splicing and two promoters, the *EcR* gene encodes three protein isoforms, EcRA, EcRB1, and EcRB2, each with different quantitative control over transcription (Mouillet, Henrich et al. 2001). Each of the three isoforms is able to dimerize with the Usp receptor, and this may explain how it is that ecdysone can initiate the large variety of responses, at various stages in the life cycle, that it does (Mouillet, Henrich et al. 2001).

The ecdysone signal activates a hierarchical response when bound to the receptor, turning on a small set of early genes that in turn activate a larger set of genes downstream. In vertebrates, activation of the homodimeric steroid receptors depends on the presence of a molecular chaperone-containing heterocomplex (MCH), which interacts with the receptors, probably to facilitate protein folding and DNA binding by their ligands. The presence of an MCH seems to be required to activate the ecdysone pathway as well (Arbeitman and Hogness 2000).

Molting is a good example to illustrate the importance under some circumstances of centrally coordinated signaling. We mentioned in Chapter 9 that some structures like limbs or teeth cannot develop until a place has been prepared for that to happen. In this case, prepatterning of cells to respond to later signals is likely to be a mechanism that enables this delayed, context-specific response to occur. An entire insect must be ready before molting can occur. At the appropriate time, a central signal is released that affects the whole body.

Neurosecretory Cells

The second insect organ system that secretes hormones consists of the neurosecretory cells of the CNS, neurons that produce small polypeptides, or neurohormones. These cells tend to be localized in the brain, although they are found in all the ganglia of the CNS; they have axons that end in *neurohemal organs* or *areas*, where the secretions are released directly into the hemolymph.

Adult insects do not undergo molting or development, so in adults the same ecdysteroids, JHs, and various neuroendocrine hormones that controlled these processes earlier in life control adult processes like diapause, migratory behavior, and reproduction—the production of yolk proteins, maturation of the ovaries, and synthesis of eggs (Hartfelder 2000; Nijhout 1994). As a generalization about hormone action in insects, a given hormone can have different effects in different target tissues and different effects on the same tissue at a different time in the life cycle. Once again we see the use and reuse of a mechanism, showing the importance of context specificity in the inducing mechanisms (which themselves may have multiple uses). As we have mentioned in Chapter 9 in context with the apparently inconsistent inductive/inhibitory effects of Dorsal protein, the specific effect of a signaling substance ultimately depends on the nature of the regulatory regions associated with a gene, and the combination of factors that must bind jointly there to activate or inhibit a gene.

Other Arthropods

Crustaceans have a number of endocrine mechanisms in common with insects, including production of *methyl farnesoate* (MF), a hormone that is the precursor of insect JH. MF seems to play a role in reproduction in crustaceans. Interestingly, although many of these substances are found in insects and crustaceans, they do not necessarily play the same role in both.

The eyestalk neurosecretory complex of decapod crustaceans is the location of the sinus gland, the source of a number of neuropeptide hormones, including those involved in pigment migration (*red pigment-concentrating hormone* and *pigment-dispersing hormone*), regulation of carbohydrate metabolism (*crustacean hyperglycemic hormone*, CHH), molting (*molt-inhibiting hormone*, MIH), and gonadal growth (*gonad-inhibiting hormone*, GIH) (Webster 1998). These neuropeptides have been found to play an inhibitory role, acting on endocrine tissues that produce the hormones that initiate reproduction and molting. DNA sequencing has shown that the genes for these hormones are structurally related. Again, they are also pleiotropic; CHH, for example, has a role in molting and in reproduction as well as in carbohydrate metabolism. Other neuroendocrine centers are located throughout the CNS, including, as well as the eyestalk, the brain and the subesophageal ganglion.

VERTEBRATE ENDOCRINE SYSTEM

As in other organisms, because hormones are a fundamental way that multicellular organisms "solve" the problem of intercellular communication and the regulation of local gene expression to serve the whole, vertebrates have evolved a substantial array of hormone-producing glands, genes, and systems that affect all aspects of life: reproduction, growth and development, maintenance of a stable internal environment, mental functions, physical activity, food seeking and satiety, many behaviors, and regulation of energy balance (Ojeda and Griffin 2000). Rudiments of these systems are found in chordate "relics" like amphioxus. As noted earlier—and as we have come to expect—at least some of these are evolutionarily related to substances found in invertebrates.

Endocrine Glands and Hormones

The vertebrate endocrine system classically consists of a number of ductless glands that produce hormones, with different functions, for release directly into the bloodstream (Table 10-3). However, a number of other glands and cells secrete hormones, and in fact some cells produce hormone for use within the cell and do not secrete it at all. These "nonclassical" hormone-secreting organs and their products are listed in Table 10-4 (Kacsoh 2000; Ojeda and Griffin 2000).

Hormones can have several effects on a target tissue, and several hormones may have the same effect. Also, as in plants and insects, hormone concentrations are controlled by feedback mechanisms, that is, they are up- or downregulated by the concentration of hormones or other compounds in the blood. Increasing levels of glucose in the blood, for example, stimulate the release of insulin, a polypeptide hormone, from the pancreas. Increased amounts of insulin in the blood in turn stimulate glucose uptake by the liver and its conversion to glycogen, and subsequent lower blood glucose levels lead to decreased secretion of insulin and slower glucose uptake.

The pancreas illustrates an interesting point about the evolution of organ systems. It is an evolutionary and developmental outcropping of the gut and also secretes digestive enzyme from a different set of cells; this indicates the general

TABLE 10-3.
Classical Endocrine Glands and their Hormone Products.

Endocrine Gland	Unit	Main secretory products
Pituitary gland (hypophysis)	Anterior lobe	Growth hormone (GH); prolactin (PRL), adrenocorticotropin (ACTH); gonadotropins (follicle-stimulating hormone (FSH); luteinizing hormone (LH)); thyroid-stimulating hormonc (TSH); β-lipotropin, β-endorphin
	Intermediate lobe	Melanocyte-stimulating hormone (MSH), β-endorphin
	Posterior lobe	Oxytocin (OT); arginine vasopressin (AVP or antidiuretic hormone (ADH))
Adrenal gland	Cortex	Aldosterone; cortisone (F); androstenedione; dehydroepiandrosterone (DHEA); DHEA-sulfate (DHEAS)
	Medulla	Adrenaline (epinephrine); noradrenaline (norepinephrine)
Testis	Leydig cells Sertoli cells	Testosterone (T); estradiol (E_2); inhibin; Müllerian inhibitory hormone. (MIH or anti-Müllerian hormone (AMH))
Ovary	Hilar cells	Testosterone
	Follicles	Estradiol (E_2); androstenedione; testosterone; inhibin
	Corpus luteum	Progesterone, estradiol (E_2); inhibin
Thyroid gland	Follicles	Thyroxine (T_4); triiodothyronine (T_3)
	Parafollicular cells	Calcitonin (CT)
Parathyroid gland	—	Parathyroid hormone (PTH)
Pancreatic islets	—	Insulin; glucagon; somatostatin (SRIF); pancreatic polypeptide
Placenta		Human chorionic gonadotropin (hCG), human placental lactogen (hPL), progesterone, estrogen
Pineal gland (epiphysis)	—	Melatonin, biogenic amines, several peptides

TABLE 10-4.
Nonclassical Endocrine Glands and their Hormone Products.[a]

Organ	Unit	Main secretory products
Brain	Especially hypothalamus	Corticotropin-releasing hormone (CRH), thyrotropin-releasing hormone (TRH), luteinizing hormone-releasing hormone (LHRH), growth hormone-releasing hormone (GHRH), somatostatin, growth factors (fibroblast growth factors), transforming growth factor-α (TGF-α), transforming growth factor-β (TGF-β), insulin-like growth factor I (IGF-I)
Heart		Atrial natriuretic peptides
Kidney		Erythropoietin, renin, 1,25-dihydroxyvitamin D fibroblasts
Liver, other organs,		IGF-I
Adipose tissue		Leptin
Gastrointestinal tract		Cholecystokinin (CCK), gastrin, secretin, vasoactive intestinal peptide (VIP), enteroglucagon, gastrin-releasing peptide
Platelets		Platelet-derived growth factor (PDGF), TGF-β
Macrophages, lymphocytes		Cytokines, TGF-β, proopiomelanocortin (POMC)-derived platelets
Various		Epidermal growth factors (EGF), TGF-α, neuregulins, neurotrophins

[a]From (Ojeda and Griffin 2000).

nature of endocrine organs as related to other developmental systems. Amphioxus, a primitive chordate thought to resemble the common ancestor of all chordates, has been found to express pancreatic hormones in the region of its gut, but without any morphologically distinct pancreatic organ.

Hormone Families

Like insects, vertebrates have directly coded hormones and others produced by enzymatic reactions. Vertebrate hormones are classified by their chemical nature into four major groups: amines, which are modified versions of the amino acid tyrosine; peptide hormones, which are short chains of amino acids; polypeptide hormones, strings of up to 200 or so amino acids; and steroids, which are lipids synthesized from cholesterol, the only class of hormones that are not direct gene products. These molecules differ in size, affinity for water, and locus of action, but, like hormones in plants and insects, and in fact all signaling molecules, they all bind to receptors on or in their target cell to effect action.

Expectedly, many hormones can be grouped into families of homologous molecules. Members of the same family have high DNA sequence homology, but they also tend to cross-react with the receptors for other members of the family, and they generally use similar mechanism or secondary messenger system to transmit their signal. The affinity with which a hormone binds to its receptor can determine the efficiency with which it transactivates a gene and thus the level at which that gene is expressed. For a given concentration of hormone, low-affinity binding will lead to the synthesis of less protein than high-affinity binding.

The amines include thyroid hormones, melatonin, and catecholamines, all derived from the amino acid tyrosine. These hormones are synthesized in the thyroid gland in the neck, are hydrophilic, and can cross cell membranes by diffusion; in fact, however, thyroid hormones tend to exit the cell by transport, packed into granules and released from the cell by exocytosis. This is the same mechanism cells use for the secretion of a number of other compounds.

The protein families of peptide and polypeptide hormones include the *insulin*, *glycoprotein*, *growth hormone*, and *secretin* families. The insulin family includes *insulin* and *relaxin*; the glycoproteins are *luteinizing hormone*, *follicle-stimulating hormone*, *thyroid-stimulating hormone*, and *chorionic gonadotrophin*; the growth hormone family includes *growth hormone*, *prolactin*, and *chorionic somatommamotropin*; and the secretin family includes *secretin*, *glucagons*, and *gastrointestinal polypeptide*.

These hormones are synthesized in the rough endoplasmic reticulum (ER) of their secreting cells. The protein as first translated is longer than the mature hormone will be and at this stage is called a *prohormone* or a *preprohormone*. The transient leading amino acid sequence, or *leader*, is hydrophobic and allows the hormone to be moved into the ER in an inactive state. The leader is cleaved when the hormone has crossed the membrane of the ER to be transported to the Golgi apparatus, where it assumes its mature conformation and then is stored in a granule that fuses with the cell membrane to be released into the perivascular extracellular space, again through exocytosis.

There are five major classes of hormone in the steroid family including the estrogens (estradiol, estriol, estrone), progesterone, androgens (DHEA, testosterone, androstenedione), glucocorticoids (cortisol, cortisone), and mineralcorticoids (aldosterone). Cholesterol is the precursor molecule, and it is transported in the bloodstream to the testes, the ovaries, and the adrenal glands, endocrine organs that convert it into steroid hormones under the direction, in turn, of hormone signals from the brain. Because they are by and large hydrophobic, steroids can readily pass through cell membranes, both to exit the cell of synthesis as well as to enter the target cell.

Inside the cell, steroids trigger gene expression by binding to nuclear receptors, which bind to DNA to transactivate gene expression. They may also have a nongenomic effect; receptors for steroid hormones also are found on cell surfaces, and receptor-hormone complex on the cell surface initiates not gene transcription but changes in intracellular concentrations of various ions by activating a second-messenger cascade.

Although these hormones are chemically related and activate gene transcription in a similar way, they have different roles in cellular physiology. The glucocorticoids and mineralcorticoids are involved in the regulation of cellular homeostasis and metabolism, whereas the estrogens, progesterones, and androgens are sex hormones, involved in the development of secondary sex characteristics as well as reproductive function.

Hormone Transport

Once in the bloodstream, hydrophilic hormones can circulate freely but hydrophobic hormones, the steroids, are generally bound to a protein carrier, which is either specific, such as *sex hormone binding protein*, or general, like *serum albumin*. As with most generalizations about the endocrine system, there are exceptions to these rules. Some hydrophilic hormones, such as IGFI, circulate bound to proteins, and this greatly extends their half-life, whereas some hydrophobic hormones, such as aldosterone, circulate unbound. Control loops and feedback mechanisms regulate the concentration of hormones in response to physiological processes and needs, and the mode of circulation determines the rate at which hormones are cleared from the bloodstream.

Hydrophilic hormones mix freely in aqueous solution that is chemically charged but cannot normally pass through the uncharged lipid cell membrane. As a result, information from hydrophilic hormones must be transduced by ligation with specific cell-surface receptors on appropriate cells that are receptive to the signal. Receptors for hydrophilic hormones are located on cell surfaces, whereas receptors to the hydrophobic hormones, the steroids, are intracellular, usually located in the nucleus. There are exceptions: Some lipophilic hormones, melatonin and the eicosanoids, like steroids, do bind to cell surface receptors to initiate cellular response.

Hormone signal transfer via extracellular ligand-receptor binding works via second-messenger cascades, mostly via single-transmembrane receptors. The basic means of information transfer was described in Chapter 7. The intracellular receptors for the hydrophobic hormones that pass through the cell membrane generally also function as TFs; the estrogen and glucocorticoid receptors are examples. When the receptor is bound with hormone, a conformational change in the receptor allows it to bind regulatory DNA sequences, which then activate gene expression. Some receptors bind to regulatory sequence in the absence of hormone but act as transcription *repressors* until bound with hormone; the thyroid hormone receptor is an example. When bound to thyroid hormone, however, the receptor-ligand configuration stimulates transcription of thyroid hormone-inducible genes (Koenig 1998).

The location of a protein within a cell is generally well regulated, and this determines the other molecules with which it interacts and thus the signaling cascade in which a signaling factor participates at any given time. In Chapter 6 and elsewhere we have referred to the sequestration of chemical cascades, themselves hierarchical and contingent based on the availability of substrates at each stage and the like. Sequestration within the cell controls the level of cross-reaction between systems, especially when they might use some of the same constituents, like amino acids or second-messenger components. The cell is also able to control the timing of the entry of transcription factors into the nucleus and thus the timing of gene expression (Downward 2001).

HORMONE EVOLUTION

Hormones cannot bind to DNA on their own but must first be bound to a receptor. How this linkage evolved is not clear. Did every mutation in a receptor have to weather the blows of natural selection until a comparable mutation arose in its ligand? This seems unlikely, but complex organisms depend extensively on such coevolution, the mechanisms of which in general are not well understood. This is another element of the "problems" of specificity, cross-reactivity, and coevolution that face systems that have multiple components evolving by modularity and duplication. However, some aspects of hormones and their receptors may help elucidate this.

A number of "orphan" nuclear receptors have been identified (by their shared amino acid sequence similarity with receptors for which ligands are known) for which a ligand has not yet been found and may not exist. Indeed, some receptors (i.e., TFs encoded by genes in a receptor gene family and having the prototypical receptor structure) seem to be able to function as TFs independently of ligand. Two possible evolutionary scenarios have been proposed: either the ancestral nuclear receptors were orphan and bound to DNA as homodimers, some of them later gaining ligand-binding capability independently (Laudet 1997), or the ancestors were liganded and some of the receptors independently lost their ligand-binding capability (Escriva et al. 2000). This supposes that the ligands—hormones—existed before ligand-receptor binding became part of the working cellular repertoire, and in fact, as we have seen, plants make steroid hormones, as do fungi, so the ancestral molecule was clearly ancient.

The affinity with which a ligand binds to its receptor varies, and indeed, some hormones bind to more than one receptor and some receptors bind more than one ligand. This kind of cross-reactivity might in principle "confuse" the organism, and there is certainly variation in nature that could reflect this (e.g., deviations from stereotypy in morphological or behavioral sexual dimorphism among individuals), but probably also provides a measure of protection via redundancy.

Finally, all of the nuclear receptors that have been identified in cnidarians are uniliganded, suggesting that this is the ancestral state (Grasso et al. 2001). The most parsimonious explanation of the evolution of hormones and receptors is perhaps that an ancestral receptor used in some other way evolved characteristics that also would bind a ligand and, after a gene duplication event and subsequent mutation, each receptor acquired the capability of binding to different ligands. This could have involved a period of cross-reactivity (which if too harmful would have been removed by selection), followed by divergence of function. We are largely handwaving here, however, because this does not explain why receptors in different subfamilies now bind ligands with no structural homology (that is, they are not themselves related to each other), how photoreceptors came to respond to light rather than chemical ligands, whether protohormones were TFs that alone could bind DNA, and why they may have lost that ability, among other provocative questions.

Perhaps the tightly paired ligands and receptors we observe and catalog today are those for which appropriate cross-reaction or coevolution happen to have occurred, whereas other signals evolved independently of ligand (e.g., some of the orphan receptors, which in fact are the majority of nuclear receptors) or ligands evolved independently of receptors (nitric oxide is an example) or have disappeared without them. The exchangeability of regulatory mechanisms means that in principle any one of many possible receptor-ligand pairs could carry out a given regulatory function and what we see today may just be those that happened to have been used. At the same time, similar systems and the use of gene family members in different species for the same job, or within species for related jobs, clearly suggests that coevolution has been important.

In this context it is interesting to ask why a truly unliganded receptor would maintain a viable ligand binding domain (that is, why would mutation not have erased the organized nature of that domain, given no selective pressure maintaining its ligand binding structure?). This could, of course, simply reflect limits on our current knowledge; ligands continue to be found for "orphan" receptors (Gustafsson 1999), thus causing the receptor to be termed "adopted" (Chawla et al. 2001). It is now recognized that some adopted receptors play a role in lipid sensing, forming heterodimers with retinoid X receptors, with low-affinity dietary lipids as ligands (Chawla, Repa et al. 2001). Also, as research goes, a gene may be long studied in the context in which it was originally found, and only later discovered to have entirely different functions. Such discoveries often tip off distant homologies among the functions, but this need not be the case.

However, orphan receptors commonly bind DNA as monomers rather than the heterodimers that steroids bind as, suggesting that this class of receptor acts in a way that is different from other receptors, whether by being truly unliganded or something else not yet understood. If any of these receptors truly have no necessary ligand, the fact that they still share the structure of liganded receptors may represent their occasional ability to cross-react in ways that have enough importance to be supported by selection, such as some undocumented specialization. On close inspection, history suggests that genes in the same family may not be as completely redundant as some experiments initially suggest.

Steroids

Steroid hormones seem to have been highly conserved until the divergence of jawed from jawless vertebrates some 450 mya. Figure 10-1 shows the structure of a number of steroid hormones. The structural diversity is produced by variation in the synthetic pathways and hence the genes that encode the synthesis of these molecules. Genes for nuclear hormone receptors form a large gene superfamily.

Peptides

Peptide hormones are typically very short strings of amino acids. Comparative sequencing suggests that after the divergence of jawed vertebrates peptide hormone genes underwent periods of sustained and rapid bursts of change, although there seems to be significant variation in the basal evolutionary rates of these genes (Liu et al. 2001; Wallis 2000). As has been suggested by studies of other genes in the vertebrate lineage, genes for protein hormones show some evidence of two whole-genome or at least large chromosomal duplication events early in the evolution of vertebrates; for example, when there is a single gene for a given hormone in *Drosophila*, four paralogous genes are found in vertebrates, creating large gene families for many hormones (Ohno 2001), and some teleost (bony) fish are tetraploid. Nonprimate mammals and the bush baby, for example, have a single-copy growth hormone gene, whereas humans and the rhesus monkey have five (Liu, Makova et al. 2001).

Coevolution at Higher Levels

Genes related to life history stages, such as different types of metamorphosis in various animal groups, must experience various forms of coevolution with other

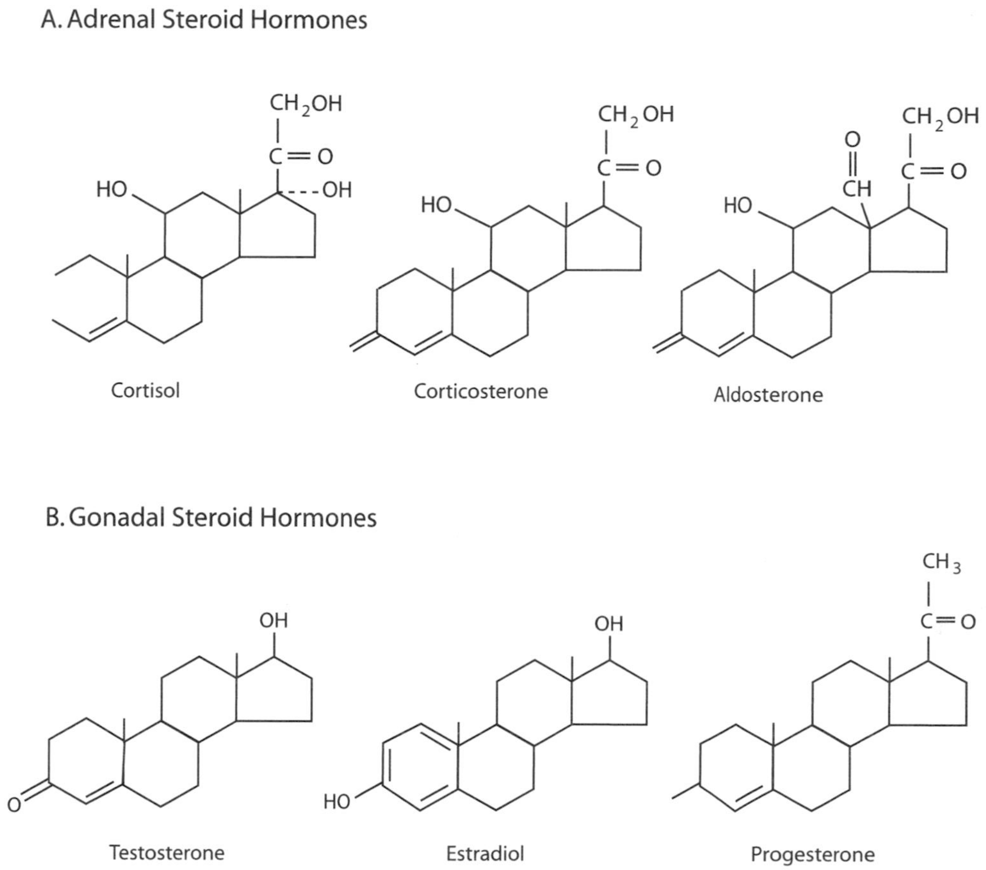

A. Adrenal Steroid Hormones

Cortisol

Corticosterone

Aldosterone

B. Gonadal Steroid Hormones

Testosterone

Estradiol

Progesterone

Figure 10-1. The chemical structure of (A) adrenal and (B) gonadal steroid hormones.

functions. These would seem to be complex, but they can evolve relatively rapidly. Thus some insects have complete metamorphosis (holometabolism) whereas others have no larval or pupal stages. Similarly, in groups of sea urchins, tunicates, or amphibians, we find that *within* some branches of the phylogeny some species experience more and others less complex development (Raff 1996). Tadpoles are found in various branches of frogs, the stage apparently having evolved independently.

All insects make ecdysone, and ecdysone controls insect metamorphosis, but metamorphosis requires many physiological changes and thus many other proteins and functions. Genes for apoptosis, for example, genes for the synthesis of new cuticle, and genes for the hormones that control other stages of metamorphosis must have all coevolved, more or less tightly, yielding the array of stages of metamorphosis that we see today.

The signaling coordination responsible for these developmental traits may involve many genes and their responses but clearly is something easy and relatively simple to evolve. Some amphibians, such as the Mexican axolotl, do not undergo metamorphosis in the normal life course, but metamorphosis is inducible with exposure to thyroid hormone. Therefore, the axolotl is still genetically capable of losing its neotenous state, but this can only happen given the provocative physiological state. The key probably lies in the quantitative or qualitative control of a small number of critical developmental, signaling, hormonal, or growth factors.

Some invertebrate hormones have marked homology to vertebrate counterparts and thus are likely to have common ancestors, which these days is no surprise. However, homologous hormones do not necessarily perform homologous functions across species, just as homologous genes are not necessarily used in homologous pathways, and other hormones seem to be found only in invertebrates, suggesting that they arose after the vertebrate-invertebrate divergence.

Like the plant story described earlier, animal hormone function (e.g., reproductive cycling and behavior in vertebrates) may involve timed, quantitative differences among a series of hormones. Not surprisingly, in some instances at least the hormones are either functionally diverged members of a gene family, or modifications of a common base molecule (steroids). Once again, we see modular evolution by duplication with divergence leading to partial sequestration of related function.

Receptors

We introduced nuclear receptors in Chapter 7 (Table 7-3). They bind ligands that make their way directly into the cell, rather than intercepting them in the extracellular space. The first nuclear receptor is likely to have arisen some 1000–800 mya, among the first metazoans. Nuclear receptors have not been found in plants. Virtually all vertebrates have the same six steroid receptors (estrogen receptors α and β, progesterone receptor, androgen receptor, glucocorticoid receptor, and mineralocorticoid receptor). Three steroid receptors have been found in agnathostomes rather than six, including an estrogen receptor, a progesterone receptor, and a corticoid receptor but no androgen receptor (Thornton 2001; Thornton and Kelley 1998).

Despite the lack of structural homology between steroid hormones, thyroid hormone, and vitamin D_3 molecules themselves, their nuclear receptors are homologous enough to be considered a steroid receptor superfamily. Nuclear receptors as a class are modular, consisting of four or five domains that function autonomously, and can be interchanged between related receptors without loss of function. This modularity presumably reflects a history of exon shuffling early in the evolution of the genes that code for these molecules (e.g., Patthy 1999).

Phylogenetic analysis of nuclear receptors led Laudet to propose six subfamilies of receptors, with members grouped by how they dimerize and bind DNA (Laudet 1997). Most of the six subfamilies are ancient and have receptors in both arthropods and vertebrates. The diversification of the superfamily may be the result of two waves of gene duplication (Escriva, Delaunay et al. 2000), which occurred before the divergence of lamprey and jawed vertebrates (Baker 1997; Thornton 2001; Thornton and Kelley 1998; Wallis 2000). There is no link between the ligand a receptor binds and the subfamily in which it is classed. That is, receptors that are closely related phylogenetically bind ligands with totally different biosynthetic pathways, suggesting that if nuclear receptors coevolved with their ligand, it was not a straightforward gain or loss of function during a gene duplication event, for example.

Phylogenetic analysis suggests that the ancestral steroid receptor was an estrogen receptor of some kind, with the androgen receptor emerging sometime after the vertebrate divergence. In lamprey, it is estrogen that regulates the development to sexual maturity in males as well as females, not androgen, suggesting that hormonal control over sexual dimorphism is relatively recent (Thornton 2001). A mineralocorticoid receptor has not been found in fishes or other lower vertebrates, indicating that it is the most recent of the steroid receptors. Agnaths apparently make most steroids, including testosterone, for which they have no receptor (personal communication, Thornton).

G Protein-Coupled Receptors

G protein-coupled receptors (GPCRs) were described in Chapter 7. As noted briefly in Table 7-4, there are several classes. The genes can be grouped into families in a number of different ways: by the molecular weight of their ligands, by the structure of their α-subunits, or by conserved amino acid structure. Five or six classes are conventionally agreed upon, and these include family I, the largest group of receptors, related to rhodopsin receptors; family II, including calcitonin-, PTH-, glucagon-receptors, and so forth; family III, containing metabotropic glutamate receptors and others, including a subgroup of vomeronasal receptors; family IV, comprised of STE2 yeast pheromone receptors; family V, yeast STE3 hormone receptors; and family VI, receptors related to slime mold cAMP receptors (Josefsson 1999). (These classes are sometimes designated by letter, A–F).

Phylogenetic analysis based on DNA sequence yields three large clades and several minor ones, all of which arose more than 800 mya (Josefsson 1999; Wess 1998). In this analysis, families I, II, V, and VI plus some unclustered receptors form one large group, family III forms a smaller clade essentially as already classified, and family IV forms a final cluster. GPCRs in these families generally bind classes of ligands with similar functions, but with much diversity between the classes (Josefsson 1999). Receptors from these families were present in the acoelomate flatworms of the Precambrian, showing that intercellular signaling is an ancient characteristic of multicellular organisms and probably of single-celled organisms.

Phenogenetic drift seems to have played a role in GPCR evolution. These genes comprise a large fraction of the genome (about three percent in mammals) and are used in a huge diversity of pathways. There is much structural diversity in the receptors as well as their ligands, even though they all operate similarly at the molecular level, so over evolutionary time there have been substitutions in the details but conservation of the mechanism and its basic logic.

GPCRs have been found in plants, but they do not seem to be as important in intercellular signaling in plants as they are in metazoans. A single gene for the α-subunit of the heterotrimeric G protein and one for the β-subunit are known in *Arabidopsis*, and homologous genes have been found in other plant species. In plants, calcium is the most important transducer of intracellular signals, as well as being involved in control of turgor pressure, cell growth, cell division, and other processes, with downstream effects including ion channel activation, gene expression, and vesicle fusion (Malhó et al. 1998). It is also an important mediator of environmental and intercellular signals. How cells interpret calcium-transduced signal is still unclear, but calcium-binding proteins are involved. One set, the *Calmodulin* genes, bind to and regulate a wide variety of signaling proteins such as kinases, which phosphorylate transcription factors, cAMP phosphodiesterase, which degrades cAMP, as well as other enzymes, ion channels and pumps, various cytoskeletal components, and the like. Calmodulin is found in all eukaryotic cells. Each Calmodulin molecule binds four Ca^{2+} ions and changes conformation on binding, which allows the complex to bind to and activate molecules downstream.

Beyond its role in inducing conformational change in binding proteins, calcium action in a cell is dependent on spatiotemporal factors, such as the timing and location of transient membrane ion channels. Distribution of Ca^{2+} is not uniform throughout the cell but, on signaling, spikes at specific intracellular locations. At rest, cytosolic calcium concentration is maintained at between 10 and 100 nM, but it can peak to between 1 and 5 μM. These transient spikes, or waves, are complex and not well understood, but it is known that the wave is initiated at a defined site by a defined signal or by passing through the plasma membrane at a single location. The wave is not transmitted by diffusion but by continuous release of calcium from subcellular stores, the ER, mitochondria, or other vesicles, followed by reuptake by cell compartments as the wave moves along. The wave can be constrained to specific locations in the cell by the factors involved in its production, and thus different parts of the cell can be separately regulated.

In Chapters 6 and 7 we described ion channels and the families of genes whose members complex to form these pores in cell membranes. These, too, have an evolutionary history. For example, there are 1 or 2 Na^+ channel genes in invertebrates but 10 in mammals. Mammalian channel genes are found on four chromosomes, suggesting multiple duplications of an ancestral chromosome, which happened before divergence of teleosts and tetrapods. Sodium channel genes are linked to *Hox* gene clusters (Lopreato et al. 2001).

CONCLUSION

We have looked at some of the specific ways that cells communicate with each other within and even between organisms. The purpose of such communication, if "purpose" is an appropriate word, is for one cell to effect changes in other cells, and this generally means changes in gene expression. Because this is so central to multicellular, organized organisms it is no surprise that many, elaborate mechanisms, with their own checks and balances, have evolved in plants and animals.

Communication of this sort is inherently four-dimensional, involving space and time (and perhaps one could suggest additional dimensions, because time and space mean different things within and between cells, and perhaps there is an additional "concentration" dimension). From immediate changes to gradual life span changes that can take decades (e.g., puberty, tooth eruption, and perhaps even some aspects of aging), or be seasonal (flowering, dormancy), cells communicate with their diversity of molecular mechanisms. It is this that produces life cycles and enables organisms to undergo development and to maintain homeostasis. In this regard, a high level of predictability is often vital. For example, stages of early development follow quickly on each other and developmental stuttering can be fatal. Similarly, mating requires that both sexes be prepared at the same time. As we mention in several contexts, the same mechanism is used *between* organisms, as in mating-related signaling, adding subtlety regarding the nature of what we call an "organism" or "being" if the reproductive behavior of one depends intrinsically on the behavior of another.

These statements apply within an organism, but we have discussed aspects of the compatible change between receptors and ligands that must occur over evolutionary time. As gene families proliferate, selection and drift must maintain at least a sufficient set of compatible ligand-receptor sets for existing function and to evolve new or modified function. In fact, additional elements of coordination are required, because the proper receptor, among its frequently numerous gene family alternatives, must be expressed in cells if they are to respond and signal-sending cells must express the appropriate ligand gene. Selection by phenotype keeps the internal chaos within bounds, but how this happens is not yet understood in detail.

But if this control is coordinated and predictable among cells, there is also a role for chance. With some exceptions such as in early development, two individuals rarely undergo changes at precisely the same time or age, and this applies to genetically near-identical animals like clones or inbred strains of mice, or plants. The similarities may be great, and we do not yet know the relative roles of somatic mutation, environment, and chance, but these factors certainly do play a role and perhaps provide opportunities for selection to act. Generally, the earlier in development, the fewer cells in an organism are committed to a particular fate and the greater the effect of a chance interruption in the nominal flow of events.

Similarly, the ability to respond to signals from the outside in itself provides organisms with a tolerance for and an ability to respond to environmental change or stress, but only within limits. Plants, for example, produce hormones to protect against drought or temperature extremes, but beyond predictable thresholds—extended drought or freezing or extreme heat, for example—they cannot survive. It is important to note, too, that the limits vary by species: tundra plants tolerate colder climates than jungle plants, flamingos tolerate warmer temperatures than penguins. Variation in response to change in environment is sometimes called the *reaction norm* (Schlichting and Pigliucci 1998). Many factors are involved, most of them unknown, but they surely will be shown to be intimately connected to cell signaling mechanisms.

Many aspects of homeostasis and life cycles are basically hardwired into the organism's makeup, with stereotyped responses sometimes rather precisely determined by genetic mechanisms. Temperature sensitivity, mating seasons, and the like may be examples. These are rather specifically predictable. Yet many vital aspects of the environment cannot be predicted specifically, and perhaps the ability of organisms to respond to things they cannot anticipate in advance is the most interesting aspect of the nature of complex organisms—even plants. Much of the remainder of this book is about what we refer to as "open-ended" sensory systems—senses such as vision and hearing, smell and taste. The organism is enabled to detect various *kinds* of externally derived input, but cannot in advance know what its details will be—yet has to be able to recognize and respond. The organism can recognize pathogens, neurons store memories of unique experiences, the heart can slow or speed its beat in response to the amount of work the organism is doing, animals can perceive a huge number of smells, we can see light within specific frequencies, and so forth. Its response involves internal communication and coordination of the sorts described here, among others.

Yet, if these abilities are crucial for the life-ways of the organisms that possess them, in fact they and their responses are not entirely open-ended or unrestricted. They typically fail above or below predictable thresholds. Primates cannot see UV light, our hearts cannot beat any faster after a certain point, we cannot defend against all types of invading pathogens, etc. The fact that a system is not perfect and has its failure limits, however, is less curious evolutionarily than is often thought: if the system is here, it has been good enough to get here. That is the only interest of evolution. It is to these systems that we now turn.

Chapter 11
Detecting and Destroying Internal Invaders

A substantial fraction of organisms that have ever lived have probably died as a result of predation, and much of evolution and of the structures of organisms can be viewed in the context of the battle to eat or be eaten. And as life would have it, many organisms alive today survive only by consuming other forms of life. Predation, broadly defined as obtaining nutrition from other living organisms, is a primary way in which organisms secure their energy needs. As a consequence, most animals and plants are meals-on-a-platter for some other organism(s). *Not* being consumed by somebody else before reproducing is a primary requisite for darwinian fitness.

Avoidance of predation by large organisms has clearly been important, and the living world has evolved many mechanisms of detection, defense, and escape. However, probably the first, and certainly still among the most dangerous, forms of predation is attack by microorganisms such as bacteria, viruses, and other parasites. These organisms are too small to be perceived by the cognitive or "organismal" sensory systems by which larger predators are perceived. Micropredators enter the body, where they can reproduce and grow and attack their prey from within. Typically they can reproduce rapidly, becoming in a sense an internal megapredator. There thus ensues a race against the attack.

Not only are microorganisms small, but they frequently attack through molecular rather than physical means, by recognizing and binding to various cell surface molecules within the prey and entering the cells, or even insinuating themselves into the host's DNA. But if their small size makes their presence difficult to detect cognitively by the host, their size means that they can be defended against by molecular means. This is convenient for complex organisms whose life is itself based on means to interact by molecular signaling and recognition. Defense against microorganisms can do double duty as defense against nonreplicating but potentially threatening exogenous microscopic or molecular entrants into the body—toxins, pollen, foreign protein, and the like. Even if not alive, such agents can be dangerous if exposure is extensive or extended over time.

Genetics and the Logic of Evolution, by Kenneth M. Weiss and Anne V. Buchanan.
ISBN 0-471-23805-8

All multicellular organisms have evolved mechanisms for protecting against molecular attack. Given that the number of pathogens and their ability to evolve rapidly greatly outstrips that of most of their prey, self-defense is no small task. It would seem, on the face of it, an almost impossible battle for big, long-living (and sometimes stationary) organisms to outcompete the swift and nimble microparasite attack. Yet this obviously happens.

The challenge in molecular self-defense is not just a matter of recognizing alien organisms within. In some way or another, an organism fighting molecules with molecules must obey the dictum "above all, know thyself." That is, organisms must be able to distinguish their own cells from pathogens, and to preferentially target the pathogens for destruction, a daunting requirement. Yet complex organisms are themselves molecularly very diverse and, due to somatic mutation and cellular differentiation, changeable during their lifetimes. Some organisms have very complex mechanisms to deal with these challenges, but others get along very well with simple mechanisms.

Some aspects of immune defenses are shared between extremely divergent organisms. In this chapter, we will discuss the immune systems of insects, plants, and animals and what is currently thought about how they evolved. Because it includes most of the components found in the immune systems of other organisms, we will begin with the most complex system, that found in vertebrates. In addition, although at first blush the necessity for an immune system at all seems to confirm the idea that life is primarily a battle to the death between competing organisms, in fact, many if not the most successful pathogens do not kill their hosts but instead have evolved a coexistence; most organisms peacefully serve as hosts to many microorganisms, and indeed cannot do without them. The challenge to live a completely aseptic life might be—or at least has proven to be—beyond the capacity or the interest of evolution.

VERTEBRATE IMMUNE RESPONSES

All vertebrates have what is traditionally termed an *innate* immune response, a rather generic, nonspecific system. Gnathostomes (jawed vertebrates) also have an *adaptive* response, that "learns" via antigen-antibody reactions to combat specific pathogens. Innate immunity does not improve with subsequent exposures to a pathogen. Adaptive immunity is specific and has *memory* in that the adaptive response to a pathogen improves with each subsequent exposure during an individual's life, but this knowledge is not transmitted to the next generation. This does not mean that jawed vertebrates have a more effective immune system in general; agnathic vertebrates are still around, and their rarity relative to gnathostomes is generally attributable to their having been outcompeted by jawed vertebrates, not pathogen attack. However, it certainly raises the question as to what forces led to the evolution of the complex adaptive system (see below).

INNATE VERTEBRATE IMMUNE RESPONSES

The overwhelming majority of multicellular organisms face the world with innate immunity alone. These systems may work against particular types of pathogen, but do not depend on recognizing specific organisms. Innate immunity begins with physical barriers, such as the skin or bark, mucous membranes, the cough response, chemical barriers such as pH of the stomach, or temperature, which produce conditions in which microorganisms cannot live. If a microorganism breaches these first lines of defense, the innate immune system detects this and launches an immediate response. Innate immunity in vertebrates is mediated by cells of a general class known as *phagocytes*, of which specific types called *macrophages* and *neutrophils* use surface receptors for common bacterial components to trap, engulf, and destroy the pathogen. Cells of the immune system are shown in Figure 11-1. The *complement system*, described below, is also activated, to *opsonize* the bacteria, that is coat

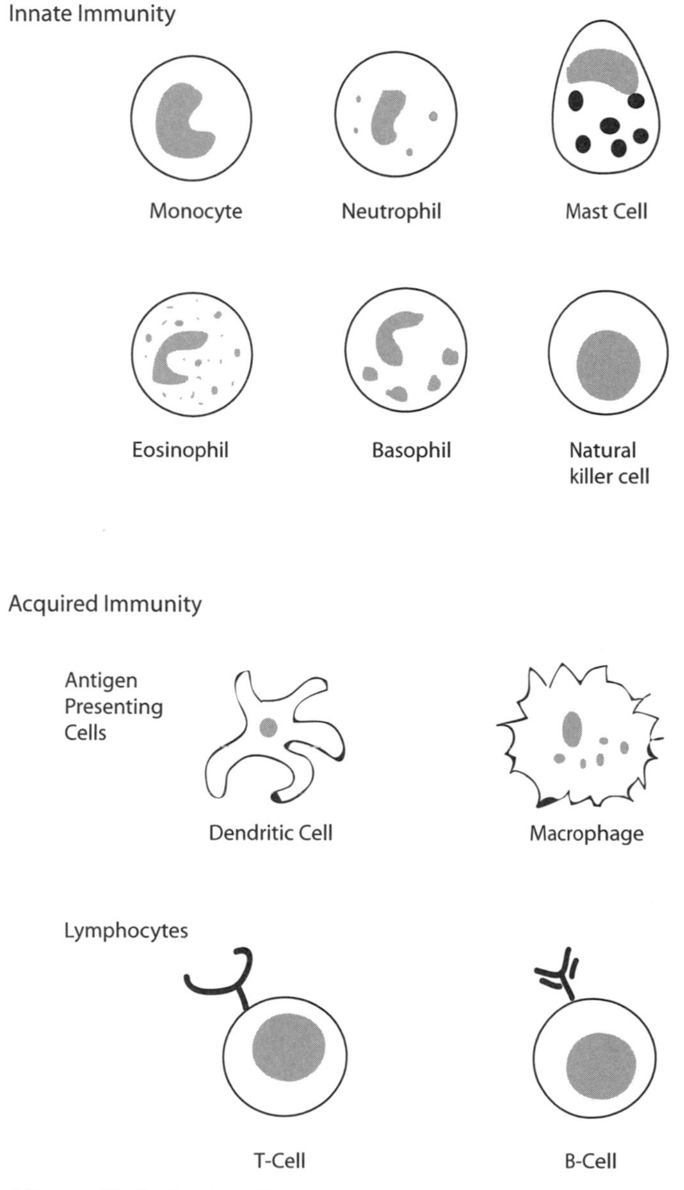

Figure 11-1. Basic cell types of vertebrate immune systems.

them for recognition and destruction by phagocytes, or to destroy it outright. These receptors are encoded in the germline, but unlike the components of the adaptive response (see below), they are inflexible. These cell lineages are part of a generalized hemopoietic differentiation process that generates the circulating blood cells.

To establish an infection, a pathogen must either avoid detection by the innate immune system or defeat or overwhelm it. To evade recognition by phagocytes, many extracellular bacteria (that is, those that do not enter the host's cells but circulate in the blood or live in mucosal tissues) develop a thick polysaccharide coating or capsule. Some pathogens have developed a way to grow inside the phagosome (inside a cell that has engulfed the pathogen).

If enough bacteria enter the body they can overwhelm the innate response, and the adaptive immune system (in jawed vertebrates) must then respond if the organism is to survive the attack. If a pathogen evades or overwhelms the innate response, the early induced responses, humoral and cell-mediated effector mechanisms to be described below, are activated. This second wave of responses may contain the infection until the adaptive responses are ready to mount a defense, and they also influence the kind of adaptive response that is mounted. Some of the same mechanisms of the innate immune response may later be enlisted to eliminate the pathogen.

On activation, phagocytes release *cytokines* into the circulation. When received by cytokine receptors, these molecules alter the behavior or induce the proliferation of the recipient cell types in the immune system. *Cytokine* genes include a family of genes called *Interleukins*, which induce the liver to produce proteins that activate complement and the opsonization of microbes. Cytokines can also trigger a fever response that can be helpful, because many bacteria do not function well at elevated body temperature.

The innate immune system fairly quickly elicits a localized *inflammatory response* at detected sites of infection. This involves the recruitment of more phagocytes and effector cells (lymphocytes that are immediately able to mediate the destruction of a pathogen without having to undergo the further differentiation that other classes of cells in the adaptive immune system require) to the infection and the release of cytokines that have local effects like inducing increased blood flow to the area, helping, for example, by improving access for more effector cells. There is also an increase in vascular permeability, which allows the local accumulation of fluid (with concomitant pain and swelling of infection) and an increase in the number of immunoglobulins and complement, among other effector cells, in the area (Janeway 2001).

Phagocytes release many other molecules as well, such as nitric oxide, toxic oxygen radicals, mediators of inflammation, and the like, that both fight the pathogen and facilitate further adaptive response. One class of molecules called *defensins* is a family of short antibacterial peptides that help to permeabilize bacterial membranes so that they can be destroyed. Similar defense mechanisms are found in plants. Viral infection induces the production and secretion of proteins known as *interferons*. Interferons, true to their name, interfere with the replication of viruses by binding to interferon receptors, initiating a signal cascade that ultimately activates the transcription of genes that degrade viral RNA or otherwise inhibit viral replication inside a cell, preventing the spread of infection to neighboring cells.

These diverse measures, each involving families of gene products that activate other cell types and thus trigger the production of many attack molecules, besiege an infected site with generic defenses. The same process induces changes in the endothelial cells (the lining of blood vessels) during inflammation to induce these cells to trigger blood clotting in small vessels near the site of infection. This reduces the ability of surviving pathogens to enter the bloodstream and travel to other sites in the body. Meanwhile, phagocytes that have engulfed pathogen are carried by the fluid that leaked into the area at the early stages of infection to nearby lymph nodes, where they trigger an adaptive immune response in which T cell receptors (TCRs) or antibodies (see below) recognize the foreign peptide on the surface of the cells and are prompted to proliferate. This leads to the destruction of the infected phagocytes (Janeway 2001).

The Complement System

This system enhances the ability of the adaptive immune system to destroy bacteria, thus the term *complement.* It is particularly important in the destruction of bacteria. It seems to be a part of the innate immune system that was partly coopted in the evolution of adaptive response. Comprised of some 20 serum proteins, the system works as a cascade of cleavage reactions, one reaction activating the next component of the system. The effector mechanisms of complement include opsonization of the surface of pathogens so that phagocytes can recognize and engulf them, direct killing of microorganisms by creating holes in their surface membrane, chemotactic attraction of leukocytes to sites of infection, and activation of leukocytes.

The system is activated via three pathways. The first is called the *classical pathway*, activated by antigen-antibody complexes and active in both innate and adaptive immunity. The second is the *mannan-binding lectin pathway* (MBLectin pathway), which responds to the binding of a serum protein called mannan-binding lectin to carbohydrates on bacteria or viruses that contain mannose, a type of sugar molecule. Finally, the *alternative pathway* is activated when the surface of a pathogen is bound by a previously activated complement component (Janeway 2001).

These "innate" systems are invoked generically, without regard to the specific pathogen involved. In that sense they are preprogrammed for a kind of blind defense. A diversity of rather crude mechanisms is used, like destroying cell walls, that are possible because innate defenses recognize generic components of broad classes of pathogen, or exploit common constituents of these organisms (such as mannose) which the pathogens cannot shed because they are so intrinsic to their survival. In that sense the innate system evolved as a response by complex organisms to constituents of bacteria in place possibly since their origin. An assault on cells that is too generic can also damage the host as well, but if the response occurs early enough and is kept localized, the damaged area can be regenerated or sloughed harmlessly, and the infection is overcome.

The Adaptive Immune Response System

Bacteria and viruses can mutate and evolve much more rapidly than animals that reproduce slowly like vertebrates, and we might expect the microbe always to be able to outevolve any specific adaptation by the host—a race that must always go to the swift. Although many slowly reproducing species nonetheless manage, immune systems in higher vertebrates include a component that is *adaptive* on a scale to match that of microbial parasites.

In vertebrates both the innate and adaptive immune responses are mediated cellularly by leukocytes. Two phagocytic classes, macrophages and neutrophils, were mentioned earlier; along with *monocytes* they are primarily involved in the innate immune response system. Phagocytes mount a first-line defense in the immune response, which is immediate but generic.

The adaptive immune response is specific but can take up to seven days (in humans) to prepare to defend against a pathogen while the innate response attempts to rid the body of it. A class of leukocytes known as *lymphocytes* are the workhorses of the adaptive response. These cells recognize specific intra- or extracellular pathogens. The two predominant types of lymphocytes are known as *B* and *T* cells, named for the location in which they have been thought mainly to develop. In mammals, B cells differentiate in the liver in the fetus, and in bone marrow postnatally whereas T cells develop in the thymus. B and T cells acquire specificity to pathogens via receptors they produce, known as *antibodies*, that recognize *antigens*, molecules on the surface of the target pathogen or a toxin that it produces.

As currently understood, the major differences between the receptors on B and T cells are that the B cell receptor has two identical antigen recognition sites and can be secreted into the circulation whereas the T cell receptor always remains anchored to the cell surface and has only one recognition site. Antibodies circulating freely in the blood or lymph are called *immunoglobulins (Ig)*. Other antibodies are anchored on the surface of B or T lymphocytes.

However, although the antibody-antigen ligation is specific at the molecular level, antibodies are not programmed in advance by specific antibody genes or alleles. Instead, lymphocytes with an essentially random assortment of antibodies, called *naïve lymphocytes* because they have not yet come into contact with antigen, circulate from the blood into the lymphoid tissues. There, an adaptive selection process unfolds in which the specific microbial antigens themselves are used to select antibodies to which they can be bound.

B Lymphocytes

A given B cell produces only a single antibody type (see below). B cells present this antibody on their surface (*surface immunoglobulin*), but once the antibody recognizes (binds) an antigen, two signals induce the lymphocyte to multiply and differentiate into immunoglobulin-secreting plasma cells. The first signal is induced by the ligated receptor and the second by a costimulatory signal in the form of a B7.1 or B7.2 molecule, coded by a member of the *B7* gene family. The first signal initiates the synthesis of one subunit of the *AP1*, or antigen-presenting, TF and the second signal directs the synthesis of the other subunit. The AP1 TF recognizes enhancers such as those for interleukin gene expression. Once the B cell differentiates it can proliferate and/or produce large quantities of immunoglobulin—its particular antibody—that bind to other circulating copies of the antigen that originally activated the cell (Medzhitov and Janeway 1998). B cells can also be induced to differentiate in the spleen into memory cells that will quickly produce antibody if the organism is subsequently rechallenged by the same pathogen.

In Chapter 7 we described a variety of reactions that involve ligand binding by cell surface receptors that triggers response via the receptors' intracellular domains. Immunoglobulins work in a logically similar way. They have two major functions, each carried out by different parts of the molecule. One region binds to antigen, and the other mediates *effector* functions, that is, the destruction of pathogens specifically based on type of infection and the pathogen's life cycle stage. There are a number of effector systems, and the same systems are used by the innate and adaptive immune systems. The simplest is neutralization, when the antibody simply binds to the pathogen and neutralizes its activity. Another is phagocytosis, where antibody activates phagocytic cells to engulf the pathogenic cell and degrade it in one of several ways. Some antibodies, as well as TCRs on T cells (see next section), when they recognize antigen, induce cytotoxic reactions that kill the pathogen outright, either by perforating the membrane of the target cell or by inducing apoptosis.

The two functions of immunoglobulins are structurally separate on the molecule. Antibodies are formed by the pairing of heavy and light polypeptide chains (see Figure 11-2). The antigen binding region, the arms of the **Y**, that include the light chains, varies enormously in several open-ended ways and thus is called the *variable (V) region*. The rest of the molecule, the leg of the **Y** composed of the heavy chains that are responsible for the effector functions, is not nearly so variable and so is called the *constant (C) region*.

There are five Ig isoforms (alternative polypeptides), each coded by a different C-region gene, and these isoforms define the functional classes of immunoglobulins in higher mammals: IgG, IgM, IgA, IgD, and IgE. IgG is the predominant immunoglobulin in humans, comprising 70–80% of the total. It is found circulating in the blood as well as interstitially (in between cells). The structure of IgGs is remarkably variable among mammals, and they cross the placenta in some but not others. Maternal IgG confers immunity to newborns for several months, until the infant is able to make its own.

IgM comprises about 10% of the body's immunoglobulins. It is generally found circulating in the blood and is the first antigen-induced response to infection. IgA accounts for 10–15% of human immunoglobulins. Secretory IgA is the primary immunoglobulin in saliva, colostrum, milk, and other secretions. IgD accounts for <1% of total immunoglobulin but exists at high levels on the surface of B cells. Its function is not known, but it may have to do with the development of the fetal immune system or with lymphocyte differentiation. IgE is usually scarce in human serum, but probably is involved with immune response to parasites as well as in allergic responses (Roitt et al. 1998).

T Lymphocytes

T cells play a variety of roles in the adaptive immune system. As currently understood, some are involved in controlling the development of B cells and their antibody production, *helper T cells* help phagocytes destroy the pathogens they have internalized, and *cytotoxic T cells* recognize and kill virally infected cells.

T cells recognize antigens that are presented on the surface of a cell by molecules of the *Major Histocompatibility Complex* (MHC) (see Figure 11-3). The TCR is specific for antigen-MHC complexes. T cells function either by releasing cytokines, which signal between cells during an immune response, indirectly by activating macrophages to destroy organisms they have engulfed, or directly by interacting with infected cells. Like B cells, once a T cell has recognized a pathogen, it multiplies to vastly increase the amount of TCR available, and because the TCR is specific this generates *clones* of the T cells appropriate to the pathogen.

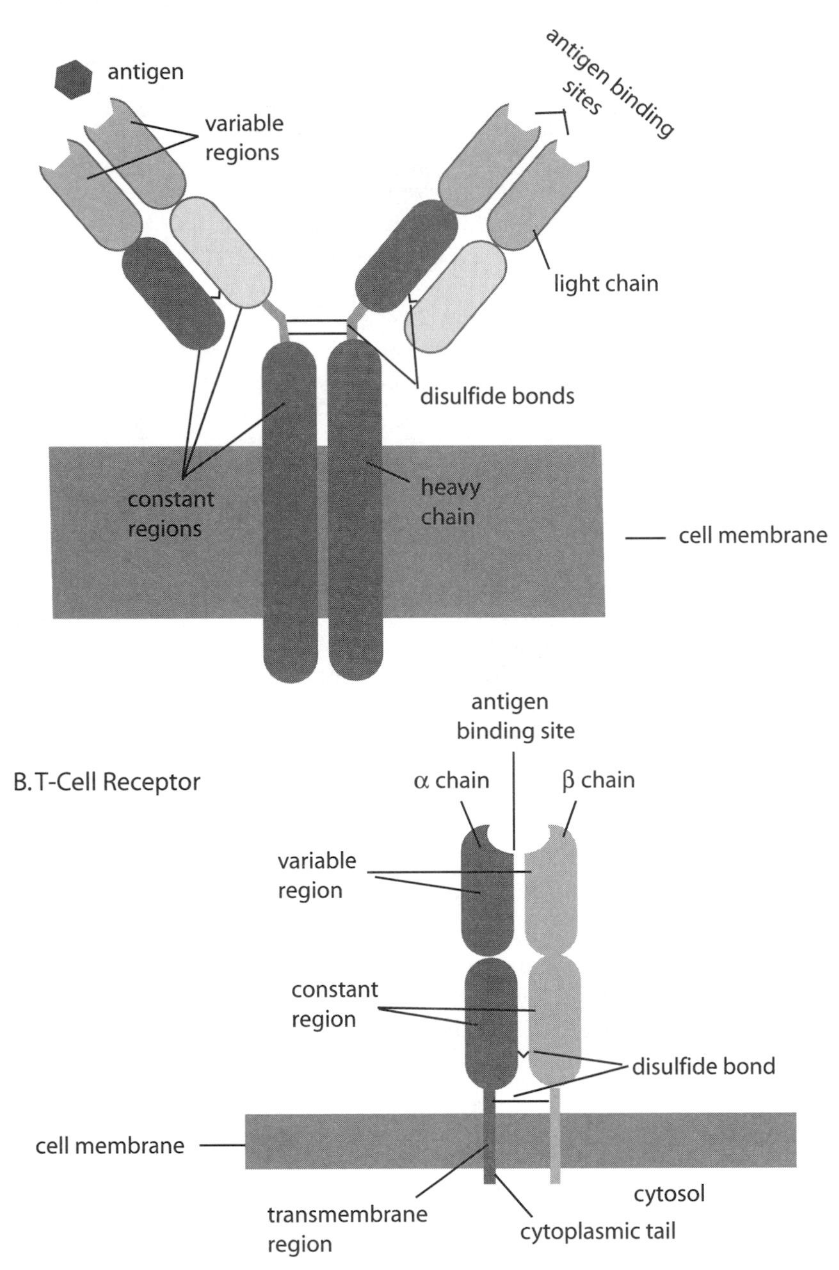

Figure 11-2. Structure of immunoglobulin proteins, showing (A) the joined light and heavy chains, constant and variable regions; (B) T cell receptors, showing their similarity to immunoglobulins.

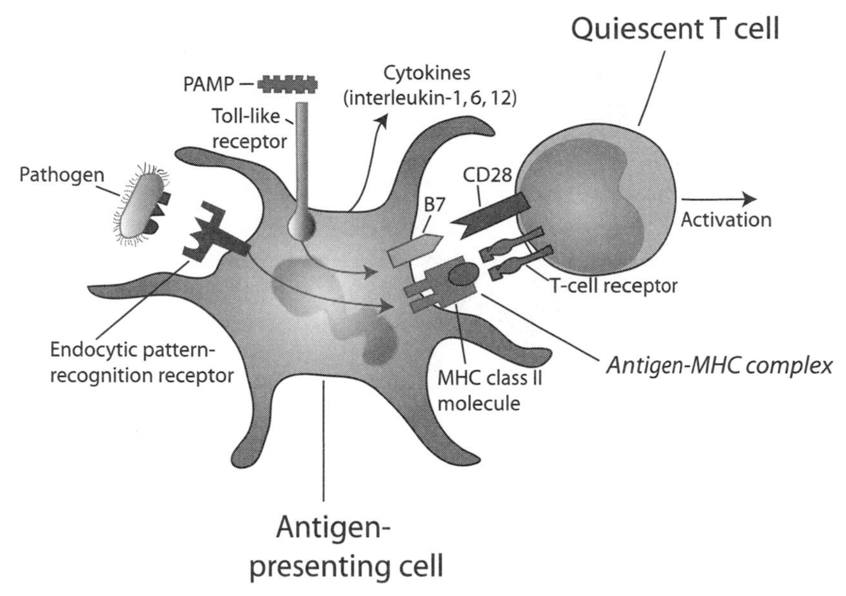

Figure 11-3. Antigen presenting cell with signals, T cell and receptors. Redrawn from Medzhitov and Janeway with permission; see that reference for details. Original figure copyright 2000, Massachusetts Medical Society, all rights reserved.

There are approximately 30,000 TCRs on the surface of every T cell. There are two types of these receptors, both formed of heterodimers composed of two polypeptide chains joined via a disulfide bond (Janeway 2001). The chain of the predominant type of receptor is made of α- and β-subunits, whereas the second, and less common, type of TCR , is a dimer of γ- and δ-chains. These receptors recognize antigen differently, and in fact, the role of the TCR is not yet fully understood. These heterodimers are physically associated on the T cell surface with five polypeptides called the CD3 TCR complex, which is required for the expression of the TCR at the cell surface, although it is not yet clear how. Ninety to ninety-five percent of T cells are αβ T cells and the rest are γδ T cells (Roitt, Brostoff et al. 1998).

Leukocytes express many different molecules on their surfaces, which are used to classify cell types by a *cluster designation* (CD) type. αβ T cells are classified as CD4+ ("helper") or CD8+ ("killer") cells. Helper T cells activate other cells to destroy pathogen, whereas CD8+ T cells kill directly. Although γδ T cells comprise a small fraction of circulating T cells they are frequent in mucosal epithelia. It is thought that these T cells may be important in protecting mucosal surfaces from some kinds of viral and bacterial infections.

The Major Histocompatibility Complex

The family of genes involved in tissue rejection comprises the MHC. This was first described in the mouse; the human analog is called the *human leukocyte antigen* (HLA) system (the organization of both systems is shown in Figure 11-4). *Histocompatibility antigens* are encoded by *histocompatibility genes*. The HLA complex is on human chromosome 6, and includes more than 200 genes, 40 of which encode leukocyte antigens. These are among the most polymorphic of all mammalian genes

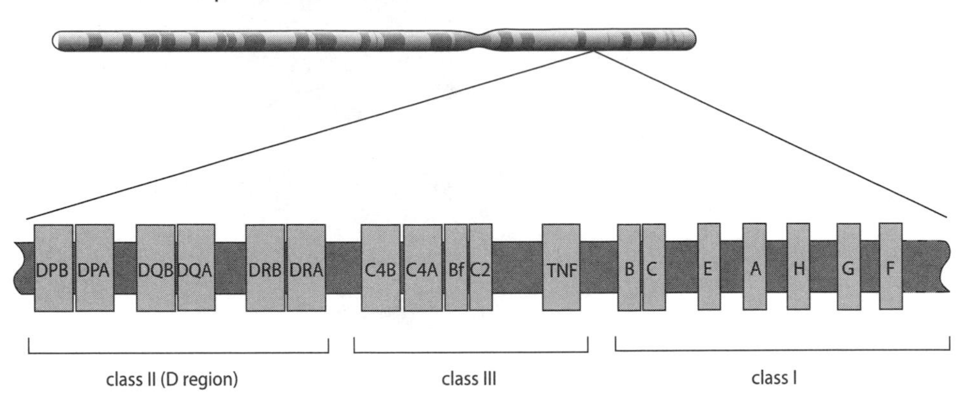

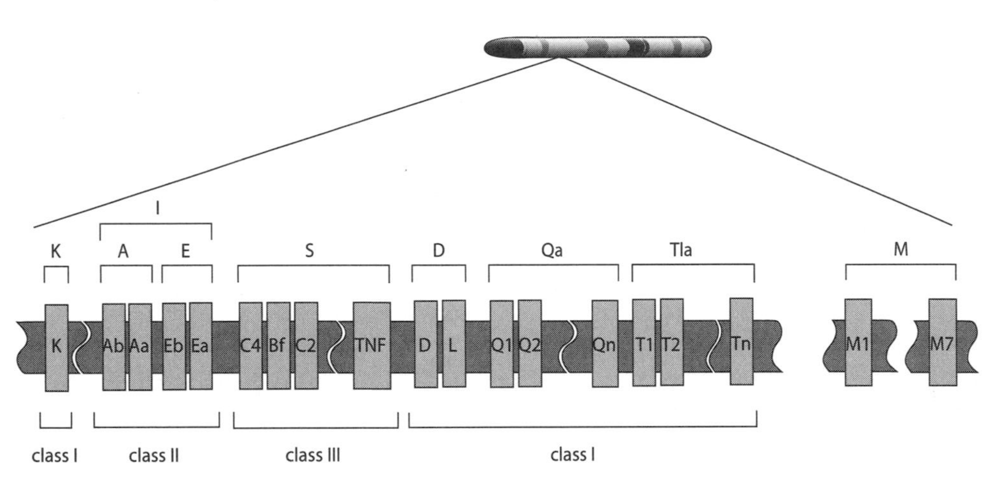

Figure 11-4. Organization of the human and homologous mouse major histocompatibility (MHC) complexes and position of major genes. The regions span three to four megabases.

(Lechler and Warrens 1999). They bear coding domains homologous to the immunoglobulin and TCR genes, all part of the large cell adhesion molecule (CAM) superfamily mentioned in Chapter 7. Other genes within the large HLA chromosomal complex encode complement, cytokines, and other molecules with a role in the immune system.

The MHC is complex incompletely understood, and varies among species. Essentially, there are two major MHC classes, encoded by *class I* and *class II genes*, although the "other" genes in the complex, physically linked but structurally unrelated to the class I and II genes, are sometimes referred to as MHC *class III* genes. The details of the structure and function of the two major classes are distinct, but the basic role of each is to present pathogen-derived peptides to the surface of cells for detection by the TCR of T cells; this happens as part of the process of cellular

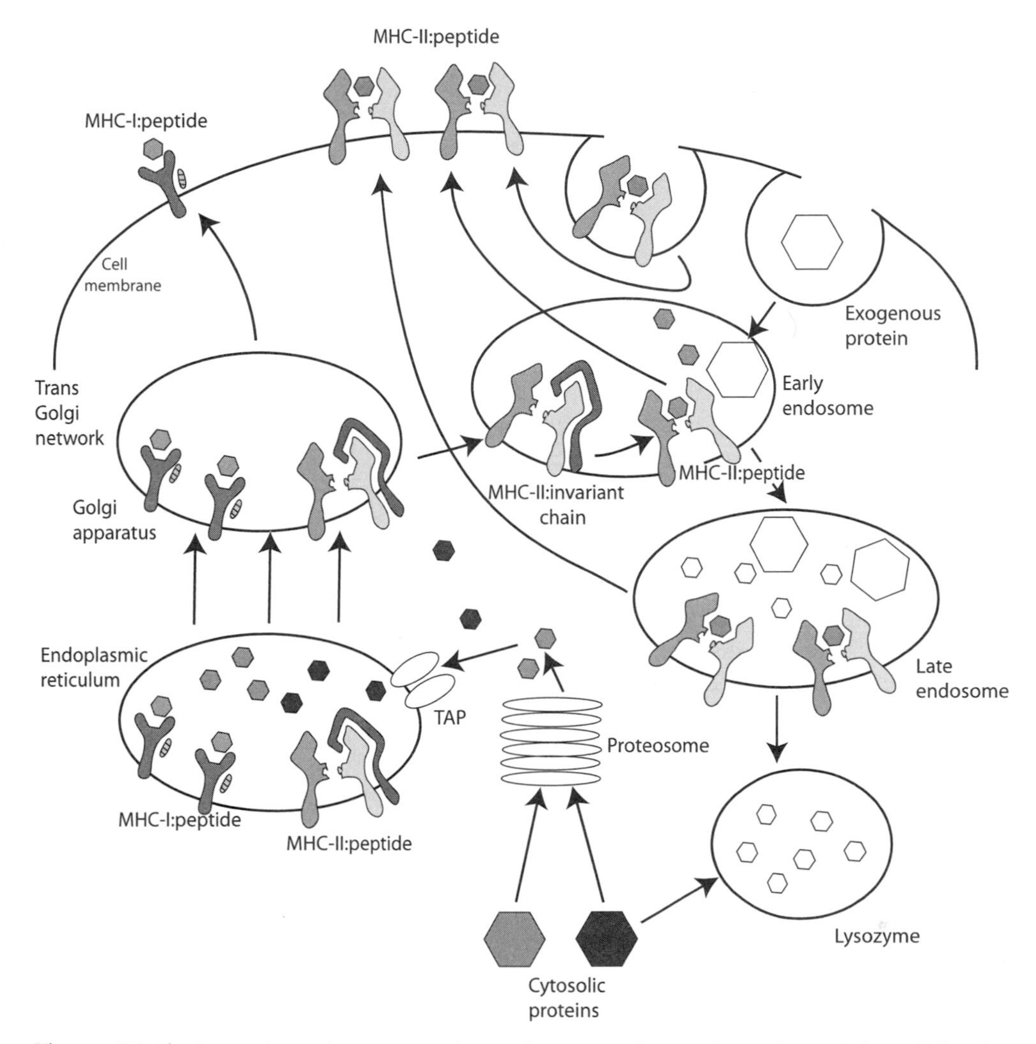

Figure 11-5. Stages in antigen processing and presentation to the surface of the cell by the MHC class I and II gene products. Redrawn after (Paul 1999) with permission. Original figure copyright Lippincott, Williams and Wilkins, 1999.

waste disposal. Class I genes are expressed on the surface of most somatic cells, whereas class II genes are expressed only on some types of immune cells, including B cells, activated T cells, macrophages, dendritic cells, and thymic epithelial cells (Klein and Sato 2000). In particular because most cells present an individual's class I molecules, the MHC has been viewed generically as a form of self/nonself recognition signaling.

Generally speaking, class I molecules carry endogenously derived peptides (intracellular, from viruses, for example) whereas class II molecules transport exogenously derived peptides (from extracellular self or foreign proteins such as bacteria or viruses that have not yet entered a cell), although this is not always true. Class I molecules plus antigen are recognized by killer T cells, whereas class II molecules and antigen are recognized by helper T cells (Klein and Sato 2000; Roitt, Brostoff et al. 1998). Thus an antigen presented by a class I molecule elicits a T cell killer response, whereas if the antigen is presented by a class II molecule, a helper T cell response follows.

Worn-out or misfolded proteins inside a cell are marked for degradation by a molecule called *ubiquitin*. The marked proteins are unfolded with the help of chaperone molecules and fed to proteasomes, which fragment them. The resulting peptides are either degraded into amino acids, which float in the cytosol of the cell and are reused, or transferred into the endoplasmic reticulum (ER) for transfer to the cell surface. This is done by *transporters associated with antigen processing*, or *TAPS*. In the ER, class I molecules pick up these peptides for transport to the surface of the cell, where they are displayed.

Extracellular proteins are treated differently. They are surrounded by invaginations from the plasma membrane of a cell, which pinch off as endocytic vesicles and fuse with lysosomes to form endosomes. Class II molecules transport these peptides to the cell surface; they do not bind to peptides in the ER, but instead the class II molecules are enclosed in membranous vesicles that fuse with endosomes to form the MHC class II compartment where exogenous proteins are degraded and subsequently transported to the cell surface.

Degradation of endogenously derived proteins, including self proteins, happens all the time in most cells, and MHC class I molecules are constantly transporting peptides to the surface of cells. The action of class II molecules on exogenously derived proteins is less ubiquitous, however, generally restricted to cells that are specialized in phagocytosis or endocytosis. In any case, most cells are studded with MHC-peptide complexes all the time, hundreds of thousands of them, and the vast majority represent self peptides (Klein and Sato 2000).

Natural Killer Cells

A third kind of lymphocyte is the *natural killer cell (NK)*. These cells originate in bone marrow. They do not express antigen receptors but have the ability to destroy some tumors by lysing tumor cells, and they play a role in innate immunity by killing cells infected with viruses.

Production of Immunoglobulin and T Cell Receptor Diversity

So much for the mechanics. To restate the evolutionary challenge presented to macroorganisms, pathogens have a much shorter life span and can rapidly outevolve the ability of large organisms to defend against them. If recognition is the challenge, then immune systems are faced with ever- and unpredictably changing diversity. One strategy for keeping up is the generation of huge amounts of diversity in the adaptive immune systems of higher vertebrates. Although somewhat different in structure and function, the fundamental tools of this system, the TCR and immunoglobulins, are both produced by a very interesting *somatic* DNA recombination mechanism that is unusual because during the development of each lymphocyte lineage, the inherited genome in the antibody coding complexes of that cell is permanently altered. This mechanism has been traced to the emergence of the ancestral jawed vertebrate about 450 million years ago.

The immunoglobulin and TCR gene complexes contain sections of coding regions known as V (variable), D (diversity), and J (joining) elements, in which there are many duplicate coding elements (especially in the V region). Active recombination-

TABLE 11-1.
Number of Functional Gene Segments in V Regions of Heavy and Light Chains, Human.

	Light chains		Heavy chain
Segment	κ	λ	*H*
Variable (V)	40	30	65
Diversity (D)	0	0	27
Joining (J)	5	4	6

Source: (Janeway 1999).

generating mechanisms randomly select one each of the V, D, and J regions during the somatic development of lymphocytes. This is what is transcribed, while the remaining alternative segments in these regions are discarded from the genome and from the cell. The repertoire from which this random selection process chooses provides a huge variety of possible VDJ combinations, and hence different lineages. Table 11-1 shows the number of functional gene segments in human V regions.

The total number of possible different receptors is enormous, and the total number of different lymphocytes circulating in an individual at any time is on the order of 10^8 (Janeway 1999). The V and C (constant) regions are connected by the short J region and, in the heavy chain, also a D region. It is the C region that determines the class of antibody and how it destroys the pathogen once it is identified and bound, whereas the variable region determines which antigen the antibody or TCR binds. Sequence variation is greater in the V region, and there are many related genes in tandem in the V region, whereas the C region has more standard genelike function. Because the former determine antigen-binding specificity of the coded molecule, this means there is a comparable diversity in the molecules.

This *somatic recombination* is triggered by the products of the recombination-activating genes *Rag1* and *Rag2*. The V, D, and J gene segments are flanked by recombination signal sequences (RSSs) that consist of conserved heptamer and nonamer sequences separated by 12- or 23-base-pair spacer sequences. The *Rag* genes encode a recombinase that catalyzes site-specific cleavage of DNA between the RSSs; two coding ends of the excised pieces are then joined in an imprecise way, with additions of random nucleotides and/or short deletions, yielding a unique new combination of V(D)J segments (Roitt, Brostoff et al. 1998) (see Figure 11-6).

Diversity in immunoglobulins can result from any one or a combination of five different processes; the different possible combinations of Ig heavy and light chains, gene conversion, recombination of V, D, and J genes (V and J for the light chains), variability in joins, or somatic hypermutation introducing mutations into the V regions of activated B cells in a way that is not yet completely understood but that involves regions of numerous point mutations, mutational hot spots, and a mechanism in which DNA breaks initiate error-prone DNA synthesis by faulty DNA polymerases in V(D)J regions (Diaz and Casali 2002). The triggering mechanisms are not yet known, but generally are well within the familiar repertoire of evolution (mutation, duplication), gene expression (context-dependent transcription, splicing), and development (combinatorial signaling).

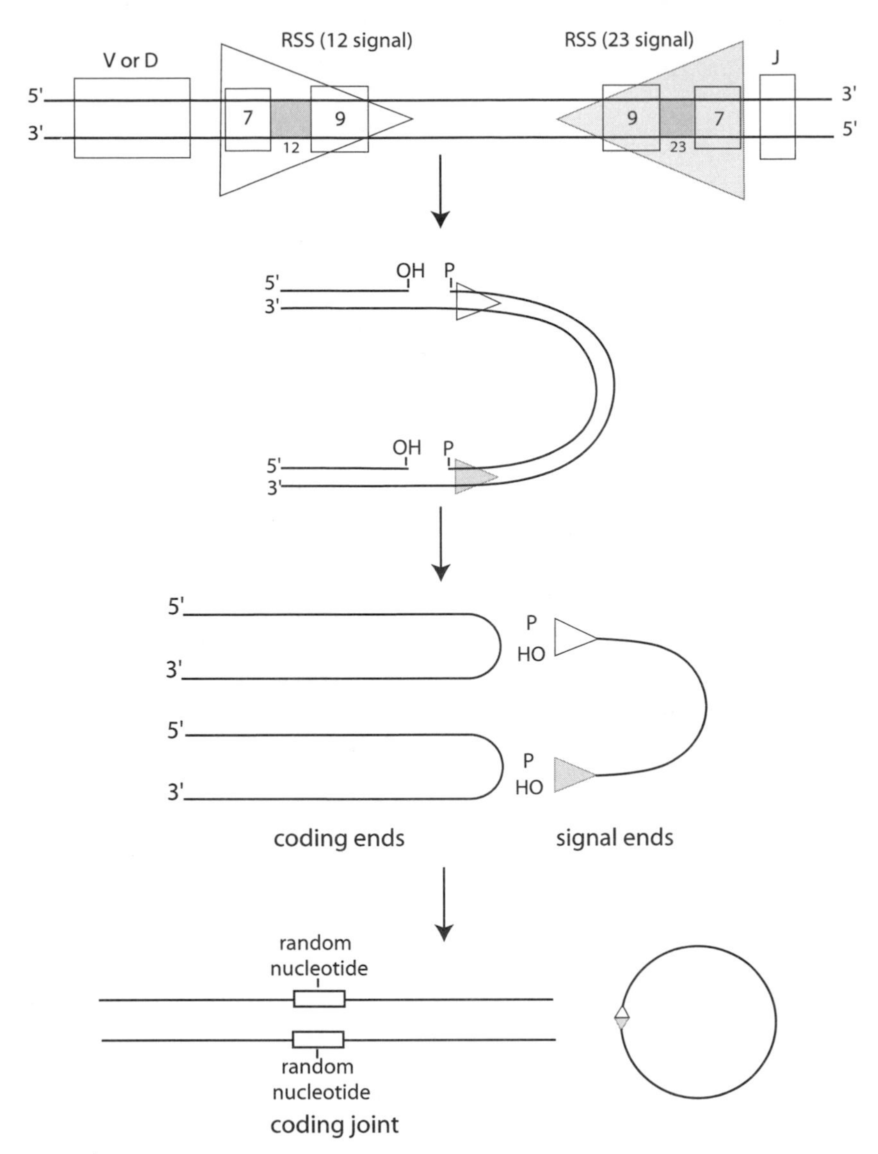

Figure 11-6. *Rag1* and *Rag2* mediated V(D)J formation (see text). Redrawn from (Ohmori and Hikida 1998) with permission.

This somatic recombination happens in *cis*, that is, along the TCR and heavy and light chain Ig gene regions. At the same time, there is a *trans* phenomenon of *allelic exclusion*, by which the same region of the homologous chromosome is inactivated, so that the cell expresses only one combination of its genes (for the TCR or its Ig, depending on the cell type) derived from only one of its chromosomes. Each of these events happens independently in each of the cell lineages. At least one mechanism of allelic exclusion seems to involve asynchronous replication in mitosis (Singh et al. 2003). The earlier-replicating chromosome can in principle produce some substances or conditions that prevent recombination from occurring on the homologous chromosome. Of course, in any evolving system, there will be diversity among species. A variety of ways in which antibody diversity is generated in vertebrates is depicted in Figure 11-7. The system in sharks depends on a large number of individual complete genes, much like that of olfactory reception (Chapter 13) and the *R*-gene system in plants (see below). We see again our common finding: in their own way, each works well enough to have evolved.

The first stages of B cell development occur in the bone marrow, where the process of gene rearrangement takes place. This stage generates an immature B cell that carries an antigen receptor. This antigen receptor is in the form of cell surface IgM, capable of interacting with antigen. ("IgM" means that in addition to its chosen V, D, and J region, the cell is using its "μ-" or M constant chain to form the immunoglobulin). Immature B cells at this stage that do not encounter antigen either die or are inactivated (become *anergic*), thereby eliminating many potentially self-reactive B cells from the repertoire of antibody-producing cells, because B cells that crosslink their own IgM are programmed to die (Janeway 2001; Roitt, Brostoff et al. 1998). Cells that survive this process are subsequently free to leave the bone marrow and enter the lymphoid system, where they may encounter foreign antigen and be activated. Activated mature B cells proliferate clonally, to become either antibody-producing plasma cells or memory cells that can respond quickly to reinfection with the same pathogen. If activated in later stages of infection, the constant chain switches from M to G, etc., depending on the context. The stages of B cell development are shown schematically in Figure 11-8.

T cells are bone marrow-derived stem cells, called *thymocytes* when they are immature lymphocytes. They differentiate and mature in the thymus, undergoing recombinational events like those that produce the differentiated antibodies of B cells. They also undergo the same kind of selection process as B cells: 98% of thymocytes die, by apoptosis, before they mature, although, unlike B cells, T cells undergo positive as well as negative selection. The process is complicated and not well understood, but it is based on the recognition by T cells of self peptide-self MHC complexes and eliminates those that are highly self-reactive, while selecting for those that have a moderate affinity for self. This negative selection thus ensures that these cells become self-tolerant. Experimental work has also shown that T cells are unable to recognize and bind to foreign peptide-foreign MHC complexes. That is, positive selection of T cells, based on moderate affinity to self MHC, ensures that T cells recognize foreign peptide-self MHC complexes (Janeway 2001; Roitt, Brostoff et al. 1998).

Several aspects of the system are worth noting. The recombinational possibilities come about because of the number of selectable regions that can be used (thanks to a history of duplication), because there are two chromosomes for each gene complex to choose from, and because recombination is an error-prone system (especially in the D and J regions). Since each activated B or T cell expresses only a single binding variant, the proliferation of just that variant represents a kind of natural selection within the organism that favors what works without proliferating a lot of antibodies that do not work against a particular pathogen (but could generate anti-self reaction). This makes the system modular in an interesting way and sequesters its specificities in controlled ways—but that do not require the organisms as a whole to "know" which of some fixed prior repertoire of antibodies to activate and which not to.

A.

1. Multiple V genes, each encoding a different V region

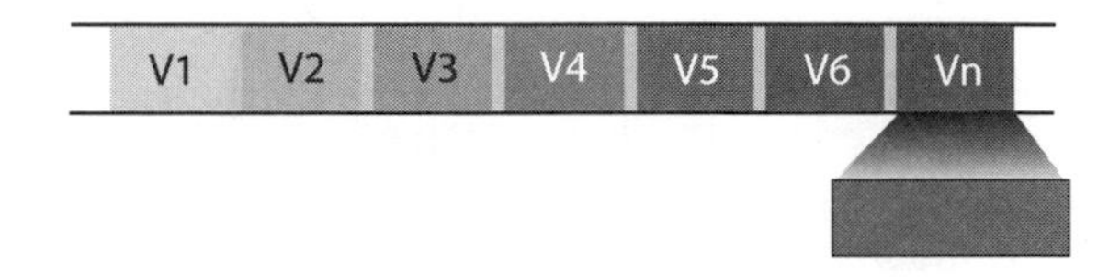

2. Somatic mutation during B cell ontogeny, producing different B cell clones

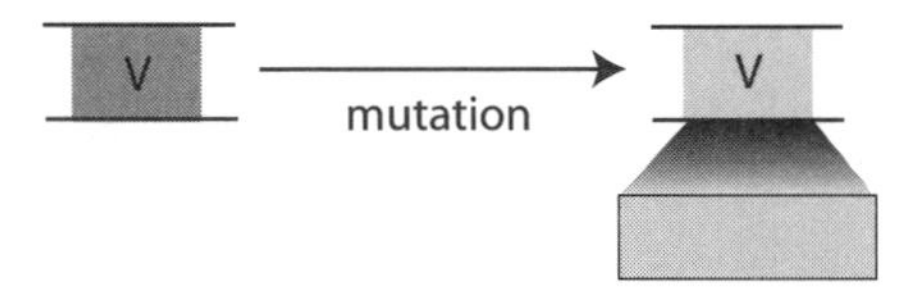

3. Somatic recombination during B cell ontogeny

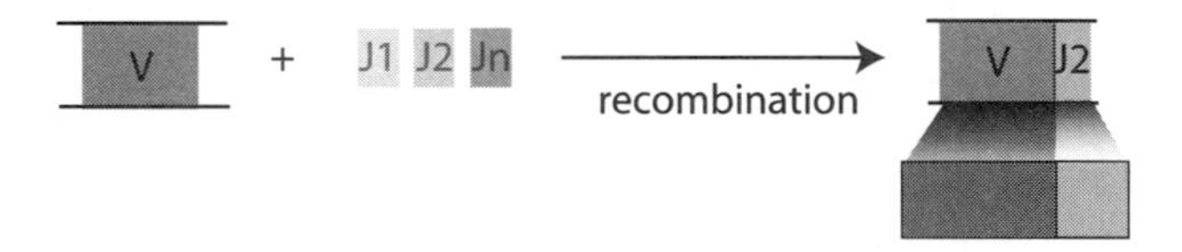

4. Gene conversion

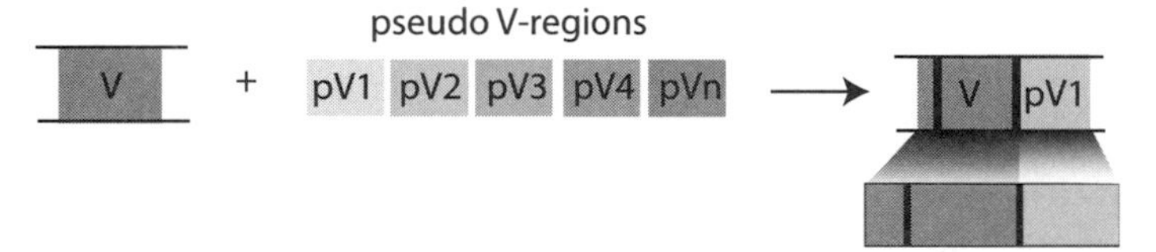

5. Nucleotide addition, when V and J regions are cut, before joining

A

Figure 11-7. Generation of antibody diversity in vertebrates (A) Various vertebrate antibody recombination mechanisms including multiple V-region genes in the germ line, somatic mutation, somatic recombination between elements forming a V-region gene, gene conversion, and nucleotide addition; (B) antibody diversity in the mouse is due to somatic recombination; in sharks to numerous antibody genes rather than somatic recombination; chickens have few antibody genes but a high level of gene conversion. (A) Redrawn from (Roitt, Brostoff et al. 1998) with permission from Elsevier, original figures copyright Elsevier 1998; (B) redrawn from (Gerhart and Kirschner 1997) with permission.

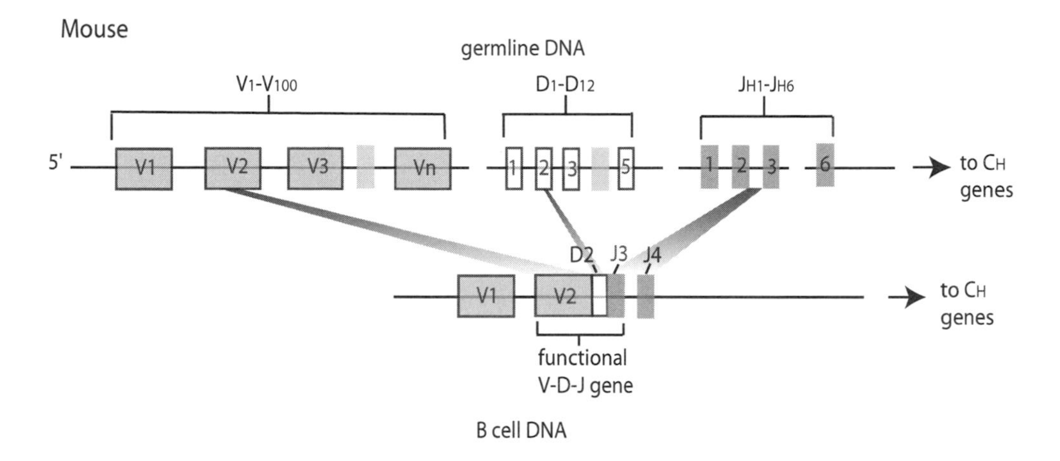

Shark

germline DNA

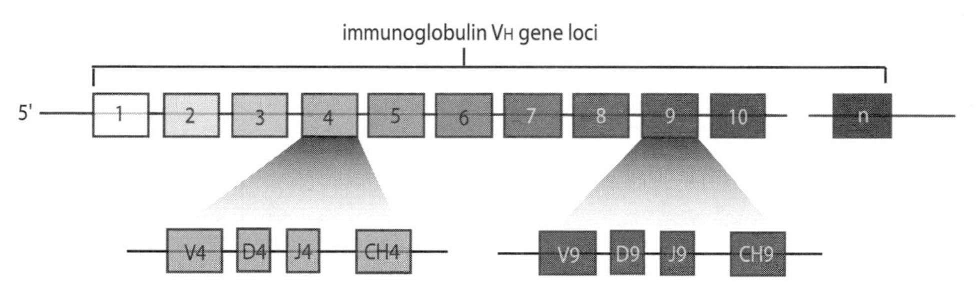

B cell DNA

Chicken

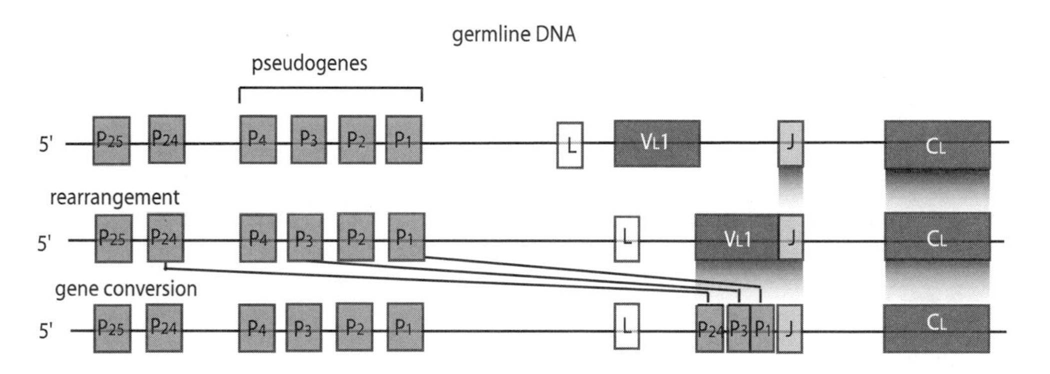

B

Figure 11-7. *Continued*

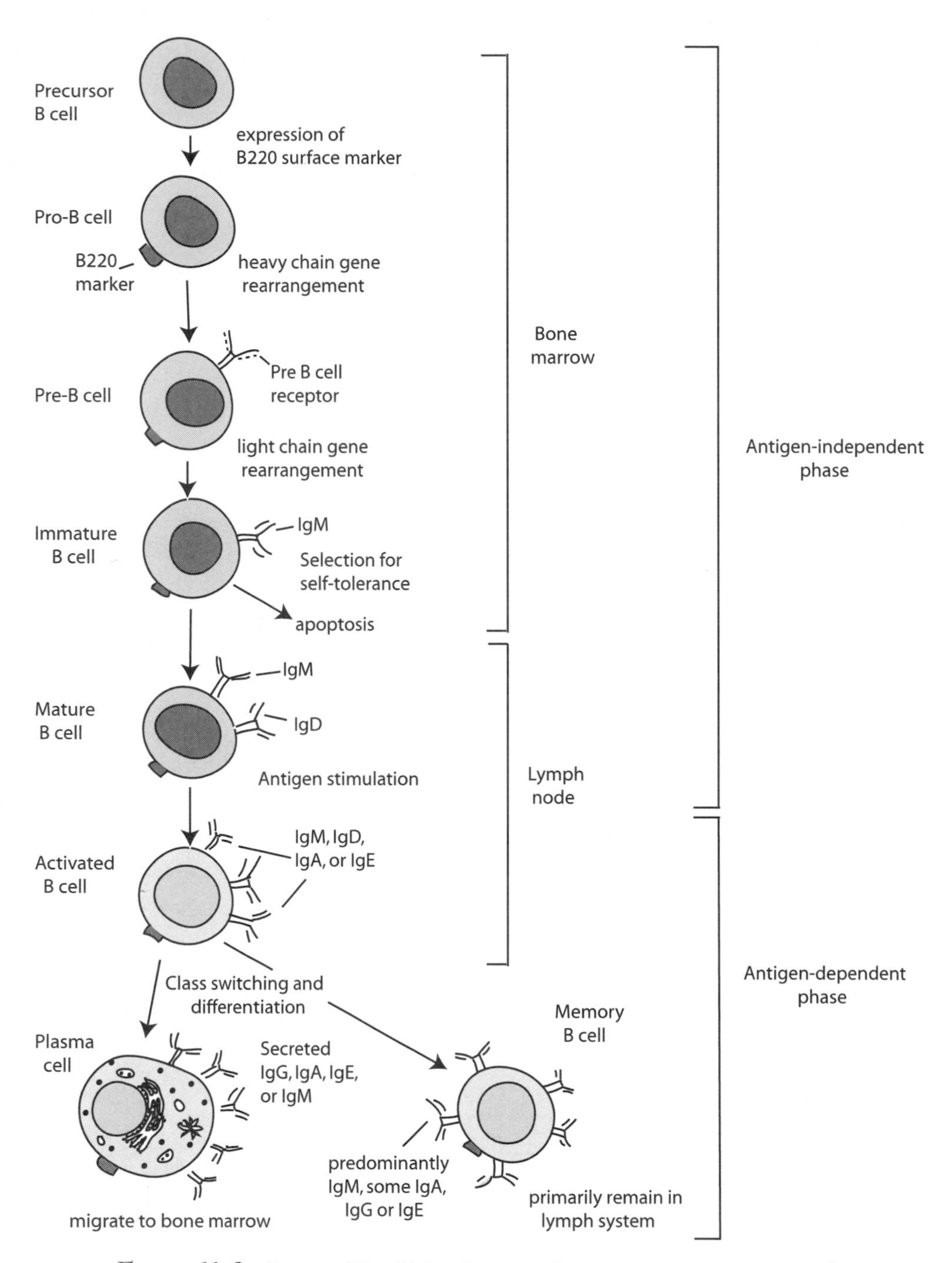

Figure 11-8. Stages of B cell development from precursor to mature cell.

Many more combinations are possible than are circulating in the blood at any one time, and these cells are being recycled throughout life. The system generates variation in a combinatorial way, but it need not be exhaustive: tiny voles and huge elephants have essentially the same immune diversity, with orders of magnitude differences in the number of lymphocytes they can circulate. There are no major differences in adaptive immunity associated with life span or body size in higher vertebrates (and fish can live very long lives). For most pathogens, the millions of antibodies present at any time appear to be sufficient to ensure that enough cells recognize the specific pathogen that the body can mount an effective defense. Thus the system is not limited by these factors (*or* it is not as important for survival as we tend to think).

Note that the immune memory that an individual develops during its lifetime is not genetically inherited (or at most is inherited in the limited extent of temporary effects of transplacental resistance from mammalian mothers to children). What is inherited is the (randomly mutated) germline repertoire and the sloppy recombination process from which any new attack is mounted. As currently understood, each new individual must generate its own set of specific antibody molecules. It starts fresh, but the mechanism is such that this is not a particular handicap relative to its benefits; and to a considerable extent, the pathogenic environment is always evolving anyway.

The system is powerful but neither necessary nor perfect; most animals do not have such a system, so it is clearly not a prerequisite for survival or "adapting" to a world filled with tiny, hungry pathogens. Also, as noted earlier, we live in commensalism with many microparasites (indeed, we could not survive without them, as illustrated by the need in vertebrates for bacterial colonization of the gut to aid in various stages of digestion; usually, these stay out of the body, where they are less vulnerable to immune attack) and, despite effective immune defense, more animals of many species may be killed by micropredation than by any other single cause.

Some Genetic Aspects of Lymphocyte Signal Transduction, Differentiation, and Response

When cell surface receptors on lymphocytes bind to lipopolysaccharide (LPS), a component of Gram-negative bacterial cell walls, peptidoglycan (PGN) from the cell wall of Gram-positive bacteria, or interleukin1 (IL1), this initiates a signal-transduction pathway, triggered by Toll-like receptors, (TLRs), which releases a TF, NFκB, from its inhibitor IκB (see Figure 11-9). This allows NFκB to enter the nucleus and bind to various promoters, activating genes involved in adaptive immunity as well as proinflammatory molecules, cytokines, that signal between cells during an immune response. Tlr4, with additional factors, is required to sense LPS, whereas the *Tlr2* gene product is required to sense PGN (Takeuchi and Akira 2001; Takeuchi et al. 1999). TLRs, a family of leucine-rich proteins, are homologous to *Drosophila* Toll protein, a membrane receptor that is also thought to function in insect microbe recognition, as well as being used in the development of dorsoventral patterning in flies as we saw in Chapter 9, among other functions.

The Innate Response Induces Adaptive Immunity and Self/Nonself Recognition

Traditional discussions of vertebrate immunity treat innate and adaptive responses as separate systems, with innate immunity being derived from the ancient response to general characteristics of pathogens and homologous to the immune systems of lower vertebrates or invertebrates. Its role was viewed as, at best, keeping pathogens

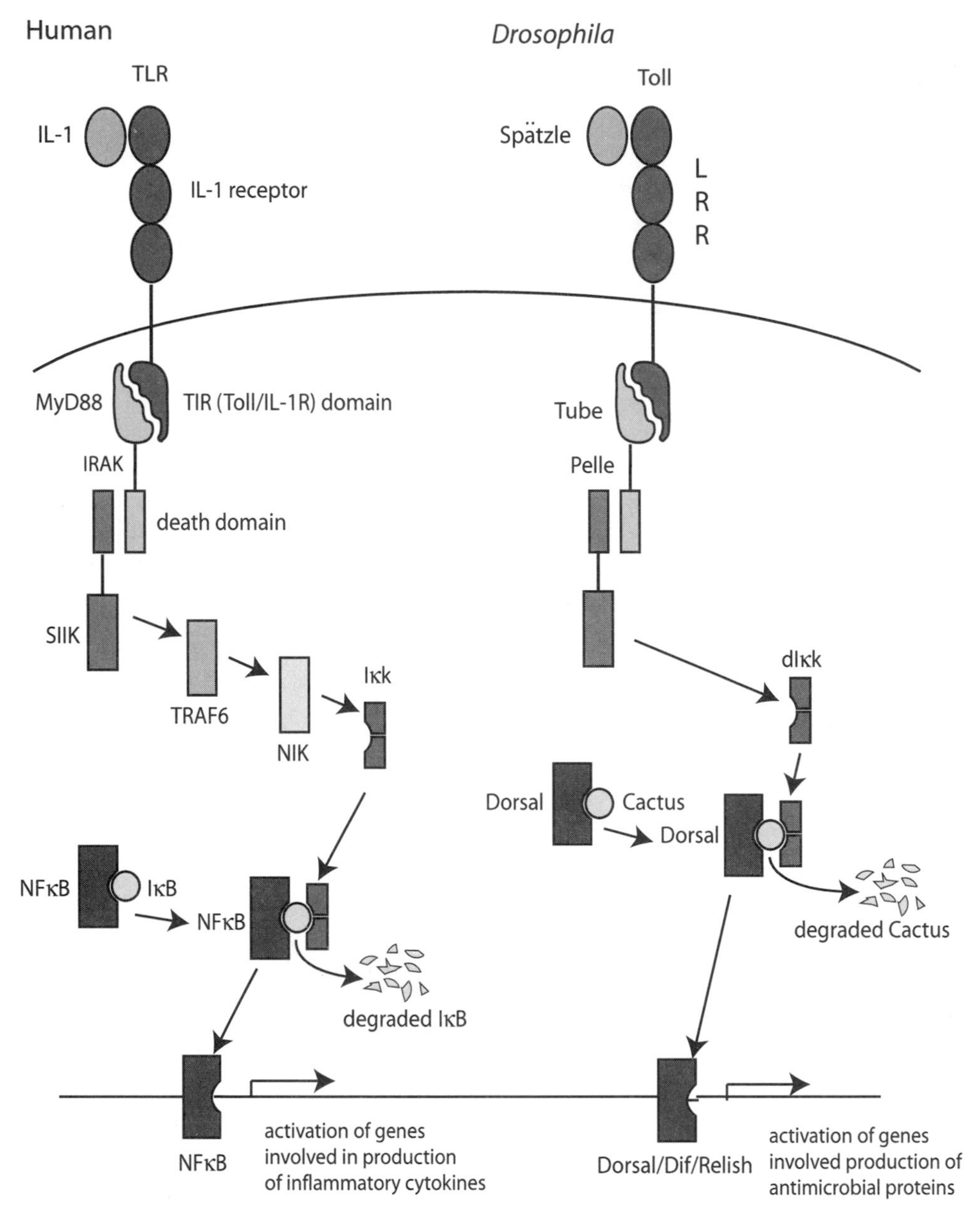

Figure 11-9. *Toll* and Toll-like signaling pathways in humans (left) and *Drosophila* (right). In human lymphocytes, the IL1 receptor phosphorylates IκB via the IRAK kinase cascade, the homolog of the *Drosophila* Pelle cascade. Once phosphorylated, the IκB protein (the *Drosophila* homolog is Cactus) is degraded, which allows the human transcription factor coded by *NFκB* (Dorsal in the fly), to enter the nucleus where it regulates the transcription of lymphocyte specific genes involved in the immune response. In flies, the same pathway effects dorsoventral patterning during development and the transcription of antimicrobial proteins in adults.

at bay while the adaptive immune system geared up to properly defend the organism. This was always a bit of a forced argument since most species keep pathogens "at bay" indefinitely by their innate system (since they have no adaptive system). The adaptive system was tailored so that it responded to characteristics of invading cells that were not found in the host's own cells (bacteria, for example, have different components of their cell walls compared with animal cells).

According to this view, the adaptive response is more recent, more highly evolved, and (to scientists) much more interesting. More recently, however, there has been discussion of the role of the innate immune system in eliciting and regulating the adaptive response (Hoffmann et al. 1999; Janeway 2001; Medzhitov and Janeway 1997; 1998). Again, we should not be surprised if a newer, but related, system is connected developmentally with its still-present predecessor. Janeway and his colleagues have proposed that the adaptive immune system is activated by a signal from the innate response. The innate immune system induces the expression of costimulator molecules on antigen-presenting cells, and without this costimulatory signal there can be no recognition of antigen by T or B cells, and thus no clonal expansion of the specific antigen-recognizing cell.

These costimulatory signals are based on the innate immune system's ability to use what Medzhitov and Janeway (Medzhitov and Janeway 2000) have called PRRs (*pattern recognition receptors*) to detect PAMPs (*pathogen-associated molecular patterns*) on invading cells, of which there are many known examples (e.g., Bauer et al. 2001; Moriwaki et al. 2001). PRRs are encoded by host genes and are specific to common molecular signals on pathogens because, these authors suggest, PRRs have coevolved with pathogens whose PAMPs are molecular structures so crucial to the pathogen that they cannot mutate to avoid detection without fatal consequences to themselves (Medzhitov and Janeway 1997). PRRs are found on the surfaces of host cells likely to be the first to encounter pathogens, such as mucosal membranes or surface epithelia as well as other types of cells of the innate immune system. We will see relevant comparisons in plant immunity below and later in Chapter 13 in relation to olfaction versus pheromone reception.

Because the V(D)J system can produce an effectively unlimited variety of antibodies, it can be expected also to produce antigens against the cells of the host's own proteins. However, if the innate induction theory of adaptive immunity is correct, then lymphocytes carrying anti-self antibodies would not destroy self proteins because a second costimulatory signal is required for lymphocyte stimulation and self proteins do not have a recognizable costimulatory molecule capable of sending this signal.

The existence of autoimmune diseases, however, perhaps weakens this idea to some extent. Human examples of autoimmunity include many life-threatening diseases: type 1 (juvenile) diabetes mellitus in which the host destroys its own pancreas, the nerve demyelination disease multiple sclerosis, systemic lupus erythematosus, which attacks a variety of organs, and many others. However, these may not simply reflect aberrant anti-self attack, because there is some evidence that they occur with various epidemic patterns in populations, including after the host has experienced an infection. An idea has been that some infections can closely mimic self antigens and the similarity triggers an autoimmune reaction as a consequence of an attack on the infectious agent. In any case, the mechanism by which purging of self-recognition molecules is controlled, once thought to be known, seems now not to be so well understood.

To some immunobiologists, the self/nonself dichotomy, however explained, does nothing to further our understanding of how the immune system works, or of its evolutionary origins. Such categorization is a human concept, not necessarily a function of the immune system as it actually evolved. The host itself is always changing depending on its developmental stage, response to environment, somatic mutation, and the like. What is "self"? How would a host be programmed to detect somatic mutation as not "self"? Should it be programmed?

In addition, the host may be large enough to withstand attack on itself so long as the latter is triggered by a genuine exogenous invasion that is cleared before too much host is self-destroyed. That the system occasionally goes awry, sometimes with lethal consequences, is not unique to immunity (and many of the known autoimmune diseases generally kill postreproductively). We do not really know why an adaptive system was favored, but clearly it was not "needed" in any ultimate sense. A system that worked well enough to combat infection better than its predecessor or competitor (in some environment that vertebrates may have had to suffer through) would tolerate the evolution of an adaptive system that simply was not too aggressive against self. In fact, given the rapidity of evolution of pathogens, the diversity of vertebrate cell types, and the often rather loose screening by selection, it may not be surprising that the system is only rather loosely discriminatory between self and nonself. If you live long enough to reproduce, evolution does not care, or even see, whether you destroy yourself subsequently or not.

Another view is that the adaptive system may have evolved to maintain the somatic integrity of the individual and only incidentally reacts against exogenous pathogens. The immune system could thus be viewed as a messy "dialogue" between the body and the cells of the immune complex (Cohen 2000). According to this view, autoimmune T cells in fact contribute to body maintenance, and play a role in maintaining the central nervous system (CNS) (Moalem et al. 1999): autoimmune T cells, reacting to myelin antigens, protect the CNS against posttrauma secondary degeneration by reducing nerve conductivity, which allows the nerve to heal. The same cells can also attack the CNS, causing disease, but the point is that autoimmune T cells are found in healthy individuals, as well as those with autoimmune disease. This, Cohen suggests, is because the business of the immune system is to maintain the body, not to discriminate between self and nonself. That such a system would, once around, come to be useful against exogenous molecular variants is not a surprising side benefit. Such discrimination may be a dispensable luxury (and was, for most of the history of evolution), if the immune reaction is timely and tempered and sufficiently localized and slows down after disposing a particular pathogen. Many vertebrate organ systems use continually renewed epithelia, and this could make the organism vulnerable to ill effects of somatic mutation, but whether this is relevant or not is wholly speculative at present.

The discussion of why an organism's immune system does not destroy the host would be a fairly arcane side branch of biology if the understanding did not have such important ramifications for how we go about developing therapies against autoimmune diseases and protecting against rejection of transplanted organs and other such medical problems. Perhaps the most important evolutionary question is why only one tiny branch of the biosphere has developed a complex adaptive system because perhaps 98–99 percent of all multicellular animals do perfectly well with innate response alone. Trees, squid, some insects, and so on are large potential pathogen dinners that often manage to live a very long time.

Detoxification Mechanisms

Other defense mechanisms also exist. One source of exogenous molecular attack is from the nonbiological world of chemicals. Many chemicals to which an organism can be exposed through the air, food, or other means may not amplify by growing within the host, but may react in damaging ways with the organism's own biochemistry. There are a number of metabolic systems that serve to inactivate or detoxify exogenous molecules that have certain properties. As an example, the cytochrome P450 (*CYP*) genes serve this function (the name comes from the history of the discovery of these enzymes).

Cytochrome P450s comprise a phylogenetically old class of heme-containing monooxygenases largely expressed in the liver that function to detoxify xenobiotic (exogenous) hydrocarbon molecules. The class of *CYP* genes appears to have evolved from related steroidogenic ancestral genes at the time of the separation of animals and plants (the latter are still found in prokaryotes and are expressed on mitochondrial and other intracellular membranes, where they help synthesize steroids and other substances used to maintain cell wall integrity).

CYP response works by catalyzing the oxidation of xenobiotic substances, which, along with other reactions, makes them more water soluble and excretable as well as inactivating them. The diversity of CYP genes in animals suggests that the system may mainly have evolved to combat substances derived from plants, in an animal-plant predator-prey competition.

CYP genes are mainly active in the liver but, like many large, old gene families, have other sites of activity and probably other functions. The genes are important in human responses to the environment but, ironically, can turn harmless human-made substances into toxic ones and can inactivate medicines or make them toxic. So the system has many facets when it comes to human affairs.

INVERTEBRATE IMMUNE RESPONSES

Immunity has, for obvious reasons of practical application, been studied most intensely in vertebrates. The vast bulk of animals must rely on their own innate, inducible, broad-spectrum, nonspecific immunity, and a few model invertebrate systems have been studied in detail, with some information available from others as well. Generally, although many invertebrates have short lives and high reproductivity (*r* strategies) as a possible defense mechanism—for the species, not the individual—even these organisms can be destroyed by pathogens and they have evolved protective defenses to combat them. In invertebrates the system relies on phagocytosis or encapsulation of the pathogen.

Insects

Most insects are *r* strategists that reproduce rapidly and have short generation times. They are also mobile. From this point of view, insect species might in principle be able to survive pathogens by reproducing fast enough to tolerate heavy loss of individuals or might be mobile enough to readily escape predation. However, some insects are relatively immobile, or large, or long-lived, so we cannot invoke generic evolutionary strategy arguments with regard to their immune capabilities.

Insect immune mechanisms begin with their hard exoskeleton, which serves as a first line of defense against pathogens or injury. The peritrophic membrane that separates the gut epithelium from the lumen (the passageway of the gut) also serves to inhibit infection by separating the contents of the lumen from the body interior, as does the lining of the trachea (Khush and Lemaitre 2000). Insects maintain a low pH and accumulate digestive enzymes and antibacterial lysozymes in their midguts as well, creating an unfavorable environment for most pathogens. However, we should note that in mammals the stomach is an acidic environment not usually hospitable to pathogens but to which a group (helicobacters) have adapted.

If an insect is wounded, or a pathogen evades the structural defenses, insects are able to mount a fast, strong response, both humoral and cellular. The system is not adaptive, nor does it have memory. It is likely that the antimicrobial response is based on the binding of receptors on circulating antibacterial proteins to general microbial epitopes (sites on these molecules that antibacterial proteins recognize and bind). The involvement of several different pathways has been described (Lee et al. 2001), but neither the receptor genes nor the mechanism that initiates insect response to wounding are as yet well understood.

Two immediate responses to wounding, infection, or the presence of other foreign objects in insects are the release of phenoloxidase and clotting of the hemolymph. Phenoloxidase induces the formation of melanin, which surrounds the wound and encapsulates parasites. Within a few hours of infection, a number of antibacterial proteins and peptides begin to build up in the hemolymph. More than 150 of these have been identified (Chung and Ourth 2000). The most significant is a family of peptides called *cecropins*, which quickly lyse bacteria. Another group of bactericidals is the *defensins* (with vertebrate homologs), which act more slowly but also permeabilize microbial or fungal cell membranes, and a third group is the *attacin-like* molecules, which disrupt dividing cells of *Escherichia coli* and other bacteria, causing the breakdown of their outer membrane. Genes coding a number of these antibacterials are known (Lee, Cho et al. 2001).

The regulatory regions for these peptides contain binding sites for transcription factors containing the Rel domain, and some of these Rel proteins are homologs of transcription factors with similar roles in immunity in mammals, the *NFκB* family. The insect gene *Toll* does not function in microbe recognition but is activated on an immune challenge. A number of TLRs, in addition to *Toll*, have been identified in *Drosophila*, and these do interact directly with microbes (Khush and Lemaitre 2000).

As in vertebrates, the immune response in insects is activated by recognition of common proteins on microbes by PRRs. Peptidoglycan recognition protein (PGRP), for example, has been identified in the moth *Trichoplusia ni* (Khush and Lemaitre 2000). It binds to Gram-positive bacteria, as does peptidoglycan recognition protein in vertebrates. Gram-negative binding proteins (GNBP) have also been isolated from insects, again with homologies to proteins in vertebrates that recognize Gram-negative bacteria and initiate an immune response (Ochiai and Ashida 1999).

Inside the body invading microbes are attacked by hemocytes, that is, phagocytic cells in the hemolymph. Antibacterials and antifungals are produced in the *fat body*, the functional equivalent of the mammalian liver. The genes that encode these peptides, including cecropins, attacins, diptericin, defensin, drosocin, drosomycin, andropin, and diptiricin-like protein, are induced by intracellular signaling cascades with homologies to the activation of *NFκB* in mammalian immunity (Lagueux et al. 2000). Other antimicrobial responses include the production of nitric oxide, which is toxic to some parasites, and perhaps sequestration of iron to limit bacterial infection (Khush and Lemaitre 2000).

Activation of the immune response in an insect, either because of microbial infection or wounding, also causes systemic and behavioral changes including reduced feeding, and developmental delay, and there is even a possible febrile response, marked by the insect's pursuit of warmer locations in its environment (Hultmark 1993). The type of immune response an insect mounts depends on the type of microbe against which it must defend. Fungi and bacteria elicit different responses in insects, as they do in vertebrates.

Mollusks and Other Invertebrates

Studies in mollusks show that at least some of the mechanisms found in insects have earlier origins. Two members of the insect defensin family of antibacterial peptides are found in mollusks (Charlet et al. 1996). PRPs specific to molecules on the surface of pathogens initiate the immune system in these animals, as in the vertebrate and invertebrate systems. Among other responses, PRPs initiate the *prophenoloxidase* (ProPO) cascade, the same melanin release system found in insects. The ProPO system recognizes nonself proteins and targets them with antimicrobial peptides and other toxic metabolites.

Some of the immune response of these animals relies on clotting, and the clotting protein has been cloned and found to be a member of the vitellogenin family. Echinoderms have a complement system, somewhat analogous to that of vertebrates.

IMMUNITY IN PLANTS

Plants Have Special Vulnerabilities

Because they have the basic characteristics of life, Aristotle hypothesized that plants have a property that we might call awareness, and we are learning that in many ways this is indeed the case. Rooted plants are sessile and many remain in the same place for a very long time, especially relative to the life spans (and mutation cycles) of microparasites. Plants can be very large and are attractive hosts for macro- as well as microparasites. One would expect that plants simply could not persist as a form of life if they were passive meals for pathogens. Although the mechanisms are different from those in animals, plants do indeed sense attack and respond defensively. But, interestingly, so far as we know, they have not evolved adaptive immune responses.

Plants have developed a number of mechanisms to defend against a wide range of bacteria, fungi, viruses, nematodes, and insects. When plants detect a pathogen they quickly launch one or more of these molecular defenses. In his seminal work on plant pathology, specifically the resistance of flax to flax rust, H. H. Flor in 1956 first suggested a *gene-for-gene* pathogen-specific response by plants to pathogens (Flor 1956), and molecular genetics has subsequently confirmed and expanded knowledge of this relationship. The gene-for-gene response involves the interaction of the products of *resistance (R) genes* in the plant and *avirulence (Avr) genes* in the pathogen. *R* genes code for receptors to products of *Avr* genes: the pathogen releases the product of the *Avr* gene into the plant at the site of invasion, and it interacts with the product of the *R* genes, whose expression is induced by the presence of Avr protein.

The *R-Avr* gene interaction activates a cascade of defensive gene expression changes. Details are just becoming known in a few model plants and pathogens. However, the signal is transduced by the *R* gene, which as well as having a ligand-binding (LRR) region to serve as a receptor, also has nucleotide-binding (NBS) regions to serve as a transcription regulator. The resulting defensive reactions include the *oxidative burst*, which can trigger rapid localized cell death at the infected site, an apoptotic reaction known as the *hypersensitive response (HR)*. This involves homologs of apoptosis-related genes in animals.

Plants have tougher cell walls than animals, but these can be penetrated by pathogens. Plants can respond by crosslinking (chemically toughening) their cell walls against further invasion, expressing at least eight classes of antimicrobial compounds including defensins, thionins and snakins, toxic secondary metabolites, and hydrolytic enzymes (Almeida et al. 2000; Garcia-Olmedo et al. 1998; Klessig et al. 2000; Segura et al. 1999).

These same responses can also be triggered by a nonspecific host response to attack. The oxidative burst, in which levels of reactive oxygen species and hydrogen peroxide rapidly increase, is the most common plant defense and is often triggered by nonspecific defense reactions. This activates the hypersensitive response, which involves an apoptotic reaction that destroys the plant's own cells in the area under attack and thus deprives the pathogen of nutrients, confines the pathogen to the site of necrosis by creating a physical barrier composed of dead plant cells, and induces the production of antimicrobial proteins (see, e.g., Mittler et al. 1999). The HR is triggered by the production of nitric oxide, which induces the expression of genes that synthesize antimicrobial compounds (Delledonne M et al. 1998).

These are localized responses that occur at detected sites of infection and serve to seal off the infected tissue and the pathogen, depriving it of the opportunity to grow or spread. However, plants have also evolved a mechanism to resist attack systemically, known as *systemic acquired resistance* (SAR), triggered by infection with a necrotizing pathogen. SAR establishes a state of heightened resistance throughout the plant to protect against subsequent attack, not only by the pathogen involved in the primary attack but also by a broad range of pathogens. This response is characterized by an accumulation of salicylic acid systemwide and the production of a set of *pathogenesis-related* (PR) proteins that confer increased nonspecific resistance to the plant, probably through antimicrobial activity (Feys and Parker 2000). SAR, once triggered, has been found to last as long as an entire season.

Another form of systemic resistance has recently been described: *induced systemic resistance* (ISR) (Pieterse et al. 1998). Independent of salicylic acid and PR proteins, plants respond to nonpathogenic colonizing bacteria, often found in the roots, by enhanced protection against pathogenic attack. ISR is triggered by signaling pathways regulated through *Npr1* that lead to the release of plant growth hormones, jasmonic acid, and ethylene. Jasmonic acid and its derivatives (known collectively as jasmonate) trigger the release of genes encoding defense-related proteins including thionins and proteinase inhibitors. Ethylene induces the expression of members of the *PR* gene family, and jasmonate and ethylene together synergistically stimulate *PR* gene expression (Xu et al. 1994) and activate the expression of genes encoding plant defensins (Penninckx et al. 1996).

The gene-for-gene response mechanisms seem to be relatively rare elicitors of plant defenses (Klessig, Durner et al. 2000) compared with the nonspecific host response. However, the gene-for-gene mechanism is phylogenetically diverse and widespread among plants, suggesting that it arose early in plant evolution. Plant resistance genes can be divided into at least five classes of related sequences, defined by the structural characteristics of the proteins for which they code (Ellis et al. 2000). They can be distributed across the genome, and there are at least 200 of these genes in *Arabidopsis* alone, comprising about 1% of its entire genome (Ellis, Dodds et al. 2000), comparable in scope to the mammalian olfactory genes (Glusman et al. 2001). Gene-for-gene immune mechanisms were thought to apply to vertebrates as well, until somatic recombination was discovered.

The majority of plant resistance genes isolated to date share NBS-LRR domains (Feys and Parker 2000; Young 2000). A major subclass also has a "TIR" region, with homology to *Toll* of *Drosophila* and interleukin receptor-like proteins in mammals, which are involved in nonspecific cellular immunity (e.g., Girardin et al. 2002). Toll-like receptors in vertebrates and insects mediate an inflammatory response that is similar to the plant defense responses: in both, the response includes production of antimicrobial active oxygen species or free radicals with bactericidal and microbicidal properties and, ultimately, death of the infected cell (Dixon et al. 2000). Some *R* genes have protein kinase intracellular domains and work through the classic message-transduction means, but others lack such an internal domain (Ellis, Dodds et al. 2000).

In the gene-for-gene model, plant and pathogen genes are under different evolutionary pressures; unlike animals, plants cannot flee their predators, and many plants have long life spans. One might expect that micropathogens that can divide (and hence mutate) in hours would easily out-evolve any defense on the part of plants. Of course, plants maintain pluripotent cells throughout their lives, they reproduce from many different places, and they reproduce by producing huge numbers of seeds (with the recombinational and mutational variation that can generate). Thus the attack of a pathogen on one branch will not necessarily destroy the reproductive capacity of other branches. However, because plants can live long lives the pathogen need not be in any hurry. Pathogens infecting seeds would cause a particular problem.

Plants must detect Avr proteins to trigger their disease resistance responses, but the pathogen benefits from evading detection. It is not likely that the role of *Avr* genes in triggering plant defenses that thwart their replication explains why these genes have been maintained by selection. Indeed, in the absence of the corresponding *R* gene, *Avr* genes play a role in the virulence of a pathogen. Virulence is recessive; if a pathogen is to become virulent in the presence of its specific receptor in the plant, its *Avr* gene must undergo a loss-of-function mutation so that the pathogen can evade detection by the plant's *R* gene (Richter and Ronald 2000).

R genes are subject to diversifying selection. To resist a new pathogen, the plant must gain a new resistance function. *R* genes are highly polymorphic, with high recombination rates and high evolvability (Ellis, Dodds et al. 2000; Richter and Ronald 2000). They typically are in clusters of many tandem duplicate genes along with a number of pseudogenes (probably the detritus of frequent recombination events), and several carry transposons, deletions, or frame shifts (Ellis and Jones 1998; Richter and Ronald 2000) on the same and on different chromosomes.

This reservoir of variation may be essential for the generation of new *R* genes in the face of evolving pathogens. However, it may not suffice to produce the amount of pathogen-specific resistance that has been found. Several instances of disease resistance conferred by the interaction of more than one *R* gene have been documented (Botella et al. 1998), which may give a combinatorial defensive capability comparable to some nonrecombining systems in animals (e.g., olfaction, discussed in Chapter 13). Additionally, some *R* genes can only elicit the hypersensitive response if the pathogen expresses an *Avr* gene and *Hrp (hypersensitive response and pathogenicity)* genes as well.

Of course, pathogens have evolved mechanisms to evade host detection. Pathogens lose avirulence (become virulent) much more quickly than plants gain resistance genes. Also, some pathogens are known to take advantage of programmed cell death to avoid the activation of host defenses. For example, necrotrophic fungi *B. cinerea* and *S. sclerotiorum* are pathogens that trigger the hypersensitive response and apoptosis to exploit necrosis to colonize the plant, in this case, *Arabidopsis* (Govrin and Levine 2000).

The genes involved in *R* gene signal transduction vary by plant and pathogen. The pathways involved have begun to be understood by the study of plant mutants; when specific genes are mutated, expression of downstream genes in the pathway may be blocked and resistance compromised or absent. Aarts (Aarts et al. 1998) has suggested that there are at least two pathways initiated by *R* genes, and the structure of the R protein in part determines which pathway it invokes. Some *R* genes belong to the leucine zipper subclass of *NBS-LRR* genes, and others belong to the class that encodes a protein with amino-terminal similarity to the cytoplasmic domains of the *Drosophila* Toll and mammalian Interleukin1 transmembrane receptors, the TIR-NBS-LRR (Aarts, Metz et al. 1998; Glazebrook 2001). Genes in the first pathway are marked by downstream activation of resistance genes that requires *Eds1* (*Enhanced Disease Susceptibility 1*), and in the other, *Ndr1* (*Non-Race Specific Disease Resistance 1*) is required (Aarts, Metz et al. 1998). There are *R* genes that require neither *Eds1* nor *Ndr1*, although these genes seem to play a role in their defense signaling, so it seems that there is at least a third pathway, the genetics of which are not yet known (Glazebrook 2001).

Some further specifics about the *Ndr1* and *Eds1* pathways are now known. For example, *R* genes that require *Eds1* also require *Pad4* (*Phytoalexin Deficient 4*). Genes that require *Ndr1* also require *Pbs2* (*avrPphB Susceptible 1*) (Glazebrook 2001) but are independent of *Eds1* (Aarts, Metz et al. 1998). A number of *R* loci are known to require *Eds1* (the *Resistance To Peronospora parasitica* genes, *Rpp2*, *Rpp4*, *Rpp5*, *Rpp21* and *Rps4*, for example). Mutant *Eds1* plants no longer have resistance to *P. parasitica* conferred by these *Rpp* genes (Aarts, Metz et al. 1998).

Plants also produce a class of steroid hormones, *phytoecdysteroids*, which are homologous to insect *ecdysteroids*. These compounds are noxious to herbivores and insects, both in flavor and effect, and can have physiological or even "xenohormonal" effects on other species, such as inducing abnormal molting or infertility in insects and clover disease in sheep, which affects the fertility of ewes by inducing histopathological changes in the uterus (Oberdörster et al. 2001). These hormones act through nongenomic pathways as well as the better-understood receptor-mediated initiation of gene expression and are often antagonistic to the predators' endogenous hormones. This shows the mixed effects of some of these plant "hormones." We would not normally use the term for substances that are toxic to predators, but as so often happens, functions even of the same substance are not neatly confined. As stated in Chapter 10, neurological, immunological, and endocrinological systems are in many ways related.

Other Molecular Antipredator Mechanisms

The role of microdefenses (gene products) against microscale internal invasion is the subject of this chapter, but the term "immune" resistance can be extended a bit to allow us to mention that microscale mechanisms exist against macro- and external predators. This is the world of toxic or bad-tasting compounds plants produce to keep predators away (it works at least to some extent, but often some predator becomes resistant and eats the plant). The same is true in some molecular ways for animals (octopus ink, skunk odor, etc.)

THE MOST PRIMITIVE IMMUNE SYSTEMS

Even simple organisms like bacteria are food for viruses and other parasites (sometimes including other bacteria). Not unexpectedly, bacteria have immune responses of their own. One response is the ability to resist natural antibiotics that can exist in their environment. Naturally occurring antibiotic resistance genes from a number of bacteria have been cloned and are routinely used in laboratory microbiology and molecular genetics research. Often, these genes are found on plasmids, simple circular DNA molecules that can be exchanged between bacteria (this exchange along with a rapid life cycle explains the ability of bacteria under antibiotic siege to develop resistance rapidly). Resistance genes on plasmids work in conjunction with other chromosomal genes in bacteria.

Bacteria can be invaded by viruses, sometimes known as *bacteriophages*. At least one major way in which they defend themselves is by the use of *restriction endonucleases* that they encode in their genome. These are a large class of genes that bind DNA at short recognition sequences and cleave it (the sequences are often palindromic in nature, taking into account the reverse complementarity of DNA, such as **GAATTC**, so that the restriction enzyme can bind equally well on both strands in opposite directions). This will destroy incoming viral DNA. The bacteria protect themselves by protecting their own DNA from cleavage, as with sequence-comparable restriction methylases that modify their DNA so the restriction enzymes will not cut it.

Another primitive defense system involves RNA interference (Carthew 2001; Sharp 1999), described in Chapter 7. Double-stranded RNA (dsRNA) complementary to a section of the mRNA from a gene can inactivate that RNA and hence prevent translation. This is a system used by various invertebrates in defense against viruses.

THE EVOLUTION OF IMMUNITY

Phylogeny of Innate Resistance Systems

The innate responses are varied and much older than the adaptive responses. Many if not most of the systems are found in vertebrates and invertebrates, and indeed often in plants as well.

For example, four *Defensin* families are found in eukaryotes; *Alpha-Defensins* and *Beta-Defensins* in mammals, insect *Defensins*, and plant *Defensins* (Hoffmann, Kafatos et al. 1999). This general mechanism arose long ago and clearly has remained useful up to the present. The production of nitric oxide by plants for use by the hypersensitive response, in inducing apoptosis and the expression of genes that induce the production of antimicrobials, is enzymatically similar to nitric oxide production in animals (Clarke et al. 2000), which is also used as an antibacterial toxin and against tumor cells (an appropriate type of autoimmunity) (Roitt, Brostoff et al. 1998). The associated cell death that occurs in the HR reaction in plants is functionally analogous between plants and mammals; it is even found in bacteria. Cytotoxic T cells can signal a targeted cell to undergo apoptosis (Roitt, Brostoff et al. 1998). We have seen many ways, including apoptosis, by which normal development is based on internally derived signals, screening, stimulus, and assault.

An Evolutionary Interface Between Innate and Adaptive Immunity?

There are similar levels of parallel in signaling pathways using *Toll/Cactus/Dorsal* in *Drosophila* and *IL1 receptor* (*IL1R/NFκB*) in mammals and *R* genes with their *NBS-LRR* domains in plants (Medzhitov and Janeway 1998). Not surprisingly, elements of the system have some correspondence with genes in bacteria, such as the NBS with that of animal apoptosis genes (Young 2000). Again, these are relatively simple or generic defenses that have stood the evolutionary test of time.

However, the *R* gene resistance mechanism seems not to be completely straightforward. Some *R* gene clusters show evidence of diversifying selection, which may be related to a general fusillade response, whereas others clearly show that classic darwinian allelic selection matching specific *R* alleles to pathogen *Avr* genes has been important. That is why the gene-for-gene system is "innate" rather than "adaptive" in the sense these terms are used here.

Although both involve a history of repeated tandem gene duplication and recombination, the *R* gene system is fundamentally different in tactic from the vertebrate adaptive immune system, which requires somatic rearrangement and mutability, rather than inherited allelic "memory," to mount a defense against a specific pathogen. However, the *R* gene system does involve combinatorial mechanisms and may be viewed as a kind of intermediate state between wholly generic chemical innate resistance and highly adaptive immunity. We do not imply that *R* genes represent an evolutionary missing link, because the common ancestor of plants and vertebrates did not have a multigene system from which *R* genes and immunoglobulins have descended. However, the vertebrate system *did* have an *R*-gene-like antecedent (see below).

Although deep homology of individual genes or gene domains shows that the basic genetic mechanisms are in some sense conserved and used for new purposes in immune defense, phenogenetic drift is always possible. In plant disease resistance responses, HR can be elicited by *R-Avr* gene interaction but also by non-*R-Avr* interaction. Some *R* genes pick up Avr signal cytoplasmically, some extracellularly, some are transmembrane transducers, some *R-Avr* interactions elicit an HR that involves apoptosis, and some do not. Some *R* genes do not transduce signal without another *R* gene involved; others do not need interaction.

Phylogeny of the Adaptive Immune Response

We tend to think of the adaptive response system as that which evolves during life, particularly the antibody generation mechanisms that we see in gnathostomes. But this system is innate in agnaths, who rely on combinational response rather than recombinational or gene-for-gene responses ("combinatorial" means that it uses multiple inherited genes, in the "hope" that at least one will work, whereas "recombinational" means using rearrangements within the organism's lifetime that are not inherited). However, it is possible in principle that natural selection in the form of a major pathogen could favor the proliferation of one of the many antibody genes possessed by an agnathic animal, as it does in the case of *R* genes in plants.

In principle, the same could still occur in gnathostomes, but this goes against the prevailing view that the adaptive system involves so much mutation and recombinational imprecision that it became more efficient and effective to generate resistance by recombination than by combinatorial attacks with inherited alleles. The gnathostome system is so mutable that classic allelic (darwinian) selection might not even work. A specific V gene allele favored in one individual would be unlikely to be sufficiently prominent among the array of somatically generated antibodies that, in other individuals, would happen to be a major aspect of the latter's defense against the pathogen. Selection would not be able to, and would not need to, single out a particular allele in a particular V region of an immunoglobulin gene cluster. Still, selection by epidemic can be severe and rapid, and it may be that more classic kinds of selection do in fact pertain to the components of antibody genes.

Some of the basic pathways of resistance are ancient and serve (and perhaps initially arose to serve) other purposes. Signaling pathways used in the vertebrate adaptive immune response, with invertebrate—and even plant—homologs, function to activate innate immunity in the absence of an adaptive response. Earlier we noted the homology of the amino terminus of other closely *NBS-LRR*-type genes to the *Drosophila* Toll or human interleukin receptor-like (TIR) region. Several human homologs of the Toll (h-Toll) protein have been isolated and shown to activate the *NFκB* pathway, which mediates lipopolysaccharide-induced cellular signaling and initiates the vertebrate adaptive response by production of inflammatory cytokines. The NBS domain of plant *R* genes is characterized by several sequence motifs found in the recently described *Ced4* and *Apaf1* animal genes. The latter genes regulate the activity of proteases that can initiate apoptotic cell death. Should we add development to the common neuro-immuno-endocrine system? Probably it is simpler to repeat that the definition of a "system" is a human artifice laid upon an evolutionary reality.

It is difficult to account in a convincing way for the evolution of the somatic recombinational adaptive immune system because so many millions of species do well in similar environments without such a system. What selective force would have given animals an advantage from producing and fine-tuning the adaptive immune system? Perhaps one major adaptive advantage of the recombinatorial response is its role in resistance to *re*infection, provided by immunological memory (Parish and O'Neill 1997). This may be a plausible argument given that the adaptive recombinational system depends on initial antigen recognition by the noncombinatorial innate system. Or, perhaps as noted earlier, the adaptive immune system evolved primarily for "body maintenance" (wound healing, tissue repair, angiogenesis, cell regeneration, and the degradation of old or abnormal cells (Cohen 2000)) and defense against pathogens is a by-product.

The immunoglobulin family with its many V, D, and J regions has had a long evolutionary history. The cell adhesion family that includes these genes (and most of the MHC genes) serves a wide variety of purposes, all mediated through similar domains (intracellular, transmembrane, and extracellular). The array of immunological *Cam* genes found in different vertebrates shows their history. In terms of immunological (and/or self-recognition or other related) functions, the MHC/T cell receptor system can be traced phylogenetically as far back as elasmobranchs (cartilaginous fishes, including sharks) and teleosts (bony fishes, including most of the more common fishes) (Nakanishi et al. 1999; Roitt, Brostoff et al. 1998). These genes generate high clonal diversity among white cells, but the mechanisms vary and have changed during the hundreds of millions of years the system has had these functions.

The repertoire of antibody diversity in vertebrates varies greatly. The genes are highly polymorphic in fish, as they are in mammals. In light of the results of allograft and other experiments, it is thought likely that fish MHC genes have an immune function similar to that in mammals (Nakanishi, Aoyagi et al. 1999; Roitt, Brostoff et al. 1998). However, potential diversity in elasmobranchs is not nearly as extensive as in higher vertebrates. This is because the heavy chain of shark Igs is constructed by the stringing together of V, D, and J segments as multiply repeated units and this eliminates the possibility of recombination between the separate gene segments (Roitt, Brostoff et al. 1998).

Chicken antibodies show considerable diversity, but this is largely due to gene conversion rather than Ig gene recombination (Roitt, Brostoff et al. 1998). In the light chain system, there is only one V, one J and one C segment gene, and in the heavy chain system, only single V and J segments. There are 16 D segments, but they do not exhibit much sequence diversity and thus do not contribute appreciably to a diverse repertoire.

Classic molecular types of response that we would describe as being an immune *system* are necessarily much older, because they are complex and hence had to evolve over some time period. On the basis of DNA sequence analysis, the origins of the adaptive immune system can be traced phylogenetically to the precursor of the recombinase activating genes, *Rag1* and *Rag2*, which arose approximately 450 million years ago (Agrawal et al. 1998). It is in part because this system seems to have arisen fairly well intact in jawed vertebrates, that the somatic recombinational machinery may have been transferred horizontally rather than evolving from endogenous precursors (Marchalonis and Schluter 1998; Marchalonis et al. 1998; Marchalonis et al. 1998), although this idea is not universally accepted (Hughes 1999) (horizontal transfer is thought to be difficult once specialized eukaryote cells evolved, as we described earlier). In any case, the use of the same set of genes for immunelike function probably antedated the recombinational mechanism for making increased clonal diversity.

An Interesting Variation on the Theme

The molecular battle between micropredator and the prey that is the object of its attention has led to recombinational as well as combinational systems. An interesting phenomenon that is conceptually related is the genetic evolution of venom proteins in a group of around 500 species of predatory *cone snails*. These diverse species have several gene families related to venom protein production. Although the sequences of their signal peptides, responsible for processing the proteins after they are coded, are relatively conserved, the venom peptide domains are hypervariable (Olivera et al. 1999). Different exons in these genes may have very different mutation rates, with the effect of generating toxin peptide diversity. These venom proteins attack ion channels involved in their victims' neurotransmission and hence act as neurotoxins. The hypervariability in these genes is a combinational strategy, like that of the *R* genes in plants and that we will revisit in Chapter 13 when we discuss olfaction. A possible explanation is that this reflects the evolution between their venom characteristics and evolution in the target molecules of their prey.

A Bit of a Misnomer

The distinction between innate and adaptive immune systems actually distinguishes between generic and somatically evolving active responses to infection. There is, however, a more direct sense in which organisms "adapt" to infection, and that is in the classic evolutionary sense, and sometimes specific to the attacker.

Mutations in genes related to the innate system or the mechanisms underlying the adaptive system can generate generalized immune deficiencies. Presumably, those aspects of immunity are maintained by darwinian selection. We also know of examples of passive resistance to, rather than active attack against, pathogens that evolve in the usual sense (alleles selected relative to specific pathogens). Selection can directly favor such host alleles, raising them to higher frequency.

The classic examples are the malaria-resistant hemoglobinopathies like sickle cell anemia. What these do is block some aspect of the pathogen's life cycle or biological needs, rather than actively attacking its cell surface antigens, etc. A stage of malarial parasites, for example, must take place within red blood cells, which are filled with hemoglobin. Mutant hemoglobin that prevents parasite entry into the red blood cells effectively resists infection. The example of sickle cell anemia is instructive also in that there is a disadvantage to having sickled hemoglobin. This leads to a balancing selection and a stable allele frequency polymorphism between "normal" hemoglobin that is a good oxygen transport molecule and "abnormal" sickle cell hemoglobin that is resistant to malaria but not a good oxygen transporter.

Many pediatric diseases in humans are recessive in the sense that a mutant allele by itself does not lead to disease. Yet the frequency of alleles associated with the disease sometimes seems too high, given the seriousness of the affected genotypes, which are often nearly lethal. Phenylketonuria, cystic fibrosis, and Tay–Sachs disease are among many examples. It is thought that the "pathogenic" alleles may have experienced a history of some as yet unknown balancing selection in these diseases, even if less dramatically than is the case with sickle cell anemia. If such explanations are made specific and prove true, it will extend the array of mechanisms against infection that are not, but probably should be, classified as part of "immune" systems. Perhaps the reason we don't now classify them as such is in a sense evolutionary: these "new" mechanisms arrived via different evolutionary pathways than the complement/antibody/MHC systems.

CONCLUSION

All organisms are under attack. If they respond well enough to go on to reproduce, they win. There is a great variety of ways that organisms respond, from simply reproducing soon and often, to having very complex, staged molecular machinery to identify, isolate, and destroy pathogens at the molecular level and, in some organisms, to remember what happened earlier so that in the event of subsequent exposure the invader can be destroyed more quickly. The organism might also just tolerate or become commensal with the pathogen or may out-reproduce or simply endure it.

It is too easy to reconstruct "had to" stories for these systems. Except for the most primitive and nonspecific innate systems, there is clearly no mandate to have an immune system of any particular type or efficiency. Either the more complex systems overlaid on simpler systems initially (and perhaps still) serve some other function, or the circumstances that called for them no longer exist, or we do not understand them. The complexity of immune systems today is "good enough" but not perfect or of the same type in different species. Slow metabolism or rapid reproductive strategies, or reproduction from diverse sequestered parts (as in plants), may allow an organism to tolerate infection rather than having to fight it.

It is not hard to see how molecular defense strategies might evolve. Any ligand-binding system in an organism might be modified by mutation (perhaps after gene duplication) to produce alleles that bind things not normally part of the system. If it binds the wrong things too aggressively it is selected against (and we might call that a "disease"), but if it does not bind self-structures too aggressively it may end up binding to other things that get inside the organism, inactivating them, and being favored by selection. It can then undergo evolutionary elaboration.

A subject beyond this book is the ecological evolution by which species adapt to commensalisms or tolerance. From a selfish evolutionary point of view, a pathogen that kills its host can survive only if transmission to a new host occurs quickly and readily enough (that is, before the original host is dead and gone). The conditions for epidemic versus endemic infection are debated among immunologists and epidemiologists (Ewald 1994). Many parasites kill their hosts only slowly or if the latter are under nutritional or other stresses. They are endemic because they are always there to infect new susceptibles born or migrating into the population. Other parasites cause little or no reaction by the host and do no harm. As we have noted above, some parasites are symbiotes: we cannot do without them, and they cannot do without us. Each case is different. The important point is that immune systems and/or parasites evolve so that the immune reaction is strong enough to protect the host but not necessarily any stronger. Of course, neither the strategy of host nor that of the pathogen is always successful. Those that have overstepped their bounds are generally no longer here to be studied.

PART IV

External Awareness

Information Transfer between Environment to Organism

The life of any organism largely consists of dealing with the outside world where things they need, like food and mates, as well as things that threaten them, are found. Organisms have evolved to receive and **perceive** *whatever aspects of the environment sufficed, in its ancestry, for survival. There is no one way to do this, and the sources of "information" in the environment are diverse.*

One important characteristic of many vital aspects of the environment is that the organism cannot know what they will be like in advance. It cannot be genetically programmed or "hard-wired" to handle unpredictable stimuli or signals. Instead, a diversity of mechanisms has evolved that enable organisms to respond to the events they need to in order to survive, seek food and mates, and so on.

Animals use some aspects of the environment that are largely predictable, such as gravity or magnetism. Other aspects are more truly open-ended. They include sounds, lights, and chemical components. Many intriguing, and sometimes closely related, means have evolved to receive input from these environmental stimuli and resources, and turn it into useful information. Examples of the responses are phototropism in plants, or vision, olfaction, and hearing in animals.

Yet, not all organisms respond to all signals that might be useful to them, and we will consider why that is so.

Genetics and the Logic of Evolution, by Kenneth M. Weiss and Anne V. Buchanan.
ISBN 0-471-23805-8

Chapter 12
Detecting Physical Variability in the Environment

Almost 2500 years ago, in his treatise *On the Soul (De Anima)*, Aristotle wrote, "There is nothing in the intellect that was not first in the senses." He enumerated five senses: sight, hearing, smell, taste, and touch. To Aristotle, the senses were formed of earth, water, air, and fire, as was all matter, and they came together in the "common sense," the heart, which brings us the awareness of sensation, allows us to distinguish between the perceptions of each sense, and yet ultimately combines them into one common experience.

Aristotle was probably the most noteworthy empiricist in the history of Western thought, at least until Bacon and perhaps Descartes in the Enlightenment period. He believed that nothing could be known without first being experienced. He had very broad interests and investigated all subjects with the same intensive data-gathering approach. His teacher, Plato, by contrast, was a rationalist. To Plato, knowledge was innate; it preceded and was not dependent upon experience.

We can argue with Aristotle's restricted list of the senses, because he should have realized that we can sense gravity, or balance, or the location of our bodies in space, or other ways in which we (and other organisms) sense our environment. Or we can disdain his primitive idea that the senses are of the four elements, because we know they are really receptors and ligands and proteins, or that the heart is the interpreter of sensation, because we know it is really the brain. We can smile at Plato's rationalism, because we, too, are empiricists in the best scientific tradition. But if we step back from these particulars, we might find that our own view of the senses, even with our extensive knowledge of genes and molecules, is in fact more similar to, or even derived from, the thought of these ancient Greeks than we care to realize. In fact, from an evolutionary perspective we should expect more connectedness, less distinction, more repetition but in other ways less "rationalism" in the traits of life.

Does knowledge precede experience, or is it the other way around? What we know about senses and learning and behavior suggests that it is not one or the other.

Genetics and the Logic of Evolution, by Kenneth M. Weiss and Anne V. Buchanan.
ISBN 0-471-23805-8

The gravitational sense, the light-seeking behavior of many organisms, and sucking in newborn mammals are all considered to be instinctive or reflexive behaviors. Animals are born "knowing" these things. Biologists would even argue that it is "coded" in the DNA, which must be as platonic a view of inherent knowledge as there can be.

On the other hand, implicit in our understanding of the senses is that they are used to acquire information—*new* and *unpredictable* information—about the world. Knowledge derives from experience, Aristotle said—so our own view of the senses has also undeniably been Aristotelian. Furthermore, many biologists assign darwinian purpose to the senses: hearing is "for" the detection of predators, taste is "for" the detection of toxic or poisonous foodstuffs, smell is "for" detecting pheromones and finding mates, etc. The knowledge we take from our senses, then, is "for" survival and reproduction.

However, in both the Aristotelian and Darwinian views of the world, knowledge—the information our senses provide us with—has two teleological components. First, it does not seem a stretch to say that the current functional purpose of our eyes is to receive light (this is not the same as saying that eyes evolved with that intended objective, which would imply that before there were eyes, an organism dreamt of seeing and strove to create a way to do so). Second, our brains use the information received from our eyes to direct us to objectives that we have in mind—pursuing that deer, searching for our lost keys. We might note that once sight evolved as a distinct function, integrated with the brain to regulate motor and other behavior, it is philosophically reasonable to say that the sense was modified by evolution (e.g., to achieve greater light and color sensitivity or focusing resolution) "for" (at least, in relation to) what today is its function. A pure mechanician might argue that our awareness that we are purposive is itself molecularly programmed, but we can leave that ultimately philosophical issue aside.

As stated earlier, we generally attribute to René Descartes the view of the body as a machine. He wrote in 1664:

> First of all, I want the reader to have a general notion of the entire machine which it is my task to describe. So I will say here that the heat in the heart is like the great spring or principle responsible for all the movements occurring in the machine. The veins are pipes which conduct the blood from all the parts of the body towards the heart, where it serves to fuel the heat there. The stomach and the intestines are another much larger pipe. . . .
>
> *(Descartes, Treatise on Man, 1664)*

Intrinsic to the Western scientific method and tradition is the purposeful dissection of a system into its constituent parts so that we might analyze them in order to understand the whole. Thus we try to understand the senses individually, in a Cartesian way, even though we know that taste without smell is not as rich and that a homing pigeon relies on smell as well as sight to find its way. What does it mean to understand the parts, then? And, even after we reduce the senses to their molecular and cellular pathways in fact this tells us nothing about their "purpose," how the brain interprets sensory signals sent to it, or how information from all the senses at once is integrated, and the like. These kinds of questions cannot be reduced, or answered sense by sense.

In this and succeeding chapters we discuss many of the ways in which organisms perceive and respond to stimuli in their external environments. The stimuli we discuss include sound, touch, movement, chemical gradients, temperature, gravity, light, taste, smell, and the electromagnetic spectrum including infrared and ultraviolet light, electricity, and magnetism. Except for the constant forces of gravity and magnetism, a key fact is that these stimuli are unpredictable and of varying intensities and ranges. This unpredictability is comparable to that of the molecular diversity confronting the immune system. An organism's ability to perceive them must be open-ended in a similar way. Light comes in many colors and levels of intensity, odors can be perceived as good or bad, seductive or noxious, sound is loud or soft, high-pitched or low.

One way or another, most if not all organisms, from bacteria to plants to invertebrates to vertebrates, have evolved receptors to at least a subset of these stimuli and are able to respond to a range of many of them. For each organism's life circumstances, the ability to discriminate among such stimuli has been at least sufficient for its survival—or, to survive, the organisms used what they had. Whether in some way it might be useful to be even more sensitive is moot. Thus whether humans would make use of better night vision or an ability to hear musical instruments producing lower sounds, or the like, these questions are irrelevant because our senses did not evolve with the *purpose* of seeing well in the dark, or capturing all vibration as sound.

A distinction is often made between the detection of internal and external stimuli, but this distinction, again in the manner of Descartes, may not be as clear-cut as it is often presented. Organisms monitor their internal as well as external state; the distension of the gastrointestinal system, blood pressure, or the levels of hormones such as insulin or glucose, salt and pH at the cellular level, and so forth. Receptors for internal events, *interoceptors*, share some principles of structure and function with receptors for external stimuli, *exteroceptors*. In all these systems, signals are received in specialized cells in the sensory organs and the information is transduced and transmitted to the central nervous system for deciphering. In addition, organisms can only respond to changes in pH and temperature and the like if they are monitoring their own internal environment so that they are able to determine when indeed there *is* a change. In fact, one could argue that the immune response is a form of monitoring and defending against threats from the external world—it just happens to be done internally. Immune detection and response is somewhat different from the usual notion of a "sense" because it happens strictly at the molecular level without being cognized; but there can be organismal response (sleeping in mammals, seeking warmth in cold-blooded species, and the like). However, for the purposes of this chapter, we will ignore these kinds of muddied distinctions and concentrate somewhat arbitrarily on responses to external physical and chemical stimuli.

WHAT ARE SENSES, AND WHY DO THEY EXIST?

If Aristotle and the other ancients defined senses in a rather restricted and perhaps anthropocentric way, here we are interested in generalities that we can draw from knowing something about these systems in very different organisms. What *is* sensation, and why do organisms have it?

As a general statement, we can characterize senses as the means by which organisms are stimulated by their external environment to react in some way. Here, the environment includes macromolecular (touch, sound), energetic (e.g., heat, magnetism, light), and chemical (pH, odors, tastes) characteristics, but signals from these kinds of stimuli may overlap. The nature of perception need not be neural and certainly not conscious, in the usual sense of awareness (but are ants "conscious" of each other when, as noted by Dante, they "touch their muzzles, each to each, perhaps to seek news of their fortunes and their journeyings" (Alighieri 1314)?). Similarly, in discussing response to environmental information we do not restrict ourselves to any particular types of response.

It is more than rhetorical to suggest that the division of sensation into discrete Cartesian categories is artificial. Such division is based on incompleteness but also to a considerable extent on how humans are structured to receive environmental signals (e.g., with distinct eyes, nose, and ears) as well the experiential aspects of how we perceive them, and thus seems "natural." For practical reasons, the treatment below classifies sensation in broad categories, but we will show how this categorization is largely a human creation and, in particular, that the genetic mechanisms involved in different modes of sensation are often very closely related and probably ancestrally identical.

SOME BASIC SENSORY PRINCIPLES

Whether of internal or external stimuli, sensory perception begins at the level of the cell. This raises two general questions: first, what kinds of signals do cells sense and what mechanisms are used? Second, what do they do with the signal? (Here, inevitably trapped by language, we use words like "information" or "signal" without implying an intentional sender or coder nor that incoming data are neatly packaged).

Some sensory mechanisms cause cells to change gene expression. Other sensory cells respond to stimuli by transmitting an impulse to another cell, as in neural perception. Much of our own sensory perception is based on central nervous system (CNS) processing of such signals, and there are several ways they are received. There are in mammals a few senses not transmitted to the CNS by neurons (e.g., light receptors in the skin), but as discussed in the introduction to this chapter, we should not be too parochial because many organisms sense their environment in wholly nonneuronal ways.

Receptor Mechanisms

Recognition of mechanical and chemical stimuli begins with the binding of a ligand to a cell surface receptor. The ligands are in the form of chemical or energy signals, and the receptors are transmembrane proteins that bind specific signals and initiate the transformation of the signal into a neural event (in organisms with an organized nervous system, at any rate). This transduction of signal into perception can take either a direct or an indirect path from the receptor cell to the CNS.

In vertebrates, the organization of the sensory cranial nerves is essentially the same for all senses. The receptor cells are either true neurons or neuronlike cells. Receptor cells that are true sensory neurons have either free sensory terminal endings or specialized sensory terminals. The axons from these cells terminate in the CNS. The senses of touch, pain, and stretch—which are perceived by the skin, muscle, and joints—generally involve this type of receptor, as does the sense of smell. Receptor cells that are not neurons are specialized cells with no axons. They relay receipt of their specific signal to the CNS by contact with neurons that transmit along their own axon. Hair cells of the vertebrate ear and lateral line of aquatic vertebrates as well as photoreceptors for light transduction are examples of this type of receptor. The basic cellular structures of different types of neurons are shown in Figure 12-1.

As with all neurons, when a sensory receptor is at rest—not activated—a *resting potential* is maintained across the cell membrane. That is, there is a difference in electrical potential across the membrane of the cell, which is generally negatively charged, and it is maintained by the sodium-potassium pump, which discharges more positive charge from the cell than it allows in. When the cell is activated by the specific stimulus to which it responds, this resting potential is disturbed and *receptor potential* results—a nerve impulse is generated. The electrochemical process that converts signals to receptor potentials is called *sensory transduction*, and generally begins with the activation of ion channels in the membrane of the receptor cell, controlling the passage of cations (positively charged ions such as sodium or potassium) into the cell.

When the flow, or *flux*, of ions into or out of the cell reaches sufficient magnitude in neuronal receptors, an *action potential* results in the axon. That is, the electric potential on the surface of the cell membrane changes, creating an electrochemical impulse that sweeps along the axon of the neuron to transmit the signal to terminals located in the brain or spinal cord. The terminals release their neuroactive chemical, which triggers *sensation*.

In a receptor cell that is not a true neuron the flow of receptor current is indirect, that is, it must be passed from the receptor to a neuron through a synapse. This often requires the intervention of a cascade of events, relayed by the second messengers, that leads to the opening of ion channels. The signal is then transmitted to a neuron to be carried along the axon to the nervous system. *Photoreceptors* and *chemoreceptors* are examples of this type of receptor. The use of secondary messengers produces a slower response than in mechanosensory transduction. Perception of sound is almost instantaneous, for example, whereas there is a delay between the stimulus and response in photo- or chemoreception.

MECHANORECEPTORS

Mechanosensory transduction is the process by which mechanical forces are converted into electrical signals. In this case, the signal is not molecular in nature, but must be translated by genetically based response mechanisms. Mechanoreceptor cells are the basis of a diversity of senses in multicelled organisms, including proprioception (the body's ability to orient itself in space and sense the movement of its own parts), balance, hearing, touch, which involves the detection of pressure on the surface of the body, the detection of stretching or twisting such as in joints or muscles or the digestive system, or the displacement of feathers or hairs or whiskers. Although they differ widely in the range of sensations they govern and the receptor that initiates the detection, the basis of each of these senses is a response to mechanical deformation—stretching, bending, displacement by air, and the like.

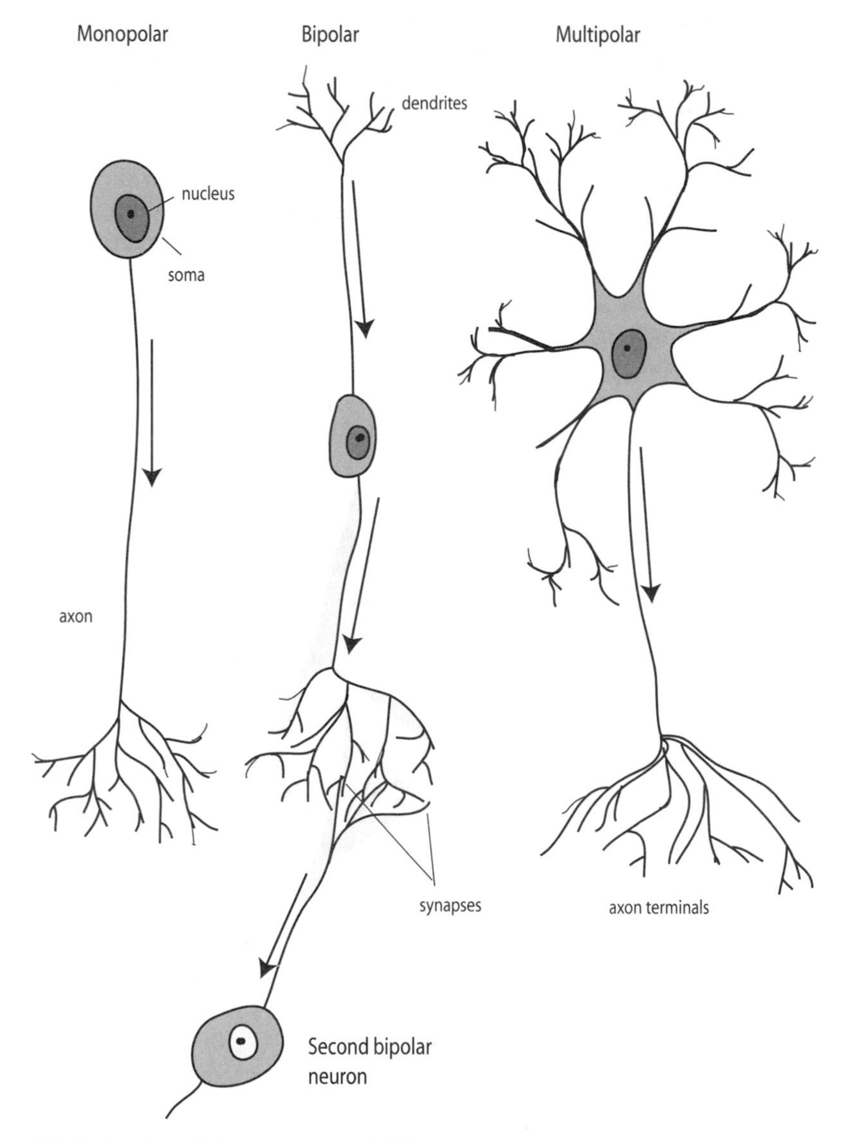

Figure 12-1. Basic cellular structure of different types of neurons. The signal travels *to* the cell body from the dendrites and *away* from the cell body, or soma, along the axon. The nucleus plays no neurological role, but functions in growth and metabolism of the cell. Multipolar neurons are the most common type in vertebrates; bipolar neurons are much less common and are found as olfactory receptors, receptors in the retina of the eye, and several cranial nerves, e.g. unipolar neurons are found in the spinal ganglia of the spinal nerves and some of the cranial nerve nuclei, where they transmit signal to the CNS.

They all depend on a mechanically gated ion channel for transduction of the mechanical force into the sensation that is ultimately perceived.

The molecular basis of the mechanical gating of ion channels is not yet well understood (Walker et al. 2000). Homologies found in this class of transduction mechanism, such as the fact that they send their message nearly instantaneously, never requiring the action of a second messenger such as a G protein cascade, suggest to some a common evolutionary origin and/or a single molecular mechanism underlying these diverse responses.

HEARING AND BALANCE

Hearing is the transduction of vibratory energy into electrical energy, which in some species is decoded in the brain into information about the pitch and magnitude of the vibration. The stimuli can come through the air for terrestrial species or through the water for aquatic species or can come from direct contact with solid objects. Because animals with hearing have bilateral symmetry, they can typically also sense the direction from which the energy comes, relative to their midline. Hearing is mainly a feature of vertebrates and arthropods, although there are suggestions that some nonarthropod invertebrates can also hear (Budelmann 1992).

Here we use the single term "hearing," but there is no one way to hear: different organisms use different aspects of sound in their own ways. Some simply need to respond to the presence of sound or, say, large, sudden sounds. Others respond to very specific patterns, frequencies, or even periodicities of sounds, such as the diverse ways that vertebrates and invertebrates respond to mating, territorial, or other such sounds from conspecifics. Organisms may need to recognize specific predator sounds. Or they may need to recognize sounds such as that of fire, moving water, and the like. Finally, of course, is the highly elaborate processing that humans do to interpret language.

Each of these uses of sound detection involves not just different levels of integration of sound impulses but different ways to detect sound. One cannot interpret elements of sound that one cannot detect and separate into its relevant elements. Nature has evolved particular mechanisms, especially perhaps in vertebrates, for discriminating sound. The variation in sound frequencies audible to a number of hearing organisms is shown in Figure 12-2.

HEARING IN VERTEBRATES

The Mammalian Model

There is great diversity in the shape of the ear in vertebrates, but the mechanism by which it transduces vibration into sound is similar across species (with similar mechanisms in some invertebrates). The classic system is that of mammals, and it illustrates the various modes of transmission. The outer ear collects the airborne sound waves and transfers them to the middle ear by vibrating the *tympanic membrane* stretched across the inner end of the ear canal; the middle ear contains three small, linked bones, the ear *ossicles*, that transfer the mechanical vibrations from the tympanic membrane to a thin, pliable membrane on the cochlea called the *oval window*, which communicates with the fluid-filled inner ear. Vibrations of the oval window are transmitted in the fluid and detected by ciliated hair cells in the spiral-shaped cochlea. Movements in the cilia are transduced into electrical energy and transmitted to the auditory portion of the brain via the acoustic nerve, which connects at the anchored end of the hair cells. The balance system, or labyrinth, is part of the same fluid-filled system and will be described below.

In all hearing vertebrates the ubiquitous and central player in this mechanosensory pathway, the sensory cell itself, is the ciliated *hair cell*. This cell type is ancient

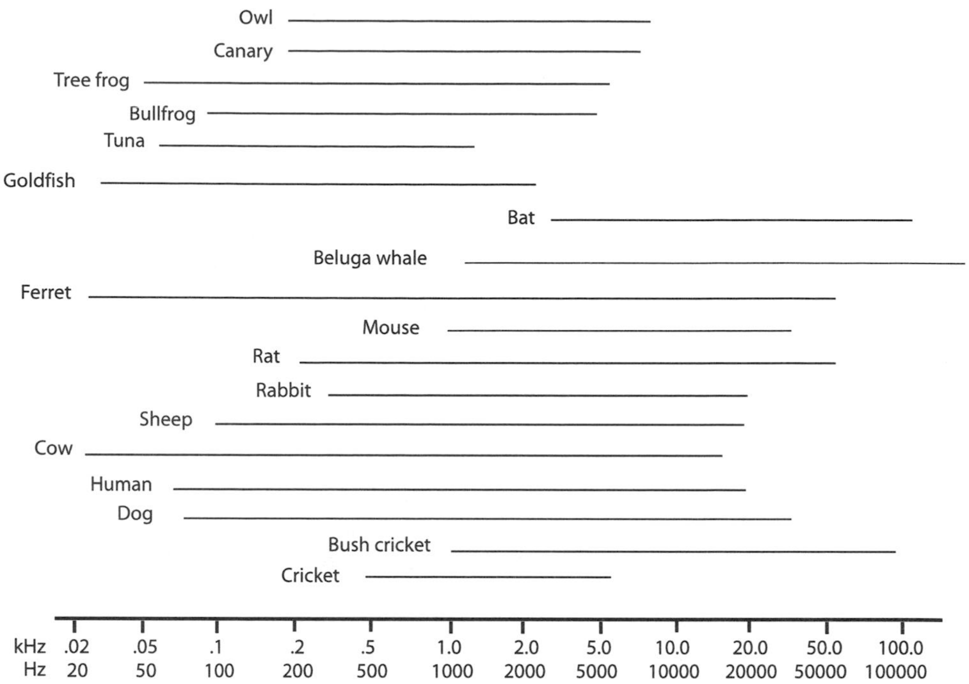

Owl
Canary
Tree frog
Bullfrog
Tuna
Goldfish
Bat
Beluga whale
Ferret
Mouse
Rat
Rabbit
Sheep
Cow
Human
Dog
Bush cricket
Cricket

kHz .02 .05 .1 .2 .5 1.0 2.0 5.0 10.0 20.0 50.0 100.0
Hz 20 50 100 200 500 1000 2000 5000 10000 20000 50000 100000

Figure 12-2. Sound frequencies (in Hz) audible to various species.

and found diversely in the living world. Cilia are tiny hairlike appendages, consisting of *microtubules* containing a protein called *tubulin*, that protrude from the surface of the cell (Alberts 1994). Ciliated cells are found in most animal species and some lower plants. Many protozoa are ciliated as well; they use the cilia to find food by washing fluid over the cell or for locomotion. Ciliated cells line the respiratory tract of many animals to move contaminants up and out of the lungs, for example, or they move eggs through the oviduct. Ciliated cells are integral to at least three sensory systems; taste buds, odorant receptors, and auditory hair cells and similar structures are used in photoreceptors.

In the hair cells of the *cochlea* and *semicircular canals* (also called the *labyrinth*) of higher vertebrates, groups of 50 or more cilia form bundles at the top of each cell and project into the fluid in the basilar membrane of the cochlea, which divides the cochlea lengthwise into an upper and a lower chamber, and in a gelatinous material, the *cupula*, in the bulging *crista ampullaris* at one end of each semicircular canal (see Figure 12-3). The cilia increase in length from one side of the bundle to the other. (See Figure 12-4 for a photograph of hair cells taken with a scanning electron microscope.) Deflection of these thin hairs by the fluid into which they protrude sends the signal to the sensory cells that connect from the basal end of the hair cell to the auditory regions of the brain. In this mechanically gated signal transduction system, if the cilia are deflected in the direction of the longest cilium, the action potential of the cell decreases, depolarizing it, and this induces the cell to release excitatory neurotransmitter at the synapse between the hair cell and the sensory neuron. Deflection away from the longest cilium, in contrast, results in hyperpolarization of the cell and decrease in the release of neurotransmitter.

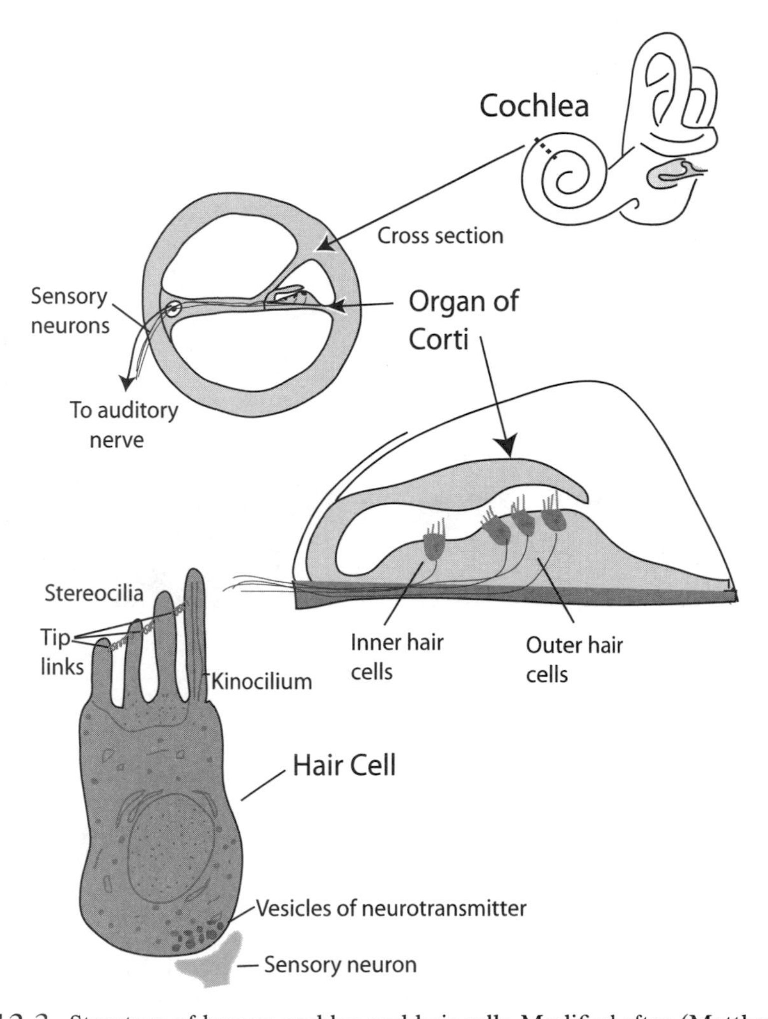

Figure 12-3. Structure of human cochlea and hair cells. Modified after (Matthews 2001).

The ear is one of the most intricately patterned structures in a vertebrate body. However, its development does not involve new principles or basically any new genes. It is beyond our scope to present ear development, but each region shown in Figure 12-3 develops through the agency of a set of signaling, transcription, and cytostructural factors (Kiernan et al. 2002; Petit 1996; Petit et al. 2001).

TFs, SFs, and signal receptor genes involved in the hind-brain region of the developing nervous system are involved in early otic patterning. A lateral otic placode invaginates to form an otic cup which closes to become the otocyst. The semicircular canals and cochlea derive from outpocketing from the otocyst. Genes from ectoderm and mesoderm are involved. Several genes related to the semicircular canals have been identified, but less is known about cochlea-specific developmental mechanisms. The hair and supporting cells become arrayed along the cochlea, probably by an activation and lateral inhibition system such as has been seen earlier, including that of the Delta/Notch system.

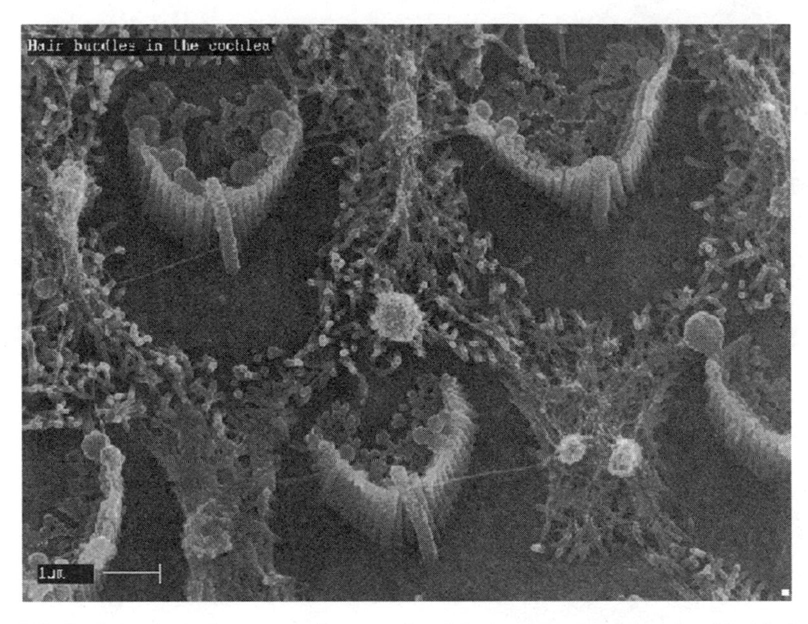

Figure 12-4. Scanning electron micrograph of hair cells. Reprinted with kind permission of Julian Thorpe and Guy Richardson, University of Sussex; see www.biols.susx.ac.uk/Home/Julian_Thorpe/coch8.htm for a series of SEM photos of hair cells at increasing magnification.

Although hair cells are not true neurons, sound detection is nearly instantaneous, as is signal transduction in all mechanically gated systems. Indeed, it is much quicker than photoreception, because the ion channel gating involved is a purely mechanical process. Each cilium in a bundle is connected to an ion channel on its neighboring cilium by a *tip link*, which is like a tiny spring (see Figure 12-3). When the bundle of cilia is deflected in the direction of the longest cilium, the tip link on the longer cilia pulls open the gate on the ion channel of the adjacent, shorter cilium. When the bundle is deflected in the opposite direction, toward the shortest cilia, the tip link relaxes, closing the ion channel of the neighboring cilium. The signal is thus transmitted through the ion channel to adjacent sensory neurons, located within the *spiral ganglion* of the cochlea. An axon extends from each spiral ganglion cell to the CNS via the auditory nerve. With the nerves that come from the vestibular apparatus of the inner ear, the semicircular canals, these axons make up cranial nerve VIII.

Frequency detection is graded along the cochlea, with the hair cells at the base nearest to the oval window responding to the highest frequencies and those at the apical end responding to the lowest frequencies. The nerve fibers that synapse with the hair cells at specific locations are "tuned" to specific frequencies. There are two competing theories as to how this happens. The first is the "place theory," which proposes that frequency is determined by the physical properties of the basilar membrane, in which the hair cells are embedded; the membrane is rigid and narrow at the basal end and wider and more flexible at the apex. The ratio of stiffness to mass determines how far sound waves will travel down the cochlea, and thus which hair cells will be excited and send their signal to the brain. A different, "temporal" or "frequency theory" is based on the idea that sound is periodic in nature; the auditory nerves fire at a rate determined by the periodicity of sound waves and the rate at which the basilar membrane vibrates, and the brain determines tone by the rate at which auditory nerves discharge their signal. Neither theory alone adequately explains all aspects of sound perception, so more research on this question is needed before we have a sufficient understanding.

Large animals (including humans) and miniscule insects detect the directionality of sound in two ways. The body itself diffracts sound waves, and the brain can detect and interpret the interaural difference in sound pressure caused by this diffraction as directionality. Second, the brain can also interpret the difference in the time of arrival of the sound at each of the ears.

In terms of genetics, the known genes involved in sending hearing and balance information to the brain are similar to those used in other membrane potential systems. Most of what we know comes from human or mouse mutations. Two classes of genetic effects are observed. One class involves syndromic hearing loss, which means that hearing is lost in association with other craniofacial developmental anomalies. In nonsyndromic hearing loss, the latter anomalies do not occur and there may be no other phenotypic effects. Over 75 genes or chromosomal regions have been identified in association with nonsyndromic hearing loss (Van Camp and Smith 2003), and this is probably an incomplete list of the many pathways that are actually involved and hence mutable.

As with so many structures, catalogs of expressed genes are being assembled by many investigators, and the lists include members of most types of genes. Perhaps the most general characteristic is that mutations in genes specifically expressed in normal cells or structures of the mature inner ear, or expressed during its development, lead to defects generally consistent with their expression pattern. The genes affect hair cell structure, cytoarchitecture, mechano- and neurotransmission, transcription factors, and genes for gap junction communication between cells, among others (see, e.g., Petit 1996; Petit, Levilliers et al. 2001); tabulated in (Van Camp and Smith 2003). Some of these genes are specific to hearing apparatus, although most if not all are members of larger gene families and some have additional sites of expression. For example, one cell-junction gene, *Connexin26*, is frequently found to affect nonsyndromic hearing loss. Many relevant genes have been discovered in surveys of cDNA (expressed genes) from embryonic ear tissue. Although mouse homologs can typically be identified, as so often happens, when comparison has been possible between human disease and effects of mutation in the mouse, the two do not always correspond closely.

The complex transmission of mechanical signal from outer to middle to inner ear might seem rather more elaborate than necessary. Indeed, some hearing is transmitted directly to the cochlear fluid through vibrations induced by sound in the skull, and other animals have more direct transduction (see below). As with so many traits, hearing evolved its elaborations in connection with a diverse set of needs (chance, developmental or genetic connectedness, or response to selection) in a way that had to be consistent with other aspects of craniofacial development.

The sensory organs that allow vertebrates to maintain their sense of balance and equilibrium are also located in the ear, next to the hearing apparatus. (See Figure 12-5 for a comparison of the anatomy of the inner ears of fish, amphibians, birds and mammals.) These are the three mutually perpendicular semicircular canals that comprise the labyrinth and detect and convey rotational movement and the two *otolith*

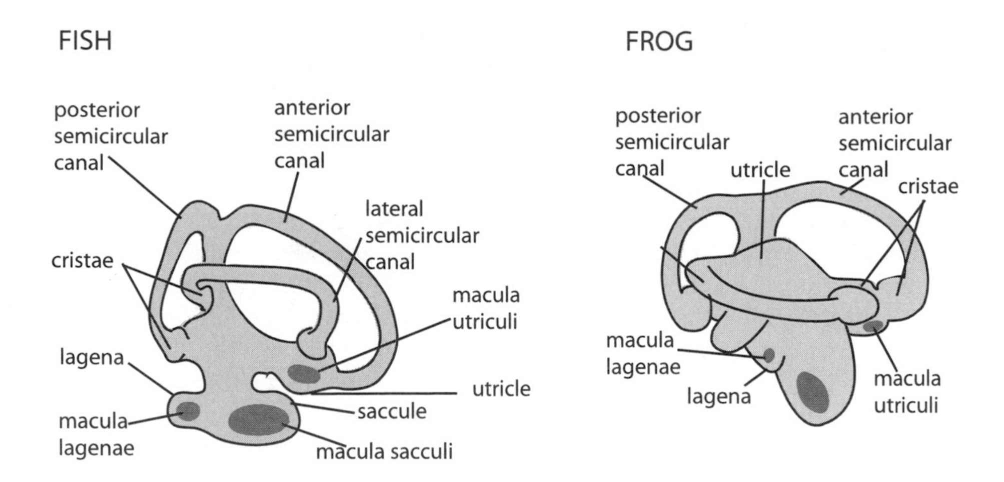

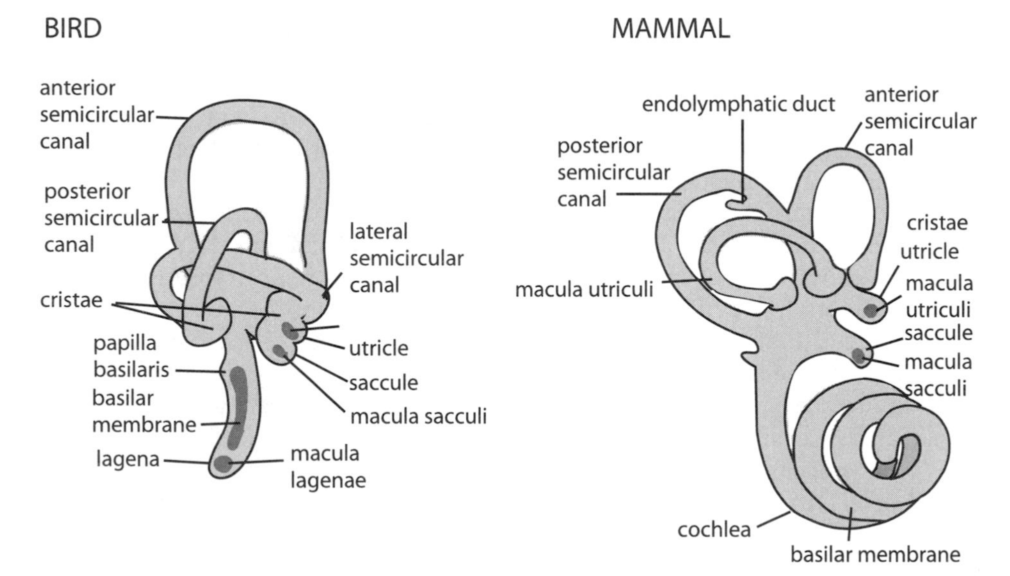

Figure 12-5. Comparative diagram of the labyrinths and inner ears of fish, amphibians, birds, and mammals. Redrawn based on version in *Encyclopedia Britannica* (Britannica 2003).

organs, the *saccule*, which detects motion in the vertical plane, and the *utricle*, which senses motion in the horizontal plane. These sacs and canals are all filled with a fluid called *endolymph*. As with hearing, the detection of motion is based on the deflection of the sensory hair cells, which occurs when the animal moves its head, causing the endolymph to lag behind. At one end of each of the semicircular canals is the knobular structure called the *ampulla* and within that, the *crista*. The upper surface of the crista contains the hair cells, which are embedded in the gelatinous cupula. As in sound reception, deflection of the hair cells stimulates the transmission of information about angular movement to the brain. When the animal moves its head the lymph within the canals moves, and this deflects the hair cells in the crista and either depolarizes or hyperpolarizes the cells to release or inhibit the release of neurotransmitter and thus transmit signal, also via the VIII nerve, to the brain.

The utricle and saccule are membranous sacs. The floor of the utricle and the wall of the saccule contain hair cells covered with gelatinous substance, as well as tiny grainlike crystals, or otoliths. At rest, the otoliths press straight down on the hair cells, but with motion, depending on the direction, the otoliths press less or more, and at different angles, and the cilia of the hair cells respond to the changes. The brain puts together signals from all of these organs to produce the sensation of motion in a three-dimensional space. People tend not to think of balance and equilibrium as one of the real "senses"—until they lose it and realize how pervasively it is used. The orientation of the semicircular canals is related to the habitual posture of a species and frequently detects motion in all three spatial planes.

Bilateral symmetry allows discrimination of direction, but many animals, especially higher vertebrates, do much more precise location of distant objects or phenomena via their sense of hearing. By turning the head or moving the outer ears, animals can intensify sound to help locate it. To do this, the brain must not only compare neural impulses from left and right cochleae but also take into account the orientation of the ears or head, thus combining macro- and microscale phenomena. There is also an ability to judge the distance of an object from the intensity as well as frequency spectrum of sound, based on experience stored in memory or in other ways. Organisms can also sometimes account for the effects of wind, echoes, and so on (but the difficulties this presents illustrate in interesting ways the limitations of the system).

Birds

The auditory structure in birds shares many of the features of mammalian ears—phylogenetic analysis suggests that hearing had its origins in reptiles in the Paleozoic with subsequent divergent evolution of the hearing apparatus in the distinct amniote lineages that then arose (Manley and Koppl 1998) (see Figure 12-6).

The auditory mechanism in birds is the *basilar papilla* in the cochlea. As in tetrapods and some insects, hearing is tympanic (based on the vibration of a membrane), and the tympanic membrane bulges outward at the surface of the head. Sound vibration is conducted to the cochlea along a single bony structure called the *columella*, which communicates with the cochlear oval window. The cochlea is a very short bony structure, although the length varies by species, and it encloses the basilar membrane and ends in the *macula lagenae* and the *lagena*, which do not have a hearing function (see Figure 12-5). In birds, hair cells are found throughout the inner ear, along the entire width of the auditory basilar papillae. Birds are capable of very acute frequency analysis, and their hearing is organized tonotopically, as is hearing in mammals.

The labyrinths of the avian inner ear are similar to those of higher vertebrates, with three canals arranged at angles that detect motion in all directions, in the same manner as the balance system of mammals.

Fish and Amphibians

The fish's ear is a membranous labyrinth that as in tetrapods serves two functions, hearing and maintenance of the fish's equilibrium. Although in the head, unlike in

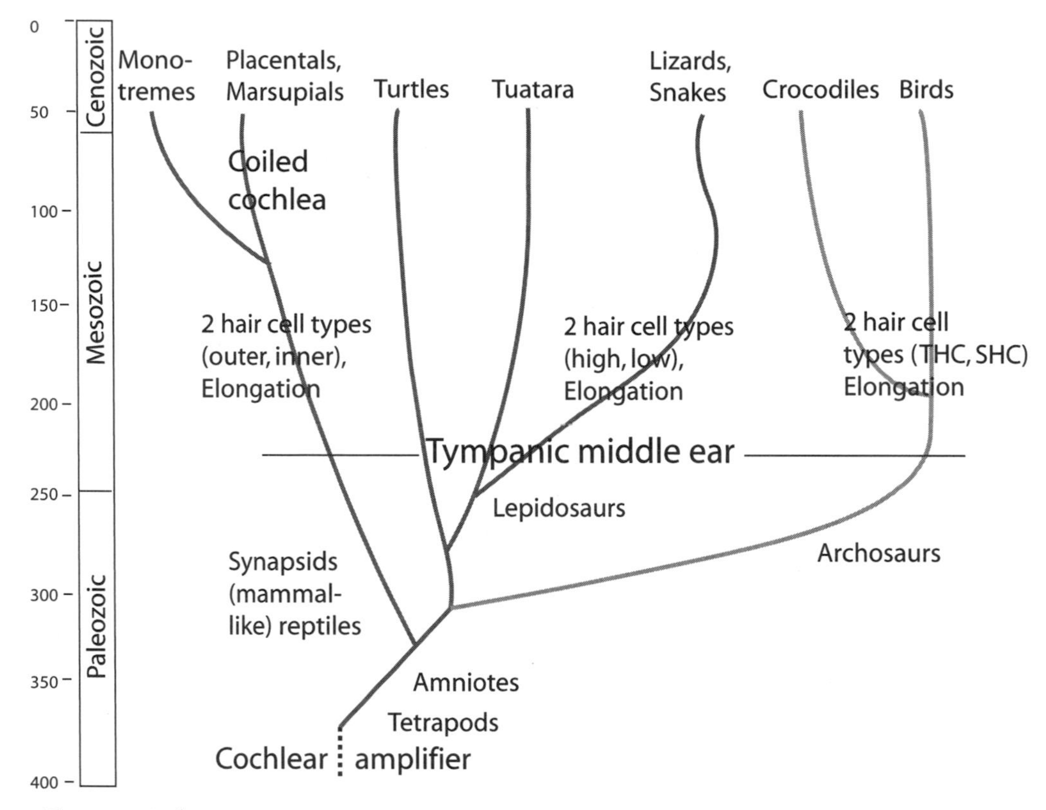

Figure 12-6. Phylogeny of cochlear types among amniote species. Redrawn from (Manley and Koppl 1998) with permission.

tetrapods, fish ears are internal organs, not open to the external environment, and there is no tympanic membrane. Sound is conducted to the hearing organs through tissue and bone. The labyrinth is made up of a series of sacs and canals, as in mammals. The semicircular canals (of which there are three in most fishes, except hagfish, which have only one, and lampreys, which have two) are oriented at right angles to each other in three different planes and are fluid filled. Each canal has an ampulla and crista, and the crista is composed of sensory cells called *neuromasts.* Neuromasts are covered by the cupula, a gelatinous membrane as in tetrapod cupulae. Fish also have the saccule and utricle and a lagena, an appendage of the saccule in fish, amphibians, and birds. In most fishes, the saccule is most likely primarily an auditory organ and the utricle primarily serves a vestibular function (Fay and Popper 1999).

Hearing in some fish also relies on an additional pathway; the swim bladder. This buoyancy organ in the body cavity of bony fishes is a gas-filled out-pocketing of the digestive tube that helps a fish maintain its depth and adjust its buoyancy. The swim bladder pathway to hearing is indirect: the gas-filled organ, or other gas bubbles near the ears, expand and contract in response to sound pressure, and this motion is transmitted to the otoliths.

Most amphibians share all the organs of the fish ear: the semicircular canals and their cristae, the saccule, lagena, and utricle. In terrestrial amphibians, reptiles, and birds, sound is conducted from the tympanic membrane to the inner ear by a single bone, rather than the three in higher vertebrates. In fish and aquatic amphibians, hair cells are found in the semicircular canals of the ear and along the lateral line. This raises the point that no one system is required for transmission of sound to the cochlea; the bony ossicles of the middle ear of mammals evolved as part of the evolution of jaws differently hinged than those of reptiles, for example. Mammalian ossicles transmit sound from the tympanic membrane to the cochlea, but this is not the only way, and it is not particularly clear what special advantage this system had. Perhaps it was just better than relying on the duller direct through-the-skull perception pathway, as bones were remodeled and recruited for the evolution of jaws and face. Maybe it did not matter, relative to the developmental constraints entailed by the importance of making jaws, so long as sound was transmitted somehow.

Fishes have evolved a large repertoire of sound-generating mechanisms that they use for attracting mates and during spawning. These include muscular vibrations of the swim bladder, pectoral girdle, and pectoral spines rubbing in the grooves of the pectoral girdle, plucking of enhanced pectoral fin tendons, or grinding of pharyngeal teeth (Ladich 2000).

Fish are probably poor at localizing sound sources; because of the way sound waves reach the hearing apparatus, interaural time and intensity differences are effectively nonexistent (Fay and Popper 1999). In some species the hearing organs are connected by perilymphatic spaces, and sound conducted by the swim bladder reaches both ears at the same time.

Echolocation

A few animals, including dolphins and bats, "see" their environment by emitting sound waves and monitoring the reflected echoes as the sound bounces off whatever it hits. The suborder of bat, *Microchiroptera* of the order *Chiroptera*, both navigates and finds its prey by *echolocating* in this way. Approximately 800 species of bats belong to this suborder, and the sound they emit ranges from biosonar pulses to clicks and other calls. These bats are able to monitor the speed of flying objects—their prey—and their size, range, and elevation among other characteristics. Bats that do not echolocate (fruit-eating bats, for example) find their prey by vision. Bats are highly speciose (richly diversified) animals as reflected in the exotic variation in their craniofacial morphology (Figure 12-6); how much of this was specifically selected for echolocation is somewhat less clear.

Bats that echolocate make laryngeal or nasal emissions (Gobbel 2002; Springer et al. 2001; Teeling et al. 2002; Teeling et al. 2000). Teeling et al. have shown that echolocating bats are probably paraphyletic, with laryngeal echolocation probably evolving first and being lost later in different lines that have modified the details of their apparatus in various ways (Teeling, Madsen et al. 2002). Dolphins have sonar, but it evolved independently of bat echolocation and uses different means. Dolphins emit clicks, receive the echo with their jawbone (*panbone*), and apparently begin to interpret it with the "melon" in their forebrain. They may actually stun their prey with sonar.

Bats, including those that echolocate, share the same auditory apparatus found in most mammals; the outer ear that captures sound waves and directs them to the

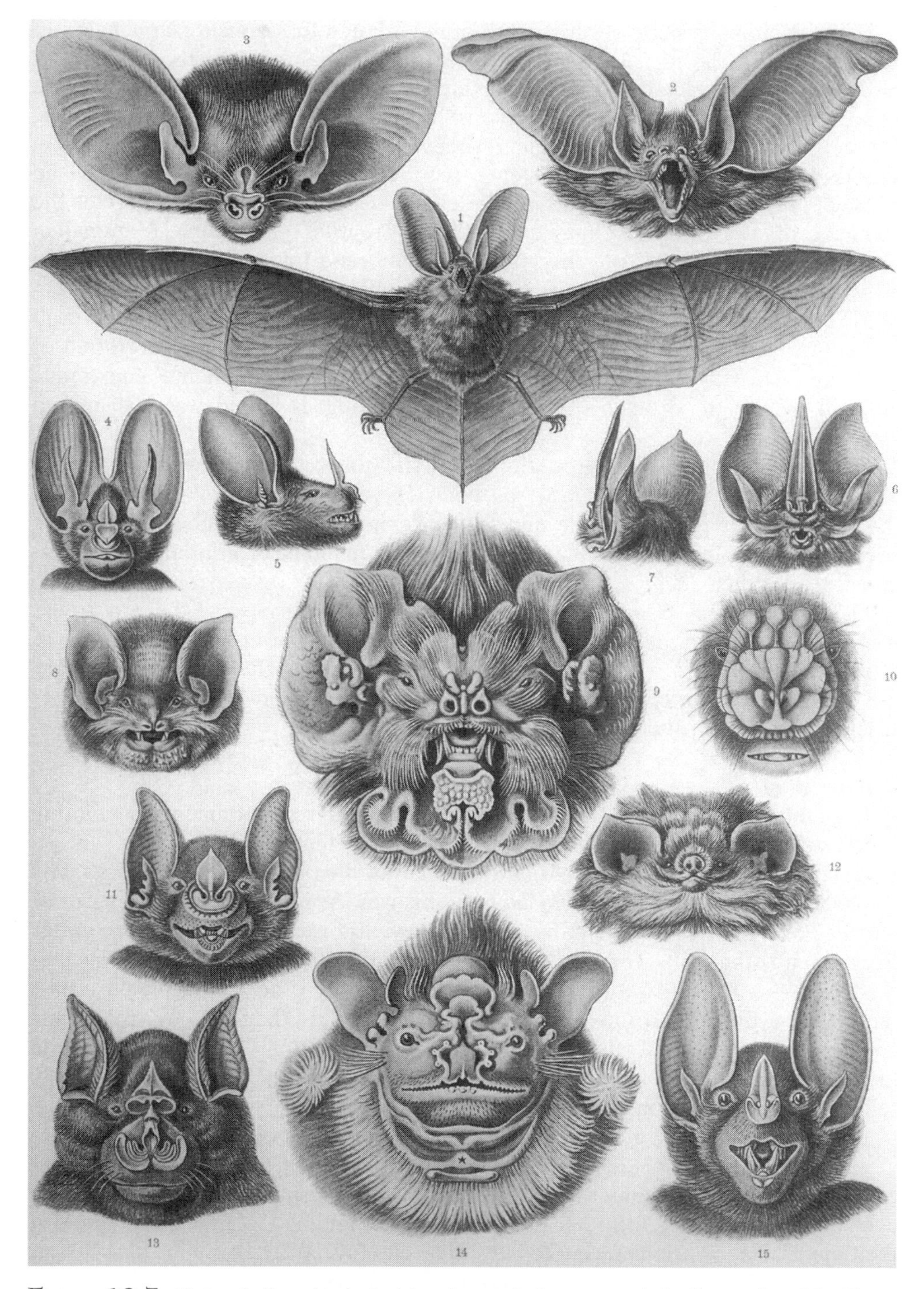

Figure 12-7. Natural diversity in bat head morphology, as artistically rendered by Ernst Haeckel in his *Art Forms in Nature* (available in reprint as Haeckel 1899).

middle ear and then the inner ear, where the sound waves are transduced from mechanical energy into electrical energy and transmitted to the brain for interpretation. However, a feature specific to echolocating bats is that the portion of the basilar membrane in the cochlea that receives sonar sound frequencies is especially sensitive and spiral ganglion cells of the brain that receive these frequencies are overrepresented relative to others.

That said, this exquisite ability to detect prey with echolocation may outrank the bat's ability to actually capture the prey once it has been detected. One study found that only about 40 percent of attempts by red bats are successful on the first try (Obrist and Wenstrup 1998). This is in part because many of the moths and other insects that bats prey upon can hear the sounds that bats emit in their echolocation calls, and in fact their hearing is most sensitive to the frequencies emitted by their most common predator and so can initiate evasive maneuvers while the bat is still fairly distant. Thus, although we think of bats as having a rather refined and exotic sensory capability (perhaps because we do not have it and hence romanticize it), echolocation is, like other systems, imperfect and highly variable among bat species, especially when the responses of moths and other prey have kept pace. If it were not so, there would be no prey—and no echolocating bats.

Hearing in Arthropods

Hearing in insects—the detection of air- or substrate-borne vibration—is used for detecting predators and for signaling to and locating mates. The mechanoreception of air- or water-borne vibration is found in seven of 27 orders of insects (Hoy et al. 1998). However, many more insects produce vibration through the branch or leaf upon which they perch by knocking against it with a leg or other body part and detect it coming from the same source with their hearing organ. Insects with body lengths less than 1 cm are generally restricted to ultrasound emissions, which are useful only at short range or in free space. Ultrasound is distorted and attenuated by most habitats; thus small insects that rely on ultrasound probably do not use sound for social communication but to detect prey or predators. Larger insects can emit sounds above about 1 kHz (Michelsen 1992), which can penetrate vegetation. Most insects that use hearing for social communication have frequency analyzers, whereas those that use it only to detect prey or predators do not seem able to detect frequency.

Interestingly, despite the bat problem, not all Lepidoptera (moths and butterflies) are able to detect sound, and the ability of those that can is highly variable, both in range and in the complexity of the actual hearing apparatus itself and its location on the body, which can be anywhere from the wings to abdomen to thorax to legs to head. Some nocturnal butterflies have "ears" on their wings that are sensitive to ultrasound (Yack and Fullard 2000); some moths hear with their mouthparts (Gopfert and Wasserthal 1999), but in all moths that can hear, the ear is tympanal and specialized to the frequency of bat ultrasonic emissions.

Insect mechanosensory organs are classified as type I or type II (Eberl et al. 2000). Type I organs have bipolar neurons with an axon extending to the CNS and a dendrite on the opposite end (see Figure 12-1). These organs are surrounded by specialized supporting cells. Type II organ cells are single multidendritic neurons with no obvious ciliary structure. Most insect sensory organs are type I (Eberl, Hardy et al. 2000).

The insects that are able to detect airborne sound have either tympanal organs, which as noted above can be almost anywhere on the body (thorax, abdomen, sternum, legs, wings, antennae), or flagellar organs. Which type of organ an insect has determines how far afield the sound it detects comes from. Tympanal organs are sound pressure detectors and as such are for detecting sound from far away. Near a sound source, on the other hand, most sound energy may be detected particle movement, and flagellar organs, which are protruding structures such as hair or antennae that function as particle velocity detectors are sufficient (Eberl 1999).

An insect's body is covered with a hard protective coating called the cuticle. Poking through at many spots are sensory bristles. These are of various types and are regularly spaced along the body, perhaps produced by reaction-diffusion types of patterning mechanisms (see, e.g., Gerhart and Kirschner 1997). Each appears to develop from a single precursor cell, and the different types of final structure are determined by combinatorial expression of numerous transcription factors (TFs) and activation-inhibition signaling not yet understood.

One of the bristle types is a sensillum known as a *chordotonal organ* (sometimes referred to as a *stretch receptor*), shown in Figure 12-8. Chordotonal organs can be specialized to perform various functions including stretch reception (of the outer chitinous shell) and a related sense of hearing. Tympanal chordotonal organs are internal structures that span two cuticle plates, where there is a thinning of the cuticle, a thin membrane that covers an air-filled sac. A four-cell structure senses

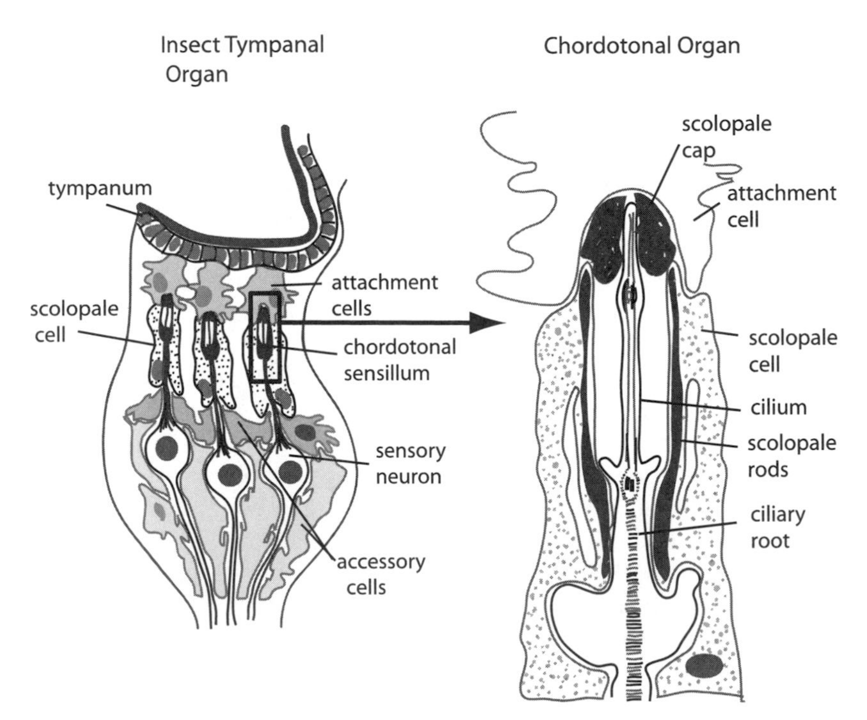

Figure 12-8. Basic structure of insect chordotonal organs. Redrawn from (Gray 1960) with permission.

vibratory movement at this location and is innervated by a single neuron. The sensory organ is sometimes called the *scolopidial organ*, in reference to the spindle-like sheaf of ciliary origin, the *scolopale cell*, into which the dendrite of the nerve cell extends. The scolopale cells and the neuron are associated with glial and support cells. These are located at basically fixed sites in a given species. Auditory chordotonal neurons project to similar sites in the CNS but what happens when the signal reaches the CNS varies among insect species, as acoustic information is extracted from the sensory input in a way specific to the sounds relevant to each insect. Although there is a wide variety of "acoustomechanical transformers" among insects, the way the mechanical signal is converted to electrochemical response in the nervous system is quite consistent across taxa (Eberl 1999).

Some of the genetics of chordotonal organs in *Drosophila* is known. The tympanal chordotonal organs are similar to the vertebrate inner ear (Eatock and Newsome 1999; Fritzsch and Beisel 2001). Both have ciliated mechanoreceptive cells and accessory or supporting cells. Homologous genes and mechanisms are involved in the development of mechanosensors in both insect chordotonal organs and vertebrate inner ears (Eberl 1999), and all ciliated mechanoreceptors share a common transduction system (Kernan and Zuker 1995).

A neurogenic TF, *Atonal*, is expressed early in embryonic development in all chordotonal organ progenitor cells (this gene is also expressed in photoreceptors). In *Atonal* mutants, all chordotonal organs are absent except for one scolopidium in the abdominal linear array of five pentascolopidial organs called *Ich5* (Lage et al. 1997). Other genes are known to be required, including *Egf receptor* signaling in precursor cells in most but not all of the eight scolopidia in each abdominal segment. Detailed studies have been done of specific gene expression in these developing organs, which essentially have to do with their patterning, number, and location.

Although most insect auditory organs are chordotonal, some also use other means of detecting air vibration. Insects such as cockroaches and crickets have, either in addition or alone, specialized bristle organs on their *cercae*, or antenna-like sensory appendages projecting from their tails, that are deflected by wind currents and can be very sensitive to sound (Eberl 1999). Bees have flagellar antennae on their heads that are thought to decipher the acoustic components of the waggle dance that is used for communicating the location of pollen sources and to detect song within the hive. *Drosophila* discriminate species-specific courtship songs at close range with their *Johnston's organ*, a collection of chordotonal organs in the antennae.

Crickets and cicadas call for mates over long distances by *stridulation*, the rubbing of ridged surfaces on their legs or wing margins, and detect sound with tympanal organs located on their front legs. Cricket songs also attract parasites (Ormiine tachinid flies, which have sternal tympanal ears for directional hearing to locate their hosts). In a defensive mechanism referred to earlier, some moths use their tympanal organs to detect the ultrasonic echolocating calls of predatory bats and respond by altering their flight patterns. Indeed, other arthropods have an even more diverse repertoire of sound-making and sensing devices. Lobsters have ridged surfaces called *plectra* on the base of their antennae that they rub against a file organ to generate sound. This seems to be for warning off predators, because the lobsters do not appear to use this for mating or defensive reactions.

Lateral Line

All the primarily aquatic vertebrates, cyclostomes (agnaths), fish and amphibians, have hair cells in the "touch organs," or mechanosensors, in their outer skin. This organ is the *lateral line*, and it is sensitive to minute water displacements from vibration as well as changes in pressure, caused both by the fish itself and by other nearby animals or fish, and so is used for various behaviors including finding prey and escaping approaching predators, social behavior such as schooling, shoaling, and avoidance of stationary objects, and hearing, among others.

The morphology of a particular lateral line system determines just what an animal is going to detect—the spatial distribution of the lateral line's mechanoreceptive organs, the neuromasts, determines the receptive field of a particular animal, the innervation patterns determine how sensitive the system will be, and the morphology of the neuromasts themselves determines which aspect of water movement the organ will respond to, velocity or acceleration (Maruska and Tricas 1998).

The lateral line and the labyrinth arise from the same embryonic placode. Together, they comprise the *acousticolateralis system*. The mechanoreceptors of the lateral line, the neuromasts, contain hair cell clusters embedded in the cupula. The organ has two kinds of sensory receptors; the *superficial neuromasts* on the skin and *canal neuromasts* recessed in fluid-filled canals beneath the skin. Superficial neuromasts detect the motion of water flow, and canal neuromasts detect its acceleration. Superficial neuromasts predominate in still water fish and canal neuromasts in fish that live in moving water (Engelmann et al. 2000).

Water enters the lateral line organ through numerous pores on the surface of the skin and flows past the neuromasts. Pressure of the water bends the cupula, and this creates an action potential in the hair cells. (See Figure 16-2.)

EVOLUTION OF THE EAR AND HEARING

In 1882, Mayser proposed that the modern ear was derived from the fish lateral line, the *acousticolateralis hypothesis*, a view that was held for most of a century (Mayser 1882). Mayser described the lateral line as an accessory hearing organ. In 1987, observing embryonic development in salmon, Wilson and Mattocks (Wilson and Mattocks 1887) suggested that the inner ear and lateral line were derived from the same embryonic placode and that the inner ear originated as part of the lateral line. Additional supporting evidence was considered to be the similarity of the hair cells in the lateral line and the inner ear.

In 1974, E. G. Wever wrote an influential paper on the evolution of vertebrate hearing, pointing out that none of the animals thought to be ancestral to vertebrates has an inner ear, so that the inner ear must not be derived from lateral line, but that the lateral line and inner ear share a common ancestor. He believed that bony fishes had the first "real ear." He suggested that the acousticolateralis hypothesis be abandoned in favor of the idea that the inner ear and the lateral line evolved from a common mechanosensory system using hair cells (Wever 1974).

Similarities in the receptor cell structure and the basic function of the ear and auditory system among vertebrate groups, and in the corresponding gene usages, suggest that the ear arose early in the evolution of vertebrates (Popper and Fay 1997). There is a consensus that the earliest inner ear structures were equilibrium receptors rather than auditory receptors. It has been proposed that cochlear amplification is ancient (Manley and Koppl 1998) and that the first auditory receptor, the basilar papilla, pinched off from the vestibular organ with the transition of vertebrates from water to land.

Manley and Koppl suggest that this first appeared almost 400 million years ago in lobe-finned fish (Manley and Koppl 1998). However, a basilar papilla seems to have evolved independently several times (Manley and Koppl 1998), after, and perhaps because of, the development of the tympanic ear in the Triassic. This may be one of many examples of the conservation of a basic common ancestral mechanism, like a mechanoreceptor cell or pathway, that has repeatedly been modified for similar purposes. An ability to detect and respond to air or physical vibration apparently evolved many times in invertebrates as well (Michelsen 1992). Animals share common ancestry more closely than was thought in the decades before genetic homologies were identifiable. The distinction between "hearing" and the detection of other forms of environmental vibration is after all a human invention.

Some but not all of these systems involve similar hair cell mechanosensory mechanisms. The near ubiquity of the sensory hair cell in vertebrate ears and the lateral line suggests that it arose early in the evolution of hearing, pressure, or vibration detection. Specialization in form and function, however, suggests that the hair cell has adapted to the specific needs of a variety of vertebrates (Fay and Popper 2000). If precedent (and the results of mouse deafness studies) is a guide, different genetic mechanisms will be found even among animals having very similar physical phenotypes.

Little is known about the evolution of hearing in insects, but it probably evolved from extant mechanoreceptors. Insect auditory receptors are all based on chordotonal sensilla—regardless of where the auditory locus is on the body—and the chordotonal system is present ontogenetically everywhere. A different part of the chordotonal system has evolved for hearing in different insect lineages (Eberl 1999). Chordotonal organs serve as proprioceptors at the appendicular joints, and at intersegmental "joints" between the thorax and abdomen and in the abdomen itself. Thus the opportunity for an ear to arise from the chordotonal primordium occurs frequently over the body of a typical insect, given its many joints and appendages.

THE SOMATOSENSORY SYSTEM

Neural pathways that process signals from receptors on the skin, muscles, and joints are called *somatosensory* pathways. There are a number of different receptor types in the somatosensory system, specific to different sensations. Some receptors perceive touch, some pressure, and others pain. There are structural differences between these receptors; some are very sensitive to touch and pressure and adapt rapidly to sustained stimulus, whereas others are less sensitive and adapt slowly.

A number of different vertebrate receptors perceive pressure, vibration, and texture. These are known as *Meissner's corpuscles*, *Pacinian corpuscles*, *Merkel's disks*, and *Ruffini endings*. The first two are rapidly *adapting*, whereas the second two adapt slowly. The act of picking up a stick, for example, is quickly perceived by Meissner's and Pacinian corpuscles, but they quickly adapt and stop sending the signal to the brain. Merkel's disks and Ruffini endings, however, continue to alert the brain to the presence of the stick in the hand. The *hair follicle receptor* is found in hairy skin, where nerve endings wrap around single hair follicles and transduce signal on deflection of the hair from that follicle.

The vertebrate pain and temperature system does not have specialized receptor organs, but instead free nerve endings throughout the skin, bone, muscle, and connective tissue perceive changes in temperature and pain peptides. Pain is sometimes the result of damage to tissue but more often is the response to substances released by damaged tissue for which free nerve endings have receptors. *Nociceptors*, receptors for pain, have axons that are either myelinated or not. Myelin is the covering of the neuron, produced by a variety of neural supporting cells, that sequesters a neuron electrically from its environment much as insulation protects electrical wires from signal leakage. Myelinated nociceptors are faster conducting and produce the immediate sensation of pain on receipt of signal. Nociceptors with nonmyelinated axons are responsible for the long-lasting pain that may follow a wound or injury.

Mechanisms for the detection of some stimuli are similar across vertebrate classes, which suggests shared evolutionary history. The somatosensory systems that originate in the spinal cord and caudal brain stem, as well as the vestibular and lateral line mechanodetection systems, are very similar across the vertebrate classes in which they are found (Hodos and Butler 1997). Other sensory systems, however, like electroreception or color vision, involve mechanisms that are different enough across species to suggest that they evolved several times. Nonetheless, although receptors for various stimuli vary across vertebrates, the basic organization of the sensory cranial nerves is remarkably alike (Hodos and Butler 1997). These nerve cells are either monopolar or pseudomonopolar neurons, such as the olfactory receptors, or bipolar neurons that innervate specialized receptor cells, such as in the octavolateralis system in fishes or the vertebrate visual system.

Invertebrates have sensory bristles on their external surface, which respond to pressure or movement. They are embedded in the cuticle in insects and are innervated by a sensory neuron that transmits the signal to the insect's central nervous system.

The Star-Nosed Mole

The star-nosed mole provides an interesting—and essentially novel—tactile sense. Its nose is splayed out (Figure 16-7) and contains physical receptors so that it can obtain a spatially arranged "map" of its underground environment. This will be described in more detail in Chapter 16.

OTHER

There is a variety of other physical sensing in the animal and plant world, probably not even easily broken into specific categories. Some plants respond quickly to touch or other physical disturbance. Temperature is a very important variable to many species. Two types of vertebrate *thermoreceptor neurons* have been described. One is the cold receptor, which reacts when the skin is colder than resting body temperature. Warm receptors respond when the skin is warmer than resting temperature. Arthropods that are blood feeders have thermoreceptors and are attracted to prey and induced to feed by heat. These receptors are usually on the antennae but can be on the legs. Many arthropods have humidity detectors. How they work is unknown.

Cold-blooded species seek places in their environment in which the temperature is lower or higher, for various reasons. The seeking of higher temperature to help fight infection in species that cannot mount a mammallike febrile (fever) response has been reported. Homeothermic (warm blooded) species have internal and external receptors that monitor temperature via the hypothalamus in the brain (see Chapter 15) and regulate metabolic responses to maintain body temperature homeostasis (see, e.g., Turner 2000).

As with the other systems we discuss in this book, clearly not all species require the same kinds of thermal sensing, because there is no one way to deal with body temperature and/or its relation to the environment. Some aspects of thermal stress seem to be more fundamental, however. Heat- and cold-shock elicit a response in many organisms, including plants, bacteria, vertebrates and invertebrates. Thermal shock can alter protein structure and function in cells, for example. The response is in the form of the production of heat- or cold-shock proteins that counteract the effect of rapid temperature change, for example, by binding to proteins to prevent their mature shape from denaturing. Heat and cold shock proteins have been found in all known cells and have many functions apart from induction of cellular response to temperature change. They can activate the immune system and generally maintain cellular homeostasis and protect cells in times of stress. The genes, the chaperonins or heat shock proteins that we have mentioned in several contexts, are an ancient family, showing the importance of this kind of protective response to thermal or other traumatic conditions.

Proprioception

In vertebrates, information about muscle length and tension is collected by sensory receptors in the joints and muscles. In invertebrates, this information is collected by stretch receptors on the outsides of muscles and chordotonal organs in the joints, which, among other things, measure tension changes. Stimulation of bristles at the joints can also be part of the proprioceptive system in arthropods.

Proprioception requires the perception of gravity for spatial orientation. Phylogenetically, structures that allow organisms to equilibrate themselves with respect to gravity and acceleration are perhaps the oldest sensory organs. As described earlier, the semicircular canals of vertebrates are based on hair cells, the same kind of sensory cell integral to the vertebrate sense of hearing. The membrane potential of hair cells changes in response to deflection of cilia that project from one end of the cell caused by vibration or the motion of fluid surrounding the cell.

In many invertebrates, the detection of motion and gravity takes place via *statocysts*, the fluid-filled chambers lined with hair cells that transduce the motion of a small granule called a *statolith*, which is free moving in some animals and loosely anchored in the hair cells in others. Statocysts are found in all major invertebrate groups, including arthropods, jellyfish, sandworms, higher crustaceans, sea cucumbers, tunicate larvae, and mollusks. In vertebrates, it is the saccule and utricle of the vestibular system of the inner ear that detect gravity.

Vertebrates and some invertebrates are also able to detect body rotation. The vertebrate ear conveys this information to the brain: when the animal moves, the inertial lag of the fluid in the semicircular canals deflects the hair cells, which transduce that information to the brain.

Many insects are sensitive to gravity, and the receptors involved tend to be found on the body surface, not internally as in statocyst-like structures. The receptors are tactile hairs, found in a variety of locations. In the honeybee, they are found between the head and thorax and between the thorax and abdomen.

The ability of plants to perceive gravity has long been recognized; in fact, Darwin wrote about the fact that roots always grow downwards. The receptor is a calcium granule, a statolith, found in statocysts, as in invertebrate animals. The statolith is pulled by gravity to the bottom of a cell in the root, and this stimulates the root to grow down.

Electroreceptors

The ability to detect electric fields is common among aquatic vertebrates, particularly fishes. Few mammals are known to have this ability, but among them are platypuses and the spiny anteater, both of which have electroreceptors in the skin of their snouts. The electroreceptors of some fish are so sensitive that they can detect the electric fields produced by the contraction of the muscles of their prey as they swim. Many fishes are also able to produce electric fields with *electric organs*, for defense as well as to stun or kill their prey.

The electroreceptors in fish are in the lateral line system. They are found on the head of all elasmobranchs and many bony fishes, electric eels, and others. They are called *ampullary lateral line organs* and differ from other sensory receptors of the lateral line system in some respects: water does not come into contact with the receptor cells, which lie within the ampulla, a vesicle that opens to the surface of the fish's body through a duct. The ampulla and the duct contain a gelatinous substance that is a good conductor of electricity, and perhaps a pH sensor as well. The environmental sequestration that was the key factor in the evolution of the earliest cells probably involved an ability to monitor and adjust to the electrical conditions in their fluid environment.

Magnetoreceptors

Many kinds of animals are known to respond to the Earth's magnetic field, including honeybees, planarians (flatworms), mice, birds, salamanders, and many bacteria. A magnetic material called *magnetite* (iron oxide) has been found in many of these animals and may be the receptor for this sense. Magnetite crystals are found, for example, in the nose of the trout, and their interaction with the earth's magnetic field is detectable by the nervous system. However, the molecular mechanism that transduces this signal to an electrical impulse perceived by the nervous system is still not known (Diebel et al. 2000).

Chemoreceptors

Chemoreceptors are cells in the olfactory and gustatory systems that initiate a neural response to molecules in the air or dissolved in liquid that come into contact with the receptors of these particular sensory systems. They function to detect the chemical structure of individual molecules, which in a logical sense is similar to immune detection but functions differently and is specialized for externally derived, nonliving molecules. These are discussed in detail in Chapter 13.

Chemesthesis

Organisms from bacteria to protists, insects, and vertebrates are able to detect chemical stimuli, and a major use is detection of noxious substances. Nerve endings responsive to irritating chemicals are a part of the general somatic sensory system, as a subset of the pain- and temperature-sensitive fibers on the skin and generally found throughout mucous membranes. The term *chemesthesis* refers to the sensations elicited by the chemical stimulation of free nerve endings (Finger and Simon 2000).

BACTERIA

Motility is an important determinant of bacterial survival. Although some only glide, many bacteria can swim by rotating their flagellum, an ability that allows them to move toward a more favorable environment when they detect toxic conditions. Bacteria sense changes in the levels of surrounding nutrients and toxic chemicals, pH, temperature, light, the magnetic field of the Earth, electron transport conditions, and the like, and they respond with *osmoregulation*, altering internal conditions to protect against the changing external environment, or by moving to a more favorable environment, *chemotaxis*. Osmoregulation is a complex "two-component system" for regulating cellular response to external conditions. Chemotaxis is also a two-component sensory system, and it is controlled by more than 40 genes and sets of genes that control the flagella, transmembrane receptors, and signal transduction involved in bacterial responses to stress. Bacterial cell membranes are loaded with *porins*, proteins that allow the diffusion of important molecules in and out to maintain homeostasis and internalize nutrients.

In a two-component system a transmembrane receptor protein transduces environmental signal into metabolic changes by phosphorylating a second protein, as described in earlier general discussions of signaling processes. The rate of phosphorylation is under the control of the receptor proteins, and in chemotaxis, for example, the second protein affects the flagella to result in a change in the direction in which the bacterium swims. Ciliated protozoa use chemical sensing to find potential mates and avoid predators. Plants and fungi also use similar two-component response systems.

Quorum Sensing

Bacteria use the information they gather from the environment for *quorum sensing*, the ability that allows them to determine the density of other bacteria around them and, depending on population size, form into biofilm. This then leads to the expression of genes that aid in the survival of the biofilm, the colonization of higher organisms, such as the roots of legumes, or finding the sites of adhesion or invasion of higher organisms to initiate infection (catheters are a common site). The environmental sensing is done by transmembrane sensory proteins, often coupled with cytoplasmic receptors, a system that has been well-characterized at the molecular level. Bacterial transmembrane receptors have common structural features: an extracellular binding domain for one or more ligand, a transmembrane region, a cytoplasmic linker, and methylation and signaling regions. The receptors form homodimers and cluster together on the cell membrane.

SLIME MOLD

Dictyostelium discoideum are amoeboid cells that live a well-characterized life cycle, many stages of which are triggered by environmental cues including depletion of the food supply (see, e.g., Bonner 2000). (See Figure 12-9.) Initially, the cells live in soil, where they feed on bacteria. While food is plentiful the cells multiply by mitotic division, but as the population grows the food supply can become depleted and the cells begin to starve. The cells are constantly monitoring population density by sensing and secreting a protein called *conditioned medium factor* (*CMF*), and when population density is large enough to threaten starvation, that is, when the concentration of CMF is high enough, cells signal other cells to aggregate. It is time to search for greener pastures.

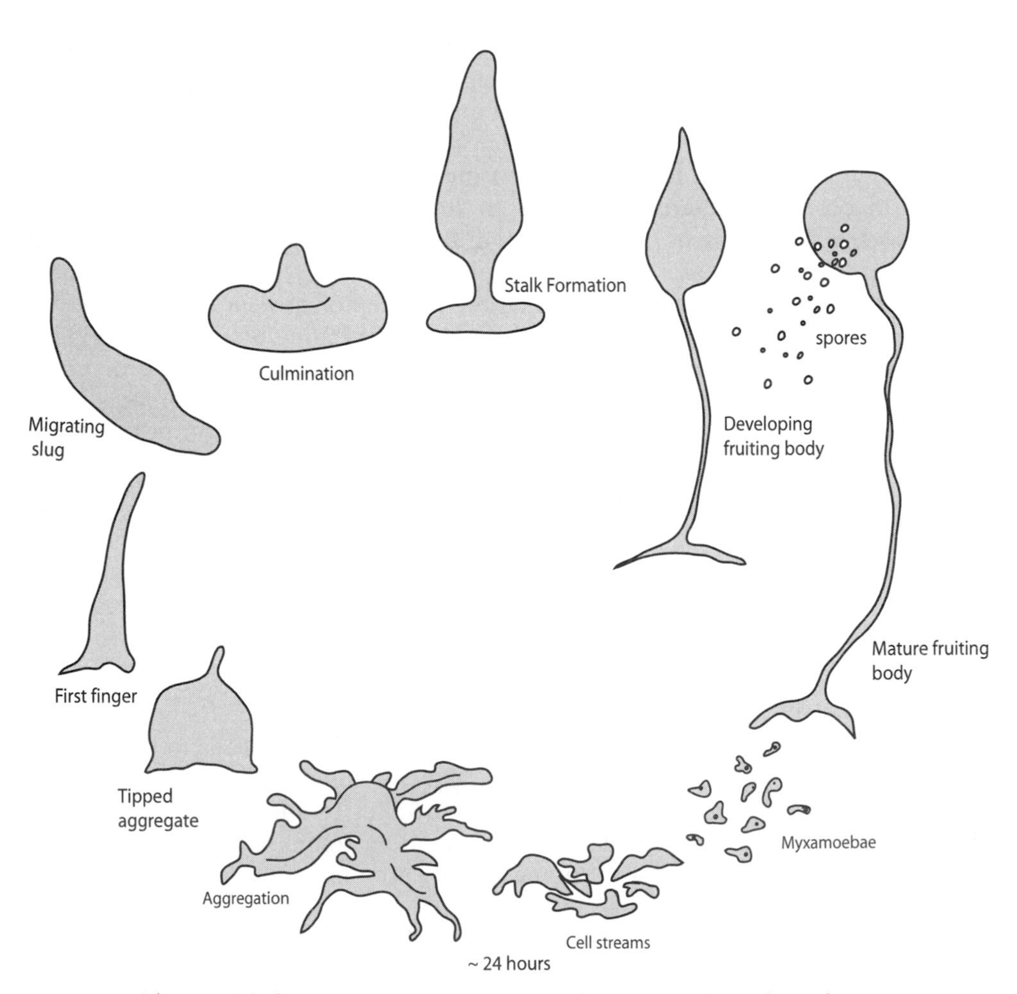

Figure 12-9. Life cycle of the slime mold *Dictyostelium discoideum*.

The signaling molecule is a substance called *acrasin*, which is cAMP. Cells adjacent to the signaling cell receive the signal via cell surface G protein-coupled receptors (GPCRs) for cAMP. The GPCRs transmit signal to the cytoplasm that initiates the pathway to the release of cAMP, which is then received by neighboring cells that, in turn, send their own signal, and so on until a colony is formed. The patterns of colony formation that result vary and are a function of the spatiotemporal pattern of signal emission. The colony is mobile, which allows the cells to migrate to a richer environment in which to produce the next generation. Within about 24 hours, the colony forms a *fruiting body* that disperses spores and begins the life cycle over again.

The colony, like a bacterial biofilm, or in fact any collection of aggregated cells, becomes more than the sum of its parts. Genes are expressed at this stage that are not expressed in independent cells. The *slug* of aggregated cells is about 1 mm long and can stay together for up to several days. The slug is polar and mobile, with its movement mediated by waves of cAMP that travel from the tip to the tail, inducing individual cells to migrate toward the tip, thus moving the slug in the direction of cAMP concentrations. The slug is extremely sensitive to environmental clues, so that small differences in light, temperature, or the concentration of ammonia gas produced by the cell mass itself will affect the direction of the slug's movement. Presumably this optimizes the location of the final fruiting body with respect to food supply. Interestingly, formation of the fruiting body depends on programmed cell death to take the slug through the stages of culmination to formation of the stalk and then to maturation, so in a very real sense this collection of single celled organisms becomes a multicellular being, with once free-standing cells sacrificing themselves for the good of the group.

A Side-Comment on Quorum Sensing, Cooperation, and Group Behavior

Quorum sensing is an interesting phenomenon, if it is being properly understood. It is one of many instances in which animals aggregate, swarm, or give display behaviors in a way that appears to reflect their population size or density and that has been interpreted as being a mutual signal among individuals that they interpret to alter their behavior. Controversy has been particularly heated over whether this kind of behavior could lead to altruistic self-sacrifice, individuals restricting their reproduction so that the group does not overpopulate relative to available resources. In a phrase, the issue is *group selection* as compared to the classic individual selection described in Chapter 3. The problem is the need to explain how alleles leading to self-sacrifice could increase in frequency. Formally, this becomes a mathematical problem for population genetics theory; informally, one can see many ways in which behavior can evolve that is good for the group so long as the sacrifice of individuals initially responsible for the behavior is not too immediate and complete.

CONCLUSION

One possible reading of the many homologies we find among the wide array of senses in all organisms alive today is that mechanoreceptors and chemoreceptors were the earliest sensory receptors, with all others being derivative, duplication of sensory modules having led to new receptors (Hodos and Butler 1997). For example, the infrared detectors of the pit organs of crotalid snakes are modified thermal detectors and electroreception has apparently evolved several times in ray-finned fishes and in sharks and monotremes (Hodos and Butler 1997). We know, however, that many of the components (or modules) of the sensory mechanisms, such as those involving transmembrane G protein-linked information transduction pathways, are used for all sorts of other processes, including developmental patterning, cellular differentiation, ion transport to maintain proper chemical conditions in the cell, and so on, as well as for signal transduction in single-celled organisms. Many others have to do with cell architecture and the moving around of materials inside the cell via the cytoskeletal tubules and molecules like myosin that use the tubules. Mechanoreception in widespread contexts and species uses these very fundamental elements of cell biology. This is also another example of at least some connections or homologies among similar systems in very dissimilar species.

Thus the general repertoire of information receipt by cells probably antedates the senses that we see as organized organ systems today. At the same time, it becomes easier to see how rudimentary systems can evolve initially and then be elaborated in diverse ways—easier than having to reinvent a suitable mechanism whenever it might do some good.

Chapter 13
Chemical Signaling and Sensation from the Outside World

The light, sound, and chemical environs of an organism are often detected at a distance from their source and, to a great extent, are varied enough that they cannot be specifically anticipated. The properties of the sensory stimuli also vary. Light is fastest, but cannot penetrate solid objects, and an object cannot be seen if the requisite energy is not available (e.g., many animals do not see well at night). Because light travels in a coherent linear way, the information it carries is precisely patterned relative to the source. Sound travels more slowly but in all directions, sometimes through solid objects, bouncing and reflecting in a less orderly way, and sounds from multiple sources mix freely. To the receiving organism, sound is less spatially patterned than a light image, but, like light, it is organized in a coherent frequency spectrum, which some organisms can detect.

Chemosensation is the ability to detect and discriminate among specific individual chemical molecules. In previous chapters we have discussed ways in which cells respond to chemical stimuli in their immediate surroundings, whether as hormones produced by neighboring cells, signaling factors that regulate gene expression, bacterial responses to chemical attractants or repellants and the like. The distinctions are clearly arbitrary and based on ways humans think of our own functional diversity, but in this chapter we focus primarily on the particular subset of chemosensory phenomena that we traditionally refer to as smell and taste, by which organisms sample their environment.

Galileo is mainly known for his physics and astronomy but in other ways he was the first modern scientist, advancing from the greats like Aristotle by the use of controlled experiments, newly invented instruments and a more critical approach to relating theory to observation. In his treatise on the philosophy of knowledge, *The Assayer* (Galilei 1622), Galileo gave a rather perceptive explanation of taste and smell: particles from the detected object wafted on air to the tongue and nose, where, after mixing with water and mucus on the tongue and nasal membranes, they were detected. The outer part of the body was assumed to be insensitive to such small

Genetics and the Logic of Evolution, by Kenneth M. Weiss and Anne V. Buchanan.
ISBN 0-471-23805-8

particles. The smell or taste of the particles depended on their concentration, the vigor of their motion, *and on their shape*. Galileo's notions of taste and smell were to some extent rooted in ancient humoral theory, but many of the basic ideas were rather prescient surmises. We could argue that our better understanding of these systems is in the detail.

Generally, *olfaction* (smell) refers to the detection of air- or water-borne chemicals that diffuse from a source to the detecting organism, whereas *gustation* (taste) and other forms of chemosensation involve direct contact with the source. Immune detection and chemosensation are molecular recognition systems, with the former generally detecting internal invaders such as living parasites and the latter detecting external factors. One might expect similar systems to be used in these two molecular detection systems, and the similarities are interesting. Genetically, however, probably the closest parallel is, perhaps surprisingly, between chemosensation and vision.

Odorants waft through the air but do not generally have a precise spatial component at their source; certainly, the mixing of odorants in the air, and their variable diffusion depending on the vagaries of air currents, mean that there is less need for a precise spatial element in odorant perception compared with that needed for light images. But as in sound perception, organisms use both direction and intensity gradients to locate olfactory sources. Organisms vary greatly in the precision with which they detect or characterize the source, as well as with which they can compensate for the effects of movement of the air.

Olfaction is used to detect many aspects of the environment, including dangers, food sources, and conspecifics. Some species also have olfactory mechanisms thought to enable them to identify specific individuals such as family or group members. Although the chemical environment can be open-ended in general, organisms may be programmed to detect some specific chemical signals, such as *pheromones* that elicit species-specific behaviors such as mating and male-male competition. Olfaction works with the endocrine system in some of these contexts, and responses may be highly ritualized, such as visual or vocal display. Recognition of pheromones requires very specific molecular detection, and both signal and receptor must coevolve in some way. Such specificity of response occurs in both vertebrates and invertebrates.

Similar to visual and sound perception, olfactory processes in the animal world share some genetic mechanisms but have evolved morphological and neurological similarities apparently independently and have evolved different ways to use similar genes for chemosensation (e.g., Strausfeld and Hildebrand 1999). Indeed, olfaction and vision have evolved from a common ancestral receptor system and rely on some similar neural mechanisms. Transcription factors, including *Pax6*, are used in the development of both systems, and the receptors are in the same family as well.

Every species is surrounded by information in its environment that it cannot or does not use. Species are able to detect what they need to detect, and vary in how they take advantage of the chemical environment. But it is worth remembering that, as with other senses or abilities, a relatively poor sense of smell generally only reflects less reliance on chemosensation, not inferior adaptation—essentially, such judgments are purely human overlays on nature.

For convenience, this chapter will discuss olfaction and gustation as separate systems. At the molecular, genetic, and neurological levels, however, these divisions are arbitrary.

OLFACTION

The ability to detect odors varies widely across the animal world. The numbers game can be misleading, but there is a general correspondence between the number of *olfactory neurons* (ONs)—cells specialized for odorant detection—that an animal has and the probability that any given odorant will come in contact with an appropriate receptor. Humans have approximately 5–10 million olfactory neurons in the *olfactory epithelium*, enabling us to detect an estimated 10,000 or more different odorants, but this is a pittance when compared with dogs, who have more than 200 million ONs. By contrast, *Drosophila* have about 1,300 ONs. For many years, the conventional wisdom was that birds do not have much of a sense of smell, but recent work suggests that the acuity of the avian sense of smell is related to the size of a bird's olfactory bulb relative to its cerebellum (Malakoff 1999). In some species, such as songbirds, it is very small, on the order of 3 percent of the size of the cerebellum, whereas in others, such as many seabirds, it is closer to 40 percent. Although olfaction in aquatic species is of water-borne chemicals, the process works in a largely similar way. It is thought that salmon, a fish that returns to its birthplace every year to spawn, may find their way home with a well-developed sense of smell and a form of imprinting (Barinaga 1999).

Mammalian odor detection begins with olfactory receptors embedded in the olfactory membrane on the roof of the nasal cavity (see Figure 13-1). The olfactory membrane is composed of three layers of cells: (1) *supporting cells*, in which (2) *olfactory receptors cells* are embedded, and (3) *basal cells*, which produce mucus and which are the source of new olfactory receptors. One end extends from the layer of supporting cells to the central nervous system (CNS) as an afferent neuron, and the other end extends to the epithelium where it forms a knob, with cilia projecting from it. This makes the olfactory epithelium the most direct connection from the outside world to the CNS. Another distinction is that, unlike most neurons, olfactory receptors are continuously replaced during life. Olfactory receptor cells are bipolar.

Odorant molecules are detected by transmembrane receptor proteins in the cilia, which are covered by mucus produced by the supporting cells. Odorants in the air or (for aquatic animals) in the water diffuse through the mucus layer to reach the receptor, although hydrophobic odorants must be transported. Among the soluble proteins found in the aqueous medium around the olfactory receptor neurons are *odorant-binding proteins* (OBPs), to be discussed below. The detection of the presence of an odorant is based on ligand-receptor binding and depends on a class of diverse receptor genes as well as the number of sensory ONs.

THE GENERALIZED VERTEBRATE OLFACTORY SYSTEM GENES

How can an organism detect (have the right ligand for) smells it could not have been preprogrammed to detect? The discovery of *odorant* (or *olfactory*) *receptor* (*OR*) genes, made first in rats in 1991 by Buck and Axel (Buck and Axel 1991), opened the way to the understanding of the molecular basis of olfaction, which has become a model system for many aspects of sensory neurobiology. *OR*s are found in the cell surface of ONs in the olfactory epithelium and are encoded by members of the 7TMR G protein-coupled receptor (*GPCR*) gene family, which also includes opsin genes for photoreception and various hormones and many other receptors. The familiar general structure of a typical olfactory receptor is shown in Figure 13-2A.

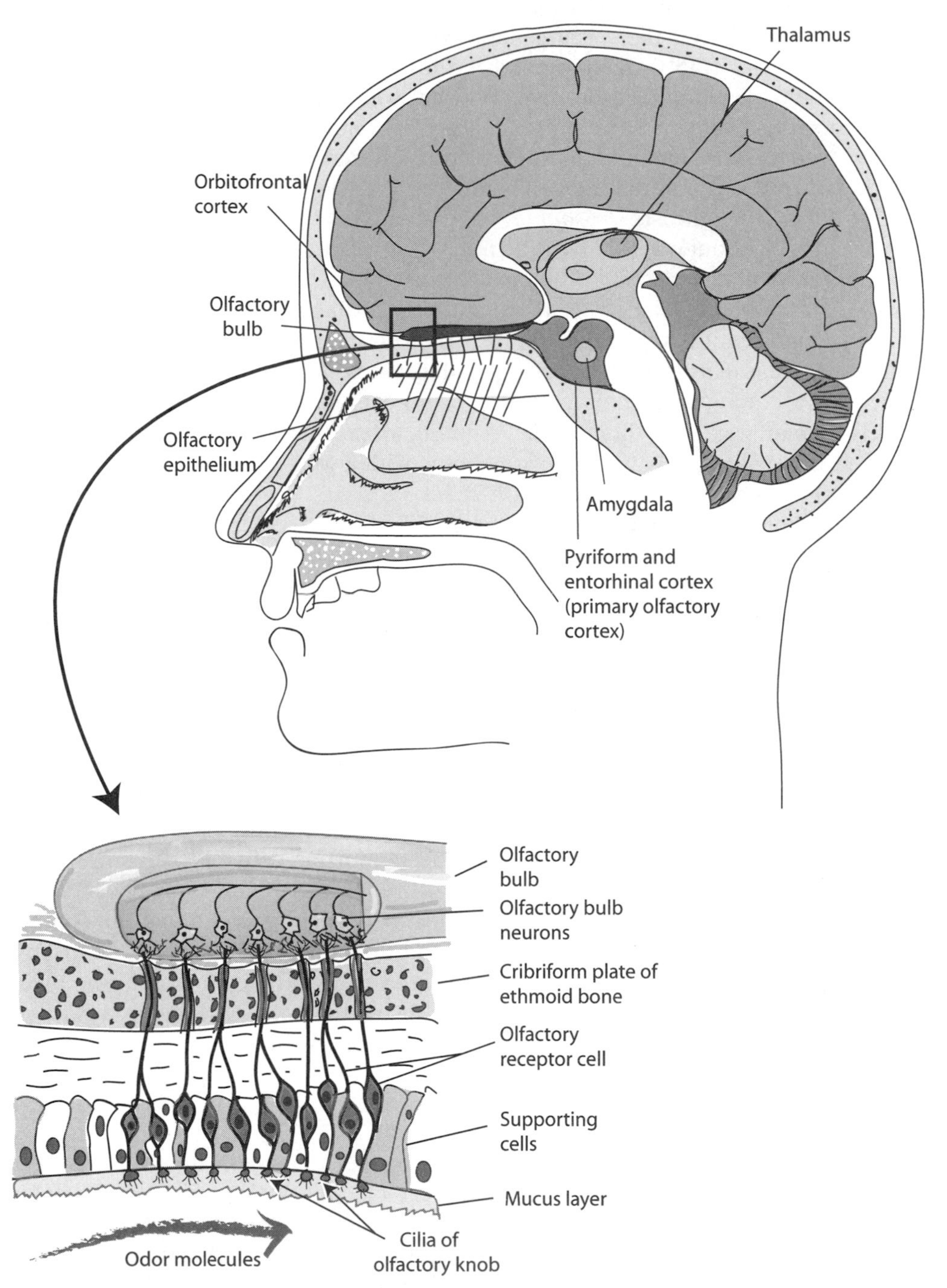

Figure 13-1. Vertebrate (human) olfactory system showing olfactory epithelium, olfactory bulb of the brain and the pathway of olfactory neurons to the olfactory regions of the brain.

OR genes are the largest subset of this gene family and comprise about 3–4 percent of the entire genomes of some species (see Crasto et al. 2003; Crasto et al. 2002; Crasto et al. 2001; Skoufos et al. 1999). Like the number of ONs, the number of *OR* genes reflects the importance of olfaction and varies greatly among species. For example, there are about 1,000 in the rat, which exceeds their number of immunoglobulin and T cell receptor genes combined (Mombaerts 2001; 2001). Among vertebrates, the number of *OR* genes varies considerably and in interesting ways that can be related to function. Some of the genes have been mutated to pseudogenes, which may relate to sloppy copying during meiosis of tandemly repeated genes—the same process that generates the useful variety of *OR*s—but may also be affected by selection. Birds typically have relatively few *OR* genes, but this varies among species. Fish have about 100 OR genes. Primates have roughly the same number of *OR* genes as other mammals (about 1000 in humans); however, among species that have been studied, about 60 percent of primate *OR* genes are pseudogenes; a typical human may have around 350 functional *OR* genes (Crasto, Singer et al. 2001; Glusman et al. 2001; Mombaerts 2001; Sosinsky et al. 2000). Interestingly, New World monkeys have few *OR* pseudogenes (Rouquier et al. 2000).

As shown below in Figure 13-4, *OR*s are distributed on almost all chromosomes in both mouse and human, and the pseudogenes have roughly similar proportional distribution. The high fraction of pseudogenes in humans thus is likely to have to do with selection or function, since the majority of mouse genes are still functional yet presumably the duplication and mutation processes are similar (but see some further thoughts at the end of this chapter).

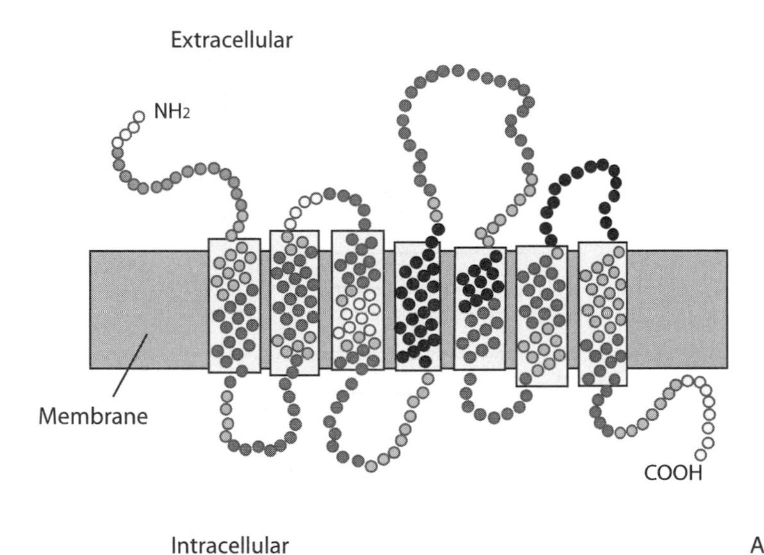

Figure 13-2. Odorant receptors and related genes. (A) Schematic of a seven transmembrane (7TMR) olfactory receptor molecule showing amino acids, specifically for receptor M71 in the mouse. The most highly conserved residues are shown in white and black, and the most variable are in shades of gray, presumably indicating the most important sites for odorant specificity. (See Firestein 2001); (B) General phylogenetic (gene sequence) relationships among major classes of chemoreceptors. The approximate numbers of genes in these classes are indicated (modified after Firestein 2001).

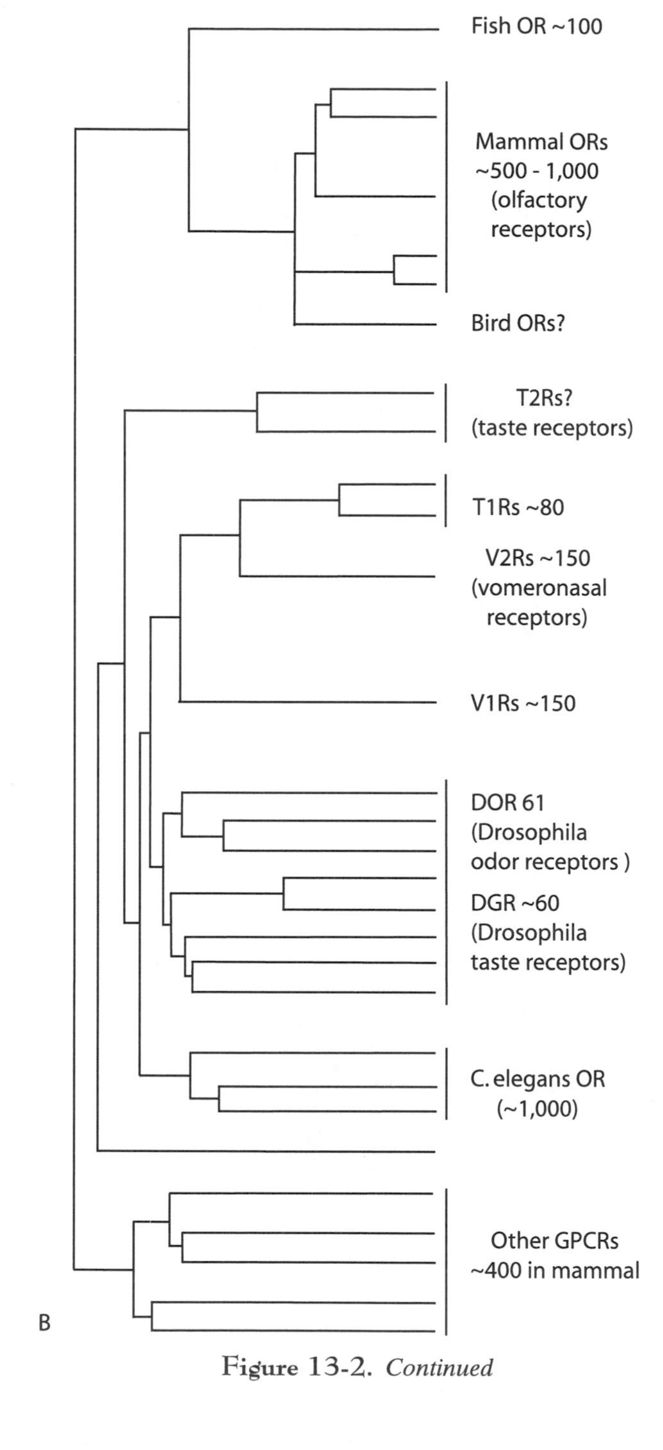

Figure 13-2. *Continued*

Vertebrate *OR*s have a roughly 1-kb intronless coding region that codes for a polypeptide with a little more than 300 amino acids. As a group, ORs can be distinguished from other seven transmembrane receptors (7TMRs) by a few conserved amino acid motifs and some conserved single residues. Seventeen hypervariable amino acid residues in transmembrane domains 3–5 are thought to define an odorant binding pocket; these domains have high diversity and little conservation among *OR* subfamilies (Mombaerts 2001). *OR* genes have variable noncoding exons and splice sites, suggesting that they are alternatively spliced (e.g., Sosinsky, Glusman et al. 2000).

The range of odorants detectable by this variety of receptors includes molecules with very different chemical structures, implying that OR ligand-binding domains must be diverse as well, and there is corresponding variability in OR structure. An OR seems to respond optimally to a particular molecular structure—for example, only one of a series of specific odorants used in experimental tests. It is often assumed that the OR has been "tuned" for this specificity by evolution (Dreyer 1998), but the use of this adjective by sensory biologists suggests an implicitly strong darwinian view of precise selection. Whether, when, or how the OR-specific response pattern was shaped by (odorant-specific?) natural selection or whether it is simply the empirical property of any given binding pocket is not clear. As a rule, odorants are not detected in nature by single receptors (a feature shared with antigen recognition by antibodies), and, as noted earlier, odorants do not have a clear or cohesive chemical "spectrum" as do light and sound.

Transduction of an olfactory signal is initiated by the binding of an odorant molecule to the receptor on the cilia of an olfactory neuron dendrite. This triggers the intracellular G protein cascade and ultimately results in the delivery of the signal to the olfactory centers in the brain where the signal is decoded. G protein linked ORs activate signal transduction in one of two ways. Olfaction-specific G proteins convert abundant ATP into cAMP to generate an action potential in the ON. Some ORs increase intracellular cAMP concentration, opening cAMP-gated cation channels and allowing an influx of sodium ions, which depolarizes the cell and initiates a nerve impulse that travels along the axon to the brain. Others activate the inositol phospholipid pathway, IP_3-gated Ca^{2+} channels, in the plasma membrane, to activate second-messenger signal transduction molecules. ONs typically produce multiple OR copies and a given cell's response may be thought of as the effect of the binding of odorants to a sufficient number of receptors on its surface over a suitably short time period, which, in aggregate, generates a threshold action potential.

ONs are the only sensory neurons whose axons connect directly to the brain (see Figure 13-3 for a schematic drawing of basic olfactory wiring to the brain). The mapping between specific odorant receptors and specific locations in the brain seems to be quite precise within an individual, as will be discussed in Chapters 15 and 16. The essential "wiring" characteristic may be that different ONs that express the same *OR* gene send axons to the same *glomerulus* (neuronal cluster), in the olfactory bulb (e.g., Kauer 2002; Mombaerts 2001; Zou et al. 2001). Replacement ONs are generated mitotically during life from cells at the base of the olfactory epithelium and express the same *OR* gene and send their axons to the same glomerulus as their predecessor, which seems to be an important means by which odor perception remains more or less constant throughout an organism's life. How ON-glomerular conservation occurs is not yet completely clear, but it appears to be ON-guided; an ON could either switch which *OR* gene it expresses or use OR-specific axonal redirection to the glomerulus appropriate for the *OR* gene it is already expressing (P. Mombaerts, personal communication).

DNA sequence phylogenies suggest that there are two main classes of vertebrate OR genes (Kratz et al. 2002; Mombaerts 2001) (see Figure 13-2B). Class I appears

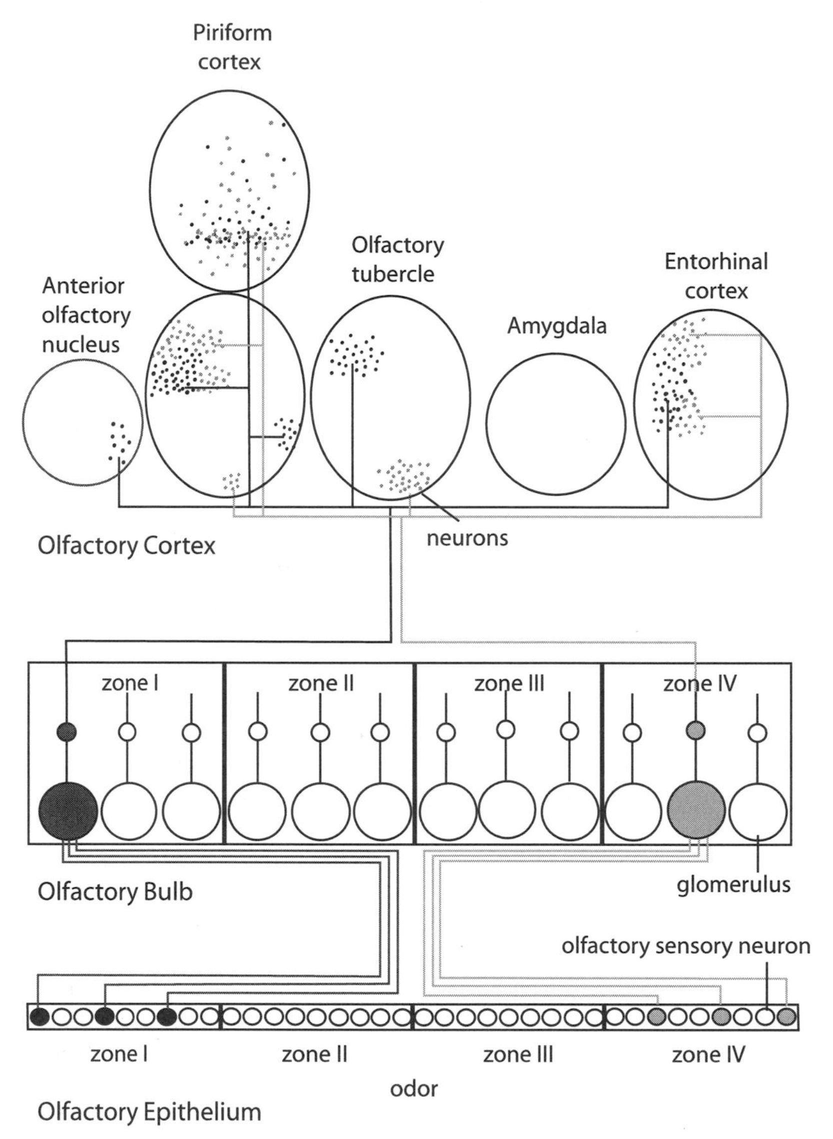

Figure 13-3. Basic olfactory wiring diagram showing pathways of two different odorant receptors. In the olfactory epithelium, sensory neurons expressing a single *OR* gene are located in the same zone. In the olfactory bulb their axons synapse with mitral cells (small circles) in the same few glomeruli (large circles). The mitral cells carrying signal of specific ORs synapse with clusters of neurons (differently shaded dots in large ovals) at stereotypical sites in olfactory areas of the cortex, creating a sensory map. Figure redrawn from (Zou, Horowitz et al. 2001) with permission. Original figure copyright 2001 by Nature Publishing.

to be the oldest and is the only group found in fish. Amphibians have class I and class II genes; at least in frogs, the class I genes are only expressed in a water-sensitive chamber and class II genes in an air-sensitive chamber. The genomic cluster arrangement of mammalian *OR*s suggests that class I genes duplicated to form the ancestor of class II genes, followed by the dispersal of duplicate *OR*s, mainly in class

II, to other chromosomes. These facts suggest that class I genes specialize for waterborne and class II for volatile odorants. However, mammals retain many class I genes scattered in their genome, and a higher fraction of human class I genes are more functional than our class II genes; thus the former seem unlikely simply to be a relic of an early aquatic life.

As with other tandemly repeated genes, *OR* genes appear to have evolved by unequal crossing over, gene conversion, and occasional chromosomal translocations or duplications, followed by subsequent additional local duplication. In mammals, genes from different subclasses are dispersed within single chromosomal clusters. There are at least two *OR* gene clusters in zebra fish, 12 in mice, and more than 25 in humans (Kratz, Dugas et al. 2002; Mombaerts 2001), in whom *OR*s are found on all chromosomes except the short chromosome 20 and the Y (Glusman, Yanai et al. 2001) (Figures 13-4 and 13-5). These genes are distributed in clusters of various sizes. Each cluster largely comprises genes from the same OR subfamily. There are two major clusters of class I *OR*s on human chromosome 11, comprising over 40 percent of our entire repertoire, pointing to the importance of this cluster in the evolution of the vertebrate *OR* genes. Another cluster is linked to the major histocompatibility complex (MHC) on chromosome 6, and it has been speculated that these *OR*s could be involved with mate identification, at least in some species like rodents (Younger et al. 2001).

There is at least some coherence in olfactory systems that may represent phylogeny as well as development. Frequent duplication, mutation, and perhaps gene conversion means that homology between pairs of *OR* genes is highly variable; however, as might be expected, the cluster organization for identifiable orthologs generally appears to be retained among mammals. *OR*s closely related in sequence are closely related in chromosomal location and regulation and with a few exceptions tend to be expressed in similar regions in the olfactory epithelium (Lane et al. 2001; Mombaerts 1999); these in turn project to localized regions in the olfactory bulb (Kratz, Dugas et al. 2002; Tsuboi et al. 1999). Thus, each zone has its own expression mechanism, and its ONs project roughly to similarly distinct regions in the olfactory bulb.

Because these epithelial regions are similar among individual animals in the same species, they may reflect temporally coordinated aspects of gene expression and embryonic neural development. Interestingly, however, little regulatory sequence sharing has been detected among paralogous genes that are expressed in the same zone of the olfactory bulb (Lane, Cutforth et al. 2001), although some potential motif sharing has been suggested (Hoppe et al. 2000; Sosinsky, Glusman et al. 2000). One problem is that recently duplicated genes may share flanking sequence, which may or may not imply that that sequence is specifically regulatory; recall that the genome is nearly saturated with potential regulatory sequences; therefore, sequence analysis alone is usually not a definitive way to identify response elements. Experiments with different *OR*s have reported differing results regarding how near the gene the relevant regulatory regions are, whether a few kilobases or hundreds (e.g., Mombaerts 2001); this may mean either that the experiments have missed something important about *OR* regulation or that that regulation varies greatly from gene to gene.

Although these spatial and chromosomal clustering aspects of organization are relevant for developmental coherence and gene expression, they may be of little relevance for odorant detection itself. The zonal organization of the olfactory

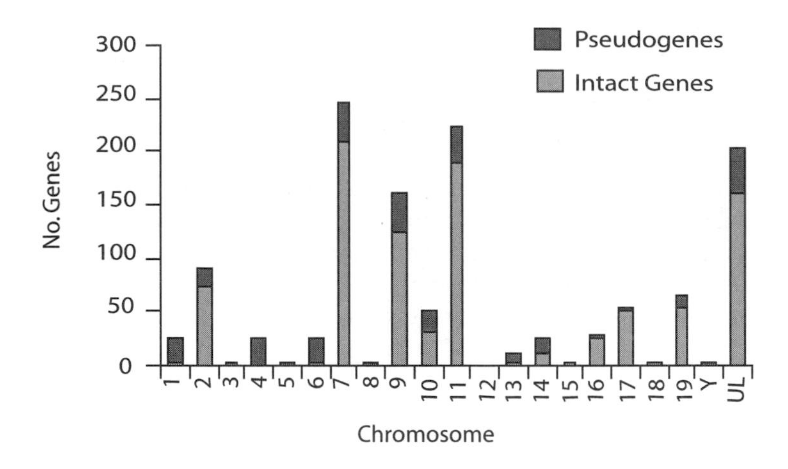

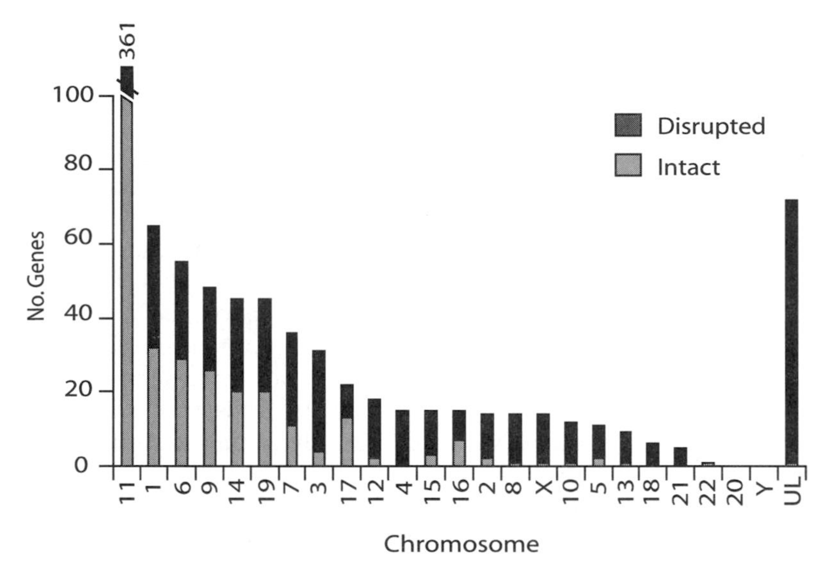

Figure 13-4. Number of functional and disrupted (pseudogene) ORs in the mouse (top), and human (bottom) genome, by chromosome. See text. Redrawn after (Zhang and Firestein 2002) and (Glusman, Yanai et al. 2001), respectively.

epithelium is widespread in general, although varying in detail, in vertebrates with very different olfactory behavior; thus, the developmental organization may relate to *OR* gene cluster and its vertebrate evolutionary history rather than clustering of ORs coding for functionally similar receptor proteins. However, an analysis of conservation of amino acid sequence motifs showed that some motifs were highly clustered within *OR* class and may have some relation to ligand-binding properties (Liu et al. 2003). While the individual motifs did not seem to these authors to be correlated with particular odorant classes or properties, some combinations of motifs did. If this interpretation bears further scrutiny, it may suggest that over evolutionary time the combinatorial shuffling of motifs among the duplicating, evolving *OR* genes is an additional means of generating olfactory diversity.

Other animals have different organizational patterns. For example, nematodes appear to have at least one chemosensory receptor that is not symmetrically expressed on left and right sides, and invertebrate receptors are located in various places, including the antennae and the separate maxillary palps (Dreyer 1998).

The regulation of *OR* expression is quite interesting, even though we do not yet understand its details. A given *OR* is expressed mainly in a single region of the olfactory epithelium. As a rule, a given vertebrate ON expresses only one *OR* gene and, at that, only one of the two alleles of that gene in the diploid individual. Similar to X-inactivation and antibody/TCR allelic exclusion, there is allelic exclusion in olfactory gene expression. Because *OR* genes are on many chromosomes (as well as the homolog of each cluster in the diploid cell) (see Figure 13-5), there must be a form of *trans-OR* expression control. Recent experiments suggest that chromosomal

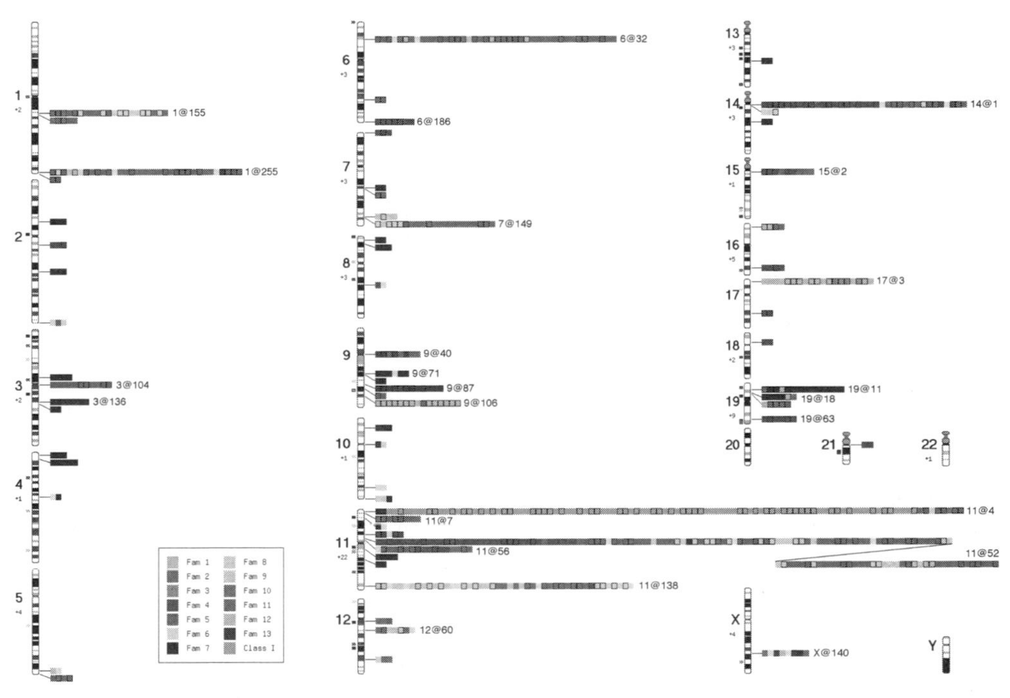

Figure 13-5. Genomic (chromosomal) distribution of *OR* classes in the human genome. This is a gray-scale version of a color original figure. Generally, the darkness shades correspond to *OR* gene subfamilies. The outlined squares are functional *OR*s (i.e., not pseudogenes). Genes shown to the left of each chromosome are singletons (isolated, not in a cluster), genes to the right are in clusters of two or more; relative order is approximate, and the large cluster on chromosome 11 is split for convenience. Even in gray-scale, the clustering of related genes and their distribution across the genome can be seen. The numbers by clusters (e.g., 3@136) refer to the chromosome and number of genes in the cluster. (Modified and reprinted with permission from Glusman, Yanai et al. 2001, and see this for details and color resolution).

exclusion may occur via asynchronous replication during mitosis, resembling a possible element of X-inactivation (Singh et al. 2003) discussed earlier. In principle, this could account for the expression of *OR* genes from only one (maternal or paternal) chromosome but does not account for the additional exclusion of expression of *OR* genes from all but the one cluster—on that same or any other chromosome. The cluster-related regionalization of expression suggests that perhaps *OR* clusters are made available for transcription (that is, the chromatin opened, and so forth) in a developmentally orderly way as the epithelium is patterned. This would reduce the exclusion problem to that of genes within the cluster(s) that are open in cells in a given epithelial region.

Within a cluster there must be some form of *cis*-exclusion. Other genes including globins and color-vision *opsins* have *cis-based* exclusion, in which only one gene within a cluster is expressed at a given time or in a given cell. Control of this type can involve enhancer competition among the genes in a cluster on the same chromosome, but *trans*-regulatory factors may also be involved (Kratz, Dugas et al. 2002; Serizawa et al. 2000). The latter possibility is suggested because single-*OR* gene expression also occurs for transgenes experimentally inserted in various places in the genome (i.e., not just in *OR* clusters). The local chromosome seems to determine the exclusion of its homolog (Singh, Ebrahimi et al. 2003), but, again, some more general *trans* mechanism(s) than asynchronous replication—perhaps the concentration of some limiting factor(s)—must exist in a cell to suppress all other *OR* expression once a first, single *OR* has been expressed anywhere in the genome. As always, there are exceptions: there is evidence that at least some rat ONs express specific pairs of *OR*s (Rawson et al. 2000) rather than only a single gene. Therefore, the system must be escapable (perhaps because the multiply expressed *OR* genes in the rat have lost their repression-control sequences).

THE VOMERONASAL ORGAN: PHEROMONES

We have described how unpredictable olfactory signals are detected. However, as we know, there are many important, indeed vital, ways in which organisms are preprogrammed for *specific* chemical signals; this is the kind of specificity found in internal communications via hormones and with developmental and regulatory signaling. How is specific *inter*organismal chemical communication achieved? In fact, many vertebrates communicate via interorganismal "hormone" systems. No new processes or mechanisms have been needed.

The nasal chemosensory organs of tetrapod vertebrates are divided into the olfactory system and the *vomeronasal* system (Johnston and Peng 2000; Wysocki 1979). The *vomeronasal organ* (VNO) is present in amphibians and reptiles which suggests that it probably emerged in early tetrapod lineages. The VNO is thought to be primarily a sensory organ for the detection of pheromones, produced in bodily secretions such as sweat, urine, and vaginal fluids, that induce stereotypical behavior (Holy et al. 2000; Keverne 1999). For example, the VNO has been shown experimentally to be vital for mating recognition and male-male aggression in mice (Stowers et al. 2002). However, at least in some animals, such as snakes, the VNO also has a role in other functions such as the detection of prey. The VNO is well developed in reptiles but less so in many mammals and has been reduced or lost in some lineages such as old-world monkeys, probably apes, and some lizards. The nature of the human VNO is still unclear. As fetuses, humans have a VNO with apparent neural connections to the olfactory bulb (Døving and Trotier 1998). Some authors suggest that there is a depression just inside the nasal opening in adults where the requisite epithelial cells can be found, but others suggest that the whole system degenerates soon after birth (Meredith 2001). In fact, evidence for an active human pheromone system is essentially nonexistent (see below).

The VNO in nonhuman vertebrates is located in a pouch off the nasal cavity, on both sides of the nasal septum dividing the nose into its right and left halves. The VNO is fluid-filled and typically not directly open to the air as is the olfactory epithelium. Pheromones must reach the VNO receptors in a way that is different from how odors reach olfactory receptor cells. Various pumplike mechanisms draw the molecules into contact with the VNO receptors. The pump can be activated by the autonomic nervous system, perhaps triggered by conventional olfactory cues (e.g., Keverne 2002). When snakes and other reptiles flick their tongues, for example, they draw in molecules from the air. When the tongue is pulled back into the mouth after each flick, the molecules pass over the duct openings that lead to the VNO, where they are detected by pheromone receptors.

Like the olfactory epithelium, the VNO employs members of the *7TMR* family. Sequence analysis has identified two *vomeronasal receptor* (*VR*) classes of 100 or more genes each, with a phylogeny separate from that of the *OR* genes. Each *VR* class uses (or at least is coincidentally expressed with) different message transduction G protein types (Bargmann 1997; Firestein 2001; Herrada and Dulac 1997; Matsunami and Buck 1997). From studies of rodents, the roughly 150 *V1R* class genes are found to be phylogenetically closer to *OR* genes, whereas the similarly large class of *V2R*s are found to be closer to a different class of *7TMRs* (Figure 13-2B, from Firestein 2001), the metabotropic glutamate receptors, the most ubiquitous neurotransmitters in the CNS and a distinct class of receptors that when activated affects internal neuronal conditions, as distinct from the direct ion-channel ionotropic receptors. The class difference has to do with the length of the extracellular N-terminus of the VR protein, that may have to do with ligand-binding (Firestein 2001). The genes in each *VR* class are expressed in a discrete part, either in the apical or basal region, of the VNO.

The VNO and olfactory epithelium appear to have a common evolutionary origin, but there are many differences between vomeronasal and olfactory neurons. The two are innervated separately. *VR*s are structurally somewhat different and share little sequence homology with *OR*s (Matsunami and Buck 1997). A vomeronasal neuron (VNN) may express multiple receptors, which converge only imprecisely to their respective region in the accessory olfactory bulb of the brain, and a given VNN sends axons to multiple glomeruli. Interestingly, although the VNO is intrinsically involved in sexual behavior, males and females express the same receptors (Holy, Dulac et al. 2000).

Because of their use in inducing stereotypical behavior, one would expect a high degree of preprogrammed sensitivity in VRs; evidence presently suggests that VRs do have more sharply restricted binding properties than ORs (e.g., Leinders-Zufall et al. 2000). Consistent with this is that specific parts of rodent olfactory bulbs are activated by mating pheromones (e.g., Keverne 1999). One might predict that at least some *VR*s will not be polymorphic because polymorphism might lead a fraction of members of a population to be unable to detect the appropriate signal. We might expect that to be at least one good example of variation easily purged by natural selection. However 200+ genes in two separate receptor groups can generate much variation in detection, and one can ask why such diversity is required if pheromones are so specific.

Important aspects of behavior in some species involve individual recognition. The highly polymorphic MHC genes play some role in individual recognition (Penn and Potts 1999). The MHC is involved in the preference of mice for mates genetically dissimilar to themselves, and there is evidence for similar mate-choice effects in other vertebrates; however, the evidence is not entirely consistent and some behaviors in some species or contexts suggest preference for MHC-similar individuals. One problem is the difficulty in evaluating behavior accurately compared with the relative ease of identifying genotypes in mates and offspring. Distinguishing among an animal's two parents by MHC mediated odors may affect mating and other behaviors. Some recent evidence suggests that, rather than simply learning by experience, this discrimination is made possible by specific genetic mechanisms that, for example, compare an organism's own alleles with those of another member of its species.

Mouse major urinary proteins (MUPs), which release small volatile pheromones, have been shown experimentally to mediate individual recognition (Hurst et al. 2001; Keverne 2002). MUPs secreted by male mice have strain specificity and pregnancy-blocking activity. MUPs also have other functions. They are produced by a diverse family of polymorphic genes with multiple tissue expression patterns, largely clustered on one chromosome in the mouse (Cavaggioni and Mucignat-Caretta 2000). The relative roles of MUPs and the MHC in individual recognition are not yet clear or consistently understood (e.g., Brennan 2001).

Adaptive evolutionary reasons have been suggested for the importance of genetically determined individual recognition. There may be competition as well as cooperation between mother and fetus in placental mammals. There are various reasons why mating behavior that favors the generation of diversity in offspring may have been selected for (e.g., Penn and Potts 1999). Heterozygosity in the immune-related MHC can improve the odds of resistance to rapidly evolving infectious organisms, and heterozygosity can protect against negative consequences of inbreeding. The allocation of cooperative behavior may also require individual identification, to distinguish relatives from nonrelatives. How important this may be in regard to natural selection is unclear, especially because many natural populations consist mainly of relatives anyway. But at least, as with most examples in this book, the mechanism to serve such functions is part of the normal repertoire of genetic mechanisms.

There is some indirect evidence for pheromonal action in humans, such as the oft-reported synchronization of menstrual cycles in women who live together (e.g., at school). However, the known human *VR* genes all appear to be pseudogenes, the corresponding accessory olfactory bulb seems to be absent, and no relevant gene expression has been detected in our VNO (Giorgi et al. 2000; Keverne 1999; 2002; Meredith 2001). Further, an ion channel gene, *Trpc2*, expressed only in VNO and thought to be required for VNO function (Liman et al. 1999) is a pseudogene in Old World monkeys and apes (including humans) (Liman and Innan 2003).

Most of the MHC evidence is equivocal in humans and has mainly been based on observed vs. expected genotype frequencies in mates or between parent and offspring. These data have generally suggested MHC-dissimilar mating preference (e.g., a reduction in homozygosity relative to expectation from Hardy-Weinberg equilibrium). Several T-shirt smelling tests have shown that humans discriminate among MHC genotypes, although sometimes preferring those like themselves and sometimes preferring those unlike themselves (e.g., Penn and Potts 1999; Potts 2002) or perhaps preferring types resembling their fathers (Jacob et al. 2002). At present, this kind of evidence is rather on the quaint side and has not been rigorously interpreted. We can, however, draw a basic conclusion: we humans seem to find our sexual way by more facultative and diverse mechanisms.

Odorant-Binding Proteins

A part of the olfactory process is getting the odorant to the detection system. A number of *odorant binding proteins* (OBPs) are found in the moist vertebrate olfactory mucosa, where they appear to bind volatile hydrophobic compounds to make them available to the ORs embedded in the hydrophobic mucosa (Finger and Simon 2000). However, the interpretation of OBPs is a complex story.

OBP genes are in the *Lipocalin* gene family (Akerstrom et al. 2000; Paine and Flower 2000; Tegoni et al. 2000). They are not as diverse as *OR*s and may be more directly relevant to the VNO than the olfactory epithelium. Lipocalins are among the MUPs that stimulate the VNOs and act as pheromones (e.g., Cavaggioni and Mucignat-Caretta 2000), but their complex expression patterns and functions go well beyond such signaling. Although lipocalins are expressed in the vertebrate nasal cavity and can bind volatile compounds, this and other lipocalin functions exist elsewhere in nature: the genes are found in invertebrates, plants, and bacteria, and there is as yet no specific sequence-based evidence that a subclass has specifically evolved to serve as OBPs (Tegoni, Pelosi et al. 2000).

Invertebrate Chemosensation

Presently, only limited data are available on chemosensation in our invertebrate relatives, and it is likely that a number of different systems remain to be identified. In this regard, invertebrates are more diverse than vertebrates, and, unlike their restriction to vertebrate respiratory intake sites, invertebrate chemosensory organs are found in various structures in different parts of the body. For example, chemosensation occurs in the *osphradium*, a sensory epithelium associated with the respiratory apparatus, in some mollusks. Nematodes rely on chemosensation as a primary sensory system, with two chemo-thermo-sensing *amphid* organs in the head. *C. elegans* has 11 pairs of individually identified chemosensory neurons, about 7 percent of their total 302 neurons; each is sensitive to different types of molecules, including at least one pheromone (Troemel 1999).

Chemosensation occurs in *sensilla* on the antennae and other loci of insects and crustaceans. Sensilla are hairlike or peglike clusters of receptor cells covered in chitinous cuticle, with dendrites that extend inside a sheath and hairshaft. They can protrude from the surface of the cuticle or they may be embedded in it.

The shape and type of olfactory sensilla varies by species, but all have a number of pores or slits in the walls of the hair that allow odorants to pass through. Inside the fluid-filled lumen of the sensilla are housed one or more bipolar olfactory receptor neurons, with an axon that extends to the CNS and a dendrite that reaches upward through the hair. The odorant passes through the pores in the wall to reach the dendrite. The number of ONs varies by species, ranging from one to 50, most commonly two to six. Like those in vertebrates, invertebrate ONs are primary receptors, each sending an axon into the CNS. They connect via glomeruli in the antennal lobes. As in fish and rodents, these can be specific, for example, for sex pheromones (see Strausfeld and Hildebrand 1999).

Also like vertebrates, and not unexpectedly, invertebrate *OR*s are in the *7TMR* gene family. Insect (*Drosophila*) *OR* genes (*DOR*s) number around 50–100, and sequence relationships show that they form a family of their own (Figure 13-2B) (Clyne et al. 1999; Vosshall 2000; 2001; Vosshall et al. 1999). About 60 of these genes are used in adults and others in larvae (Firestein 2001). Unlike vertebrates, however, insect *OR* coding regions are interrupted by introns and have no specific homology to the *OR* subclasses in their vertebrate or nematode olfactory counterparts (Strausfeld and Hildebrand 1999; Vosshall, Amrein et al. 1999).

Current evidence suggests that each *OR* is expressed in a nonoverlapping set of neurons; that is, neurons express one or distinct subsets of a few members of the *OR* repertoire. This is bilaterally symmetric and conserved among flies (of the same species tested); the pattern appears to be more stereotypical than the vertebrate pattern in which, although restricted to a general region of the olfactory epithelium, receptors are more or less randomly expressed among the ONs. The pattern is shared among individuals. As in vertebrates, however, neurons expressing similar *OR* genes are wired together through the same glomerulus. The location of olfactory and gustatory receptors in a stereotypical insect is shown in Figure 13-6.

Insects also appear to have a diversity of odorant binding proteins to transport odorants within the chemosensory sensilla on their antenna, and these include pheromone binding proteins. However, insect OBPs appear to have no homology to the lipocalins used as vertebrate OBPs (Galindo and Smith 2001; Robertson et al. 1999). As with lipocalins in vertebrates, it may be that insect OBPs function to protect against sensory overload or to speed up sensory recovery by removing odorants from the organism.

Nematodes have about 500 active *OR* genes plus many pseudogenes, but it is not clear whether all the "active" genes are actually used for chemosensation. Nematode *OR*s are *7TMRs* unrelated to the *OR* subfamilies found in other species. The presence of so many genes and pseudogenes suggests that sloppy meiosis generates diversity, and one can predict a certain amount of point-mutational variation as well. Unlike insects and vertebrates, each nematode chemosensory neuron expresses multiple (15–25) receptor genes rather randomly distributed among the subclasses. These elicit two major responses, either attraction or repulsion, a restricted discriminatory power relative to the other systems that we have discussed. This innervation pattern raises questions about how discriminating their perception is (Troemel 1999); obviously, it's good enough.

Perhaps rather surprising to our vertebrate perspective, insects can be trained by classical punishment-reward experiments to respond positively or negatively to odors (e.g., see Waddell and Quinn 2001). Mutations have been identified that affect the ability of flies to respond to odorants, as evaluated by training experiments, and this fact has enabled some of the neural signaling mechanisms involved in olfaction to be understood. For instance, training involves interpreting coincident stimuli, and the detection of temporal signal pairing appears to occur in structures called *mushroom bodies* in the fly brain. Perhaps more important than neurological details, these studies have shown clearly that "flies are not automata. Their tiny brains are capable of much more than hard-wired reactions" (Waddell and Quinn 2001). This is a lesson for our human-centric world that conceives of thought or mind or information

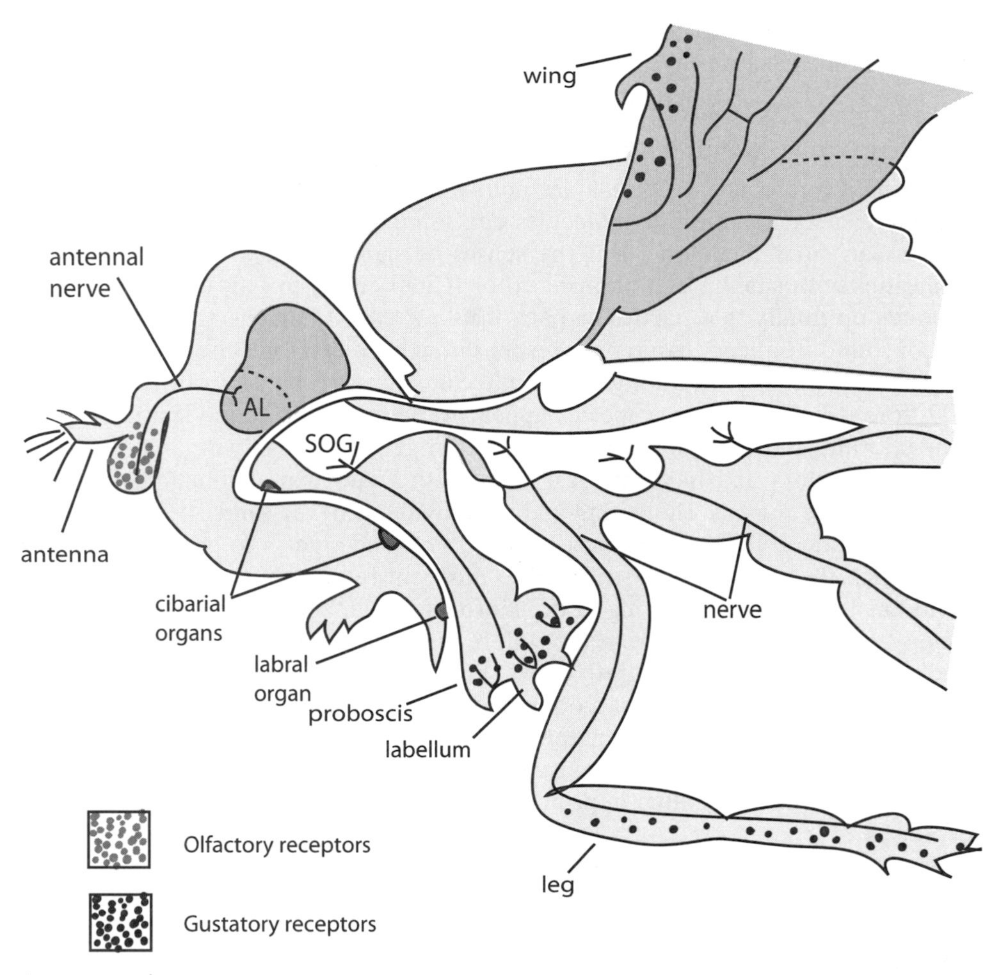

Figure 13-6. Olfactory and gustatory receptors are located on many parts of an insect's body, here showing *Drosophila.* Redrawn from (Stocker 1994), with permission. The SOG (*subesophageal ganglion*) and AL (*antennal lobe*) are parts of the insect brain that receive signals from ORs and GRs. *Labral* and *cibarial organs* are sense organs in the mouth, each with chemosensory receptors that project to the CNS.

processing in human terms, that we generally associate with our experience with consciousness.

Olfaction appears to be another example of the evolution of systems that involve shared genetic mechanisms used in independently evolved organs, as also seen in mechanoreception mechanisms in hearing, and will be seen for vision. The location and morphology of invertebrate olfactory organs is highly variable, but there are shared neural similarities. It will be interesting to see the degree to which the signaling mechanisms that induce the development of chemosensory structures are shared. Yet, although chemosensation involving the basic *7TMRs* probably existed very early in metazoan life, the particulars do not seem to have been shared since these diverse animal groups' common ancestor. For example, crustaceans don't have olfactory glomeruli or the characteristic mushroom bodies. Instead, the ORs of decapod crustaceans are found in *antennules*, appendages on the heads of these arthropods, and their axons terminate in the *antennular lobes* of the brain, suggesting that these structures evolved multiple times (Strausfeld and Hildebrand 1999).

HOW DOES IT WORK?

Unlike light and sound, odorants are not sampled from a continuous or coherent frequency spectrum. Similar molecules can "smell" different, and different compounds can smell the same. But the senses of sight and hearing do have some similarities with smell. Each photoreceptor (Chapter 14) or hair cell (Chapter 12) responds optimally to a particular part of its corresponding spectrum, but a given light or sound frequency can trigger responses in more than one photoreceptor gene or hair cell. Similarly, an odorant molecule can trigger response from various ORs.

Different individuals in a species appear to detect and respond similarly to the same odorant, at least to some extent. Receptor genes are highly organized in chromosomal clusters and have at least some coordinate expression in segmentally arranged tissue regions. The wiring among individuals of the same species is similar but not identical. Different classes of receptors send signals to similar glomeruli, from which the signals are distributed to different regions of the brain, still maintaining at least some clustering (Zou, Horowitz et al. 2001). What the brain then does is *compare* the signal from the different classes of receptors. The set of ONs sending signals can be integrated in a kind of binary algebra, resulting in an "address" or signature of a given odor, and signature can be remembered.

This description does not enlighten in regard to how a given odorant may control behavior, however. Behavior response is somewhat clearer with pheromones, which bind to preset receptors; thus, in principle, they can have developmentally prewired responses.

Vertebrate and invertebrate olfaction are different in many ways, but similarities do exist. Both groups of animals actively sample their olfactory environment: vertebrates repeatedly sniff to sample ambient odors, whereas insects may flick their olfactory appendages (e.g., Laurent 1999). Both groups also track the source of an odor in searching for mates or food from which a signal emanates. Regardless of how the information is processed cognitively, as an animal moves it compares the changing signal strength (and right-left differences). Thus, an insect may fly in and out of a signal plume, adjusting its direction as the signal changes. Olfactory pursuit requires that a signal be cleared and the system refreshed quickly enough to detect small changes in intensity or direction (e.g., Laurent 1999; MacLeod et al. 1998).

Molecules of a given odorant may have low concentration or spatial density. By chance, any given OR might not be "hit" by a ligand molecule in a given sniff or might not be hit from sniff to sniff; the aroma would seem to the animal to come and go. This could be a problem if each odorant could only be detected by a single ON or if all the ONs expressing a given receptor were tightly arranged in the olfactory epithelia. Instead, like-expressing ONs are multiple and scattered at least around a region of the olfactory epithelium. In this way, approximately the same *number* of OR-specific hits may occur from sniff to sniff, yielding roughly the same signal strength; of course, the brain then must be able to do its sums. The specificity of an odorant can be perceived because ONs expressing the same OR send their signal to similar glomeruli, where a sufficient integration of signal strength that does not depend on a single hit can occur; but again, the brain must be able to do its bookkeeping.

These comments of course relate to detecting a signal and not to its cognitive perception as an "odor" with a behavioral message. That is a separate topic, which, although important, is still very poorly understood (but we review it briefly in Chapter 16).

GUSTATION

Taste is the detection of soluble chemicals to elicit feeding and perhaps other behaviors. Gustatory receptors (GRs or, alternatively, TRs for "taste receptors") are found in the epithelium of vertebrate body parts used for the ingestion of food: lips, oral cavity, tongue, and pharynx (see Figure 13-7 for location of GRs on the mammalian tongue). In invertebrates, they are located in various appendages of the head (especially the proboscis), where they would be expected, but also in the wings, feet, and genitalia (Scott et al. 2001) (see Figure 13-6); insects can thus explore potential food sources in more varied ways than vertebrates. Insect larvae also make use of GRs. Aquatic animals, both vertebrate and invertebrate, can have chemoreceptors on their body surfaces, which are used for locating food. These chemoreceptors are the basis of a sense that is only somewhat analogous to the sense of taste, however, and are not structurally the same as "true" taste receptors. Of course, we cannot really guess what the sea "tastes" like to a fish in this respect.

Reflecting the common evolutionary origins of chemosensation, *GR* genes are *7TMR*s, but they again form a subfamily different from the *OR*s. *GR*s have introns. Mammals have 50–100 *GR* genes in at least two subfamilies, far fewer than their number of *OR* genes. The *T2R* family of genes are used for bitter taste reception and may have distant homology to the *V1R* family of rodent VRs; putative sweet or other taste receptors in the *T1R* family have homology to the *V2R*s (Mombaerts 2001). A given taste receptor cell expresses multiple, related GRs (Mombaerts 2001).

There are about 50 *GR* genes in flies, roughly corresponding to their *OR* repertoire, suggesting that soluble and volatile chemosensation are of comparable importance to insects. Nematodes (*C. elegans*) use chemoreceptors in a way that resembles taste in vertebrates, where, as noted earlier, multiple receptors are expressed in a few sensory neurons (Firestein 2001).

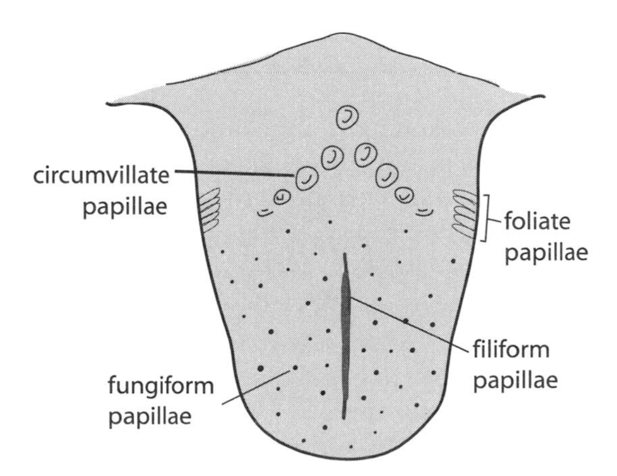

Figure 13-7. Location of gustatory receptors on a mammalian tongue; four major classes.

The taste mechanism in vertebrates is patterned on the tongue and develops separately from the olfactory and vomeronasal systems. Taste receptor cells have voltage-gated ion channels, which in vertebrates synapse into three of the primary cranial nerves (facial, glosopharyngeal, and vagus); remarkably, this is true even of distal taste buds on the body surface of fish (Finger 1997). Taste neurons are secondary receptors, which is in contrast to olfactory neurons, which are primary receptors; that is, taste receptors have no axon but synapse with sensory nerves that go to the brain. Different taste regions use different molecular reception mechanisms because of the chemical properties of the molecules being detected. Salt and acids can pass directly through ion channels, whereas sugars and bitter substances must activate a second messenger to be detected. In insects, the neurons are bipolar with a distal process that extends to the surface of the epithelium, usually ending at an opening in the cuticle or exoskeleton, which chemicals can penetrate, and a central process extending into the CNS (Finger and Simon 2000).

Gustation and olfaction are related in interesting ways, not least of course being their use of related chemoreceptor genes. The organization of taste receptors resembles that of the olfactory epithelium, but this is due to developmentally different patterning events. The classic senses of salt, sweet, sour, and bitter (and umami, the taste elicited by glutamate, found naturally in some foods and added to others as monosodium glutamate, or MSG) are detected by receptors for specific chemicals, and the long-held idea (going back to the classics) that these are located exclusively in different parts of the tongue has been shown to be incorrect. The regional differentiation of the olfactory epithelium and VNO is not highly correlated (if at all) with particular odorant properties; similar ORs are located similarly, but odorant perception is not obviously regionalized in the nose. "Taste" as we usually refer to it, really is an integral use of taste and smell receptors. However, taste receptors map to distinct regions of the vertebrate brain even though the senses lead to related and integrated percepts. A separation of smell and taste in the brain occurs also in insects, but the distinction is less clear and at least some insect *GR* genes also function in olfactory circuits.

OTHER VERTEBRATE CHEMOSENSORY MECHANISMS

Chemosensation is a very general phenomenon, and two mechanisms unrelated to immune detection and olfaction merit mention (e.g., Finger 1997). One is the human "common chemical sense," which is a property of free somatosensory nerve endings in epithelia such as on the exposed surfaces of eyes, nose, mouth, and throat. These respond to substances such as ammonia, mint, and pepper and provide sensations of burning and coolness, often leading to avoidance reactions. A second and perhaps related system involves the solitary chemoreceptor cell (SCC). Cells of this type are secondary sensory neurons, as in taste cells, and are located on external surfaces of nonamniotic aquatic craniates and used in feeding and predator avoidance. The SCC may be related evolutionarily to taste but is connected to the somatosensory system.

CHEMOSENSATION IN PLANTS AND SINGLE CELLED ORGANISMS

Although we concentrate in this chapter on olfaction and gestation in multicellular animals, other organisms detect external chemical signals in other ways that we can briefly mention (again acknowledging that our singling out of animal taste and smell as separate senses is somewhat arbitrary and artificial relative to chemosensation in the world as a whole). Plants do not have sensory systems equivalent to taste or smell but they do respond to external chemical signals including by differential growth corresponding to chemical gradients in the environment, ripening induced by ethylene, responses to herbicides, and the like.

Single-celled organisms respond to chemicals as well. As mentioned in Chapter 12, planktonic *Dictyostelium discoideum* induce aggregation among their peers by emitting a chemical signal and as aggregates they respond to environmental chemicals, particularly ammonia, in various ways. *E. coli* and *Salmonella*, and many other single celled organisms, navigate toward and away from chemical attractants or repellants, for example, toward nutrients or away from toxins.

SOME EVOLUTIONARY THOUGHTS

Chemosensory evolution presents various challenges. For example, there is a potential conflict of interest between mechanisms that preserve specificity and those that generate diversity. The immune system provides interesting contrasts and similarities.

Chemosensory receptor clusters have a history of frequent gene duplication, with the subsequent accumulation of diversity between *OR* genes. That is typical of tandemly repeated clusters of related genes, including the immunoglobulin and MHC clusters. Indeed, the pressure for olfactory diversity is shown by the fact that the *OR* family is even larger. The similarity between *OR* and immune allelic exclusion provides a kind of perceptive specificity, within diversity, sequestered in specific cells to keep things orderly and coherent. This tempts the speculation that *OR*s are also rearranged somatically during development of the olfactory epithelium (e.g., Mombaerts 1999). However, if this is going on it has not yet been demonstrated.

A partial similarity between immune and chemodetection is that both systems bind their target molecules combinatorially. Many antibody molecules may bind the same or different haptens of a circulating antigen, and a given odorant will typically be a ligand for several different ORs. Unlike the vertebrate somatic recombination-generated diversity, however, the strategy for detection of an odorant appears to involve a combinatorial process. The brain senses a signal from a particular set of simultaneous signals from multiple, presumably replicable and inherited, ORs. Once an antibody molecule has been generated, the combinatorial aspect of somatic rearrangement is over for that particular cell; combinatorial immune attacks involve multiple cells each with rearranged genomes, but there needs to be no centralized accounting of which cells are at work. By contrast, such accounting is vital for organismal response to chemosensation.

Another difference is that in mammalian immunity a response gradually becomes "focused" by selecting for the preferential amplification of cells producing the best among the diversity of antibodies that recognize an intruding molecule. Immune focusing can continually occur, so as to track mutational changes in the pathogen. To some extent, each generation of vertebrates faces a different diversity of pathogenic organisms. The olfactory environment may change from moment to moment just as the immune environment can, but changes occur much less rapidly and unpredictably across generations. Although immune recognition sharpens, olfactory perception becomes dulled with prolonged exposure; the evidence to date doesn't seem to suggest that there is olfactory focusing. A far as we know, however, our immune system does not require any form of cognitive integration—so what ensures that different individuals, who smell the same thing in combinatorally different ways, will react appropriately?

We should remember that somatic recombination is not a requisite for effective immune resistance, and in fact even long-lived plants do perfectly well with their olfaction-like *R*-gene system (which uses genomic combinatorial rather than somatically recombinatorially generated diversity). The large number of *OR* genes and their intergenic diversity has typically been viewed (as in the *R*-gene system in plants) as having been selected to generate high amounts of odorant binding diversity. The general lack of pseudogenes has reinforced this idea of stringent natural selection for odorant-specific *OR* genes. Plants do appear to have pathogen-specific *R* genes. There seems to be a consistent pattern of a large number of ORs, with few pseudogenes, in species with high reliance on olfaction. The many *OR* genes in dogs and few if any in dolphins are good examples, but our own olfactory epithelium is more patchy than continuous, and we have only around 350 functional *OR* genes, with the rest of our *OR* genes being pseudogenes.

There is a high degree of polymorphism *within* human OR genes (see Mombaerts 2001), which might seem consistent with this evolutionary story, for example, if there has been relaxed selection in our ancestry, which is the usual inference. One upshot of this level of variation is that an *OR* may be "pseudo" in some people and functional in others. Among other changes, many human *OR* pseudogenes have had stop codons created by mutations in the coding region; these are either not fixed in our species or mutation may have recreated open reading frames by converting such stops back into amino acid codons. With a high degree of heterozygosity, each person may bear two different alleles at their roughly 350 functional *OR* genes, which effectively doubles the available repertoire of diversity, even if no two people have the same set of alleles or even the same set of functional genes.

Rather than viewing human olfaction as degenerate, one might make an alternative darwinian interpretation that, as in immunology and MHC specification, selection has favored olfactory diversity even in humans. This may sound fine in principle, but, unlike the immune system, we have to react cognitively to an odorant. If there is too much variation, we might not be able to detect any given odorant, and indeed odorant-specific anosmias are common and the perfume industry is kept busy because we more. Yet, with the amount and chaotic organization of variation, we should be than we are different vary in what we perceive or how we react.

The pattern of variation in the major *OR* gene cluster on human chromosome 17 is interesting in this regard (Gilad et al. 2000). As seen in 20 sequenced individuals, the functional *OR* genes in this cluster vary less than the pseudogenes, and there is evidence from comparison with orthologs in chimps that there has been weak positive selection at the genes (but not the pseudogenes), which may maintain *OR* diversity. This may be too weak to be attributed to odorant specificity, although for some odorants most people do react in a similar positive or negative way, as if there is some form of specificity. These facts need to be reconciled with the observation that at least many human *OR* genes are not pseudo in everyone.

At the same time, a comparison between humans and other primates found that humans have accumulated pseudogene-producing mutations (that is, that disrupt coding relative to functional orthologs) at a rate roughly four times that in other primates (Gilad et al. 2003; Rouquier, Blancher et al. 2000). This suggests excess loss of genes in the human lineage, though the other primates also have considerably higher fractions of pseudogenes than does the mouse. Whether what has been favored is a kind of variable pseudogene pattern remains to be shown.

Important information will come from the analysis of variation in the *OR* genes in animals that have few pseudogenes, which is assumed to be due to selection for function. Interestingly, based on the early indications, mice do have *OR* polymorphisms (P. Mombaerts, personal communication). What is preventing pseudogenes in these species? The question is cogent because even in mice the evidence suggests that odorant detection is combinatorial and open-ended rather than prescriptively specifying one set of receptors for each possible odorant. Why wouldn't mice benefit from a high mutational repertoire, even at the expense of making some genes pseudogenes in some individuals? Possibly, the determination of olfactory genetics from inbred mice (in most cases, probably from studies of a single strain) could obscure some of these questions until wild mice, or mice from multiple independently-derived strains, are examined.

A cautionary note in any such functional evolutionary speculation is that ORs are expressed in nonolfactory tissues like testis and heart and thus may have pleiotropic functions. One possibility is that there have been various types of balancing selection, but a simpler explanation may be that this just reflects "leaky" nonspecific gene expression; the testis expresses many genes that have no obvious germline function (Mombaerts 1999). Alternatively, it has been rather loosely speculated that distributed expression of highly variable OR genes can be used in development as a kind of "area code" to identify tissue-specific codes during development (Dryer 2000).

The answers to these many questions will be interesting. We need to remember also that it is after the fact that we evaluate the nature or importance of chemosensation to a given species. Chemosensation is a generic need of cellular life, but a given species uses what it has and has what it uses. Why, for example shouldn't birds or humans have a better sense of smell? There is no one chemosensory "need" for an organism or a chemosensory problem to "solve."

CONCLUSION

Chemosensation is widespread in the living world. As so typifies evolution, there is no single way to detect chemicals in the environment. Communication among cells is an extension of interaction within cells and is about chemical information exchange. The division of chemosensation into internal and external systems is a somewhat arbitrary distinction (as pheromones show). The various chemical senses employ the widespread *7TMR* class of signal receptors. Within clusters, the genes seem generally to be of similar origin, but between *types* (*OR*s, *VR*s, *GR*s, and other chemosensory receptors) there is only distant homology. This may indicate that, as in so many other systems, chemosensory functions have arisen multiple times independently and/or that the different aspects of chemosensation have evolved in their own ways. But this has happened from a common general starting mechanism.

The various sensory cells are located differently in different organisms, which may in part reflect the relatively less organized nature of chemical information in the world, compared with light and sound. Unlike the programmed ability to detect specific molecules such as hormones or pheromones, olfaction appears to be designed, like the immune system, as a general open-ended molecular detection system.

Olfaction is, however, more than the reception of a molecular "signal." It should come as no surprise that many other genes involved in the processing and use of the signal, locally as well as to and in the brain, as well as the *OR* genes themselves, are involved. A series of experiments in *Drosophila* that involved crosses between olfactory mutant flies have shown that olfactory response is a kind of complex trait, with allelic variation having quantitative effects on the trait, epistatic effects as well as complementarity or other interactions among the genes (Anholt et al. 2001; Anholt and Mackay 2001; Fedorowicz et al. 1998). This elegant experimental result shows what we know in so many ways to be generally true of complex traits in nature. It is in a sense also reassuring: not only are traits assembled over evolutionary time to involve many elements, but selection does not stipulate that there is one "wild type" way to make a trait. Instead, what we get is variation and to some extent complexity that buffers a multigenic system against mutation.

We have outlined the basic mechanical means by which the molecular detection is done, but the most important question perhaps has not been touched. It appears sufficient for the immune system to recognize and inactivate invading organisms. Odors, however, seem to require cognitive, information-integrating responses, which should, at least in some instances, depend on the nature of the compound. To a great extent, reactions are learned: avoidance of a substance is based on experience. Many bitter-tasting butterflies have to be sampled before being shunned. But is reaction all learned by experience, or are there undiscovered mechanisms by which the *type* of molecule being detected is interpreted? Do the different classes of odorant receptors carry, historically, some information in their ligand-binding properties that is not simply due to chance?

We know this is true for specific types of molecules: taste for sweet and bitter, for example, and pheromones. Are there others? Do these exist in mammals as well as invertebrates (which have typically been assumed to be "hard-wired" automotons)? Or are there cognitive processes to do this in ways unknown to us at present? These questions probably will be answered in the near future.

Chapter 14
Detecting Light

Light is the ultimate resource for all forms of life. Organisms use light in three basic ways, the most ubiquitous being in direct physiological processes, the capture of energy via photosynthesis. This is of course what plants do and, in terms of the ecology of the Earth, photosynthesis is probably the most active important use of light energy by living organisms. Second, many organisms simply detect the location of a light source, moving toward or away from it. Third is *vision*, the interpretive use of spatially arranged patterns of light by animals as a source of information about objects in their environment. Mobile organisms do this to avoid collision, avoid predators, pursue prey and other food or water sources, detect conspecifics, associate with mates, and otherwise participate in social behavior, among many other purposes. Information can also be *sent* by organisms through light transmission, as by flowers to attract pollinating insects or by behavioral displays (like leks, in which clouds of males aggregate to attract females). Many attributes of light are used in this way, including wavelength or frequency and intensity.

However, as important as light is in this latter sense, organisms do not need to perceive light or do so in any particular way, even for many of the uses just listed. This is shown, as we will see, by the great variation in what animals, even closely related ones, can see. Simple detection of light without regard for pattern can be sufficient for some, whereas others depend on fine resolution of specific objects (prey) or aspects of objects (berries on a tree). Some use light only to detect motion.

As we have asked in other contexts, how can organisms be genetically programmed to use light to resolve the huge diversity of conditions that they cannot specifically anticipate in a hard-wiring sense, unique conditions that have never occurred—not even once—in the entire 3 billion years of their ancestry?

LIGHT RECEPTION

Energy organized as electromagnetic waves, and/or streams of photons, is called radiant energy. Visible light comprises a small subset of the total spectrum of radiant energy (Figure 14-1).

At one end of the spectrum are cosmic rays and at the other are (for humans) electrical power waves (even lower frequency radiation is possible). The different kinds of radiant energy can be characterized by their wavelength (inversely related

Genetics and the Logic of Evolution, by Kenneth M. Weiss and Anne V. Buchanan.
ISBN 0-471-23805-8

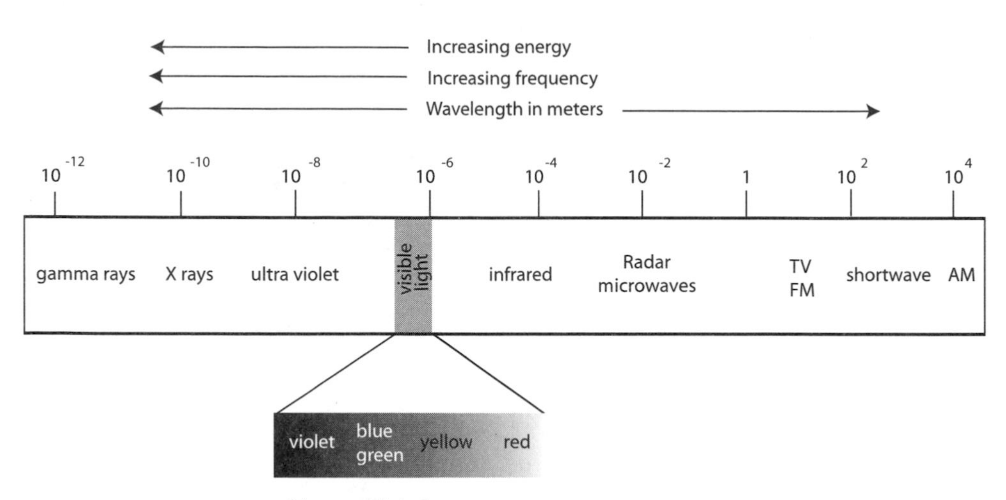

Figure 14-1. Radiant-energy spectrum.

to frequency because the energy travels at the speed of light). The average wavelength of cosmic rays, the shortest cycles of radiant energy known, is in the range 10^{-12} meters (million millionths of a meter). The wavelength of electric power waves is on the order of 10^6 (million) meters (50–60 Hz, or cycles per second). The radiant energy visible to humans comprises a narrow portion of this spectrum, ranging from 380 to 760 nm (nanometers or billionths of a meter). Other organisms can see somewhat more, or less, of this same range. Snakes, for example, have infrared (heat) receptors, and bees and birds have receptors for ultraviolet light.

Visible light is detected by *photoreceptors*. These can take many forms, and we can only describe some of them. For example, the entire cell of some single-celled organisms, such as amoebas, may be sensitive to light, moving toward or away from it. They use a membrane-bound photoreceptor that induces the release of cAMP, which in turn induces changes in the beating of their cilia and cellular motion. Some worms have photoreceptive cells, or eyespots, scattered throughout the epithelium on the surface of their bodies. In the earthworm, these serve to orient the organism directionally, as they prefer to live underground in darkness.

Direct sunlight includes energy in the ultraviolet (UV) part of the spectrum, but reflected light is dimmer and retains little UV, and many species are sensitive to both. Vertebrates have evolved two basic kinds of photoreceptor cells, the *rod* and *cone* cells, which are embedded in the retina at the back of the eye and attached to the axons of neurons in the optic nerve (Figure 14-2A shows the structure of the human eye and retina, and 14-2B depicts the structure of rods and cones with a retinal molecule bound to a 7TMR photoreceptor protein embedded in the surface of a rhodopsin disc). Based on its properties, a photoreceptor molecule responds to light of particular energy and frequency. Rods perceive light and dark (i.e., black, white, and shades of gray), and organisms use these mainly for perception in dim light; they are maximally sensitive in the middle of the visual spectrum and able to respond to a very small amount of light. Cones are more specialized for color perception and acuity of vision and are used to detect form and motion but do require more light than rods to be activated, hence do not function well in dim light (humans lose color

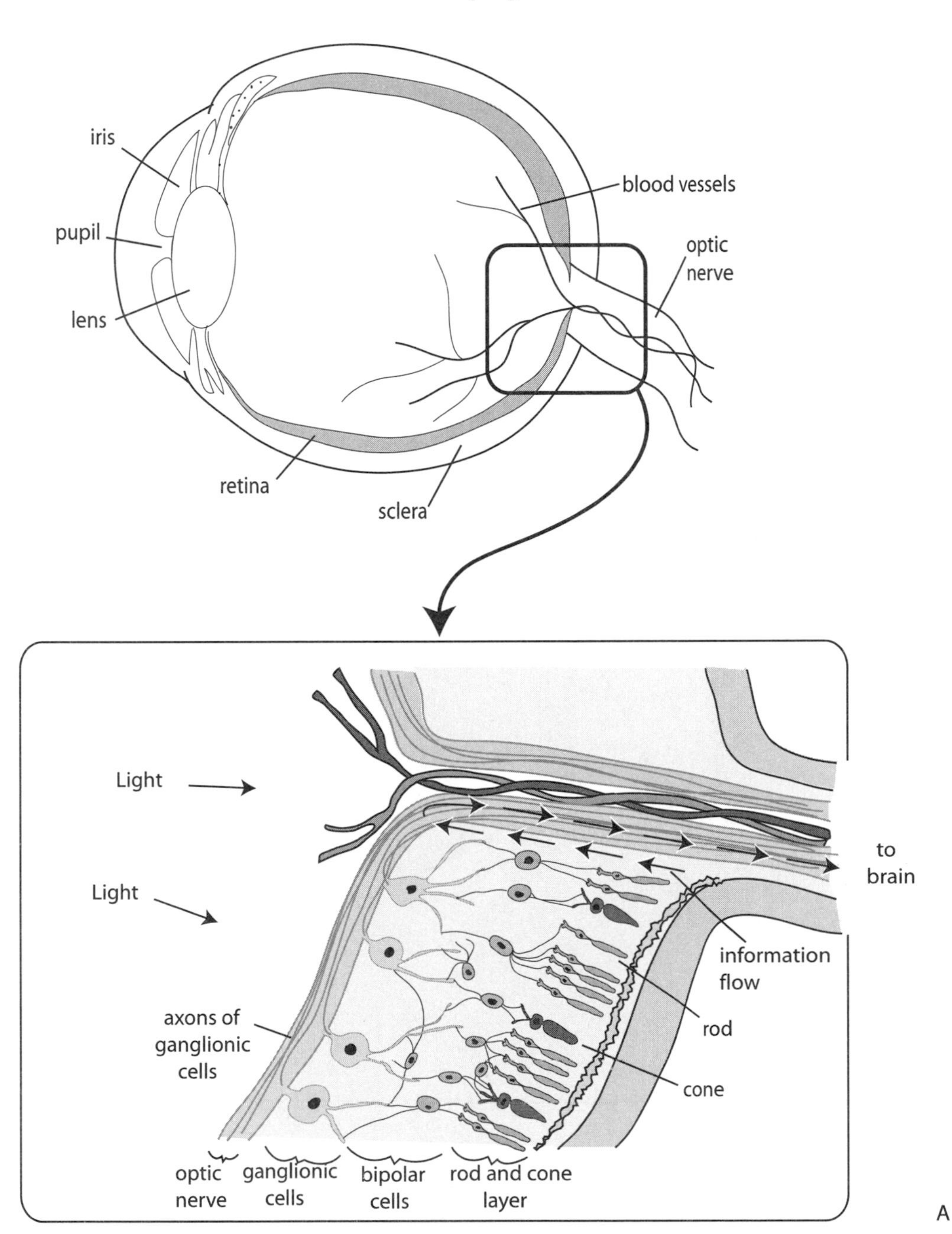

Figure 14-2. Parts of vertebrate visual system (human). (A) Structure of the eye and retina; (B) cellular structure of rods and cones, including a diagram of an opsin and its chromophore.

perception as light dims). The ratio of rods to cones varies across species as well as in different areas of their respective retinas and is related to their adaptations, but the cells function in a similar manner from species to species. UV-sensitive vision in vertebrates is generally a separate phenomenon, although it involves related genes and mechanism (Yokoyama 2000; Yokoyama and Yokoyama 2000).

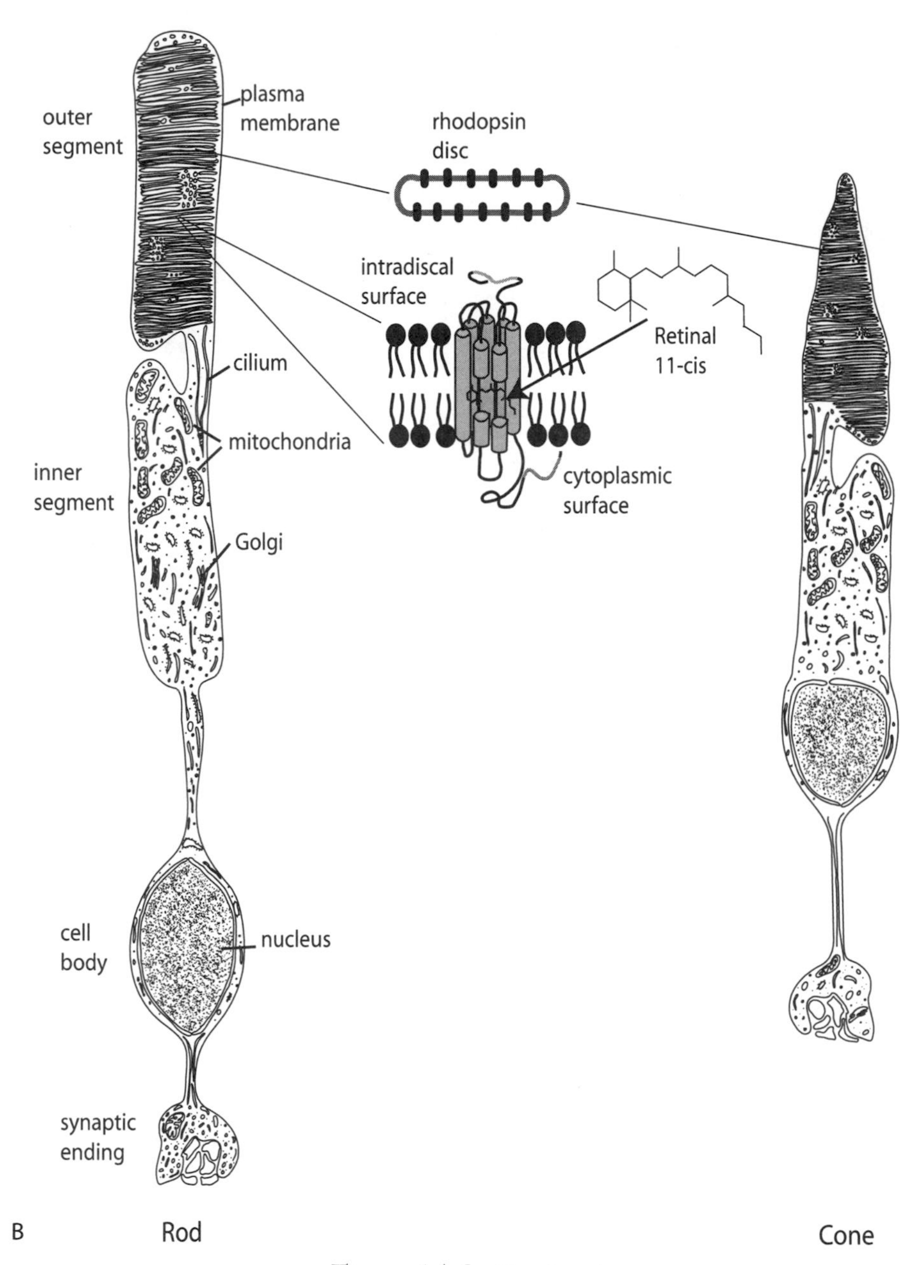

Figure 14-2. *Continued*

Rods and cones have inner and outer segments. The outer segments are composed of a series of stacked bilipid membranous disks that contain visual pigment called *photopigment*. There are two main ways in which these disks can be presented: at the apical end of the cell or in cilia formed on that end. These cellular arrangements have been thought to divide protostomes and deuterostomes, respectively, but the photoreceptor phylogeny turns out to be rather more complex and even raises questions about the nature of chordate-nonchordate eye homologies, or the ease with which similar arrangements of photoreceptors can re-evolve (Arendt and Wittbrodt 2001), a problem we have visited several times (e.g., sea urchin larval stages, Figure 8-9). Based on differences between message transduction pathways that are used, Arendt and Willbrodt suggest that the early bilateran ancestor species had both cellular types.

The photopigment is composed of an *opsin* protein and a *chromophore* called *retinal*, derived from vitamin A. The opsin and chromophore are bound together and embedded in transmembrane receptors on the outer disks. Light energy is captured in these disks and, through a series of chemical reactions that take place in the photopigment, is transduced into receptor potential and sent, via the optic nerve, to the brain where it is interpreted as light and color. Alternative chromophores used by some taxa modify the spectral response. Bird, amphibian, and reptile photoreceptor pigments have an overlying oil droplet that filters light and modifies their spectral sensitivities. As a result, different taxa can use different pigment and chromophore combinations to achieve a similar spectral response.

When light is received by a photoreceptor, the chromophore all-*trans* retinal converts to 11-*cis* retinal, which changes the conformation of the opsin protein. This in turn modifies its intracellular domain (the opsin must then be recharged by a new chromophore molecule).

In vertebrates this "bleaching" releases the chromophore from the opsin; the chromophore is then regenerated by neighboring cells reversing its conformation to the all-*trans* form so it can be reused. In invertebrates, the chromophore does not leave the opsin before being reversed.

Invertebrates and vertebrates use homologous *opsins*. These are coded by yet another branch of the *7TMR* gene family, although the two major animal groups use unrelated second messengers to relay the signal. Vertebrate photoreceptors are coupled to the G protein *transducin*. Absorption of light causes a conformational change in the receptor, leading to increased binding of GTP by the α-subunit of transducin. This activates cGMP phosphodiesterase, which degrades cGMP, causing cGMP-gated Na^+ channels to close, hyperpolarizing the cell and inhibiting neurotransmitter release. This in turn propagates signal to the brain. Thus, neurons are inhibited in the dark, when neurotransmitter is released at a high rate. By contrast, the invertebrate second messenger is inositol triphosphate, and their photoreceptors respond by depolarizing rather than hyperpolarizing (Hardie and Raghu 2001; Nilsson 1996; Ranganathan et al. 1995).

Invertebrate photoreceptor cells also have layers of membranes filled with *rhodopsin*, a photoreceptor pigment, and the photoreceptor collects light in the same way as in vertebrates (Yarfitz and Hurley 1994). Each photoreceptor cell in the *Drosophila* eye has a structure called a *rhabdomere* that contains ~60,000 microvilli. Millions of rhodopsin molecules associated with the downstream light transduction cascade reside in the microvilli.

Opsin Evolution and Function

The number of *opsin* genes varies among species (Pichaud et al. 1999). Although the basic structure of opsins is conserved, each gene has a distribution of wavelengths to which it responds, a distribution with a peak wavelength sensitivity, and diminishing sensitivity at surrounding wavelengths (Figure 14-3C). The spectral sensitivity can be evaluated experimentally and is determined by the amino acid

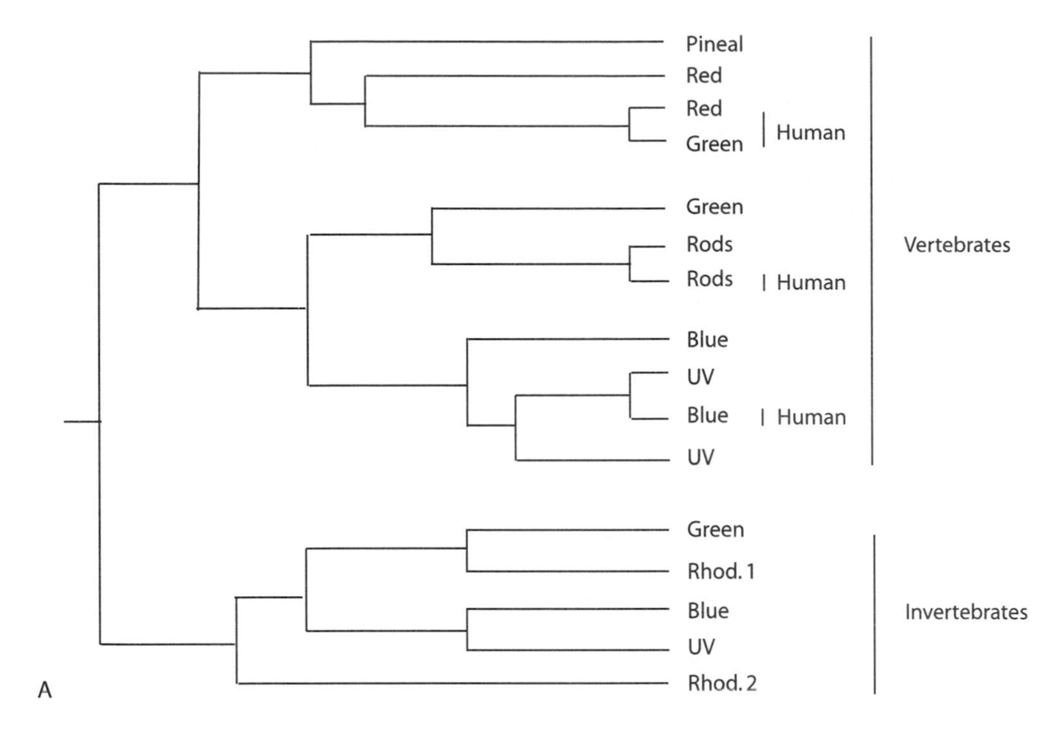

Figure 14-3. *Opsin* genes. (A) Simplified *opsin* gene tree; in many instances similar but not identical *opsins* (based on gene sequence and estimated peak sensitivity) are found within the groups tested, but shown here as one branch of similar genes; (B) location of the amino acids in the opsin protein structure that have the greatest effect in red-green spectral cone sensitivity among naturally observed opsins, showing their likely important relationship to the retinal molecule bound to the photoreceptor (e.g., Yokoyama 2000; Yokoyama and Yokoyama 2000); (C) distribution of wavelengths to which opsins respond (rhodopsins have peak sensitivities around the 500 nm range).

sequence of the opsin protein and its interaction with the chromophore. Modifications in the photoreceptor itself, such as the use of oil droplets by some species, mean that the actual response of the photoreceptor may not correspond to that of the visual pigment alone. Examples are the chicken *Rh2* gene that is paired with a green oil droplet and the goldfish *Rh2* that is green-shifted because of the modified chromophore it uses (Yokoyama and Yokoyama 2000). Because its different opsins respond differently, an organism can compare the signal strength of adjacent opsins to assess "color" and other attributes of incident light coming from the same part of the environment.

Insects have dual dim-bright and color vision capabilities that vary among the species that have been tested (Pichaud, Briscoe et al. 1999). Flies have independent receptors for the two types of vision, but the ability of other insects to distinguish between the two is less clear because the neural impulses to the brain from the two systems seem to merge. Honey bee and *Drosophila* eyes express three *opsins* that absorb maximally in the UV, blue, and green ranges. Some butterflies appear to have

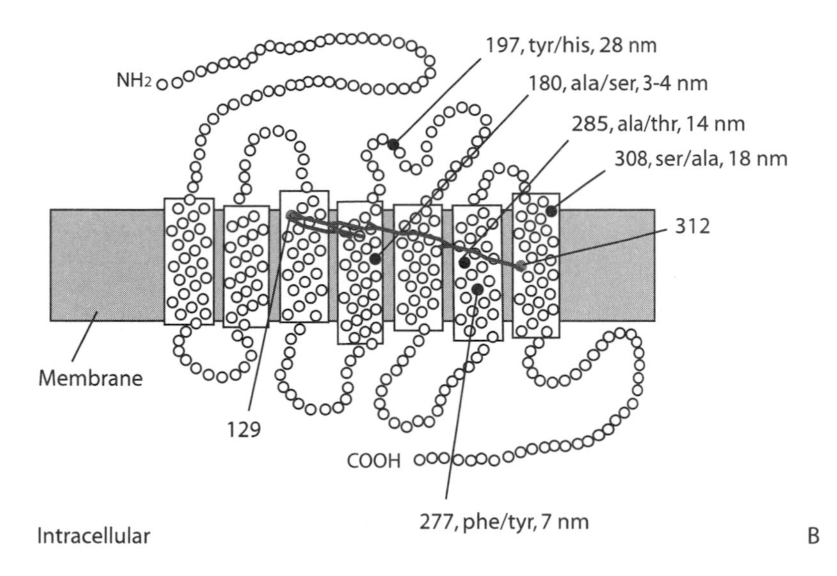

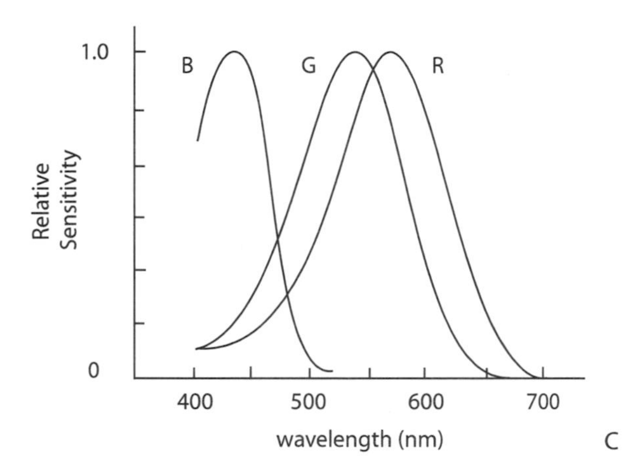

Figure 14-3. *Continued*

combinations of opsins and filtering pigments that allow for six classes of spectral peak sensitivities (Pichaud, Briscoe et al. 1999).

Opsins responding to similar light frequencies are typically closely related in protein structure and, presumably, evolutionary history (Figure 14-3A) (Yokoyama 2000). However, through gene duplication, a number of cone *opsins* evolved, and today there is great variety among vertebrates and invertebrates in their cone and rod cell arrangements, relative number, type number, and wiring to the central nervous system. Both vertebrate and invertebrate *opsin* sequence relationships suggest that rhodopsin evolved from green-sensitive ancestry.

Many amino acids vary among opsins, but a relatively small number of key amino acids appear to account for most of the variation in spectral sensitivity, at least in the repeated evolution of red-green sensitivity differences that have been studied in detail (Yokoyama 2000; Yokoyama and Yokoyama 2000). Current reconstructions based on teleost fish, amphibian, reptile, bird, and mammalian data suggest that the ancestral vertebrate had five *opsins*, *rhodopsin* plus four cone types, probably indicating tetrachromatic vision (Bowmaker 1998). The latter include, *opsins* sensitive to red, green, blue, and ultraviolet parts of the spectrum, along with a pineal *opsin* (see below) (Bowmaker 1998; Yokoyama 2000; Yokoyama and Yokoyama 2000). The gene phylogeny suggests that only the long (red-green) and the shortest (blue) wavelength *opsins* were present in stem mammals, giving them dichromatic vision, with a relative sparseness of cones in their retinas that may indicate that their early evolution was as nocturnal species. Invertebrates have rhodopsin-green and blue-UV gene classes and may only have had the two in their stem ancestors. Clearly, terms such as "tetrachromatic" are misnomers in that with four different peak sensitivities the organism can actually parse a wide range of colors, not just the four—just as we can see the entire color range with our three (red, blue, green) *opsins*.

For foraging and social signaling, some vertebrates such as birds use UV light as well as the broad spectrum of visible light. This is a potentially important fact in assessing the value of protective coloration as an adaptation against bird predation because the usual hypotheses have been based on what human investigators can see. A famous example is the case of industrial melanism, in which the rapid evolution of protection in peppered moths was said to have occurred because moths that matched the color of lichens on tree trunks on which the moths rested were not seen (or eaten!) by bird predators. Visible moths were eaten, and their unfortunate wing-color alleles disappeared along with them. However, the UV sensitivity of bird vision may have rendered moths less effectively disguised than they appear to our human eyes (Grant 1999; Weiss 2002c).

In the retina, cones tend to cluster in hexagonal groupings, surrounded by rods, but with a greater overall concentration of cones near the center and of rods near the periphery of the retina. Rod cells predominate in vertebrates that live in dim light. Interestingly, it appears that a given cone cell as a rule expresses only one type of *opsin*, although this may be less tightly regulated in species other than primates (Wang et al. 1999). Some primates, including humans, have two closely linked *opsins* on their X chromosome, that arose by gene duplication. Today, these respond to red and green ranges of the spectrum. The expression of only a single gene in a cluster is another example of *cis* allelic exclusion previously seen in immune, globin, and olfactory genes. Which of the tandem X-linked *opsin* genes is expressed in a given cone cell may involve competitive binding of regulatory proteins (Wang, Smallwood et al. 1999). This selection occurs probabilistically, so that both genes are expressed in sufficient numbers of cone cells.

X-inactivation in females ensures that only one chromosome's genes are used for red and green in any given cell and introduces some variation between each retina and among areas within the same retina. Exclusion does not occur in blue cones, which express both copies of the gene (blue *opsin* is autosomal and hence diploid). How blue vs. red-green *trans* exclusion occurs so that a given cone expresses only one or the other, when the two are on different chromosomes, is not known. In fact, some species do express both blue and green *opsins* in the same receptor cell (Glosmann and Ahnelt 2002).

Other genes are involved in a variety of "nonvisual" animal light perception phenomena. Among these are light-sensitive melanopsins found in frog skin and mammal retinal ganglia, which help regulate circadian (day-night) rhythm and pupil reflex and others. These genes are widespread in nature, and at least some are part of the *opsin* gene family. However, melanopsins in vertebrates appear more closely related to those in their invertebrate relatives; for example, in situ, their chromophore conformation is reversible and does not require helper cells as in vertebrate retinal photoreceptors. Vertebrates use genes in the *P* (*pineal*) subgroup to regulate diurnal cycling mechanism through the *retinohypothalamic tract*. A net of *melanopsin*-expressing cells in the inner mammalian retina may serve this function independent of the outer retinal photoreceptor cells (rods and cones) (e.g., Grant 1999; Provencio et al. 2000; Provencio et al. 2002). Blind subterranean mole rats appear to use this mechanism for circadian rhythms despite having degenerate eyes (Hannibal et al. 2002). Blind humans and experimentally blinded rodents may retain light-responsive nonvisual functionality. Insects with eyes removed can also perceive light intensity through a rudimentary "dermal" light sense (Steven 1963).

The Evolution of Color Vision

It is tempting to relate color sensitivities to aspects of the environment that may have provided the adaptive darwinian basis of organisms by selection; environmental lighting conditions, the color of food sources, mates and conspecifics, and the like have been suggested as the selective forces (e.g., Mollon 1989; Treisman 1999; Yokoyama 2000). For example, vertebrates living in dim light often mainly have long-wave (blue) sensitive cone pigments as well as rods. Selection based on visual cues, both to favor seeing ability and to favor being seen or not, depending on the circumstances, can be strong and rapid. Adaptation to wavelength sensitivity is often referred to as spectral "tuning" by natural selection (we also saw this notion in regard to olfaction). It is, however, more difficult to evaluate specifically the range and nature of what an organism can actually see, than to evaluate the specific nature of an opsin protein.

The tandem nature of human red and green pigment genes provides an interesting test case in regard to color vision. A male has only a single X, and all red- or green-expressing opsins use the respective alleles on that single chromosome. Because of X-inactivation, even a female, who has two X chromosomes, will only express one of her two red or green opsins in any given cone. This adds an element of stochastic variation among females (and between the left and right eyes of a given female) in their red-green sensitivity. Generally, this has subtle effects at most.

Mutant opsins are reasonably common. Because blue opsin is autosomal (chromosome 7), both males and females have two copies of the blue opsin, and blue color blindness is rare in either sex: even if one allele is defective, the chances are small that the other will be as well. However, red or green color blindness is not unusual in males; they only have a single red and green *opsin*, so that if either gene is defective there is no normal allele to "cover" for it. It is relatively unlikely that a female will inherit two dysfunctional red or green *opsins*. (In Hardy-Weinberg terms, if p is the defective allele frequency, her chance of having two such alleles is p^2, typically a very small value, and similar to the situation for the autosomal blue *opsin*. Even though a female only expresses genes from one of her X chromosomes in any specific retinal cell, roughly half of her cones will express her functional allele. This may affect her color sensitivity somewhat, but she will not be color blind.). By contrast, a fraction p of males will inherit the defective allele, but having only that single X have no covering protection.

Color blindness is considered a kind of disorder but that is a human subjective judgment. There is generally considerable variation in the color sensitivities of *opsins*. The red and green *opsin* genes are closely linked, and mutation, recombination, and/or inaccurate replication produces a variety of copied, disrupted, or fused *opsins*, with substantial variation in the resulting peak sensitivities that can be explained by the nature of the mutation (Figure 14-4) (Neitz et al. 1996; Neitz et al.

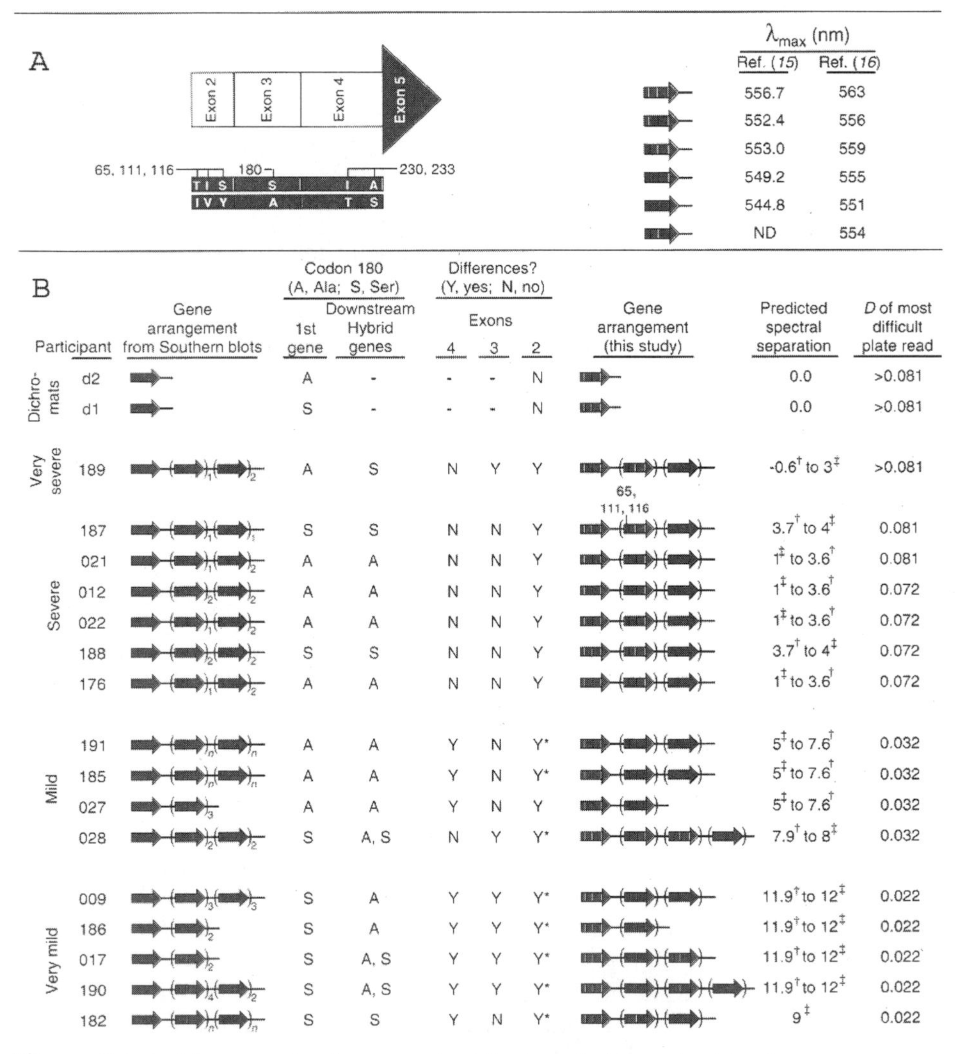

	Participant	Gene arrangement from Southern blots	Codon 180 (A, Ala; S, Ser): 1st gene	Codon 180: Downstream Hybrid genes	Differences? (Y, yes; N, no) Exons: 4	3	2	Gene arrangement (this study)	Predicted spectral separation	*D* of most difficult plate read
Dichromats	d2		A	-	-	-	N		0.0	>0.081
	d1		S	-	-	-	N		0.0	>0.081
Very severe	189		A	S	N	Y	Y		-0.6† to 3‡	>0.081
Severe	187		S	S	N	N	Y		3.7† to 4‡	0.081
	021		A	A	N	N	Y		1‡ to 3.6†	0.081
	012		A	A	N	N	Y		1‡ to 3.6†	0.072
	022		A	A	N	N	Y		1‡ to 3.6†	0.072
	188		S	S	N	N	Y		3.7† to 4‡	0.072
	176		A	A	N	N	Y		1‡ to 3.6†	0.072
Mild	191		A	A	Y	N	Y*		5‡ to 7.6†	0.032
	185		A	A	Y	N	Y*		5‡ to 7.6†	0.032
	027		A	A	Y	N	Y		5‡ to 7.6†	0.032
	028		S	A, S	N	Y	Y*		7.9† to 8‡	0.032
Very mild	009		S	A	Y	Y	Y*		11.9† to 12‡	0.022
	186		S	A	Y	Y	Y*		11.9† to 12‡	0.022
	017		S	A, S	Y	Y	Y*		11.9† to 12‡	0.022
	190		S	A, S	Y	Y	Y*		11.9† to 12‡	0.022
	182		S	S	Y	N	Y*		9‡	0.022

Figure 14-4. The type of red-green color blindness is determined by the arrangement and mutation of the X-linked *opsin* genes. This table shows phenotypes associated with mutant, deleted, or fused genes that have been observed. Spectral sensitivities were measured in two ways. Reprinted from Neitz et al. (Neitz et al.) with permission; see the original for details.

1995). The amount of variation probably should temper what we consider "normal" and is also relevant to the age-old philosophical question of whether all people see the same colors. Not only is there variation in the genes themselves, but there is a stochastic element in the *opsins* expressed by any given cone, meaning that not even the left and right retinas of the same person are exactly the same. Of course, even if they were identical, this cannot answer the question of whether people who detect the same thing *perceive* things similarly, which probably remains as philosophical as ever.

The patterns of red-green X-linked *opsins* found among primates has been used to infer whether their common ancestor had di- or trichromatic vision. Old World monkeys and apes have trichromatic color vision as we do, with both red- and green-sensitive *opsins*, suggesting that this was probably the condition or our shared ancestor. However, most New World monkeys have only one X-linked *opsin*, and there's a twist. The single X-linked *opsin* is polymorphic in many species that have been studied, and the observed alleles are variously sensitive to red- or green-range light. Depending on the alleles' frequencies, a female's genotype may be **RR**, **RG**, or **GG**, whereas a male is either **R** or **G** sensitive. The relative frequencies of these genotypes will be determined by the allele frequencies (e.g., p_R^2, $2p_R(1 - p_R)$, $(1 - p_R)^2$ in Hardy Weinberg Equilibrium, using p_R for the frequency of the **R** gene).

What accounts for this widespread variation in the peak sensitivities and polymorphism among monkeys with but a single X-linked *opsin*? The phenomenon provides an opportunity to examine the nature of darwinian explanations (e.g., Mollon 1989; Nathans 1999; Yokoyama 2000). The typical suggested scenario is that there has been selection for ability to detect fruit or other plant foods (for example, leaves whose color indicates taste or toxicity) in the dappled background of tropical forests. Animals who could see colors appropriate to their dietary staples had higher fitness, over time spectrally "tuning" their opsins' peak sensitivity frequencies. The polymorphic nature of the single X-linked opsin cannot be quite so easily explained. Why was it not "tuned" to an optimal frequency? Instead, as it is now, members of the same population are variably sensitive to color in a strange way: males are differently sensitive and to only one of the available colors, in proportion to the **R** and **G** allele frequencies. Females vary in proportion to the respective genotype frequencies (as above), the **RR**'s and **GG**'s being dichromatic and only the **RG**'s being trichromatic. How would selection produce this?

Assuming that trichromacy must be evolutionarily better and hence favored by selection (from a human-centered perspective), it has been suggested that the (at most) few trichromatic females in any small local monkey troop may have led their peers to food sources. Otherwise, only a single best allele would have been favored by selection. If the selective scenario is correct, one might expect a kind of balancing selection to have optimized the frequency of female heterozygotes, which will occur with allele frequencies close to 0.5 (which maximizes $2p_R(1 - p_R)$); this does not seem generally to be the case (Cropp et al. 2002; Heesy and Ross 2001), although available samples are inadequate for a definitive answer.

Perhaps there is a simpler and more natural explanation that is also more parsimonious (though, as noted early in this book, there is no reason that evolution needs to have followed the simplest path). It is useful first to note that the placement of cones in the retina may be related to spatial as well as, or even rather than, color perception. Also, even fully trichromatic individuals use other features of light such as shading and lightness in image discrimination perception. Figure 14-5 shows the

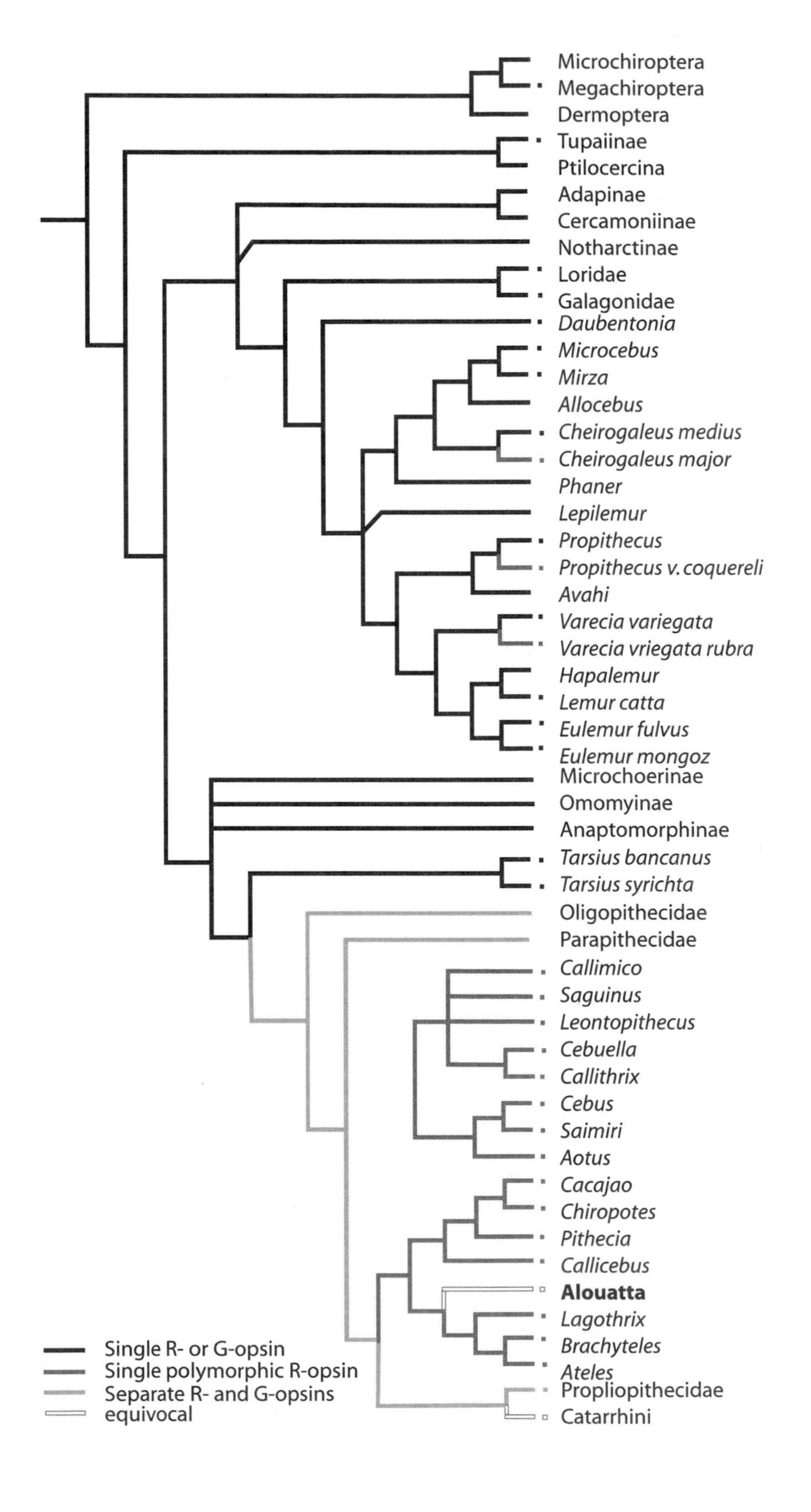

Microchiroptera
Megachiroptera
Dermoptera
Tupaiinae
Ptilocercina
Adapinae
Cercamoniinae
Notharctinae
Loridae
Galagonidae
Daubentonia
Microcebus
Mirza
Allocebus
Cheirogaleus medius
Cheirogaleus major
Phaner
Lepilemur
Propithecus
Propithecus v. coquereli
Avahi
Varecia variegata
Varecia vriegata rubra
Hapalemur
Lemur catta
Eulemur fulvus
Eulemur mongoz
Microchoerinae
Omomyinae
Anaptomorphinae
Tarsius bancanus
Tarsius syrichta
Oligopithecidae
Parapithecidae
Callimico
Saguinus
Leontopithecus
Cebuella
Callithrix
Cebus
Saimiri
Aotus
Cacajao
Chiropotes
Pithecia
Callicebus
Alouatta
Lagothrix
Brachyteles
Ateles
Propliopithecidae
Catarrhini
Single R- or G-opsin
Single polymorphic R-opsin
Separate R- and G-opsins
equivocal

color sensitivities of the red/green *opsin* genes in primates (Heesy and Ross 2001), but the distribution of peak sensitivities among these alleles (Figure 14-6) does not suggest tight selection for red-centered and green-centered alleles but rather suggests that the alleles are spread across the entire red-green part of the spectrum. Selection might only have ensured that an animal would have opsins sensitive to *some* part of the red-green range rather than any *particular* one.

Because of codon redundancy, the polymorphisms in the key amino acids that have mainly been responsible for peak sensitivities in this range (Nathans 1999; Yokoyama 2000) require only a single nucleotide change; therefore, recurrent mutation to similar alleles in different species is not implausible on an evolutionary time scale. However, the appearance of similar mutations in multiple lineages might just reflect genetic drift on an ancient polymorphism in the ancestral lineage. Consistent with this, in both squirrel monkeys and humans, there is less variation in the single blue opsin than in the X-linked genes (Shimmin et al. 1998), suggesting that the blue-sensitive gene has been under stronger selective constraint than the polymorphic X-linked gene in squirrel monkeys (Cropp, Boinski et al. 2002).

The presumed need for precise spectral tuning may be less than is typically thought if one considers the many ways in which visual signals are interpreted, that opsins have sensitivity at wavelengths around their peak sensitivity, and that colors are not perceived strictly on the basis of peak sensitivities. For example, vertebrate retinas compare the relative strength of signal detected by different parts of the system (e.g., from two types of nearby cone cells) in various ways, even before signal is sent to the brain (e.g., Nathans 1999). In addition, what an animal "sees" is affected by its past experience.

In a species with *two* X-linked opsins, it is plausible that selection could keep their sensitivities separated as part of their normal repertoire (gene duplication of X-linked opsins occurred at least twice in primate lineages; New World howler monkeys have trichromacy similar to that in Old World monkeys and apes). But these systems, too, have considerable normal variation, suggesting that selection has, at most, not been too stringent.

The evolution of color vision can also illustrate the potential importance of organismal as opposed to classical natural selection. Some fish live deep in the sea where there is little light and what gets through is of short wavelength. Coelocanths use a combination of two rhodopsin-related opsins that are sensitive only to such light (Yokoyama et al. 1999). The fish are adapted to life in the depths. A fish will

◄

Figure 14-5. Phylogeny of primate X-linked color vision capabilities. Phylogenetic tree of primates showing species that have alleles conferring only red (R) or green (G) sensitivity, a single gene with observed polymorphisms conferring alleles with R and G sensitivity, two X-linked genes (R- and G-sensitive), or some less-clear sensitivity. One can infer that the original primates may have been "dichromats" but with a single gene that was polymorphic for peak sensitivities in the red-green range. Gene duplication led to two X-linked genes, one specialized for red and the other for the green ends of this part of the spectrum, enabling "true" (human-like) trichromacy in the ancestor of Old World monkeys and apes. Gene phylogenies suggest that this occurred independently in the lineage of howler monkeys (*Alouatta*) in the New World. Taxa with no symbol indicate no data available. Redrawn from (Heesy and Ross 2001) with permission.

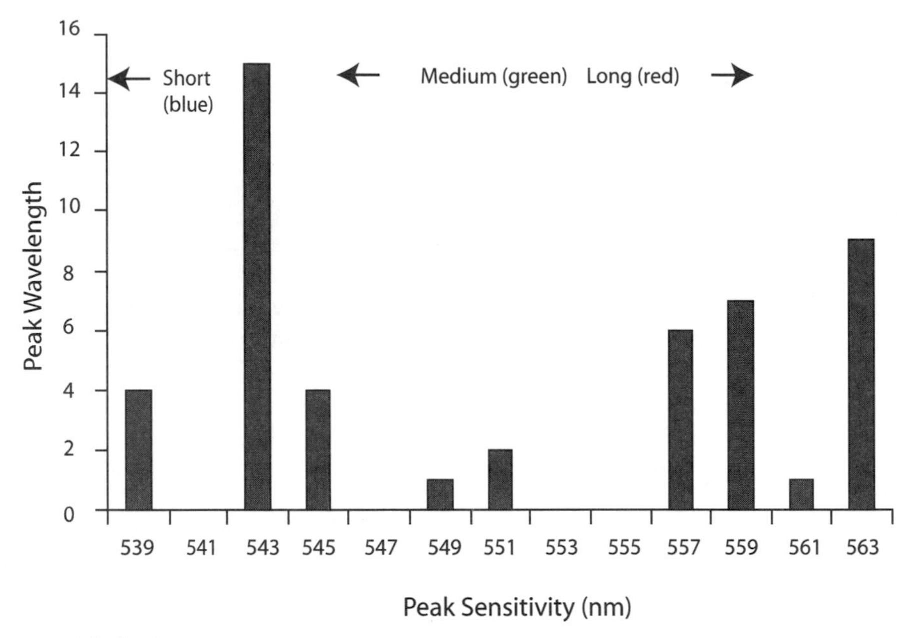

Figure 14-6. Peak spectral sensitivities in X-linked primate *opsin* genes. Height of each bar is the count of alleles of that peak sensitivity observed in tested primate species (that is, these are not allele frequencies: each observed allele of that sensitivity category incremented the category count by 1, regardless of the frequency of that allele in the species). Reprinted from (Weiss 2002c) with permission.

move around and seek food wherever it might be detected. Fish that by mutational chance were able to see in dim low wavelength light might gravitate toward greater sea depths and mate with others they find there. This scenario would require choice of available subenvironments, but not classical darwinian selection within a single population, even though color vision is important in animal ways of life (Weiss 2002c).

Cichlid fish have in some instances become highly speciose (numerous highly diverse species, perhaps including incipient species), even within what appears to be a continuous environment, including the large Lake Victoria in Africa (Terai et al. 2002) and crater lakes in Central America (Wilson et al. 2000). The various species are characterized largely by color variation and in some instances trophic variation (e.g., structural variation in their eating apparatus that suits their local subenvironment within the lake). This has presented the problem of sympatric speciation (Chapter 2), that is, how species can form without a physical mating barrier. One explanation is organismal selection, perhaps related to sexual selection; fish choose regions of the lake suited to their traits and mate with others in the local subenvironment.

Organismal selection can reflect existing genetically determined traits like color and color vision (on which, for example, color-based sexual selection may depend). Variation in the long-wavelength sensitive *opsin* gene in the Victoria cichlids is associated with fish color and/or local food species. The locations of at least some of the amino acid variants observed are those shown to have spectral sensitivity effects as described earlier (Figure 14-3B). This is at least suggestive evidence for color-based sympatric speciation (Terai, Mayer et al. 2002), although the pattern of polymorphisms, specific *opsin*-environment correlations, and means other than direct color sensation for picking mates and finding food might alter the explanation. In any case, there are many similarities between this variation and that in primates, that may provide an excellent "laboratory" for the challenge to interpret genetic patterns to the nature and extent of selection events that may be responsible.

The Origins of Photoreception

Some unicellular organisms, such as Euglena or algae such as *Chlamydomonas*, *Volvox*, or *Haematococcus*, have a light-sensitive *eyespot*, a region with a higher sensitivity to light than the rest of the cell. The eyespot is located between the cell's two flagella and its equator; it sends information about the quality and intensity of surrounding light to the flagella, enabling the organism to orient vis-à-vis the light, in a process called *phototaxis*. These eyes have optics, photoreceptors, and a primitive signal transduction mechanism, with layers of hexagonally packed lipid globules that reflect more or less light, depending on the number of layers (Hegemann 1997). Signal transduction uses retinal-containing photoreceptors, two rhodopsins in *Chlamydomonas* (Ebrey 2002; Sineshchekov et al. 2002).

Bacteria have rhodopsin-like molecules with seven transmembrane domains but little sequence resemblance to the corresponding eukaryote genes. Eubacteria and cyanobacteria are photosensitive or photosynthetic and produce phytochromes that work via phosphorylation signal transduction. At least some of these appear to be ancestral and related to mechanisms found in plants. Several classes of genes are involved, or possibly involved, some of which employ attached chromophore molecules (Herdman et al. 2000). This suggests that eukaryote and prokaryote photosensitivities have at least some common origins, although new mechanisms seem to have evolved in multicellular organisms.

PHOTORECEPTION IN PLANTS

Plants are always "seeking the light desiring the sunshine, . . . unconscious of either" (Joseph Conrad, *Almayer's Folly*, 1895). Plants evolved from organisms like simple photosynthesizing cyanobacteria into large organisms with many leaves that allow the plant to maximize photosynthetic activity. They depend on light for many aspects of their life cycle, including *photosynthesis*, *photoperiodism*, *photomovement*, and development, flowering, and seed germination. Although they are sessile, they have become exquisitely adept at responding to and exploiting variation in light duration, quality, quantity, and direction (Neff et al. 2000), and in the kind of intraorganismal independence mentioned in Chapter 8, different parts of a plant can respond independently (only some branches need move in a given direction).

Plants can respond with both positive phototropism, or movement toward the light by the leaves and stems, and negative *phototropism*, or movement away from light by the roots, responses that are mediated by the hormone auxin (light is only one factor in the direction of growth of a seedling however; shoots grow upward even when a plant is growing in complete darkness and roots downward, due to positive and negative *gravitropism*). Chloroplasts in the leaves move toward light to maximize the capture of photons, and the *stomata* (pores in leaves) respond to light by opening (see Figure 14-7A for a drawing of a chloroplast). Plants respond to more than simply the presence or absence of sunlight, but also to wavelength, intensity, and directionality of light as well as length of day. They use these cues to modulate their growth and development, in a process called *photomorphogenesis*.

A number of photoreceptors and photopigments have evolved to exploit whatever light they can get. These are coded by genes in three classes related to cryptochromes and phototropins (UV-B, UV-A, blue), and phytochromes (red/far-red), inducing expression of the appropriate response genes (e.g., Quail 2002).

Homologies have been found between some plant and animal photoreceptors. Among other light-induced responses in plants, cryptochromes are involved in the regulation of circadian rhythms and the timing of flowering. A diverse family of photoreceptors, *cryptochromes*, are found throughout higher eukaryotes, both plants and animals, including humans (Lin 2002), where they are involved in regulation of circadian rhythms. Cryptochromes share protein structure and composition of the chromophore with microbial DNA photolyases, or DNA repair enzymes. Phytochromes homologous to those in plants have been found in cyanobacteria (Quail 2002), and the "photosensory domain" is highly conserved between phyla. *Opsin*-like photoreceptors may also be found in plants; they are in green algae, and retinal has been purified from plants.

Plants build carbohydrates from light energy and raw materials from water and CO_2. In the natural cycle of the biosphere, they obtain hydrogen from water and carbon and oxygen from the CO_2 expelled into the atmosphere by animals. The carbon and O_2 are taken into leaves through stomata, and water is taken up by roots and circulated up through the vasculature to the leaves, where the light energy is used to break down the raw materials into hydrogen, carbon, and oxygen, which are then recombined into carbohydrates. The plant subsequently uses these carbohydrates as energy, stores them, or builds them into more complex molecules such as oils or proteins.

Photosynthesis is initiated by "visible" light in the 400–700 nm range, primarily red and blue light. This is a complicated electrochemical process and will only be sketched out lightly here. Light is absorbed by *chlorophyll a* and *b* and *carotenoids*, which are pigment molecules in the chloroplasts of green plants. These pigments absorb green light poorly—it is the reflection of the light at the green wavelengths that causes chlorophyll-laden plants to appear green.

There are 200 to 300 pigment molecules bound to protein complexes in the photosynthetic, or *thylakoid*, membrane of the chloroplast (Figure 14-7B), and they form an *antenna system* that absorbs light energy and transfers it in the form of excited electrons to a chlorophyll molecule that serves as a *reaction center*. The reaction center then transfers its high-energy electrons to an acceptor molecule in the electron transport chain. High-energy electrons are then transferred through several membranes, for synthesis of ATP and the electron-carrier NADPH, which is used later for the synthesis of carbohydrates from CO_2 and water.

Photoreceptors to UV-B light are important in the development of seedlings (Fankhauser and Chory 1999). *Cryptochromes* (CRY1 and CRY2) and *nonphototropic hypocotyls1* (NPH1) detect UV-A and blue light, respectively. The cryptochromes are important in seedling development as well as the transition to flowering, whereas NPH1 and other factors are important for phototropism. Phototropins, of which there are at least two in the laboratory model plant (the mustard family member *Arabidopsis*), control growth and movement toward the light source.

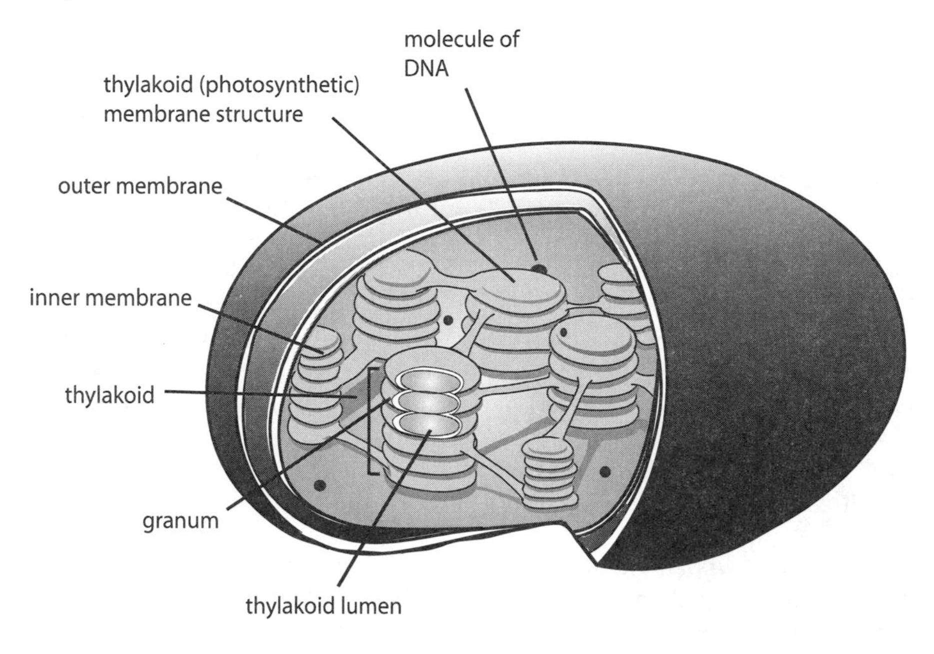

B.

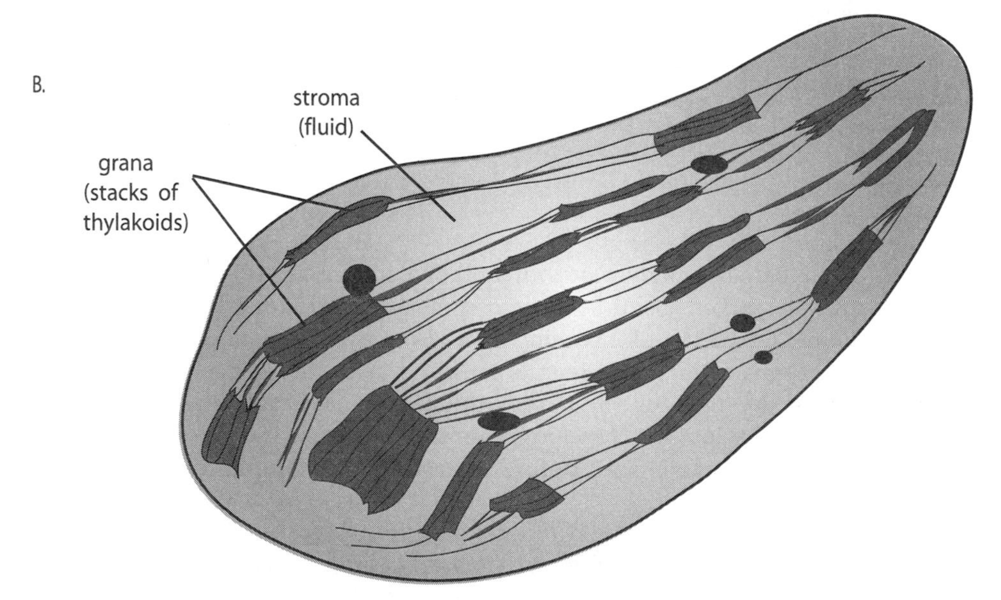

Figure 14-7. Plant chloroplasts. (A) Chloroplast structure; (B) cross section.

Phytochromes are important in many aspects of plant development (Fankhauser and Chory 1999). These are used in the detection of neighboring vegetation and the initiation of shade avoidance, the measurement of day length and light exposure and the consequent initiation of reactions to day length and duration of light exposure so that flowering occurs at the appropriate time and the like, and the control of the timing of reproduction.

Phytochromes are encoded by small and evolutionarily very old multigene families (Quail 2002), with structural similarity to prokaryotic histidine kinases, although the plant proteins phosphorylate the amino acids serine and threonine rather than histidine or aspartate. *Arabidopsis* has five known phytochromes, A to E, which play a role in plant development at all stages. Photoreceptors for light essential to growth apparently have multiple interactions that provide complementary or redundant ways through common transcriptional pathways. Phytochromes are pigmented proteins that can convert between two spectrally distinct forms, Pr, a red-absorbing form, and Pfr, a far-red-absorbing form. Photoconversion of Pr to Pfr induces a range of developmental responses in the plant, which have been shown to halt when Pfr converts back to Pr, showing that Pfr is the active form of the receptor.

Biological pathways in plant light response involve familiar elements of gene regulation (Quail 2002). For example, although plants have far fewer than animals, some G proteins have been identified as components in the phytochrome signaling pathway. They induce chlorophyll biosynthesis, among other functions, as well as the expression of many light-regulated genes. Light-induced changes in ion transport across the cell membrane, and other kinds of signal transduction, are apparent, but they have not yet been characterized in detail (Fankhauser and Chory 1999; Quail 2002). Essentially, many aspects of plant metabolism, growth and reproduction rely on light and plants have elaborate gene-regulation signal transduction mechanisms that respond to different aspects of received light, which involve several gene families specific to plants.

THE EVOLUTION OF EYES

Light can be detected in many ways, but light detection is not exactly what we mean when we think of *vision*. Vision is more complex because it implies the processing of intricate *spatial* patterns of incoming light along with its intensity and aspects of its spectral characteristics. In higher organisms, vision involves interpretation of complex signals by the central nervous system and translation of these into decision-making that usually includes a directed response, such as by muscular instructions directing movement. Before the response, the patterned aspect of the signal must be *detected*.

Because light travels in straight lines, an organism receives spatially organized light energy that is a map of the objects from which the energy comes. To see pattern, the organism must have some corresponding form of neurological spatial "map". This is in contrast to the much less spatially organized nature of olfaction, sound, or (sometimes) heat, the essentially nonspatial nature of immunological information, and the temporal arrangement of daylight or air temperature. For spatial processing, there must be a receptor map. Interestingly, as we will discuss in Chapter 16, even senses that do not depend on spatial perception use spatial maps in the brain to decode signal. This is probably a function of the fairly uniform histological, laminar, and synaptic organization of the segments of the brain that decode sensory input.

The hundreds of ommatidia that comprise the compound eyes of insects provide a fixed receptor array, each with its own neuronal connection to the brain. In camera-type eyes, such as those in mammals, the rods and cones in the retina are fixed in location and their corresponding neurons can be connected to the brain in a way that directly conserves spatial relationships, or at least the brain can reconstruct them because the signals from the same matrix position in the retina (or compound eye) are fixed. In mammals, unified stereoscopic vision further integrates the slightly different images from right and left eyes. An important part of this is that there need be no specific a priori aspect of each ommatidium or retinal cell, relative to images that are going to be detected: the organism is set up to detect *any* spatial light pattern. It does not need, for example, built-in diagrams of its predators, food, or landscape panoramas.

The Multiple Evolutionary Histories of Eyes

There are many ways in which eyes that are used for spatial inference can be constructed. Figure 14-8 shows some of them. The arrangement and number of eyes vary extensively, even when considering just bilateral cerebral eyes (e.g., see Arendt and Wittbrodt 2001). These include relatively simple larval eye spots in many branches of animal taxa, to more elaborate eyes formed by an array, usually a cup, of retinal cells. In some eyes, this has deepened and been closed except for a small opening, which like a pinhole camera allows a single image to be received in an array of photoreceptor cells on the inside. More precise (in human terms, at least), focused images are formed by eyes with lenses in this aperture, as found in mammals. Each ommatidium in an insect compound eye is a simple eye in itself, with spatially distributed neurons from its individual receptor cells receiving similar signals that are sent to similar regions of the brain.

Comparative anatomy and taxonomy have shown that eyes have evolved independently many times (a classic analysis suggests 40–65) (Arendt and Wittbrodt 2001; Salvini-Plawin and Mayr 1961). For that reason, eyes have long been used as exemplars of parallel evolution whose distribution is in the phylogeny of animals; that is, eyes are examples of analogy rather than homology because similar eyes appeared intermingled among taxa that do not appear to have shared one type of eye in their common ancestor. The evolutionary complexity of eyes perplexed Darwin (*Origin of Species*), who confessed that he could hardly imagine something as structurally complex as an eye evolving and reevolving independently so many times. A simpler explanation would be that animal eyes were homologous rather than analogous, but to rescue such a notion required a common ancestral eye that could have evolved into the diverse descendant forms seen today. Darwin hypothesized a primitive animal eye consisting of two cell types, a photoreceptor and a nonphotosensitive pigment cell with a substance that shades the light so that its direction could be detected, all perhaps encased in some kind of translucent covering. Primitive eyes resembling this structure exist in various taxa, but it was the remarkable discovery that genes fundamental to vision were widely shared among animals that made the idea credible.

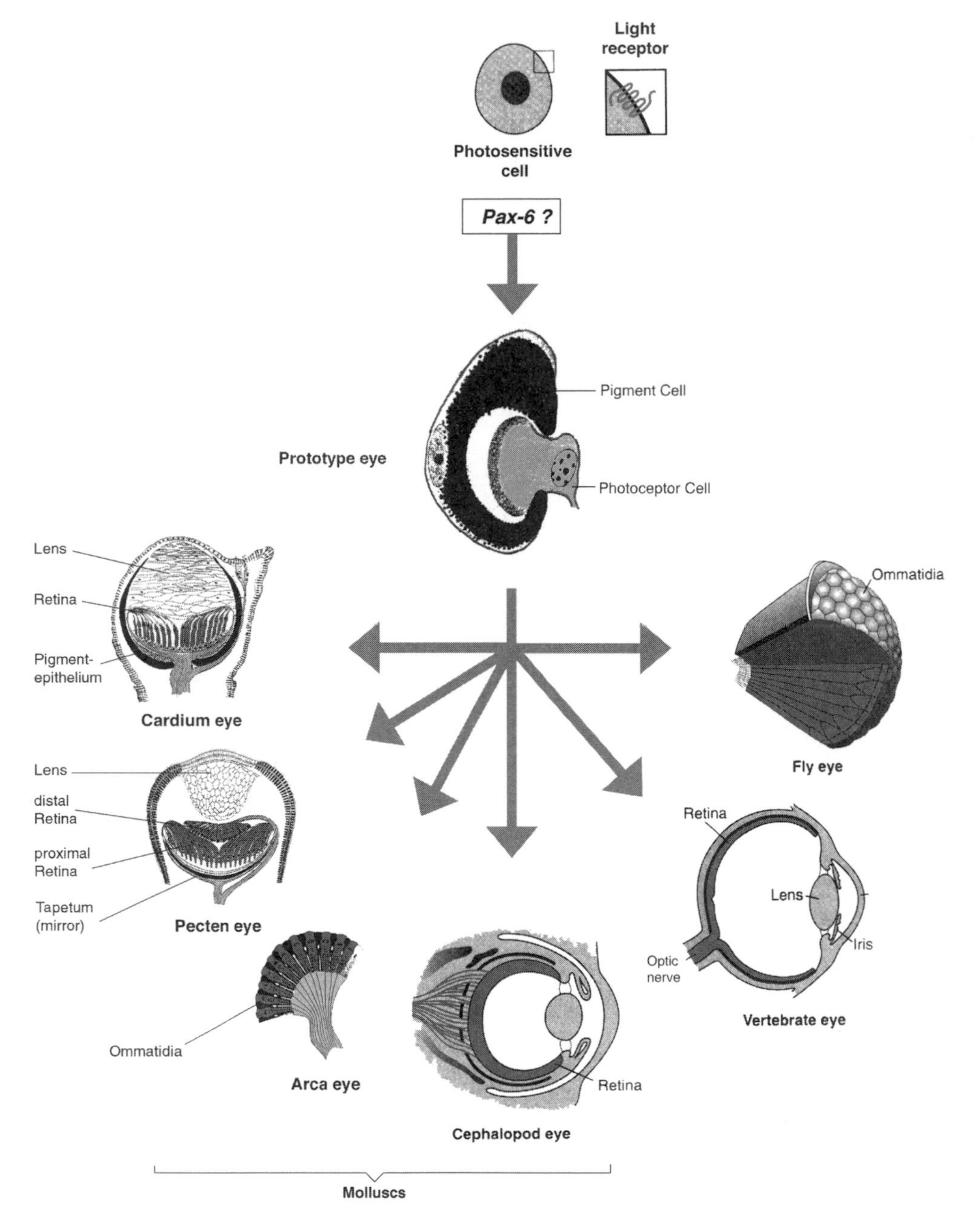

Figure 14-8. Evolution of diverse eyes from a common rudimentary photoreceptor, following Darwin's original notions; variation even within mollusks is shown, as is some of the recurrent evolution of eye types. From (Gehring 2002) with permission.

PAX6 AND CONSERVED HOMOLOGY AMONG EYES

The transcription factor (TF) *Pax6* was one of these genes. Expressed early in development, *Pax6* appears to serve as a selector for the initiation of differentiation cascades involved in eye development in most metazoans tested. Evidence showing this has come from expression experiments, as well as from natural or artificial mutational studies in mouse, humans, flies, and other animal species (e.g., Gehring 2002; Gehring and Ikeo 1999; Punzo et al. 2001). Rhodopsins and intracellular light-sensitive mechanisms exist in bacteria, suggesting that light sensitivity evolved before multicellular life. Furthermore, *Pax6* is expressed in several other later-stage eye structures and even regulates the expression of at least some opsin genes, like insect rhodopsin, expressed in mature photoreceptors.

The phylogenetically deep sharing of *Pax6* and *rhodopsins* can be seen as developmentally bracketing the morphological and sensory aspects of light reception among animals. This might seem to provide Darwin with the elements of his suggested common Precambrian origin for animal eyes (Figure 14-8), (Gehring 2002; Gehring and Ikeo 1999). But the sharing of a couple of genes does not provide obvious ease for his concern as to how similar types of complex eye *morphologies* have evolved several times independently or how similar eyes can evolve through different *developmental* pathways even in related species and from very different tissue contexts (e.g., Hall 1999). It is these morphologies that account for how light is used in various taxa. One explanation is that developmental patterning mechanisms were recruited, or *intercalated*, between the initiation of visual systems by *Pax6* and the later induction of *opsin* gene expression (e.g., Gehring 2002; Gehring and Ikeo 1999).

Different types of eyes involve different morphogenic processes, such as placode formation, invagination (e.g., of an optic cup), and periodic patterning (e.g., ommatidia). As reviewed in Chapter 9, these are standard parts of the developmental repertoire of complex organisms, which during evolution could have been invoked in new contexts in between the first expression of *Pax6* and the later expression of photoreceptors. The various forms of eye could have evolved sometimes inserting shared but other times different morphogenic mechanisms. Because the developmental mechanisms are resident parts of the animal developmental toolkit, such evolution could be relatively rapid (Pichaud et al. 2001).

Relevant to this is that eyelike structures can be induced experimentally in insects and vertebrates by ectopic application of Pax6 protein (Gehring and Ikeo 1999), in locations where relevant signaling molecules are already expressed (e.g., Kumar and Moses 2001; Pichaud, Treisman et al. 2001), but in contexts that normally do not develop into eyes (e.g., other imaginal disks in insects). Pax6 from one species can even induce such development in another.

These results suggest that different eyes share genetic homologies and that they did not entirely evolve independently as analogous organs. This is interesting because the intercalated mechanisms are also used in other structures, so that eyes would in part be homologous to antennae, feathers, and teeth (which also use Hedgehog, Bmps, and the like). The basic elements of the logic of the rapid evolution of diverse eyes are also found in other systems, including olfaction and various aspects of developmental polarity, repeated structures, and the like that we have seen are so characteristic of evolution.

Crystallins and the Nonconserved "Homology"

Another interesting aspect of eye evolution concerns the *crystallins*, major proteins found in vertebrate lens and corneal tissue. An important characteristic of crystallins is their light transparency properties when desiccated and compacted. However, genes coding crystallins did not evolve to serve a visual purpose. Instead, different species have opportunistically recruited different, often wholly unrelated proteins to use as their lens or corneal proteins (Tomarev and Piatigorsky 1996). These proteins typically also serve other functions in the same organism, such as heat shock response or housekeeping enzymes. Lens and cornea develop embryologically from the same tissue, but in a given species the corneal crystallins are different from those used in the lens. Many crystallins include *Pax6* REs in their 5′ flanking regulatory regions, and experiments show that the enhancers drive appropriate eye-specific expression, often across species (e.g., the enhancer from a chick can induce lens expression in a mouse). *Sox2* and retinoic acid receptor pathways may be alternative or additional parts of the shared regulatory mechanism (see below). However, unlike the conservation of the *Pax6* regulatory pathway itself, in the case of crystallins, the *function* rather than the specific gene is conserved. Phenogenetic drift has replaced one gene with another in different lineages that have shared the form of their eyes since a common ancestor.

EYES EXEMPLIFY IMPORTANT LESSONS ABOUT RECONSTRUCTING THE EVOLUTION OF COMPLEX STRUCTURES

There has been enthusiastic acceptance of the *Pax6/opsin* solution to Darwin's problem about the repeated, apparently independent evolution of complex eyes. The intercalary evolution scenario has seemed quite convincing, and similar relationships are being identified for other traits—indeed, in some ways that is becoming the rule rather than the exception, as we have seen. However, the picture is more complex and shows some of the dangers inherent in accepting evolutionary reconstructions that are too tidy. It is instructive to see what some of these problems are.

The "Master"-y of *Pax6*

It is clear that *Pax6* is important in eye development. No one thinks that genes act in isolation, but the idea that *Pax6* is *the* master selector for vision has gained common currency. There are many reasons why we like simple, grand explanations and metaphors (and usually they fit other things in our culture—this may account for the reliance on competition as a mechanism and the notion of "master" genes—the CEOs of development; we have noted similar caveats in regard to terms like "organizer" and "selector"). However, oversimplification gives a misrepresentative picture of eye biology and of evolution in general. Remember the complexity of developmental and signal-transducing pathways, and the ways that they are often incompletely documented (Chapter 7). Few if any pathways act alone.

The original *Eyeless* mutant was one of the classic traits of early fly genetics and ultimately led to the isolation of the homeobox/paired-box *Eyeless* (*Ey*) TF—a homolog of *Pax6* in vertebrates. However, *Pax6* is only one of several (known) genes that are roughly comparable in their importance in eye development (*Ey*, *Toy*, *So*,

Eya, *Dac*, *Eyg*, *Opt*, *Hth*, *Tsh*). These genes can induce ectopic eyes much as *Pax6* can (e.g., Bessa et al. 2002; Goudreau et al. 2002; Treisman 1999). The gene *Twin of eyeless (Toy)* is a *Pax6*-type gene closely related to *Ey* and perhaps evolutionarily closer to vertebrate *Pax6* than *Ey* is itself. Experimental ectopic expression of *Ey*, or of *Toy*, results in the formation of ommatidia on the legs, wings, and antennae, but *Toy* is expressed developmentally upstream of *Ey* and directly regulates the latter's eye-specific enhancer.

Other TFs, including *Sine oculis* (*So*), *Eyes absent* (*Eya*), and *Dachshund* (*Dac*), interact and are essential for development of compound eyes (Cutforth and Gaul 1997; Czerny et al. 1999; Punzo et al. 2002). Loss of *So* or *Dac* results in loss of eye structures. Ectopic expression of *Dac* induces expression of *Ey*, and the subsequent formation of an eye, although at lower efficiency than with ectopic expression of *Ey* alone. *So* probably acts downstream of both *Dac* and *Ey*, as it is not sufficient for eye formation. *So* is one of the class of TFs called *Six* genes (Pineda et al. 2000) whose members interact with *Pax6* at different stages of eye development in different species, including planarians, insects, and vertebrates, and can compensate to some extent for each other. The fly paralog *Optix* can generate eyes in an *Ey*-independent way (Seimiya and Gehring 2000).

Several homologs of these genes are also expressed in vertebrate eye development, but comparison is difficult because vertebrate eyes develop differently, and vertebrates also often have more members of the same gene families than do flies. For example, eyes develop from a single imaginal disk primordial tissue in flies and through interaction of two tissues (an ectodermal lens placode and the underlying optic vesicle of the brain) in mammals. *Pax6* is higher in the eye pathways in flies than in vertebrates; for example, optic vesicles form in *Pax6*-mutant (*Sey*) mice (Harris 1997). *Pax6* and *Six3* interact in a quantitative, mutually inductive way in eye development, suggesting that neither acts as "the" control gene.

In mice, Pax6 and Sox2 protein bind together to induce lens development, and cooperative enhancers for these genes are found flanking the *δ-crystallin* gene, which is subsequently expressed in the lens in mice and chickens (Furuta and Hogan 1998; Kamachi et al. 2001). Similar interactions between *Pax* and *Six* homologs are involved early in fly eye development. *Pax*-coded protein may provide a DNA-binding domain and Six a domain that facilitates transcriptional activation. Expression of the signaling factor *Bmp4* in the optic vesicle is also involved, but this appears to function independently of *Pax6* (Furuta and Hogan 1998). *Pax6* also interacts with *Pax2* and at least one other related gene in eye development (Papatsenko et al. 2001).

The homeobox gene *Rx* is neurally expressed earlier than the eye in vertebrates, and experiments show that the development of retinal tissue (a late stage) depends on it (Mathers et al. 1997). In at least some fish, a TF *Rx3* is needed to initiate the early stages of eye development (Loosli et al. 2001). There is a homolog in flies (*DRx*), but it is not expressed in eye imaginal disks.

It is probably better to view these various genes, including *Pax6* but perhaps others not yet known, as forming a horizontal network of control rather than a vertical Master-driven hierarchy (Kumar and Moses 2001; Pichaud, Briscoe et al. 1999; Pichaud, Treisman et al. 2001). Inactivating this gene can leave eyes unimpaired. This is consistent with the general nature of gene activation, which requires the binding of complexes of multiple regulatory proteins. A selector ("master" regulatory) gene needs a tissue environment previously made ready, which in the case of eyes is due to expression of growth factors such as *Bmp*s (in flies, the homolog *Dpp*), *Egf*s, and *Hedgehog*, or *Notch* signaling, and the absence of inhibiting signals such as *Wnt* (in flies, the homolog *Wg*), among other factors. Similar factors are needed to specify eye vs. antenna fate in flies, for example, from a common starting primordium, and their expression is not restricted to eye primordia. Experimentally induced ectopic *Ey* expression only induces eyes in prepared soil, so to speak. *Pax6* is an early neural gene in many ways that would be atop more than just eyes in any case. A gene *Tsh* (*Teashirt*) is not needed in normal fly eye development but can trigger ectopic eyes experimentally, presumably by interacting with other genes.

THE NECESSITY FOR *PAX6*

There are other ways in which the universality or necessity for genes like *Pax6* has perhaps been overstated or oversimplified (Gehring 2002; Gehring and Ikeo 1999; Kumar and Moses 1997; Kumar 2001). *Pax6* is clearly important to eyes, but it is important to *us* that we have a realistic understanding of the role of genes in evolution, which often requires sorting out dauntingly complex or seemingly contradictory facts. Even in the classical fly *Ey* mutations, eyes were not entirely or always missing, and over some generations regeneration could occur, and this involved different genes in different lines. Although the fly is used as the classic example of the universal master nature of the *Pax6* pathway in visual systems, even in the fly the ocelli (central, simple eyes) do not require *Ey* but *Toy* instead (Punzo, Seimiya et al. 2002); ocelli use *opsin* genes, sometimes different from those used in ommatidia, as the *Rh2* gene is in *Drosophila* (Pollock and Benzer 1988; Smith et al. 1993). *Pax6* knockout flies have ocelli, but at least some knockouts of the related gene *Eya* lose both compound and medial eyes (T. Oakley, personal communication). The apparent independence of *melanopsin* pathways from the classical ones used in eyes also suggests that there is more to this story.

Pax6 is a member of the "paired" class of TF, and like some of the members of that family has two DNA binding domains, a homeodomain related to that in the homeobox TFs and another called the paired domain. *Pax6* binding sites have been found flanking crystallin and perhaps other eye-related genes, a fact used to support the master gene idea, but the story is not so simple. The paired domain seems to be needed, but evidence about the homeodomain is to date unclear. For example, it has been shown that *Pax6* can regulate the expression of invertebrate rhodopsins, but this regulation is complex (Papatsenko, Nazina et al. 2001), and eyes can develop normally, including rhodopsin expression, in the absence of the homeodomain (Punzo, Kurata et al. 2001).

Pax6 has complex binding domain and splice variants, so that the natural pattern of eye-related effects is not simple (Duncan et al. 2000) and may not be easily inferable from a knowledge of RE sequences alone. How this relates to the other genes in the eye-related regulatory hierarchies is unclear. The fact that *Ey* regulates *Rh1* in flies, a late-stage event, shows that the role of *Pax6* is not just as the master that sets off the eye cascade (this illustrates the potentially misleading metaphoric use of the word "master"—CEOs are not found on the assembly line, and generals no longer lead the charge into battle). There are occasions when the "master" gene has been substituted over evolutionary time and occasions when the later events came first, which were then regulated by a higher-level "master" gene (e.g., Graham et al. 2003).

The presence of *Pax6* homeodomain enhancers in opsin genes has been used to support the master-gene argument, but the gene is not expressed in differentiated vertebrate rod or cone cells that do express *opsins*. Perhaps, *Pax6* is involved only earlier in photoreceptor development. Planarians have two *Pax6* homologs, but experimentally neither is needed for the regeneration or maintenance of eyes, although a *Six* homolog is essential (Goudreau, Petrou et al. 2002; Harris 1997; Pineda et al. 2002). A comparable role has been demonstrated for *Six* class genes, including *Sine oculis*, in flies; as noted above, *Six* and *Pax* genes seem to interact.

The Nature and Role of Crystallins

Many lens and corneal crystallin genes have *Pax6* enhancers, suggesting that this is how they were intercalated opportunistically by natural selection into the eye development cascade in different lineages. *Pax6* seems to be generally involved in early neural development; however, crystallins are nonneural in nature. Crystallin genes have acquired *Pax6* enhancers, but this gene is not required by all crystallins, for example, duck crystallin (which may also not require *Sox2*) (Brunekreef et al. 1996). In some vertebrates (e.g., chick and mouse), the optic vesicle in the forebrain plays a necessary role in lens induction, and one study has shown that *Bmp4* is necessary for this to occur, but this pathway may be independent of *Pax6* (Furuta and Hogan 1998).

With regard to crystallins, one argument includes the notion that *some* such transparent lens and corneal proteins are needed—and recruited, for example, by the selection-favored appearance of *Pax* enhancers in the future crystallin genes' regulatory regions—in eye assembly. The idea is that this was an opportunistic pathway in that it did not matter to natural selection *which* genes were recruited. However, gene knockout experiments have generally found that there is little if anything wrong with eye development when the crystallins are missing. Experimental overexpression of Pax6 protein in mice has induced cataract, but apparently due to cytostructure genes rather than crystallin anomalies (Duncan, Kozmik et al. 2000).

One possible interpretation that requires less invocation of directional adaptive selection is that *Pax6* and other eye-related regulatory genes bind enhancers flanking hundreds of genes that become expressed in eyes, and hence in lens and cornea. Among proteins expressed in a tissue, there will be a distribution of abundance, and distributions of that sort are usually skewed—many elements will be relatively rare, only a few will be common. Many factors, including chance or the presence of other REs, can affect the relative amount of the gene product found in any tissue, including eyes. One can always extract proteins from lens or cornea, rank them by expression level, and *define* those with the highest level to be "crystallins." Crystallins are water soluble and densely packed in these tissues, but this is true of many other proteins with lower expression levels. Nothing prevents us from crediting the presence of these relatively abundant proteins to their *Pax6* enhancers as if driven by selection for visual purposes. There may also be other important related genes that serve as, or with, the known crystallins (Chauhan et al. 2002; Chauhan et al. 2002, and A. Cvekl, personal communication).

As far as selection goes, the key fact may simply be the compaction of sufficient water-soluble proteins, and among the hundreds of proteins expressed in a lens or cornea, there may be little in the way of selection involved in their evolution to "be" crystallins. It is important to be aware of the possible arbitrariness of our labeling based on strong darwinian assumptions and to be sure that we have evidence that selection is in fact responsible. Other mechanisms can also result in abundant lens expression. Duck crystallin appears not to be regulated by *Pax6* in the lens; instead, the ubiquitous protein Sp1 seems to suffice (Brunekreef, Kraft et al. 1996).

Blind mole rats use their eyes for circadian rhythm. They, and moles, have little if any real vision, yet crystallins are still expressed in their lenses (Avivi et al. 2001; Hendriks et al. 1987; Quax-Jeuken et al. 1985; Smulders et al. 2002). Mutations have accumulated in the mole rat lens crystallins—but not as rapidly as in a pseudogene. These "crystallins" may still have some function. But are they really crystallins? There is also a report of *alphaA-crystallin* that is expressed in the eyes of a species of blind cave fish (Behrens et al. 1997; Behrens et al. 1998) in the expected developmental stages and places based on closely related sighted fish.

Some of these data may turn out to be experimental artifacts, and selection explanations can of course be correct. These examples do illustrate the general point that evolutionary stories are usually not simple, and phylogenetically deep conservation is usually not complete. Nonetheless, there *are* clear genetic homologies among eyes that *do* suggest the unity of animal evolution. Thus, the differences among animal eyes may show that the common origins were at a much simpler, older molecular level than that of a common "eye" (e.g., Arendt and Wittbrodt 2001), just as Darwin suggested. His speculation that this could not have happened by chance was prescient. That we can now see this in genetic terms is not surprising, when we think of the highly conserved developmental processes that produce complex structures in multicellular organisms.

The discovery of the genes added new and convincing evidence, but it should not be surprising that paired anterior cerebral light-sensing structures found in diverse animals would not be entirely unrelated. An ironic note is that, despite playing a major role in showing these connections, and their prediction by Darwin, Gehring (like Darwin) suggests that the original eye evolved by *chance* (Gehring and Ikeo 1999). However, that is a strange invocation of chance from a darwinian point of view because the first start toward eyes in animals may *not* have required a "hopeful monster" generated by chance, given the primitive photosensors in single-celled organisms.

A number of "natural experiments" with eyes show that mutation can inactivate vision in many ways. Blind fish have evolved in various cave lakes in Mexico, although this process has been relatively recent in evolutionary terms, and vision can sometimes be restored in crosses between species (or subspecies). The reason is that different genes have been mutated in different species of fish, and crosses allow the functional genes in each blind parental strain to complement each other in the offspring, restoring function. Other animals not using vision, such as those in areas where it would be of no use, have lost much or most of the sight that was presumably found in their ancestors (moles are a favorite example). The star-nosed mole has an unusual variant on these themes—its nose provides a special kind of tactile rather than light-based spatial sense, as will be described in Chapter 16.

Ants and bees vary greatly in the size, components, and acuity of their eyes, and this is generally correlated with their role in life and the amount of time they spend outside their hive. In an historically well-known observation, Henry Bates' (Bates 1862) attention was drawn to Amazonian leaf-cutter, fungus-culturing *Sauba* ants (genus *Atta*). He observed three main groups of worker ants, one being a large-headed class that another famous Amazonian adventurer H. M. Tomlinson noted keep "moving hither and thither about the main body; having an eye on matters generally, I suppose" (Tomlinson 1912). Ironically, Bates identified a third group with "an eye"—a single, central eye—but he only observed this Cyclopean caste deep underground where there is no light. Actually, there seems to be a continuum of size in these workers, who vary in the jobs that they can do (Hölldobler and Wilson 1990); there are overall correlations between head and eye size, so whether the variable nature of these particular eyes represents an anomaly of any kind is unclear.

CONCLUSION: "IF THE ÆGLE WERE JUDGE"

Organisms use light directly as an energy source or as a source of information about their surroundings. The ability to detect and characterize objects at a distance is important in the way of life of many or even most animals. A variety of attributes of light are used, including direction, brightness, image, distance, and motion of objects. The evolution of vision in turn affected the evolution of the organisms being *detected* by their light images. This occurs in many ways, not least being the evolution of mating rituals and display, protective camouflage and deceptive coloration (e.g., caudal eye spots that disguise a tail as a head to trick predators, and Batesian mimicry in which tasty species mimic bitter ones), light-based attractants in flowers and fruit, motion behavior to escape predators. Visual cues are also used to set traps, by disguising a predator. Fireflies have species-specific flash signals to attract each other (and their predators who mimic them to lure a lusty but careless fly to its death). These phenomena variously use color alone and/or coloration pattern.

As important as it has been, however, seeing is not a specific problem to be solved by organisms, and no environment demands any particular type of vision. Indeed, the presence of various primitive photoreception mechanisms shows that organisms can use light information without having a brain (e.g., primitive organellar vision in unicellular organisms; jellyfish). Vision probably also illustrates the role organismal selection may play in nature. Organisms use what they have, and this may sort them out as well as classical darwinian selection does, but without fitness differences.

The homologies indicated by *Pax6* were dramatic findings. But this was only among the first of many similar findings of deeply conserved genetic mechanisms involving regulatory, enzymatic, and structural genes in many different structures and organ systems. The eye story, along with a few others like the *Hox* axial patterning system, led the way in these discoveries, but findings like this are rather general. Indeed, we have now come to expect to find such conservation. This unifies the animal world and may suggest that life is younger than we thought, relative to the evolution of genetic mechanisms—there has not been enough time for the sharing of these genetic mechanisms to be erased, replaced by selection, or subjected to phenogenetic drift, as sequestration among species almost guarantees will eventually occur if the Earth remains hospitable to life for a long enough time. However, the sharing that has persisted so far has perhaps tempted overstatements of homology because the conservation has been far from complete or simple, as eyes and vision illustrate.

Photoreception is evolutionarily related in interesting ways to olfaction and has diverged from a common ancestral mechanism for cells to sense external conditions. They use related members of the *7TMR* family (and *Pax6* is involved in olfactory development). This homology can be overstated because there are so many genes in this family and they serve such diverse purposes. However, like olfactory receptors, photoreceptors are G protein-linked receptors that activate cyclic nucleotide-gated channels using cGMP to affect cAMP concentration (it is synthesized by guanylyl cyclase and degraded by cGMP phosphodiesterase). A major step was the evolution of the different binding properties of these related genes—one for chemical shape and the other for light energy.

As is true of hearing and olfaction, seeing is not just a molecular trick played by the brain with opsin firing. The eyes and head can be moved relative to the environment, and the brain must integrate that movement with the neural information it receives from the eyes or other sensory neurons. Although this book is about genetic mechanisms and how they have been used to receive diverse information from the environment—especially unique or unpredictable information—we should not forget that morphology is also a major aspect of organisms. Morphology (and its coordinated use) is responsible for the mobility of external ears, eyes, nostrils, heads, and bodies, and this is all an integral part of how organisms detect and respond to their environment.

Most animals can respond to the world in ways that are not prespecified. Some merely need to know where a light source is. Others need a spatially arranged map of some aspects of the external environment and have evolved complex means to integrate spatially arrayed input, as we will see in Chapter 16. There is a correspondence between the properties of optics and the structures found in the eyes in nature, strongly suggesting that adaptation by natural selection has been at work.

But in nature, there are many ways to see, and no organism has them all. Animal brains have evolved to be able to receive organized light signals and resolve them in one way or another. There isn't one species that might not do better with additional visual ability, but all make do with what they have, and that has been good enough.

In Greek mythology, Juno jealously suspected that her husband Jupiter had a mistress who took the form of a heifer to disguise her identity (Juno was right). She hired the herdsman Argus to keep an eye—actually, to keep his *hundred* eyes—on the heifer to prevent further mischief. Argus never slept with more than two of his eyes closed and so was ever-vigilant. This evolutionary experiment ended suddenly, however, when Jupiter commissioned Mercury to get rid of Argus. Mercury lulled Argus to sleep with story-telling and then killed him in a stroke. Juno's revenge included bedecking the peacock's tail with Argus' eyes, which are now seen but no longer see. Being covered with 100 eyes always on the alert might have been selected for, but it wasn't (except, perhaps, in the scallop, whose hundred or so eyes are arrayed, beadlike, along its mantle).

Aristotle and his peers through Classical times had many ideas about what species could see. Aristotle, for example, thought that moles were blind. But in a huge compendium to debunk long-held mistaken ideas published in 1646, Thomas Browne (Browne 1646) reported that he observed mole embryos to develop eyes and that in fact moles could see. That a species could not see as well as others do, he notes, is a value judgment of which we need beware: "if the Ægle were judge, wee might be blinde our selves."

Chapter 15
The Development and Structure of Nervous Systems

In previous chapters, we have discussed the open-ended environmental variability that organisms encounter and *detect* with sight, taste, smell, feel, and so on. In Chapters 15 and 16, we will describe in basic terms what is known about the central problem of how organisms *perceive* this infinite and constantly changing variety of sensory signals.

As humans, we tend to relate these issues to our own personal experience, leading us to equate perception with consciousness. But the two are highly or perhaps even entirely different. In fact, we only have indirect indicators—at best—of how different the mental experience of other organisms may be from those that we understand. The nature and even the definition of consciousness are still debated and unclear, as is whether nonhuman organisms experience it at all. Many organisms to whom we would not attribute consciousness clearly do have integrated, organized responses to complex environmental input. Is that so different from what humans experience? How would we know one way or the other?

An organism does not have to be complex and multicellular to respond to light, vibration, temperature change, and other environmental signals. Some do so locally, with no centralized processing, and we noted that immune response was like this and that in plants each part can in many ways respond independently of the rest of the plant. Other species, however, have a hierarchical and highly specialized central nervous system (CNS) that serves to create an internal representation of environmental cues and to organize responses. Even plants and single-celled organisms integrate and respond to multiple environmental cues; therefore, a neural mechanism for receiving and responding to unpredictable environmental signals is clearly not required for survival, or even for adapting to changing external conditions. Only a small subset of living organisms does so through a CNS.

Charles Darwin himself, in his 1872 *Descent of Man*, goes to great lengths to connect human behavioral components with those of other species, to show that in fact we share common origins with other animals. He was appropriately

Genetics and the Logic of Evolution, by Kenneth M. Weiss and Anne V. Buchanan.
ISBN 0-471-23805-8

impressed even by ants: "the wonderfully diversified instincts, mental powers, and affections of ants are notorious, yet their cerebral ganglia are not so large as the quarter of a small pin's head. Under this view, the brain of an ant is one of the most marvelous atoms of matter in the world, perhaps more so than the brain of a man." Do we have a rational basis for asserting that simple organisms do not have a form of "awareness" or organismal level percept and that their reactions to inputs are as unknowing as, for example, that of a motion detector that turns on lights in a house?

In organisms with a CNS, sensory input may undergo some processing locally at the site where it is received, but ultimately raw light, sound, taste, touch, and other signals all proceed to the nervous system in essentially the same way, via action potentials, and the signals are decoded by modality in the brain. The pathway the signal travels in the brain determines how it is ultimately perceived, and the pathway is determined by the destination of the particular neuron through which the signal has traveled to reach the brain. Light signals are sent from the eye to the visual centers in the brain, where they are translated into perception of color, shape, motion, and the like. Sound waves are collected in the ear and shunted along to the auditory centers in the brain to be translated into perceived danger signals, music, communication, and so forth. However, *percept*—whatever it actually turns out to be—is not entirely location dependent because, as we will see, the same kinds of input (e.g., sound, light) can occur in different parts of the brains in different people or even in the same person at different times in his/her life.

In most sensory systems, neurons project their axons *topographically*, that is, in an orderly fashion that provides a precise spatial "map" or representation of the location of a particular class of receptors on the surface of the body, whether the retina, the olfactory epithelium, the cochlea, or the skin. The axons terminate in distinct sensory centers of the brain, their parametric representation of the physical world still conserved. That is, neurons that are adjacent in the receptor surface terminate adjacent to each other in the sensory area of the brain to which they project. The central processing system detects in detail where the signal came from, and this may represent the external world in a literal sense, as in vision and touch, but need not as in olfaction or hearing. These ensembles are called *neural maps*.

Signals are processed through multiple levels in the CNS before final perception. In most sensory systems, both parallel and hierarchical (serial, with one step preceding the next in order) processing are involved in the anatomic connections that send signals to the appropriate centers of the brain for interpretation. Auditory systems, for example, can simultaneously process many sounds as well as many aspects of each sound. Light is processed into increasing complexity by higher cortical areas, and different qualitative aspects of light are processed simultaneously. The various sensory areas in the brain have many functional and structural commonalities, and, despite segregation of the sensory areas of the CNS, in the end the brain integrates the different signals to create what is experienced as a unitary but multidimensional representation of the external world.

It might be rather subjective to discuss which organ system is the most complex in its structure or function, but the nervous system would certainly be a candidate, especially in vertebrates where the developmental dynamics are far from a simple hierarchy and the connection between structure and function not always clear. In Chapter 15, we will describe in broad terms the various types of nervous systems known, their architecture, how they work, and some of what is known about the genetics of their development. In Chapter 16, we will discuss different sensory modalities and how they are decoded and perceived in the brain.

INVERTEBRATE NERVOUS SYSTEMS

Action potentials probably existed in the first single-celled organisms, predating neuronal activity and the evolution of the nervous system, evolved for some other purpose, perhaps having to do with response to external chemical conditions. Some modern single-celled organisms can effect a kind of preneuronal electrical signaling and are capable of producing action potentials that regulate, for example, the direction of the beating of cilia and thus movement of the cell.

Colonies of multicelled organisms that do not have a central nervous system, such as the cnidarian *Obelia*, may also produce action potentials. *Obelia* colonies share a digestive system, and electrical signals can spread through the common epithelial lining of the digestive cavity. Other cnidarians have nervous systems of a different sort. Hydra, for example, have *nerve nets*, collections of nonpolar neurons that send electrical impulses in any direction such that the whole body responds to any stimulus, although there can be a concentration of these nerves around the mouth for enhanced detection of food. There is no hierarchy of control, nor any specialization of neuronal function. The nerve nets also control the organism's movement (Matthews 2001) as well as acquisition of food.

Echinoderms like starfish typically have a locally organized nervous system, with no central *ganglion* (see Figure 15-1). A ganglion is a knotlike cluster of nerve cells. The neurons of these radially symmetric organisms connect into a *nerve ring*, with branches extending into each of the five arms to innervate them. Jelly fishes have a neural net with ganglia, which allows some organization of movement such as swimming by pulsation. Wormlike hemichordates have a nerve net in the epidermis, which thickens to form several solid nerve cords, and a neurocord, a collection of giant nerve cells, which connects the nerve cords formed from the nerve net. The neurocord probably plays a role in rapid motor responses (Butler and Hodos 1996). Flatworms have a collection of enlarged anterior ganglia, analogous to a simple brain. Longitudinal nerve cords allow for some control of body movements.

Although insects have brains, their nervous systems are much less centralized than those of vertebrates. Generally, an insect CNS is organized into a series of ganglia, strung together along a *commissure* (a junction at which corresponding structures join), that runs along the ventral nerve cord, the latter being linked to the brain (see Figure 15-2A). The brain itself is a collection of six fused ganglia (three pairs) that each control specifically circumscribed activities. The first pair, the *protocerebrum*, innervates the compound eyes and the ocelli. The second, the *deutocerebrum*, integrates signals from the antennae. The third, the *tritocerebrum*, innervates the *labrum* (roof of the mouth) and integrates input from the protocerebrum and the deutocerebrum. It also links the brain to the rest of the ventral nerve cord and the ganglia along the cord that control other organs and behaviors, such as feeding, mating, locomotion, and sensory reception. As in the vertebrate CNS, there are three kinds of neurons in the insect nervous system: *sensory afferents*, *motoneurons*, and *interneurons*, nerve cells in the CNS that link the sensory and motoneurons.

The major structures of the insect brain include the *mushroom bodies*, *central body*, the *optic lobes*, and *the antennal lobes* (Figure 15-2B). The mushroom bodies

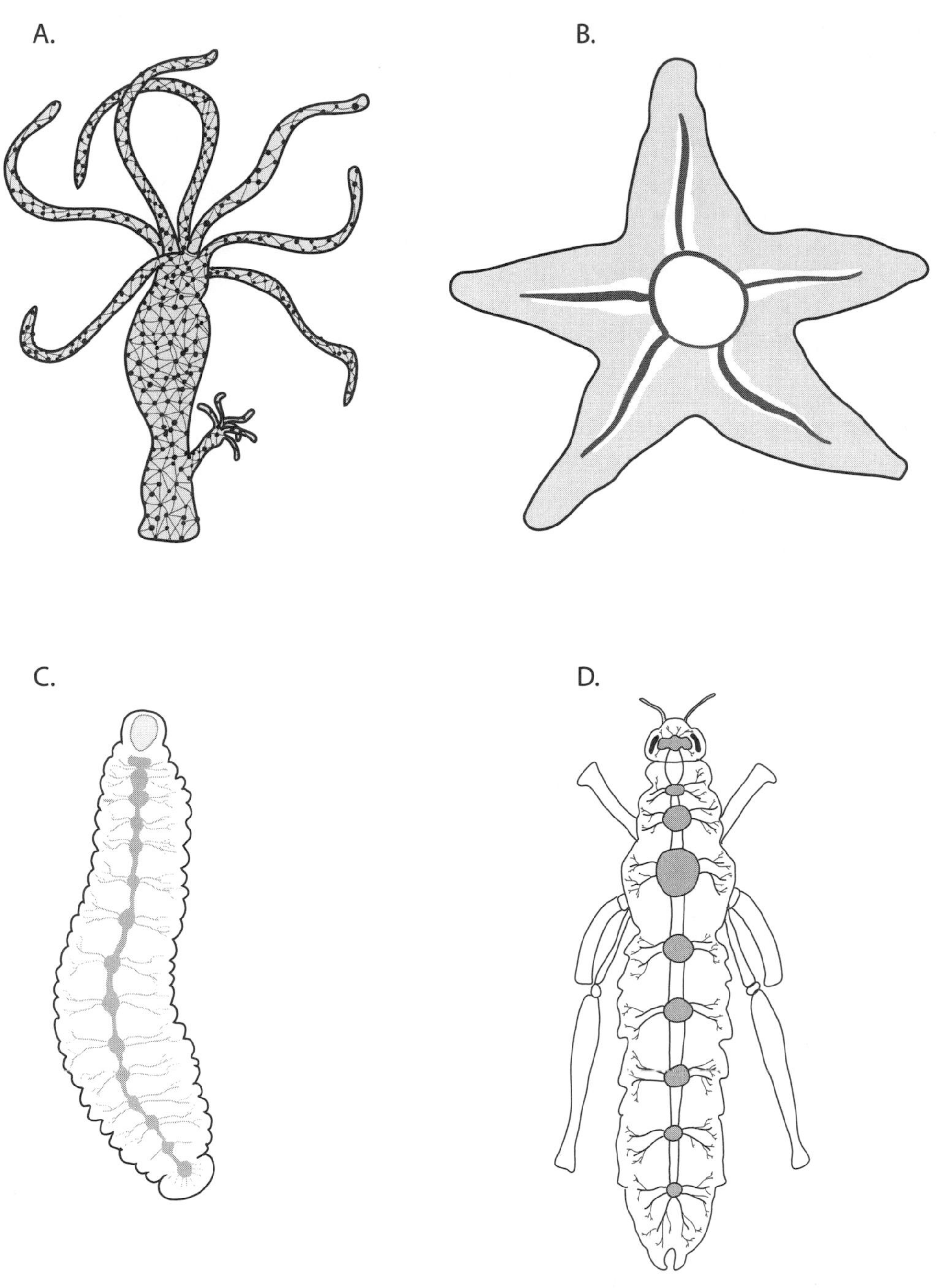

Figure 15-1. Organization of the nervous system. (A) The nerve net of the hydra; (B) the starfish nerve ring; (C) the leech ganglia and; (D) the cephalized nervous system of the grasshopper. Redrawn from (Matthews 2001).

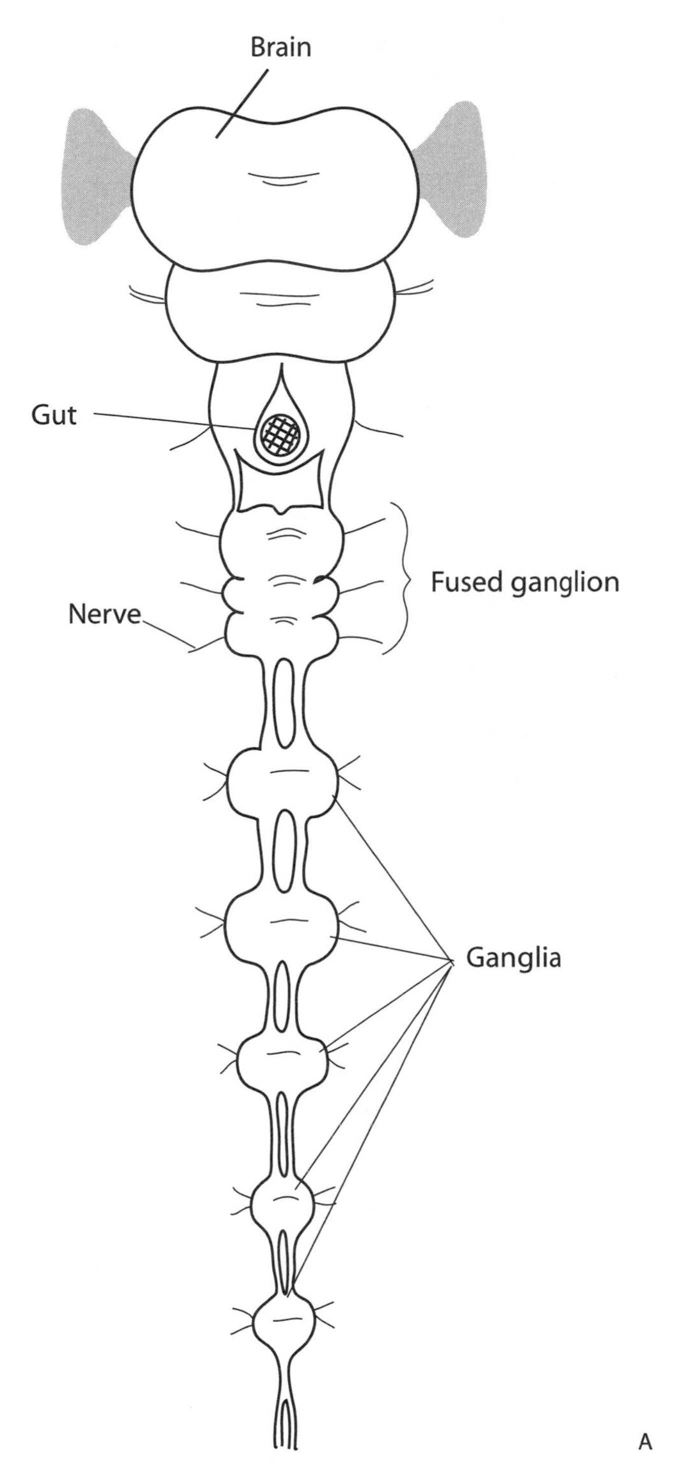

Figure 15-2. Schematics of the insect nervous system (A) and major structures of the insect brain (B).

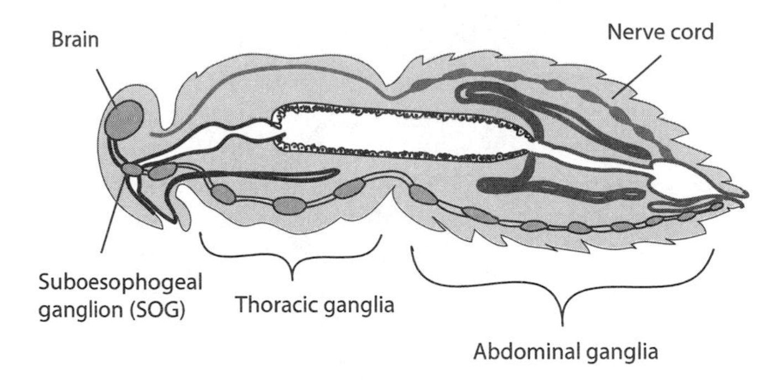

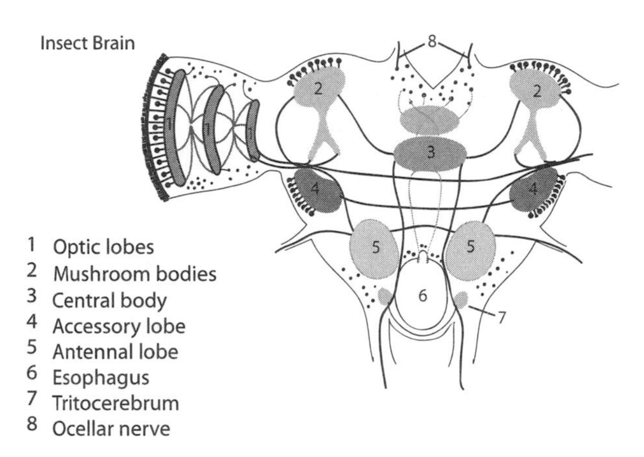

Figure 15-2. *Continued*

are involved in olfaction. The central body is a way-station through which nerve fibers pass from one hemisphere to the other. The nature of the signal processing that takes place here is unclear; it may be involved in visual processing (Burrows 1996). The optic lobes are laminar, that is, organized in layers, and process information from the insect's compound eyes through the layers in a way that, as in vertebrates, retains the retinotopic (spatial) map of the image. The antennal lobes process olfactory signals coming from the antennae.

This may be an appropriate place for us again to note the many parallels between vertebrate and invertebrate nervous systems. Biologists long noted the similarity of organization (eyes, mouth, digestive system, limbs, and so forth) between invertebrates and vertebrates, although they were considered analogous rather than homologous. As we have already seen in many ways, recent studies have shown similarities—sometimes close similarities—system by system, at the gene, developmental, and cytological levels. We noted in Chapter 9 the suggestion by Geoffroy St. Hilaire in the early 1800s that insects were inverted vertebrates, in that the dorsal nerve cord in vertebrates is homologous to the ventral dorsal nerve in invertebrates (and the digestive systems are in the corresponding inverted locations). There are some corresponding genetic similarities as well, but it is important to resist the temptation to extrapolate too far or to expect even true homology to be too precise at the trait level.

Overall, invertebrate nervous systems range from very simple and diffuse to relatively complex and centralized. As a result, the organisms react in ways that seem interpretable to us. Darwin went through a list of ways. Insects exhibit what appears to be fear, which can be elicited by various signals, panic (when they are upside down, for example), and exploratory behaviors, not to mention courtship (including males fighting for mates). There are many examples of learned behavior in insects as well, and they can be trained in the laboratory (Waddell and Quinn 2001). This may seem surprising, but their brains do have room for the integration of new experiences and behaviors (Meinertzhagen 2001). None of this explains how information is integrated, however, much less how the organism experiences the information. Another thing to consider is whether we are anthropomorphizing with terms like "panic" and "courtship," especially if consciousness is not involved (but is it?).

VERTEBRATE NERVOUS SYSTEMS

As chordate body plans evolved, neuronal connections tended to become centralized and cephalized with a knot of neurons organized into a ganglion at the head of the animal. Neuronal function became more specialized and hierarchically organized. This has been much more elaborated in vertebrates than in the somewhat similar path taken by some arthropod lineages. Vertebrate CNS development involves most of the signaling, transduction, transcription, receptor, patterning, and combinatorial expression phenomena that we have seen many times, as well as mechanisms for highly controlled migration of cells through areas already inhabited by other cells.

Early in the evolution of the vertebrate CNS, the brain stem was the only "cerebral ganglion" that could receive and process sensory information, and this is still the way that the nervous systems of some vertebrates are organized. The lancelet, or *amphioxus*, is an example of a modern "lower" chordate with a notochord and a nerve cord above it, but no brain, no eyes, and, in fact, no head. Over evolutionary time, although the more "advanced" anterior parts of the brain grew larger, much of the synaptic organization of the brain stem was conserved and modified. Some sensory nerves still make synapses (connections) in the brain stem on their way to processing centers in higher, more recently evolved regions of the brain (Matthews 2001).

Why did an organized nervous system that is centralized at one end evolve, and why at the front? One obvious answer is that "front" is basically defined by the direction the animal moves. This centralized end confronts the environment first—if the main sensors responsible for finding food and mates and avoiding predators are close together and localized in an area that first contacts the approaching environment neural connections can be short and the response can be quick and more easily integrated. This is a rather circular explanation, in part because of how we define "front" (and we ignore as exceptions due to phylogenetic inertia those animals, like some crabs, that generally move "backward"). The evolutionary explanations could be countered by saying that the tail end should have the sensors because this is the direction from which the greatest dangers may come or that animals should have eyes in the back *and* front of their heads (or, like Argus, 100 eyes). Perhaps the brain should be closest to the endocrine organs, to facilitate quick response, or buried deep in the middle of the organism to protect the vital "nerve center," much as military nerve centers are protected in deep underground bunkers. After all, this is how we explain the fact that veins are more superficial than arteries (which, when severed, cause more immediate threat to survival than veins that are severed).

Vertebrate nervous systems are segmented and modular at morphological, histological, cytological, and physiological levels (e.g., Carlson 1999; Redies and Puelles 2001). In some ways, the brain can be thought of as a set of Matryoshka dolls, with major segments composed of smaller units, themselves formed of still smaller ones, in turn made of even smaller ones, and so on, defined in different ways at each step, down to the level of the neuron itself. The brain is physically as well as "virtually" (functionally) segmented, and the distinction is important because under some conditions the physical location of a function can vary or move (and hence, percept, whatever it is experientially, need not be physically localized in the brain). However, not all segments in an adult vertebrate brain correspond to the initial segment structure, and indeed the segments are not always neatly nested; they are generally interdigitated structurally and even functionally.

Segmented Development and Neural Differentiation in Vertebrates

The nervous system is segmented along its anterior-posterior axis from the rostral (face) to the caudal (tail) ends and transversely, that is, dorsoventrally, and also from medial or mesial (inside, central) to lateral or distal (outside, peripheral). In a hierarchical or nested way, segments at an early stage give rise to segmental substages later on. The system initially forms along the dorsal midline, from a layer of cells called the *neural plate* that overlies an anterior-posterior supporting rod called the *notochord* (e.g., Carlson 1999; Gilbert 2003; Matthews 2001). The notochord is a source of Shh (*Hedgehog* class) SF (Gavalas and Krumlauf 2000; Lumsden and Krumlauf 1996; Wurst and Bally-Cuif 2001), which also induces *Shh* expression in the adjacent overlying neural plate cells. Laterally (on either side of the Shh source) cells express *Bmp* SF proteins (*Bmp4* and *Bmp7*). In a process called *neurulation*, this *Bmp*-expressing flanking tissue grows up and around on either side, closing dorsally to form the top, or *roof plate*, of a hollow *neural tube*. The ventral and medial part, the *Shh*-expressing cells that overlie the notochord, is known as the *floor plate*. (Recall that Bmp proteins have dorsal effects.)

The *Bmp*-signaling area is now dorsal, atop the closed tube, and diffuses down both sides, inducing various transcription factors (TFs), including *Pax3* and *7* and *Msx1* and *2*, whose expression defines the dorsolateral *alar plate*. Shh diffuses upward on either side from the floor plate, inducing genes including *Pax6* and *Nkx* TFs to define the ventrolateral *basal plate*. These expression patterns extend to the spinal cord caudally and are important in establishing regional identity in the forebrain.

The Bmp signal diffusing from lateral ectoderm also induces expression of TFs, including the zinc-finger gene *Slug*, in cells at the crest of the juncture that formed the roof plate. These cells take on a special role generally considered to be a defining characteristic of vertebrates—perhaps even qualifying as a fourth primary germ layer (Hall 2000)—by becoming the migrating cells of the neural crest (NC), described in Chapter 8. NC cells are involved in peripheral nervous system and cranial neural crest (CNC) in craniofacial bone, tooth, and other tissues.

The extent to which CNC cells are regionally prepatterned before migration away from their source at the neural tube is unclear. CNC cells have plasticity such that to some extent their role is determined by response to signals they meet, possibly including from endoderm, on their migratory way (e.g., Trainor and Krumlauf 2000). In Chapter 8, we described their interaction with overlying epithelial cells in the process of epithelial-mesenchymal interaction that generates various structures. There is evidence that the overlying tissue has already been prepatterned to produce signaling molecules that induce a response from in-migrating CNC cells, and postcranial NC cells cannot be induced to form head structures (Graveson et al. 1997).

The spinal cord predominates in the control of functions in lower vertebrates, with the brain gaining dominance with the evolution of increasing size of the forebrain in later vertebrates. Figure 15-3 shows the relative sizes of the major sections of various vertebrate brains. This evolutionary differentiation with its complex behavioral and sensory consequences can thus be viewed as having come about, in part at least, through relatively simple allometric modifications of existing structures that were easy to achieve. Some of the evolutionarily "early" functions remain largely controlled by their original parts of the CNS.

The spinal cord is a collection of nerve fibers and nerve cells in vertebrates that extends from the *medulla oblongata* at the base of the brain through the spinal column. The number of nerves comprising the spinal cord differs somewhat among vertebrate species, but there is remarkable stability in their position and function. In primates, 31 pairs of nerves, both afferent (incoming sensory) and efferent (outgoing motor), travel along the length of the cord, to emerge at various points between the vertebrae to relay information to and from the brain and the rest of the body. Twelve nerve pairs, the *cranial nerves*, extend directly from the ventral surface of the brain itself to affect functions in the head and face, as well as some functions in the trunk (e.g., diaphragm). Most of these are both sensory and motor, although the cranial nerves involved in olfaction, vision, and hearing are exclusively sensory.

The brain itself develops as a complex convoluted layer of cells called the *cortex* that surrounds an inner fluid space (the core of the original neural tube) known as the *ventricle*. Longitudinally, the brain forms three main regions: the *hindbrain* (*rhombencephalon*), the *midbrain* (*mesencephalon*), and the *forebrain* (*prosencephalon*). The hindbrain functionally divides into the *metencephalon*, which includes the *cerebellum*, and the *pons* and the myelencephalon, which includes the medulla oblongata. A constriction called an *isthmus* separates the midbrain and hindbrain.

Neurons from the medulla oblongata help to control functions such as breathing, swallowing, cardiovascular function, digestion, and some body movement. The pons, at the junction between the hindbrain and the midbrain, is also involved in regulating breathing. The cerebellum receives sensory information from many parts of the body as well as signals from motor control areas of the forebrain, which it helps to coordinate into motor commands. In popular vernacular, we do not have to "think" about these functions for them to work. They work during sleep and even during coma, and hence do not require consciousness. The rest of the brain *is* aware

A.

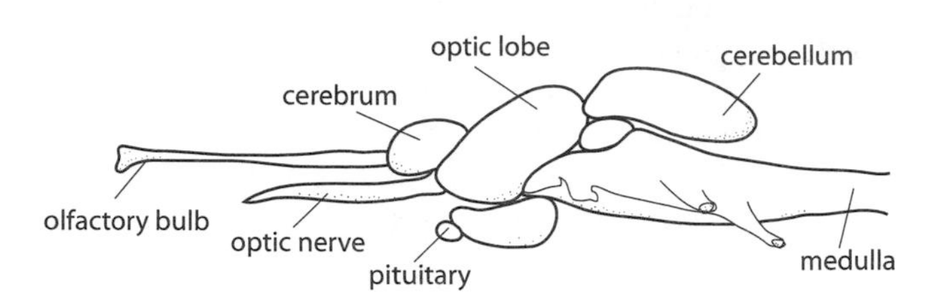

B.

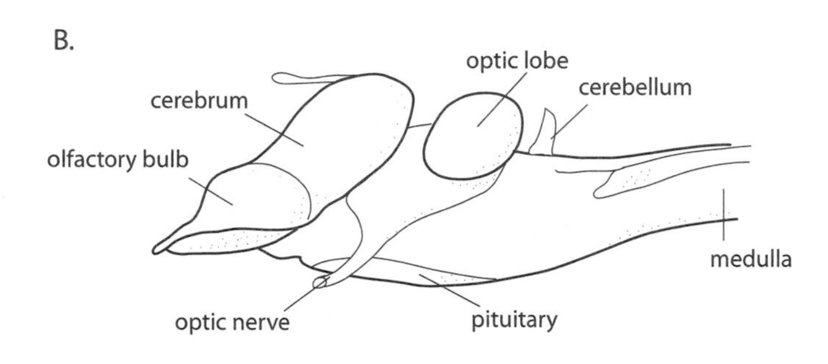

C.

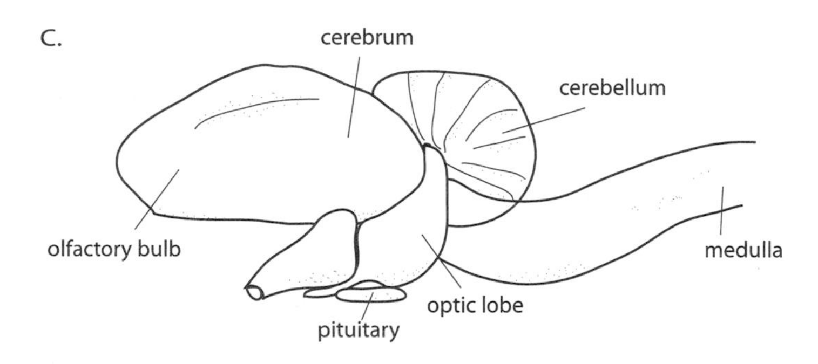

D.

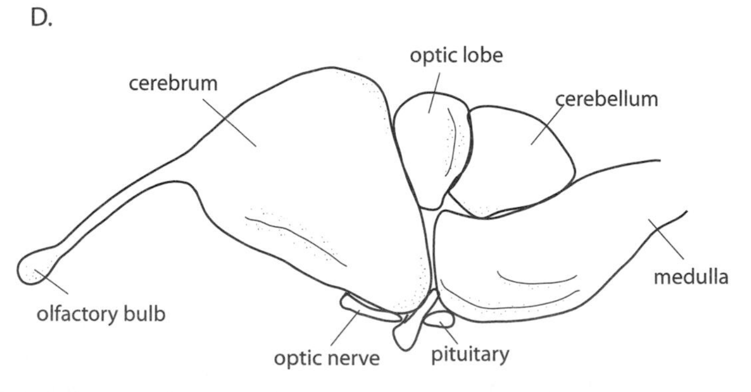

Figure 15-3. Major sections of selected vertebrate brains. (A) fish; (B) amphibian (frog); (C) bird; (D) reptile (alligator).

of these functions; for example, they can change circumstantially and can even be influenced by thought, as the racing heart rate triggered by fear.

The hindbrain and midbrain constitute the *brain stem*. The midbrain contributes to the control of movement and receives sensory information that it passes along to higher areas for interpretation and response. Its major subdivisions are the *inferior* and the *superior colliculus*, the first a relay and processing center for auditory signals and the second for visual signals. The *reticular formation* comprises some of the rest of the midbrain. This section controls the organism's state of arousal and plays a role in various sensory and motor systems.

Figure 15-4 shows the expression boundaries in the hindbrain-midbrain region of many genes that have been examined. Keep in mind the mixed and incomplete dartboard nature of such expression or pathway catalogs. Regional differentiation establishes the isthmus, which then becomes a boundary relative to expression of signaling factors *Fgf8* posteriorly and *Wnt1* anteriorly. The isthmus is a similarly sharp expression boundary of two TFs: *Otx2* anteriorly and *Gbx2* posteriorly. Not all genes important in patterning in this area are so restricted, however; *En1* and *En2* are expressed on either side of the isthmus.

The hindbrain develops into a series of segments called *rhombomeres* that are reflected in expression boundaries. For example, *Krox20* expression differentiates odd from even rhombomeres, and combinatorial expression patterns of genes from the four *Hox* clusters along with *Krox20*, *Follistatin*, *Kreisler*, and others identify each rhombomere (Voiculescu et al. 2001). This is the anterior-most part of the famous AP combinatorial patterning effect of the *Hox* clusters presented in Chapter 9 and elsewhere. Consistent with this, experiments that individually inactivate these genes have homeotic (segment-shifting) effects to varying degrees. Functional studies have also shown that these various segment-defining genes interact. Signaling by *Fgf8* and diffusible retinoic acid (a morphogen related to vitamin A, not a gene product) and its receptors and binding proteins (which are gene products), affect *Hox* expression domains and hence rhombomere identity (e.g., Trainor and Krumlauf 2000). Rhombomere-specific gene expression patterns make it possible to trace the migration of NC cells from specific rhombomeres into structures like the pharyngeal (gill) arches (Gavalas and Krumlauf 2000).

A series of genes and their receptors (*Eph* and *Ephrin* in Figure 15-4) affect cellular differentiation in the region, and most of the genes shown, which are important to hindbrain specification, are also involved in other developmental processes; for example, Follistatin is a secreted molecule involved later in life in controlling reproductive hormone levels.

The isthmus is frequently referred to as an "organizer" because, like Spemann's, the apical ectodermal ridge (AER) in the developing limb, enamel knots in teeth, the focal Dll spot on butterfly wings, and numerous others, it is a source of signaling molecules that help induce subsequent development in receiving cells around it. Thus, Fgf8 signal emanating from the isthmus inhibits expression of *HoxA2* that helps define the anterior hindbrain subregion and in turn is part of the specification of NC cells migrating from the hindbrain to affect skeletal development in the jaws. *Hox* genes are expressed in the mandibular arch of embryonic agnaths, suggesting that the loss of this expression enabled the anterior pharyngeal arch to evolve into a jaw (Cohn 2002).

However, this is a useful place to restate the caveat that "organizer," like "master" and "selector," is a concept with complex meanings in the culture of the scientists

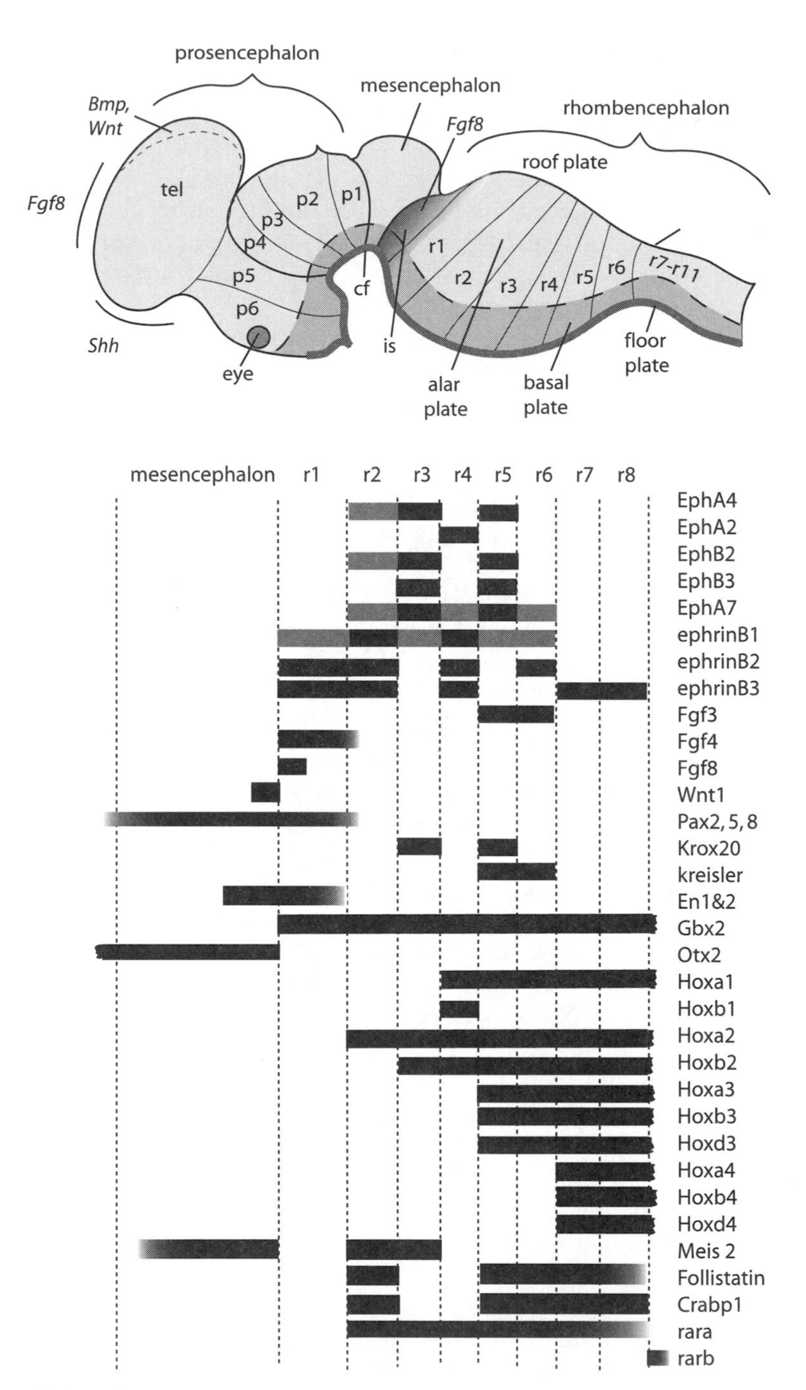

Figure 15-4. Longitudinal segmentation and associated gene expression in the early vertebrate brain. Schematic and selected (and many genes are not yet known). "Is" denotes the isthmus; p1-p6 prosomeres, and r1-411 the rhombomeres. Genes shown are *Hox* cluster genes, other TFs (*Drox20*, *Kreisler*, *En1&2*, *Gbx2*, *Otx2*, *Pax*, *Meis2*); SFs and receptors (*Eph*, *Ephrin*, *Fgf*, *Wnt*); factors related to retinoic acid and its receptors and binding proteins (*rara*, *rarb*, *Crabp1*). Composite based on (Gavalas and Krumlauf 2000; Lumsden and Krumlauf 1996; Marin and Rubenstein 2002; Pasini and Wilkinson 2002; Simeone 2000; Wilson and Rubenstein 2000; Wurst and Bally-Cuif 2001).

coining such usages in genetics. As such, the term can be misperceived to suggest a discrete, externally imposed purposive entity, rather than a basically quantitative region vs. qualitative entity, which itself arises from prior signaling interactions (Wilson and Rubenstein 2000). It is called an organizer because it is a regional source of signal that affects the development—sometimes but not always with sharp boundaries—of subsequent structures. In fact, things are not so simple and the defining genes are not the only important factors in phenomena affected by their production in the isthmus (e.g., Chambers and McGonnell 2002), nor is the isthmus the only place they are important even, even in neural development.

Vertebrate Forebrain Development

The forebrain, or prosencephalon, is generally thought of (anthropocentrically) as the place where the really important biological traits reside: where vertebrates interpret, integrate, and respond to sensory and somatic input and motor commands (e.g., Matthews 2001). The relative size of the forebrain varies greatly in vertebrates, from being among the smallest sections of the brain of fish and reptiles, to constituting most of the brain in primates. Many of the functions carried out in this part of more complex brains, such as visual processing, occur in the midbrain of animals with small forebrains, although some of the "higher" functions are missing from the latter. (Does this mean they "perceive" vision differently?) It is perhaps worth considering that, from the point of view of members of species without forebrains—species that are doing perfectly well in the world—the forebrain might seem a grotesquely exaggerated tissue that requires a lot of extra DNA, metabolic energy, and maintenance, and thus a lot of food. This belies any notion that evolution is an energetically parsimonious phenomenon.

Each part of the brain is further subdivided into sections with their own specialized functions. Our understanding of the degree to which specific functions, other than the major obvious ones like vision and olfaction, are precise and replicable is still incomplete but scans of brain activity under various conditions and in experimental animals whose brains have been modified, or affected by certain diseases, show that functions can be remarkably tightly located. That is, brain activity changes only in a very localized region of the forebrain during the experimental or test experience.

The forebrain develops from an outpocketing of anterior neural tissues (first to form the optic vesicle) and is affected by signaling from several major regions (for a detailed discussion of these complex processes see Marin and Rubenstein 2002); these are generally indicated in Figure 15-4. Early dorsal signals involve *Bmp* and *Wnt* genes. A second signaling area of importance in forebrain development is the *anterior neural plate* (ANP). This most rostral area is another source of Fgf8 signal in the developing brain and is also referred to as a forebrain organizer. Ventrally, forebrain development is affected by *Shh* to which we have already referred.

The outgrowth forms a hollow structure on each side, with a dorsal groove or sulcus between them that separates the future left and right cerebral hemispheres. Also forming are areas of thickening, including the ventral septum and the medial and lateral ganglionic eminences (MGE and LGE), as indicated by the transverse sections shown in Figure 15-5. Ventral to these areas are the *anterior entopeduncular area* and associated *anterior preoptic area*, where major tracts of neural fibers go into or come out of the telencephalon. Darker shading indicates the *ventricular zone*

(VZ) in which neural stem cells produce differentiated glial and neuronal cells that move into the adjoining *subventricular zones* (lighter shading). The dorsal region (dorsal and medial pallium in the figure) becomes the cortex.

The prosencephalon can be viewed as being divided into six segments called *prosomeres*, by analogy to rhombomeres, though the most anterior three of these are less clearly defined. Prosomeres are characterized by differential gene expression patterns (e.g., Marin and Rubenstein 2002). The *diencephalon* and the *telencephalon* are the major subsequent divisions of the forebrain. These, however, are further divided—the diencephalon into the *thalamus* and the *hypothalamus* and the telen-

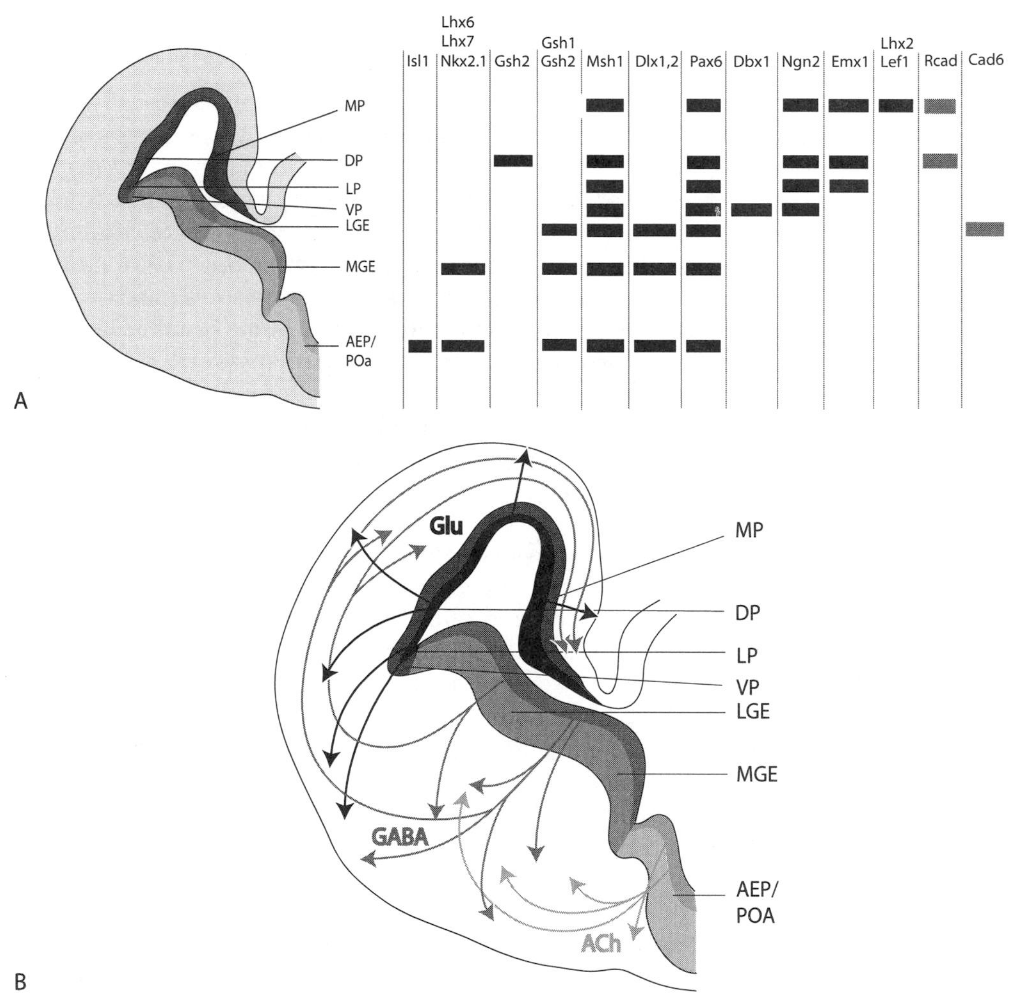

Figure 15-5. Gene expression in forebrain development. (A) Schematic of genes expressed in different regions. Bars show zones of expression for each gene. These are TFs except the two Cadherins on the right; (B) neurons migrate from the same areas but are at least partly prespecified by the time of migration, as shown here in regard to type of neurotransmitter expressed (Glutamate, GABA, and Acetylcholine). Highly schematic, roughly at embryonic day 14.5 or midgestation in the mouse. For anatomical terms, see text. Greyshade: darker, ventricular zone; lighter, subventricular zone. Composite modified after (Marin and Rubenstein 2002; O'Leary and Nakagawa 2002; Pasini and Wilkinson 2002; Redies and Puelles 2001; Schuurmans and Guillemot 2002; Wurst and Bally-Cuif 2001).

cephalon into the *cerebrum* and the *basal ganglia*, which are involved in control of movement. As will be seen below, the thalamus is central to lower processing and relay of signals from sensory organs to higher centers of the cerebrum, where they are further processed and integrated.

Figure 15-5A indicates some of the genes whose regional expression is correlated with these morphological areas. Overall dorsoventral differentiation is established or affected by mutually inhibitory interactions between dorsal Bmp4 and ventral Foxg1 (also called BF1), whereas the cortical LGE-MGE distinction reflects mutual inhibitory interactions between genes activated via *Nkx2.1* and *Pax6*; *Dlx*, *Msh*, *Gsh*, and *Lhx* gene expression also differentiates these regions. These expression patterns also reflect and are markers for the histological, and cytological differentiation of the region, and are generally conserved among vertebrates that have been tested (e.g., see Monuki et al. 2001; Puelles et al. 1999; Redies and Puelles 2001). The regional expression pattern is also consistent with known signal-transduction interactions, in that the interacting genes are expressed in the same region.

Although these expression patterns still mainly serve as regional markers, over- and under-expression studies have confirmed their general functional importance and relevance, and some of the temporal and regulatory hierarchies are being identified (Schuurmans and Guillemot 2002). Further, their importance in indicating the orderly and segmental, hierarchical nature of brain development is that it makes the evolutionary origin and development of an organ of this complex nature easier to understand.

As the size and complexity of vertebrates increased, the cerebrum began to dominate the forebrain. The cerebrum is filled with neurons and is where complex processing of sensory information (for example, auditory, visual, olfactory), the intentional planning and initiation of movement, and many forms of learning, among other activities, take place. The *cerebral cortex* is the 1- to 4-mm outer layer of the cerebrum and formed of three subdivisions, sometimes called the *neocortex* (the new cortex, unique to mammals), the *paleocortex* (old cortex), and the *archicortex*. (These names are intended to reflect the phylogenetic origins of the regions, but may not in fact do so, as there is not a clear consensus about the origins of the neocortex (Northcutt and Kaas 1995)). The paleocortex is basically the olfactory region of the brain and receives direct input from olfactory neurons, as specified in Chapter 13. The archicortex is primarily the *hippocampus*, or memory area, an inner fold of the temporal lobe. These regions differ cytoarchitecturally—the archicortex is a single cell layer, the paleocortex is two layers of cells, and the neocortex is formed of six layers, described below.

Several ideas regarding the origins of the neocortex have been proposed. It may be an enlargement of the reptilian dorsal *pallium* (roof, literally "cloak" or outermost layer) or the migration of two sources of cells from lower parts of the neuraxis (see Karten 1997 and Northcutt and Kaas 1995 for discussion of primitive mammalian cortexes). Since the evolved function is what counts, this might be largely an academic question. But it could turn out to have implications if, for example, there are as yet unknown vestiges of the original functions that may constrain the current ones in some way, be related to particular diseases, and so on. Searches for expression of cortex-specific gene combinations across the spectrum of living vertebrates will undoubtedly provide many clues about how this evolution occurred. Full documentation of the patterning processes will help.

The neocortex comprises about 95 percent of the cerebral cortex and is somewhat arbitrarily subdivided into four or five major regions (Buxhoeveden and Casanova 2002): the *frontal*, *parietal*, *temporal* and *occipital lobes* and sometimes including the *limbic system* (Figure 15-6). Although sensory processing is integrated throughout the brain, each of these lobes has specialized functions. The frontal lobes are involved in emotion, thinking, planning, personality, language, and sexual and social behavior, among other functions. Of course, some of these functions—as we humans choose to label them—are specific to humans (e.g., language, planning), and some (e.g., pheromone responses) to nonhuman (including some Old World monkeys and apes) vertebrates. But we must again guard against unwarranted anthropocentrism. Squirrels plan for the distant future when they bury nuts in the summertime; can we say they do not know why, nor envision a remembered meal on a snow-covered branch?

The cortical lobes are not symmetric, and, as a general rule, the human left side predominates in control of language, whereas the right lobe controls nonverbal activities. This is by no means invariant, nor is it clear how these separate functions are established genetically, although some of the molecular mechanisms are beginning to be elucidated (Essner et al. 2000; Morgan et al. 2002). The parietal lobe is involved in mechanosensory perception; touch, temperature, pain, pressure, as well as taste. It also helps integrate visual input. The temporal lobe is involved in auditory perception and integration, memory, and language abilities, among other functions such as the sense of "self" and "other," generally on the left and rights sides, respectively (this perceptual self-nonself distinction is not to be confused with a circulating, molecular, immunologically based distinction). The occipital lobe is responsible for some aspects of the processing and interpretation of visual input. The limbic system controls aspects of the autonomic and endocrine systems, including feeding behavior, homeostasis, and emotion.

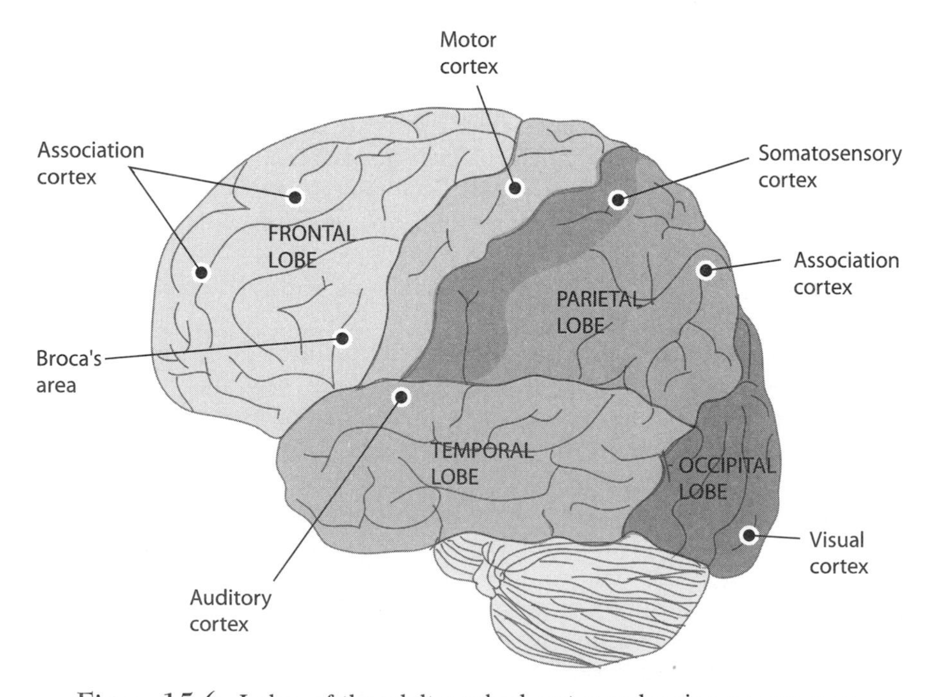

Figure 15-6. Lobes of the adult cerebral cortex and major sensory areas.

The external surface of the forebrain is characterized by species-specific patterns of fissures, *sulci* (grooves), and *gyri* (ridges) of tissue. The human brain is densely covered with such fissures. The degree of fissuring varies among species, however, with some like rodents having little such structure and primates much more. The deepest sulci reflect the early, major, primary division of the forebrain into its various lobes, which are not only phylogenetically conserved but reflect the kind of basic segmenting developmental processes we have discussed. However, the meaning of the rest of the fissuring is less clear, but is interesting from developmental, evolutionary, and functional points of view. Genetic evidence is currently weak and mainly suggests that overall pattern, size, and shape (plus the major fissural structures) are genetically influenced, at least to some general extent.

We have earlier seen drawings by the anatomist Andreas Vesalius. In his 1543 work, he also depicted the fissures of the human brain only *schematically* to indicate their nature. Today, we tend to assume things should be more literally represented to be "scientific." However, sulcal and gyral patterns vary even among twins and from left to right side in the same individual, and it is not yet clear how much

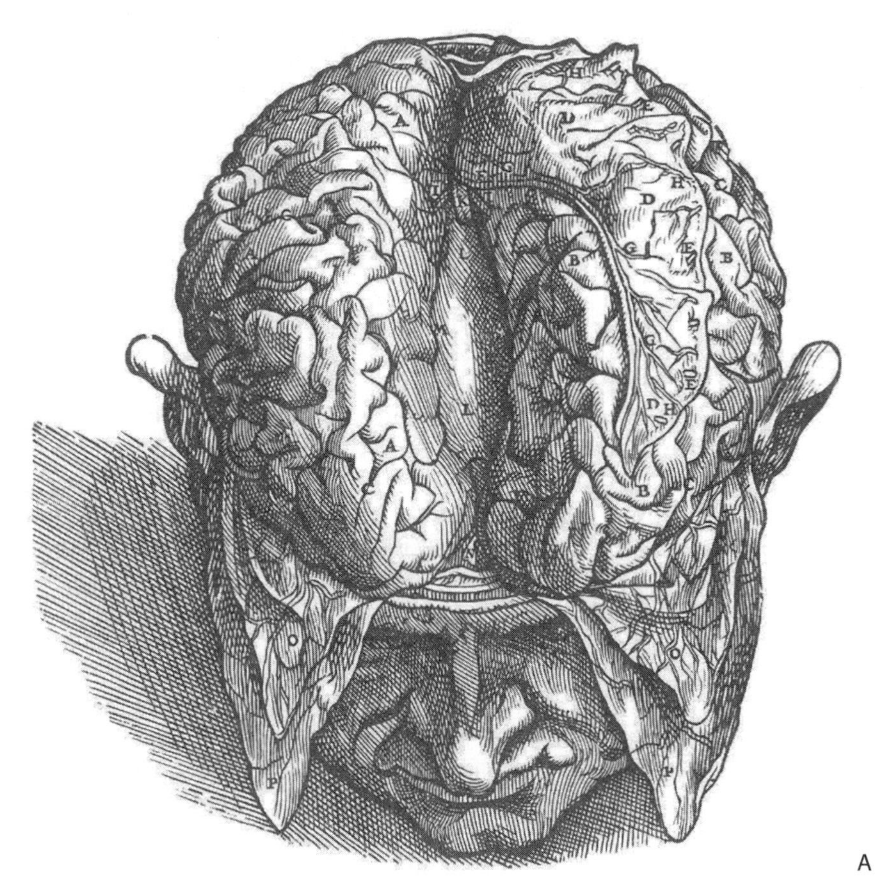

Figure 15-7. Brain surface patterns in humans. (A) Vesalius' drawing in 1543. (B) sulcal patterns vary even between left and right sides of the same brain. (A) Vesalius reprints are from (Vesalius 1543); (B) MRI data from Pearlson and Bata compiled into a figure by K. Aldridge and reprinted with her kind permission.

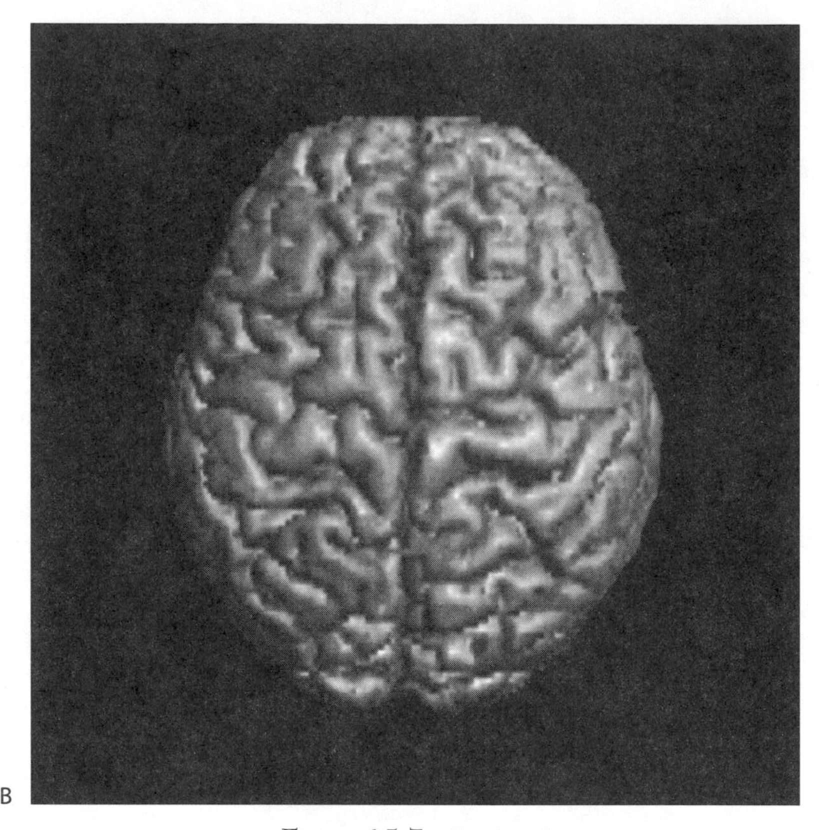

Figure 15-7. *Continued*

of this is purely random (and we don't understand the processes responsible). How much of the pattern is genetic, in what way it is genetic (specific sulci? the overall complexity of sulci?), and what functions are associated? For example, to what extent does the asymmetry reflect—or determine—the existence of hemispheric dominance, handedness, and so forth (Weiss and Aldridge 2003)?

These questions are not easy to answer but could be important to our understanding of brain evolution, particularly of the "higher" primates whose brains are particularly characterized by such patterns. If the fissure pattern has a large chance component, and if this affects function because, say, of the relative amount of cortex available for some particular function like literary or athletic ability—or Einstein's unique physics ability—then evolution might be viewed as providing that randomness as a kind of generator of variation, much as we see in our immune system. Alternatively, the details of fissure patterning could have no particular meaning. It is a very difficult question to answer, or even to state *how* to answer it.

The cortex is internally organized in a segmental way. The prevailing view is that the basic structure is columnar, with each column (or "minicolumn") being a fundamental functional processing unit about 50 μm wide and containing 80–100 neurons. All neurons within a column may serve the same sensory modality, although this is not a universally accepted interpretation (Nieuwenhuys et al. 1998).

The human cortex has six basic transverse cellular layers, each with distinct neuronal organization. The number of layers in the cortex differs among vertebrates, and the primordial number was probably two or three, as evinced by the number in contemporary lower vertebrates as well as in the evolutionarily earlier sections of the mammalian brain. Some of the layers contain neurons that project to subcortical structures; others contain neurons that extend only locally within the column. Groups of 60–80 minicolumns are, in turn, bound together by short neuronal connections into "macrocolumns," (Buxhoeveden and Casanova 2002). The laminar and columnar organization of nonmammalian brains is less visibly obvious, however. Columns are essentially the same size in most brains; apparently, the cortex expanded in size over evolutionary time with increases in column number, not column width or basic structure.

The six radial layers that form the primate neocortex are depicted in Figure 15-8. The cellular composition of the layers differs somewhat even among mammals (Northcutt and Kaas 1995), but, in primates, layer I is acellular, primarily an area in which incoming axons form synapses with dendrites from neurons from deeper layers. Layers II and III contain numerous pyramidally shaped neurons, *pyramid cells*, and these cells make synapses with many of the incoming axons from layer I. Layer IV contains star-shaped *stellate* cells, which have axons that remain local and receive input from axons carrying sensory signals from other regions of the cortex. Layer V is full of pyramidal cells with the largest cell bodies in the cortex. The axons from these neurons extend long distances, to the brain stem and spinal cord. Axons from pyramidal cells in layer VI extend to noncortical regions of the CNS, including the thalamus.

Neuronal Development and Migration

The cortex is thus a complex structure highly differentiated at the cell level and much more "refined" than the general regional segmentation reflected in the expression patterns described above. The ventricular zone is the source of differentiation of a variety of neurons and their associated *glial* (supporting) cells. In cortical development in mammals, neurons arise some distance away from their eventual destination. Neurons that will migrate to the cerebral cortex are formed early in gestation deep within the brain, in the ventricular zone. The VZ is developmentally transient; neurons form in this region, and when cells in the VZ stop dividing, neurons generally no longer proliferate.

In the forebrain, neurons migrate outward along transverse ("radial," inside-out, or ventricular-pallial) glial supporting cells, perhaps the way ivy climbs a trellis, though relying more on molecular than physical motive force, to form first a *preplate* above the ventricular zone. The next set of cells to migrate form a *subplate*, and these separate to form a region called the *cortical plate*. The outer layer of cells is then known as the *marginal zone*, and this differentiates into *Cajal-Retzius* cells. Neurons migrate into the cortical plate where they begin to differentiate, some becoming excitatory neurotransmitters, for example, and others inhibitory (the two provide a balance to activate and inactivate signal transmission at a given juncture, policing the neurotransmitter's ion channel states).

Neurons migrate along radial glial cells as guides. They enter the cortical plate and keep moving past other cells until they reach the Cajal-Retzius cells. Each new generation ends up passing previous generations to come to rest on the then-outermost layer. In this way, all six layers form, with layer VI deeper but forming earlier than layer II.

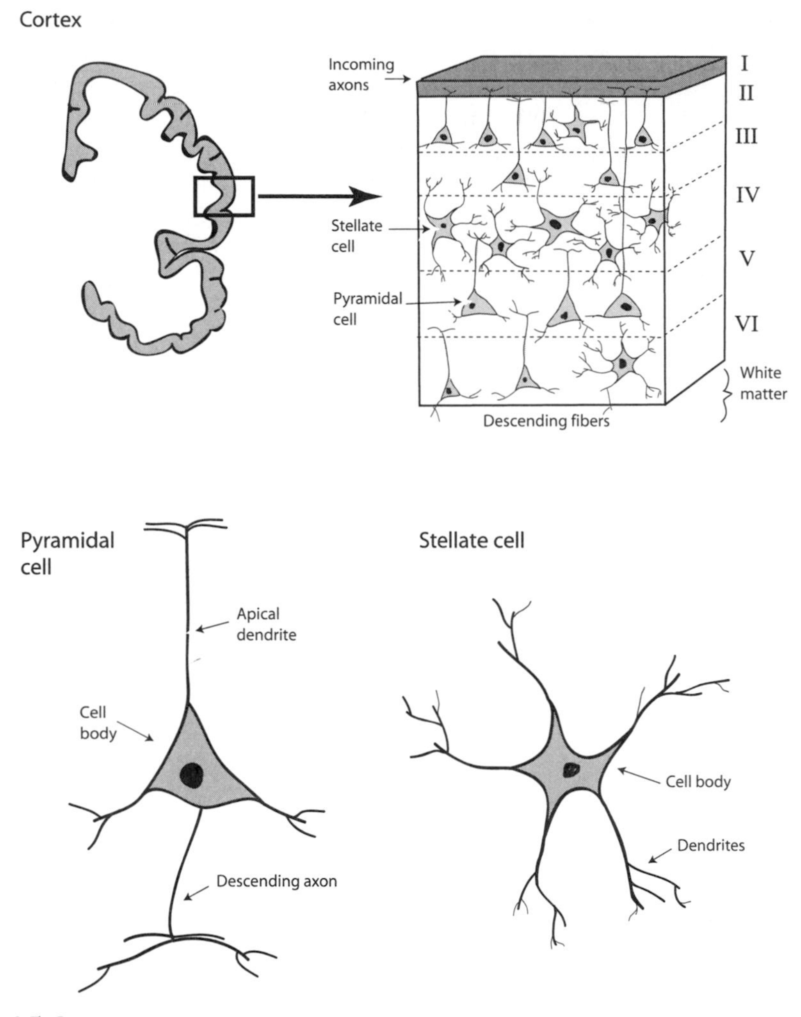

Figure 15-8. Cellular layers of the adult cortex; pyramidal and stellate cells, the principal neuronal cells of the cerebral cortex.

Some of these aspects can change over evolutionary time, perhaps by phenotypic substitution in which essentially the same functional result is achieved in developmentally different ways, even in related animals. In reptiles, the cortex develops in the opposite direction, from the outside in (Butler and Hodos 1996). The dorsal cortex is considered by some to be a homolog of the mammalian neocortex, and, unlike the neocortex, the dorsal cortex develops in such a way that the cells that form later are located below the earlier cells.

In an adult mammalian cortex, the cells on each layer share characteristic properties, such as the extent of axonal projection beyond or beneath the cortical column they form. But cellular identity seems to be plastic at least in part, determined by environmental cues, not genetic programming at first mitosis. Transplant experiments have shown that the cell's laminar position, when it undergoes its final differentiation, determines its fate. Arealization, however, does seem to be genetically programmed by mechanisms that maintain architecture or segmental boundaries, although neurons can also be redifferentiated to a new functional context.

The minicolumns of the cortex are homogeneous in appearance, despite the diversity of the information they process. This is in contrast with the specialized neurons in the different areas of the subcortical regions, which have an appearance specific to their task. The staged development of the mammalian cortex, with the cells that proliferate first forming the deepest cortical layer, and subsequent generations layering above them to form the layers closest to the surface, is called the "radial unit hypothesis" because it is believed to explain the development of the columnar organization of the cortex; related neurons develop by layering on each other and subsequently developing synaptic connections among themselves and in neighboring columns, to form the specialized regions of the cortex (Mountcastle 1978; Rakic 1988; Rakic and Lombroso 1998). Again, the functional significance of the columnar organization of the cortex has still not been substantiated, and, in fact, it is still not completely accepted that the entire mammalian neocortex indeed *is* structurally columnar (Nieuwenhuys, Donkelaar et al. 1998).

The molecular and genetic basis of these migrating pathways is beginning to be understood (Gupta et al. 2002; Hatten 2002; Lambert de Rouvroit and Goffinet 2001; Marin and Rubenstein 2002; Rice and Curran 2001). Much of the knowledge derives originally from naturally occurring human or mouse diseases, which led to the discovery of the genes involved. Here, we present a simplified account of what is currently understood. The Cajal-Retzius cells are a source of the *Reln* (or *Reelin*) gene product. A TF called *Tbr1* seems to be important in the formation of Cajal-Retzius cells and hence of Reln production. This secreted Reln protein attaches to the extracellular matrix and forms a concentration gradient from the marginal zone downward. Migrating neurons express various receptors on their surface, including the VLDR and ApoER2 lipoprotein receptors. These bind to Reln, triggering an intracellular messenger cascade that involves various factors (e.g., *Dab1*, *Cdk5*, *Cdk5r*) that, in ways not yet fully understood, may cause the neuron to lose cell adhesion cell-surface properties required to stay attached to guide glial cells. In any case, the neurons appear to drop off the glia at this point—right under the Cajal-Retzius layer that produced the Reln gradient to come to rest.

Radial migration retains the relative spatial location between the precursor cells in the VZ and their corresponding cells in the cortex; therefore, the cells involved have been called *projection neurons*. By contrast, other neurons migrate tangentially (laterally to the columns) and travel considerable distance within a layer to become *interneurons* and are thought to be involved in local circuitry integration (Anderson et al. 2001; Marin and Rubenstein 2001; 2003; Monuki, Porter et al. 2001; Monuki and Walsh 2001). In terms of what is known, the two types of neurons arise from different precursor populations and can be distinguished in part by expression of TFs that characterize the source and destination regions (Figure 15-5B).

Neurons can also be characterized by their major neurotransmitter expression, the particular ion channels, or other mechanisms used to receive or trigger the transmission of a signal impulse. The three main classes are functionally relevant as well as useful in identifying developmental lineages and aspects of brain regionalization. Their migration patterns are shown schematically in Figure 15-5B. Radially developing projection neurons generally are in the glutamatergic class, and tangentially developing interneurons are in the GABAergic class. Cholinergic neurons mainly use acetylcholine. Other neurotransmitters, such as dopamine, are used by other types of neurons.

Glutamatergic neurons predominate in the cortex, but the LGE and MGE send a substantial number of GABAergic neurons into the cortex (Marin and Rubenstein 2001; 2003; Puelles, Kuwana et al. 1999) (Figure 15-5B). GABAergic neurons migrate tangentially out of the MGE and LGE (e.g., Hatten 2002; Marin and Rubenstein 2001; 2002). The mechanisms of the extensive and highly patterned neural migration are varied. Some of the migrating cells express *neuropilins*, which are receptors for semaphorin 3A and 3F. The latter are expressed in the *striatum* region (below the cortex, including the LGE), and this appears to guide or trap these interneurons for a striatal destination, whereas other cells that do not express neuorpilins migrate to the cortex. Guidance of this migration seems to be affected by extracellular matrix molecules, including gene products *Slit* and its ligand *Robo*, and *Netrin* and its ligand *Dcc*, but details are not well known.

Olfactory neurons are a special class because they are continually replaced by new olfactory neurons, which originate in the subventricular zones and move to the olfactory bulb in a chain of cells ensheathed by protective, guiding glial cells. Two axon guidance repellents include *EphB* and *EphA* RTK receptors and their *Ephrin* ligands, and surface cell adhesion gene products are also important (see below).

How Are Boundaries and Segments Maintained?

The migration of neurons is but one of several cell migrations in vertebrate development (and such things occur to a lesser extent in other animal groups); NC cells are a prominent example. Cells that migrate need to find a path, to have a mechanism for following it, and to associate only with the appropriate cells in the neighborhood as they pass through. Migration involves both intracellular mechanisms and external chemotactic (attractant or repellant) clues and involves the production of signaling molecules that may diffuse or may be attached, as in the Reln mechanism for projection neuron migration. Also, the migrating cells must express the appropriate receptor.

Experimentally mingled neural cells may aggregate if they are genetically similar or repel if they are different. Alternatively, introduced cells may redifferentiate to take on the expression patterns of their surrounding cells.

Several cytological mechanisms are involved in boundary maintenance. Cells may bundle physically together based on identity of cell surface adhesion molecules. One family of such molecules is the *Cadherins* (e.g., Pasini and Wilkinson 2002). Cells that express the same set of Cadherins on their surface adhere to each other but will not adhere to cells expressing other members of the gene family. In this way, Cadherin expression "codes" can subdivide the developing brain either regionally early in development and/or functionally later on. Fiber tracts, or bundles of cells of common origin, can stay together, retaining morphological or functional coherence, while crossing through regions.

For example, as indicated in Figure 15-5A, the cortex expresses *Rcad*, whereas the adjacent LGE expresses *Cad6*. That the two sets of cells are distinct is also reflected by their expression of *Pax6* and *Dlx1*, respectively. Combinatorial *Cadherin* expression codes appear to play a role in maintaining prosomere integrity from the early to the mature brain (e.g., Redies and Puelles 2001). Fiber tracts from the motor area of the cortex, or to the cortex from auditory and visual sensory organs, pass through several brain regions and must thus maintain coherence. Thus, areas in the cortex express *Cad6*, *8*, and *11*, as do the incoming sensory thalamic nuclei (see below).

Once appropriately attached, neurons are physically guided by glial cells, hence keeping the right neurons going to the right place even when surrounded by other types of cells. Olfactory neurons may retain (or obtain) their specificity in terms of the olfactory receptor gene they express. This may serve as a kind of guide mechanism between the parts of brain and olfactory epithelium appropriate for that olfactory receptor gene (of course, this does not address the question of what determines spatial expression of different types of genes in the first place). Two subsets of *Ephrin* genes for membrane-bound tyrosine kinase receptors and their ligands, designated *EphA* and *B*, appear to be involved. Class specificity appears to be involved in axon guidance by separating A-expressing from B-expressing cells. *Ephrin* expression differences help separate the hindbrain into odd- and even-numbered rhombomeres. Other cell adhesion molecules like *Integrins* and diffusible gradients (*Netrins*, *Slit*, and others) also help guide axon migration (Redies and Puelles 2001).

The familiar *Notch* receptors, their ligands *Delta* and *Serrate*, and the modifying gene *Fringe* play compartmentalizing roles in *Drosophila* patterning, and the homologous mechanism appears to be involved in vertebrate neural development. Fringe modifies Notch receptor structure, suppressing its activation by Serrate, and together these genes can establish patterns of affinity or repulsion. The same system seems to be involved in the specification of certain neuronal precursor cells in the VZ to differentiate into neurons, while surrounded by inhibition zones of nondeveloping cells. *Notch*-related patterning mechanisms also appear to be involved in thalamus development and in dendrite morphology in cortical neurons and perhaps gliogenesis.

Neurons can also be reprogrammed when they contact differently programmed neurons. Reprogramming of intermingling neurons by their local context has been shown experimentally in the hindbrain (Pasini and Wilkinson 2002). As noted earlier, the isthmus separates anterior and posterior cell types, which we can identify by their morphology and expression of indicator genes *Fgf8* posteriorly and *Wnt1* anteriorly. Transplanted cells become reprogrammed to express indicator genes appropriate to their new context. The initial boundaries are sometimes "fuzzy" but subsequently sharpen. A model suggested by Pasini and Wilkinson is that cells respond to quantitative gradient thresholds of SF molecules like Fgf8. This is not a precise mechanism, but response to SF levels exceeding the threshold induces appropriate *Hox* and associated *Krox20* and *Kr* expression. In turn, this induces *Ephrin*, which, being a qualitative cell-binding system, sharpens boundaries between cell types. Consistent with these ideas is that inactivation of genes mentioned in this chapter often has quantitative rather than clear, simple, or qualitative effects.

Not only must a cell have a path, it has to have a means for moving along it, and there are several (Gupta, Tsai et al. 2002). In one, the cell creeps along following a projection that detects extracellular signals or works its way along a glial guide cell. In another, the cell sends out a filamental projection (*filopodium*) toward a guiding signal, which fills with cytoskeletal tubules. The nucleus and rest of the cell are then pulled into the projection, and the process begins anew. Appropriately, cytoskeletal and adhesion genes (*Lis1*, *Dcx*, *Filamin1*, *Cdk5*, and *Astn1*, *Integrin3*, respectively) have been found to be mutant in natural disorders of cortical layering (or the latter produced by experimental mutation in these genes (Gupta, Tsai et al. 2002; Hatten 2002; Marin and Rubenstein 2002).

Function and Cortical Areas

We can relate these various developmental process notions to the way that the neocortex becomes functionally organized, peri- or postnatally depending on the species, into *areas* of the brain specific to different sensory inputs, such as optical, auditory, or olfactory areas. The thalamus is a relay center for most incoming sensory signals as we will see in Chapter 16 in regard to particular sensory systems. At present, it is not readily apparent whether the role of thalamocortical axons (TCAs) in arealization is molecular or due to external cues like sensory input relayed to the brain, and there has been longstanding debate about this issue.

The *protomap* hypothesis suggests that the arealization is controlled intrinsically, when the neurons are generated (Rakic 1988). This would imply a kind of prepatterning in the brain, which would not be without precedent in vertebrate development, as we have seen. The alternative *protocortex* hypothesis suggests that arealization instead takes place when the neurons are in place and receiving incoming sensory signal (O'Leary 1989). For example, "rewiring" experiments in which afferents are redirected into areas to which they do not normally project—for example, optic afferents sent to the olfactory area—result in functional remapping of the reassigned cortical area (e.g., Scalia et al. 1995), suggest that the arealization of the cortex is driven by experience rather than by preprogramming. This view would be consistent with the brain as a flexible instrument of organismal survival rather than a form of response automaton. It would also be consistent with examples we have seen, including in neurally relevant situations, such as epithelial-mesenchymal interaction involving migrating NC cells.

Both kinds of patterning are probably important and act separately and synergistically, at different developmental stages (e.g., O'Leary and Nakagawa 2002; Sur and Leamey 2001). There are ample mechanisms available in the developmental repertoire. Before their differentiation, prenatally, differential expression of a number of familiar gene families and pathways is seen in numerous neocortical regions, as discussed above. The areas are then differentiated cytoarchitecturally and chemoarchitecturally, as well as by their distinct efferent and afferent pathways.

The borders between many functional areas in adults are distinct, and easily discernible, but the borders of these regions of gene expression during development are not. The borders subsequently become distinct at about the time that TCAs have extended to cortical layer IV, around the time of birth. The temporal patterning, coupled with many negative results of searches for genes confined to functional cortical areas before thalamic innervation, suggests that perhaps prenatal regionalization is genetically controlled but that arealization is controlled by thalamic inputs (Pallas 2001).

A FEW STRAY THOUGHTS

Much remains to be learned about the role of specific genes in brain development, but what we know so far leaves us with a kind of paradox. The brain and its intricate connections with the entire rest of the body might seem to be a system of a different kind from others. Yet we have seen most of the genes and signaling mechanisms in brain development, like *Fgf* or *Bmp* or *Shh* signaling, or *Hox* combinations, elsewhere in the body.

Patterns of gene expression during brain development generally suggest roles for genes that are confirmed by gene inactivation or mutation experiments (e.g., Gupta, Tsai et al. 2002; Marin and Rubenstein 2002), and these genes seem to be working in ways expected of them. For example, an experiment in which expression of *Fgf8* at embryonic day 11 in mice was altered or blocked in the developing neocortex before the formation of thalamocortical connections, yielded animals with dramatically reorganized cortical maps including cortical pattern shifts (Fukuchi-Shimogori and Grove 2001). *Bmp4* overexpression had segmental cytodifferentiation effects (Gomes et al. 2003). In both cases, the effects were confined to the expected expression of these interacting SFs (consistent, for example, with their playing roles in activation-inhibition kinds of patterning). Knockouts of the regulatory genes *Emx2* or *Pax6* have also altered cortical organization, with loss of portions of the cortical epithelium (Pallas 2001). *Holoprosencephaly* is the failure of the two hemispheres to separate, and several genes are known that can produce this; these have relevant expression domains. *Reelin* was discovered in a behaviorally anomalous animal whose cortical layers are aberrantly developed.

These experiments and many others suggest that the brain is regionally patterned through familiar mechanisms. However, even after extensive work, only a few genes have been found to be expressed solely within defined cortical areas *before* thalamocortical innervation, suggesting that definition of areal borders requires input from thalamocortical neurons relaying signal from sensory receptors themselves (Pallas 2001). Perhaps there is a kind of reverse prepatterning, in which the central location awaits information returning there from distal locations.

The nervous system is not only highly integrated at morphological and cytohistological levels itself, but must be integrated in relatively precise ways with all other body systems. This has to occur in real time, because of the intricate innervation of all structures. It may be that neurological development involves larger or more complex sets of genes than other systems, and the brain is an organ that expresses a higher fraction of all genes than most (however, the liver is another expression-rich organ).

Presently, experimental gene manipulation involves only one or at most two or three genes. Such experiments are vulnerable to oversimplified interpretation. For example, *Shh* mutations lead to holoprosencephaly, but many other genes or chromosomal regions have also been implicated in this trait, showing that a set or network of pathways rather than a single gene is responsible. Tens if not hundreds of genes, when mutated, can affect very specific aspects of human brain function or cognition—learning, language, or personality—or very general traits like intelligence or motor control. Genes may affect such traits through morphological development or subsequent function (e.g., neurotransmission).

Perhaps it is a human conceit by which we view the brain as more intricate than other systems or having uniquely complex higher-level organization. A lesson in humility is to dissect a limb, or hand. Still, the higher integrative functions *are* interesting in their own right. Major success in understanding them will come when we learn how expressing particular networks of genes leads to the end phenomena—smell, vision, problem solving, and the like—but this may have to await better methods for understanding emergent traits.

There are other interesting differences from nonneural traits that may be important here. In structures like ribs, eyes, and gut, the differentiated cells stay where they are and function locally. In the nervous system, local differentiation results in distant function. An electrochemical impulse in the nose leads to a "smell" in the brain. Indeed, neurons develop not only from local signals that start them off on their journey, but they respond to local context as they move or extend into different parts of the brain, or extend into distant parts of the body. Cognitive and integrative functions seem to be different sorts of phenomena from digestion, muscle contraction, and making bone. We have very few ideas about the ultimate nature of these fundamental brain functions.

Nonetheless, a prediction borne of experience is that these systems are not as unique as they seem to be but once again involve familiar elements. For example, the traditional notion of a prepattern is that prepared cells remain developmentally quiescent until they encounter an inductive signal. An example may be the role of neural crest cells in epithelial-mesenchymal interactions that develop structures like feathers. However, these cells follow signals along the way and there may be more direct contact or connection as both the NC cells and the tissues through which they move both grow out from the midline. Neurons also follow signals as they grow to innervate developing tissues like skin, muscle, and gut. This kind of prepared-track may be responsible for the migration of olfactory neurons, and perhaps in a way not yet known, for that of interneurons into the cortical layers (as shown in Figure 15-5B). Neurons also follow the growing structures they will innervate.

Some if not most of the signaling and cell-movement mechanisms also involve familiar cytoarchitectural and cytoskeletal genes and interactions. *Reelin*, for example, is expressed in cells elsewhere than the brain. *Tbr1* that may help induce *Reelin* expression may be largely confined to the forebrain, but is part of a larger family (the *T-box* TFs) that is, like other such families, old and has more diverse expression. Most of the architectural and extracellular matrix genes referred to in this chapter, that help guide axon migration, have multiple expression sites.

There are homologies between invertebrates and vertebrates in many of the pathways and mechanisms for neural cell-type specification and brain regionalization (Hirth and Reichert 1999). *Hox* genes are expressed in the embryonic vertebrate hindbrains and caudal nervous system of insects (based on studies in *Drosophila*) in a spatiotemporally conserved way and are involved in neuronal identity. *Otd/Otx* genes are expressed anteriorly, in rostral brain development in insects, and are required for vertebrate mid- and forebrain formation. A mechanism identified in vertebrates—for example, by a human disease—can often, perhaps typically, be found to have a homolog in *Drosophila* and many other animal species (even yeast or fungi). This is useful for learning about the mechanics and origins, but does not tell us what the brain is actually doing.

Nonetheless, we have at least a general sense of the overall organization of central nervous systems in different classes of animals. This makes it possible to make more sense of the way information received from an external source by the various sensory systems is processed, to which we now turn.

Chapter 16
Perceiving: Integrating Signals from the Environment

> . . . every sense has something peculiar, and also something common; peculiar, as, e.g., seeing is to the sense of sight, hearing to the auditory sense, and so on with the other senses severally; while all are accompanied by a common power, in virtue whereof a person perceives that he sees or hears (for, assuredly, it is not by the special sense of sight that one sees that he sees; and it is not by mere taste, or sight, or both together that one discerns, and has the faculty of discerning, that sweet things are different from white things, but by a faculty connected in common with all the organs of sense; for there is one sensory function, and the controlling sensory faculty is one . . .)
>
> (Aristotle, *On Sleep and Sleeplessness*)

In Chapter 16, we will provide a general overview of the ways organisms collect and make use of different signals from their environment. We will describe the peculiarities of each sense, as well as what is known of the "common power" that the senses share. Intriguingly, the answer to the question of how "a person perceives that he sees or hears" may be as elusive to us today as it was to Aristotle. We know many more details about the senses and how they function than did Aristotle, particularly at the molecular and cellular levels where he had no knowledge at all; however, we are still pondering how "that faculty connected in common with all the organs of sense" interprets the information the senses collect.

For some biological functions, there is a relatively direct correspondence between genes and function. For example, the protein code specified by the genes for the hemoglobin molecule relates to that molecule's ability to bind heme (an iron group) that relates to its capacity to bind oxygen. And similarly, the protein structure of collagen is related directly to its physical role as a connective tissue. But is there any such connection between protein structure and the "common power" of percept? If so, it seems totally to have eluded us. Instead, what seems much

Genetics and the Logic of Evolution, by Kenneth M. Weiss and Anne V. Buchanan.
ISBN 0-471-23805-8

more likely is that the power lies strictly in the *organization* of neural processing systems.

How this works is not clearly related to genes per se but to their expression patterns (including during development). Similarly, as Darwin's and Bates' comments on insect behavior mentioned in previous chapters show, we cannot attribute function just to a numbers of cells game. Once again we can cite Aristotle. Although Francis Bacon and others who came along nearly two millennia later are usually credited with inaugurating the reductionist worldview of science, in his *Physics*, Aristotle asserts that "we do not think that we know a thing until we are acquainted with its primary causes or first principles, and have carried our analysis as far as its elements." This was not exactly modern molecular reductionism, but it is curious to ponder how apt this view may be in regard to understanding cognitive function, an "emergent" trait if there ever was one. Once again, and especially for traits determined by interaction among elements rather than the inherent properties of the elements, the specific genes used to lay down the wiring network are functionally arbitrary when it is within that network itself that the function lies. As a result, the survey in this chapter is not strictly "genetic," even if perception ultimately depends on gene expression.

In Chapter 15, we laid the basic groundwork for understanding how "higher" organisms integrate environmental cues by describing the development and structure of central nervous systems (CNS). Again, it is important to remember that only a small subset of life collects and makes sense of environmental signals in this way. Despite our own very understandable romance with the importance of the head in "higher" organisms, many organisms with noncentralized signal collection and integration systems, from single-celled bacteria to bacterial colonies to cnidarians and echinoderms and plants, have been supremely successful with no head.

LIGHT

As we saw in Chapter 14, most organisms in this world, single-celled bacteria, plants, insects, mammals and fish, make use of light energy in some way. Some convert solar into chemical energy, whereas others have receptors that trigger responses ranging from movement toward or away from a light source to conversion of the light into a perception of the surroundings.

Plant Light Reception

Plants use light in more ways than do animals and have evolved more pigments to collect and transduce it. They use light for photosynthesis, photoperiodism, photomovement, and photomorphogenesis. We might say plants have organismal responses that require some forms of perception because they respond to light in multiple ways (e.g., positive phototropism in stems and leaves and negative phototropism in roots), and these responses are based on signals beyond light itself (e.g., gravity, hormones). Furthermore, responses vary from intracellular ones, such as chloroplast movement, to higher level ones, such as stomata opening or differential growth of different sides of branches in the process of photomorphogenesis. Plants are also responding to more than simply the presence or absence of sunlight: they also respond to wavelength, intensity, and directionality of the light, as well as diurnal length.

The plant's local, noncentralized photoreception means that, although the photosynthesizing leaf cannot survive without the roots and circulatory system of the plant, the plant itself can survive without a specific leaf or even without many of its leaves. Unlike an organism with a cephalized nervous system, which cannot perceive light without its central ganglion (the head), a plant, with its many loci of efficient photosynthesis, can lose a lot of leaves before it isn't taking in enough solar energy to survive or to make local or organismal responses to light. We can think of this as automatic or purely molecular rather than "intelligent" response—so far as we know (reputed response to music and kind words notwithstanding), plants are unaware that they are reaching toward the skies—but, unless we agree to reserve the term for brain activity, that distinction seems somewhat artificial. Indeed, it may be rather anthropocentric and consciousness-focused to make such a distinction.

We also noted in Chapter 14 that algae can sense light and respond with phototactic motion directed in relation to its source. This involves complex signaling and response mechanisms, although of course less complex than in, say, organisms with a CNS that must respond by coordinated locomotion involving multiple parts. It is important not to forget that the result is not just the activation of a molecular pathway but an organismal response, even in the lowly algae.

Insect Vision

Drosophila have two visual systems: the compound eye for image formation and ocelli for light detection. The compound eye is formed of 700 or so ommatidia, each of which contains photoreceptor cells. The ommatidia are arranged in a honeycomb-like fashion, and the entire eye connects to three relay points in the optic lobes of the fly's CNS: the *lamina, medulla,* and *lobula complex.* The simple central *ocelli*, of which *Drosophila* have three, synapse with a single point in the CNS, the ocellar ganglion.

The ommatidial neurons terminate in one of the three layers. The projections from a single ommatidium extend to the lamina, and from there some axons continue on to the medulla. As in vertebrate vision, these projections are topographic. That is, the image is relayed to the brain in a way that preserves its physical layout or "map" in the eye which, because of the linear propagation of light itself, comes directly from the perceived image.

Some insects have an extraocular or *dermal light sense.* This has been shown with experiments in which the eyes have been made nonfunctional. This sense involves single neurons in the brain and/or ventral nerve cord responding to light.

In Chapter 14, we described some key aspects of the cascade of regulatory gene expression that is used to form the insect eyes themselves (Cutforth and Gaul 1997; Czerny et al. 1999; Punzo et al. 2002). Some of the regulatory genes that are involved in development of the optic lobes of the adult *Drosophila* nervous system have been identified and include *Wingless* (*Wg*) and *Decapentaplegic* (*Dpp*), genes already familiar to us because of their involvement in many other structures. *Minibrain,* a protein serine/threonine kinase gene, and *Division abnormally delayed* (*Dally*) are also involved in cell proliferation in the visual lobes of the fly brain. The final differentiation of cells in the developing lamina depends on the arrival of axons from the ommatidial photoreceptors, in a way homologous to the development of vertebrate olfaction (Cutforth and Gaul 1997), which will be described below. The process involves differential signal transduction, in this case among adjacent cells.

How an insect "perceives" light images is, however, still unknown, especially in terms of its experience. For example, how does a particular image generate a response such as flight or avoidance? Can these be called "emotional" responses, and if so how are they "felt" by the fly?

Vertebrate Vision

Light sensation typically requires detection of direction and strength of the signal. Some organisms do more with light, and it is for this that we use the term "vision." Except perhaps for touch, vision is the one sense that specifically requires the (sometimes detailed) *spatial* characterization of the signal. It is this that allows us to interpret an *image* that comes essentially in the form of pixels, to give an integrated assessment of the spatial relationships of objects distant from us, and we might say this is the value of vision. Eyes such as ours perceive many other aspects of light in addition to spatial ones, providing an even more nuanced sense of our environment.

We tend to think of the vertebrate retina as a simple analog of a camera's passive film or other kind of pixel sensor. Rods detect brightness and cones the various primary colors. But the interpretation of light absorbed by the rods and cones begins before the signal leaves the eye. Retinal neurons are specialized for different aspects of light, and the image that is sent to the brain is not that of absolute levels of illumination but instead a retinal map of spatial patterns of regions of relative light and dark, color, intensity, depth, and so forth.

The spatial orientation of the light that hits the retina is precisely maintained, although somewhat distorted, in the form of a *retinotopic map*, sent to each visual processing center in the brain; the map is intact in that neighboring cells in the retina project to groups of neighboring cells in the visual area of the thalamus, which in turn project to neighboring regions of the *striate cortex* (Kandel et al. 2000). "Somewhat distorted" means that the *relative* position of the signals is retained but the translation of a three-dimensional image received by a curved retina into a two-dimensional one necessarily alters it somewhat, and the absolute distances between receipt centers of two signals in the brain and between their receiving photoreceptors is not maintained. The *percept* provided by the brain compensates for this kind of "stretch" distortion with mechanisms at least partially understood (see below). In other words, the correction appears to be such that the percept more closely maps to the image striking the retina than to the way that image enters the brain. In fact, most sensory systems project their receptive surface to the appropriate brain centers in a similar way, but the spatial correction is most important in visual and tactile perception.

Two predominant pathways carry the signal from eye to brain in vertebrates: the *retinotectal* and the *retinogeniculate* pathways. In vertebrates with a small forebrain, such as birds, amphibians, and reptiles, the retinotectal projection is the most prominent. Here, the photoreceptors in the retina project onto bipolar nerve cells (cells with two processes extending from the cell body, the axon and the dendrite), with limited dendritic branching on one end and a short axon on the other, a structure that allows fast and precise conductance of the signal. These neurons in turn connect with retinal ganglion cells that form the optic nerve and go primarily to a part of the midbrain called the *optic tectum* ("roof"). This structure coordinates orienting responses—turning toward light or sensing prey or danger—rather than analysis of form or shape or the like, as in mammals. That is, retinotectal vision is predominantly linked with motor control and is representationally rather crude.

In mammals, the homolog to the optic tectum is the *superior colliculus* (a small moundlike region in the midbrain), which receives light signals, as well as other kinds of sensory input. The superior colliculus helps orient the head and eyes in relation to these other kinds of sensory information.

The second pathway, the retinogeniculate, is the predominant visual signal relay pathway in mammals but is only barely evident in vertebrates with small forebrains. As in the retinotectal pathway, light signals are conveyed along retinal ganglion cells to the visual processing centers of the brain. The retinal ganglion cells are called X or Y cells (M or P cells in primates) and comprise two different major routes for visual information to reach the brain. The X and Y cells transmit slightly different visual information to slightly different areas of the brain, with the X cells projecting to the magnocellular, or large-celled layers, and the Y cells projecting to the parvocellular, or small-celled layers of the laminar *lateral geniculate nucleus* (LGN) in the thalamus.

The most important difference between the X and Y cells is their response to color contrasts; X cells are essential for color vision. But X cells are also important in distinguishing images that require high spatial and low temporal resolution (i.e., they specialize in sustained responses and are best at analysis of stationary objects), whereas Y cells are important in vision that requires the opposite: low spatial and high temporal resolution (i.e., they have fast and transient responses and detect movement, basic shape, depth, brightness, texture, etc., of objects but are poor at analysis when objects are stationary) (Kandel, Schwartz et al. 2000).

Retinal ganglion cells exit each eye bunched together into the *optic nerve.* The optic nerve from each eye meets in the *optic chiasma* (see Figure 16-1), where signals from each side cross to the other brain hemisphere, to travel on to the lateral geniculate to be relayed from there to the primary cortical visual center. This crossover is essential for coordination of the images from both eyes to create stereoscopic vision.

Each neuron in the LGN is responsive to a single spot of light in a region of the visual field, and each layer of the LGN is monocular, that is, responsive to only one eye. The retinotopic map is maintained by the neurons of the LGN, which are primarily projection or *relay* neurons, that in turn project to different layers of the *primary visual cortex* (also called *Brodmann's area 17*, *V1*, or the *striate cortex*) of the forebrain. The X and Y pathways still project to different sublayers of the cortical layer IV, maintaining the separation of input from each eye.

The primary visual cortex, as much of the nervous system, is modularly organized into sets of columns. This "multiple columnar system," with hypercolumns for different tasks, includes *orientation* and *ocular dominance columns.* The neurons in the ocular dominance columns receive input from a small spot on the retina of a particular eye, and the orientation columns receive input from light that hits the retina on a particular plane, that is, of lines and edges all tilted at essentially the same angle to the vertical. To represent all orientations for both eyes, 18–20 columns are required. Neighboring hypercolumns represent neighboring sections of the retina. The neighboring hypercolumns communicate through horizontal connections that link cells with the same specific tasks in the same layers and in this way integrate information over many millimeters of cortex. Information from outside a cell's

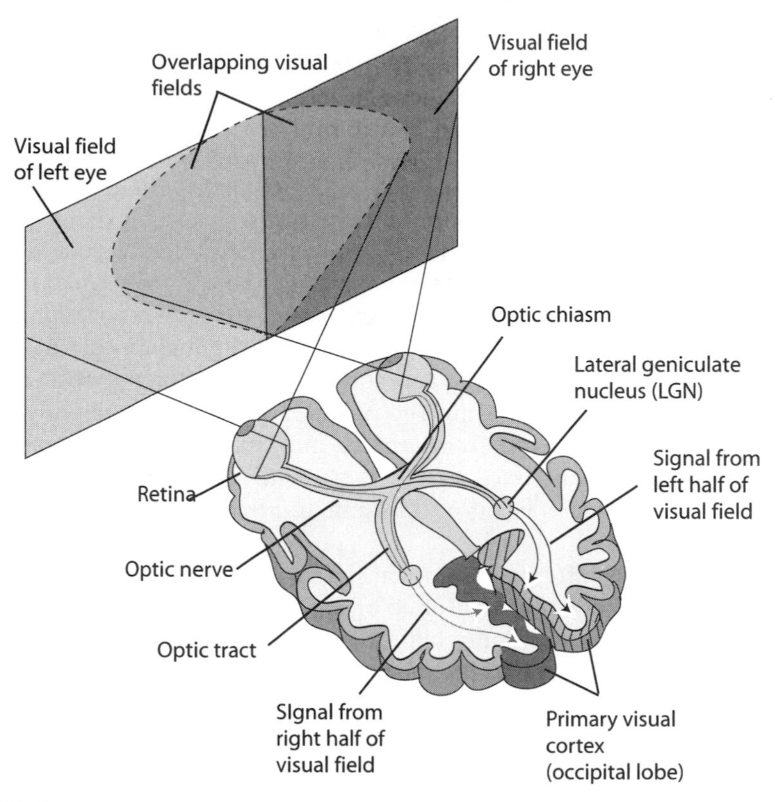

Figure 16-1. Transverse section showing pathway of visual signal to the human primary visual cortex. Redrawn from (Driesen 2003).

immediate environs may influence the way it processes information and thus influence the way we evaluate in context what we see.

After visual information leaves the primary visual cortex (V1), the signals go to 30 or more secondary visual processing centers in the occipital lobe and parts of the parietal and temporal lobes (Figure 15-6). In the end, perhaps 50 percent of the neocortex is involved in visual processing. Beyond the striate cortex, signal is processed with respect to color, motion, intensity, depth, and form.

The X and Y pathways remain segregated as the information leaves the primary visual cortex. The X pathway extends into the inferior temporal cortex, as the ventral cortical pathway and the Y pathway extend to the posterior parietal cortex. The dorsal pathway is largely responsible for the perceptions of motion, depth, and form, and the perception of contrast and contours takes place in the ventral pathway, although there is a good deal of overlap. Put simply, the dorsal pathway determines *where* an object is, and the ventral pathway is involved in recognizing *what* the object is (Kandel, Schwartz et al. 2000).

Visual images are passed to our brain inverted and backward, as they are represented on the retina, a curved but essentially two-dimensional surface. Bilateral symmetry allows organisms to see on both sides of themselves. One eye detects what is to the left of the organism, the other what is to the right. In some organisms, both eyes face forward enough that the fields of vision (images) of the two eyes overlap;

it then becomes possible to characterize the outside world more precisely in three dimensions, that is, additional information about relative distances from the eyes is available. The three dimensional aspects of vision are reproduced in the higher cortical centers, using both monocular and binocular cues. Binocular resolution allows the kinds of "triangulation" of differences that enable us to perceive depth or distance. Our brains apparently learn from experience to integrate the many cues to determine an object's distance, its size relative to other objects in the frame, motion, and other characteristics that give an image its three dimensionality, as well as to correct the inversion and reversal of the image as it enters the brain.

The retinal image is represented numerous times in the cortex, with many cells processing the image in different ways at once, in at least two different pathways. How do all the separate representations of what we see come together into a single image? This is known as the Binding Problem and is one of the central questions in cognitive psychology. The answer is still not clear. The brain apparently constructs a visual image in layers, by putting together the numerous interpretations produced at each level of processing, but whether there is a common pathway that integrates all this or whether the various afferent pathways interact in some way is not yet known.

Extraocular Vertebrate Photoreception

We referred briefly in Chapter 14 to the fact that organisms perceive light in many ways that do not involve eyes. We referred to vertebrate extraretinal photoreceptors that are important in regulation of circadian rhythms and photoperiodicity. The pineal and parapineal glands within the brain are the most important such photoreceptors. Brains are more or less permeable to light, depending on the circumference of the head: although the light becomes somewhat refracted and filtered, in many animals sufficient light penetrates the skull and brain to reach photoreceptors in the pineal and parapineal glands. In larger animals, light input to the pineal gland is part of the visual pathway. The pineal gland regulates synthesis of the hormone melatonin, which is derived from the neurotransmitter serotonin and is secreted at night. The neurochemical basis for its regulation of circadian rhythms is not known. Although this response to light is in some senses behavioral, it appears to be simple in that it does not involve detecting other aspects of light such as direction or, especially, image.

Many animals also use light in nonperceptive or certainly nonbehavioral ways. Some require sunlight in the ultraviolet range, for example, for the conversion of cholesterol into vitamin D or to induce the production of melanin by melanocytes to protect cells from the damage ultraviolet light can cause to DNA. Interestingly, melanocytes are derived from NC cells and in that sense are neural structures.

VIBRATION SENSING

Vibration, the wavelike movement of air or other objects in which an organism comes into contact, can be created by traveling sound waves or by mechanical disturbance (such as water currents or wind). By itself, vibration detection is a form of the sense of touch. However, like light, there is additional information carried by the frequency, amplitude, complexity, and location of vibratory motion. It is this that we refer to as hearing, but as usual there are subtleties and gradations. And it is not only animals that sense and respond to vibration.

Plants

Some plants, such as mimosa, are able to perceive vibration and respond by folding the leaves that have sensed the motion. This is presumably a protective response, to shield delicate leaves. Response to vibration is called *seismonasty*; the plant may have the same response to touch, and this is called *thigmonasty*. The response is caused by changes in turgor pressure, which is driven electrochemically, through depolarization of cell membranes. The cell membranes of specialized *motor cells* become more permeable to potassium ions, resulting in the outflow of ions and osmotic loss of water from the cell. This results in the shrinking of the cells that sensed the vibration, which causes the leaves to droop. A whole plant may respond to loud sound or shaking, but this is not thought to be an organismal response, but one that takes place independently in each leaf.

Lateral Line

The lateral line is a vibration detection mechanism of fish and aquatic amphibians (see Chapter 12 and Figure 16-2) and is used for many vital individual as well as social behaviors. The spatial distribution of its mechanoreceptive organs, known as neuromasts, determines the receptive field, which can vary among species, and the innervation patterns determine how sensitive the system is going to be. Neuromast morphology also varies and is related to the types of fluid motion that will be detected, whether velocity or acceleration (Maruska and Tricas 1998).

The lateral line system comprises a series of neuromasts located in tunnel-like canals in the dermis of the head and along the midlateral flank of the fish and in pit organs throughout the body. The tunnels open at intervals at the body surface to expose the mechanosensory cells to the exterior. Neuromasts are the basic functional organ of the lateral line and are composed of groups of about 30 sensory hair cells and 60 supporting cells and are covered by a gelatinous cupula. The hair cells of the lateral line are essentially identical morphologically and functionally to those in the vertebrate inner ear and detect movement via the displacement of stereocilia in the same way. Most fish have both canal neuromasts and superficial neuromasts, which detect water current. Neuromasts are innervated by branches of the posterior (PLL) and anterior (ALL) lateral line nerves—those on the head by the ALL and those on the sides and tail by the PLL. Projections from the ALL and PLL extend to two locations in the hindbrain in two neighboring columns, preserving a somatotopic representation of neuromast order and thus the location in the body where a vibration was perceived.

The vertebrate homologs of the *Drosophila* proneural gene *Atonal* (*Ato*) include *Math1* in mouse, *Zath1* in zebra fish, and *Atoh* in humans; these bHLH-class TFs are all essential for the development of inner ear hair cells and promote neuroblast differentiation and subsequent differentiation of the peripheral and CNS, including the brain (Itoh and Chitnis 2001).

There is great interspecific diversity in brain morphology among fish, with the area or size of a particular functional part of the brain correlated with modal specialization of the species. Deep-water fish have poor color vision, for example, and

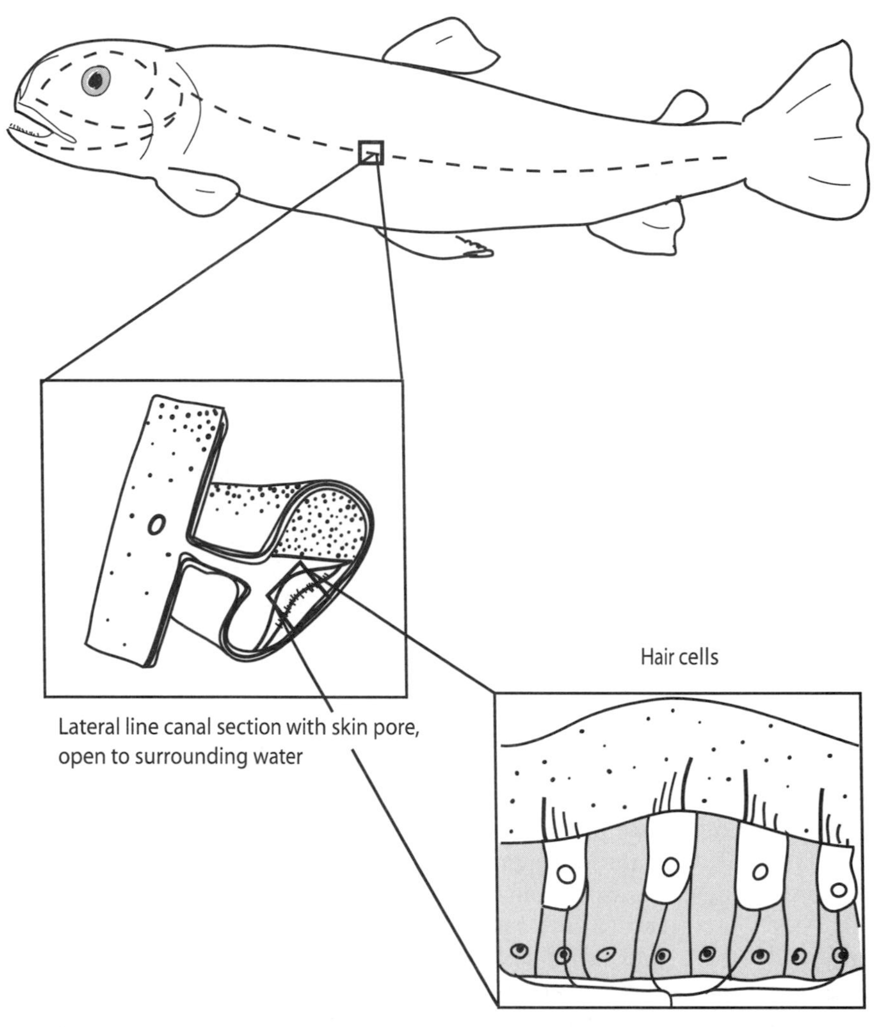

Figure 16-2. Fish lateral line system; schematic layout on the body; lateral line canal section with skin pore, open to surrounding water; and hair cells embedded in the lateral line.

may have poor vision in general or are even blind, whereas shallow water fish can differentiate colors; thus the size of the brain areas associated with color vision and vision in general are differentially enhanced.

Vertebrate Hearing

Hearing is the specialized reception and interpretation of another kind of vibration—sound waves, the mechanical displacement of the medium in which an animal lives, whether water or air. As with all sensory processes, the energy of the vibration must be transduced into an electrical signal. In the auditory system, this is done by the hair cells of the inner ear, as described in Chapter 12. As the term is generally used, "sound" refers to more than just general strength of vibratory signal—it also refers to the details and complexities of its frequency and "hearing" refers more specifically to its interpretation. The obvious reason is that this provides much more information about the source than just its presence.

The hair cells in the cochlea synapse with neurons whose cell bodies lie in the cochlear or spiral ganglion at the base of the *vestibulocochlear* or *auditory* nerve from each ear, which carry the signal to the *cochlear nucleus* in the brain stem. The vestibulocochlear nerves have two branches that project to the brain—one from the cochlea and the other from the vestibular system. The cochlear nerve contains about 30,000 afferent axons, most of them from inner hair cells in the cochlea, but 5–10 percent from the outer hair cells, which seem to play a role in tuning sound as a cochlear amplifier, among other functions.

The axon of each neuron in the cochlear nerve connects to an area of the cochlea that is most responsive to a particular *characteristic frequency*, and sound signals are carried to the brain in an orderly way that represents the spatial frequency tuning along the cochlea—the apical end receiving low frequencies and the basal end high frequencies. Moving from the beginning to the smaller inside end of the coiled cochlea, sounds of increasing frequency are detected.

The means by which each region along the cochlea responds most to a given frequency range is a complex matter of acoustical physics and hydrodynamics, and is not yet completely understood. Generally, the length and shape of the cochlear canal, and the differential stiffness of the basilar membrane along it, determine the vibration characteristics of the fluid in the cochlea. There are also differences in hair cell structure and firing behavior that vary along the cochlea. Sound energy applied to the basilar end sets up waves that travel up the cochlea through this fluid. The changing shape and stiffness of the basilar membrane along the canal dampen the passing waves, which as a result end up establishing standing peaks centered at a frequency-dependent distance from the basilar end.

In this way, a given sound frequency triggers hair cells in a specific region along the cochlea. Because this is a replicable property of fluid dynamics, the brain can recognize each occurrence of the same sound frequency by the location triggered. Further, while the relationship between frequency and position varies among species, and within individuals, and is not perfectly linear along the cochlea (it may be log-linear), the *relative* positions of activation can be used by the brain to work out complex sound characteristics, such as (at least for humans) harmonies, octaves, consonance, and the like. Thus, in a general way, successive octave frequencies activate regularly spaced locations.

Variation in gene expression leading to allometric variation in growth dynamics of the size, shape, length, and stiffness of the cochlea will produce a chamber with particular response characteristics, and these differ by species, as is shown in Figure 12-2. As a result, organisms are suited to particular hearing behavior. There is a danger that this will invoke an adaptive illusion, in this case of fine-tuning by selection. Such selection could in principle be what happened, but there are alternative potential explanations.

The basic mechanisms of mechanoreception and ion channel-based signal transmission were available early in evolution, and were cobbled together into the wide range of hearing mechanisms we see today. Rather than the sensitive touch of selection this could be explained as sloppy jerry-rigging of the pieces that evolution had to work with at the time, yielding hearing mechanisms that fell fairly randomly all over the map with regard to what organisms could hear. Organisms then made do with what they had, "tuning" themselves by organismal selection, regardless of the fact that the ability to hear a wider range of frequencies, or softer sounds, or from further afield would serve them even better. Evolution has built a basic vibration-response mechanism with wide interspecific, and even intraspecific, variation in how it works, with the only measure of importance being that it does work.

In fact, as with other senses, the sound wave peaks are spread over an area within the cochlea, so that there is overlapping sensitivity among hair cells. This and other aspects of interaction among cells along the cochlea give the brain the wherewithal to integrate the information and compensate for imprecision.

Relative position along the cochlea is conserved by relative position of fibers of the auditory nerve, with apical afferent fibers in the center and basilar ones around the outside. They thus arrive in the brain with locational information intact. This *tonotopic map* is analogous to the retinotopic map of the visual system in that an orderly spatial representation of the cochlea is sent to the brain but differs fundamentally in that the tonotopic map is totally unrelated to any spatial aspect of sound. The tonotopic map translates frequency into cochlear position, whereas the retinotopic map conserves the relative spatial positions of incoming photons. Perhaps this is related to the reason that, although sound is topographically represented in the brain, we don't perceive it as an image the way we do light (there is no reason, in principle, why the brain could not present sound to us that way; in fact, we make oscilloscopes to do just that). But the orderly map is in a sense a de facto result of the structural means by which sound is parsed so that individual frequencies can be detected. That structure takes advantage of the relative propagation properties of vibration of different frequencies in an enclosed fluid.

Auditory neurons are specific to different aspects of sound; some, for example, are responsive to the onset of a sound, transmitting information about the initiation of the sound but quickly dampening, whereas others don't begin to fire until the sound has been sustained for some time, thereby transmitting information about the sound's intensity and duration.

Neurons in the cochlear nucleus project to other auditory areas in the brain via three different pathways: the *dorsal acoustic stria*, the *intermediate acoustic stria*, and the *trapezoid body*. As with vision, sound is processed in parallel pathways that finally converge as complex acoustical information about sound source, intensity, frequency, and duration. Signals from both ears join in the *superior olivary nucleus*, which receives input from the trapezoid body. Postsynaptic axons from the superior olivary nucleus and axons from the cochlear nuclei project to the *inferior colliculus* in the midbrain via the *lateral lemniscus*, both of which, again, receive binaural input. Efferents from the superior olive project back to the cochlea to control sensitivity. The primary function of the superior olive is sound localization. Cells in the colliculus project to the *medial geniculate nucleus* of the thalamus, and axons from the geniculate nucleus terminate in the *primary auditory cortex* in the temporal lobe and in the *superior temporal gyrus.* See Figure 16-3 for a diagram of the central auditory pathways.

The primary auditory cortex is segmented in several ways, as is most of the neocortex. In particular, its segmentation is similar to that of the visual cortex in being laminar, with each layer receiving input from neurons from different places. It is also organized into functional columns, as is the visual cortex, with all neurons in a given column responding optimally to sounds with the same frequency. The dorsal portion of the auditory cortex responds to lower frequencies and the anterior portion to higher frequencies, with a gradient of frequency responses in the columns

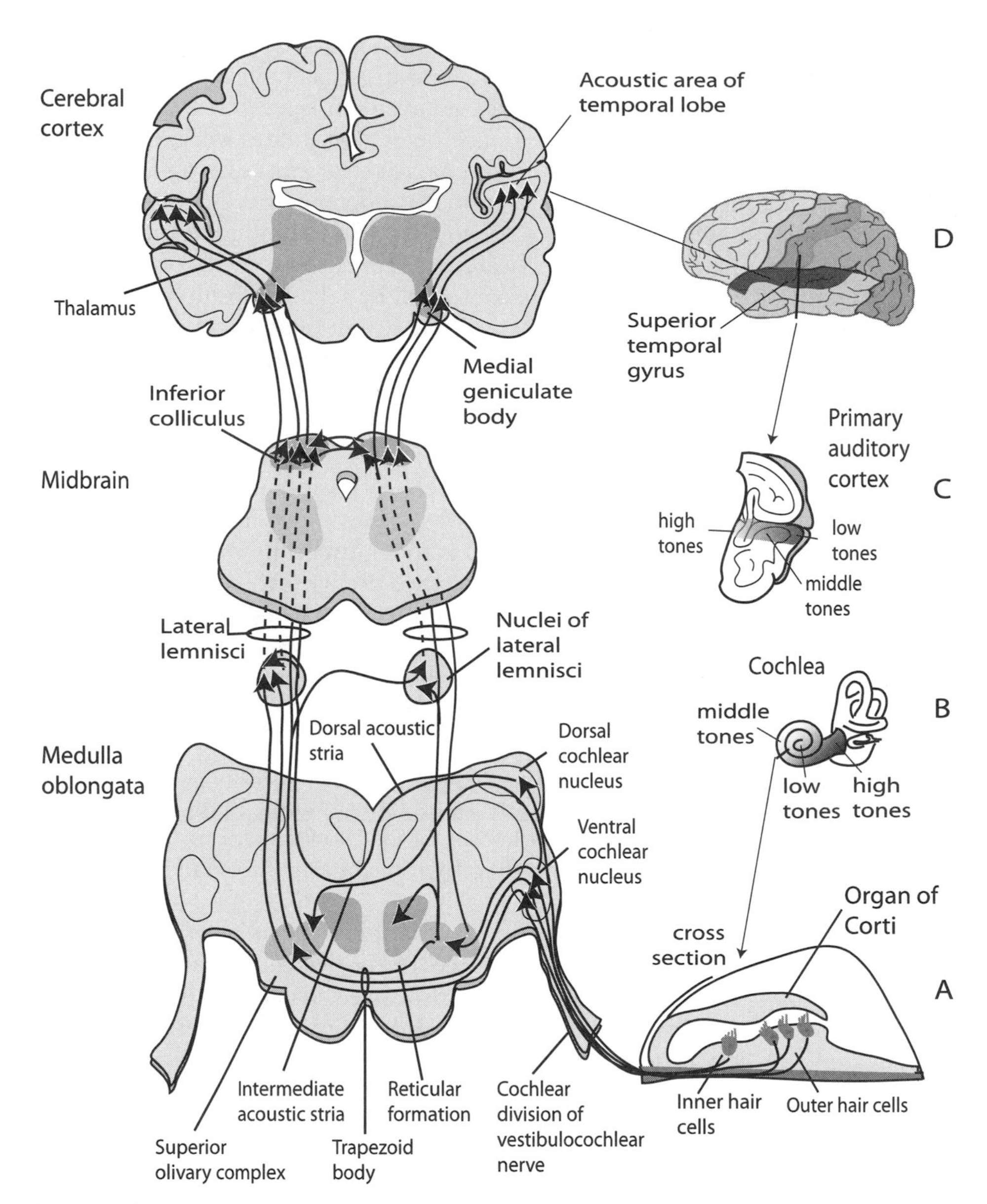

Figure 16-3. Central auditory pathways. Sound is transmitted from **A**, the organ of Corti in **B**, the cochlea, where different frequencies stimulate the hair cells in stereotypical areas; signal is transmitted from the cochlea through the brainstem to higher auditory areas, with the tonotopic map maintained, and ultimately to **C**, the primary auditory cortex, where signal is perceived in the acoustic area of the temporal lobe, **D**, still with the tonotopic map intact.

in between. The auditory cortex is also divided into two types of alternating "zones"—*summation columns* and *suppression columns.* The summation columns are composed of neurons that are excited by stimulation from either ear, and the suppression columns are composed of neurons that are stimulated by input from only one ear and inhibited by stimulation from the opposite ear. The spatial organization of these columns relative to the axis of tonotopic mapping enables the primary auditory cortex to respond to every audible frequency and interaural interaction (Kandel, Schwartz et al. 2000).

In mammals, sound is generally sent for further processing to a number of other areas beyond the primary auditory cortex, in a parallel and hierarchical way similar to the visual system, but the process is not nearly as well understood. In marsupials, sound processing seems to be confined to the primary auditory cortex; however, in eutherian mammals, there are at least nine regions beyond the primary auditory cortex, although the number seems to vary by species and/or depending on the method of data collection. In all cases, sound continues to be organized according to the same tonotopic map received and processed by the primary auditory cortex. The number of areas in insectivores seems to be three or four, in rodents four to seven, and in carnivores and primates six to eight or nine (Ehret 1997).

Again like vision, bilateral symmetry makes it possible to locate a sound in a more three-dimensional way when integrated in a stereophonic rather than simple left-right way. Unlike light, sound reaches both ears no matter which direction its source. This means that it is possible to detect the slight delay that results from signal coming from one side or the other and even to assess distance to some extent.

Radios rely on devices to filter out signals at all frequencies other than the one to which they are tuned. In that way, a signal is detected out of a background that can contain energy of all frequencies. Auditory systems in the brain also tease out single sounds or sound frequencies from the generally broad panoply of incoming sounds, so, for example, a person can hear the melody of the violins in an orchestra or attend to a single conversation in a crowded room. How we do this *auditory scene analysis* is an important but still not well-understood phenomenon that probably involves learning from experience combined with accurate representation and localization of sound by the ear and brain.

Sound interpretation is more than just the detection of characteristic vibration signatures, the way a mass spectrophotometer detects the characteristic spectrum of each type of molecule. Signature detection is certainly part of this, but, at least in higher vertebrates, many more subtleties and complexities are resolved. Also as occurs in vision, this is basically done in the brain, after the information has left the detector itself.

Human speech processing is an important higher area of sound processing in humans, but understanding of the neural pathways and processes involved is still fairly rudimentary, in part because there are no laboratory animals in which it can be studied.

Invertebrate Hearing

It appears that only a minority of insects have the ability to hear, but insect hearing may have evolved at least 19 times (Yager 1999). Like vision, we will probably learn that rudimentary elements of the system, perhaps shared with vertebrates, were present in stem animals (with whom animals on two divergent branches share ancestry) with morphological details independently intercalated in between an early induction signal and later receptor-system differentiation. Cytoarchitectural processes recruited for mechanoreception are likely to be its basis.

Paired hearing organs are generally peripherally located on almost any part of the body, including various abdominal or thoracic body parts, legs, wings, or mouth.

Almost all insect ears share a tympanum covering a tracheal sac and tympanal organ. The somites and dendrites of insect auditory receptors are in the hearing organs themselves; sound frequency is first analyzed there. The axons of the auditory receptors enter the nearby segmental ganglion or ganglia and carry auditory input in a tonotopically organized way, as in the vertebrate hearing system.

Thus, insects also interpret a signal that has been given an orderly translation from a frequency pattern into a neurological space-based "map." The signal is then sent to the brain for decoding of the sound's direction, pattern, frequency, and subsequent localization (Pollack 1998; Stumpner and von Helversen 2001). The specifics of the neuronal pathways differ by species; some, for example, use parallel independent processing to recognize and localize signals and some use a process in which localization depends on recognition (Pollack 1998).

As in vertebrates, sound direction is determined by comparison of the auditory input from each of the two ears. The vertebrate brain uses interaural arrival time difference and intensity difference to make that determination. The insect brain uses intensity difference as well but whether there is enough difference in time of arrival of a sound at two ears located so close to each other to be useful in determining sound directionality is still open to question (Pollack 1998). (Again, we must be careful not to view the world from our human frame of reference: some mammals that can detect directionality of sound, like voles and bats, are certainly also very small.) Inhibiting or destroying sound reception of one ear does result in an insect no longer being able to distinguish the location of a sound source, such as the call of an echolocating bat.

SOMATIC SENSATION

Plant Thigmotropy

We referred earlier to thigmonasty or vibration sense in plants. Plants also have *thigmotropy* or a sense of mechanical disturbance. The leaves of some plants, such as some mimosa or the venus fly trap, close when touched, in the same way that they close when sensing vibration. Roots often exhibit *negative thigmotropy* by turning away from objects in the soil that they touch as they grow. Perhaps the best example of thigmotropy in plants is the manner in which tendrils curl around objects. When they come into sustained contact with an object, the opposite side of the stem begins to grow more rapidly, thus elongating relative to the contact side. At the same time, the contact cells lose their turgor pressure and become flaccid and smaller. Sustained differential growth and turgid pressure cause the tendril to curl around the object it is touching. How the touch stimulus is communicated throughout the plant is not known and may work differently in different plants.

Biochemically, mechanical stimulation has been found to elicit the release of *alamethicin*, *jasmonic acid*, and *12-oxo-phytodienoic acid*, which induce coiling in some plants. Alamethicin is a peptide that forms voltage-gated ion channels and elicits the synthesis of a number of volatile compounds. 12-oxo-phytodienoic acid is a powerful inhibitor of growth and is differentially found on the contact side of a tendril (Engelberth et al. 2000); if the contact side slows its growth but the outer side continues to grow, the tendril will curl around the contacted object.

Plants growing in conditions of sustained exposure to touch, wind, water spray, and other similar physical stimuli are shorter and stockier than plants grown without exposure to these stimuli. This is called *thigmomorphogenesis.* At least four touch-induced (TCH) genes are known in *Arabidopsis thaliana*; they belong to a family related to *Calmodulin* genes, which code for calcium-binding proteins. TCH mRNAs accumulate rapidly on touch, within 10 minutes (Johnson et al. 1998), but the pathway by which these genes induce thigmomorphogenesis is not understood.

Vertebrate Somatosensation

As discussed in Chapter 12, many sensory receptors are located in the skin: mechanoreceptors that sense touch, pressure, vibration, and hair displacement; thermoreceptors that sense temperature; and nociceptors for damage and pain. These receptors are not uniformly distributed throughout the skin but instead are concentrated in some areas more than others. Proprioceptors relay information about joint and limb position.

The cell bodies for these receptors cluster to form the dorsal root ganglia along the spinal cord. The location of the ganglia provides a relative positional map of the location of the stimulating signal on the body. The axons of these neurons ascend to the brain in two separate pathways in the spinal cord—one that transmits information about highly localized sensations (fine touch and the proprioceptive system) in the *dorsal column medial-lemniscal* system and the other system, the interneurons in the spinal cord that synapse with receptors of poorly localized sensations, pain and temperature, in the lateral sensory tract of the spinal cord (the *anterolateral system*).

Each of these pathways projects to different areas of the brain. The fast-acting interneurons in the anterolateral system project directly into the thalamus without synapsing first in the brain stem. This is the pathway that is responsible for immediate pain perception after injury. The nerve fibers in this system are myelinated and large to transmit signal rapidly (diameter of the neuron affects the speed of conductance). The prolonged pain of injury is the product of the slower-acting dorsal column medial-lemniscal pathway. This system for perception of touch and proprioception ascends the spinal cord ipsilaterally (on the side of the spinal cord that corresponds to the side of the body that was touched), and axons synapse in the *dorsal column nuclei* of the medulla in the brain stem before moving on to the thalamus. The receptors in this pathway are unmyelinated small fibers.

In the dorsal column medial-lemniscal pathway, the first axons to enter the spinal cord, from the sacral (lower back) region, are centrally located in the dorsal column; axons that enter from successively higher areas of the body are progressively more lateral. At the upper region of the spinal cord, the dorsal columns divide into two bundles of axons—the *gracile fascicle* and the *cuneate fascicle.* The nerve fibers in these bundles are organized somatotopically—those in the gracile fascicle from the ipsilateral sacral, lumbar, and lower thoracic segments and those in the cuneate fascicle from the upper thoracic and cervical segments.

As in the visual and auditory systems, the somatotopic organization of the ascending axons is maintained as a map at each step of the processing of somatosensory information in the brain (Kandel, Schwartz et al. 2000). Touch and pain, however, require rapid reaction capabilities and in general (in mammals, or at least

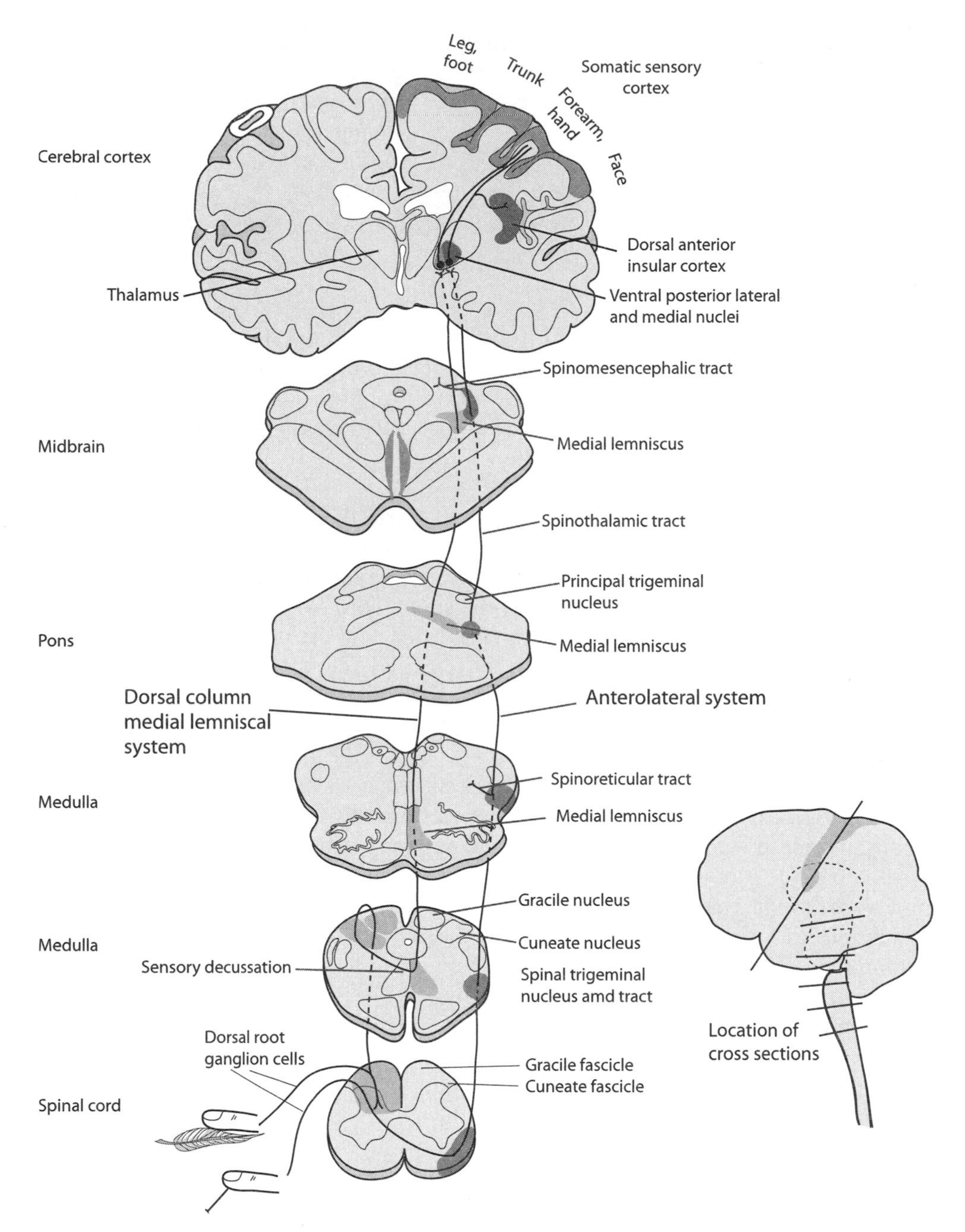

Figure 16-4. Anterolateral and dorsal column medial-lemniscal pathways: touch. Redrawn from (Kandel, Schwartz et al. 2000).

humans) less detailed understanding of the source of the sensation. We need to know the location of a cut or contact with flame, not the shape of the thorn or fire.

Because the brain stem was the first cerebral ganglion to form in the evolution of the cephalized CNS of vertebrates, most (but not all) sensory axons ascending the spinal cord, even in vertebrates with an extensive forebrain, still make synapses in the brain stem before traveling to higher parts of the brain. Axons from the gracile and cuneate bundles terminate in the gracile nucleus and the cuneate nucleus in the lower medulla. Somatosensory signals from the face and scalp are sent to the *principal trigeminal nucleus*, which is also in the brain stem. Axons that originate in the dorsal column nuclei decussate, or cross, the midline of the spinal cord and ascend in the medial lemniscus to the *ventral posterior thalamic nucleus* in the thalamus. As these fibers decussate, the body map becomes its mirror image, and the sacral segments are now located most laterally and the higher segments are located medially. Signals from the face, once they reach the principal trigeminal nucleus, are transmitted in the *trigeminal lemniscus*, joining axons from the arm and back of the head in the medial lemniscus, to the ventral posterior medial nucleus of the thalamus.

The information reaching the thalamus in the right side of the brain comes from the left side of the body and vice versa. In the anterolateral system, the crossover takes place in the spinal cord, whereas in the dorsal column medial-lemniscal system it takes place in the medulla. The decussation always occurs with the first interneurons to synapse with the primary sensory neurons (Matthews 2001).

From the ventral posterior thalamus, signal is relayed to the cortex. Most neurons from this area project to the *primary somatosensory cortex* (*SI*), with a minority projecting to the *secondary somatosensory cortex* (*SII*). Again, the body surface is represented in an orderly way. Some areas of the body are overrepresented in the brain with respect to relative body size, however, so that more brain area, for example, is devoted to afferent connections from a human's fingertips or lips than the legs or toes. More cortical area, in fact, is devoted to the palm and fingers than to the leg, trunk, and arm combined. These areas themselves also have a higher density of sensory nerve endings. The somatotopic map of the correspondence between skin surface and cortical area was first depicted as a *homunculus*, a human figure distorted to represent the amount of cortical area devoted to specific body parts, by neurosurgeon Wilder Penfield in the 1950s (see Figure 16-5). He determined, from testing individuals with epilepsy during brain surgery, the areas in the cortex in which afferent connections from different parts of the body terminated. The growth of individual nerves to the periphery of the body innervates the skin in a modular, segmented way, establishing *dermatomes* or areas on the skin served by branches of a given nerve. Since the branching is developmentally hierarchical, the brain is enabled to recognize regions and subregions in an orderly way.

This is something of a different kind of topographic map. Here, relative position in the body is represented, but there is differential distortion based on use or importance. Vision and hearing are not so distorted in this way, although retinal images have different strengths and sensitivities in the center and peripheral regions of the retina. There are also species-specific differences that, at least to some extent, correspond to use and need. In these senses, relative function *is* preprogrammed into even the brain (and innervation patterns of the body).

Like other cortical areas that we have discussed, the somatosensory cortex is laminar. Again, axons of afferents coming from the ventral posterior thalamus synapse with the stellate interneuron cells of layer IV, which in turn connect with the pyramidal neurons of the other layers of the cortex, including motor regions, and back to the thalamus in a feedback loop that supplies sensory information to the cortex. The motor regions of the brain can then act on proprioceptive information or cues about the body's contact with the environment.

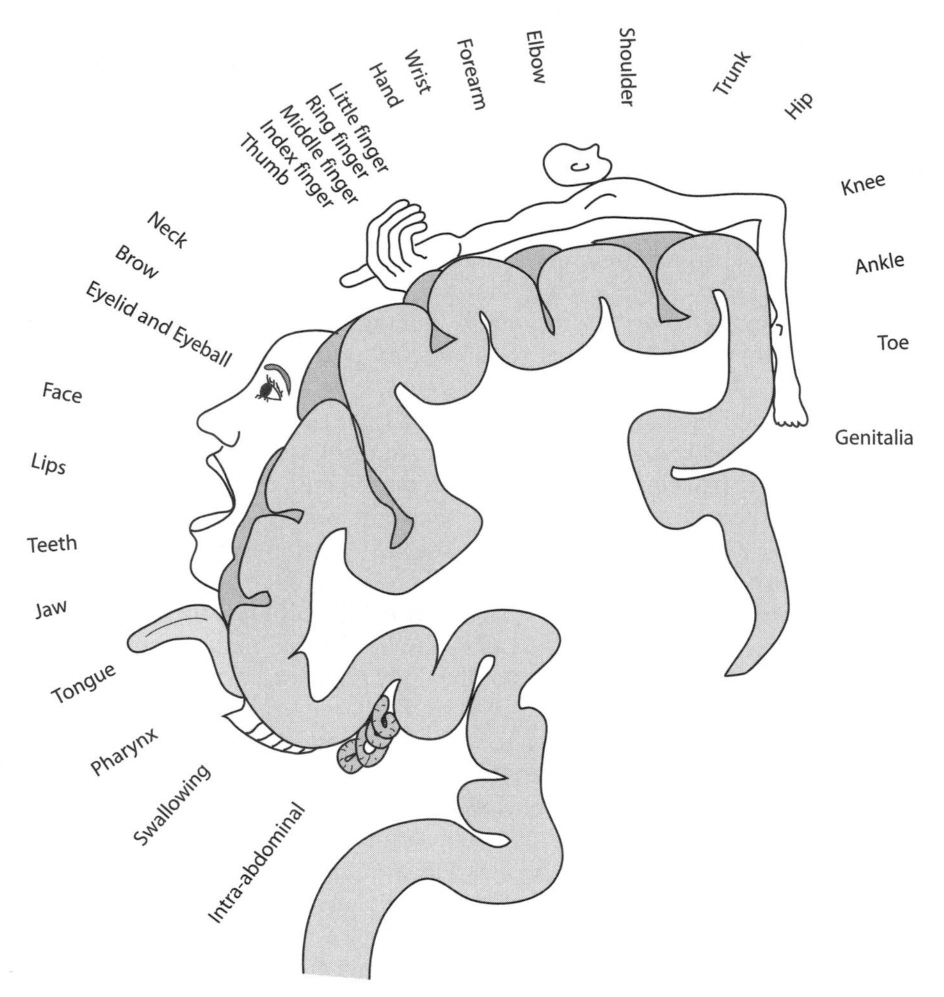

Figure 16-5. Homunculus, representing the topographic organization of the human sensory cortex, and relative size of brain areas dedicated to perception by various body surfaces.

Sensory modalities are separated in SI into four distinct functional areas, each of which contains its own representation of the whole body. Two of these areas respond primarily to deep pressure and movement, and the other two respond primarily to surface touch. Some neurons from these areas project to the secondary somatosensory cortex. Both the primary and secondary somatosensory cortical areas are organized in columns as well as layers, just as are the visual and auditory areas of the cortex. Neurons in each column respond to stimulation from a specific region of the body, and there is extensive interconnection between the four areas. This manner of morphological organization allows the somatosensory system to function via both parallel and hierarchical processing, again as with other areas of the brain.

The somatosensory cortex not only enables perception of touch and temperature but also is responsible for localization of the source and intensity of a stimulus, as well as further dissection of somatosensory signals. The secondary somatosensory cortex is innervated by neurons from SI. SII requires projections from SI to function. Neurons in SII project to the *insular cortex,* which in turn relays signal to portions of the temporal lobe thought to be involved in creating and storing tactile memory. Areas in the *posterior parietal cortex* receive input from SI and other areas and are involved in integrating tactile signals with proprioceptive signals and with integrating information from the two hands. Other areas integrate visual with somatosensory information and initiate sensing and movement as a response to tactile and proprioceptive information (Kandel, Schwartz et al. 2000).

Rodent Vibrissae

A much-studied example of topographic mapping of a somatosensory system is that of *vibrissae,* or whiskers, in rodents. The vibrissae are the principle tactile receptors in these animals. They form along a line by a repetitive periodic patterning process that generates them one by one, roman-candle fashion along the line. The number is generally invariant in the mouse (Dun and Fraser 1959), although artificial selection experiments and some naturally occurring mutation (in the *Ta*, or *Tabby* gene) can expose underlying variation. This is thought to be a cell-surface ligand gene product whose extracellular domain is a ligand for a TNF-class cytokine receptor (related to IL1 and NFκB pathways we have seen earlier), and is used in epithelial-mesenchymal patterning of structures like hair and teeth.

As the developing face projects outward, presumably each new vibrissal primordium produces its sensory and transmitting neurological elements when the vibrissa is initiated. As it develops, each whisker maps to a specific site in layer IV of the rodent primary somatosensory cortex. The somatotopic map represents precisely the position of each whisker on the rodent snout. A single whisker is innervated by about 100 neurons, which convey the directional movements of the whiskers to the brain. The afferents project from the whiskers to the SI through the thalamus.

In the SI, the neurons are arranged in functional units called *barrels*, one per whisker, corresponding to the array of the whiskers on the snout (see Figure 16-6). (The nomenclature is due to the shape of the cell bodies of these neurons because they appear to form barrel-shaped arrays when the cortex is cut parallel to the cortical surface.) As in other synaptic networks in the brain, there is a critical developmental period for the growth of the optimal network of connections among barrels in the cortex of the rat. When vibrissae are removed from a young rat at an early age, the cortical connections to the different cortical layers are much less dense than in normal rats, suggesting that experience "trains" the synapses (Fox 1994).

Interestingly, not all species that have whiskers have corresponding cortical barrels. Barrels can be found in, for example, the mouse, rat, squirrel, porcupine, and walrus but not in the dog, cat, or raccoon. In addition, barrels are found in some animals that have minimal use for their whiskers, such as the guinea pig and chinchilla (Nieuwenhuys et al. 1998). However our understanding of the brain is still incomplete if not perhaps only rudimentary, so caution is needed in generalizations about brain structure and function.

Star-Nosed Mole

An unusual example of an organ for touch that may give this animal the keenest sense of touch known in the animal kingdom is the star on the nose of the star-

Figure 16-6. Topographic map of mouse vibrissae. (A) Vibrissae; (B) corresponding spatial organization of barrels in the cerebral cortex. Reprinted with kind permission of T. A. Woolsey, originally published in (Woolsey and Van der Loos 1970).

nosed mole. The snout of this animal extends in a splayed protrusion of 22 finger-like appendages that are used to explore the animal's immediate surroundings. This is shown in Figure 16-7. Despite its location, the organ is not olfactory, nor used for manipulating food or other objects, but instead for touch. The appendages are covered with tiny *Eimer's organs*, hairlike touch receptors, 1,000 or so on each of the 22 rays of the star. More than 100,000 nerves project from the nose to the brain, compared with the 17,000 or so from the human hand (Catania 1999). When the nose comes into contact with an object, it engorges with blood, and this pushes the receptors closer to the surface of the nose, thereby increasing its sensitivity. This provides a kind of light-less spatial representation of the environment in front of this burrowing animal, a kind of tactile retina, whose spatial aspect is vital to interpretation. Does the star-nose mole "see" its environment or just "feel" it?

Given what we know about the organization of the somatosensory areas of the cortex vis-à-vis touch sensitivity, it is not surprising that the star-nose is highly over-represented in this animal's brain. The star is represented three times by a stripe of tissue in the cortex of each hemisphere, once in the SI, again in the SII, and yet again as a smaller representation in an area termed SIII (Catania and Kaas 2001). This is unusual; the somatotopic map of vibrissae, for example, is represented only once, in SI. The amount of cortical surface devoted to the star-nose reflects the importance of this tactile organ to the mole. Furthermore, it is possible that distributing the cortical connections among three small areas rather than one large one may speed interneural connections, and this could be important for an organ upon which an animal relies so heavily.

TASTE

As we saw in Chapter 13, the receptors for the sense of taste in vertebrates are grouped together in taste buds, each of which contains about 100 *taste receptor cells*. Taste buds in mammals are located primarily on the tongue, but also on the pharynx, the laryngeal epiglottis, and at the entrance of the esophagus. Some fish have taste buds over the entire surface of the body (Butler and Hodos 1996).

Figure 16-7. The star-nosed mole. The splayed nose provides a spatial tactile sense of a light-less environment. Source: http://www.moleplace.com/photo_page.htm.

The surface of each taste cell is covered with microvilli, which protrude through a pore in the cell (the taste pore) to be exposed to the oral cavity. They are specialized to detect chemicals dissolved in the saliva on the tongue. The microvilli in turn are clustered into papillae that are embedded in the epithelial layer of the tongue. Afferent nerve axons enter the taste bud at the base and each one synapses with multiple receptor cells in the taste bud.

The taste cells can detect five basic stimuli: bitter, salty, sour, sweet, and umami. The ability to taste amino acids seems to be particularly acute in fishes, but it is not obvious why this particular sense developed. Combinations of the five basic tastant stimuli lead to perceptions of complex tastes. Each stimulus is transduced by a different mechanism, and the same taste may be elicited by different stimuli and different mechanisms. In addition, the mechanism used to sense the same stimulus may differ between vertebrate species.

Sour and salty tastants do not have a receptor molecule in the receptor cell in the usual sense; these tastes are due to the hydrogen or sodium ions in the substance. They are not detected with the usual receptors but by changing the membrane potential of ion channels in the taste receptor cell. Sweet, bitter, and umami reception are more complex, and transduction of these tastes is more similar to phototransduction and olfactory transduction: the tastant is bound by a G protein membrane receptor, which in turn activates intracellular signals that affect the ionic permeability of the taste receptor cell. Because many molecules can be perceived as sweet or bitter, receptors for these tastants are many and varied.

Two nerves carry input from the tongue to the brain: the facial nerve (cranial nerve VII) conveys information from the front of the tongue and the glossopharyngeal nerve (cranial nerve IX) from the back. These two nerves also contain the sensory fibers for transduction of touch, pressure, and temperature. Axons from these nerve bundles enter the brain stem in the medulla and synapse in a thin line of cells called the *nucleus* of the *solitary tract.* As with other sensory systems, the somatotopic organization of taste is maintained in the solitary, or gustatory, nucleus, such that gustatory nerves from different regions of the head, and skin in animals with taste buds on the body, enter this area in the order in which they are located on the body. From there, depending on the extent of development of the animal's thalamocortical system, input is conveyed to the gustatory region of the somatosensory cortex, the primary gustatory cortex, where the perception of taste is formed, and, in animals with more fully developed forebrains, to the hypothalamus, amygdala, and insula, the limbic area, where behavioral responses to the perception of taste—aversion, salivation, gastric secretion, pleasure, etc.—are triggered.

Fish have a very well developed sense of taste in general, with taste receptors on their lips and in their mouth and pharynx, but several suborders have evolved taste receptors over the entire surface of their bodies. In these fish, the gustatory nervous system is very complex and supplies them with a detailed topographical taste map of their surroundings. These *silurids* (catfishes, of which there are more than 1,000 species) and *cyprinids* (carps, minnows, chubs, and goldfish) are bottom feeders and have evolved an elaborate system for separating inedible particles from food as they pick up mouthfuls of sediment from the bottom of the stream or ocean (Butler and Hodos 1996).

Insect Taste

In insects, detection of soluble chemicals by gustatory neurons can elicit feeding behavior, but mating behavior as well (Scott et al. 2001). *Drosophila*, for example, have chemosensory hairs on their legs and proboscis that activate proboscis extension and feeding when they detect sweet compounds. Female *Drosophila* have bristles on their genitalia that elicit ovipositing upon detection of nutrients, which probably maximizes the probability that eggs are laid in an environment in which the newly hatched can feed.

Taste receptors in *Drosophila* are found in sensory sensilla located on the fly wing, legs, proboscis, and genitalia. Two types of sensilla have been characterized: *taste bristles*, which are located on the legs, wings, ovipositor, and mouthparts, and *taste pegs*, on the oral surface and in the pharynx. Bristles are hollow hairlike lymph-filled structures with a pore at the terminal end, through which nutrients enter to dissolve in the lymph. Each bristle contains neurons that are specific for sugar, salt, or water, as well as a single mechanoreceptor neuron (Shanbhag et al. 2001). Taste pegs, in contrast, do not have obvious pores in the termini or side-walls, but their role in gustation has been shown behaviorally in blowflies, and electrophysiological assays suggest that they have sugar and salt receptors, as well as a mechanoreceptor. Stimulation of these cells induces feeding (Shanbhag et al. 2001).

Adult *Drosophila* have about 2,000 chemosensory neurons in the sensilla. Up to four neurons innervate each sensillum. Taste pegs are innervated by two receptor cells: a chemoreceptor and a mechanoreceptor (Shanbhag et al. 2001). Sensory neurons from the proboscis extend to the subesophogeal ganglion (SOG) in the brain and then are relayed to other gustatory areas for further processing. Gustatory neurons from other parts of the body project locally to peripheral ganglia. How taste is represented in the *Drosophila* brain is not yet known.

Differential taste responses could arise from different locations on the body or tongue as a consequence of the developmental patterning mechanism that generates the taste buds or receptors. That there is a topographic map of taste receptors in terms of their location of the body is no surprise; it provides a sense of where a given tastant is being detected, but this need not have any *particular* evolutionary function and instead may just be a useful result of developmental patterning. But how different and variable locations on insect bodies are all interpreted as *taste*, or whether something "tastes" different depending on which sensors it activates, are more intriguing questions.

Odorant detection is odorant specific at the level of the cell in many animal species. This has to do with the developmental allelic and gene exclusion in generating the olfactory neuron. Taste is generally a different kind of detection, more general, and it is easy to understand why each taste bud might be sensitive to multiple tastant characters: it could enable the organism to determine the generic "flavor" of the stimulating substance, rather than its specifics. At the same time, it could also simply be the consequence of the way taste-sensing units develop.

OLFACTION

Odorants traveling in the air do not maintain a precise relative location and may mix and mingle. Most interesting objects have complex odorants that have no spatial relevance to a detecting organism. Except for strength and general direction, olfaction neither has nor requires a spatial map that faithfully represents details of the emitting source.

For these reasons there would have been little selective pressure to detect odors with a spatial orientation. However, olfactory signals *do* map spatially in terms of the relationship between the position of the detecting cells in the olfactory epithelia and where the associated neurons synapse in the brain. This relationship is probably used by the brain to keep track of what is coming in, because a given receptor will be triggered by the same odor whenever it occurs. The regionalization of the olfactory epithelium and the sets of related olfactory receptor genes expressed in each region, as described in Chapter 13, are easy to account for in principle by the generally modular nature of developmental patterning processes.

VERTEBRATE OLFACTION

The olfactory sensory neurons in vertebrates are in the nasal cavity, in the olfactory epithelium, an area of about $5\,cm^2$ in humans. Several million olfactory neurons are embedded in this small region, interspersed among supporting cells. A single dendrite extends from the olfactory sensory neuron to the epithelial surface of the nasal cavity, where it swells into a knob. Five to 20 cilia protrude from this knob into the mucus-covered epithelium. These cilia have receptors capable of binding specific chemical characteristics and hence able to bind specific odorants, which initiates the steps that transduce the olfactory signal to the sensory neuron and on to the brain. In Chapter 13, we described the genetics of odorant detection itself and the olfactory receptor gene families.

Each vertebrate olfactory neuron seems to express only one odorant receptor gene and thus transmits information about only one odorant to the brain, although one odorant may be detected by several different receptors and a receptor may respond to several different odorants. Animals can detect many more odors than they have odorant receptor genes because of this combinatorial approach to the open-ended nature of chemical detection. The perception is related to the combination a given odorant triggers.

The olfactory epithelium (Figure 16-8) is organized into four different zones. Groups of neurons with the same receptors cluster together in these zones, and axons from the different zones project to distinct regions of the olfactory bulb, thereby preserving the topographical organization of the receptors in the olfactory epithelium. The spatial map of these zones is nearly identical in both hemispheres of the brain and between individuals (Zou et al. 2001).

From the basal end of the olfactory neuron, a single axon projects to the olfactory bulb in the brain, above the nasal cavity, to synapse with olfactory bulb neurons. These neurons are organized into small glomeruli, of which there are about 2,000 per bulb in mice (Kandel, Schwartz et al. 2000). Several thousand sensory neurons project to each glomerulus, each of which apparently receives input from only one type of receptor (that is, one class within the odorant receptor (*OR*) gene family). Glomeruli that receive input from specific types of receptors are located in the same place in the olfactory bulb in different individuals. Each sensory neuron synapses in only one glomerulus. In each glomerulus, there are 20–50 different relay neurons, and axons synapse with three different types: the *mitral* and *tufted relay* neurons that send the signal to the olfactory cortex and the *periglomerular interneurons* that encircle the glomerulus and make inhibitory synapses with mitral cell dendrites.

Olfactory signals seem to be extensively processed before they are relayed on. The inhibitory connections made by the periglomerular interneurons may be part of this preprocessing, as are feedback connections from the olfactory cortex and parts of the forebrain back to the olfactory bulb. An odorant's effect on an animal's behavior may depend on the animal's physiological state; an odor may heighten an animal's hunger, for example (Kandel, Schwartz et al. 2000), or elicit sexual behavior only during receptive periods.

The mitral and tufted relay neurons project directly to the cerebral cortex for further processing and to the limbic system, which is involved in emotions and mediates emotional responses to smells. Olfaction is a primitive or early-evolving sense, and this may be why the olfactory nerve is the only cranial nerve that sends signals directly to the cerebrum, bypassing the thalamus. This probably reflects the importance of chemodetection in early vertebrate evolution, which would not be surprising given the aquatic environment in which chemicals more than light or sound might have been the primary environmental stimulant. The primary olfactory cortex is actually in the paleocortex, a part of the brain evolutionarily older than the neocortex with somewhat different laminar structure.

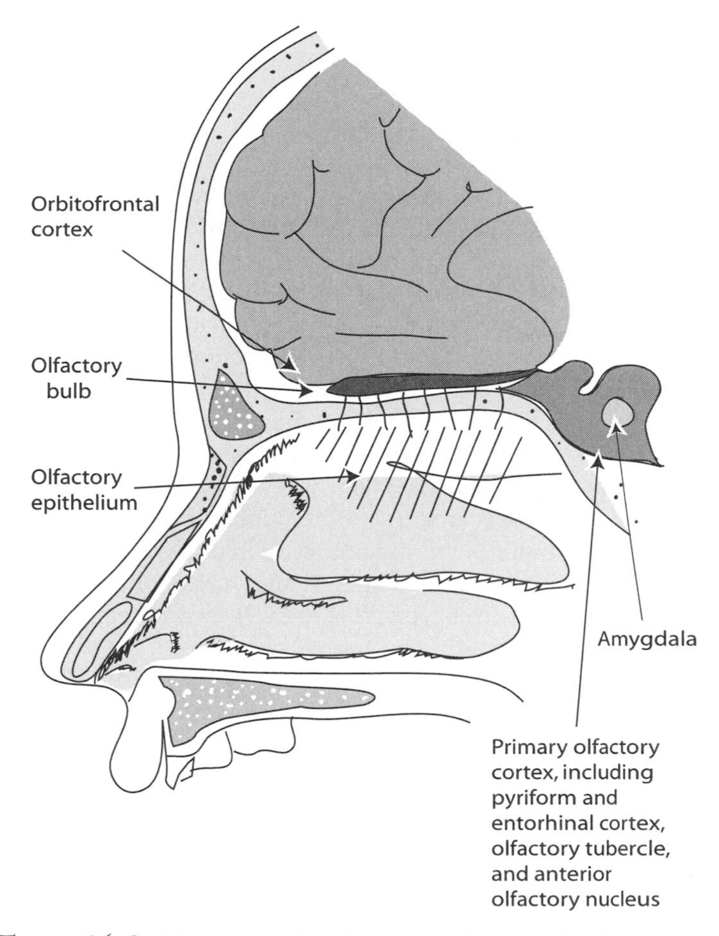

Figure 16-8. Diagram of the olfactory pathway to the human CNS.

The olfactory cortex is divided into five regions: (1) the *piriform cortex* (the largest olfactory area), (2) *olfactory tubercle*, (3) *anterior olfactory nucleus*, and parts of the (4) *amygdala* and (5) *entorhinal cortex*. Mitral cells project to all parts of the olfactory paleocortex, whereas tufted cells project only to the most anterior regions (Zou, Horowitz et al. 2001). From the primary olfactory cortex, signal is relayed to secondary and tertiary olfactory regions, including the hippocampus, ventral striatum and pallidum, hypothalamus, thalamus, orbitofrontal cortex, agranular insular cortex, and cingulate gyrus (Kandel, Schwartz et al. 2000; Weismann et al. 2001). Studies of people with brain lesions suggest that the pathway to the orbitofrontal cortex from the thalamus regulates perception and discrimination of odors, whereas the amygdala regulates emotional responses to odor.

Olfactory input seems to be relayed to the same brain areas across individuals, suggesting the existence of a stereotyped map of axonal connections to the olfactory cortex (Zou, Horowitz et al. 2001). This is what would be expected of hierarchical or regional patterning mechanisms, as seen in the other systems we have described. However, a stereotyped map of connections does not mean that this replicable organization is related to anything inherent in the objects emitting odors or the odors themselves that are being detected. Instead, today it mainly reflects the evolutionary history of gene duplication in the *OR* clusters, the developmental patterning process, and perhaps the nature and timing of the process by which each cell "chooses" which *OR* to express. As we have seen, this may affect the localized migratory pattern of that neuron into the brain.

Still, consider that to at least some extent, odorants with somewhat similar characteristics are today detected by receptors in similar *OR* classes, that are expressed in generally similar regions of the olfactory epithelium (Liu et al. 2003). Given the activation-inhibition nature of repetitive patterning mechanisms, it is likely that when there were but few *OR* genes just accumulating binding differences (and hence, the class differences we find today), the cell-patterning mechanism would leave them regionalized with a de facto function-location correlation. Whether that was important to the *nature* of the odorant classes that were detected by the simple early system for any particular functional reason, is a separate question.

As in other sensory systems, olfactory input is processed hierarchically as well as in parallel in different regions of the brain, when, as with other senses, olfactory input is sent to more than one area at a time. Unlike other senses, the olfactory topographic map in the brain does not maintain spatial information about odors, or retain the bulb map, but instead it may encode the quality of an odor (Mombaerts et al. 1996; Wong et al. 2002; Zou, Horowitz et al. 2001), with organized input coming from different glomeruli.

Information from different odorant receptors is segregated until it reaches the olfactory cortex, where input from many glomeruli clusters into overlapping neuronal groupings and the olfactory cortex integrates these signals to produce the perception of many different and complex odors in a way that is not yet well understood. Some olfactory signals are then relayed to the limbic system, where they may affect emotional states or instinctive behaviors, and to the neocortex, where they are further processed.

Insect Olfaction

We described insect olfactory reception in Chapter 13. Odor receptors in insects are found in sensilla, usually sensory hairs, which project from the cuticle. The sensilla have tiny pores on their surfaces through which odorants pass, and they stimulate the dendrites of the odorant receptors inside, which are covered in lymph. Most sensilla in most insects are found on the antennae.

Briefly, adult *Drosophila* have about 1,300 odorant receptors (Vosshall 2001), compartmentalized in sensory hairs on the surface of the third antennal segment and on the maxillary palp, an olfactory organ on the proboscis. This compartmentalization is a major difference between vertebrate and invertebrate olfactory systems. Each antenna holds about 600 sensilla of three different morphological types, whereas each maxillary palp has 60 of a single type. The *OR*s code for G protein-coupled receptors (GPCRs), of which there are three classes (Figure 13-2B), and each of the sensilla in turn contains two neurons, arranged in stereotyped pairs. Thus, there are six types of neurons, each with its own particular response. If each gene is expressed with the same probability, the relative frequency of the types will depend on the relative numbers of genes in each class. Whether this has functional or evolutionary relevance is not known and would probably require categorizing these genes in multiple species.

The sensilla are situated on the palp in a fixed configuration, and each exhibits a different specific response. A certain odor may excite one OR and inhibit another, and a single OR may be excited by one odor but inhibited by another. The ORs all behaving in this fixed way generate an olfactory code. There appears to be no sexual dimorphism in the kinds of sensilla present in male or female flies, in contrast to a marked sexual dimorphism found in the sensilla and olfactory glomeruli of some moths (e.g., de Bruyne et al. 1999).

As noted earlier, the *OR* gene class in *Drosophila* is quite divergent from the corresponding genes in vertebrates (Scott, Brady et al. 2001; Vosshall 2001). Each olfactory neuron expresses only one receptor gene, as in mammals (with the exception noted earlier of one *OR*, *Or83b,* that seems to be expressed in every olfactory neuron) (Scott, Brady et al. 2001; Vosshall 2001).

In arthropods, as in vertebrates, afferents from ONs project to glomeruli in the CNS, maintaining the topographic organization of the expressed ORs in the sensory projections (Wong, Wang et al. 2002). There are 43 olfactory glomeruli in the *Drosophila* antennal lobe, the fly equivalent of vertebrate olfactory bulbs. As in vertebrates, functional imaging of brain activity in insects shows that different odors elicit activity in different glomeruli (Wong, Wang et al. 2002). Projection neurons then relay input from the antennal lobe to the mushroom body and the lateral horn of the fly protocerebrum in a stereotyped set of axon branching patterns (Marin et al. 2002), for higher processing. If, as in vertebrates, a combinatorial code defines an odor, the specifics of how input from different ORs is integrated in the higher areas of the insect brain remain to be elucidated. It is known that projection neurons connect to a stereotyped set of third-order neurons, and it is at the level of these third-order neurons that the integration appears to take place.

MULTISENSORY PERCEPTION

We have largely discussed sensory perception in a Cartesian way, as though each of the senses were an independent function. Although we see orderly ways in which these things are organized, and we can relate that to the known kinds of developmental processes, the emergent nature of perception in each type of sense is still rather elusive. In fact, although many of these sensory systems do function without input from others, higher sensory processing often involves their interaction.

Morphological, visual, olfactory, auditory information is routinely integrated in evaluating threat postures, prey or predator behavior, and the like. Perception of smell and taste interact to produce a more complex sensation, and spoken language comprehension can be more precise when complemented with visual processing of lip movements. Significant interaction takes place between visual and auditory input in vertebrate development, and visual input can influence tonotopic mapping in the auditory cortex (King 1999). Indeed, it is this *intersystem* integration that constitutes behavioral responses to the environment, a fact reflected at the neural level as well. These statements are true of invertebrates as well as vertebrates and indeed, in their way, of plants and single-celled organisms, too.

In many vertebrates, some processing actually takes place in multisensory neurons. Cortical and subcortical afferents from a variety of different modalities converge in these neurons and allow enhanced (or sometimes depressed) perception of an event. That is, neurons that respond to visual cues may also receive auditory and touch cues and then issue multiple motor commands, controlling the orientation of the head, eyes, ears, and so forth, in response.

Located in different regions of the neuraxis, the most well-understood being the superior colliculus of the midbrain, where sensory cues are received and motor commands initiated, these neurons have overlapping receptive fields for each separate modality from which they receive signal (e.g., Kadunce et al. 2001; Meredith and Stein 1986; Stein et al. 2001). Interestingly, although they already have multiple receptive fields and act as fully developed multisensory neurons in their signal reception and sending capacities, these neurons are not able to process multiple modalities at birth. Their multisensory processing capabilities only develop with increasing sensory experience (Wallace and Stein 2001).

This fact is not a surprise, however, as it is known that if one sensory modality is lost, the parts of the brain that perceived and controlled responses to that sense can be recruited by other sensory modalities. This rewiring or remapping does not need to happen only during development; the brain is plastic throughout life. The visual cortex of someone who loses vision as an adult, for example, can be partially recruited for touch or hearing. As with many aspects of perception, sensory and multisensory processing in the brain is the product of hard-wiring *and* experience.

THINKING: WHAT PERCEPTION "FEELS" LIKE AND ORGANISMAL RESPONSES

It is difficult for us to assess the relationship between detection and perception without involving our own experience and notions that connect awareness and consciousness. Indeed, that experience places limits on our attempts to understand perception generally, especially in other species. Thus, a discussion of consciousness itself, speculative though it must necessarily be, is in order.

Nerve nets and ganglia allow some forms of coordinated or integrated responses. Plants, however, do not have any kind of organized nervous system, and, as we have seen, they manage to have coordinated and integrated responses to environmental signals nonetheless; therefore, a centralized perception of the environment or a higher-level organization of response is not necessary for organized multicellular life. With respect to plants and the simplest nerve nets, we may say that "all" this is is reflex triggering of neural impulses by input impulses—a purely mechanical process. In the case of ganglia and more complex systems, we might say that this is "simply" a more highly organized form of the same thing. Cognition—the *real thing*?—involves similar chemical reactions. What is the difference, and, especially, what is the difference when it comes to consciousness?

One standard if nearly metaphysical answer is that these higher-order processes are analogous to electromagnetic fields in physics: when many neurons fire at once, a higher-order emergent phenomenon occurs, say a "field" of "electro-chemo-magnetic" energy, and if enough neurons fire at once, that becomes the phenomenon that we, at least, experience as *consciousness*. As humans, we tend to equate this with the organized unifying phenomenon we refer to as perception.

How this would work or what it means in scientific rather than metaphysical terms are difficult questions. But let us take the ideas as generically correct; if perception *is* an emergent phenomenon, can a simple organism with a neural net, or even a ganglion itself, also generate some form of emergent "aura" of awareness that we would recognize relevant to our own human experiences? For that matter, because plants also organize reactions in a complex way, is there any sense in which we could refer to this as plant "perception"?

We have several times referred to the complex social organization and problem solving abilities of ants and other social insects, who have the tiniest of brains. We can list neural wiring patterns, name some lumps and bumps and connection zones in the brain, and identify neurotransmitters, ion channel genes, and adhesion and signaling factors and receptors that are expressed in brain development. But do we understand what this means any better now than Aristotle did 2,400 years ago?

Natura non Facit Saltum: Perception is Not an All-or-Nothing Phenomenon

Humans with severe forms of epilepsy have sometimes been treated by severing the *corpus callosum*, one of the major neural throughways that connect the left and right sides of the brain. In these "split-brain" individuals, who behave and report still feeling their normal identity, the evidence suggests that only one hemisphere of the brain (usually, the left) is the seat of consciousness. The other hemisphere is a fully functional, problem–solving entity that can even be shown to be self-aware. But it may not have consciousness (an important part of this may be aspects of language and verbal expression, usually controlled by the left side). Perception and sophisticated problem solving are possible without explicit consciousness.

It is clear that there is no one such thing as consciousness or the experience of awareness. We sleep, dream, and can drive cars without "thinking" about it. Each person experiences things in different ways at different times and differently from other people. Some people have a "feel" for science, music, personal interactions, or the flow of a hockey game. Some smell or see things others completely miss, facts that in this case we know directly, from studies of opsin spectral analysis or odorant-specific sniff-testing. Some people even claim to have direct contact with the immaterial world, whereas others who lack that experience strongly declare that to be delusional. Fanciful as such contact experience may seem, areas of the brain that are responsible (for the experience, whether or not the reality) are identified by studies of brain activity scans and the association of some such activities with epileptic episodes and so on.

The most central of all observations of evolution is that systems in related organisms are similar because of shared descent with modification. There is considerable freedom in evolution, but not complete freedom; thus, this kind of relationship has an organic reality. There is no reason whatever not to extend the same principles to perception and consciousness itself.

Darwin was committed to the notion *natura non facit saltum*—nature does not take leaps. To him, this principle gave evolution by natural selection its plausibility, and he used this principle in defending many attacks on his theory (especially in expanded discussion in the sixth edition of *Origin of Species*). His inference was that evolution and selection worked gradually over time. Traits did not emerge de novo. This in a sense is the basis of the modern view by which the step-by-step evolution of complex traits occurs.

Evolutionary biologists have come, perhaps somewhat reluctantly, to accept "punctuated" events in evolution in two senses. First, it is generally accepted that there can be times of acceleration in the rate of change relative to longer times of slower change. This can be brought about by things like the invasion by a species of a new territory or by rapid climate change (or by intense use of antibiotics). Secondly, homeotic change is recognized as a mode of evolutionary "jumps" in which the number of segments like digits or vertebrae can change as a result of changes in meristic patterning processes. In a sense, the latter is just a quantitative change, requiring no sudden novel mechanisms. But beyond that, we have no good examples to persuade us that a new trait can suddenly appear.

It is thus interesting that a darwinian (gradualistic) view of consciousness should be controversial, but it seems to be so. Many scientists are reluctant even to acknowledge that chimpanzees share the human experience of consciousness in more than a rudimentary way. In what can be characterized as at least a bit anthropocentric, since humans and chimpanzees have evolved numerous important differences in other traits, efforts to explain our evolutionary difference from apes has often been focused on the brain. The classic example is the famous but vain struggle by Richard Owen in the 1800s to find a trait unique to the human brain. He thought he had done so with the hippocampus major but was famously embarrassed in that notion by the ever-combative Thomas Huxley. Recent invocations of the brain-centered bias have demonstrated genetic differences between humans and chimpanzees that involve language or differential gene expression in the brain (Enard et al. 2002a; Enard et al. 2002b).

Any application of human experience to "lesser" species—even dogs and their emotions, not to mention the social behavior of ants—is typically denigrated as anthropomorphizing (or worse). However, some authors have tried to justify the position of humans as part of the natural world (rather than above it, by virtue of our unique traits, like consciousness) by insisting that consciousness has arisen out of a phenomenon that somehow pertains to *everything*. These views have typically come out of philosophical or religious perspectives under notions that might generally be referred to as *animism*. Animism asserts some type of universal internal awareness and drive, of which the human mind is the logical, inevitable, or creational acme. Famous biologists and writers on biology including Lamarck, Henri Bergson, and Teilhard de Chardin (to name some of the most prominent) have held such views, often including rudiments in atoms themselves to make the system universal (e.g., Chardin's *Phenomenon of Man*). Even one of the cofounders of population genetics theory, J. B. S. Haldane, made this kind of point (Haldane 1932).

We can look at this subject from an entirely rationalistic point of view, with no mysticism attached. All we need do is apply the same evolutionary notions that we apply to other traits, asserting the gradual or quasi-gradual origins of new traits (no saltations). *Natura non facit saltum.* It follows almost automatically that other species would at the very least have identifiable run-ups to what we experience as consciousness, and these should involve the same sensory neural-integrating processes in other species that they do in humans. We may not know what context-dependent, intermittently flickering, or partial consciousness feels like. Nor are there unambiguous criteria for speculating how far into the range of species this phenomenon may reach. But by any consistent standard of evolution, this important trait must have evolved over time and rudiments of some sort may still characterize much of animal life.

It is hard to accept that even insects might have complex perceptions to go with the behavior that has impressed so many. Henry W. Bates, one-time exploring companion with Wallace, remarked of Amazonian sand wasps: "The action of the wasp [in building a nest for its young and stocking it with paralyzed insects] would be said to be instinctive; but it seems plain that the instinct is no mysterious and unintelligible agent, but a mental process in each individual, differing from the same in man only by its unerring certainty" (Bates 1863). Anyone watching social insects can easily see that their behavior is to some extent open-ended in the sense we have used the term in this book: they may or may not have a fixed or limited repertoire of basic interactions, but they use them in nontrivial, context-dependent ways that we should not diminish by a facile assumption that what appears so complex and well organized must be "just" instinct instead. One day we may develop some way to understand what the experience is like to be an ant or wasp.

The dramatic difference seen today between humans and others seems to rest clearly on language and/or whatever symboling abilities that entails and entailed evolutionarily. If we compare electrical engineers or Shakespeare, with chimpanzees, the difference of course seems qualitative. However, several species of human-like ancestors have existed in the past, and we can trace through the fossils the progressive increase in cortical brain size. It seems hard to envision that this was not accompanied by progressive increase in "consciousness." Not even *Homo erectus* suddenly emerged with consciousness.

To assert that consciousness is a totally new phenomenon or even represented solely by ourselves, would verge uncomfortably on an essentially unprecedented view of the sudden appearance of what Richard Goldschmidt, in opposition to the Darwinian assertion of graduation, famously termed a "hopeful monster" (a baby who could speak language to parents who couldn't?). It would be like invoking spontaneous generation or a form of special creationism—the very last thing most biologists would want to be accused of.

WHAT IS PERCEPTION ANYWAY?

It is easy to say that perception is an organism's way to develop an internal map of the external world, but, given that different species thrive with the external world internalized in such diverse ways and that even individuals within a species do not completely share internal maps, no specific internal map can be said to be a "true" representation of the outside world, but only one that suffices for the particular individual or species. In that sense, the subset of cues an animal takes from the environment are what its ancestors required to live and reproduce; however, in a sort of darwinian feedback mechanism, these cues also drive the evolution of the pathways that allow the animal to collect the information that it needs. And, in an echo of Kant, an organism's world only *is* those aspects it can perceive.

There are many other fundamental questions about sensory perception that we don't yet know how to answer. For example, given that there are critical periods for synapse formation, and if external stimuli and experience are crucial to their formation, how can stereotyped patterns arise? How much of what appears to be pattern merely reflects the shared experience (e.g., uterine environment) of all members of a species? If there are stereotyped patterns, brain areas "for" different senses, how can these areas also be so plastic, so readily recruited by other senses, in the event of, say, a brain lesion or the interruption of a neural pathway or its development or the loss of the relevant neural input? Is brain plasticity a generic phenomenon, so that the synaptic reorganization involved in acquisition of knowledge or new memories is the same as that involved in remapping of brain areas? How does the brain reintegrate sensory input into a single image or sound or smell after it is processed by a number of different areas? How does the brain integrate perceptions from separate modalities into one? How does a smell evoke emotion?

It is in this area, more than perhaps any other area in biology, that differences between molecular and organismal biologists' interests most diverge. There has been incredibly rapid progress in identifying genetic and molecular aspects of neural development and arrangement and of the genetic and neural wiring of various senses that must integrate information from the environment. But lists of transmitters, receptors, and gene expression cascades do not bring us much closer to understanding the phenomenon, or perhaps better the *experience* of perception, than philosophers do.

CONCLUSIONS

We referred earlier to Aristotle's defining of the classical separate senses. Here we have noted that not only is each sense a complex problem of perception of detection, but the senses are not really entirely independent in that their perception is often integrative. Indeed, the poet Dante Alighieri even suggested that shades (souls) of the departed were equipped with organs for each of the senses (Alighieri 1314). How else could he have spoken with them on his tour through the post-mortem worlds of Heaven, Purgatory, and Hell?

We need not delve into that particular issue, but as this and the previous chapter suggest, our current repertoire of analytical techniques is good at uncovering neural pathways and visualizing brain activity given different sensory exposures. We have shown in a physical sense what topographic sensory maps are like, that is, how physical relationships among receptors are maintained by physical relationships in the brain. We have not discussed many specific genes in Chapter 16. Genes such as for olfactory receptors have made it possible to trace individual neurons, and transcription factor families have made it possible to characterize segmental development. Genes associated with apoptosis have been informative about the remodeling processes in brain development. And genetic disease and experimental mutations, along with informatic methods and expression profiling that allow unknown genes to be found that are related to genes of known function, have all aided in the understanding of the wiring and developmental processes.

The processes by which the brain develops and by which connections from the periphery to the CNS are built involve the same kinds of processes we have seen again and again in surveying the way organisms get through life. We have learned much in recent years about neural mapping, especially of olfaction and other senses. In brain development and sensory mapping, division, segmentation, differentiation, cell adhesion, signaling, receptor-ligand information transfer, and apoptotic signals are involved. We have seen these processes used numerous times in the development of plants and animals and have seen that they also involve many of the same genes.

In a way this reveals how the somatotopic mapping of inputs to the CNS is achieved, almost as the requisite spin-off of standard, relatively simple and straightforward patterning and developmental processes. No specific natural selection for topographic maps per se is needed. The organization of sensation of open-ended systems into topographic maps from sensor to brain is sufficiently accounted for by the directness of the processes by which segmentally or regionally organized tissues are laid down—progressively during development, and ramifying in nested, modular form from simpler beginnings over evolutionary time. Development of the tissues at both the detecting and the receiving end will have these modular characteristics, as they are the general characteristics of tissue development. Once the detection-reception connection was made in the rudimentary system, that it proliferates and retains spatial relationships as the system evolves could basically be predicted.

For visual imaging and some aspects of touch, the topographic map was of particular importance because it represented the actual spatial arrangement of the outside world providing the stimulus. For sound, the cochlear tonotopic map is a byproduct of the developmental process generating the physical structure by which wavelength detection occurs. For odor detection, the map is a useful matter of bookkeeping. In all these cases the nature of the developmental process involved in making the isomorphic maps did not require fundamental new mechanisms.

Generally, we have done a lot of hand-waving in this chapter. The genetic basis of these neural wiring patterns is only partially understood. More importantly, the perceptive side, the side that we may perhaps be most interested in as human beings and as scientists, remains largely a black box. Why is it that we have maps of sound, light, and touch in the brain, but these perceptions *feel* different? Why do we see an object as a kind of information panel but not hear sound as a straight line as on an oscilloscope? Understanding these aspects of perception is one of the most fascinating questions in biology.

An important fact is that the neurochemistry and cell architecture are essentially the same in the visual and auditory parts of brains, and indeed one area can take on functions of the other, as we have noted. This helps provide an answer to the question as to how such diverse and complex systems can possibly evolve and develop. Each sensory system uses its own means to detect signal. But *all* the systems are translated into the *same* lingua franca of perception: the *interaction* of signals in networks of neurons. Perception in this sense has nothing to do with the nature of neurons, any more than **AGA** has to do with the functional characteristics of arginine for which it codes, or Fgf8 has to do with the morphological structure of teeth. Unless a truly new principle is discovered, this fact would have to stand as an additional profound way in which the traits of life are achieved by functionally arbitrary means.

PART V

Finale

Evolutionary Order and Disorder between Phenotypes and Genotypes

We have reviewed many aspects of the lives and times of complex organisms. It can be seen both in terms of traits and their associated genes that life is a universally connected phenomenon, in many simple but elegant ways. Genes are complicated, but the process by which they have been strung together and used to make the Great Chain of Beings that live today, and that have lived, is a mesh of chemical interactions that follow some simple general processes that lead us to a unified view of life.

Indeed, even the traditional definition of what constitutes a "being" relative to the systems of life, has been somewhat arbitrary and restrictive. The relationships of genes within cells, of cells within organisms, of organisms within species, and of species within ecosystems have remarkably similar features. In many ways, it is the emergent properties of sometimes arbitrary interactions, not the chemicals themselves, which have the most profound meaning in life.

Genetics and the Logic of Evolution, by Kenneth M. Weiss and Anne V. Buchanan.
ISBN 0-471-23805-8

Chapter 17
A Great Chain of Beings

It doesn't take a brilliant mind to discern that there are extensive and orderly relationships among the creatures on this Earth. But it has taken a number of brilliant minds to try to explain *why* that is. Charles Darwin and Alfred Wallace were the first to articulate clearly the basics of what is still the prevailing view, that natural selection is responsible for the diversity and connectedness of life as we know it, giving a new kind of meaning to the order of life classically developed by Aristotle. We use these thinkers as representative of many others, whose work also contributed to our evolving modern understanding. However, their explanation didn't satisfy everyone—many naturalists thought it was incomplete and it challenged the fundamental world-view (not to mention long-held pronouncements) of many others. Beginning almost immediately after the publication of the first edition of *Origin of Species*, there has been a steady stream of resistance to the idea of natural selection, and many of the same arguments are still raised by skeptics nearly 150 years later. There are still religious and political agendas that lead to some opposition, but some unease continues within professional biology as well, and the basic problem has not really been resolved.

To biologists satisfied with the classical darwinian explanations, a blanket invocation of evolution by a persistent, consistent directive *force* of selection seems to suffice, though the long geologic time scale means this must generally be surmised rather than observed directly. To those who are not satisfied with so generic an assumption, the major problem is to understand more specifically how a simple process like blind (not teleological) natural selection could bring the diversity and apparent high degree of specificity of complex adaptations about, especially in terms of the genetic mechanisms that are responsible.

Here, we will try to present an integrated summary and overview of the major points of this book, much of which deals with this basic question. A modest number of basically simple general principles show how evolutionary processes can indeed have achieved what we see. Some of the generalizations are the classics that go back to Darwin and before. Others are not usually thought of as general evolutionary principles, although we believe they deserve that status and have tremendous

Genetics and the Logic of Evolution, by Kenneth M. Weiss and Anne V. Buchanan.
ISBN 0-471-23805-8

Figure 17-1. (A) Aristotle; (B) young Darwin; (C) young Wallace. (A) Statue in Vienna art museum, copied from (Bowder 1982), (B) 1840 painting by George Richmond.

explanatory power. Among our points is that one can accept Darwin and Wallace's deep insight, yet need not *overly* or uncritically invoke the single force of systematic natural selection as we try to understand the evolution of life.

Before there was much formalized history and essentially no paleontology—no sense of great time depth in the affairs of our planet—a static view of the nature of the relationships among the different forms of life on Earth was not just natural but probably the best supportable one (although there were dynamic cosmologies even among the ancient Greeks). In a static worldview, it was easy to group similar-looking animals, like different types of fish or birds or butterflies, and to infer that they have a relationship with each other.

Aristotle is widely credited with expressing the first systematized biology, which included the notion of a qualitative *scala naturae* (natural ladder, scale, or order of all life), the natural ranking of beings according to their relationships, and considering them in order. In a classic book, Arthur Lovejoy (Lovejoy 1936) traced the history by which this idea, which came to be called the Great Chain of Being, was established in Western culture; mainly, it was fitted into our biblically based cosmology with a point of origin at the Creation. In most Western versions, of course, this is a linear hierarchy with humans at the top (but under God) and the rest of the chain in service to us.

However, evidence was accumulating into the 19th century that the Earth was older and organic beings were not so static as had once been thought. Discoveries in geology, agricultural breeding, geographic distribution of forms found by world exploration, and a growing appreciation of the meaning of fossils all led to the burgeoning realization that species had changed over time. Darwin and Wallace provided a very general *process* that could account in principle for this dynamic history and connect the entire living world—plant and animal—into a single phenomenon. They stressed selection, but probably as important was that *time* and connectedness through *common ancestry* became key variables in biological explanation.

Biology has not abandoned a great chain of being, but we account for it in a new and less arbitrary, materially more testable way. And, taking organisms past and present, the chain has been rearranged. The evolutionary process leads to diversifying rather than linear connectedness, and does not imply that the tips of one branch are qualitatively better than those of another, only that their ancestors were successful *in their own times and relative to their local situations*. Despite this, however, an informal notion of progress, basically of a qualitative hierarchy in nature, is surprisingly persistent in our culture and in biology itself (Ruse 1996), even though it is manifestly true that simple organisms still thrive today. The notion of progress is more than a human conceit; it leads to a kind of tacit general inference of perfection and tightness of adaptation in the world.

The new evolutionary worldview was wondrously reaffirmed during the 20th century in a totally unpredicted way. Biologists adopted the molecular reductionism of the physical sciences and used it as the framework for looking inside organisms, to show that their parts, including their genes, had their own reality as biological entities, and were also connected in an historical way and that this was related to the general pattern of connectedness of whole organisms. Rather than a great chain of static being, what we see now is all organisms *and* their multilayered internal constituents, interdependent today and linked through a three- or four-billion-year-old continuum of interwoven connections: a Great Chain of Beings.

It is worth quoting Darwin's famous reflection on the grandeur of life by which he closed *Origin of Species* in 1859:

> It is interesting to contemplate an entangled bank, clothed with many plants of many kinds, with birds singing on the bushes, with various insects flitting about, and with worms crawling through the damp earth, and to reflect that these elaborately constructed forms, so different from each other, and dependent on each other in so complex a manner, have all been produced by laws acting around us. These laws, taken in the largest sense, being Growth with Reproduction; inheritance which is almost implied by reproduction; Variability from the indirect and direct action of the external conditions of life, and from use and disuse; a Ratio of Increase so high as to lead to a Struggle for Life, and as a consequence to Natural Selection, entailing Divergence of Character and the Extinction of less-improved forms. Thus, from the war of nature, from famine and death, the most exalted object which we are capable of conceiving, namely, the production of the higher animals, directly follows. There is grandeur in this view of life, with its several powers, having been originally breathed into a few forms or into one; and that, whilst this planet has gone cycling on according to the fixed law of gravity, from so simple a beginning endless forms most beautiful and most wonderful have been, and are being, evolved.

This "vision statement" is interesting in ways that are perhaps unappreciated. Much food for thought is here, including mistakes relative to the currently prevailing view. Darwin uses the image of an "entangled" bank, of life interdependent, brought about through common "laws" that we still see "acting around us." He states the fundamental core principles we outlined in Chapter 1. He keeps open the possibility that life may have derived from more than one founding form—and we have seen some evidence of a reticulated rather than simple Tree of Life, perhaps particularly early on before the highly structured cell that we know today stabilized as a basic form (Chapter 2). Darwin stresses the divergent nature of the process, but his view is somewhat lamarckian ("variability from . . . use and disuse"). His evolutionary theory was so powerful and represented so deep a truth that it withstood incorrect notions even about core elements such as the nature of heredity and the age of the Earth.

Darwin's is largely a metaphor of the world as a Hobbesian war of all against all, driven by overpopulation relative to resources. Historians have suggested that his view of life was colored by his being a wealthy member of the world's Imperial power. This predisposition is said to be reflected in Darwin's regular allusion to the struggle for survival as an essentially deterministic and gradual view of life driven by laws as ineluctable as those of physics. Systematically, the better and more powerful are destined to prevail; the law of the jungle is the law of nature. Whether or not Darwin's view is based on his wealth and privilege in the British Empire, that view is the prevailing one today.

When Mendel demonstrated the discrete nonblending nature of heredity—that heredity was a natural "atomic" process of some kind—he opened an important door to further the understanding of evolutionary mechanisms. A century of intense work identified the molecules involved and demonstrated that they were inherited and modified in ways that could be fitted to the general processes leading to accumulated biological diversity that Darwin and Wallace had invoked. The result has been a set of broad, general principles, formalized as population genetics, which are applied universally to the evolution of life. That is what makes the work of Darwin, Mendel, Wallace, and other founders of modern biology so powerful.

In this book, we have looked at how evolution and genetics apply to a selection of different kinds of biological processes that exemplify the basic aspects of biological complexity. These include:

Figure 17-2. An entangled bank. Central Pennsylvania, USA. Photograph by E. Weiss.

1. how the largely prespecified complex forms of a differentiated organism start from simple beginnings like a single cell;
2. how complex differentiated entities can reproduce;
3. how the components within an organism communicate to bring about a unity of coordinated functions;
4. how organisms detect a variety of external conditions that cannot be prespecified but that are important to their life ways; and
5. how the latter information is interpreted and translated into responses.

We have discussed how a set of simple principles stated by Darwin and Wallace, a few others that supplement their theory, and a modest toolkit of genetic mechanisms can account for the diverse world of biological complexity and achievement. Some of the principles are not typically included as formal premises of biological theory, but including them helps integrate a unified and more complete understanding of the great chain of beings in the world today.

Darwin's Basic Postulates

For most purposes, all living forms can be viewed as descendants of a single origin on Earth. A "single origin" for all life does not mean a single original molecule, species, trait or gene. All have their own individual, partially independent nested origins over time. Rather, biology posits a single set of starting conditions and that the essence of the system, once begun, did not receive meaningful extraterrestrial contributions (even if, say, amino acids continue to rain down on Earth) and did not keep originating (and in particular, required no spontaneous creation of complex organisms). In many ways including ancient or even occasional modern horizontal gene transfer, sexual reproduction, and recombination, life is as interconnected today by biological processes as it is by a unitary ancestry, and the former is ultimately because of the latter. If other forms of biological activity—for example, those not relying on DNA or RNA or protein coding—existed back at the beginning, they have become extinct without leaving a trace that we recognize today. They need not be considered to understand how biology works today, but it is worth noting that if such other orders did exist, our *assumption* that the coalescent structure of all life today reconstructs the *origin* of life, will blind us to them.

Biological information that is replicated across the generations of reproduction can accumulate a trace of its past, as we see in DNA. This is due to the modular, slowly mutating nature of DNA but is not a formal necessity of darwinian evolution per se. This is clearly so, as Darwin's ideas of genes were largely wrong. Many aspects of a cell are specific to each organism and are inherited but do not bear such a trace; examples are the particular mix of minerals, salts, pH, vitamins, minerals, and so on. These are inherited in the fertilized cell by which life begins but do not retain a permanent kind of information the way genes do. Still this illustrates a legitimate and important point about primary versus secondary causation in life: is it DNA or cytoplasm? Chickens or eggs? This is not a conundrum at all: the answer is *both*. Eggs are continuations of chickens, a chain of cell division going back billions of years.

If we do not force ourselves to be constrained by an overly rigid definition of inheritance as strictly applying to genes, similar statements are true of aspects of life that are inherited in other ways (Chapter 3). Indeed, some are inherited in a lamarckian fashion; for example, behaviors learned or traits acquired in life that are transmitted to offspring, like local microenvironments, birds' nests, and aspects of verbalization. Of course, cultural inheritance is vital to human survival, and its accumulated sophistication makes possible our *social* and *technical* complexity—that makes us *seem* qualitatively different and superior to all other organisms. But a human stripped naked and dumped alone deep in a wilderness would be a laughable King of Life. If John Milton (*Paradise Lost*) is any guide, Adam and Eve were created with a house, plantation, and cookware. Culture is inherent in what differentiates us from other animals, not a trivial add-on unrelated to our evolution, and something similar is true of most if not all animals. Of course, the inheritance of acquired characters we are referring to has no implication of inner drive toward some long-term goal, although human culture is probably more truly Lamarckian in that respect than almost anything else in the living world.

Malthusian population pressure was an important stimulus for Darwin and Wallace, and is important in life but not necessary to evolution. Phenotypic change could in principle occur in a population that did not suffer from overreproduction, for example, if all individuals had equal chances of reproducing (that is, change by *drift*). But *if* more individuals exist at any time than their source of nutrients or propagation can support, and *if* relative success depends on heritable information, and *if* that information relates to form, then whatever the cause, there will be change of form. Darwin and Wallace had the great insight that *if* there was a systematic favoring of a particular subset of competing forms, that could produce particular kinds of adaptation, and natural selection became a transforming concept. But population pressure does not guarantee by itself that there will be this kind of adaptation, and we know that differential reproduction is heavily affected by chance. We also have seen in Chapter 2 and elsewhere ways in which the fact or nature of adaptation, always viewed after the fact, can be illusory. *All* organisms are the descendents of an unbroken chain of nearly four billion years of successfully *adapted* ancestors—whether or not they had the highest fitness, as measured by population genetics, among their respective peers.

The basic darwinian principles not only provide a logically coherent explanation of how complex evolution can occur, but *predictions* from this reasoning are borne out regularly in new data that were not used to develop the theory. Examples are the statistical correlation of DNA sequence differences with times of separation among species, and the lower level of variation in DNA sequence affected by selection. Although the theory does not enable us to predict the specifics, the general kind of predictive power of evolutionary biology constitutes a convincing demonstration that the notions are compatible with a large body of data, and that is the most we can ask of any science.

Evolutionary theory allows fewer "bits" of information to account for at least some aspects of life than their complete description and enumeration. However, there is a danger that a convincing theory becomes a constraining ideology, with vested interests that resist dissent just as in any other ideology, and in some hands this has occurred in biology. Ideology fetters thinking, and this is important to avoid as we try to understand how life has evolved because there really are a few problems, of incompleteness if not also of stress, with the elements of the theory of Darwinian evolution.

Some Additional Principles: Modularity and Sequestration

Darwin's notion of divergence of character is empirical rather than logically necessary. Competition by itself does not imply systematic or deterministic change of an adaptive nature. Even when it has a genetic basis, evolutionary theory does not claim to provide any guide as to *how* adaptation will occur, only *that* it will. Nor does the theory say anything about the mechanisms by which complex traits come about in organisms. A few additional principles help fill those holes.

The DNA/RNA coding system, based on complementary base pairing between nucleotides, and the nature of protein-protein interaction specificity are principles of evolution as it has happened on our planet. DNA base pairing provides a universal, modifiable system with historical memory that, once evolved, makes the rest of divergent evolution and adaptation relatively easy and plausible to understand. As has been said of cells, after they evolved the rest was easy—and is explicable in cellular terms. In many ways, after the RNA/DNA system, the rest can be explained in terms that relate to this system.

Everywhere in life the importance of *modularity* or segmentation is clear. From the modular nature of proteins and DNA, through cells, through perhaps the vast majority of the structures and systems of complex organisms, nature is composed of modules. This is immediately apparent as a fundamental characteristic of genes themselves. New genes could in principle arise by the incremental accretion of nucleotides to the ends of chromosomes until, by chance, they formed a valid coding unit that could be expressed, but they don't. The genome can be viewed as an assemblage of modular elements, including gene family members (each with internally replicated structures like exons, splice signals, and the like), replicated telomeres and centromeres, and regulatory elements.

This reflects billions of years of duplication events plus the evolution of short elements like enhancers by mutation in otherwise noncoding DNA. These processes result in genes connected together on chromosomes along with regulatory sequence, which allows for differentiation among cells, and which in turn allows for multicellular organisms. Indeed, life works the way it does fundamentally because genes are concatenated on chromosomes and thus fundamentally through modular organization that was made possible by this duplication history (something presciently seen, with relatively little data, by Ohno (Ohno 1970)). In retrospect, we can see the chemical ways in which mutation and erroneous replication lead to duplication of structures. Thus, duplication is a commitment made so early that it is now as fundamental as any other of the postulates of biological evolution as it happened on Earth.

Modular physical structures occur from cells on up and across all of life. They include direct repeats, as in hair or leaves, or modified repeats as in regionally differentiated vertebrae, digits, or limb segments or butterfly color spots. Physiological systems are similarly organized. They involve the interaction and differentiated use of the products of duplicated, differentiated gene families. From lipid and oxygen transport to neurotransmission, transcription regulation, olfaction, immune response, selective sampling of frequencies in the light spectrum, hormonal signaling, ion channel formation, and most else, this is the molecular nature of cells and organisms. In fact, the core constituents of life themselves (amino acids, nucleic acids, lipids) are modular branches of the core biochemical energy-processing system of life (Morowitz et al. 2000).

Gene duplication provides some of the wherewithal for repetitive structures, but modular structures themselves consist of spatial or temporal repeats of patterns of similar gene expression rather than literal duplicates of a separate set of genes for each module in the system. Here, replication, sometimes with modification, occurs via interacting combinations of products of gene families and results in spaced replicated zones of structure activation surrounded by inhibition zones. In the sense of scientific understanding, but also in the sense of the phenomenon itself, repetitive patterning as a *process* involves fewer "bits" to make something complex.

None of this works without extensive *sequestration*. Darwin knew that there were problems with his blending theory of inheritance, but the fundamental importance of nonblending applies at all levels of life. Life may have begun in a fluid soup of some sort, in which simple chemical reactions took place in public. But at some point billions of years ago, this changed, which explains differentiated life as we have seen it from the earliest fossils to the present. From the nonblending nature of DNA sequences (including codons, enhancers, and so on) to the origin of cells as ways to protect an accumulating repertoire of vital internal reactions in controlled isolation from the external environment, sequestration of biological elements allows them to differentiate, specialize, and interact.

Sequestration makes possible the discrimination of immune, visual, auditory, and olfactory signals. To the extent that signals from different receptors activate the same neuron, the organism cannot make specific discrimination as to the source (e.g., in nematode olfaction each cell expresses multiple receptor genes and, in some insects, brightness and color signals may merge in their path to the brain). Diffusible signals in development are by definition not sequestered for the very reason that their relative strength across a tissue field is presumed to carry the developmental information for local cells within the field. But the reception of these signals *is* sequestered in that there are distinct, separate receptors or binding proteins for each signal to be received and specifically recognized. The receiving cells must be sequestered so that they can develop independently, which is how such a uniform tissue field becomes differentiated.

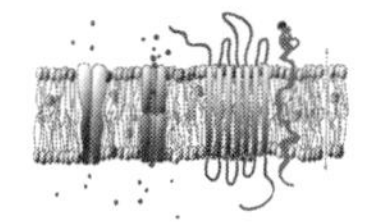

Actually, sequestration has led to life being a network of interactions among otherwise isolated elements. Divergence is important, but horizontal interaction (that is, among contemporary molecules, cells, organisms, and even species) is completely essential to organized life. Sequestration in many if not most instances is not or perhaps cannot be complete. Interactions are fundamental to development, homeostasis, reproduction, and even gene evolution (recombination, gene conversion, and gene regulation by *trans* factors acting on *cis* gene-regulatory elements). Complete sequestration would prevent that. And if hormones are diffusible signals within an organism, pheromones, flower color or odor, and ritual display or sound are examples of diffusible signals among organisms and species. In each case, the recipient has to be enabled to receive the signal, and we have seen many ways in which this happens; much of life depends on ligand-receptor binding to transfer information of diverse kinds in diverse ways.

Related to the notion of sequestration is the fact that most biological processes are *contingent*. This is implied by modularity, sequestration, and nested hierarchical organization because each new step in a process or stage of development is based on conditions at the time. However, it is the *partial* nature of sequestration that enables divergence to occur but not to isolate components completely. This is an important way in which the notion of cooperation in life is extremely important, to keep systems, organisms, species, and ecosystems integrated to varying degrees.

An Expanded View of the Role of Chance

Because we can always observe the current adaptation of a species, nothing prevents us from expressing (and, if we wish, believing) the idea that the adaptation is "remarkable" and "had" to have been molded by natural selection. But the after-the-fact nature of our observations means that this is a human judgment, and we are necessarily blinded to the nature of time periods far longer than our direct experience; we have to model them with some sort of law-like theory, often using the mathematics of population genetics because that captures general ideas in somewhat tractable form—abstract, but digestible.

A modified view actually makes Darwin's central idea of divergence from shared ancestry even more plausible: present-day adaptation does not in principle require a systematic, steady, or prescriptive selective environment. Chance is a frustration treated as a source of measurement noise in a science seeking deterministic highly predictive laws of nature, but chance has much more of a direct influence on life itself than is generally credited.

The difference between chance and selection largely depends on the parameters of the process, including population size and structure, intensity of selection and the like. The question as to whether there has been "enough" time for chance to have brought about given adaptation really rests on the general property that change happens faster when directed than when meandering. In this sense, as the estimated age of life moves backward (getting older), the role of chance via phenotypic drift and phenotypic substitution becomes more tenable in principle.

But since what is here is here because it has worked, and no specific thing *had*, a priori, to be here, the time question is a somewhat moot point, as explained in Chapter 2. Has there been enough time for *R*-genes or olfactory genes to evolve sufficient diversity? Is this even a meaningful scientific question? Of course there has been enough time, unless our most fundamental idea about a terrestrial, unitary origin of life is wrong. The more important issue is what mix of factors may have been responsible.

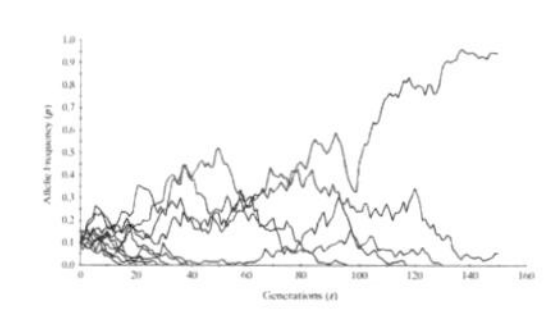

A highly deterministic view that assumes that selection is intolerant of variation, and highly prescriptive in nature is not necessarily accurate, and we know it is often inaccurate. In most of the examples we have covered in this book—from the constituents of cells on up—it would be difficult to argue, other than post hoc, that what exists today is actually optimized. Would one be able to say that about sex-determining mechanisms? If so, which ones? What about vision or olfaction or immunity? If these have been optimized, in what sense, and why are they so different, both among species and between individuals? What about people who type with two fingers rather than all ten? Daily life all around us is manifestly inefficient. Perhaps a better question is *how* prescriptive can we expect selection to be?

The answer seems clearly to be contextual. Optimizing or any other kind of tightly determining selection can certainly occur in principle, in the laboratory, or under the appropriately strong limiting conditions (like nutritional stress or the imposition of antibiotics or pesticides, but breeding experiments show that there are limits to what even artificial selection can accomplish). Nonetheless, although conditions for strong selection *might* occur, chance is *always* a factor in change from one generation to another. Phenotypic drift is a legitimate and logically plausible means of change. Humans are builders, and it seems difficult to understand that chance can play a role in what from our perspective seems clearly designed for its present adaptation. But accepting a greater role of chance relieves some of the need to construct specific post hoc selective scenarios that, no matter how attractive, often show exceptions and complications that must then have to be accounted for by caveats and additional explanations.

For most traits that we have discussed in this book, there is a diversity of standing variation. This applies to perceiving light or chemicals, defending against microbial attack, and even the essential phenomenon of sexual reproduction. These traits seem to have been shared by common ancestors of animals and plants, so their diverse forms today are one kind of evidence that *non*specificity of mechanism is important, not just a curious observation, and is a widespread if not fundamental characteristic of evolution.

Also as noted in Chapter 3 and elsewhere, change in the genetic basis of phenotypes can occur even when classical natural selection is taking place. Selection screens on phenotypes and does not "care" or detect how those are brought about. It works only indirectly on genes. This is a powerful protective mechanism, in that it allows for redundancy and alternative genetic pathways that may make survival and persistence more likely. Some core metabolic processes and highly specific communication via pheromones seem to have been rather tightly controlled even at the gene level, and may place deep constraints on evolution. However, as frequently demonstrated by gene knockout experiments in mice that fail to show the expected phenotypic effect or show it only in some background strains, even basic developmental systems that produce rather invariant phenotypes have buffered or alternative mechanisms. And some important functions, like olfaction, vision, and immunity, have specifically *imprecise* mechanisms that increase the chance of success: their sensitivity does not depend on the presence of one specific allele or gene.

Tolerance of variation is itself a survival mechanism, and every redundancy and imprecision is an opportunity for drift. It is in this sense wrong to think of variation as "noise" around some true signal (often called the "wild type") specified by selection. It is true that rabbits breed rabbits rather than mice, and this is certainly the result of genes. But the appearance of specificity in comparing species is at least in part an artifact of reproductive isolation. After long time periods, the variances around trait modes among related species may no longer overlap.

But within species there is always variation, and gene mapping studies and studies of allelic effects on most complex traits that have been looked at carefully bear this out. This is true of human and mammalian disease mapping, studies of yeast and bacteria, and natural and experimental observations in fruit flies, agricultural breeding, and so on. Many genes contribute allelic variation, generally of low individual penetrance, and there is phenogenetic equivalence. One cannot accurately predict the phenotype from knowledge of the genotype (except under favorable or highly constrained circumstances, or in a probabilistic sense). Similarly, and for related reasons, we usually cannot infer the underlying genotype from the observed phenotype, and if we can't do that, natural selection—much blinder than we are because it only uses the one criterion of fitness—can't do it either.

An important lesson we have learned from experimental and observational studies of genotype-phenotype relationships is that they often depend on the population of inference; they are not universals (Lewontin 2000; Schlichting and Pigliucci 1998; Weiss and Buchanan 2003). Related to the presence of redundancy at any given time is the evolutionary consequence that selection can keep a trait or function around over evolutionary time while its underlying genetic basis or even its physical basis changes. We have described examples of phenogenetic as well as phenotypic drift. Phenotypes, by and large, are not inherent properties of genotypes, and this is even truer of fitness, and as we tried to stress early on, it can be a mistake to treat selective coefficients in that way.

One gets a rather different view of evolution from the perspective of the trait or organism, not the gene. However, this is more in line with what motivated Darwin and Wallace in the first place, and it has ironic implications. Genes may bear the information trace of the past, and organisms may typically develop from single cells and hence be genetically driven; however, if the trait rather than the gene is what selection maintains, the ephemeral trait may be more "real" or lasting in that sense than its underlying genetic basis.

This certainly does not imply that traits evolve without underlying genes. Nor does it imply that there is no conservation of mechanism. Indeed, we have seen throughout this book exceedingly deep conservation of at least some aspects of phenogenetic mechanisms. The role of *Pax6* and opsins in vision across the animal world and extending even to algae and of genes inducing dorsoventral patterning between vertebrates and invertebrates are examples. But phenogenetic drift does occur and provides a view of evolution that is less imprisoned than reasoning that demands more causal genetic precision by selection need be.

Adaptive Evolution: Selection and Drift are a Continuum of Effects

There has long been debate about the relative prevalence of selection *versus* drift in evolution. In fact, these constitute a continuum. We discussed the key elements in the early chapters of this book. Genetic drift occurs when selective coefficients, s, equal zero (that is, the genotypes being considered have equal fitness). But this parameter is fundamentally context-dependent (it is *defined* as pertaining to peers within a population of inference). Unfortunately, this leads to a practical problem, because of the notorious difficulty in detecting selection in nature.

Even some of the most classic examples of selection such as related to the beaks of Darwin's famous Galapagos finches, or protective coloration associated with industrial melanism in peppered moths over the past two centuries, are not as clear as had been thought (Grant 1999; Grant and Grant 2002; Weiss 2002a). These are cases where the supposed selection probably *is* at least one of the important factors involved, and is probably a relatively strong force.

Instead of direct observation, we usually have to detect the effects of selection in a general way by comparing genomic sequence in various ways, and inferring that selection is responsible for the more stable aspects of the data. Thus, we infer that selection is responsible for the systematically lower observed variation in coding than in intronic DNA, or in regions of a gene showing more sequence conservation in one species than in another. Equating function with reduced variation is at least a little bit circular, but it generally corresponds with what we know of DNA function. It is this kind of analysis that led to the genetic load problem we discussed in Chapter 4, because while selection is so difficult to document in regard to genes on the ground, the statistical evidence for it is pervasive across the genome. It is not easy to account for how daily life can support the amount of selective loss required to maintain so many variation-constrained regions across the genome.

The most persuasive resulting generalization from this kind of population genetic approach to sequence variation is known as the "nearly neutral" view of evolution. As mentioned in Chapter 3, it seems that most of the time most selective coefficients are so small that the future fate of an allele is as much or more affected by drift than by the effects of selection (Ohta 1992).

Population genetic data have been showing things like this for a long time, but it is nonetheless common, if not usual, to see evolutionary explanations that equate present-day function with selective history. This often assumes steady, gradual selection as the causative agent and accepts the *net* result of selection as an adequate way to account for the local, day-to-day events that were actually responsible over evolutionary time and space. That verges on a kind of determinism that implies that what is here was destined in advance, and is in part a product of the adaptive illusion.

Despite this tendency, biologists are in unanimous agreement that this is an incorrectly teleological view. To account for complex evolution in a nonteleological way, we take from Darwin himself the assumption that complex traits got here through a series of intermediate precursors (sometimes called "exaptations") that had their own evolutionary reason for being. The usual reconstruction is to attribute selection to each such stage. The fossil record or comparative biology sometimes shows us what these earlier stages were, and sometimes we guess at what they may have been. Evolution is contingent in that changes from one stage to the next depended on selection among variation and by conditions that existed at the time, having nothing directly to do with what might happen in the future, and in that sense the model is one of "chance" evolution, even if selection is responsible all along the way.

However, as we noted in Chapter 3, our everyday experience trying to demonstrate selection in action today, tells us that these local stages probably were, in their time, typically not under intense selection. Slow evolution with small selective coefficients at any given point in turn means stepwise nearly-neutral evolution. This in turn implies that drift can be important if not predominant at many or even most stages of the process. Of course, there is no way or reason to rule out occasional bursts of more stringent selection, nor that at all points on the way selection may have truncated phenotypes that were out of bounds for their local conditions. But small s means the bounds were broad, and tolerant of variation, for example, culling only at the extremes and shaping variation, but only weakly.

This scenario allows phenotypic drift to apply to the incremental changes, allowing a much greater role for chance than in the usual view. Indeed, pushed to its extreme, it allows what we judge retrospectively to be highly molded and focused to be due instead largely to chance, both in the contingency and the drift sense. This general model of complex trait evolution is consistent with the combined action of natural and organismal selection, various levels of drift, the extensive evidence for nearly neutral evolution inferred directly from patterns of DNA variation, phenogenetic equivalence due to many loci affecting complex traits, the existence of standing variation in populations today in essentially all traits (indeed, the fuel for future evolution), the imprecision and "imperfection" of biological traits, the widely varying forms of traits even among related species (that the examples in this book clearly show), and the correlation between separation times and phenotypic differences.

This view is completely consistent with the genetic theory of evolution, though it does not over-invoke classical, gradual, steady selection. Its elements are all familiar. The most difficult thing is to shake the anthropic illusion and to allow something that is functional to have evolved largely by chance in a much deeper sense than we usually accept. But because selection and drift are themselves in many ways different points on a continuum of effects, between this view and one invoking selection more strongly, the differences are matters of scale and perspective, as is so often the case in science.

Some Additional Points

Watson and Crick characterized the basic nature of genes, which shows how the replication of chromosomes makes it possible for every cell in an organism to contain the entire inherited genome (except for somatic mutations). But this did not explain how cells with the same genome could produce a differentiated organism. We now know quite a lot about the way in which specific subsets of genes are activated in specific cellular contexts via *cis*-regulation using modular response elements in and around them. We have learned of other ways gene expression level is adjusted, including RNA interference, and quantitative effects of different numbers of copies of given enhancers, tolerance of variation in enhancer binding sequences, and the packaging or chemical modification of DNA near a gene that may affect transcription factor binding.

We have learned in recent decades of other sequence-based functions in DNA, including chromosome protection (telomeres), separation during cell division (centromeres), packaging (histone binding sites), and multiple RNA splicing to affect protein structure. These aspects of DNA sequence expanded the traditional meaning of the word "gene" from just protein coding to include these many additional functions. Others likely remain to be discovered. This changing understanding of the nature of the gene does not challenge the essential ideas in the genetic theory of evolution, which is based on the change in frequency of heritable variants, whatever their nature. But the diverse functions of DNA add richness to our understanding of evolution's mechanisms, and contribute to a more complete theory of evolution.

The function of many genes (perhaps most genes in complex organisms) involves regulating the expression of other genes, signaling, or other similar kinds of indirect function. Genes responsible for the final aspects of the traits—that is, the physical phenotypes of traditional evolutionary interest like morphology or behavior—are buffered from direct detection by selection by layers of causal interactions. Unlike specific proteins such as hemoglobin, most of these indirectly acting genes have multiple uses and interact with different genes in different circumstances. Epistasis and pleiotropy can constrain the freedom of action of selection because tinkering with these genes via selection on one of the traits they affect can have negative effects on the other traits.

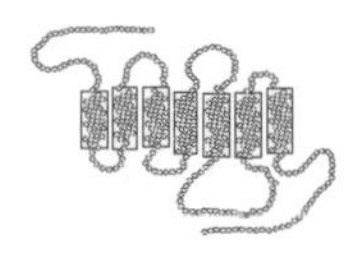

However, the system that evolved uncoupled these pleiotropic regulatory proteins from what they regulate, because the protein can remain protected from mutational damage by selection, so it can do its work whenever needed, while mutational change can add, delete, or modify enhancer sequences the regulatory protein recognizes, allowing each *use* to evolve more independently. This allows considerable flexibility for selection while preserving a relatively stable toolkit, and is another way in which it has not been necessary for each new function to be built de novo by entirely new genes. The indirect, sequestered control of gene function by regulatory elements is another fundamental characteristic of life.

Some of the regulatory processes are subtle, including the way that mammalian X chromosomes are inactivated by being covered with *Xist* RNA; the competitive binding (or other) mechanism by which the members of *globin* or *Hox* gene clusters are sequentially activated; the means by which only one X-linked red or green *opsin* is expressed in a given retinal photoreceptor; the rearrangement of *immunoglobulin* or *T cell receptor* genes and the sequential use of the constant regions as response to infection proceeds; the allelic exclusion by inactivation of the other chromosome for these genes that leads to unique antibodies being produced by a given lymphocyte; and the near total exclusion of all but one *olfactory receptor* gene from expression in a given olfactory neuron. Posttranscriptional regulation by antisense RNA interference is an additional subtlety that we have mentioned.

A consequence of this layered nature of causation is that a major fraction of biological activity arises from the action of genes that have nothing specific to do with the nature of the final trait to which they contribute. We have referred to this as logically necessary (the trait must be produced) but functionally arbitrary (it doesn't matter how it is produced). Signaling ties the living world together and is another of the general characteristics of life not specifically implied by the fundamental darwinian postulates.

Yet diffusible signals are a kind of arbitrary information-by-agreement. Nothing about an *Fgf* gene or its receptor need have anything at all to do with the nature of the final trait being made. The same is true with the many transcription factors, second messengers, and the like. Because they are arbitrary they, like codons and enhancer sequences, can form a general-purpose toolkit. However, this is not the same kind of "code," in that there is nothing in *Fgf* itself that is a stand–in for some trait the way a codon is for an amino acid.

In the last couple of chapters we have taken the notion of functional arbitrariness even further. If a generic toolkit of regulatory factors can in a sense account for all morphologies, it seems true that all of the physically diverse sensory inputs can be *perceived* by a single set of tools. The diverse functions of perception and central control are achieved by the interaction arrangements among a few types of neural cells. But the cellular properties of neural cells *per se* have little if anything to do with the nature of an image or smell, with whether one is looking at, touching, or tasting ice or ice cream.

The use of functionally arbitrary processes is deeply a part of life. Neural behavior most compellingly forces attention on the importance of learning how to understand "emergent" traits—that are fundamentally due to the *interaction* of elements rather than the sum or nature of those elements. How this will relate to or can be achieved by reductionist approaches, with their centuries' head start and long and successful record, or whether entirely new conceptual approaches will be developed, only the future will tell. But functional arbitrariness does seem to imply that we will have to understand many of the traits in life somewhere "above" the level of the gene.

As we saw in Chapters 4 and 5, a problem with genetic mechanisms being so utterly modular is that the genome is saturated with potential functional sequence. As scientists, we understand that function because of decades of experimental investigation carried on from without the organism. Organisms, however, have to figure this out for themselves, and from within. The DNA sequence of a new organism can be interpreted only because it arrives in an appropriate interpretive environment (e.g., mRNA, pH, and so forth in the cell). Again, this is why the distinction between a chicken and an egg is largely a false one: *life is continuity*. (Various proposed attempts to generate life spontaneously in a test–tube of ingredients may ultimately succeed but will not invalidate this last assertion, because the recipe that will be used will be derived from a knowledge of the nature of current life, that took billions of years to produce from the inside out).

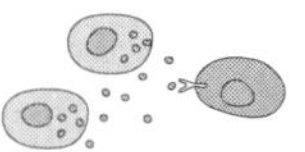

There has been tremendous recent progress in identifying genes and processes involved in regulatory aspects of complex traits. But gene lists are not of themselves particularly more informative than was the classificational analysis of beetles in the 19th century, which is often denigrated from our modern perspective as having been "just" descriptive natural history. Dartboard-like today, regulatory pathway diagrams may ultimately become complete, via tools like expression profiling. Experts in the area are eager to deal with such data (e.g., Davidson 2001; Davidson et al. 2002a; Davidson et al. 2002b), but precedent and even the existing difficulties of understanding known pathway stereotypes suggest to us that how we will deal with the impending information inundation to move from even longer lists to better understanding, is by no means clear. For one thing, each new element adds another source of inter-species, inter-strain, quantitative, or stochastic variation. However one thing even incomplete gene lists *have* already shown is, again, the ubiquitous use of a few pathways, like *Wnt* and *Fgf* signaling and the like. In a sense, this shows how life is a combinatorial molecular phenomenon, another aspect of its fundamentally modular nature—and another challenge to understand emergence.

COMPLEX PHENOTYPES ARISE OUT OF A FEW BASIC PROCESSES

Even the most complex of phenotypes are produced to a great extent by a few basic developmental principles, a fact that makes it easier to understand the evolution of the diverse complexities of life. Not surprisingly, this directly reflects the underlying modular genetic toolkit and shares many of its organizational properties.

Complexity is an illusive term, but getting big is one kind of complexity. Size can be achieved simply by mitosis of adherent cells that function together as a physical entity. Spatially varied scaling (allometry) or temporally varying growth rates (heterochrony) among the parts of an organism are simple kinds of processes repeatedly used in different ways among organisms. But to be more than a structureless lump, such growth needs to be regionally differentiated, and temporal or spatial asymmetries that locate function are vital to many complex phenomena. We see this even in single-celled organisms, the classic example being the syncytium of a fly egg that establishes polarity via the interaction of factors with different concentration gradients within the cell.

Given the redundancy and diversity of function occurring even within a single cell, there would perhaps be too much chaotic cross-reaction had internal organization not already evolved early in the history of cells (and there is plenty of molecular chaos in cells as it is!). For example, cell membranes are loaded with ion channels and signal receptors responding to a host of different conditions, but they use similar sets of internal mechanisms such as protein kinase reactions. Intracellular membranes, lipid molecules, organelles, cytostructural molecules, transport and packaging proteins, nuclear membrane structures, and chromosome modification produce extensive intracellular sequestration.

As noted earlier, morphological and physiological systems are typically characterized by repetition of modular units and/or segmentation. Rather than gene-for-unit specification, quantitative patterning *processes* bring repetitive structures about and as we have seen in several chapters, such processes may only require a modest number of interacting critical factors.

In some of the most well-understood examples, the factors are diffusible signals and their receptors that activate or inhibit selective gene expression. In principle, a two-component reaction-diffusion-like process can generate complex periodic patterning. We do not yet know how many factors are critical for the kinds of repetitive patterning we have described, from ommatidia in flies to hair to the arrangement of regions in vomeronasal or olfactory epithelia and the like. We know that there are some simple components, such as interactions between Bmps and Fgfs, or the *Delta-Notch* system, but we know that many if not most of the systems studied so far have redundancy or alternative pathways, that organisms can compensate for missing components, or, as revealed by many mouse knockout experiments, that the effects of a pathway are variable depending on naturally occurring variation even within a species.

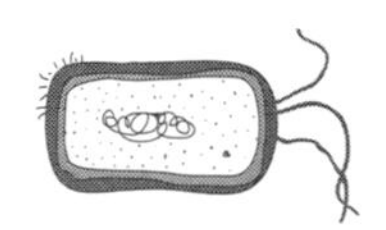

Simple patterning processes can work because of the partial sequestration among the cells in a tissue field, which enables each cell to interpret and respond to the relative concentration of diffusing signaling factors in its particular extracellular environment. This allows repetitions of the same structure to occur, surrounded by inhibition zones. Making this even easier to understand is that these patterning processes are nested from the first stage of an organism. One process sets up fields, for example, basic polarity, anatomic axes, or organizing centers, which provide partial isolation or differentiation that can initiate subsequent more regionalized patterning, generating a hierarchy of ever more localized differentiation. This seems at least in part to be how limbs, vertebral columns, and invertebrate segmentation work. Because of the evolution of families of signaling and transcription factors, and their receptors and enhancers, initial pathways can diversify by specializing the various gene family members, another aspect of sequestration (that is often only partial because there can be cross-reactions).

Morphological modularity is facilitated by, but also facilitates, the reusable nature of regulatory and signaling genes. Because these genes can be used in different combinations at different times or contexts in the same organism; prior stages in which a gene has been used lead to sequestered, differentiated descendant cells that then can respond to the same signaling factor in a different way. Some signaling pathways are used multiple times in different ways, even during the development of a single structure; an example is *Fgf* and *Hedgehog* signaling in dental patterning and tooth development (Jernvall and Thesleff 2000). It is the *cis*-regulation of gene expression that makes this combinatorial phenomenon possible, and these ubiquitous facts constitute another major fundamental commitment to a particular way of life that was made billions of years ago.

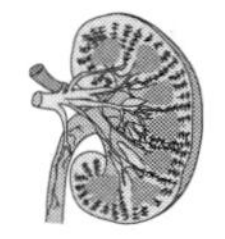

With the use of this set of mechanisms, complex traits develop via a few simple processes in addition to periodic patterning. Budding out or invaginating inward is brought about by local asymmetric cell growth, a signaled process. Branching is another simple process that is widespread in nature and can be repeated to generate a nested hierarchy of components. Among the widespread examples of complexity produced by such processes are the lung and bronchial trees of animals, the distribution of blood vessels, and, of course, *real* branches on plants. Plants avoid the need for layered repetitive patterning by retaining the potential for generating diversified structures that can be repetitively invoked in the sequestered environments produced by branching.

Differentiation by combinatorial expression of regulatory genes also applies to physiological traits and sometimes involves related or identical genes. This is the case, for example, for ion channel function, cell adhesion genes in neural development, the control of neuronal firing, osmotic function, lipid transport, the differentiation of cells from common precursors in the blood system, and many others.

We haven't covered all biological systems in this book by any means. We've concentrated on particular phenomena that are important to understanding how complex organisms work and how they got that way. The principles and often even the specific genetic phenomena are, however, similar for systems that we have not mentioned. For example, digestion involves the breakdown, absorption, and so on of proteins, fats, sugars, and other carbohydrates, using gene products just like the other systems (e.g., proteases, binding factors). Kidney filtration rests on ion channels and similar structures. Many of these are pure "chemical" processes; that is, they involve genes to the extent of synthesizing chemicals (e.g., HCl, pancreatic enzymes) and secreting them from cells but are mainly not "informational" in the sense of most systems and phenomena in which we have been interested here.

We can illustrate the kind of evolution that leads to diverse complex traits via a single set of mechanisms by an example from our own work. Among the most important characteristics of vertebrates are their mineralized tissues. As vertebrates evolved, the initial calcified tissue of external scales expanded to include teeth and bones (actually, it is not entirely clear which of these came first or whether the pattern was the same in all early vertebrate lineages). A class of secretory calcium-binding phosphoprotein (SCPP) genes, almost all still linked in a single chromosomal cluster (and that appear to have evolved through a series of gene duplication events in ray-finned and tetrapod lineages), is involved in the formation of different mineralized tissues in different species (Kawasaki and Weiss 2003). Some of these genes are expressed in bone development, others in forming the mineralized parts of teeth, and still others in lactation and salivary secretion (calcium binding in saliva can secure the mineral and probably has antibacterial function), tissues that arise from different parts of the embryo's developmental tree. Most or all of these tissues involve epithelial-mesenchymal interaction of the type described in Chapter 9.

These SCPP genes are physiological and not informational; they directly serve a final structural function rather than a developmental or patterning one. But the nature and arrangement of the gene family shows how step-by-step a diversity of complex traits can evolve through or taking advantage of gene duplication. The first SCPP gene was present before vertebrates, serving calcium-related metabolic function(s). In vertebrates, this trait could be used in the evolution of anterior feeding mechanisms (teeth), body protection (scales), and internal body support (bone). Subsequently, new functions took advantage of these genes and their epithelial expression pattern in the evolution of salivary and lactational functions that are basically unrelated to bones (except perhaps in the indirect sense like the use of casein to make calcium available to mammalian infants). Thus, metabolic function evolves through the same kind of gene duplication and subsequent divergence, resulting in the modular function "strategies" that we see in complex morphological structures.

Branching: A Common Metaphor with Differences

This is a point to note that some of the same metaphors apply at many different levels, as a reflection of correlated effects of the basic nature of evolution as revealed by the principles we have been describing. The idea of branching and nested hierarchy is a prime example. This is indicated in Figure 17-3. Karl Ernst von Baer was one of the 19th century founders of modern embryology. In pre-evolutionary times, he noted the way the embryos of collections of species, like vertebrates, begin life looking very similar. But as the embryos age, they diverge in form in the various species, ending up with adult variations on the theme of their shared overall body plan.

Figure 17-3A shows an attempt by Charles Darwin to understand von Baer's notions. Evolution in its original sense of developmental "unfolding" from a shared body plan might be due to a Creator's design, but Darwin saw that the embryological data were highly relevant to the problem of the evolution of species. He was among the first to use a similar branching metaphor for the divergence of species from a common ancestor (Figure 17-3B). Again there was a shared form, but that reflected the state of the ancestor. Development and evolution relate morphological similarities to very different time scales. There was both confusion and connection between the two, accounting for the famous Biogenic Law of Ernst Haeckel (shared to some extent by Darwin) that during embryogenesis (ontology) species sequentially recapitulate their ancestral forms (phylogeny).

Figure 17-3C schematically shows the somatic divergence of organ systems within the body of a vertebrate, starting from a single cell, the fertilized egg at the top, and ending up with the shedding of another single cell to form the next generation (sperm cells at the bottom). Finally, Figure 17-3D shows a fractal simulation of branching that, as we saw in Chapter 9, has been likened schematically to the structure *within* organs like the lung.

Metaphors are only so useful, and the phenomena in Figure 17-3 are different in important ways. But to a very real extent they are all the same, and for similar reasons: they all relate to the descent with modification, by duplication with variation, of partially sequestered lineages of genetically differentiated modules of cells. We have tried to show in this book how parallels like these are found throughout the living world, from genomes to species.

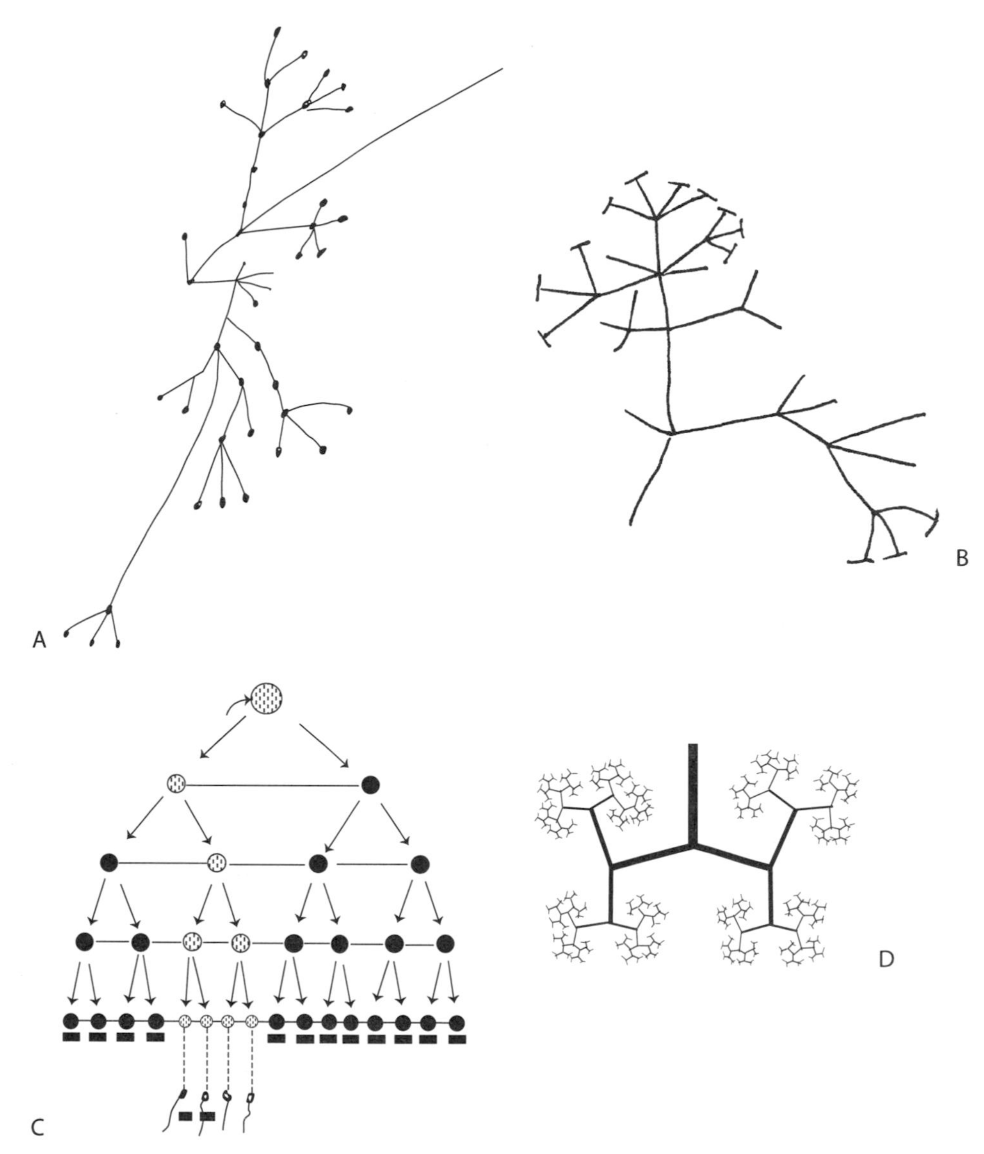

Figure 17-3. (A) Darwin's drawing of developmental divergence of basic body plan components from a common core body plan as proposed by von Baer; (B) Darwin's early sketch of the divergent nature of his emerging view of species evolution; (C) a somatic tree of development of organs within an organism; the sequestered germline cells are shown in gray; (D) a fractal pattern of branching as roughly approximates much of organ branching seen in veins, nerves, bronchial trees and real trees. (A,B,D) redrawn by the authors (Darwin's original figures can be seen in Richards 1992); (C) from Wilder (Wilder 1909).

HIGHER LEVELS OF EVOLUTIONARY ORDER

In Chapter 8, we alluded to various concepts of organization above the level of what is typically considered to be the organism. Such organizational structures have been referred to as "superorganisms," but for reasons discussed in that chapter, the definition of what constitutes an organism or "being" is more arbitrary than is often thought.

Bacterial biofilms and slime molds are among the most simply organized coalitions of individual organisms, although social insects are probably the most often-used example of clearcut organization above the level of what is traditionally considered an organism. Of course, human societies and hundreds of less elaborate animal social structures can be thought of in the same way.

From a darwinian point of view, of ever-present competition for reproductive gain, social organization should occur only if it confers a benefit that outweighs to the risks of group life (like potential exhaustion of food supplies). A major way this has been put is the need to explain the widespread occurrence of altruism, or how genes can evolve whose bearers sacrifice some or all of their own reproductive potential for that of others. This is an inflammatory dispute within biology, fought with ideological vigor. The argument is often engaged under the term "group selection." Strict genetic darwinism is individual-focused, strongly selectionistic, and will at most grudgingly accept sacrifice for the good of the group because of a lack of an obvious mechanism by which that can evolve. (Group selection was important in Wallace's view of evolution, however.)

The most common explanation for altruism is that its beneficiaries are close relatives who share similar genotypes with the individual making the sacrifice. Formally, in the general case, what an individual should be willing to sacrifice for another individual depends on the precise degree of relationship between them. As a general principle this seems reasonable, though it fails to account without contortions verging on implausibility for many of the social behaviors found in the world.

The individual cells in some bacterial biofilms are clones of each other, but many biofilms are aggregates of different kinds of bacteria. Why would different strains of bacteria be "willing" to enhance the survival of others? Is it that they are closely related *enough*? In *Dictyostelium* the slug forms from a collection of (former) individuals and it appears to be only a matter of chance which of them end up producing spores for the next generation. But to a considerable extent the issue is artificial. Most groups of organisms are collections of relatives to varying degrees, simply because individuals of most species do not disperse very widely from their place of birth or, if born into a horde or school, disperse *as* a reproductive group. Giving your all for the group usually means giving your all for your relatives in one way or another. Your peers are extensions of yourself. Precision in evaluating degrees of kinship before acting, or precise screening of individuals by selection in this regard, for the many reasons we have discussed, is neither required nor to be expected.

More interesting than the fine points of the altruism debate is a broader view of the nature of communities and the concept of community itself. An entire insect colony is produced by one or a few queens, and hence the entire progeny shares the half of the genome they inherit from her, with the other half coming from reproductive males. This means a peculiarly close kind of genetic kinship among members of the colony, as discussed in Chapter 8. This very close connection is why the term "superorganism" has been used to describe insect colonies.

But we can extend the concept. Even a cell can be viewed as a superorganism, of individual genes, and an individual organism a superorganism of cells. In each case, there is internal connectedness, communication via signaling and response elements, with associated specialization of function and coordination of action. The immediate life history as well as evolutionary fate of one is dependent on that of others. There is active and sometimes externally instructed self-sacrifice, as seen in the widespread apoptosis among the cells in the life of a complex organism (or controlled mRNA degradation within a cell). Gene inactivation is to a cell, and apoptosis to an organism, what altruism is to a member of a society.

These internal characteristics of organisms—genealogical connections among genes, signaling systems that alter gene expression, and responses by one component to conditions of another—also apply to relationships between members of the same species and, indeed even between members of different species. Social insects represent but one way in which this occurs; in fact, it is a phenomenon found from biofilms to ecosystems.

Organisms communicate and coordinate by pheromones, visual, auditory, tactile, olfactory, and other means of message transfer, reception, interpretation, and response. Even under the most classical of darwinian scenarios, flowers evolve odors that insects can smell and vice versa. Signal and ligand, reception and response. The genomes are intimately connected and literally interdependent.

As noted in Chapter 8, one distinction used in the definition of what we refer to as an "organism" is that its cells are connected together to form a single body, and derive from a common cellular ancestor. But this is a somewhat false distinction. Cells and other elements within an organism are not always physically attached (for example, circulating blood cells), whereas in many species distinct organisms *are* physically connected when they reproduce—usually an essential part of reproduction. And members of a population are clearly connected by their unbroken physical chain of shared cellular ancestry.

Social and ecological interactions affect the collective genome(s) involved, which of course is thus of direct evolutionary import. Like any other set of interacting factors, the relative frequency and spatiotemporal relationships among the factors depends on the dynamics of their interactions. Just as in development, they can generate gradients of location in space (such as microenvironments), wave-like oscillations (such as predator-prey cycles), or relatively stable frequencies (such as population size). These are "emergent" properties of communities in the same way that similar patterned traits within individual organisms are properties of interactions among signaling factors.

At the embryological, organismal, and ecological levels, the interacting system involves the partial sequestration of its modular components. Indeed, though we think of organisms as entirely different kinds of "being," their interactions, as in pheromone signaling or sepual reproduction, involve exactly the same molecular processes—even genes and types of genes—as found in developmental signaling. If biofilms are any indicator, it may be that interactions among single-celled organisms were exaptations for the evolution of multicellular organisms in the first place.

In this sense, it is rather arbitrary what we should call an "organism." For appropriate purposes, even the entire biosphere can be viewed, truly, as composed of comparable entities interacting in complex, hierarchically nested and networked ways, by related genes and common molecular mechanisms. The distinctions between kinds of biological entities blur because they are all interconnected, in similar ways, now and by common ancestry back through the entirety of life. This is the Great Chain of Beings.

A BIG BLACK BOX STILL REMAINS

Discussion of the nature of emergent traits draws attention to at least one box that remains rather completely black. Inside that black box is *perception*. Perception is made possible by patterning processes, activation/inhibition, cell adhesion, ligand binding to alter gene expression, and so on—simple developmental processes that in natural and understandable ways provide spatiotemporal maps from sense to sensor. The developmental processes generating sensoritopic maps account for how individual incoming signal properties, like wavelength, sound frequency, or specific odorant molecules, can (in some species and for some senses) be distinct in an orderly and sufficiently replicable way. This is what we have seen for other organ systems.

Perhaps the most interesting aspect of sensory and cognitive systems, however, is not their molecular components but what these perceptive systems "feel" like to the organism, their emergent property. For example, how and why does hearing feel different from seeing when both involve structurally and chemically very similar neural systems (indeed, that can remap from one type of sensation to another)? This immediately leads to the issue of most interest to humans: the relationships between perception and consciousness.

As far as keeping track of incoming signals, one might liken the brain to a building guard sitting in the guardroom of a bank, watching the halls, teller windows, and so on through television monitors (it is in fact debated, but immaterial to our point here, whether the brain actually works by having a monitoring center—sometimes also referred to as a "homunculus"—or not). The guard does not see anything directly and must respond on the basis of video images provided by the monitors. This is inherently a limited amount of information compared with what is really going on out there in the bank. However, because television monitors are *designed* specifically to mimic human visual perceptive experience, the guard does have a rather natural, if one step removed, sense of at least that slice of external reality.

A somewhat more apt image might be to submerge the monitor more deeply, and think of a sailor in a submarine working with sonar and radio information. Sonar does not represent a natural human information gathering system. Why? Because undersea sounds are not the kind we are used to or evolved for interpreting. Undersea sound has to be translated into something interpretable, and we choose mainly to do that on a video screen, as blips identified as to distance, direction, and azimuth (or as audible beeps produced in earphones). These are all arbitrary ways to represent the information. Electronic detection (e.g., of radio signals) may be presented in some other computer-digested form. Indeed, sensors are made to detect things humans have no sensory means to detect (e.g., very low frequency radiation), and this information may be presented as spectral frequency pattern data, such as on an oscilloscope. In some cases information is presented in the even more abstract and arbitrary form as text on a screen, representing a totally black box analysis by a computer. Again, the data are translated into a form interpretable by humans, and the person has essentially only limited data, indirect contact, and must rely on integrated experience to interpret what the sensors provide. Every organism lives in a Kantian world, and can only know aspects of reality for which it has some form of receptors.

In this imagery, what counts is that some aspect of the signal, which we select for our own purposes such as to thwart burglars or survive attack, is chosen for detection. We design our interpretive machinery from the outside, perhaps using cost or other considerations to decide what aspects of the signal to bother about. Evolution has had to do this by trial and error from the inside, and the only known criterion is reproductive success. Each natural sensory system is built today to detect a subset of environmental information in a way that was evolutionarily sufficient in the past. But the neurons themselves, like the electrons used in the submarine's systems, are not directly related to the nature of what is being detected by them.

It is not surprising that in many systems the topographic relationship between the signal and the interpretation is maintained. This helps the brain keep track of important aspects of the signal, although what counts as important varies. Above all, it seems to mean replication: the same signal will fire the same distinct set of neurons. In some systems like visual image perception and touch, the map directly reflects the structure of the outside world, while for others like olfaction it seems mainly to serve a bookkeeping function. Vertebrate hearing uses a physical trick of forming a fluid-filled tube to decompose a complex wavelike signal. None of these particular kinds of sensory maps are essential, however; some species ignore aspects of these information sources while detecting others.

We have not really considered other "higher" aspects of the integration of environmental information, like prey tracking, locomotion, mating, eating, and the like, which integrate environmental sensing, physical activity, proprioception, and so on. These are internal supersystems, integrating muscular, neural, sensory, and motor functions, each of which is an organized system in itself. Emergence upon emergence. The overall ways these complex integrative functions are orchestrated are not understood (although some of the wiring is known).

We have suggested that these functions, including consciousness itself, evolved gradually and may constitute traits of which we have not much more direct sense of the reality than a sonar operator does of things detected from the murky deep. But this intriguing subject has to remain for the future. Even attempts to specify what the phenomenon of consciousness actually is are confessedly highly speculative (Crick and Koch 2003). Many biologists would argue that, with present knowledge and tools, these questions are currently beyond the reach of science.

SOME CAUTIONS ABOUT EVOLUTIONARY AND GENETIC INTERPRETATIONS

One problem in reconstructing life is that adaptation—whatever its cause—is not perfect, and there is no way to define what "perfect" might mean. If the only criterion is that the fit of the organism to the environment be "good enough," and this depends on the changing landscape of local environments, competitors, colleagues, and the genotypes available, then we cannot expect definitive answers. Rather, the generalization seems to be that there is no one way and no need to be better than good enough under the circumstances—and lucky.

Members of some species are largely safe from predation (elephants, lions, humans, giant turtles, probably many viruses and bacteria, and so on). Some seem not just essentially safe but to have rather open life spans, such as venerable olives, the famous Tule tree in Oaxaca, Mexico, and some giant sequoias in California that are thousands of years old. Other species live but a fleeting hour or so before reproducing (e.g., bacteria, adult mayflies). Some produce young that are almost all immediately devoured. Every creature gets through life differently.

Interestingly, however, each must face the same kinds of challenges, of escaping from predators and microbial attack and finding food sources. The differences are mostly of scale and circumstance. There is nothing inherent in mammaldom that should make them vulnerable to a greater diversity of parasites, for example, than a tree that has to just stand there and take it. This means that there is little in the way of a priori criteria by which to predict how or what a given organism will do or be like. It is in this important sense that the organism-as-machine metaphor can be highly misleading; it is thinking of organisms in human terms, as being designed to solve problems laid before them by nature. An individual organism may try to get "there" from "here" (e.g., to catch that rabbit), but evolutionarily the only problem it is solving is to persist and reproduce (to catch a rabbit *today*). And this is why it can be misleading to think of there being a fixed number of senses, or ways to hunt, or a single kind of "immune" system, and so on.

Not only is chance a pervasive factor in the environments inside and outside of organisms, but imperfection, if we can use the term, is an essential factor that makes evolution possible and ongoing. Energetic *inefficiency* is what enables evolution and in that sense it is non-sense to ask whether a trait has evolved to be energetically efficient. The puffer fish makes do with a genome only about 1/8th the size of its vertebrate relatives who, nonetheless, have roughly the same genes. This shows that such baggage can in practice be off-loaded to save on the substantial metabolic demands it must make. Yet selection is often invoked to account even for what seem to be the fine points of nonfunctional DNA (for example, it is said that there can *be* no truly nonfunctional DNA or selection would have eliminated it). But if energetic considerations of that kind applied as often as they are invoked, most individuals in most species would be on the brink of starvation, and thus need to shed every ounce of needless base pairs. In energetic terms themselves, purging has clearly not been worth the cost. If selection were typically too stringent, evolution as we know it might not have been possible—nobody would have survived.

Again, it is necessary to think contingently and that means separately for each case, and one result is a substitution of description for scientific generality. But that itself seems to be one of the realities. For example, one can ask whether evolution has made some trait more efficient than it used to be for the same use, but the truth is only that the trait is as efficient as it has had to be and the uses are always changing. A classic example is the argument that mammalian legs are oriented to work more for-and-aft than the arc-sweeping of reptile limbs, and that this evolved because it produced more efficient locomotion in mammals. But think of alligators in motion. Whether they use more energy per foot traveled than mammals, many a mammal has paid the ultimate price despite having "more efficient" locomotion. Or are the winners among competing alligators those with more efficient inefficient locomotion?

This exemplifies the problem of making evolutionary reconstruction stories based on how any particular trait "must" have come about because the post hoc nature of evolutionary inference means that multiple explanations can have comparable plausibility. This is known as a *nonidentifiability* problem, and it is a ubiquitous fact of evolutionary biology.

This is relevant to the persistent division of opinion about the roles of competition and cooperation in evolution. If a genetic variant becomes more common over time, one can always *assume* or *define* its increased frequency as due to adaptational competition and can in principle then look back into the specific history of the gene(s) and infer that what happened was deterministic. From this point of view, persons who see (or desire to see) cooperation as being as important in nature as competition can always be refuted: if social organisms, organelles, genes, or individuals cooperate and this leads to differential proliferation of associated genetic variation, then that cooperation can be expressed in a consistent way in terms of competition (some alleles do win, after all, if that's how we define winning).

But if the World According to Hobbes is a consistent one, the same outcome—what we observe today—can be accounted in a different way. Chance and organismal selection can lead to genetic change without the kind of darwinian warfare that makes for good television viewing (and has so often been used to justify social inequity). Organisms build their own environments and choose their own niches when they can. If it looks like cooperation, or *feels* like cooperation, then for all practical purposes it *is* cooperation on the level of organization at which it actually occurs. Cooperation *is* fundamental to organized life, from genes on up.

Focusing on how or why cooperation, or cultural inheritance, or social behavior, or organismal selection are "really" just wolves in sheep's clothing may help explain some aspects of life; however, this focus can lead to tunnel vision, drawing attention away from important or even pervasive aspects of how life works. If evolution is the meandering contingent process that it seems to be, the ultimate explanation that cooperation really represents a past history of competition may be true but too generic to explain very much. It is true that our house is made of nails, paint, and boards, but that does not explain our house.

In many ways, the competition-cooperation distinction is one more instance of a false distinction of perspective. Sequestration by itself almost implies cooperation if complex organization is to evolve because the interaction of isolated units like cells is the essence of such complexity. Without "cooperation" between various biological molecules, nothing in life happens. The catalyzing of biological reactions was for much of the 20th century taken to be a definition of the difference between life and nonlife. DNA does not even replicate itself without help, despite that often being stated as a biological fundamental.

Chance is always present as is the potential for natural selection, but the same can be said of organismal selection, a potentially less combative source of adaptation. Organisms sort themselves into local environments depending on what their capabilities are, and over time genetic changes that need not have to do with classical adaptation (e.g., chromosome rearrangements) can produce a species barrier. Indeed, that they are mobile and seeking is one of the traditional definitions of what it means to be an "animal." Facultative searching of environments is also done by single-celled organisms and even by plants.

Like sexual selection, self-sorting by organismal selection can be faster and more precisely related to function than classical natural selection, because organisms know better what they can do best than the crude screen of selection may be able to detect. There is nothing unnatural about this form of sorting and proliferating of genetic and phenotypic variation, perhaps supplemented by genetic assimilation, though it has been treated as a kind of backwater of biological theory. Instead, the search *for* such mechanisms as a way to account for adaptations should be more active.

Not Laws, But Principles, and Why There is no Argus

We have identified a variety of principles that we believe apply generally to evolution as it has occurred, and we think these make a useful addition to the usual darwinian principles. They are not new to us, nor to biological thinking, but they are not usually treated as part of the theory of evolution itself. If our view is justified, we might ask if these principles have any predictive power, one of the criteria for generalizations in science. In fact, we think this is indeed the case.

The pattern of the traits that we have described in this book is largely a very orderly one. Traits evolve differences, and new traits arise, out of earlier stages and by reusing processes that already exist. Traits of more ancient origin are more persistent, and derived traits are constrained by that fact. These general statements are true at the gene, morphology, and biochemical levels. As a result we can explain in a fairly formal and rigorous way, why certain patterns occur and others do not. There is much flexibility in these constraints as shown, for example, by the intercalative nature of the apparent re-evolution of traits like eyes. But because of the constraints we have described, we would agree with another observation of Thomas Browne (whom we quoted in Chapter 14 in regard to eyes and Ægles (Browne 1646)), that in "sanguineous" (vertebrate) animals, there are but two eyes, and they are in the head. There can be no Argus. In 1646 he was surmising, but evolutionary biology can explain the reason why.

There are undoubtedly errors in this book, although we hope they are not too many. Errors arise first from our own misunderstanding of existing knowledge in areas we have tried to represent that are beyond our own prior expertise, and then from the incompleteness and rapidly changing nature of that knowledge even if we have interpreted it accurately. However, the broad picture presented in these pages is likely to be robust even in the face of those kinds of error. It is always possible that fundamental new properties of genetic life will be discovered, and new fossil or living species are sure to be found. We cannot know what they might be, but from what we *do* know, it will be very surprising if they do not have the same general properties we have seen so pervasively so far: genealogy, divergence, duplication, reticulation, modularity, sequestration, interaction, functional use and reuse, and chance.

Early in their lives, Darwin and Wallace described the world as a self-directed phenomenon of change driven by competition, unfolding to the present panoply of complex organisms. Later in life, both men had their doubts. Darwin persisted in thinking of a Creator responsible for the initial start who perhaps then left things to go their own way. Wallace studied spiritualism and could never believe that the human mind could be the product of natural selection. Were these the bedside conversions of men facing death or the wishful thinking soft-headedness of age? Unlikely. It is clear from the work of both that these thoughts were present from the beginning. And their brilliant minds are not alone in this experience.

However it works, after billions of years in the making, evolution has produced such grandeur, so difficult to reconstruct in retrospect that it is difficult even for many evolutionary biologists today to accept completely mechanical explanations.

Figure 17-4. (A) Old Wallace; (B) Old Darwin (painting by John Collier, in Down House).

The most reductionist of molecular reductionists have sometimes struggled not to invoke teleological notions but to substitute what comes within a hair's breadth of doing so, for example, via a very determinative view of selection, or Jacques Monod's use of "teleonomy" and the "project" of life to explain a kind of inherent property, if not drive, in the very molecules of life (Monod 1971). René Descartes, who is often credited with starting reductionist science in the first place, said that even if an organism is a machine, it was driven by the spirit. Whatever that is.

Darwin suggested in the *Descent of Man* (Darwin 1871) that despite its importance he may have placed too much stress on the role of natural selection but that he did so to show that natural processes would suffice to explain the living world without the need for special divine intervention. In this book, we have pointed out ways in which selection may be somewhat less necessary than Darwin stressed because other natural processes contribute to plausible explanation. It is remarkable that so modest a number of basic principles can account for both the production *and* evolution of the diversity of life we see on Earth today.

One of the most consistent findings, and a continual source of doubt, is that there are exceptions to these principles. Indeed, viewing them *as* exceptions shows a danger in scientific inference. The exceptions can only be viewed that way if we take our rules too seriously, thinking of them as classic "laws of nature." That there will be exceptions probably is itself one of the fundamental laws of the nature of life.

References

Aarts, N., M. Metz, et al. (1998). "Different requirements for EDS1 and NDR1 by disease resistance genes define at least two R gene-mediated signaling pathways in Arabidopsis." *Proc Natl Acad Sci U S A* **95**(17): 10306–11.

Agathon, A., C. Thisse, et al. (2003). "The molecular nature of the zebrafish tail organizer." *Nature* **424**(6947): 448–52.

Agrawal, A., Q. M. Eastman, et al. (1998). "Transposition mediated by RAG1 and RAG2 and its implications for the evolution of the immune system." *Nature* **394**(6695): 744–51.

Akerstrom, B., D. R. Flower, et al. (2000). "Lipocalins: unity in diversity." *Biochim Biophys Acta* **1482**(1–2): 1–8.

Alberts, B. (1994). *Molecular biology of the cell.* New York, Garland Pub.

Alighieri, D. (1314). *The Divine Comedy: Purgatorio*. New York, Knopf Everyman's Library.

Almeida, M. S., K. M. Cabral, et al. (2000). "Characterization of two novel defense peptides from pea (Pisum sativum) seeds." *Arch Biochem Biophys* **378**(2): 278–86.

Alstad, D. (2003). Populus 5.2. Minneapolis, D. Alstad.

Altman, S. (1990). "Nobel lecture. Enzymatic cleavage of RNA by RNA." *Biosci Rep* **10**(4): 317–37.

Amor, D. J. and K. H. Choo (2002). "Neocentromeres: role in human disease, evolution, and centromere study." *Am J Hum Genet* **71**(4): 695–714.

Ancel, L. W. (1999). "A quantitative model of the Simpson-Baldwin effect." *Journal of Theoretical Biology* **196**: 197–209.

Ancel, L. W. and W. Fontana (2000). "Plasticity, evolvability, and modularity in RNA." *J Exp Zool* **288**(3): 242–83.

Anderson, S. A., O. Marin, et al. (2001). "Distinct cortical migrations from the medial and lateral ganglionic eminences." *Development* **128**(3): 353–63.

Andersson, J. O., A. M. Sjogren, et al. (2003). "Phylogenetic analyses of diplomonad genes reveal frequent lateral gene transfers affecting eukaryotes." *Curr Biol* **13**(2): 94–104.

Andersson, S. G., A. Zomorodipour, et al. (1998). "The genome sequence of Rickettsia prowazekii and the origin of mitochondria." *Nature* **396**(6707): 133–40.

Anholt, R. R., J. J. Fanara, et al. (2001). "Functional genomics of odor-guided behavior in Drosophila melanogaster." *Chem Senses* **26**(2): 215–21.

Anholt, R. R. and T. F. Mackay (2001). "The genetic architecture of odor-guided behavior in Drosophila melanogaster." *Behav Genet* **31**(1): 17–27.

Appel, T. A. (1987). *The Cuvier-Geoffory Debate: French Biology in the Decades Before Darwin*. New York, Oxford University Press.

Aranda, A. and A. Pascual (2001). "Nuclear hormone receptors and gene expression." *Physiol Rev* **81**(3): 1269–304.

Arbeitman, M. N. and D. S. Hogness (2000). "Molecular chaperones activate the Drosophila ecdysone receptor, an RXR heterodimer." *Cell* **101**(1): 67–77.

Archibald, J. M., M. B. Rogers, et al. (2003). "Lateral gene transfer and the evolution of plastid-targeted proteins in the secondary plastid-containing alga Bigelowiella natans." *Proc Natl Acad Sci U S A* **100**(13): 7678–83.

Genetics and the Logic of Evolution, by Kenneth M. Weiss and Anne V. Buchanan.
ISBN 0-471-23805-8

Arendt, D. and K. Nubler-Jung (1994). "Inversion of dorsoventral axis?" *Nature* **371**(6492): 26.

Arendt, D. and J. Wittbrodt (2001). "Reconstructing the eyes of Urbilateria." *Philos Trans R Soc Lond B Biol Sci* **356**(1414): 1545–63.

Armitage, J. P. (1999). "Bacterial tactic responses." *Adv Microb Physiol* **41**: 229–89.

Arthur, W. (2002). "The emerging conceptual framework of evolutionary developmental biology." *Nature* **415**(6873): 757–64.

Avivi, A., A. Joel, et al. (2001). "The lens protein alpha-B-crystallin of the blind subterranean mole-rat: high homology with sighted mammals." *Gene* **264**(1): 45–9.

Bailey, J. A., L. Carrel, et al. (2000). "Molecular evidence for a relationship between LINE-1 elements and X chromosome inactivation: the Lyon repeat hypothesis." *Proc Natl Acad Sci U S A* **97**(12): 6634–9.

Baker, M. E. (1997). "Steroid receptor phylogeny and vertebrate origins." *Mol Cell Endocrinol* **135**(2): 101–7.

Banerjee, D. and F. Slack (2002). "Control of developmental timing by small temporal RNAs: a paradigm for RNA-mediated regulation of gene expression." *Bioessays* **24**(2): 119–29.

Banfield, J. F. and C. R. Marshall (2000). "Perspectives: earth science and evolution. Genomics and the geosciences." *Science* **287**(5453): 605–6.

Bargmann, C. I. (1997). "Olfactory receptors, vomeronasal receptors, and the organization of olfactory information." *Cell* **90**(4): 585–7.

Barinaga, M. (1999). "Salmon follow watery odors home." *Science* **286**(5440): 705–6.

Barlow, L. A. and R. G. Northcutt (1998). "The role of innervation in the development of taste buds: insights from studies of amphibian embryos." *Ann N Y Acad Sci* **855**: 58–69.

Bar-Yam, Y. (1997). *Dynamics of Complex Systems*. Reading, Massachusetts, Addison-Wesley.

Bates, H. W. (1862). "Contributions to an insect fauna of the Amazon Valley." *Transactions of the Linnean Society of London* **23**: 495–566.

Bates, H. W. (1863). *The Naturalist on the River Amazons*. London, John Murray.

Bateson, W. (1886). "On the ancestry of the Chordata." *Quarterly Journal of Microscopic Science* **26**: 535–71.

Bateson, W. (1894). *Materials for the Study of Variation, Treated with Special Regard to Discontinuity in the Origin of Species*. London, Macmillan.

Bateson, W. (1913). *Problems of Genetics*. New Haven, Yale University Press.

Bauer, F., K. Schweimer, et al. (2001). "Structure determination of human and murine beta-defensins reveals structural conservation in the absence of significant sequence similarity." *Protein Sci* **10**(12): 2470–9.

Behrens, M., T. G. Langecker, et al. (1997). "Comparative analysis of Pax-6 sequence and expression in the eye development of the blind cave fish Astyanax fasciatus and its epigean conspecific." *Molecular Biology and Evolution* **14**: 299–308.

Behrens, M., H. Wilkens, et al. (1998). "Cloning of the alphaA-crystallin genes of a blind cave form and the epigean form of Astyanax fasciatus: a comparataive analysis of structure, expression and evolutionary conservation." *Gene* **216**: 319–26.

Beldade, P. and P. M. Brakefield (2003). "Concerted evolution and developmental integration in modular butterfly wing patterns." *Evolution & Development* **5**: 169–72.

Benfey, P. N. (1999). "Stem cells: A tale of two kingdoms." *Curr Biol* **9**(5): R171–2.

Bergson, H. (1907). *Lévolution créatrice*. Paris, F. Alcan.

Bessa, J., B. Gebelein, et al. (2002). "Combinatorial control of Drosophila eye development by eyeless, homothorax, and teashirt." *Genes Dev* **16**(18): 2415–27.

Bharathan, G., T. E. Goliber, et al. (2002). "Homologies in leaf form inferred from KNOXI gene expression during development." *Science* **296**(5574): 1858–60.

Bininda-Emonds, O. R., J. E. Jeffery, et al. (2003). "Inverting the hourglass: quantitative evidence against the phylotypic stage in vertebrate development." *Proc R Soc Lond B Biol Sci* **270**(1513): 341–6.

Bird, A. (2001). "Molecular biology. Methylation talk between histones and DNA." *Science* **294**(5549): 2113–15.

Black, B. E., J. M. Holaska, et al. (2001). "DNA binding domains in diverse nuclear receptors function as nuclear export signals." *Curr Biol* **11**(22): 1749–58.

Blanchette, M., B. Schwikowski, et al. (2002). "Algorithms for phylogenetic footprinting." *J Comput Biol* **9**(2): 211–23.

Blanchette, M. and M. Tompa (2002). "Discovery of regulatory elements by a computational method for phylogenetic footprinting." *Genome Res* **12**(5): 739–48.

Bockaert, J. and J. P. Pin (1999). "Molecular tinkering of G protein-coupled receptors: an evolutionary success." *Embo J* **18**(7): 1723–9.

Bonifer, C. (2000). "Developmental regulation of eukaryotic gene loci: which cis-regulatory information is required?" *Trends Genet* **16**(7): 310–15.

Bonner, J. T. (1988). *The Evolution of Complexity by Means of Natural Selection*. Princeton, NJ, Princeton University Press.

Bonner, J. T. (1998). "The origins of multicellularity." *Integrative Biology* **1**: 27–36.

Bonner, J. T. (2000). *First Signals: the Evolution of Multicellular Development*. Princeton, Princeton University Press.

Botella, M. A., J. E. Parker, et al. (1998). "Three genes of the Arabidopsis RPP1 complex resistance locus recognize distinct Peronospora parasitica avirulence determinants." *Plant Cell* **10**(11): 1847–60.

Bouadloun, F., D. Donner, et al. (1983). "Codon-specific missense errors in vivo." *Embo J* **2**(8): 1351–6.

Bowder, D. (1982). *Who was who in the Greek world, 776 BC-30 BC*. Ithaca, N.Y., Cornell University Press.

Bowmaker, J. K. (1998). "Evolution of colour vision in vertebrates." *Eye* **12**((Pt 3b)): 541–7.

Brennan, P. (2001). "How mice make their mark." *Nature* **414**(6864): 590–1.

Britannica, E. (2003). Ear, human.

Brockdorff, N. (2002). "X-chromosome inactivation: closing in on proteins that bind Xist RNA." *Trends Genet* **18**(7): 352–8.

Browne, T. (1646). *Pseudodoxia Epidemica (Reprinted, 1981)*. Oxford, Clarendon.

Brunekreef, G. A., H. J. Kraft, et al. (1996). "The mechanism of recruitment of the lactate dehydrogenase-B/epsilon-crystallin gene by the duck lens." *J Mol Biol* **262**(5): 629–39.

Buck, L. and R. Axel (1991). "A novel multigene family may encode odorant receptors: a molecular basis for odor recognition." *Cell* **65**(1): 175–87.

Budelmann, B. U. (1992). Hearing in Crustacea. *Evolutionary biology of hearing*. D. B. Webster, R. Fay and A. Popper. Berlin Heidelberg New York, Springer: 131–9.

Burke, A. C., C. E. Nelson, et al. (1995). "Hox genes and the evolution of vertebrate axial morphology." *Development* **121**: 333–46.

Burrows, M. (1996). *The neurobiology of an insect brain*. New York, Oxford University Press.

Buss, L. W. (1987). *The Evolution of Individuality*. Princeton, NJ, Princeton University Press.

Bussemaker, H. J., H. Li, et al. (2000). "Building a dictionary for genomes: identification of presumptive regulatory sites by statistical analysis." *Proc Natl Acad Sci U S A* **97**(18): 10096–100.

Butler, A. B. and W. Hodos (1996). *Comparative vertebrate neuroanatomy: evolution and adaptation*. New York, Wiley-Liss.

Buxhoeveden, D. P. and M. F. Casanova (2002). "The minicolumn hypothesis in neuroscience." *Brain* **125**(Pt 5): 935–51.

Calder, W. A. (1984). *Size, Function, and Life History*. Cambridge, MA, Harvard University Press.

Callaerts, P., G. Halder, et al. (1997). "PAX-6 in development and evolution." *Annu Rev Neurosci* **20**: 483–532.

Campbell, G. (2002). "Distalization of the Drosophila leg by graded EGF-receptor activity." *Nature* **418**(6899): 781–5.

Caporale, L. H. (1999). *Molecular Strategies in Biological Evolution*. New York, New York Academy of Sciences.

Carlson, B. M. (1999). *Human Embryology & Developmental Biology*. St Louis, Mosby.

Caron, H., B. van Schaik, et al. (2001). "The human transcriptome map: clustering of highly expressed genes in chromosomal domains." *Science* **291**(5507): 1289–92.

Carroll, S., J. Grenier, et al. (2001). *From DNA to Diversity: Molecular Genetics and the Evolution of Animal Design*. Malden, MA, Blackwell.

Carroll, S. B. (2001). "Chance and necessity: the evolution of morphological complexity and diversity." *Nature* **409**(6823): 1102–9.

Carthew, R. W. (2001). "Gene silencing by double-stranded RNA." *Curr Opin Cell Biol* **13**(2): 244–8.

Catania, K. C. (1999). "A nose that looks like a hand and acts like an eye: the unusual mechanosensory system of the star-nosed mole." *J Comp Physiol [A]* **185**(4): 367–72.

Catania, K. C. and J. H. Kaas (2001). "Areal and callosal connections in the somatosensory cortex of the star-nosed mole." *Somatosens Mot Res* **18**(4): 303–11.

Cavaggioni, A. and C. Mucignat-Caretta (2000). "Major urinary proteins, alpha(2U)-globulins and aphrodisin." *Biochim Biophys Acta* **1482**(1–2): 218–28.

Cavalli-Sforza, L. L. and W. F. Bodmer (1971). *The Genetics of Human Populations*. San Francisco, CA, WH Freeman.

Cech, T. R. (1986). "A model for the RNA-catalyzed replication of RNA." *Proc Natl Acad Sci U S A* **83**(12): 4360–3.

Cech, T. R. (1990). "Nobel lecture. Self-splicing and enzymatic activity of an intervening sequence RNA from Tetrahymena." *Biosci Rep* **10**(3): 239–61.

Cerutti, H. (2003). "RNA interference: traveling in the cell and gaining functions?" *Trends Genet* **19**(1): 39–46.

Chambers, D. and I. M. McGonnell (2002). "Neural crest: facing the facts of head development." *Trends Genet* **18**(8): 381–4.

Charlesworth, B. (1994). *Evolution in age-structured populations*. Cambridge [England]; New York, Cambridge University Press.

Charlet, M., S. Chernysh, et al. (1996). "Innate immunity. Isolation of several cysteine-rich antimicrobial peptides from the blood of a mollusc, Mytilus edulis." *J Biol Chem* **271**(36): 21808–13.

Chauhan, B. K., N. A. Reed, et al. (2002). "A comparative cDNA microarray analysis reveals a spectrum of genes regulated by Pax6 in mouse lens." *Genes Cells* **7**(12): 1267–83.

Chauhan, B. K., W. Zhang, et al. (2002). "Identification of differentially expressed genes in mouse Pax6 heterozygous lenses." *Invest Ophthalmol Vis Sci* **43**(6): 1884–90.

Chawla, A., J. J. Repa, et al. (2001). "Nuclear Receptors and Lipid Physiology: Opening the X-Files." *Science* **294**(5548): 1866–1870.

Chiang, C., Y. Litingtung, et al. (2001). "Manifestation of the limb prepattern: limb development in the absence of sonic hedgehog function." *Dev Biol* **236**(2): 421–35.

Chipman, A. D., A. Haas, et al. (2000). "Variation in anuran embryogenesis: differences in sequence and timing of early developmental events." *J Exp Zool* **288**(4): 352–65.

Chiu, C. H., C. Amemiya, et al. (2002). "Molecular evolution of the HoxA cluster in the three major gnathostome lineages." *Proc Natl Acad Sci U S A* **99**(8): 5492–7.

Christensen, S. K., N. Dagenais, et al. (2000). "Regulation of auxin response by the protein kinase PINOID." *Cell* **100**(4): 469–78.

Chung, K. T. and D. D. Ourth (2000). "Viresin. A novel antibacterial protein from immune hemolymph of Heliothis virescens pupae." *Eur J Biochem* **267**(3): 677–83.

Chuong, C. M., N. Patel, et al. (2000). "Sonic hedgehog signaling pathway in vertebrate epithelial appendage morphogenesis: perspectives in development and evolution." *Cell Mol Life Sci* **57**(12): 1672–81.

Clark, A. G. (1998). "Mutation-selection balance with multiple alleles." *Genetica* **102–103** (1–6): 41–7.

Clarke, A., R. Desikan, et al. (2000). "NO way back: nitric oxide and programmed cell death in Arabidopsis thaliana suspension cultures." *Plant J* **24**(5): 667–77.

Cleland, R. (1999). Introduction: Nature, occurrence and functioning of plant hormones. *Biochemistry and Molecular Biology of Plant Hormones*. P. J. J. Hooykaas, M. A. Hall and K. R. Libbenga. Amsterdam, Elsevier. **33**: 541.

Clyne, P. J., C. G. Warr, et al. (1999). "A novel family of divergent seven-transmembrane proteins: candidate odorant receptors in Drosophila." *Neuron* **22**(2): 327–38.

Cohen, I. R. (2000). "Discrimination and dialogue in the immune system." *Semin Immunol* **12**(3): 215–9; discussion 257–344.

Cohn, M. J. (2002). "Evolutionary biology: lamprey Hox genes and the origin of jaws." *Nature* **416**(6879): 386–7.

Conn, P. M. and A. R. Means (2000). *Principles of molecular regulation.* Totowa, N.J., Humana Press.

Cooper, T. F., D. E. Rozen, et al. (2003). "Parallel changes in gene expression after 20,000 generations of evolution in Escherichiacoli." *Proc Natl Acad Sci U S A* **100**(3): 1072–7.

Couvier, G. (1830). "Considerations sur les Mollusuqes, et en particulier sur les Cephalopodes." *Annales des Sciences Naturelles* **19**: 241–59.

Crasto, C., M. L., et al. (2003). The Olfactory Receptor Database. **2003**.

Crasto, C., L. Marenco, et al. (2002). "Olfactory Receptor Database: a metadata-driven automated population from sources of gene and protein sequences." *Nucleic Acids Research* **1**: 354–60.

Crasto, C., M. S. Singer, et al. (2001). "The olfactory receptor family album." *Genome Biol* **2**(10): REVIEWS1027.

Crick, F. and C. Koch (2003). "A framework for consciousness." *Nat Neurosci* **6**(2): 119–26.

Cropp, S., S. Boinski, et al. (2002). "Allelic variation in the squirrel monkey x-linked color vision gene: biogeographical and behavioral correlates." *J Mol Evol* **54**(6): 734–45.

Crow, J. F. (1958). "Some possibilities for measuring selection intensities in man." *Human Biology* **30**: 1–13.

Crow, J. F. (2000). "The origins, patterns and implications of human spontaneous mutation." *Nat Rev Genet* **1**(1): 40–7.

Crow, J. F. and M. Kimura (1971). *An Introduction to Population Genetics Theory*. New York, Harper & Row.

Cutforth, T. and U. Gaul (1997). "The genetics of visual system development in Drosophila: specification, connectivity and asymmetry." *Curr Opin Neurobiol* **7**(1): 48–54.

Cvekl, A. and J. Piatigorsky (1996). "Lens development and crystallin gene expression: many roles for Pax-6." *Bioessays* **18**(8): 621–30.

Czerny, T., G. Halder, et al. (1999). "twin of eyeless, a second Pax-6 gene of Drosophila, acts upstream of eyeless in the control of eye development." *Mol Cell* **3**(3): 297–307.

Daniels, R., M. Zuccotti, et al. (1997). "XIST expression in human oocytes and preimplantation embryos." *Am J Hum Genet* **61**(1): 33–9.

Darwin, C. (1862). *On the various contrivances by which British and foreign orchids are fertilised by insects*. London, J. Murray.

Darwin, C. (1871). *The descent of man and selection in relation to sex*. London, J. Murray.

Darwin, C. D. (1859). *On the Origin of Species*. London, John Murray.

Davey, K. G. (2000). "The modes of action of juvenile hormones: some questions we ought to ask." *Insect Biochem Mol Biol* **30**(8–9): 663–9.

Davidson, E. H. (2001). *Genomic Regulatory Systems: Development and Evolution*. San Diego, CA, Academic Press.

Davidson, E. H., J. P. Rast, et al. (2002a). "A genomic regulatory network for development." *Science* **295**(5560): 1669–78.

Davidson, E. H., J. P. Rast, et al. (2002b). "A provisional regulatory gene network for specification of endomesoderm in the sea urchin embryo." *Dev Biol* **246**(1): 162–90.

Dawkins, R. (1981). *The Selfish Gene*. Oxford, Oxford University Press.

de Bruyne, M., P. J. Clyne, et al. (1999). "Odor coding in a model olfactory organ: the Drosophila maxillary palp." *J Neurosci* **19**(11): 4520–32.

de Duve, C. (1991). *Blueprint for a Cell.* Burlington, North Carolina, Neil Patterson.
De Robertis, E. M., J. Larrain, et al. (2000). "The establishment of Spemann's organizer and patterning of the vertebrate embryo." *Nat Rev Genet* **1**(3): 171–81.
De Robertis, E. M. and Y. Sasai (1996). "A common plan for dorsoventral patterning in Bilateria." *Nature* **380**(6569): 37–40.
Delledonne M, Xia Yiji, et al. (1998). "Nitric oxide functions as a signal in plant disease resistance." *Nature* **394**(6 August): 585–8.
Dengler, N. and J. Kang (2001). "Vascular patterning and leaf shape." *Curr Opin Plant Biol* **4**(1): 50–6.
Dermitzakis, E. T. and A. G. Clark (2002). "Evolution of transcription factor-binding sites in mammalian gene regulatory regions: conservation and turnover." *Molecular Biology and Evolution* **19**(6): in press.
DeSilva, U., L. Elnitski, et al. (2002). "Generation and comparative analysis of approximately 3.3 Mb of mouse genomic sequence orthologous to the region of human chromosome 7q11.23 implicated in Williams syndrome." *Genome Res* **12**(1): 3–15.
Desmond, A. (1994). *Huxley: The Devil's Disciple.* London, Michael Joseph.
Diaz, M. and P. Casali (2002). "Somatic immunoglobulin hypermutation." *Curr Opin Immunol* **14**(2): 235–40.
Diebel, C. E., R. Proksch, et al. (2000). "Magnetite defines a vertebrate magnetoreceptor." *Nature* **406**(6793): 299–302.
Dixon, M. S., C. Golstein, et al. (2000). "Genetic complexity of pathogen perception by plants: the example of Rcr3, a tomato gene required specifically by Cf-2." *Proc Natl Acad Sci U S A* **97**(16): 8807–14.
Dolch, R. and T. Tscharntke (2000). "Defoliation of alders (Alnus glutinosa) affects herbivory by leaf beetles on undamaged neighbours." *Oecologia* **125**(4): 504–11.
Doolittle, R. F. (1998). "Microbial genomes opened up." *Nature* **392**(6674): 339–42.
Doolittle, W. F. (1998). "A paradigm gets shifty." *Nature* **392**(6671): 15–16.
Doolittle, W. F. (1999). "Phylogenetic classification and the universal tree." *Science* **284**(5423): 2124–9.
Doolittle, W. F. and J. M. Logsdon, Jr. (1998). "Archaeal genomics: do archaea have a mixed heritage?" *Curr Biol* **8**(6): R209–11.
Doudna, J. A. and T. R. Cech (2002). "The chemical repertoire of natural ribozymes." *Nature* **418**(6894): 222–8.
Døving, K. B. and D. Trotier (1998). "Structure and function of the vomeronasal organ." *J Exp Biol* **201**(Pt 21): 2913–25.
Downward, J. (2001). "The ins and outs of signalling." *Nature* **411**(6839): 759–62.
Dreyer, W. J. (1998). "The area code hypothesis revisited: olfactory receptors and other related transmembrane receptors may function as the last digits in a cell surface code for assembling embryos." *Proc Natl Acad Sci U S A* **95**(16): 9072–7.
Driesen, J. L. (2003). Primary visual pathway.
Driever, W. and C. Nusslein-Volhard (1988). "The bicoid protein determines position in the Drosophila embryo in a concentration-dependent manner." *Cell* **54**(1): 95–104.
Dryer, I. (2000). "Evolution of odorant receptors." *Bioessays* **22**: 803–10.
Dubendorfer, A., M. Hediger, et al. (2002). "Musca domestica, a window on the evolution of sex-determining mechanisms in insects." *Int J Dev Biol* **46**(1): 75–9.
Duboule, D. (2002). "Making progress with limb models." *Nature* **418**(6897): 492–3.
Dudley, A. T., M. A. Ros, et al. (2002). "A re-examination of proximodistal patterning during vertebrate limb development." *Nature* **418**(6897): 539–44.
Dun, R. B. and A. S. Fraser (1959). "Selection for an invariant character, vibrissa number, in the house mouse." *Australian Journal of Biological Sciences* **21**: 506–23.
Duncan, M. K., Z. Kozmik, et al. (2000). "Overexpression of PAX6(5a) in lens fiber cells results in cataract and upregulation of (alpha)5(beta)1 integrin expression." *J Cell Sci* **113**(Pt 18): 3173–85.

Eatock, R. A. and W. T. Newsome (1999). "Sensory systems [editorial]." *Curr Opin Neurobiol* **9**(4): 385–8.
Eberl, D. F. (1999). "Feeling the vibes: chordotonal mechanisms in insect hearing." *Curr Opin Neurobiol* **9**(4): 389–93.
Eberl, D. F., R. W. Hardy, et al. (2000). "Genetically similar transduction mechanisms for touch and hearing in Drosophila." *J Neurosci* **20**(16): 5981–8.
Ebrey, T. G. (2002). "A new type of photoreceptor in algae." *Proc Natl Acad Sci U S A* **99**(13): 8463–4.
Eckstein, F. (2003). RNA function.
Edgell, D. R., H. P. Klenk, et al. (1997). "Gene duplications in evolution of archaeal family B DNA polymerases." *J Bacteriol* **179**(8): 2632–40.
Ehret, G. (1997). "The auditory cortex." *J Comp Physiol [A]* **181**(6): 547–57.
Eisen, M. B., P. T. Spellman, et al. (1998). "Cluster analysis and display of genome-wide expression patterns." *Proc Natl Acad Sci U S A* **95**(25): 14863–8.
Ellis, C. and J. G. Turner (2001). "The Arabidopsis mutant cev1 has constitutively active jasmonate and ethylene signal pathways and enhanced resistance to pathogens." *Plant Cell* **13**(5): 1025–33.
Ellis, J., P. Dodds, et al. (2000). "Structure, function and evolution of plant disease resistance genes." *Current Opinion in Plant BIology* **3**(4): 278–84.
Ellis, J. and D. Jones (1998). "Structure and function of proteins controlling strain-specific pathogen resistance in plants." *Curr Opin Plant Biol* **1**(4): 288–93.
Enard, W., P. Khaitovich, et al. (2002a). "Intra- and interspecific variation in gene expression patterns." *Science* **296**: 340–3.
Enard, W., M. Przeworski, et al. (2002b). "Molecular evolution of FOXP2, a gene involved in speech and language." *Nature* **418**: 869–72.
Engelberth, J., T. Koch, et al. (2000). "Channel-Forming Peptaibols Are Potent Elicitors of Plant Secondary Metabolism and Tendril Coiling We gratefully acknowledge the gift of ampullosporin A, bergofungins A-C, and the chrysospermins from Prof. U. Grafe (Hans-Knoll Institute for Natural Products Research, Jena), and we thank Dr. T. Nurnberger (Institute for Plant Biochemistry, Halle) for a sample of systemin." *Angew Chem Int Ed Engl* **39**(10): 1860–2.
Engelmann, J., W. Hanke, et al. (2000). "Hydrodynamic stimuli and the fish lateral line." *Nature* **408**(6808): 51–2.
Escriva, H., F. Delaunay, et al. (2000). "Ligand binding and nuclear receptor evolution." *Bioessays* **22**(8): 717–27.
Essner, J. J., W. W. Branford, et al. (2000). "Mesendoderm and left-right brain, heart and gut development are differentially regulated by pitx2 isoforms." *Development* **127**(5): 1081–93.
Ewald, P. W. (1994). *Evolution of infectious disease*. Oxford; New York, Oxford University Press.
Eyre-Walker, A., P. D. Keightley, et al. (2002). "Quantifying the slightly deleterious mutation model of molecular evolution." *Mol Biol Evol* **19**(12): 2142–9.
Fankhauser, C. and J. Chory (1999). "Light receptor kinases in plants!" *Curr Biol* **9**(4): R123–6.
Fay, R. R. and A. N. Popper (1999). *Comparative hearing:fish and amphibians*. New York, Springer.
Fay, R. R. and A. N. Popper (2000). "Evolution of hearing in vertebrates: the inner ears and processing [In Process Citation]." *Hear Res* **149**(1–2): 1–10.
Fedorowicz, G. M., J. D. Fry, et al. (1998). "Epistatic interactions between smell-impaired loci in Drosophila melanogaster." *Genetics* **148**(4): 1885–91.
Feys, B. J. and J. E. Parker (2000). "Interplay of signaling pathways in plant disease resistance [In Process Citation]." *Trends Genet* **16**(10): 449–55.
Finch, C. E. and T. B. L. Kirkwood (2000). *Chance, Development, and Aging*. New York, Oxford University Press.
Finger, T. and S. Simon (2000). Cell Biology of Taste Epithelium. *The Neurobiology of Taste and Smell*. T. Finger, W. Silver and D. Restrepo. New York, Wiley-Liss: 287–314.

Finger, T. E. (1997). "Evolution of taste and solitary chemoreceptor cell systems." *Brain Behav Evol* **50**(4): 234–43.
Firestein, S. (2001). "How the olfactory system makes sense of scents." *Nature* **413**(6852): 211–18.
Fisher, R. A. (1930). *The Genetical Theory of Natural Selection*. Oxford, Clarendon.
Fitch, D. H. and W. Sudhaus (2002). "One small step for worms, one giant leap for 'Bauplan'?" *Evol Dev* **4**(4): 243–6.
Fitz-Gibbon, S. T. and C. H. House (1999). "Whole genome-based phylogenetic analysis of free-living microorganisms." *Nucleic Acids Res* **27**(21): 4218–22.
Fleck, L. (1979). *Genesis and development of a scientific fact (orig. 1930)*. Chicago, University of Chicago Press.
Flor, H. H. (1956). "The complementary genetic systems in flax and flax rust." *Adv Genet* **8**: 29–54.
Fontana, W. (2002). "Modelling 'evo-devo' with RNA." *Bioessays* **24**(12): 1164–77.
Fontana, W. and P. Schuster (1998). "Continuity in evolution: on the nature of transitions." *Science* **280**(5368): 1451–5.
Fox, K. (1994). "The cortical component of experience-dependent synaptic plasticity in the rat barrel cortex." *J Neurosci* **14**(12): 7665–79.
Freeland, S. J. and L. D. Hurst (1998). "The genetic code is one in a million." *J Mol Evol* **47**(3): 238–48.
Freeland, S. J., R. D. Knight, et al. (2000). "Early fixation of an optimal genetic code." *Mol Biol Evol* **17**(4): 511–18.
Fritzsch, B. and K. W. Beisel (2001). "Evolution and development of the vertebrate ear." *Brain Res Bull* **55**(6): 711–21.
Fry, J. D., K. A. de Ronde, et al. (1995). "Polygenic mutation in Drosophila melanogaster: genetic analysis of selection lines." *Genetics* **139**(3): 1293–307.
Fukuchi-Shimogori, T. and E. A. Grove (2001). "Neocortex patterning by the secreted signaling molecule FGF8." *Science* **294**(5544): 1071–4.
Furuta, Y. and B. L. Hogan (1998). "BMP4 is essential for lens induction in the mouse embryo." *Genes Dev* **12**(23): 3764–75.
Galilei, G. (1622). *The Assayer*. Rome, The Academy of the Lynxes (reprinted in Drake, S. Discoveries and Opnions of Galileo. Garden City, NY, Doubleday, 1957).
Galindo, K. and D. P. Smith (2001). "A large family of divergent Drosophila odorant-binding proteins expressed in gustatory and olfactory sensilla." *Genetics* **159**(3): 1059–72.
Gampala, S. S., R. R. Finkelstein, et al. (2001). "ABA INSENSITIVE-5 interacts with ABA signaling effectors in rice protoplasts." *J Biol Chem*.
Garcia-Olmedo, F., A. Molina, et al. (1998). "Plant defense peptides." *Biopolymers* **47**(6): 479–91.
Gavalas, A. and R. Krumlauf (2000). "Retinoid signalling and hindbrain patterning." *Curr Opin Genet Dev* **10**(4): 380–6.
Gehring, W. J. (2002). "The genetic control of eye development and its implications for the evolution of the various eye-types." *Int J Dev Biol* **46**(1 Spec No): 65–73.
Gehring, W. J. and K. Ikeo (1999). "Pax 6: mastering eye morphogenesis and eye evolution." *Trends Genet* **15**(9): 371–7.
Gellon, G. and W. McGinnis (1998). "Shaping animal body plans in development and evolution by modulation of Hox expression patterns." *Bioessays* **20**(2): 116–25.
Gene Ontology Consortium (2003). The gene ontology database.
Gerhart, J. (2000). "Inversion of the chordate body axis: are there alternatives?" *Proc Natl Acad Sci U S A* **97**(9): 4445–8.
Gerhart, J. and M. Kirschner (1997). *Cells, Embryos, and Evolution: Toward a Cellular and Developmental Understanding of Phenotypic Variation and Evolutionary Adaptability.* Malden, MA, Blackwell Scientific.

Ghbeish, N., C. C. Tsai, et al. (2001). "The dual role of ultraspiracle, the Drosophila retinoid X receptor, in the ecdysone response." *Proc Natl Acad Sci U S A* **98**(7): 3867–72.

Gilad, Y., O. Man, et al. (2003). "Human specific loss of olfactory receptor genes." *PNAS* **100**: 3324–7.

Gilad, Y., D. Segre, et al. (2000). "Dichotomy of single-nucleotide polymorphism haplotypes in olfactory receptor genes and pseudogenes." *Nat Genet* **26**(2): 221–4.

Gilbert, S. (2003). *Developmental Biology*. Sunderland, MA, Sinauer.

Gilbert, S. F., J. M. Opitz, et al. (1996). "Resynthesizing evolutionary and developmental biology." *Dev Biol* **173**(2): 357–72.

Gilbert, W. (1986). "The RNA world." *Nature* **319**: 618.

Gillespie, J. H. (1998). *Population Genetics: a Concise Guide*. Baltimore, MD, Johns Hopkins Press.

Giorgi, D., C. Friedman, et al. (2000). "Characterization of nonfunctional V1R-like pheromone receptor sequences in human." *Genome research* **10**: 1979–5.

Girardin, S. E., P. J. Sansonetti, et al. (2002). "Intracellular vs extracellular recognition of pathogens—common concepts in mammals and flies." *Trends Microbiol* **10**(4): 193–9.

Glazebrook, J. (2001). "Genes controlling expression of defense responses in Arabidopsis—2001 status." *Curr Opin Plant Biol* **4**(4): 301–8.

Glosmann, M. and P. K. Ahnelt (2002). "A mouse-like retinal cone phenotype in the Syrian hamster: S opsin coexpressed with M opsin in a common cone photoreceptor." *Brain Res* **929**(1): 139–46.

Glusman, G., I. Yanai, et al. (2001). "The complete human olfactory subgenome." *Genome Res* **11**(5): 685–702.

Gobbel, L. (2002). "Morphology of the external nose in Hipposideros diadema and Lavia frons with comments on its diversity and evolution among leaf-nosed Microchiroptera." *Cells Tissues Organs* **170**(1): 39–60.

Gomes, W. A., M. F. Mehler, et al. (2003). "Transgenic overexpression of BMP4 increases astroglial and decreases oligodendroglial lineage commitment." *Dev Biol* **255**(1): 164–77.

Gopfert, M. C. and L. T. Wasserthal (1999). "Hearing with the mouthparts: behavioural responses and the structural basis of ultrasound perception in acherontiine hawkmoths." *J Exp Biol* **202**(Pt 8): 909–18.

Gottlieb, G. (1997). *Synthesizing Nature-Nurture: Prenatal Roots of Instinctive Behavior*. Mahwah, N.J., Erlbaum.

Gottlieb, T. M., M. J. Wade, et al. (2002). " Potential genetic variance and the domestication of maize." *Bioessays* **24**: 685–9.

Goudreau, G., P. Petrou, et al. (2002). "Mutually regulated expression of Pax6 and Six3 and its implications for the Pax6 haploinsufficient lens phenotype." *Proc Natl Acad Sci U S A* **99**(13): 8719–24.

Gould, S. J. (1977). *Ontogeny and Phylogeny*. Cambridge, MA, Belknap Press.

Gould, S. J. and E. Vrba (1982). "Exaptation—A Missing Term in the Science of Form." *Paleobiology* **8**: 4–15.

Govrin, E. M. and A. Levine (2000). "The hypersensitive response facilitates plant infection by the necrotrophic pathogen Botrytis cinerea." *Curr Biol* **10**(13): 751–7.

GPCR (2003). The evolution and classification of G-protein-coupled receptors.

Graham, P., J. K. Penn, et al. (2003). "Masters change, slaves remain." *Bioessays* **25**(1): 1–4.

Grant, B. S. (1999). "Fine tuning the peppered moth paradigm." *Evolution* **53**: 980–4.

Grant, P. R. and B. R. Grant (2002). "Unpredictable evolution in a 30-year study of Darwin's finches." *Science* **296**(5568): 707–11.

Grasso, L. C., D. C. Hayward, et al. (2001). "The evolution of nuclear receptors: evidence from the coral acropora." *Mol Phylogenet Evol* **21**(1): 93–102.

Graur, D. and W-H. Li (2000). *Fundamentals of molecular evolution*. Sunderland, Mass., Sinauer.

Graveson, A. C., M. M. Smith, et al. (1997). "Neural crest potential for tooth development in a urodele amphibian: developmental and evolutionary significance." *Dev Biol* **188**(1): 34–42.

Gray, E. (1960). "The fine structure of the insect ear." *Phil Trans R Soc B* **243**: 75–94.

Greenspan, R. J. (2001). "The flexible genome." *Nat Rev Genet* **2**(5): 383–7.

Griffiths, A. J. E., J. H. Miller, et al. (1996). *An Introduction of Genetic Analysis*. New York, Freeman.

Grosshans, H. and F. J. Slack (2002). "Micro-RNAs: small is plentiful." *J Cell Biol* **156**(1): 17–21.

Gulick, J. T. (1872). "On diversity of evolution under one set of external conditions." *Linnean Society Journal of Zoology* **11**: 496–505.

Gupta, A., L. H. Tsai, et al. (2002). "Life is a journey: a genetic look at neocortical development." *Nat Rev Genet* **3**(5): 342–55.

Gustafsson, J.-A. (1999). "Seeking Ligands for Lonely Orphan Receptors." *Science* **284**(5418): 1285–6.

Haber, J. E. (1998). "Mating-type gene switching in Saccharomyces cerevisiae." *Annu Rev Genet* **32**: 561–99.

Haeckel, E. (1899). *Art Forms in Nature*. New York, Dover.

Haeckel, E. (1906). *The Evolution of Man*. London, Watts.

Haig, D. and L. D. Hurst (1991). "A quantitative measure of error minimization in the genetic code." *J Mol Evol* **33**(5): 412–17.

Haldane, J. B. S. (1932). *The Inequality of Man*. London, Penguin.

Hall, B. K. (1999). *Evolutionary Developmental Biology*. Dordrecht, Holland, Kluwer Academic.

Hall, B. K. (2000). "The neural crest as a fourth germ layer and vertebrates as quadroblastic not triploblastic." *Evol Dev* **2**(1): 3–5.

Halle, F. (1999). Ecology of reiteration in tropical trees. *The Evolution of Plant Architecture*. M. H. Kurmann and A. R. Hemsley. London, Kew Botanic Gardens.

Hannibal, J., P. Hindersson, et al. (2002). "The circadian photopigment melanopsin is expressed in the blind subterranean mole rat, Spalax." *Neuroreport* **13**(11): 1411–14.

Hannon, G. J. (2002). "RNA interference." *Nature* **418**(6894): 244–51.

Hardie, R. C. and P. Raghu (2001). "Visual transduction in Drosophila." *Nature* **413**(6852): 186–93.

Hardison, R. C. (2000). "Conserved noncoding sequences are reliable guides to regulatory elements." *Trends Genet* **16**(9): 369–72.

Hardison, R. C. (2001). "New views of evolution and regulation of vertebrate beta-like globin gene clusters from an orphaned gene in marsupials." *Proc Natl Acad Sci U S A* **98**(4): 1327–9.

Harris, W. A. (1997). "Pax-6: where to be conserved is not conservative." *Proc Natl Acad Sci U S A* **94**(6): 2098–100.

Hartfelder, K. (2000). "Insect juvenile hormone: from 'status quo' to high society." *Braz J Med Biol Res* **33**(2): 157–77.

Hartl, D. and A. G. Clark (1997). *Principles of Population Genetics*. Sunderland, MA, Sinauer.

Hartl, D. L. and R. B. Campbell (1982). "Allele multiplicity in simple Mendelian disorders." *Am J Hum Genet* **34**(6): 866–73.

Hartman, H. and A. Fedorov (2002). "The origin of the eukaryotic cell: a genomic investigation." *Proc Natl Acad Sci U S A* **99**(3): 1420–5.

Hatten, M. E. (2002). "New directions in neuronal migration." *Science* **297**(5587): 1660–3.

Hedrick, P. (2000). *Genetics of Populations*. Boston, Jones and Bartlett.

Heesy, C. P. and C. F. Ross (2001). "Evolution of activity patterns and chromatic vision in primates: morphometrics, genetics and cladistics." *J Hum Evol* **40**(2): 111–49.

Hegemann, P. (1997). "Vision in microalgae." *Planta* **203**(3): 265–74.

Hendriks, W., J. Leunissen, et al. (1987). "The lens protein alpha A-crystallin of the blind mole rat, Spalax ehrenbergi: evolutionary change and functional constraints." *Proc Natl Acad Sci U S A* **84**(15): 5320–4.

Herdman, M., T. Coursin, et al. (2000). "A new appraisal of the prokaryotic origin of eukaryotic phytochromes." *J Mol Evol* **51**(3): 205–13.

Herrada, G. and C. Dulac (1997). "A novel family of putative pheromone receptors in mammals with a topographically organized and sexually dimorphic distribution." *Cell* **90**(4): 763–73.

Hirth, F. and H. Reichert (1999). "Conserved genetic programs in insect and mammalian brain development." *Bioessays* **21**(8): 677–84.

Hodos, W. and A. B. Butler (1997). "Evolution of sensory pathways in vertebrates." *Brain Behav Evol* **50**(4): 189–97.

Hoffmann, J. A., F. C. Kafatos, et al. (1999). "Phylogenetic perspectives in innate immunity." *Science* **284**(5418): 1313–18.

Hogan, B. L. (1999). "Morphogenesis." *Cell* **96**(2): 225–33.

Hölldobler, B. and E. O. Wilson (1990). *The ants*. Cambridge, Mass., Belknap Press of Harvard University Press.

Holley, S. A., P. D. Jackson, et al. (1995). "A conserved system for dorsal-ventral patterning in insects and vertebrates involving sog and chordin." *Nature* **376**(6537): 249–53.

Hollick, J. B., J. E. Dorweiler, et al. (1997). "Paramutation and related allelic interactions." *Trends Genet* **13**(8): 302–8.

Holy, T. E., C. Dulac, et al. (2000). "Responses of vomeronasal neurons to natural stimuli." *Science* **289**(5484): 1569–72.

Hoppe, R., M. Weimer, et al. (2000). "Sequence analyses of the olfactory receptor gene cluster mOR37 on mouse chromosome 4." *Genomics* **66**(3): 284–95.

Houchmandzadeh, B., E. Wieschaus, et al. (2002). "Establishment of developmental precision and proportions in the early Drosophila embryo." *Nature* **415**(6873): 798–802.

Hoy, R. R., A. N. Popper, et al. (1998). *Comparative Hearing: Insects*. New York, Springer.

Hoyle, B. D. and J. W. Costerton (1991). "Bacterial resistance to antibiotics: the role of biofilms." *Prog Drug Res* **37**: 91–105.

Hughes, A. L. (1999). "Genomic catastrophism and the origin of vertebrate immunity." *Arch Immunol Ther Exp (Warsz)* **47**(6): 347–53.

Hultmark, D. (1993). "Immune reactions in Drosophila and other insects: a model for innate immunity." *Trends in Genetics* **9**(5): 178–83.

Hume, D. A. (2000). "Probability in transcriptional regulation and its implications for leukocyte differentiation and inducible gene expression." *Blood* **96**(7): 2323–8.

Hunter, T. (1995). "Protein kinases and phosphatases: the yin and yang of protein phosphorylation and signaling." *Cell* **80**(2): 225–36.

Hurst, J. L., C. E. Payne, et al. (2001). "Individual recognition in mice mediated by major urinary proteins." *Nature* **414**(6864): 631–4.

Hurst, L. D. (1995). "Selfish genetic elements and their role in evolution: the evolution of sex and some of what that entails." *Philos Trans R Soc Lond B Biol Sci* **349**(1329): 321–32.

Hurst, L. D., A. Atlan, et al. (1996). "Genetic conflicts." *Q Rev Biol* **71**(3): 317–64.

Hutchinson, G. E. and S. Rachootin (1979). Historical introduction. *Problems of Genetics*. W. Bateson. New Haven, Connecticut, Yale University Press: vii–xxii.

Hutton, J. (1788). "Theory of the earth; or an investigation of the laws discernible in the composition, dissolution and restoration of land upon the globe." *Transactions of the Royal Society of Edinburgh* **1**: 209–304.

Hwang, I. and J. Sheen (2001). "Two-component circuitry in Arabidopsis cytokinin signal transduction." *Nature* **413**(6854): 383–9.

Inoue, T., M. Higuchi, et al. (2001). "Identification of CRE1 as a cytokinin receptor from Arabidopsis." *Nature* **409**(6823): 1060–3.

Itoh, M. and A. B. Chitnis (2001). "Expression of proneural and neurogenic genes in the zebrafish lateral line primordium correlates with selection of hair cell fate in neuromasts." *Mech Dev* **102**(1–2): 263–6.

Jacob, S., M. K. McClintock, et al. (2002). "Paternally inherited HLA alleles are associated with women's choice of male odor." *Nat Genet* **30**(2): 175–9.

Jain, R., M. C. Rivera, et al. (1999). "Horizontal gene transfer among genomes: the complexity hypothesis." *Proc Natl Acad Sci U S A* **96**(7): 3801–6.

Janeway, C. (1999). *Immunobiology : the immune system in health and disease*. London New York, NY, US, Current Biology Publications; Garland Pub.

Janeway, C. A., Jr. (2001). "How the immune system works to protect the host from infection: a personal view." *Proc Natl Acad Sci U S A* **98**(13): 7461–8.

Jernvall, J. and H. S. Jung (2000). "Genotype, phenotype, and developmental biology of molar tooth characters." *Am J Phys Anthropol* **Suppl 31**: 171–90.

Jernvall, J. and I. Thesleff (2000). "Reiterative signaling and patterning during mammalian tooth morphogenesis." *Mech Dev* **92**(1): 19–29.

Jiang, T. X., H. S. Jung, et al. (1999). "Self-organization of periodic patterns by dissociated feather mesenchymal cells and the regulation of size, number and spacing of primordia." *Development* **126**(22): 4997–5009.

Johnson, K. A., M. L. Sistrunk, et al. (1998). "Arabidopsis thaliana responses to mechanical stimulation do not require ETR1 or EIN2." *Plant Physiol* **116**(2): 643–9.

Johnson, P. R. and J. R. Ecker (1998). "The ethylene gas signal transduction pathway: a molecular perspective." *Annu Rev Genet* **32**: 227–54.

Johnston, R. E. and M. Peng (2000). "The vomeronasal organ is involved in discrimination of individual odors by males but not by females in golden hamsters." *Physiol Behav* **70**(5): 537–49.

Josefsson, L. G. (1999). "Evidence for kinship between diverse G-protein coupled receptors." *Gene* **239**(2): 333–40.

Joyce, G. F. (2002). "The antiquity of RNA-based evolution." *Nature* **418**(6894): 214–21.

Judson, O. P. and B. B. Normark (2000). "Evolutionary genetics. Sinless originals." *Science* **288**(5469): 1185–6.

Jung, H. S., P. H. Francis-West, et al. (1998). "Local inhibitory action of BMPs and their relationships with activators in feather formation: implications for periodic patterning." *Dev Biol* **196**(1): 11–23.

Jung, H. S., L. Wolpert, et al. (1999). "The formation of the feather pattern in chick skin after a proportion of cells have been killed by X-irradiation." *Int J Dev Biol* **43**(2): 117–23.

Jurgens, G. (1994). Pattern formation in the embryo. *Arabidopsis*. E. M. Meyerowitz and C. R. Somerville. Cold Spring Harbor, New York, Cold Spring Harbor Laboratory Press. **Monograph 27**.

Kacsoh, B. (2000). *Endocrine physiology*. New York, McGraw-Hill Health Professions Division.

Kadunce, D. C., J. W. Vaughan, et al. (2001). "The influence of visual and auditory receptive field organization on multisensory integration in the superior colliculus." *Exp Brain Res* **139**(3): 303–10.

Kamachi, Y., M. Uchikawa, et al. (2001). "Pax6 and SOX2 form a co-DNA-binding partner complex that regulates initiation of lens development." *Genes Dev* **15**(10): 1272–86.

Kandel, E. R., J. H. Schwartz, et al. (2000). *Principles of neural science*. New York, McGraw-Hill Health Professions Division.

Karban, R., I. Baldwin, et al. (2000). "Communication between plants: induced resistance in wild tobacco plants following clipping of neighboring sagebrush." *Oecologia* **121**(1): 66–71.

Karten, H. J. (1997). "Evolutionary developmental biology meets the brain: the origins of mammalian cortex." *Proc Natl Acad Sci U S A* **94**(7): 2800–4.

Katz, L. A. (1999). "The Tangled Web: Gene Genealogies and the Origin of Eukaryotes." *Am Nat* **154**(S4): S137–45.

Kauer, J. S. (2002). "On the scents of smell in the salamander." *Nature* **417**(6886): 336–42.

Kawasaki, K. and K. M. Weiss (2003). "Mineralized tissue and vertebrate evolution: The secretory calcium-binding phosphoprotein gene cluster." *Proc Natl Acad Sci U S A*.

Kay, L. (2000). *Who Wrote the Book of Life?* Stanford, CA, Stanford University Press.

Keefe, A. D. and J. W. Szostak (2001). "Functional proteins from a random-sequence library." *Nature* **410**(6829): 715–18.

Keightley, P. D. and W. G. Hill (1983). "Effects of linkage on response to directional selection from new mutations." *Genet Res* **42**(2): 193–206.

Keller, E. F. (2002). *Making Sense of Life*. Cambridge, Massachusetts, Harvard University Press.

Kernan, M. and C. Zuker (1995). "Genetic approaches to mechanosensory transduction." *Curr Opin Neurobiol* **5**(4): 443–8.

Kettunen, P., J. Laurikkala, et al. (2000). "Associations of FGF-3 and FGF-10 with signaling networks regulating tooth morphogenesis." *Dev Dyn* **219**(3): 322–32.

Keverne, E. B. (1999). "The vomeronasal organ." *Science* **286**(5440): 716–20.

Keverne, E. B. (2002). "Pheromones, vomeronasal function, and gender-specific behavior." *Cell* **108**(6): 735–8.

Keys, D. N., D. L. Lewis, et al. (1999). "Recruitment of a hedgehog regulatory circuit in butterfly eyespot evolution." *Science* **283**(5401): 532–4.

Khush, R. S. and B. Lemaitre (2000). "Genes that fight infection:what the drosophila genome says about animal immunity [In Process Citation]." *Trends Genet* **16**(10): 442–9.

Kiernan, A. E., K. P. Steel, et al. (2002). Development of the mouse inner ear. *Mouse Development: Patterning, Morphogenesis, and Organogenesis*. J. Rossant and P. P. L. Tam. San Diego, California, Academic Press: 539–66.

King, A. J. (1999). "Sensory experience and the formation of a computational map of auditory space in the brain." *Bioessays* **21**(11): 900–11.

King, I. and R. Kingston (2001). "Specifying transcription." *Nature* **414**: 858–61.

Kirschner, M., J. Gerhart, et al. (2000). "Molecular "vitalism"." *Cell* **100**(1): 79–88.

Kissinger, J. C. and R. A. Raff (1998). "Evolutionary changes in sites and timing of actin gene expression in embryos of the direct- and indirect-developing sea urchins, Heliocidaris erythrogramma and H. tuberculata." *Dev Genes Evol* **208**(2): 82–93.

Klein, J. and A. Sato (2000). "The HLA system. First of two parts." *N Engl J Med* **343**(10): 702–9.

Klessig, D. F., J. Durner, et al. (2000). "Nitric oxide and salicylic acid signaling in plant defense." *Proc Natl Acad Sci U S A* **97**(16): 8849–55.

Kmita, M. and D. Duboule (2003). "Organizing axes in time and space; 25 years of colinear tinkering." *Science* **301**(5631): 331–3.

Koenig, R. J. (1998). "Thyroid hormone receptor coactivators and corepressors." *Thyroid* **8**(8): 703–13.

Koonin, E. V., L. Aravind, et al. (2000). "The impact of comparative genomics on our understanding of evolution." *Cell* **101**(6): 573–6.

Kramer, E. M. and V. F. Irish (1999). "Evolution of genetic mechanisms controlling petal development." *Nature* **399**(6732): 144–8.

Kratz, E., J. C. Dugas, et al. (2002). "Odorant receptor gene regulation: implications from genomic organization." *Trends Genet* **18**(1): 29–34.

Kroiher, M., M. A. Miller, et al. (2001). "Deceiving appearances: signaling by "dead" and "fractured" receptor protein-tyrosine kinases." *Bioessays* **23**(1): 69–76.

Kuhn, T. (1962). *The Structure of Scientific Revolutions*. Chicago, University of Chicago Press.

Kulesa, P. M., G. C. Cruywagen, et al. (1996). "On a model mechanism for the spatial patterning of teeth primordia in the alligator." *Journal of Theoretical Biology* **180**: 287–96.

Kumar, J. and K. Moses (1997). "Transcription factors in eye development: a gorgeous mosaic?" *Genes Dev* **11**(16): 2023–8.

Kumar, J. P. (2001). "Signalling pathways in Drosophila and vertebrate retinal development." *Nat Rev Genet* **2**(11): 846–57.

Kumar, J. P. and K. Moses (2001). "EGF receptor and Notch signaling act upstream of Eyeless/Pax6 to control eye specification." *Cell* **104**(5): 687–97.

Kumar, N. M. and N. B. Gilula (1996). "The gap junction communication channel." *Cell* **84**(3): 381–8.

Kurland, C. and J. Gallant (1996). "Errors of heterologous protein expression." *Curr Opin Biotechnol* **7**(5): 489–93.

Kurland, C. G. (1992). "Translational accuracy and the fitness of bacteria." *Annu Rev Genet* **26**: 29–50.

Ladich, F. (2000). "Acoustic communication and the evolution of hearing in fishes [In Process Citation]." *Philos Trans R Soc Lond B Biol Sci* **355**(1401): 1285–8.

Lage, P., Y. N. Jan, et al. (1997). "Requirement for EGF receptor signalling in neural recruitment during formation of Drosophila chordotonal sense organ clusters." *Curr Biol* **7**(3): 166–75.

Lagueux, M., E. Perrodou, et al. (2000). "Constitutive expression of a complement-like protein in toll and JAK gain-of-function mutants of drosophila [In Process Citation]." *Proc Natl Acad Sci U S A* **97**(21): 11427–32.

Lambert de Rouvroit, C. and A. M. Goffinet (2001). "Neuronal migration." *Mech Dev* **105**(1–2): 47–56.

Lander, E. S., L. M. Linton, et al. (2001). "Initial sequencing and analysis of the human genome." *Nature* **409**(6822): 860–921.

Lane, R. P., T. Cutforth, et al. (2001). "Genomic analysis of orthologous mouse and human olfactory receptor loci." *Proc Natl Acad Sci U S A* **98**(13): 7390–5.

Laudet, V. (1997). "Evolution of the nuclear receptor superfamily: early diversification from an ancestral orphan receptor." *J Mol Endocrinol* **19**(3): 207–26.

Laughlin, R. B., D. Pines, et al. (2000). "The middle way." *Proc Natl Acad Sci U S A* **97**(1): 32–7.

Laurent, G. (1999). "A systems perspective on early olfactory coding." *Science* **286**(5440): 723–8.

Lauter, N. and J. Doebley (2002). "Genetic variation for phenotypically invariant traits detected in teosinte: implications for the evolution of novel forms." *Genetics* **160**(1): 333–42.

Laux, T. (2003). "The stem cell concept in plants: a matter of debate." *Cell* **113**(3): 281–3.

Lawrence, P. A. (1992). *The making of a fly : the genetics of animal design*. Oxford [England]; Cambridge, Mass., USA, Blackwell Science.

Lechler, R. and A. N. Warrens (1999). *HLA in health and disease*. San Diego, Calif. London, Academic.

Lee, J. H., K. S. Cho, et al. (2001). "Diptericin-like protein: an immune response gene regulated by the anti-bacterial gene induction pathway in Drosophila." *Gene* **271**(2): 233–8.

Lehner, B., G. Williams, et al. (2002). "Antisense transcripts in the human genome." *Trends Genet* **18**(2): 63–5.

Leinders-Zufall, T., A. P. Lane, et al. (2000). "Ultrasensitive pheromone detection by mammalian vomeronasal neurons." *Nature* **405**(6788): 792–6.

Lemon, B., C. Inouye, et al. (2001). "Selectivity of chromatin-remodelling cofactors for ligand-activated transcription." *Nature* **414**: 924–8.

Lemon, B. and R. Tjian (2000). "Orchestrated response: a symphony of transcription factors for gene control." *Genes Dev* **14**(20): 2551–69.

Lenski, R. E., C. L. Winkworth, et al. (2003). "Rates of DNA Sequence Evolution in Experimental Populations of Escherichia coli During 20,000 Generations." *J Mol Evol* **56**(4): 498–508.

Lercher, M. J., A. O. Urrutia, et al. (2002). "Clustering of housekeeping genes provides a unified model of gene order in the human genome." *Nat Genet* **31**(2): 180–3.

Levine, S. S., A. Weiss, et al. (2002). "The core of the polycomb repressive complex is compositionally and functionally conserved in flies and humans." *Mol Cell Biol* **22**(17): 6070–8.

Lewin, B. (2000). *Genes VII*. New York, Oxford University Press.

Lewis, E. B. (1978). "A gene complex controlling segmentation in Drosophila." *Nature* **276**(5688): 565–70.

Lewontin, R. C. (1996). Evolution as engineering. *Integrative Approaches to Molecular Biology*. J. Collado-Vides, B. Magasanik and T. F. Smith. Cambridge, MA, MIT Press: 1–10.

Lewontin, R. C. (2000). *The Triple Helix*. Cambridge, MA, Harvard University Press.

Li, J., K. A. Lease, et al. (2001). "BRS1, a serine carboxypeptidase, regulates BRI1 signaling in Arabidopsis thaliana." *Proc Natl Acad Sci U S A* **98**(10): 5916–21.

Li, W.-H. (1997). *Molecular Evolution*. Sunderland, MA, Sinauer.

Liang, Z. and M. D. Biggin (1998). "Eve and ftz regulate a wide array of genes in blastoderm embryos: the selector homeoproteins directly or indirectly regulate most genes in Drosophila." *Development* **125**(22): 4471–82.

Lieb, J. D., X. Liu, et al. (2001). "Promoter-specific binding of Rap1 revealed by genome-wide maps of protein-DNA association." *Nat Genet* **28**(4): 327–34.

Liman, E. R., D. P. Corey, et al. (1999). "TRP2: a candidate transduction channel for mammalian pheromone sensory signaling." *Proc Natl Acad Sci U S A* **96**(10): 5791–6.

Liman, E. R. and H. Innan (2003). "Relaxed selective pressure on an essential component of pheromone transduction in primate evolution." *PNAS* **100**: 3328–32.

Lin, C. (2002). "Blue light receptors and signal transduction." *Plant Cell* **14 Suppl**: S207–25.

Liu, A. H., X. Zhang, et al. (2003). "Motif-based construction of a functional map for mammalian olfactory receptors." *Genomics* **81**(5): 443–56.

Liu, J. C., K. D. Makova, et al. (2001). "Episodic evolution of growth hormone in primates and emergence of the species specificity of human growth hormone receptor." *Mol Biol Evol* **18**(6): 945–53.

Loosli, F., S. Winkler, et al. (2001). "Medaka eyeless is the key factor linking retinal determination and eye growth." *Development* **128**(20): 4035–44.

Lopreato, G. F., Y. Lu, et al. (2001). "Evolution and divergence of sodium channel genes in vertebrates." *Proc Natl Acad Sci U S A* **98**(13): 7588–92.

Losick, R. and A. L. Sonenshein (2001). "Molecular biology. Turning gene regulation on its head." *Science* **293**(5537): 2018–19.

Lovejoy, A. O. (1936). *The great chain of being: a study of the history of an idea. The William James lectures delivered at Harvard university, 1933*. Cambridge, Mass., Harvard university press.

Lovelock, J. E. (1979). *Gaia: A New Look at Life on Earth*. Oxford, Oxford University Press.

Lovelock, J. E. (1988). *The Ages of Gaia: A Biography of our Living Earth*. New York, W.W. Norton.

Lovelock, J. E. and L. Margulis (1974). "Homeostatic tendencies of the earth's atmosphere." *Orig Life* **5**(1): 93–103.

Ludwig, M. Z., C. Bergman, et al. (2000). "Evidence for stabilizing selection in a eukaryotic enhancer element." *Nature* **403**(6769): 564–7.

Ludwig, M. Z. and M. Kreitman (1995). "Evolutionary dynamics of the enhancer region of even-skipped in Drosophila." *Mol Biol Evol* **12**(6): 1002–11.

Ludwig, M. Z., N. H. Patel, et al. (1998). "Functional analysis of eve stripe 2 enhancer evolution in Drosophila: rules governing conservation and change." *Development* **125**(5): 949–58.

Lumsden, A. and R. Krumlauf (1996). "Patterning the vertebrate neuraxis." *Science* **274**(5290): 1109–15.

Luo, C. C., W. H. Li, et al. (1986). "Structure and evolution of the apolipoprotein multigene family." *J Mol Biol* **187**(3): 325–40.

Lynch, M. and B. Walsh (1998). *Genetics and Analysis of Quantitative Traits*. Sunderland, MA, Sinauer.

Ma, H. and C. dePamphilis (2000). "The ABCs of floral evolution." *Cell* **101**(1): 5–8.

Mackay, T. F. (1995). "The genetic basis of quantitative variation: numbers of sensory bristles of Drosophila melanogaster as a model system." *Trends Genet* **11**(12): 464–70.

Mackay, T. F. (1996). "The nature of quantitativc genetic variation revisited: lessons from Drosophila bristles." *Bioessays* **18**(2): 113–21.

Mackay, T. F. (2001). "The genetic architecture of quantitative traits." *Annu Rev Genet* **35**: 303–39.

Mackay, T. F. (2001). "Quantitative trait loci in Drosophila." *Nat Rev Genet* **2**(1): 11–20.

MacLeod, K., A. Backer, et al. (1998). "Who reads temporal information contained across synchronized and oscillatory spike trains?" *Nature* **395**(6703): 693–8.

Maddison, D. (2003). Web of Life.

Magner, L. N. (1994). *A history of the life sciences*. New York, M. Dekker.

Malakoff, D. (1999). "Following the scent of avian olfaction." *Science* **286**(5440): 704–5.

Malhó, R., A. Moutinho, et al. (1998). "Spatial characteristics of calcium signalling: the calcium wave as a basic unit in plant cell calcium signalling." *Philosophical Transactions of the Royal Society of London B Biological Sciences* **353**(1374): 1463–73.

Manley, G. A. and C. Koppl (1998). "Phylogenetic development of the cochlea and its innervation." *Curr Opin Neurobiol* **8**(4): 468–74.

Marchalonis, J. J. and S. F. Schluter (1998). "A stochastic model for the rapid emergence of specific vertebrate immunity incorporating horizontal transfer of systems enabling duplication and combinational diversification." *J Theor Biol* **193**(3): 429–44.

Marchalonis, J. J., S. F. Schluter, et al. (1998). "Antibodies of sharks: revolution and evolution." *Immunol Rev* **166**: 103–22.

Marchalonis, J. J., S. F. Schluter, et al. (1998). "Phylogenetic emergence and molecular evolution of the immunoglobulin family." *Adv Immunol* **70**: 417–506.

Margulis, L. (1970). *Origin of eukaryotic cells : evidence and research implications for a theory of the origin and evolution of microbial, plant, and animal cells on the Precambrian earth*. New Haven, Yale University Press.

Marin, E. C., G. S. Jefferis, et al. (2002). "Representation of the glomerular olfactory map in the Drosophila brain." *Cell* **109**(2): 243–55.

Marin, O. and J. L. Rubenstein (2001). "A long, remarkable journey: tangential migration in the telencephalon." *Nat Rev Neurosci* **2**(11): 780–90.

Marin, O. and J. L. Rubenstein (2002). Patterning, regionalization, and cell differentiation in the forebrain. *Mouse Development: Patterning, Morphogenesis, and Organogenesis*. J. Rossant and P. P. L. Tam. San Diego, California, Academic Press: 75–106.

Marin, O. and J. L. Rubenstein (2003). "Cell Migration in the Forebrain." *Annu Rev Neurosci*.

Marshall, C. R., E. C. Raff, et al. (1994). "Dollo's law and the death and resurrection of genes." *Proc Natl Acad Sci U S A* **91**(25): 12283–7.

Maruska, K. and T. Tricas (1998). "Morphology of the mechanosensory lateral line system in the Atlantic stingray, *Dasyatis sabina:* the mechanotactile hypothesis." *Journal of Morphology* **238**: 1–22.

Mathers, P. H., A. Grinberg, et al. (1997). "The Rx homeobox gene is essential for vertebrate eye development." *Nature* **387**(6633): 603–7.

Matsunami, H. and L. B. Buck (1997). "A multigene family encoding a diverse array of putative pheromone receptors in mammals." *Cell* **90**(4): 775–84.

Matthews, G. G. (2001). *Neurobiology : molecules, cells, and systems*. Malden, MA, Blackwell Science.

Maynard Smith, J. and E. Szathmary (1995). *The Major Transitions in Evolution*. Oxford, Freeman.

Mayr, E. (1982). *The Growth of Biological Thought: Diversity, Evolution, and Inheritance*. Cambridge, MA, Belknap Press.

Mayser, P. (1882). "Vergleichend anatomische Studien uber das Gehirn der Knochenfisch besonderer Berucksichtigung der Cyprinoiden." *Zeitschrift fur wissenschaftliche zoologie* **36**: 259–64.

McAdams, H. H. and A. Arkin (1999). "It's a noisy business! Genetic regulation at the nanomolar scale." *Trends Genet* **15**(2): 65–9.
McCarty, D. R. and J. Chory (2000). "Conservation and innovation in plant signaling pathways." *Cell* **103**(2): 201–9.
McMillan, W. O., A. Monteiro, et al. (2002). "Development and evolution on the wing." *Trends in Ecology & Evolution* **17**: 125–33.
McPhee, J. (2002). *The Founding Fish*. New York, Farrar, Straus, Giroux.
Medzhitov, R. and C. A. Janeway, Jr. (1997). "Innate immunity: impact on the adaptive immune response." *Curr Opin Immunol* **9**(1): 4–9.
Medzhitov, R. and C. A. Janeway, Jr. (1998). "Innate immune recognition and control of adaptive immune responses." *Semin Immunol* **10**(5): 351–3.
Medzhitov, R. and C. A. Janeway, Jr. (2000). "How does the immune system distinguish self from nonself?" *Semin Immunol* **12**(3): 185–8; discussion 257–344.
Meinertzhagen, I. A. (2001). Plasticity in the insect nervous system. *Advances in Insect Physiology*. P. D. Evans. San Diego, CA, Academic Press. **28**: 85–••.
Meinhardt, H. (1996). "Models of biological pattern formation: common mechanism in plant and animal development." *Int J Dev Biol* **40**(1): 123–34.
Meinhardt, H. (2000). "Models for organizer and notochord formation." *C R Acad Sci III* **323**(1): 23–30.
Meinhardt, H. (2001). "Organizer and axes formation as a self-organizing process." *Int J Dev Biol* **45**(1 Spec No): 177–88.
Meinhardt, H. (2003). Theoretical aspects of pattern formation and neuronal development.
Meredith, M. (2001). "Human vomeronasal organ function: a critical review of best and worst cases." *Chem Senses* **26**(4): 433–45.
Meredith, M. A. and B. E. Stein (1986). "Visual, auditory, and somatosensory convergence on cells in superior colliculus results in multisensory integration." *J Neurophysiol* **56**(3): 640–62.
Metzger, R. J. and M. A. Krasnow (1999). "Genetic control of branching morphogenesis." *Science* **284**(5420): 1635–9.
Michelsen, A. (1992). Hearing and sound communication in small animals: evolutionary adaptations to the laws of physics. *The Evolutionary Biology of Hearing*. D. B. Webster, Fay, R. R. & Popper, A. N. New York, Springer: 61–78.
Michelson, A. M. (2002). "Deciphering genetic regulatory codes: A challenge for functional genomics." *Proc Natl Acad Sci U S A* **99**(2): 546–8.
Mighell, A. J., A. F. Markham, et al. (1997). "Alu sequences." *FEBS Lett* **417**(1): 1–5.
Miller, M. B. and B. L. Bassler (2001). "Quorum sensing in bacteria." *Annu Rev Microbiol* **55**: 165–99.
Millikan, R. (2002). "The changing face of epidemiology in the genomics era." *Epidemiology* **13**(4): in press.
Misteli, T. (2001). "The concept of self-organization in cellular architecture." *J Cell Biol* **155**(2): 181–5.
Misteli, T. (2001). "Protein dynamics: implications for nuclear architecture and gene expression." *Science* **291**(5505): 843–7.
Mittler, R., E. H. Herr, et al. (1999). "Transgenic tobacco plants with reduced capability to detoxify reactive oxygen intermediates are hyperresponsive to pathogen infection." *Proc Natl Acad Sci U S A* **96**(24): 14165–70.
Moalem, G., R. Leibowitz-Amit, et al. (1999). "Autoimmune T cells protect neurons from secondary degeneration after central nervous system axotomy." *Nat Med* **5**(1): 49–55.
Moller, S. G. and N. H. Chua (1999). "Interactions and intersections of plant signaling pathways." *J Mol Biol* **293**(2): 219–34.
Mollon, J. D. (1989). " 'Tho' she kneel'd in that place where they grew . . .' The uses and origins of primate colour vision." *J Exp Biol* **146**: 21–38.
Mombaerts, P. (1999). "Odorant receptor genes in humans." *Curr Opin Genet Dev* **9**(3): 315–20.

Mombaerts, P. (1999). "Seven-transmembrane proteins as odorant and chemosensory receptors." *Science* **286**(5440): 707–11.

Mombaerts, P. (2001). "How smell develops." *Nat Neurosci* **4 Suppl**: 1192–8.

Mombaerts, P. (2001). "The human repertoire of odorant receptor genes and pseudogenes." *Annu Rev Genomics Hum Genet* **2**: 493–510.

Mombaerts, P., F. Wang, et al. (1996). "Visualizing an olfactory sensory map." *Cell* **87**(4): 675–86.

Monod, J. (1971). *Chance and Necessity*. New York, Knopf.

Monteiro, A., J. Prijs, et al. (2003). "Mutants highlight the modular control of butterfly eyespot patterns." *Evolution & Development* **5**: 180–7.

Monteiro, A. F., P. M. Brakefield, et al. (1994). "The evolutionary genetics and developmental basis of wing pattern variation in the butterfly *Bicyclus anyana*." *Evolution* **48**: 1147–57.

Monuki, E. S., F. D. Porter, et al. (2001). "Patterning of the dorsal telencephalon and cerebral cortex by a roof plate-Lhx2 pathway." *Neuron* **32**(4): 591–604.

Monuki, E. S. and C. A. Walsh (2001). "Mechanisms of cerebral cortical patterning in mice and humans." *Nat Neurosci* **4 Suppl**: 1199–206.

Moore, J. A. (1987). *Form and function*. Science as a Way of Knowing, Boston, MA, American Society of Zoologists.

Moore, J. A. (1989). *A conceptual framework for biology. Part II*. Science as a Way of Knowing, Boston, MA, American Society of Zoologists.

Morgan, D., J. Goodship, et al. (2002). "The left-right determinant inversin has highly conserved ankyrin repeat and IQ domains and interacts with calmodulin." *Hum Genet* **110**(4): 377–84.

Moriwaki, Y., N. A. Begum, et al. (2001). "Mycobacterium bovis Bacillus Calmette-Guerin and its cell wall complex induce a novel lysosomal membrane protein, SIMPLE, that bridges the missing link between lipopolysaccharide and p53-inducible gene, LITAF(PIG7), and estrogen-inducible gene, EET-1." *J Biol Chem* **276**(25): 23065–76.

Morowitz, H. J., J. D. Kostelnik, et al. (2000). "The origin of intermediary metabolism." *Proc Natl Acad Sci U S A* **97**(14): 7704–8.

Mouillet, J. F., V. C. Henrich, et al. (2001). "Differential control of gene activity by isoforms A, B1 and B2 of the Drosophila ecdysone receptor." *Eur J Biochem* **268**(6): 1811–19.

Mountcastle, V. (1978). An organizing principle for cerebral function. *The Mindful Brain*. G. Edelman and V. Mountcastle. Cambridge (MA), MIT Press: 7–50.

Murray, J. (1993). *Mathematical Biology*. Berlin, Springer Verlag.

Mussig, C. and T. Altmann (2001). "Brassinosteroid signaling in plants." *Trends Endocrinol Metab* **12**(9): 398–402.

Nakanishi, T., K. Aoyagi, et al. (1999). "Specific cell-mediated immunity in fish." *Vet Immunol Immunopathol* **72**(1–2): 101–9.

Nakashita, H., M. Yasuda, et al. (2003). "Brassinosteroid functions in a broad range of disease resistance in tobacco and rice." *Plant J* **33**(5): 887–98.

Namba, R., T. M. Pazdera, et al. (1997). "Drosophila embryonic pattern repair: how embryos respond to bicoid dosage alteration." *Development* **124**(7): 1393–403.

Nathans, J. (1999). "The evolution and physiology of human color vision: insights from molecular genetic studies of visual pigments." *Neuron* **24**(2): 299–312.

Navarro, A. and N. H. Barton (2003). "Chromosomal speciation and molecular divergence–accelerated evolution in rearranged chromosomes." *Science* **300**(5617): 321–4.

Navarro, M. and K. Gull (2001). "A pol I transcriptional body associated with VSG monoallelic expression in Trypanosoma brucei." *Nature* **414**(6865): 759–63.

Neff, M. M., C. Fankhauser, et al. (2000). "Light: an indicator of time and place." *Genes Dev* **14**(3): 257–71.

Nei, M. (1987). *Molecular Evolutionary Genetics*. New York, Columbia University Press.

Nei, M. and S. Kumar (2000). *Molecular Evolution and Phylogenetics*. Oxford, Oxford University Press.

Neitz, J., M. Neitz, et al. (1996). "Visual pigment gene structure and the severity of color vision defects." *Science* **274**(5288): 801–4.
Neitz, M., J. Neitz, et al. (1995). "Polymorphism in the number of genes encoding long-wavelength-sensitive cone pigments among males with normal color vision." *Vision Res* **35**(17): 2395–407.
Ng, M. and M. F. Yanofsky (2000). "Three ways to learn the ABCs." *Curr Opin Plant Biol* **3**(1): 47–52.
Niehrs, C. (2001). "Developmental biology. Solving a sticky problem." *Nature* **413**(6858): 787–8.
Niehrs, C. and H. Meinhardt (2002). "Modular feedback." *Nature* **417**(6884): 35–6.
Nielsen, C. (1999). "Origin of the chordate central nervous system—and the origin of chordates." *Dev Genes Evol* **209**(3): 198–205.
Nielsen, M. G., E. Popodi, et al. (2003). "Evolutionary convergence in Otx expression in the pentameral adult rudiment in direct-developing sea urchins." *Dev Genes Evol* **213**(2): 73–82.
Nieuwenhuys, R., H. J. T. Donkelaar, et al. (1998). *The central nervous system of vertebrates*. New York, Springer.
Nijhout, H. F. (1991). *The Development and Evolution of Butterfly Wing Patterns*. Washington, Smithsonian Institution Press.
Nijhout, H. F. (1994). "Genes on the wing." *Science* **265**(5168): 44–5.
Nijhout, H. F. (1994). *Insect hormones*. Princeton, N.J., Princeton University Press.
Nilsson, D. E. (1996). "Eye ancestry: old genes for new eyes." *Curr Biol* **6**(1): 39–42.
Noncher, S., M. Maconochie, et al. (1996). "The conserved role of krox-zo in directing Hox gene expression during vertebrate hindbrain segmentation." *Proc Natl Acad Sci USA* **93**(18): 9339–45.
Nordborg, M. and S. Tavare (2002). "Linkage disequilibrium: what history has to tell us." *Trends Genet* **18**(2): 83–90.
Northcutt, R. G. and L. A. Barlow (1998). "Amphibians provide new insights into taste-bud development." *Trends Neurosci* **21**(1): 38–43.
Northcutt, R. G. and J. H. Kaas (1995). "The emergence and evolution of mammalian neocortex." *Trends Neurosci* **18**(9): 373–9.
Oberdörster, E., M. A. Clay, et al. (2001). "Common phytochemicals are ecdysteroid agonists and antagonists: a possible evolutionary link between vertebrate and invertebrate steroid hormones." *J Steroid Biochem Mol Biol* **77**(4–5): 229–38.
Obrist, M. K. and J. J. Wenstrup (1998). "Hearing and hunting in red bats (Lasiurus borealis, Vespertilionidae): audiogram and ear properties." *J Exp Biol* **201**(Pt 1): 143–54.
Ochiai, M. and M. Ashida (1999). "A pattern recognition protein for peptidoglycan. Cloning the cDNA and the gene of the silkworm, Bombyx mori." *J Biol Chem* **274**(17): 11854–8.
Ochman, H. (2001). "Lateral and oblique gene transfer." *Curr Opin Genet Dev* **11**(6): 616–19.
Ochman, H., J. G. Lawrence, et al. (2000). "Lateral gene transfer and the nature of bacterial innovation." *Nature* **405**(6784): 299–304.
O'Day, D. H. (2003). Hox gene expression web page.
Ohmori, H. and M. Hikida (1998). "Expression and function of recombination activating genes in mature B cells." *Crit Rev Immunol* **18**(3): 221–35.
Ohno, S. (1970). *Evolution by gene duplication*. Berlin; New York, Springer-Verlag.
Ohno, S. (2001). "The one-to-four rule and paralogues of sex-determining genes." ***Exs***(91): 1–10.
Ohta, T. (1992). "The nearly neutral theory of molecular evolution." *Annual Review of Ecology and Systematics* **23**: 263–86.
Ojeda, S. R. and J. E. Griffin (2000). Organization of the Endocrine System. *Textbook of Endocrine Physiology*. J. E. Griffin and S. R. Ojeda. Oxford, Oxford University Press: 3–18.
O'Leary, D. D. (1989). "Do cortical areas emerge from a protocortex?" *Trends Neurosci* **12**(10): 400–6.
O'Leary, D. D. and Y. Nakagawa (2002). "Patterning centers, regulatory genes and extrinsic mechanisms controlling arealization of the neocortex." *Curr Opin Neurobiol* **12**(1): 14–25.

Olivera, B. M., C. H. Walker, et al. (1999). Speciation of cone snails and interspecirfic hyperdivergence of their venom peptides. Potential evolutionary significance of introns. *Molecular Strategies in Biological Evolution*. L. H. Caporale. New York, New York Academy of Sciences: 223–37.

Oro, A. E. and M. P. Scott (1998). "Splitting hairs: dissecting roles of signaling systems in epidermal development." *Cell* **95**(5): 575–8.

Ostrer, H. (2001). "Invited review: sex-based differences in gene expression." *J Appl Physiol* **91**(5): 2384–8.

Ostrer, H. (2001). "Sex determination: lessons from families and embryos." *Clin Genet* **59**(4): 207–15.

Ouaked, F., W. Rozhon, et al. (2003). "A MAPK pathway mediates ethylene signaling in plants." *Embo J* **22**(6): 1282–8.

Ouellet, F., P. J. Overvoorde, et al. (2001). "IAA17/AXR3. Biochemical insight into an auxin mutant phenotype." *Plant Cell* **13**(4): 829–42.

Oyama, S. (2000). *The ontogeny of information: developmental systems and evolution*. Durham, Duke University Press.

Oyama, S., P. E. Griffiths, et al., Eds. (2001). *Cycles of Contingency: Developmental Systems and Evolution*. Cambridge, MA, MIT Press.

Paine, K. and D. R. Flower (2000). "The lipocalin website." *Biochim Biophys Acta* **1482**(1–2): 351–2.

Pallas, S. L. (2001). "Intrinsic and extrinsic factors that shape neocortical specification." *Trends Neurosci* **24**(7): 417–23.

Papadopoulos, D., D. Schneider, et al. (1999). "Genomic evolution during a 10,000-generation experiment with bacteria." *Proc Natl Acad Sci U S A* **96**(7): 3807–12.

Papatsenko, D., A. Nazina, et al. (2001). "A conserved regulatory element present in all Drosophila rhodopsin genes mediates Pax6 functions and participates in the fine-tuning of cell-specific expression." *Mech Dev* **101**(1–2): 143–53.

Parish, C. R. and E. R. O'Neill (1997). "Dependence of the adaptive immune response on innate immunity: some questions answered but new paradoxes emerge." *Immunol Cell Biol* **75**(6): 523–7.

Parker, G. A. and L. Partridge (1998). "Sexual conflict and speciation." *Philos Trans R Soc Lond B Biol Sci* **353**(1366): 261–74.

Parsek, M. R. and E. P. Greenberg (2000). "Acyl-homoserine lactone quorum sensing in gram-negative bacteria: a signaling mechanism involved in associations with higher organisms." *Proc Natl Acad Sci U S A* **97**(16): 8789–93.

Partridge, L. and L. D. Hurst (1998). "Sex and conflict." *Science* **281**(5385): 2003–8.

Pasini, A. and D. G. Wilkinson (2002). "Stabilizing the regionalisation of the developing vertebrate central nervous system." *Bioessays* **24**(5): 427–38.

Patel, K., H. Makarenkova, et al. (1999). "The role of long range, local and direct signalling molecules during chick feather bud development involving the BMPs, follistatin and the Eph receptor tyrosine kinase Eph-A4." *Mech Dev* **86**(1–2): 51–62.

Patthy, L. (1999). "Genome evolution and the evolution of exon-shuffling—a review." *Gene* **238**(1): 103–14.

Paul, W. (1999). *Fundamental Immunology*. Philadelphia, Lippincott-Raven.

Penn, D. and W. Potts (1999). "The evolution of mating preferences and major histocompatibility complex genes." *American Naturalist* **153**(2): 145–64.

Pennacchio, L. A. and E. M. Rubin (2001). "Genomic strategies to identify mammalian regulatory sequences." *Nat Rev Genet* **2**(2): 100–9.

Penninckx, I. A., K. Eggermont, et al. (1996). "Pathogen-induced systemic activation of a plant defensin gene in Arabidopsis follows a salicylic acid-independent pathway." *Plant Cell* **8**(12): 2309–23.

Petit, C. (1996). "Genes responsible for human hereditary deafness: symphony of a thousand." *Nat Genet* **14**(4): 385–91.

Petit, C., J. Levilliers, et al. (2001). "Molecular genetics of hearing loss." *Annu Rev Genet* **35**: 589–646.

Pichaud, F., A. Briscoe, et al. (1999). "Evolution of color vision." *Curr Opin Neurobiol* **9**(5): 622–7.

Pichaud, F., J. Treisman, et al. (2001). "Reinventing a common strategy for patterning the eye." *Cell* **105**(1): 9–12.

Pieterse, C. M., S. C. van Wees, et al. (1998). "A novel signaling pathway controlling induced systemic resistance in Arabidopsis." *Plant Cell* **10**(9): 1571–80.

Pineda, D., J. Gonzalez, et al. (2000). "Searching for the prototypic eye genetic network: Sine oculis is essential for eye regeneration in planarians." *Proc Natl Acad Sci U S A* **97**(9): 4525–9.

Pineda, D., L. Rossi, et al. (2002). "The genetic network of prototypic planarian eye regeneration is Pax6 independent." *Development* **129**(6): 1423–34.

Plath, K., S. Mlynarczyk-Evans, et al. (2002). "Xist RNA and the mechanism of x chromosome inactivation." *Annu Rev Genet* **36**: 233–78.

Pollack, G. (1998). Neural Processing of Acoustic Sound. *Comparative Hearing: Insects*. R. R. Hoy, Popper, A.N., Fay, R.R. New York, Springer: 139–96.

Pollock, J. A. and S. Benzer (1988). "Transcript localization of four opsin genes in the three visual organs of Drosophila; RH2 is ocellus specific." *Nature* **333**(6175): 779–82.

Popodi, E. and R. A. Raff (2001). "Hox genes in a pentameral animal." *Bioessays* **23**(3): 211–14.

Popper, A. N. and R. R. Fay (1997). "Evolution of the ear and hearing: issues and questions." *Brain Behav Evol* **50**(4): 213–21.

Potts, W. K. (2002). "Wisdom through immunogenetics." *Nature Genetics* **30**: 130–1.

Pratt, L. A. and R. Kolter (1999). "Genetic analyses of bacterial biofilm formation." *Curr Opin Microbiol* **2**(6): 598–603.

Preston, CA, Laue G, et al. (2001). "Methyl jasmonate is blowing in the wind, but can it act as a plant-plant airborne signal?" *Biochemical Systematics and Ecology* **29**: 1007–3.

Prigent-Combaret, C., O. Vidal, et al. (1999). "Abiotic surface sensing and biofilm-dependent regulation of gene expression in Escherichia coli." *J Bacteriol* **181**(19): 5993–6002.

Proteases (2003). Introduction to the proteases.

Provencio, I., I. R. Rodriguez, et al. (2000). "A novel human opsin in the inner retina." *J Neurosci* **20**(2): 600–5.

Provencio, I., M. D. Rollag, et al. (2002). "Photoreceptive net in the mammalian retina. This mesh of cells may explain how some blind mice can still tell day from night." *Nature* **415**(6871): 493.

Puelles, L., E. Kuwana, et al. (1999). "Comparison of the mammalian and avian telencephalon from the perspective of gene expression data." *Eur J Morphol* **37**(2–3): 139–50.

Punzo, C., S. Kurata, et al. (2001). "The eyeless homeodomain is dispensable for eye development in Drosophila." *Genes Dev* **15**(13): 1716–23.

Punzo, C., M. Seimiya, et al. (2002). "Differential interactions of eyeless and twin of eyeless with the sine oculis enhancer." *Development* **129**(3): 625–34.

Quail, P. H. (2002). "Phytochrome photosensory signalling networks." *Nat Rev Mol Cell Biol* **3**(2): 85–93.

Quax-Jeuken, Y., S. Bruisten, et al. (1985). "Evolution of crystallins: expression of lens-specific proteins in the blind mammals mole (Talpa europaea) and mole rat (Spalax ehrenbergi)." *Mol Biol Evol* **2**(4): 279–88.

Raff, R. A. (1996). *The Shape of Life: Genes, Development, and the Evolution of Animal Form*. Chicago, University of Chicago Press.

Raff, R. A. (1999). "Larval homologies and radical evolutionary changes in early development." *Novartis Found Symp* **222**: 110–21; discussion 121–4.

Rajavashisth, T. B., J. S. Kaptein, et al. (1985). "Evolution of apolipoprotein E: mouse sequence and evidence for an 11-nucleotide ancestral unit." *Proc Natl Acad Sci U S A* **82**(23): 8085–9.

Rakic, P. (1988). "Specification of cerebral cortical areas." *Science* **241**(4862): 170–6.
Rakic, P. and P. J. Lombroso (1998). "Development of the cerebral cortex: I. Forming the cortical structure." *J Am Acad Child Adolesc Psychiatry* **37**(1): 116–17.
Ranganathan, R., D. M. Malicki, et al. (1995). "Signal transduction in Drosophila photoreceptors." *Annu Rev Neurosci* **18**: 283–317.
Rawson, N. E., J. Eberwine, et al. (2000). "Expression of mRNAs encoding for two different olfactory receptors in a subset of olfactory receptor neurons." *J Neurochem* **75**(1): 185–95.
Redies, C. and L. Puelles (2001). "Modularity in vertebrate brain development and evolution." *Bioessays* **23**(12): 1100–11.
Reenan, R. A. (2001). "The RNA world meets behavior: AfiI pre-mRNA editing in animals." *Trends Genet* **17**(2): 53–6.
Renfree, M. B. and G. Shaw (2001). "Germ cells, gonads and sex reversal in marsupials." *Int J Dev Biol* **45**(3 Spec No): 557–67.
Reymond, P., H. Weber, et al. (2000). "Differential gene expression in response to mechanical wounding and insect feeding in Arabidopsis." *Plant Cell* **12**(5): 707–20.
Reynolds, S. J. (1913). *The Vertebrate Skeleton*. Cambridge, England, Cambridge University Press.
Rice, D. S. and T. Curran (2001). "Role of the reelin signaling pathway in central nervous system development." *Annu Rev Neurosci* **24**: 1005–39.
Richards, R. J. (1992). *The Meaning of Evolution: The Morphological Construction and Ideological Reconstruction of Darwin's Theory*. Chicago, University of Chicago Press.
Richardson, M. K., J. Hanken, et al. (1997). "There is no highly conserved embryonic stage in the vertebrates: implications for current theories of evolution and development." *Anat Embryol (Berl)* **196**(2): 91–106.
Richardson, M. K. and G. Keuck (2001). "A question of intent: when is a 'schematic' illustration a fraud?" *Nature* **410**(6825): 144.
Richardson, M. K. and G. Keuck (2002). "Haeckel's ABC of evolution and development." *Biol Rev Camb Philos Soc* **77**(4): 495–528.
Richter, T. E. and P. C. Ronald (2000). "The evolution of disease resistance genes." *Plant Mol Biol* **42**(1): 195–204.
Rieseberg, L. H. and K. Livingstone (2003). "Evolution. Chromosomal speciation in primates." *Science* **300**(5617): 267–8.
Rivera-Pomar, R., D. Niessing, et al. (1996). "RNA binding and translational suppression by bicoid." *Nature* **379**(6567): 746–9.
Robert, J. S. (2001). "Interpreting the homeobox: metaphors of gene action and activation in development and evolution." *Evol Dev* **3**(4): 287–95.
Robertson, H. M., R. Martos, et al. (1999). "Diversity of odourant binding proteins revealed by an expressed sequence tag project on male Manduca sexta moth antennae." *Insect Mol Biol* **8**(4): 501–18.
Roitt, I. M., J. Brostoff, et al. (1998). *Immunology*. London; Philadelphia, Mosby.
Romer, A. S. and L. I. Price (1940). "Review of the Pelycosauria." *Geological Society of America, Special Papers* **28**: 1–538.
Ronneberg, T. A., L. F. Landweber, et al. (2000). "Testing a biosynthetic theory of the genetic code: fact or artifact?" *Proc Natl Acad Sci U S A* **97**(25): 13690–5.
Rosenberg, N. A. and M. Nordborg (2002). "Genealogical trees, coalescent theory and the analysis of genetic polymorphisms." *Nat Rev Genet* **3**(5): 380–90.
Rouquier, S., A. Blancher, et al. (2000). "The olfactory receptor gene repertoire in primates and mouse: evidence for reduction of the functional fraction in primates." *Proc Natl Acad Sci U S A* **97**(6): 2870–4.
Rouse, D., P. Mackay, et al. (1998). "Changes in auxin response from mutations in an AUX/IAA gene." *Science* **279**(5355): 1371–3.
Royle, S. J. and R. D. Murrell-Lagnado (2003). "Constitutive cycling: a general mechanism to regulate cell surface proteins." *Bioessays* **25**(1): 39–46.

Ruse, M. (1996). *Monad to man: the concept of progress in evolutionary biology*. Cambridge, Mass., Harvard University Press.

Rutherford, S. L. and S. Henikoff (2003). "Quantitative epigenetics." *Nat Genet* **33**(1): 6–8.

Sadikov, Z. T., Z. Saatov, et al. (2000). "Sileneoside H, a new phytoecdysteroid from Silene brahuica." *J Nat Prod* **63**(7): 987–8.

Salazar-Ciudad, I. and J. Jernvall (2002). "A gene network model accounting for development and evolution of mammalian teeth." *Proc Natl Acad Sci U S A* **99**(12): 8116–20.

Salvini-Plawin, L. and E. Mayr (1961). "On the evolution of photo receptors and eyes." *Evolutionary Biology* **10**: 207–63.

Samakovlis, C., N. Hacohen, et al. (1996). "Development of the Drosophila tracheal system occurs by a series of morphologically distinct but genetically coupled branching events." *Development* **122**(5): 1395–407.

Sartre, J. P. and S. Gilbert (1947). *No exit (Huis clos): a play in one act, & The flies (Les mouches) a play in three acts*. New York, A. A. Knopf.

Sauer, F., R. Rivera-Pomar, et al. (1996). "Gene regulation in the Drosophila embryo." *Philos Trans R Soc Lond B Biol Sci* **351**(1339): 579–87.

Scalia, F., A. C. Grant, et al. (1995). "Functional properties of regenerated optic axons terminating in the primary olfactory cortex." *Brain Res* **685**(1–2): 187–97.

Schlichting, C. and M. Pigliucci (1998). *Phenotypic evolution : a reaction norm perspective*. Sunderland, Mass., Sinauer.

Schopf, J. W. (1994). "Disparate rates, differing fates: tempo and mode of evolution changed from the Precambrian to the Phanerozoic." *Proc Natl Acad Sci U S A* **91**(15): 6735–42.

Schopf, J. W. (2000). "Solution to Darwin's dilemma: discovery of the missing Precambrian record of life." *Proc Natl Acad Sci U S A* **97**(13): 6947–53.

Schopf, J. W. (1992). *Major events in the history of life*. Boston, Jones and Bartlett Publishers.

Schroder, R. (2003). "The genes orthodenticle and hunchback substitute for bicoid in the beetle Tribolium." *Nature* **422**(6932): 621–5.

Schumaker, K. and M. Dietrich (1998). "Hormone-inducedd signaling during moss development." *Annu. Rev. Plant Physiol. Plant Mol. Biol.* **49**: 501–23.

Schuster, P. (2000). "Taming combinatorial explosion." *Proc Natl Acad Sci U S A* **97**(14): 7678–80.

Schutt, C. and R. Nothiger (2000). "Structure, function and evolution of sex-determining systems in Dipteran insects." *Development* **127**(4): 667–77.

Schuurmans, C. and F. Guillemot (2002). "Molecular mechanisms underlying cell fate specification in the developing telencephalon." *Curr Opin Neurobiol* **12**(1): 26–34.

Schwartz, S., Z. Zhang, et al. (2000). "PipMaker—a web server for aligning two genomic DNA sequences." *Genome Res* **10**(4): 577–86.

Scott, K., R. Brady, Jr., et al. (2001). "A chemosensory gene family encoding candidate gustatory and olfactory receptors in Drosophila." *Cell* **104**(5): 661–73.

Scott, M. P. (1994). "Intimations of a creature." *Cell* **79**(7): 1121–4.

Scriver, C. R. (2002). "Does hereditary metabolic disease modulate senescence and ageing?" *J Inherit Metab Dis* **25**(3): 235–51.

Segura, A., M. Moreno, et al. (1999). "Snakin-1, a peptide from potato that is active against plant pathogens." *Mol Plant Microbe Interact* **12**(1): 16–23.

Seimiya, M. and W. J. Gehring (2000). "The Drosophila homeobox gene optix is capable of inducing ectopic eyes by an eyeless-independent mechanism." *Development* **127**(9): 1879–86.

Sempere, L. F., E. B. Dubrovsky, et al. (2002). "The expression of the let-7 small regulatory RNA is controlled by ecdysone during metamorphosis in Drosophila melanogaster." *Dev Biol* **244**(1): 170–9.

Sengupta, A. M., M. Djordjevic, et al. (2002). "Specificity and robustness in transcription control networks." *Proc Natl Acad Sci U S A* **99**(4): 2072–7.

Serizawa, S., T. Ishii, et al. (2000). "Mutually exclusive expression of odorant receptor transgenes." *Nat Neurosci* **3**(7): 687–93.

Sham, P. (1997). *Statistics in Human Genetics*. Oxford, Oxford University Press.

Shanbhag, S. R., S. K. Park, et al. (2001). "Gustatory organs of Drosophila melanogaster: fine structure and expression of the putative odorant-binding protein PBPRP2." *Cell Tissue Res* **304**(3): 423–37.

Shapiro, J. A. (1998). "Thinking about bacterial populations as multicellular organisms." *Annu Rev Microbiol* **52**: 81–104.

Sharp, P. A. (1999). "RNAi and double-strand RNA." *Genes & Development* **13**: 139–41.

Shashikant, C. S., C. B. Kim, et al. (1998). "Comparative studies on mammalian Hoxc8 early enhancer sequence reveal a baleen whale-specific deletion of a cis-acting element." *Proc Natl Acad Sci U S A* **95**(26): 15446–51.

Sheng, M. (2001). "Molecular organization of the postsynaptic specialization." *Proc Natl Acad Sci U S A* **98**(13): 7058–61.

Sheng, M. and S. H. Lee (2001). "AMPA receptor trafficking and the control of synaptic transmission." *Cell* **105**(7): 825–8.

Sheng, M. and C. Sala (2001). "PDZ domains and the organization of supramolecular complexes." *Annu Rev Neurosci* **24**: 1–29.

Shimmin, L. C., J. Miller, et al. (1998). "Contrasting levels of DNA polymorphism at the autosomal and X-linked visual color pigment loci in humans and squirrel monkeys." *Mol Biol Evol* **15**(4): 449–55.

Siegal, M. L. and A. Bergman (2002). "Waddington's canalization revisited: developmental stability and evolution." *Proc Natl Acad Sci U S A* **99**(16): 10528–32.

Simeone, A. (2000). "Positioning the isthmic organizer where Otx2 and Gbx2meet." *Trends Genet* **16**(6): 237–40.

Sineshchekov, O. A., K. H. Jung, et al. (2002). "Two rhodopsins mediate phototaxis to low- and high-intensity light in Chlamydomonas reinhardtii." *Proc Natl Acad Sci U S A* **99**(13): 8689–94.

Singh, N., F. A. Ebrahimi, et al. (2003). "Coordination of the random asynchronous replication of autosomal loci." *Nat Genet* **33**(3): 339–41.

Skoufos, E., M. D. Healy, et al. (1999). "Olfactory Receptor Database: a database of the largest eukaryotic gene family." *Nucleic Acids Res* **27**(1): 343–5.

Smith, W. C., D. A. Price, et al. (1993). "Opsins from the lateral eyes and ocelli of the horseshoe crab, Limulus polyphemus." *Proc Natl Acad Sci U S A* **90**(13): 6150–4.

Smulders, R. H., M. A. van Dijk, et al. (2002). "The eye lens protein alphaA-crystallin of the blind mole rat Spalax ehrenbergi: effects of altered functional constraints." *Exp Eye Res* **74**(2): 285–91.

Sollars, V., X. Lu, et al. (2003). "Evidence for an epigenetic mechanism by which Hsp90 acts as a capacitor for morphological evolution." *Nat Genet* **33**(1): 70–4.

Sosinsky, A., G. Glusman, et al. (2000). "The genomic structure of human olfactory receptor genes." *Genomics* **70**(1): 49–61.

Spellman, P. T. and G. M. Rubin (2002). "Evidence for large domains of similarly expressed genes in the Drosophila genome." *J Biol* **1**(1): 5.

Sponges (2002). "Introduction to porifera."

Springer, M. S., E. C. Teeling, et al. (2001). "Integrated fossil and molecular data reconstruct bat echolocation." *Proc Natl Acad Sci U S A* **98**(11): 6241–6.

Srinivasan, S., K. E. Rashka, et al. (2002). "Creation of a Sog morphogen gradient in the Drosophila embryo." *Dev Cell* **2**(1): 91–101.

Stein, B. E., W. Jiang, et al. (2001). "Nonvisual influences on visual-information processing in the superior colliculus." *Prog Brain Res* **134**: 143–56.

Steinberg, M. (1998). "Goal-directedness in embryonic development." *Integrative Biology* **1**: 49–59.

Stenoien, D. L., S. Simeoni, et al. (2000). "Subnuclear dynamics and transcription factor function." *J Cell Biochem* **Suppl 35**: 99–106.

Stepanova, A. N. and J. R. Ecker (2000). "Ethylene signaling: from mutants to molecules." *Curr Opin Plant Biol* **3**(5): 353–60.

Steven, D. M. (1963). "The dermal light sense." *Biol Rev Camb Philos Soc* **38**(2): 204–40.

Stewart, P. S. (2002). "Mechanisms of antibiotic resistance in bacterial biofilms." *Int J Med Microbiol* **292**(2): 107–13.

Stewart, P. S. and J. W. Costerton (2001). "Antibiotic resistance of bacteria in biofilms." *Lancet* **358**(9276): 135–8.

Stocker, R. F. (1994). "The organization of the chemosensory system in Drosophila melanogaster: a review." *Cell Tissue Res* **275**(1): 3–26.

Stowers, L., T. E. Holy, et al. (2002). "Loss of sex discrimination and male-male aggression in mice deficient for TRP2." *Science* **295**(5559): 1493–500.

Strausfeld, N. J. and J. G. Hildebrand (1999). "Olfactory systems: common design, uncommon origins?" *Curr Opin Neurobiol* **9**(5): 634–9.

Struhl, K. (2001). "Gene regulation. A paradigm for precision." *Science* **293**(5532): 1054–5.

Stumpner, A. and D. von Helversen (2001). "Evolution and function of auditory systems in insects." *Naturwissenschaften* **88**(4): 159–70.

Suemori, H. and S. Noguchi (2000). "Hox C cluster genes are dispensable for overall body plan of mouse embryonic development." *Dev Biol* **220**(2): 333–42.

Sun, X., H. D. Le, et al. (2003). "Sequence analysis of a functional Drosophila centromere." *Genome Res* **13**(2): 182–94.

Sun, X., F. V. Mariani, et al. (2002). "Functions of FGF signalling from the apical ectodermal ridge in limb development." *Nature* **418**(6897): 501–8.

Sur, M. and C. A. Leamey (2001). "Development and plasticity of cortical areas and networks." *Nat Rev Neurosci* **2**(4): 251–62.

Sussex, I. M. and N. M. Kerk (2001). "The evolution of plant architecture." *Current Opinion in Plant BIology* **4**: 33–7.

Sutherland, D., C. Samakovlis, et al. (1996). "branchless encodes a Drosophila FGF homolog that controls tracheal cell migration and the pattern of branching." *Cell* **87**(6): 1091–101.

Suzuki, M. G., T. Shimada, et al. (1998). "Absence of dosage compensation at the transcription level of a sex-linked gene in a female heterogametic insect, Bombyx mori." *Heredity* **81**(Pt 3): 275–83.

Tabata, T. (2001). "Genetics of morphogen gradients." *Nat Rev Genet* **2**(8): 620–30.

Takeuchi, O. and S. Akira (2001). "Toll-like receptors; their physiological role and signal transduction system." *Int Immunopharmacol* **1**(4): 625–35.

Takeuchi, O., K. Hoshino, et al. (1999). "Differential roles of TLR2 and TLR4 in recognition of gram-negative and gram-positive bacterial cell wall components." *Immunity* **11**(4): 443–51.

Tamayo, P., D. Slonim, et al. (1999). "Interpreting patterns of gene expression with self-organizing maps: methods and application to hematopoietic differentiation." *Proc Natl Acad Sci U S A* **96**(6): 2907–12.

Tautz, D. (2000). "A genetic uncertainty problem." *Trends Genet* **16**(11): 475–7.

Teeling, E. C., O. Madsen, et al. (2002). "Microbat paraphyly and the convergent evolution of a key innovation in Old World rhinolophoid microbats." *Proc Natl Acad Sci U S A* **99**(3): 1431–6.

Teeling, E. C., M. Scally, et al. (2000). "Molecular evidence regarding the origin of echolocation and flight in bats." *Nature* **403**(6766): 188–92.

Tegoni, M., P. Pelosi, et al. (2000). "Mammalian odorant binding proteins." *Biochim Biophys Acta* **1482**(1–2): 229–40.

Teleman, A. A., M. Strigini, et al. (2001). "Shaping morphogen gradients." *Cell* **105**(5): 559–62.

Terai, Y., W. E. Mayer, et al. (2002). "The effect of selection on a long wavelength-sensitive (LWS) opsin gene of Lake Victoria cichlid fishes." *Proc Natl Acad Sci U S A* **99**(24): 15501–6.

Thesleff, I., S. Keranen, et al. (2001). "Enamel knots as signaling centers linking tooth morphogenesis and odontoblast differentiation." *Adv Dent Res* **15**: 14–18.

Thomas, M. A., K. A. Walsh, et al. (2000). "Molecular phylogenetic analysis of evolutionary trends in stonefly wing structure and locomotor behavior." *Proc Natl Acad Sci U S A* **97**(24): 13178–83.

Thompson, D. A. W. (1917). *On Growth and Form*. Cambridge, England, Cambridge University Press.

Thornton, J. W. (2001). "Evolution of vertebrate steroid receptors from an ancestral estrogen receptor by ligand exploitation and serial genome expansions." *Proc Natl Acad Sci U S A* **98**(10): 5671–6.

Thornton, J. W. and D. B. Kelley (1998). "Evolution of the androgen receptor: structure-function implications." *Bioessays* **20**(10): 860–9.

Tijsterman, M., R. F. Ketting, et al. (2002). "The genetics of RNA silencing." *Annu Rev Genet* **36**: 489–519.

Tomarev, S. I. and J. Piatigorsky (1996). "Lens crystallins of invertebrates—diversity and recruitment from detoxification enzymes and novel proteins." *Eur J Biochem* **235**(3): 449–65.

Tomlinson, H. M. (1912). *The Sea and the Jungle*. New York, Dutton.

Trainor, P. A. and R. Krumlauf (2000). "Patterning the cranial neural crest: hindbrain segmentation and Hox gene plasticity." *Nat Rev Neurosci* **1**(2): 116–24.

Transfac (2003). The transcription factor database.

Travisano, M., J. A. Mongold, et al. (1995). "Experimental tests of the roles of adaptation, chance, and history in evolution." *Science* **267**(5194): 87–90.

Treisman, J. E. (1999). "A conserved blueprint for the eye?" *Bioessays* **21**(10): 843–50.

Trifonov, E. N. (1999). Elucidating sequence codes: three codes for evolution. *Molecular Strategies in Biological Evolution*. L. H. Caporale. New York, New York Academy of Sciences: 330–8.

Troemel, E. R. (1999). "Chemosensory signaling in C. elegans." *Bioessays* **21**(12): 1011–20.

True, J. R. and E. S. Haag (2001). "Developmental system drift and flexibility in evolutionary trajectories." *Evol Dev* **3**(2): 109–19.

Tsuboi, A., S. Yoshihara, et al. (1999). "Olfactory neurons expressing closely linked and homologous odorant receptor genes tend to project their axons to neighboring glomeruli on the olfactory bulb." *J Neurosci* **19**(19): 8409–18.

Tudge, C. (2000). *The Variety of Life: A Survey and a Celebration of all the Creatures that Have Ever Lived*. Oxford, Oxford University Press.

Turing, A. (1952). "The chemical basis of morphogenesis." *Philosophical Transactions of the Royal Society of London, Series B* **237**: 37–72.

Turner, J. S. (2000). *The Extended Organism: The Physiology of Animal-Built Structures*. Cambridge, MA, Harvard University Press.

Tyler-Smith, C. and G. Floridia (2000). "Many paths to the top of the mountain: diverse evolutionary solutions to centromere structure." *Cell* **102**(1): 5–8.

Urrutia, A. O. and L. D. Hurst (2001). "Codon usage bias covaries with expression breadth and the rate of synonymous evolution in humans, but this is not evidence for selection." *Genetics* **159**(3): 1191–9.

Van Camp, G. and R. Smith (2003). Heridtary Hearing Loss Homepage. **2003**.

van den Biggelaar, J. A. M., E. Edinger-Gonzales, et al. (2002). "The improbability of dorso-ventral axis inversion during animal evolution, as presumed by Geoffroy Saint Hilaire." *Contributions to Zoology* **71**: 29–36.

Veitch, E., J. Begbie, et al. (1999). "Pharyngeal arch patterning in the absence of neural crest." *Curr Biol* **9**(24): 1481–4.

Vellai, T. and G. Vida (1999). "The origin of eukaryotes: the difference between prokaryotic and eukaryotic cells." *Proc R Soc Lond B Biol Sci* **266**(1428): 1571–7.

Vesalius, A. (1543). *De Humani corporis fabrica liborum, epitome*. Basel, Oporinus (Copied from Dover reprint edition).

Via, S. (2001). "Sympatric speciation in animals: the ugly duckling grows up." *Trends Ecol Evol* **16**(7): 381–90.
Voiculescu, O., E. Taillebourg, et al. (2001). "Hindbrain patterning: Krox20 couples segmentation and specification of regional identity." *Development* **128**(24): 4967–78.
von Bubnoff, A. and K. W. Cho (2001). "Intracellular BMP signaling regulation in vertebrates: pathway or network?" *Dev Biol* **239**(1): 1–14.
Vosshall, L. B. (2000). "Olfaction in Drosophila." *Curr Opin Neurobiol* **10**(4): 498–503.
Vosshall, L. B. (2001). "The molecular logic of olfaction in Drosophila." *Chem Senses* **26**(2): 207–13.
Vosshall, L. B., H. Amrein, et al. (1999). "A spatial map of olfactory receptor expression in the Drosophila antenna." *Cell* **96**(5): 725–36.
Vossilenko, K. (2003). tRNA sequences.
Waddell, S. and W. G. Quinn (2001). "What can we teach Drosophila? What can they teach us?" *Trends Genet* **17**(12): 719–26.
Waddington, C. H. (1942). "Canalization of development and the inheritance of acquired characters." *Nature* **150**: 563–5.
Waddington, C. H. (1953). "Genetic assimilation of an acquired character." *Evolution* **7**: 118–26.
Waddington, C. H. (1956). "Geneticv assimilationof bithorax phenotype." *Evolution* **10**: 1–13.
Waddington, C. H. (1957). *The Strategy of the Genes: A Discussion of Some Aspects of Theoretical Biology*. London, George Allen & Unwin.
Wagner, A. (2000). "Robustness against mutations in genetic networks of yeast." *Nat Genet* **24**(4): 355–61.
Wagner, G., G. Booth, et al. (1997). "A population genetics thoery of canalization." *Evolution* **51**: 329–47.
Wagner, G. P. (1996). "Homologues, natural kinds and the evolution of modularity." *American Zoologist* **36**: 36–43.
Wagner, G. P. (1999). "A research programme for testing the biological homology concept." *Novartis Found Symp* **222**: 125–34; discussion 134–40.
Wagner, G. P. and J. A. Gauthier (1999). "1,2,3 = 2,3,4: a solution to the problem of the homology of the digits in the avian hand." *Proc Natl Acad Sci U S A* **96**(9): 5111–16.
Wagner, G. P. and B. Y. Misof (1993). "How can a character be developmentally constrained despite variation in developmental pathways?" *Journal of Evolutionary Biology* (449–455).
Wake, M. H. (1979). *Hyman's Comparative Vertebrate Anatomy*. Chicago, University of Chicago.
Walcott, C. D. (1918). "Cambrian geology and paleontology. IV. No. 4. Appendages of trilobites." *Smithsonian Miscellaneous Collections* **67**(4).
Walker, R. G., A. T. Willingham, et al. (2000). "A Drosophila mechanosensory transduction channel." *Science* **287**(5461): 2229–34.
Wallace, M. T. and B. E. Stein (2001). "Sensory and multisensory responses in the newborn monkey superior colliculus." *J Neurosci* **21**(22): 8886–94.
Wallis, M. (2000). "Episodic evolution of protein hormones: molecular evolution of pituitary prolactin." *J Mol Evol* **50**(5): 465–73.
Wang, Y., P. M. Smallwood, et al. (1999). "Mutually exclusive expression of human red and green visual pigment-reporter transgenes occurs at high frequency in murine cone photoreceptors." *Proc Natl Acad Sci U S A* **96**(9): 5251–6.
Wang, Z. Y., H. Seto, et al. (2001). "BRI1 is a critical component of a plasma-membrane receptor for plant steroids." *Nature* **410**(6826): 380–3.
Webster, D. B., R. R. Fay, et al. (1992). *The Evolutionary biology of hearing*. New York, Springer-Verlag.

Webster, G. (1992). William Bateson and the science of form. *Materials for the Study of Variation: Treated with especial regard to discontinuity int he origin of species*. W. Bateson. Baltimore, Johns Hopkins University Press: xxix–lix.

Webster, S. G. (1998). Neuropeptides inhibiting growth and reproduction in crustaceans. *Recent Advances in Arthropod Endocrinology*. G. M. Coast and S. G. Webster. Cambridge, Cambridge University Press: 32–52.

Weigel, D. and G. Jurgens (2002). "Stem cells that make stems." *Nature* **415**(6873): 751–4.

Weigel, D. and E. M. Meyerowitz (1994). "The ABCs of floral homeotic genes." *Cell* **78**(2): 203–9.

Weismann, M., I. Yousry, et al. (2001). "Functional magnetic resonance imaging of human olfaction." *Neuroimaging Clin N Am* **11**(2): 237–50, viii.

Weiss, K. M. (2002a). "Come to me my melancholic baby!" *Evolutionary Anthropology* **12**: 3–6.

Weiss, K. M. (2002b). "Goings on in Mendel's garden." *Evolutionary Anthropology* **11**: 40–4.

Weiss, K. M. (2002c). "How the eye got its brain." *Evolutionary Anthropology* **11**: 215–19.

Weiss, K. M. (2002d). "Is the message the medium: biological traits and their regulation." *Evolutionary Anthropology* **11**(3): 88–93.

Weiss, K. M. (2002e). Phenotype and genotype. *Keywords and Concepts in Evolutionary Developmental Biology*. B. H. Hall and W. Olson. Cambridge, MA, Harvard University Press: 279–88.

Weiss, K. M. (2003a). "Good vibrations: the silent symphony of life." *Evolutionary Anthropology* **11**: 176–82.

Weiss, K. M. (2003b). "Ludwik Fleck and the art-of-fact." *Evolutionary Anthropology* **12**: in press.

Weiss, K. M. (2003c). Phenotype and genotype. *Keywords in Evolution and Development*. B. Hall, Olson, W. Cambridge, MA, Harvard University Press: 279–88.

Weiss, K. M. and K. Aldridge (2003). "What stamps the wrinkle deeper on the brow?" *Evolutionary Anthropology* **12**: 205–10.

Weiss, K. M. and A. Buchanan (2003). "Evolution by phenotype." *Perspectives in Biology and Medicine* **46**: 159–82.

Weiss, K. M. and S. M. Fullerton (2000). "Phenogenetic drift and the evolution of genotype-phenotype relationships." *Theor Popul Biol* **57**(3): 187–95.

Weiss, K. M. and J. D. Terwilliger (2000). "How many diseases does it take to map a gene with SNPs?" *Nat Genet* **26**(2): 151–7.

Weiss, K. M., Z. Zhao, et al. (1998). "Dynamic interactions and dental patterning." *Critical Reviews in Oral Biology and Medicine* **9**: 369–98.

Welch, D. M. and M. Meselson (2000). "Evidence for the evolution of bdelloid rotifers without sexual reproduction or genetic exchange." *Science* **288**(5469): 1211–15.

Wess, J. (1998). "Molecular basis of receptor/G-protein-coupling selectivity." *Pharmacol Ther* **80**(3): 231–64.

West-Eberhard, M. J. (2003). *Developmental and Evolutionary Plasticity*. New York, Oxford University Press.

Wever, E. (1974). The evolution of vertebrate hearing. *Handbook of Sensory Physiology: Auditory System*, Springer-Verlag. **5**: 234–454.

White, D. (1899). *Fossil flora of the Lower Coal Measures of Missouri*. Washington, U.S. Government Printing Office.

Wickner, S., M. R. Maurizi, et al. (1999). "Posttranslational quality control: folding, refolding, and degrading proteins." *Science* **286**(5446): 1888–93.

Wilder, H. H. (1909). *History of the Human Body*. New York, Henry Holt.

Wilkins, A. S. (2002). *The Evolution of Developmental Pathways*. Sunderland, MA, Sinauer.

Wilson, A. B., K. Noack-Kunnmann, et al. (2000). "Incipient speciation in sympatric Nicaraguan crater lake cichlid fishes: sexual selection versus ecological diversification." *Proceedings of the Royal Society. Biological Sciences* **267**: 2133–41.

Wilson, E. O. (1998). *Consilience : the unity of knowledge*. New York, Knopf.

Wilson, H. and J. Mattocks (1887). "The lateral sensory anlage in the salmon." *Anat Anz* **13**: 658–60.

Wilson, M. R., T. W. Close, et al. (2000). "Cell population dynamics (apoptosis, mitosis, and cell-cell communication) during disruption of homeostasis." *Exp Cell Res* **254**(2): 257–68.

Wilson, S. W. and J. L. Rubenstein (2000). "Induction and dorsoventral patterning of the telencephalon." *Neuron* **28**(3): 641–51.

Woese, C. (1998). "The universal ancestor." *Proc Natl Acad Sci U S A* **95**(12): 6854–9.

Woese, C. R. (2000). "Interpreting the universal phylogenetic tree." *Proc Natl Acad Sci U S A* **97**(15): 8392–6.

Woese, C. R. (2002). "On the evolution of cells." *Proc Natl Acad Sci U S A* **99**(13): 8742–7.

Wolpert, L. (1969). "Positional information and the spatial patttern of cellular differentiation." *Journal of Theoretical Biology* **25**: 1–47.

Wolpert, L. (1981). "Positional information and pattern formation." *Philosophical Transactions of the Royal Society of London* **295**: 441–50.

Wolpert, L., R. Beddington, et al. (1998). *Principles of Development.* London, Current Biology Ltd.

Wong, A. M., J. W. Wang, et al. (2002). "Spatial representation of the glomerular map in the Drosophila protocerebrum." *Cell* **109**(2): 229–41.

Woolsey, T. A. and H. Van der Loos (1970). "The structural organization of layer IV in the somatosensory region (SI) of mouse cerebral cortex. The description of a cortical field composed of discrete cytoarchitectonic units." *Brain Res* **17**(2): 205–42.

Wray, G. A., M. W. Hahn, et al. (2003). "The Evolution of Transcriptional Regulation in Eukaryotes." *Mol Biol Evol* **9**: 1377–419.

Wurst, W. and L. Bally-Cuif (2001). "Neural plate patterning: upstream and downstream of the isthmic organizer." *Nat Rev Neurosci* **2**(2): 99–108.

Wysocki, C. J. (1979). "Neurobehavioral evidence for the involvement of the vomeronasal system in mammalian reproduction." *Neurosci Biobehav Rev* **3**(4): 301–41.

Xu, Y., P.-F. L. Chang, et al. (1994). "Plant defense genes are synergistically induced by ethylene and methyl jasmonate." *Plant Cell* **6**: 1077–85.

Yack, J. E. and J. H. Fullard (2000). "Ultrasonic hearing in nocturnal butterflies." *Nature* **403**(6767): 265–6.

Yager, D. D. (1999). "Structure, development, and evolution of insect auditory systems." *Microsc Res Tech* **47**(6): 380–400.

Yamada, H., T. Suzuki, et al. (2001). "The Arabidopsis AHK4 histidine kinase is a cytokinin-binding receptor that transduces cytokinin signals across the membrane." *Plant Cell Physiol* **42**(9): 1017–23.

Yarfitz, S. and J. B. Hurley (1994). "Transduction mechanisms of vertebrate and invertebrate photoreceptors." *J Biol Chem* **269**(20): 14329–32.

Yedid, G. and G. Bell (2002). "Macroevolution simulated with autonomously replicating computer programs." *Nature* **420**(6917): 810–12.

Yokoyama, S. (2000). "Molecular evolution of vertebrate visual pigments." *Prog Retin Eye Res* **19**(4): 385–419.

Yokoyama, S. and R. Yokoyama (2000). Comparative molecular biology of visual pigments. *Handbook of Biological Physics*. D. G. Stavenga, W. J. DeGrip and E. N. Puch, Jr. St. Louis, MO, Elsevier.

Yokoyama, S., H. Zhang, et al. (1999). "Adaptive evolution of color vision of the Comoran coelacanth (Latimeria chalumnae)." *Proc Natl Acad Sci U S A* **96**(11): 6279–84.

Young, N. (2000). "The genetic architecture of resistance." *Current Opinion in Plant Biology* **3**(4): 285–90.

Younger, R. M., C. Amadou, et al. (2001). "Characterization of clustered MHC-linked olfactory receptor genes in human and mouse." *Genome Res* **11**(4): 519–30.

Yu, M., P. Wu, et al. (2002). "The morphogenesis of feathers." *Nature* **420**(6913): 308–12.

Zamore, P. D. (2002). "Ancient pathways programmed by small RNAs." *Science* **296**(5571): 1265–9.
Zhang, X. and S. Firestein (2002). "The olfactory receptor gene superfamily of the mouse." *Nat Neurosci* **5**(2): 124–33.
Zhao, Y., S. K. Christensen, et al. (2001). "A role for flavin monooxygenase-like enzymes in auxin biosynthesis." *Science* **291**(5502): 306–9.
Zou, Z., L. F. Horowitz, et al. (2001). "Genetic tracing reveals a stereotyped sensory map in the olfactory cortex." *Nature* **414**(6860): 173–9.

Index

References to Tables are in **Bold** face, Figures in *Italics*

Genetics and the Logic of Evolution, by Kenneth M. Weiss and Anne V. Buchanan.
ISBN 0-471-23805-8